Handbook of Research on AI and ML for Intelligent Machines and Systems

Brij B. Gupta
Asia University, Taiwan

Francesco Colace
University of Salerno, Italy

A volume in the Advances in Computational Intelligence and Robotics (ACIR) Book Series

Published in the United States of America by
IGI Global
Engineering Science Reference (an imprint of IGI Global)
701 E. Chocolate Avenue
Hershey PA, USA 17033
Tel: 717-533-8845
Fax: 717-533-8661
E-mail: cust@igi-global.com
Web site: http://www.igi-global.com

Library of Congress Cataloging-in-Publication Data

Names: Gupta, Brij, 1982- editor. | Colace, Francesco, 1972- editor.
Title: Handbook of research on AI and ML for intelligent machines and systems / edted by: Brij B. Gupta, and Francesco Colace.
Description: Hershey PA : Engineering Science Reference, [2024] | Includes bibliographical references. | Summary: "By compiling recent advancements in intelligent machines that rely on machine learning and deep learning technologies, this book serves as a vital resource for researchers, graduate students, PhD scholars, faculty members, scientists, and software developers. It offers valuable insights into the key concepts of AI and ML, covering essential security aspects, current trends, and often overlooked perspectives that are crucial for achieving comprehensive understanding. It not only explores the theoretical foundations of AI and ML but also provides guidance on applying these techniques to solve real-world problems. Unlike traditional texts, it offers flexibility through its distinctive module-based structure, allowing readers to follow their own learning paths"-- Provided by publisher.
Identifiers: LCCN 2023025890 (print) | LCCN 2023025891 (ebook) | ISBN 9781668499993 (hardcover) | ISBN 9798369300008 (ebook)
Subjects: LCSH: Artificial intelligence--Industrial applications. | Machine learning--Industrial applications.
Classification: LCC TA347.A78 H356 2024 (print) | LCC TA347.A78 (ebook) | DDC 006.3/1--dc23/eng/20230810
LC record available at https://lccn.loc.gov/2023025890
LC ebook record available at https://lccn.loc.gov/2023025891

This book is published in the IGI Global book series Advances in Computational Intelligence and Robotics (ACIR) (ISSN: 2327-0411; eISSN: 2327-042X)

British Cataloguing in Publication Data
A Cataloguing in Publication record for this book is available from the British Library.

For electronic access to this publication, please contact: eresources@igi-global.com.

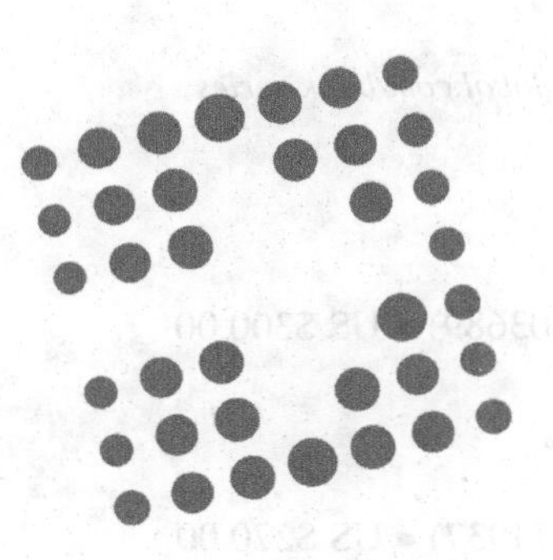

Advances in Computational Intelligence and Robotics (ACIR) Book Series

Ivan Giannoccaro
University of Salento, Italy

ISSN:2327-0411
EISSN:2327-042X

Mission

While intelligence is traditionally a term applied to humans and human cognition, technology has progressed in such a way to allow for the development of intelligent systems able to simulate many human traits. With this new era of simulated and artificial intelligence, much research is needed in order to continue to advance the field and also to evaluate the ethical and societal concerns of the existence of artificial life and machine learning.

The **Advances in Computational Intelligence and Robotics (ACIR) Book Series** encourages scholarly discourse on all topics pertaining to evolutionary computing, artificial life, computational intelligence, machine learning, and robotics. ACIR presents the latest research being conducted on diverse topics in intelligence technologies with the goal of advancing knowledge and applications in this rapidly evolving field.

Coverage

- Intelligent Control
- Fuzzy Systems
- Brain Simulation
- Cyborgs
- Evolutionary Computing
- Artificial Intelligence
- Neural Networks
- Artificial Life
- Robotics
- Natural Language Processing

IGI Global is currently accepting manuscripts for publication within this series. To submit a proposal for a volume in this series, please contact our Acquisition Editors at Acquisitions@igi-global.com or visit: http://www.igi-global.com/publish/.

The Advances in Computational Intelligence and Robotics (ACIR) Book Series (ISSN 2327-0411) is published by IGI Global, 701 E. Chocolate Avenue, Hershey, PA 17033-1240, USA, www.igi-global.com. This series is composed of titles available for purchase individually; each title is edited to be contextually exclusive from any other title within the series. For pricing and ordering information please visit http://www.igi-global.com/book-series/advances-computational-intelligence-robotics/73674. Postmaster: Send all address changes to above address.

Titles in this Series

For a list of additional titles in this series, please visit: www.igi-global.com/book-series

Artificial Intelligence in the Age of Nanotechnology
Wassim Jaber (ESPCI Paris - PSL, Fance)
Engineering Science Reference • © 2024 • 330pp • H/C (ISBN: 9798369303689) • US $300.00

Application and Adoption of Robotic Process Automation for Smart Cities
R.K. Tailor (Manipal University Jaipur, ndia)
Engineering Science Reference • © 2023 • 226pp • H/C (ISBN: 9781668471937) • US $270.00

Deterministic and Stochastic Approaches in Computer Modeling and Simulation
Radi Petrov Romansky (Technical University of Sofia, Bulgaria) and Nikolay Lyuboslavov Hinov (Technical University of Sofia, Bulgaria)
Engineering Science Reference • © 2023 • 513pp • H/C (ISBN: 9781668489475) • US $265.00

Technological Tools for Predicting Pregnancy Complications
D. Satishkumar (Nehru Institute of Technology, India) and P. Maniiarasan (Nehru Institute of Engineering and Technology, India)
Engineering Science Reference • © 2023 • 392pp • H/C (ISBN: 9798369317181) • US $365.00

Meta-Learning Frameworks for Imaging Applications
Ashok Sharma (University of Jammu, India) Sandeep Singh Sengar (Cardiff Metropolitan University, UK) and Parveen Singh (Cluster University, Jammu, India)
Engineering Science Reference • © 2023 • 253pp • H/C (ISBN: 9781668476598) • US $270.00

Predicting Pregnancy Complications Through Artificial Intelligence and Machine Learning
D. Satishkumar (Nehru Institute of Technology, India) and P. Maniiarasan (Nehru Institute of Engineering and Technology, India)
Medical Information Science Reference • © 2023 • 350pp • H/C (ISBN: 9781668489741) • US $350.00

Effective AI, Blockchain, and E-Governance Applications for Knowledge Discovery and Management
Rajeev Kumar (Moradabad Institute of Technology, India) Abu Bakar Abdul Hamid (Infrastructure University, Kuala Lumpur, Malaysia) and Noor Inayah Binti Ya'akub (Infrastructure University, Kuala Lumpur, Malaysia)
Engineering Science Reference • © 2023 • 403pp • H/C (ISBN: 9781668491515) • US $270.00

701 East Chocolate Avenue, Hershey, PA 17033, USA
Tel: 717-533-8845 x100 • Fax: 717-533-8661
E-Mail: cust@igi-global.com • www.igi-global.com

Handbook of Research on AI and ML for Intelligent Machines and Systems explores the forefront of technological innovation, offering insights and applications in a concise and insightful manner. Ideal for researchers and enthusiasts, this handbook navigates the evolving world of AI and ML.

- Kwok Tai Chui, Hong Kong Metropolitan University (HKMU), Hong Kong

Explore the frontier of technological evolution with the Handbook of Research on AI and ML for Intelligent Machines and Systems. This concise yet profound guide unveils the transformative power of artificial intelligence and machine learning, making it an indispensable resource for researchers, practitioners, and visionaries shaping the future of intelligent systems.

- Priyanka Chaurasia, University of Ulster, UK

With deep appreciation and love, this book is dedicated to my family for their enduring support and encouragement.

\- Brij B. Gupta

With deep appreciation and love, this book is dedicated to my family for their enduring support and encouragement.

\- Francesco Colace

List of Contributors

Anitha Jebamani, S. / *Department of Information Technology, Sri Sai Ram Engineering College, Chennai, India* 251
Anitha, K. / *Department of Computing Technologies, School of Computing, SRM Institute of Science and Technology, India* 374
Arunarani, A. R. / *Department of Computational Intelligence, SRM Institute of Science and Technology, India* 223
Baidoo, Charity Y. M. / *University of Ghana, Ghana* 27
Boopathi, Sampath / *Mechanical Engineering, Muthayammal Engineering College, India* 195, 223, 348, 374
Casillo, Mario / *University of Salerno, Italy* 50, 107
Colace, Francesco / *University of Salerno, Italy* 50, 107, 305
Das, Pratyay / *Maulana Abul Kalam Azad University of Technology, West Bengal, India* 137
G. Nadakinamani, Rajkumar / *Badr Al Samaa Hospital, Oman* 276
G. S., Hari Priya / *Department of Computer Science, M.S. Ramaiah College of Arts Science and Commerce, Bengaluru, India* 169
Gao, Xinyi / *Auckland University of Technology, New Zealand* 94
Ghosh, Ahona / *Maulana Abul Kalam Azad University of Technology, West Bengal, India* 137
Gupta, Brij B. / *Asia University, Taichung, Taiwan & Lebanese American University, Beirut, Lebanon* 50, 107, 305
Halder, Shubhajit / *Department of Chemistry, Hislop College, Nagpur, India* 348
Kumar, K. Pradeep Mohan / *Department of Computing Technologies, SRM Institute of Science and Technology, Chennai, India* 348
Lorusso, Angelo / *University of Salerno, Italy* 107, 305
Maheshwari, A. / *Department of Computational Intelligence, SRM Institute of Science and Technology, India* 223
Malathi, J. / *Department of Computer Science and Business Systems, Sri Sai Ram Engineering College, Chennai, India* 169
Marongiu, Francesco / *University of Salerno, Italy* 50
Mohanty, Akash / *School of Mechanical Engineering, Vellore Institute of Technology, India* 251
Myilsamy, Sureshkumar / *Mechanical Engineering, Bannari Amman Institute of Technology, India* 251, 276
Nguyen, Minh / *Auckland University of Technology, New Zealand* 94, 400
Owusu, Ebenezer / *University of Ghana, Ghana* 27
Pachiappan, Krishnagandhi / *Department of Electrical and Electronics Engineering, Nandha Engineering College, India* 374

Padmini, S. / *Department of Computing Technologies, SRM Institute of Science and Technology, Chennai, India* 348
Pappachan, Princy / *Asia University, Taiwan* 1
Pitchai, R. / *Department of Computer Science and Engineering, B.V. Raju Institute of Technology, India* 374
Rahaman, Mosiur / *Asia University, Taiwan* 1
Rahamathunnisa, U. / *School of Computer Science Engineering and Information Systems, Vellore Institute of Technology, India* 251
Rebecca, B. / *Department of Computer Science and Engineering (Data Science), Marri Laxman Reddy Institute of Technology and Management, Hyderabad, India* 348
S., Murugan / *Sona College of Technology, India* 169
Saha, Sriparna / *Maulana Abul Kalam Azad University of Technology, West Bengal, India* 137
Sangeetha, S. / *Department of Computer Science Engineering, Karpagam College of Engineering, India* 374
Santaniello, Domenico / *University of Salerno, Italy* 50, 107, 305
Satapathy, Suchismita / *KIIT University, India* 75
Satyanarayana, T. V. V. / *Department of Electronics and Communication Engineering, Mohan Babu University, India* 374
Sethuramalingam, T. K. / *Department of Electronics and Communication Engineering, Karpagam College of Engineering, Coimbatore, India* 276
Shankar, Amit Kumar / *Maulana Abul Kalam Azad University of Technology, West Bengal, India* 137
Sharma, Manju / *Department of Computer Science and Engineering, University Institute of Engineering and Technology, India & Maharshi Dayanand University, India* 195
Sharma, Monika / *Department of Computer Science and Engineering, The Technological Institute of Textile and Sciences, Bhiwani, India* 195
Sharma, Neerav / *Department of Computer Science and Engineering, BITS College, Bhiwani, India* 195
Singh, Ajeet / *School of Computing Science and Engineering, VIT Bhopal University, India* 223
Singh, Vikash / *Department of Civil Engineering, Institute of Engineering and Technology, Lucknow, India* 169
Sreerakuvandana / *Jain University, India* 1
Srivastava, Bipin Kumar / *Department of Applied Sciences, Galgotias College of Engineering and Technology, India* 348
Sudhakar, K. / *Department of Computer Science and Engineering, Madanapalle Institute of Technology and Science, Madanapalle, India* 251
Sumathy, G. / *Department of Computational Intelligence, SRM Institute of Science and Technology, India* 223, 276
Sundaramoorthy, K. / *Department of Information Technology, Jerusalem College of Engineering, India* 223
Syamala, Maganti / *Department of Computer Science and Engineering, Koneru Lakshmaiah Education Foundation, Vaddeswaram, India* 169
Troiano, Alfredo / *University of Salerno, Italy* 305
Udendhran, R. / *Department of Computational Intelligence, SRM Institute of Science and Technology, India* 251

Uma Maheswari, B. / *Department of Computer Science and Engineering, St. Joseph's College of Engineering, Chennai, India* 169
Valentino, Carmine / *University of Salerno, Italy* 107, 305
Xiao, Bingjie / *Auckland University of Technology, New Zealand* 400
Xue, Yinzhe / *Auckland University of Technology, New Zealand* 421
Yan, Wei Qi / *Auckland University of Technology, New Zealand* 94, 400, 421
Yaokumah, Winfred / *University of Ghana, Ghana* 27

Table of Contents

Preface xxii

Acknowledgement xxvii

Chapter 1
Conceptualising the Role of Intellectual Property and Ethical Behaviour in Artificial Intelligence...... 1
Princy Pappachan, Asia University, Taiwan
Sreerakuvandana, Jain University, India
Mosiur Rahaman, Asia University, Taiwan

Chapter 2
Measuring Throughput and Latency of Machine Learning Techniques for Intrusion Detection 27
Winfred Yaokumah, University of Ghana, Ghana
Charity Y. M. Baidoo, University of Ghana, Ghana
Ebenezer Owusu, University of Ghana, Ghana

Chapter 3
Securing Digital Ecosystems: Harnessing the Power of Intelligent Machines in a Secure and Sustainable Environment 50
Mario Casillo, University of Salerno, Italy
Francesco Colace, University of Salerno, Italy
Brij B. Gupta, Asia University, Taichung, Taiwan & Lebanese American University, Beirut, Lebanon
Francesco Marongiu, University of Salerno, Italy
Domenico Santaniello, University of Salerno, Italy

Chapter 4
IoT-Based Economic Flame Detection Device for Safety........ 75
Suchismita Satapathy, KIIT University, India

Chapter 5
Human Face Mask Detection Using YOLOv7+CBAM in Deep Learning 94
Xinyi Gao, Auckland University of Technology, New Zealand
Minh Nguyen, Auckland University of Technology, New Zealand
Wei Qi Yan, Auckland University of Technology, New Zealand

Chapter 6
The Role of AI in Improving Interaction With Cultural Heritage: An Overview............................ 107
Mario Casillo, University of Salerno, Italy
Francesco Colace, University of Salerno, Italy
Brij B. Gupta, Asia University, Taichung, Taiwan & Lebanese American University, Beirut, Lebanon
Angelo Lorusso, University of Salerno, Italy
Domenico Santaniello, University of Salerno, Italy
Carmine Valentino, University of Salerno, Italy

Chapter 7
Machine Learning Approach for Robot Navigation Using Motor Imagery Signals 137
Pratyay Das, Maulana Abul Kalam Azad University of Technology, West Bengal, India
Amit Kumar Shankar, Maulana Abul Kalam Azad University of Technology, West Bengal, India
Ahona Ghosh, Maulana Abul Kalam Azad University of Technology, West Bengal, India
Sriparna Saha, Maulana Abul Kalam Azad University of Technology, West Bengal, India

Chapter 8
Cloud Solutions for Smart Parking and Traffic Control in Smart Cities 169
Maganti Syamala, Department of Computer Science and Engineering, Koneru Lakshmaiah Education Foundation, Vaddeswaram, India
J. Malathi, Department of Computer Science and Business Systems, Sri Sai Ram Engineering College, Chennai, India
Vikash Singh, Department of Civil Engineering, Institute of Engineering and Technology, Lucknow, India
Hari Priya G. S., Department of Computer Science, M.S. Ramaiah College of Arts Science and Commerce, Bengaluru, India
B. Uma Maheswari, Department of Computer Science and Engineering, St. Joseph's College of Engineering, Chennai, India
Murugan S., Sona College of Technology, India

Chapter 9
Building Sustainable Smart Cities Through Cloud and Intelligent Parking System 195
Monika Sharma, Department of Computer Science and Engineering, The Technological Institute of Textile and Sciences, Bhiwani, India
Manju Sharma, Department of Computer Science and Engineering, University Institute of Engineering and Technology, India & Maharshi Dayanand University, India
Neerav Sharma, Department of Computer Science and Engineering, BITS College, Bhiwani, India
Sampath Boopathi, Mechanical Engineering, Muthayammal Engineering College, India

Chapter 10
A Study on AI and Blockchain-Powered Smart Parking Models for Urban Mobility........................ 223
K. Sundaramoorthy, Department of Information Technology, Jerusalem College of Engineering, India
Ajeet Singh, School of Computing Science and Engineering, VIT Bhopal University, India
G. Sumathy, Department of Computational Intelligence, SRM Institute of Science and Technology, India
A. Maheshwari, Department of Computational Intelligence, SRM Institute of Science and Technology, India
A. R. Arunarani, Department of Computational Intelligence, SRM Institute of Science and Technology, India
Sampath Boopathi, Mechanical Engineering, Muthayammal Engineering College, India

Chapter 11
Machine Learning and Deep Learning for Intelligent Systems in Small Aircraft Applications......... 251
U. Rahamathunnisa, School of Computer Science Engineering and Information Systems, Vellore Institute of Technology, India
Akash Mohanty, School of Mechanical Engineering, Vellore Institute of Technology, India
K. Sudhakar, Department of Computer Science and Engineering, Madanapalle Institute of Technology and Science, Madanapalle, India
S. Anitha Jebamani, Department of Information Technology, Sri Sai Ram Engineering College, Chennai, India
R. Udendhran, Department of Computational Intelligence, SRM Institute of Science and Technology, India
Sureshkumar Myilsamy, Mechanical Engineering, Bannari Amman Institute of Technology, India

Chapter 12
Machine Learning in E-Health and Digital Healthcare: Practical Strategies for Transformation 276
T. K. Sethuramalingam, Department of Electronics and Communication Engineering, Karpagam College of Engineering, Coimbatore, India
Rajkumar G. Nadakinamani, Badr Al Samaa Hospital, Oman
G. Sumathy, Department of Computational Intelligence, SRM Institute of Science and Technology, India
Sureshkumar Myilsamy, Mechanical Engineering, Bannari Amman Institute of Technology, India

Chapter 13
Unsupervised Learning Techniques for Vibration-Based Structural Health Monitoring Systems Driven by Data: A General Overview.. 305
Francesco Colace, University of Salerno, Italy
Brij B. Gupta, Asia University, Taichung, Taiwan & Lebanese American University, Beirut, Lebanon
Angelo Lorusso, University of Salerno, Italy
Alfredo Troiano, University of Salerno, Italy
Domenico Santaniello, University of Salerno, Italy
Carmine Valentino, University of Salerno, Italy

Chapter 14
Convergence of Data Science-AI-Green Chemistry-Affordable Medicine: Transforming Drug Discovery 348
B. Rebecca, Department of Computer Science and Engineering (Data Science), Marri Laxman Reddy Institute of Technology and Management, Hyderabad, India
K. Pradeep Mohan Kumar, Department of Computing Technologies, SRM Institute of Science and Technology, Chennai, India
S. Padmini, Department of Computing Technologies, SRM Institute of Science and Technology, Chennai, India
Bipin Kumar Srivastava, Department of Applied Sciences, Galgotias College of Engineering and Technology, India
Shubhajit Halder, Department of Chemistry, Hislop College, Nagpur, India
Sampath Boopathi, Mechanical Engineering, Muthayammal Engineering College, India

Chapter 15
Intelligent Machines, IoT, and AI in Revolutionizing Agriculture for Water Processing 374
Krishnagandhi Pachiappan, Department of Electrical and Electronics Engineering, Nandha Engineering College, India
K. Anitha, Department of Computing Technologies, School of Computing, SRM Institute of Science and Technology, India
R. Pitchai, Department of Computer Science and Engineering, B.V. Raju Institute of Technology, India
S. Sangeetha, Department of Computer Science Engineering, Karpagam College of Engineering, India
T. V. V. Satyanarayana, Department of Electronics and Communication Engineering, Mohan Babu University, India
Sampath Boopathi, Mechanical Engineering, Muthayammal Engineering College, India

Chapter 16
A Mixture Model for Fruit Ripeness Identification in Deep Learning 400
Bingjie Xiao, Auckland University of Technology, New Zealand
Minh Nguyen, Auckland University of Technology, New Zealand
Wei Qi Yan, Auckland University of Technology, New Zealand

Chapter 17
YOLO Models for Fresh Fruit Classification From Digital Videos 421
Yinzhe Xue, Auckland University of Technology, New Zealand
Wei Qi Yan, Auckland University of Technology, New Zealand

Compilation of References 436

About the Contributors 493

Index 500

Detailed Table of Contents

Preface...xxii

Acknowledgement xxvii

Chapter 1
Conceptualising the Role of Intellectual Property and Ethical Behaviour in Artificial Intelligence...... 1
Princy Pappachan, Asia University, Taiwan
Sreerakuvandana, Jain University, India
Mosiur Rahaman, Asia University, Taiwan

The development of artificial intelligence has significantly affected all facets of human life, leading to many ethical and legal questions that have positive and negative consequences for individuals, organisations, and society. Amidst this lies the interaction between intellectual property rights and ethical behaviour in the development and use of artificial intelligence. While intellectual property acts as a catalyst for innovation, ethical behaviour ensures responsible and accountable artificial intelligence consistent with social standards. Accordingly, the beneficial effects of artificial intelligence can only be guaranteed by balancing intellectual property rights and ethical behaviour. This chapter thus discusses the notion of intellectual property and ethical behaviour in the context of artificial intelligence by providing a comprehensive historical review and reflecting on the creation and implementation of ethical artificial intelligence while preserving intellectual property rights.

Chapter 2
Measuring Throughput and Latency of Machine Learning Techniques for Intrusion Detection 27
Winfred Yaokumah, University of Ghana, Ghana
Charity Y. M. Baidoo, University of Ghana, Ghana
Ebenezer Owusu, University of Ghana, Ghana

When evaluating the effectiveness of machine learning algorithms for intrusion detection, it is insufficient to only focus on their performance metrics. One must also focus on the overhead metrics of the models. In this study, the performance accuracy, latency, and throughput of seven supervised machine learning algorithms and a proposed ensemble model were measured. The study performs a series of experiments using two recent datasets, and two filter-based feature selection methods were employed. The results show that, on average, the naive bayes achieved the lowest latency, highest throughput, and lowest accuracy on both datasets. The logistics regression had the maximum throughput. The proposed ensemble method recorded the highest latency for both feature selection methods. Overall, the Spearman feature selection technique increased throughput for almost all the models, whereas the Pearson feature selection approach maximized performance accuracies for both datasets.

Chapter 3
Securing Digital Ecosystems: Harnessing the Power of Intelligent Machines in a Secure and Sustainable Environment 50
Mario Casillo, University of Salerno, Italy
Francesco Colace, University of Salerno, Italy
Brij B. Gupta, Asia University, Taichung, Taiwan & Lebanese American University, Beirut, Lebanon
Francesco Marongiu, University of Salerno, Italy
Domenico Santaniello, University of Salerno, Italy

Industries are evolving towards an integral digitisation of their processes. In the face of ever-faster market demands and ever-increasing quality, information technology (IT) progress represents the only solution to these needs. Industry 4.0 was born with this focus, where cybernetic systems interact with each other to achieve, efficiently, a predetermined goal. The whole process takes place with minimal, or in some cases total, absence of human intervention, leaving the systems to interact in full autonomy. This approach commonly falls under the internet of things (IoT) paradigm, in which all objects, regardless of size and functionality, are connected in a standard network exchanging information. In this sense, objects acquire intelligence because they can modify their behaviours based on the data they receive and transmit.

Chapter 4
IoT-Based Economic Flame Detection Device for Safety 75
Suchismita Satapathy, KIIT University, India

Mainly fires are of three types (i.e., ground, surface, and crown fire), which occur as wild land/forest fire, residential fire, building fire, and others. The number and effect of fires are an outcome of global warming, extinction of species, and climate change. To battle against these parameters/disasters, it is important to take on an exhaustive and complex methodology that empowers nonstop situational mindfulness and moment responsiveness. The outcome is unrecoverable and dangerous to the climate, environment, people's lives, and causes economic losses. The issues/barriers in the detection of fire are discussed here. For that, the authors have identified and ranked those challenges by using the best worst method. They have designed an automatic fire alarm detector at sensitive sites as one of the preventive steps to avoid the hazard. It can detect heat in a specific environment, raise an alert, turn off the building's mains, and even spray water to minimize the intensity of the fire.

Chapter 5
Human Face Mask Detection Using YOLOv7+CBAM in Deep Learning 94
Xinyi Gao, Auckland University of Technology, New Zealand
Minh Nguyen, Auckland University of Technology, New Zealand
Wei Qi Yan, Auckland University of Technology, New Zealand

COVID-19 and its variants have affected millions of people around the world. Wearing a mask is an effective way to reduce the spread of the epidemic. While wearing masks is a proven strategy to mitigate the spread, monitoring compliance remains a challenge. In this chapter, the authors propose a mask detection method based on deep learning and convolutional block attention module (CBAM). In this chapter, they extract representative features from input images through supervised learning. In order to improve the recognition accuracy under limited computing resources. They choose YOLOv7 network model and incorporate CBAM into its network structure. Compared with the original version of YOLOv7, the proposed network model improves the mean average precision (mAP) up to 0.3% in face mask detection process. Meanwhile, the method improves the detection speed of each frame 73ms. These advancements have significant implications for real-time, large-scale monitoring systems, thereby contributing to public health and safety.

Chapter 6
The Role of AI in Improving Interaction With Cultural Heritage: An Overview............................. 107
Mario Casillo, University of Salerno, Italy
Francesco Colace, University of Salerno, Italy
Brij B. Gupta, Asia University, Taichung, Taiwan & Lebanese American University, Beirut, Lebanon
Angelo Lorusso, University of Salerno, Italy
Domenico Santaniello, University of Salerno, Italy
Carmine Valentino, University of Salerno, Italy

Over the years, artificial intelligence techniques have been applied in several application fields, exploiting data to execute different tasks and achieve disparate objectives. Therefore, the cultural heritage field can utilise AI techniques to improve the interaction between visitors and cultural assets. Then, this work aims to present the background related to the principal AI techniques and provides an overview of the literature aimed at improving the user cultural experience. This overview focuses on AI integration with tools aimed to enhance the interaction among visitors and cultural sites, such as recommender systems, context-aware recommender systems, and chatbots. Finally, the most common measure used for estimating the accuracy of the AI methodologies will be introduced.

Chapter 7
Machine Learning Approach for Robot Navigation Using Motor Imagery Signals.......................... 137
Pratyay Das, Maulana Abul Kalam Azad University of Technology, West Bengal, India
Amit Kumar Shankar, Maulana Abul Kalam Azad University of Technology, West Bengal, India
Ahona Ghosh, Maulana Abul Kalam Azad University of Technology, West Bengal, India
Sriparna Saha, Maulana Abul Kalam Azad University of Technology, West Bengal, India

Electroencephalography (EEG) signals have been used for different healthcare applications like motor and cognitive rehabilitation. In this study, motor imagery data of different subjects' rest vs. movement and different movements is categorized from a publicly available dataset. The authors have first applied a lowpass filter to the EEG signals to reduce noise and a fast fourier transform analysis to extract features from the filtered data. Utilizing principal component analysis, relevant features are selected. With an accuracy of 95.02%, they have classified rest vs. movement using the k-nearest neighbor algorithm. Using the random forest algorithm, they have classified various movement types with an accuracy of 96.45%. The success in differentiating between movement and rest raises the possibility that EEG signals can recognize a user's intention to move. Accurately classifying different movement types opens the possibility of navigating robots accordingly in the real-time scenario for people with motor disabilities to assist them with robotic arms and prosthetic limbs.

Chapter 8
Cloud Solutions for Smart Parking and Traffic Control in Smart Cities 169
Maganti Syamala, Department of Computer Science and Engineering, Koneru Lakshmaiah Education Foundation, Vaddeswaram, India
J. Malathi, Department of Computer Science and Business Systems, Sri Sai Ram Engineering College, Chennai, India
Vikash Singh, Department of Civil Engineering, Institute of Engineering and Technology, Lucknow, India
Hari Priya G. S., Department of Computer Science, M.S. Ramaiah College of Arts Science and Commerce, Bengaluru, India
B. Uma Maheswari, Department of Computer Science and Engineering, St. Joseph's College of Engineering, Chennai, India
Murugan S., Sona College of Technology, India

Urban mobility trends include 5G connectivity, autonomous vehicles, electric and sustainable modes, AI and machine learning, drones, and air mobility. These technologies enable real-time data exchange, reduce congestion, enhance safety, optimize road capacity, and optimize infrastructure planning. AI and machine learning algorithms provide accurate predictive analytics, adaptive traffic control, and personalized services. Cloud computing, IoT, and data analytics enable predictive modeling for mobility planning, traffic flow forecasting, demand forecasting, and behavioral analysis. MaaS platforms facilitate seamless integration of modes, while shared mobility services like car-sharing and ride-hailing grow, reducing private vehicle ownership and promoting efficient resource use. Mobility data transforms urban planning, infrastructure optimization, mixed-use development, and smart city integration, guiding transportation layouts, traffic signal placements, parking facilities, and neighborhood design.

Chapter 9
Building Sustainable Smart Cities Through Cloud and Intelligent Parking System 195
Monika Sharma, Department of Computer Science and Engineering, The Technological Institute of Textile and Sciences, Bhiwani, India
Manju Sharma, Department of Computer Science and Engineering, University Institute of Engineering and Technology, India & Maharshi Dayanand University, India
Neerav Sharma, Department of Computer Science and Engineering, BITS College, Bhiwani, India
Sampath Boopathi, Mechanical Engineering, Muthayammal Engineering College, India

This chapter discusses the role of cloud computing and intelligent parking systems in sustainable smart cities, addressing challenges like traffic congestion, pollution, and resource inefficiency. These technologies enhance urban mobility, reduce environmental impact, and improve quality of life in cities facing rapid urbanization worldwide. This chapter offers a thorough analysis of the integration of cloud computing and intelligent parking systems in sustainable urban development, highlighting successful implementations and lessons learned. It also explores potential future developments and policy considerations to facilitate widespread adoption of these technologies, highlighting the importance of global best practices.

Chapter 10
A Study on AI and Blockchain-Powered Smart Parking Models for Urban Mobility........................ 223
K. Sundaramoorthy, Department of Information Technology, Jerusalem College of Engineering, India
Ajeet Singh, School of Computing Science and Engineering, VIT Bhopal University, India
G. Sumathy, Department of Computational Intelligence, SRM Institute of Science and Technology, India
A. Maheshwari, Department of Computational Intelligence, SRM Institute of Science and Technology, India
A. R. Arunarani, Department of Computational Intelligence, SRM Institute of Science and Technology, India
Sampath Boopathi, Mechanical Engineering, Muthayammal Engineering College, India

Urban problems like traffic jams and a lack of parking spaces can be solved in an innovative way with the help of smart parking models powered by AI and blockchain technology. These models enhance user experience, optimise space allocation, and shorten search times. Predictive analytics and real-time data from IoT sensors direct drivers to available parking spaces, minimising traffic and environmental impact. By protecting user privacy, controlling access, and securing transactions, blockchain technology improves AI. Users are empowered by blockchain-based decentralised digital identities, which also guarantee data privacy and transparent business dealings. With less traffic, more user happiness, and significant cost savings, this combination produces user-centric, environmentally friendly, and cost-effective smart parking solutions. The cost-benefit analysis for AI and blockchain-powered smart parking demonstrates a favourable return on investment, paving the way for smarter, greener cities and more interconnected urban settings.

Chapter 11
Machine Learning and Deep Learning for Intelligent Systems in Small Aircraft Applications......... 251
U. Rahamathunnisa, School of Computer Science Engineering and Information Systems, Vellore Institute of Technology, India
Akash Mohanty, School of Mechanical Engineering, Vellore Institute of Technology, India
K. Sudhakar, Department of Computer Science and Engineering, Madanapalle Institute of Technology and Science, Madanapalle, India
S. Anitha Jebamani, Department of Information Technology, Sri Sai Ram Engineering College, Chennai, India
R. Udendhran, Department of Computational Intelligence, SRM Institute of Science and Technology, India
Sureshkumar Myilsamy, Mechanical Engineering, Bannari Amman Institute of Technology, India

This chapter explores the integration of machine learning and deep learning techniques in small aircraft applications. The aviation industry is exploring innovative solutions to improve safety, efficiency, and performance in these operations. The chapter explores the advantages, challenges, and future prospects of implementing intelligent systems in small aircraft, including autopilot systems, navigation assistance, fault detection, and pilot support systems. Real-world case studies and applications demonstrate the transformative impact of these technologies on small aircraft operations. The chapter provides a comprehensive overview of the latest advancements in machine learning and deep learning, highlighting their pivotal role in improving small aircraft intelligence, safety, and efficiency.

Chapter 12
Machine Learning in E-Health and Digital Healthcare: Practical Strategies for Transformation...... 276
T. K. Sethuramalingam, Department of Electronics and Communication Engineering, Karpagam College of Engineering, Coimbatore, India
Rajkumar G. Nadakinamani, Badr Al Samaa Hospital, Oman
G. Sumathy, Department of Computational Intelligence, SRM Institute of Science and Technology, India
Sureshkumar Myilsamy, Mechanical Engineering, Bannari Amman Institute of Technology, India

Machine learning is revolutionizing healthcare by offering innovative solutions to complex challenges. This chapter explores the practical strategies, ethical considerations, and real-world applications of machine learning in the healthcare domain. It delves into data collection and management, model development, integration with existing systems, and the importance of interdisciplinary collaboration. The chapter also discusses the ethical dimensions of healthcare AI, such as data privacy, bias mitigation, and regulatory compliance. Real-world case studies highlight the impact of machine learning on early disease detection, drug discovery, and precision medicine. The chapter concludes by examining future trends, including emerging technologies like quantum computing, nanomedicine, and the growing role of AI in drug discovery and genomic medicine. As machine learning continues to reshape healthcare, understanding these practical strategies and ethical considerations is essential for optimizing patient care and advancing the healthcare industry.

Chapter 13
Unsupervised Learning Techniques for Vibration-Based Structural Health Monitoring Systems Driven by Data: A General Overview... 305
Francesco Colace, University of Salerno, Italy
Brij B. Gupta, Asia University, Taichung, Taiwan & Lebanese American University, Beirut, Lebanon
Angelo Lorusso, University of Salerno, Italy
Alfredo Troiano, University of Salerno, Italy
Domenico Santaniello, University of Salerno, Italy
Carmine Valentino, University of Salerno, Italy

Structural damage detection is a crucial issue for the safety of civil buildings, which are subject to gradual deterioration over time and at risk from sudden seismic events. To prevent irreparable damage, the scientific community has directed its attention toward developing innovative methods for structural health monitoring (SHM), which can provide a timely and reliable assessment of structural conditions. In this domain, the significance of unsupervised learning approaches has grown considerably, as they enable the identification of structural irregularities solely based on data obtained from intact structures to train statistical models. Despite the importance of studies on unsupervised learning methods for structural health monitoring, no reviews are specifically dedicated to this topic, considering the application part. The review of studies, therefore, made it possible to highlight the progress achieved in this field and identify areas where improvements could still be made to develop increasingly accurate and effective methods for structural damage detection.

Chapter 14
Convergence of Data Science-AI-Green Chemistry-Affordable Medicine: Transforming Drug Discovery 348
B. Rebecca, Department of Computer Science and Engineering (Data Science), Marri Laxman Reddy Institute of Technology and Management, Hyderabad, India
K. Pradeep Mohan Kumar, Department of Computing Technologies, SRM Institute of Science and Technology, Chennai, India
S. Padmini, Department of Computing Technologies, SRM Institute of Science and Technology, Chennai, India
Bipin Kumar Srivastava, Department of Applied Sciences, Galgotias College of Engineering and Technology, India
Shubhajit Halder, Department of Chemistry, Hislop College, Nagpur, India
Sampath Boopathi, Mechanical Engineering, Muthayammal Engineering College, India

The drug discovery and design process has been significantly transformed by the integration of data science, artificial intelligence (AI), green chemistry principles, and affordable medicine. AI techniques enable rapid analysis of vast datasets, predicting molecular interactions, optimizing drug candidates, and identifying potential therapeutics. Green chemistry practices promote sustainability and efficiency, resulting in environmentally friendly and cost-effective production processes. The goal is to develop affordable medicines that are not only efficacious but also accessible to a wider population. This chapter explores case studies and emerging trends to highlight the transformation of the pharmaceutical industry and innovation in drug discovery.

Chapter 15
Intelligent Machines, IoT, and AI in Revolutionizing Agriculture for Water Processing 374
Krishnagandhi Pachiappan, Department of Electrical and Electronics Engineering, Nandha Engineering College, India
K. Anitha, Department of Computing Technologies, School of Computing, SRM Institute of Science and Technology, India
R. Pitchai, Department of Computer Science and Engineering, B.V. Raju Institute of Technology, India
S. Sangeetha, Department of Computer Science Engineering, Karpagam College of Engineering, India
T. V. V. Satyanarayana, Department of Electronics and Communication Engineering, Mohan Babu University, India
Sampath Boopathi, Mechanical Engineering, Muthayammal Engineering College, India

Modern agriculture faces numerous challenges, ranging from rising global food demand to water scarcity. To address these issues, the incorporation of intelligent machines, the Internet of Things (IoT), and artificial intelligence (AI) in agricultural water processing has become critical. This chapter investigates these technologies' transformative potential for optimizing water usage, increasing crop yields, and ensuring sustainable agricultural practices. It delves into the key concepts and applications, emphasizing the advantages and disadvantages of this novel approach. Farmers can make data-driven decisions, automate irrigation processes, and adapt to changing environmental conditions by leveraging AI and IoT-enabled systems, ultimately contributing to a more efficient and environmentally friendly agricultural sector.

Chapter 16
A Mixture Model for Fruit Ripeness Identification in Deep Learning 400
Bingjie Xiao, Auckland University of Technology, New Zealand
Minh Nguyen, Auckland University of Technology, New Zealand
Wei Qi Yan, Auckland University of Technology, New Zealand

Visual object detection is a foundation in the field of computer vision. Since the size of visual objects in an images is various, the speed and accuracy of object detection are the focus of current research projects in computer vision. In this book chapter, the datasets consist of fruit images with various maturity. Different types of fruit are divided into the classes "ripe" and "overripe" according to the degree of skin folds. Then the object detection model is employed to automatically classify different ripeness of fruits. A family of YOLO models are representative algorithms for visual object detection. The authors make use of ConvNeXt and YOLOv7, which belong to the CNN network, to locate and detect fruits, respectively. YOLOv7 employs the bag-of-freebies training method to achieve its objectives, which reduces training costs and enhances detection accuracy. An extended E-ELAN module, based on the original ELAN, is proposed within YOLOv7 to increase group convolution and improve visual feature extraction. In contrast, ConvNeXt makes use of a standard neural network architecture, with ResNet-50 serving as the baseline. The authors compare the proposed models, which result in an optimal classification model with best precision of 98.9%.

Chapter 17
YOLO Models for Fresh Fruit Classification From Digital Videos... 421
Yinzhe Xue, Auckland University of Technology, New Zealand
Wei Qi Yan, Auckland University of Technology, New Zealand

Identifying food freshness is a very important; it is a part of a long historical actions by humans, because fruit freshness can tell us the information about the quality of foods. With the advancement of machine learning and computer science, which will be broadly employed in factories and markets, instead of manual classification. Recognition of the freshness of food is rapidly being replaced by computers or robots. In this book chapter, the authors conduct the research work on fruit freshness detection, we make use of YOLOv6, YOLOv7, and YOLOv8 in this project to implement fruit classifications based on a variety of digital images, which can improve the efficiency and accuracy of the classification incredibly; after the classification, the output will showcase the result of fruit fresheness classification, namely, fresh, or rotten, etc. They also compare the results of different deep learning models to discover which architecture is the best one in terms of speed and accuracy. At the end of this book chapter, the authors made use of the majority vote method to combine the results of different models to get better accuracy and recall scores. To generate the final result, the authors trained the three models individually, and also propose a majority vote to get a better performance for fresh fruit detection. Compared with the previous work, this method has higher accuracy and a much faster speed. Because this one uses the clustering method to generate the final result, it will be easy for researchers to change the backbone and get a better result in the future.

Compilation of References 436

About the Contributors 493

Index 500

Preface

As editors of *the Handbook of Research on AI and ML for Intelligent Machines and Systems*, we are delighted to present this comprehensive reference work that delves into the ever-evolving realm of intelligent machines and the pivotal role that artificial intelligence (AI) and machine learning (ML) play in their development and transformation.

Intelligent machines represent a revolutionary paradigm in technology, characterized by their ability to autonomously interact with their environment and adapt to new situations. At the heart of this technological evolution lies the fusion of AI and ML, two dynamic fields that have proven their worth in creating robust intelligent machine systems and services. The need of the hour, however, is to bring these foundational technologies under a common roof to harness their collective power, thereby ushering in a new era of transformation for intelligent machines.

The fusion of AI and ML serves as the cornerstone for the development of intelligent machines. These technologies, on their own, have already demonstrated their efficacy in empowering machines to perform complex tasks with precision and efficiency. Yet, it is their synergy, their harmonious interplay, that holds the potential to propel intelligent machines to unparalleled heights of capability and interactivity. This book is dedicated to realizing this transformation.

Today, the demand for intelligent machines is on the rise across various sectors, from traffic monitoring and speech recognition to face recognition and automated manufacturing. These machines are instrumental in enhancing the efficiency of complex tasks, making them indispensable in the modern world. To comprehend the intricacies of intelligent machines, one must explore the integration of machine learning and deep learning technologies.

This handbook seeks to unravel the core concepts of AI and ML, offering a wealth of valuable insights that extend beyond the typical boundaries of other texts in this field. By addressing vital aspects of security and current trends often overlooked elsewhere, it provides a comprehensive survey designed with both students and professionals in mind.

Furthermore, this book serves as a practical guide for those eager to apply AI and ML techniques to real-world challenges. Its unique modular structure allows readers to tailor their learning experience to their individual needs, making it an ideal resource for students, professionals, and anyone interested in the domains of big data and sentiment analysis in online social media.

Our primary audience comprises graduate students, PhD scholars, faculty members, scientists, and software developers seeking to embark on research in the fields of data mining for big data and sentiment analysis in online social media, among other related areas. This handbook not only presents advances in these domains but also paves the way for future research directions.

The content of this book covers a wide spectrum of topics, including human-machine interaction, the role of intelligent machines in enhancing the cyber economy, intelligent network traffic monitoring, healthcare applications, diagnostic systems, smart city integration, wireless sensing, image enhancement, edge devices, speech and language recognition, manufacturing, agriculture, food processing, big data analytics, structural engineering, optimization techniques, performance comparisons, ethical considerations, and the integration of intelligent machines with conventional devices.

ORGANIZATION OF BOOK

Chapter 1

The development of artificial intelligence has transformed nearly every aspect of human life, from individual experiences to the operations of organizations and society at large. Yet, this transformation brings forth a myriad of ethical and legal questions that carry both positive and negative consequences. This chapter delves into the intricate relationship between intellectual property rights and ethical behavior in the realm of artificial intelligence. Intellectual property rights act as catalysts for innovation, while ethical behavior ensures responsible and accountable AI development in line with societal standards. By exploring the historical context and addressing the creation and implementation of ethical AI within the framework of intellectual property rights, this chapter strives to strike a balance that guarantees the positive impacts of artificial intelligence.

Chapter 2

When it comes to assessing the effectiveness of machine learning algorithms for intrusion detection, performance metrics alone are not sufficient. This chapter delves into the intricacies of evaluating seven supervised machine learning algorithms and an ensemble model in terms of performance accuracy, latency, and throughput. A series of experiments, utilizing recent datasets and feature selection methods, offers valuable insights. The results highlight the Naive Bayes algorithm's low latency and high throughput, along with the Logistic Regression algorithm's maximum throughput. Additionally, the chapter demonstrates how different feature selection methods impact the performance of these algorithms.

Chapter 3

Industries are evolving rapidly, and the digitization of processes has become paramount to meet the ever-accelerating market demands and quality standards. Industry 4.0, driven by cyber-physical systems that interact autonomously, is at the forefront of this transformation. This chapter explores the integration of Industry 4.0 and the Internet of Things (IoT), where objects of all types and functionalities are interconnected, exchanging data and gaining intelligence. The autonomy of these systems minimizes human intervention, allowing them to adapt based on the information they receive and transmit. The chapter elucidates how this paradigm shift is shaping industries and our world.

Chapter 4

The increasing incidence of various types of fires, exacerbated by factors like global warming and climate change, necessitates a comprehensive approach to prevent and mitigate disasters. This chapter addresses the challenges in fire detection and ranks them using the Best Worst Method. It introduces an IoT-based fire alarm detection and control system that can detect heat, raise alerts, shut down building mains, and even dispense water to minimize fire intensity. The system is a crucial step in achieving continuous situational awareness and rapid responsiveness, thereby reducing the devastating impact of fires on the environment, human lives, and economic losses.

Chapter 5

In the midst of the COVID-19 pandemic, the use of face masks has become a critical measure to prevent the spread of the virus. This chapter introduces a face mask detection method based on deep learning and Convolutional Block Attention Module (CBAM). It leverages supervised learning to extract visually representative features from images, enhancing recognition accuracy within limited computing resources. By integrating the CBAM model into the YOLOv7 structure, the chapter achieves improved accuracy and reduced detection time, making a valuable contribution to surveillance in crowded public spaces.

Chapter 6

Artificial Intelligence is now enhancing the interaction between visitors and cultural assets in the field of cultural heritage. This chapter provides an overview of AI techniques and their integration with tools designed to enrich cultural experiences. It delves into Recommender Systems, Context-Aware Recommender Systems, and Chatbots as means to improve the visitor's cultural journey. The chapter also explores key measures for evaluating the accuracy of AI methodologies in this context.

Chapter 7

Electroencephalography (EEG) signals play a crucial role in healthcare applications, particularly for motor and cognitive rehabilitation. This chapter focuses on the classification of motor imagery data using EEG signals, applying low-pass filters and Fast Fourier Transform analysis. By employing the k-nearest neighbor algorithm, it successfully distinguishes between rest and movement with an accuracy of 95.02%. Additionally, it uses the Random Forest algorithm to classify different types of movements with an accuracy of 96.45%, offering potential assistance for individuals with motor disabilities.

Chapter 8

Urban mobility is undergoing a revolution, with the integration of technologies like 5G connectivity, autonomous vehicles, AI, and more. This chapter delves into the transformative potential of these technologies, offering predictive analytics, adaptive traffic control, and personalized services. Cloud computing, IoT, and data analytics enable predictive modeling for mobility planning and traffic flow forecasting, paving the way for smarter, more efficient cities.

Chapter 9

This chapter explores the role of cloud computing and intelligent parking systems in sustainable smart cities, addressing challenges like traffic congestion and pollution. It offers insights into successful implementations and lessons learned, as well as potential future developments and policy considerations to promote the widespread adoption of these technologies, ultimately leading to more sustainable urban living.

Chapter 10

Innovative solutions for urban traffic issues, such as traffic congestion and parking shortages, are explored in this chapter. It introduces the concept of smart parking models powered by AI and blockchain technology, offering efficient space allocation and reducing search times. The technology enables predictive analytics for mobility planning and minimizes the environmental impact, all while preserving user privacy and security.

Chapter 11

The aviation industry is embracing machine learning and deep learning techniques to enhance the safety, efficiency, and performance of small aircraft. This chapter delves into the advantages, challenges, and future prospects of implementing intelligent systems in small aircraft. It covers autopilot systems, navigation assistance, fault detection, and pilot support systems, showcasing real-world case studies and applications.

Chapter 12

Machine learning is revolutionizing healthcare, offering solutions to complex challenges. This chapter explores practical strategies, ethical considerations, and real-world applications of machine learning in healthcare. It emphasizes data management, model development, interdisciplinary collaboration, and ethical dimensions. The chapter concludes by examining emerging trends and the growing role of AI in drug discovery and genomic medicine.

Chapter 13

Structural damage detection is vital for the safety of civil buildings, especially in the face of gradual deterioration and seismic events. This chapter highlights the significance of unsupervised learning methods for structural health monitoring. It reviews the progress in this field and identifies areas where improvements can be made for more accurate and effective structural damage detection.

Chapter 14

The drug discovery and design process is undergoing a significant transformation, leveraging data science, artificial intelligence, green chemistry, and affordability. This chapter explores the integration of these elements in pharmaceutical research, enabling the rapid analysis of datasets, predicting molecular interactions, and optimizing drug candidates. The goal is to develop affordable and accessible medicines that are both efficacious and sustainable.

Chapter 15

Modern agriculture faces challenges like rising global food demand and water scarcity. This chapter investigates how intelligent machines, the Internet of Things (IoT), and artificial intelligence (AI) are transforming agricultural water processing. It explores key concepts, applications, and the advantages of this approach. By leveraging AI and IoT-enabled systems, farmers can make data-driven decisions, automate irrigation, and adapt to changing environmental conditions for more efficient and sustainable agriculture.

Chapter 16

Visual object detection is a fundamental aspect of computer vision. This chapter explores the use of advanced YOLO models for detecting objects in images of varying object sizes. By employing ConvNeXt and YOLOv7, the chapter enhances object detection accuracy and efficiency, ultimately achieving a precision of 98.9%.

Chapter 17

Detecting the freshness of fruits is crucial for quality control in food industries and markets. This chapter focuses on using YOLOv6, YOLOv7, and YOLOv8 to classify fruit images based on their freshness. These models improve classification efficiency and accuracy, providing an innovative solution for food quality assessment.

As editors, we have strived to compile a diverse and insightful collection of contributions from experts in the field. We believe that this handbook will serve as a valuable resource for those seeking to explore, understand, and innovate in the dynamic world of AI and ML for intelligent machines and systems.

Brij B. Gupta
Asia University, Taichung, Taiwan & Lebanese American University, Beirut, Lebanon

Francesco Colace
University of Salerno, Italy

Acknowledgement

Many people have contributed greatly to this book on "Handbook of Research on AI and ML for Intelligent Machines and Systems". We, the editors, would like to acknowledge all of them for their valuable help and generous ideas in improving the quality of this Handbook. With our feelings of gratitude, we would like to introduce them in turn. The first mention is the authors and reviewers of each chapter of this Handbook. Without their outstanding expertise, constructive reviews and devoted effort, this comprehensive book would become something without contents. The second mention is the IGI Global publisher staffs for their constant encouragement, continuous assistance, and untiring support. Without their technical support, this Handbook would not be completed. The third mention is the editor's family for being the source of continuous love, unconditional support, and prayers not only for this work, but throughout our life. Last but far from least, we express our heartfelt thanks to the Almighty for bestowing over us the courage to face the complexities of life and complete this work.

Brij B. Gupta

Francesco Colace

Chapter 1
Conceptualising the Role of Intellectual Property and Ethical Behaviour in Artificial Intelligence

Princy Pappachan
Asia University, Taiwan

Sreerakuvandana
Jain University, India

Mosiur Rahaman
Asia University, Taiwan

ABSTRACT

The development of artificial intelligence has significantly affected all facets of human life, leading to many ethical and legal questions that have positive and negative consequences for individuals, organisations, and society. Amidst this lies the interaction between intellectual property rights and ethical behaviour in the development and use of artificial intelligence. While intellectual property acts as a catalyst for innovation, ethical behaviour ensures responsible and accountable artificial intelligence consistent with social standards. Accordingly, the beneficial effects of artificial intelligence can only be guaranteed by balancing intellectual property rights and ethical behaviour. This chapter thus discusses the notion of intellectual property and ethical behaviour in the context of artificial intelligence by providing a comprehensive historical review and reflecting on the creation and implementation of ethical artificial intelligence while preserving intellectual property rights.

DOI: 10.4018/978-1-6684-9999-3.ch001

INTRODUCTION

Artificial Intelligence or AI, even today, continues to emerge as a technology for general purposes, with its applications expanding throughout society. From self-driving cars to generative chatbots, artificial intelligence has revolutionised how we interact with technology. The advent of generative artificial intelligence, however poses a two-edged sword challenge in which one side strives to promote scientific development and economic success while addressing public ethical concerns with issues on the intellectual property rights of citizens. Over the past few years, there has been an enormous rise in the importance of intellectual property law. Intellectual property law has gained new prominence as one of the most crucial aspects promoting innovation and economic growth due to digitisation, electronic records, and the advent of post-industrial information-based sectors. The comprehension of artificial intelligence is imperative to understand the importance of intellectual property rights and the need to consider ethical concerns in artificial intelligence.

Artificial intelligence moved away from computer programs or software with its ability to learn and replicate "human-level intelligence" (Jordan, 2019), as machine learning techniques allowed artificial intelligence systems to learn using deep, supervised, and unsupervised learning (Wang & Siau, 2019). Additionally, with the interface of machine learning and natural language processing, Generative Pre-Trained Transformer could read and generate text from vast training datasets.

With respect to intellectual property rights and ethical concerns, there are two kinds of artificial intelligence; Weak artificial intelligence and Strong artificial intelligence. Weak artificial intelligence recognises the need for humans to carry out tasks. On the other hand, strong artificial intelligence is equipped with strong reasoning and problem-solving skills, making its responses indistinguishable from human-generated responses (Bechmann & Bowker, 2019). At present, artificial intelligence has entered the arena of strong artificial intelligence. This expeditious revolutionary transition from weak artificial intelligence to strong artificial intelligence poses serious questions on intellectual property rights concerning authorship and data protection and ethical concerns concerning data privacy and transparency in artificial intelligence.

This chapter thus investigates the different types of intellectual property rights and ethical concerns in artificial intelligence. Reviewing the challenges of intellectual property rights and ethical concerns in artificial intelligence, it discusses the legal perspective, existing policies and current strategies adopted to address its effect on present and future artificial intelligence by presenting case studies of the artificial intelligence-driven healthcare sector.

The chapter argues that current strategies and policies are still lacking in addressing the protection of intellectual property rights. Though copyright ownership and patent authorship have been addressed in artificial intelligence, it still requires modification to address what happens in generative artificial intelligence. Additionally, with respect to ethical challenges and concerns, legal resolutions to address these issues fall short of guaranteeing ethical artificial intelligence. The chapter thus argues that the current policies that talk about ethical concerns have not been able to provide a comprehensive framework on how ethical concerns can be addressed.

1. UNDERSTANDING INTELLECTUAL PROPERTY

Intellectual Property (IP) rights are understood to be one of the most quintessential assets to any business, especially small business. The law governing intellectual property encompasses a wide range of topics, from books, and computer programmes to even the genetically modified versions of plants and animals (Baan, Allo & Patak, 2022). Essentially, human intelligence is the major source of the creation of intellectual property. These are the ideas of the human mind which are transformed to creations, manifestations of creativity like inventions, works that are literary and artistry, they also can be symbols and images used in any form of commerce ("What Is Intellectual Property?"). The key forms of intellectual property are patents, trademarks and copyrights; all of these share the characteristics of an asset's properties, enabling the rights of selling, buying and even licensing. There are also intellectual property laws that enable inventors and owners to protect their property from unauthorised access or use.

Safeguarding these property rights is imperative to foster economic growth. Besides, it will act as an incentive for technological innovation, thereby attracting investments that can create new job opportunities for the citizens. Given this stature that intellectual property holds, Intellectual Property Rights (IPR), was introduced, with its roots in Europe. Intellectual property rights are the legal rights conferred on the inventor or creator for a specific period of time (Singh, 2008). This means the inventor, the creator, or even the assignee gets an exclusive right to maximally utilise his/her invention or creation for a given period. It also recognises the need for intellectual labour associated with the invention or creation to be given due importance owing to the many opportunities that emanate from it. Intellectual property rights are a powerful tool that helps protect investments, time, money and mainly the effort put in by the inventor or creator of an intellectual property. This way, intellectual property rights directly help in the economic development of a country and promote healthy competition, thus encouraging industrial development and economic growth.

The many laws and administrative procedures surrounding intellectual property rights have been traced and identified to have their roots in Europe. The idea of granting patents to creators or inventors started in the fourteenth century, with England being technologically more advanced than any other European countries, attracting artisans from all over. Venice is generally regarded as the hub of an established intellectual property system, with laws and legal systems being made for the first time. Other countries eventually followed (Bainbridge, 2010). The very first Indian 1856 Patent Act, based on the British Patent System Patent, exceeds 150 years in existence.

1.1 Types of Intellectual Property

Traditionally, trademarks, industrial design and patents were the only ones protected under 'Industrial Property'. The term 'Intellectual Property' has a wider coverage. Intellectual property is typically classified into two main areas;

1.1.1. Industrial Property

Industrial Property is an area that covers the protection of distinctive signs, such as trademarks (trademarks distinguish goods and services of one undertaking from that of others) and geographical indications (where a good is characterised as belonging to one place and the characteristic of the good also can be attributed to its geographical location). Such a protection of distinctive signs is mandated to stimulate and ensure fair competition. It also helps customers make an informed choice between the various goods and services.

There is also yet another kind of industrial property right that ensures protection in areas such as stimulating innovation design and also in the creation of technology. This category captures what is known as patents, trade secrets and also industrial designs. A patent is usually awarded for an invention provided it satisfies the criteria of being a novel idea, of being non-obvious and has an industrial or commercial application. However, intellectual property rights can be transferred, gifted or even sold, just like any other property (New Delhi, 2005).

1.1.2. Copyright

The second kind of industrial property is the copyright policy and the rights related to it. Copyright covers the rights of authors of literary work and also artistic works such as books, music, paintings, films and even computer programs. They are protected by copyright for a minimum period of 50 years after the death of the creator or inventor. The sole intention behind copyrights and any rights related to it is to encourage creative work.

A strong system of intellectual property rights emerged when the pressure of globalisation and internationalisation started becoming intense. Until then, the need for intellectual property rights was not recognised. Pharmaceutical, chemical, electronic and information technology industries are the drivers of globalisation, resulting in a lot of investment into the Research and Development (R&D) sector. A majority of the industries also realised that mere trade secrets and safeguarding the trade secrets were not enough to protect technology. It became increasingly difficult to fully leverage the benefits that innovation created unless there was a set of uniform laws and rules for the protection of patents, copyright or even trademarks. This is how intellectual property rights became a significant constituent of the World Trade Organisation (WTO) (Watal, 2001).

1.2. Challenges of IP Protection in AI

Today, artificial intelligence is a prevalent tool in almost all domains, from the industrial sector to academics. Artificial intelligence involves simulating human intelligence processes by machines, specifically, computer systems. Some of the specific applications of artificial intelligence are Natural Language Processing, Expert Systems and Speech Recognition (Burns et al., 2023). Broadly, artificial intelligence is capable of performing tasks commonly associated with cognitive functions that humans do, such as identifying patterns, playing games and even interpreting speech. The system learns this by processing large amounts of data; it identifies patterns in the presented data so as to model its own decision-making. This branch of artificial intelligence and computer science that emphasises using data and algorithms to mimic the way humans learn, eventually increasing its accuracy, is known as Machine Learning. Machine learning is already playing a pivotal role in how organisations and services function. It excels in areas such as healthcare, finance, social media platforms and even academics. However, the steps needed to train the model and to deploy it will depend on the data available at hand.

Humans also have a role in supervising an artificial intelligence's learning process by reinforcing good and bad decisions. This kind of learning is known as supervised learning. Supervised learning requires input and output data during its training phase. The training data is labelled by a data scientist during the preparation stage, just before allowing the testing and training on it. Once the model learns the connection between the output and the input, the model learns to classify and even predict datasets it has never encountered before. On the other hand, artificial intelligence is also set to learn on its own without any kind of human interference and this is known as unsupervised learning; for example, allowing the artificial intelligence to play a game on its own over and over, until it figures out the rules of the game by its own. In unsupervised learning, the model runs on raw and unlabeled data. This is often used in the exploratory stage of training to understand the datasets better. In recent years, artificial intelligence has evolved significantly, with many advances seen in data storage, algorithms and computing power. In contrast to human labour, an artificial intelligence system does not need any incentive to perform or improve. All it needs is data.

Developments like these have also made artificial intelligence more sophisticated and powerful, thus leading to an increased usage of artificial intelligence in various industries such as entertainment, healthcare and finance. While this may bring numerous opportunities, it raises some significant questions regarding intellectual property. The intersection of artificial intelligence and intellectual property is a presumptuous one. Despite such advancements made to protect the creations of the human mind and its associated products, there is no regulation regulating artificial intelligence as an intellectual property object. Essentially, this means, artificial intelligence is not subject to any of the intellectual property laws. One of the key challenges here is providing intellectual property protection for artificial intelligence concerning ownership. As per the traditional intellectual property law, the creator or inventor of any work owns the associated intellectual property rights. However, with artificial intelligence-generated inventions, this line of citizenship has become blurred. For instance, when an artificial intelligence 'creates' something new, who gets to keep the patent rights? Is it the developer who developed the artificial intelligence system or the user who actually trained the system? Alternatively, could it be the artificial intelligence system itself?

The other major challenge here is the issue of biases and discrimination in artificial intelligence systems. When an artificial intelligence system is trained on biased data, the output of that data will be biased. In this case, some concerning questions include who takes responsibility for this biased output and who should be charged for any subsequent harm. One side of the debate insists that creators of artificial intelligence systems and the ones who provide the data for training should have greater control over the product and can claim the profit from their innovation. This is presumably so because creations such as these involve a significant amount of time and effort; creators should be rewarded for their efforts. The opponents from the other side opine that the existing laws are not adequately laid out to address the challenges posed by artificial intelligence-generated work. Their contention is that artificial intelligence models can also be developed from data drawn from the public domain, thereby making it difficult to figure out who is the actual owner of that data. Also, most of the time, the artificial intelligence-generated output is a result of multiple collaborators, making it challenging to identify the actual owner of the artificial intelligence-generated output or product (AI IP Protection Debate, n.d.). The third challenge is the kind of output an artificial intelligence model will produce. Since artificial intelligence models learn and evolve over time, it is possible that the same algorithm can produce a range of outputs. This will make it difficult to decide what out of this, should be protected under intellectual property law.

In order to efficiently address these concerns, some have proposed intellectual property protection laws specifically suited for artificial intelligence-generated work. "Machine Learning Patents" is the name suggested for the same. This will be awarded to artificial intelligence systems that can demonstrate any significant breakthrough.

Nonetheless, there are also potential benefits in providing intellectual property protection for artificial intelligence. It will help boost investment in the artificial intelligence research and development sector, provide incentives for the creation of more advanced and innovative systems within artificial intelligence. In order to effectively address some of these challenges, lawmakers and experts have suggested other alternative forms of intellectual property protection such as trade secrets. Yet another set of experts propose creating separate intellectual property rights that are specific for artificial intelligence-generated innovations. Intellectual property rights are usually claimed by a legal entity such as a company or a human. It is also not possible to ask artificial intelligence to protect its own invention as this may have consequences under the infringement of intellectual property. Therefore, the major question is, if the subject content was indeed generated by an artificial intelligence machine, who actually becomes the first owner of the intellectual property? If artificial intelligence can create a subject matter, it should also be held responsible in some situations. That is, it could be subjected to allegations of violation of copyright.

There is also a growing body of concern over artificial intelligence creating inventions completely against the betterment of humankind (*What Is Intellectual Property?*, n.d.) . In cases like these, where artificial intelligence users are responsible for caring for and overseeing the results and the outcomes it produces, they could be held liable. On the other hand, if an artificial intelligence machine functions autonomously, without the intervention of a human, or develops anything via self-learning, then the liability of producing anything wrong can fall upon artificial intelligence itself. In particular, the World Intellectual Property Organization (WIPO) highlights in its patent document policy the patentability guidelines. These patentability guidelines may or may not be modified to regulate artificial intelligence-based inventions. WIPO poses three unsolved questions surrounding the artificial intelligence-based invention property regulation. Firstly, if the artificial intelligence-based inventions are excluded from patent eligibility altogether. Secondly, should artificial intelligence-based inventions be regulated by whatever regulations that exist with respect to computer-assisted inventions? And lastly, should there be an addition of specific guidelines with respect to the regulation of artificial intelligence?

2. ETHICAL CONSIDERATIONS AND CHALLENGES IN AI

The technological advancements in artificial intelligence brought forth critical questions surrounding human-oriented aspects. The early beginnings of artificial intelligence (1940s-1950s) only catered around the potential consequences that may arise from mimicking the humane part (Rai, 2022). By the later years, 1980s-1990s, artificial intelligence witnessed its emergence of using datasets to answer questions and solve problems in specific domains of knowledge. Examples of such expert systems are Dendral developed in 1965, MYCIN developed in 1972, and XCON created in 1980 (Marquis, Papini, & Prade, 2020). These expert systems began to assist and replace human-decision decision-making processes in domains like medicine and law. This brought forth the initial ethical concerns that dealt with accountability, bias and errors. Artificial intelligence gradually evolved from simple rule-based expert systems to systems that relied on data-based machine learning (Slavina, 2023). The integration of machine learning algorithms into artificial intelligence helped analyse vast amounts of data, deduce patterns and predict results (Taricani & Saris, 2020). Accordingly, vast amounts of data were gathered to develop data sets, raising ethical concerns about how data is collected and processed and the bias within the algorithms is used to predict and forecast human behaviour.

In general, there is a confound of selection bias within artificial intelligence in the datasets it uses in order to create an algorithm. Buolamwini and Gebru (2018) demonstrated the artificial intelligence bias in facial recognition, resulting in decreased accuracy when having to recognise dark-skinned faces, particularly in women. The data sets for machine learning need to be large and also be the ones that are often used in clinical trials, they need to be derived from the majority of the population; failing which can lead to resulting algorithms being more biased and will fail to represent underserved and also the underrepresented population.

2.1. Ethical Framework and Principles in AI

Ethical framework is an attempt to build consensus surrounding the values and norms that can be safely adopted by a community. Many organisations are known to have participated in developing an artificial intelligence ethical framework. A recent study by Jobin et al. (2019) states that artificial intelligence ethics has now converged on five principles, namely, non-maleficence, responsibility and accountability, transparency and explainability, justice and fairness, and respect for privacy and security.

Richard O. Mason, in the year 1986, proposed four ethical issues of the information age; privacy being the primary one, followed by accuracy, property and accessibility. While these were not proposed under the context of artificial intelligence or data analytics, they were discussed in the light of information age, which involved the production of intellectual property. He proposes that information systems should not invade privacy, must protect intellectual property, be accurate and be accessible for all (Mason, 1986). Mason's theory is understood to be the basic one, but still holds the elementary structure required that IT systems should follow. The frameworks that are available for artificial intelligence at present intend to identify the ethical challenges and also go on to suggest some remedies in order to mitigate the risks. These frameworks provide a ground for discussing the concepts of the ethical aspects of artificial intelligence and their impact. Additionally, it also lists the potential principles and concerns and provides remedies for addressing the concerns. These remedies are more like recommendations on how to best design an artificial intelligence system. On the other hand, the conceptual angle focuses on describing the concepts that underlie these ethical principles. These principles are stated in the form of desirable properties of an artificial intelligence system, like the transparency of these systems, and the privacy of data used in the development of artificial intelligence. Morely et al. (2020) consider the principle of explicability as one of the requirements for understanding the artificial intelligence system (Morley et al., 2020). The concern, on the other hand is that artificial intelligence systems should behave in a transparent manner and there should be accountability. Traditionally, principles are identified as critical instruments for defining ethicality. They are intended to guide a person on developing and using artificial intelligence.

The ethics of artificial intelligence is primarily concerned with the "concerns" surrounding it. Privacy and surveillance have been a general concern in the information age (Macnish 2017; Roessler 2017). It mainly refers to access to private data and any other data that is identifiable personally. Some areas of artificial intelligence ethics are:

a. AI and privacy: Artificial intelligence relies heavily on information to learn, and a significant portion of this information comes from users. The privacy concern here is that the users are unaware of the information being gathered from them. The users are also not aware of how this information is being used to make decisions that may or may not affect them. Internet searches and online purchases are tracked efficiently to provide a personalised experience for users. While on one hand, this can be positive, it can also have negative consequences such as unexpected bias in offers being provided to some customers and not others.

b. Avoiding AI bias: Poorly constructed AIs can demonstrate bias against data that is poorly demonstrated. Since AIs learn from data, it is necessary that artificial intelligence data represents equality, devoid of any sort of bias such as bias against minorities and underrepresented groups.
c. Addressing AI environmental impact: Since the artificial intelligence models are expanding and getting larger daily, these large models take significant energy to train. Therefore, researchers, today are developing techniques to keep artificial intelligence models energy efficient so as to balance the performance and also to keep it energy efficient.

Some of the pressing issues within the ethical framework are outlined below.

2.1.1. Bias and Discrimination

Like discussed above, artificial intelligence systems are trained on a massive amount of data and often embedded in these data is some amount of societal bias. Subsequently, these biases will become ingrained in the artificial intelligence algorithms, perpetuating unfair or discriminatory results in areas such as hiring, medicine, resource allocation, etc. Even without human prompting, these artificial intelligence systems can generate stereotyped and racist content on their own, without being able to charge anyone or anything in particular. While developers are doing their best to eliminate such biases, the task is incredibly nuanced as datasets include a vast amount of images and words.

2.1.2. Ownership and Creativity

AI-generated art is very prevalent now. When a human generates a piece of art simply with the help of text prompts into an artificial intelligence system and when the system generates an art as per the prompt, then who owns this generated artificial intelligence-generated art is the major question. There is also no clear consensus on who can commercialise it and who may be possibly be at the risk of infringement. Therefore, this area needs lawmakers who can clarify ownership rights and also possibly provide some guidelines so as to navigate any potential infringements.

2.1.3. Privacy and Security

The effectiveness of artificial intelligence solely relies on the availability of large volumes of personal data. With the expansion in the usage of artificial intelligence, concerns primarily revolve around how this information is collected and stored.

2.1.4. Misinformation and Manipulation

In competitive fields like business and politics, fake news, false information and manipulation of information is common. Artificial intelligence algorithms can be exploited to to manipulate any public opinion and therefore amplify social division.

2.1.5. Transparency

Transparency and accuracy is another concerning area. Artificial intelligence systems are often known to operate in a "black box", where there is limited interpretability of how they work and how do they arrive at certain kinds of decisions. Transparency becomes extremely vital in areas like medicine, autonomous vehicles so that we can ascertain who bears responsibility for the decisions made. Clarifying accountability also becomes important, especially when artificial intelligence systems make an error or pose any harmful output. Thus, clarifying will ensure some appropriate measures be taken into account so that it can be combatted with corrective measures. Combatting black box challanges is one area that researchers are working in by way of developing explainable artificial intelligence that will help to accurately characterise the fairness of the model, its accuracy and also the potential bias it holds.

2.2. Ethical Challenges and Concerns in AI Development and Deployment

One of the ways to minimise or combat such biases is to involve diverse teams when developing and testing the algorithms, ranging from varied ethnicity, gender and even socioeconomic status to educational background with respect to knowledge, beliefs and morals. This may help make biases a little more less obvious. In addition, developers also need to promise the security and fairness of these artificial intelligence systems through creating a sense of accountability. The European Union's General Data Protection Regulation (GDPR) provides for the legal consequences of non-compliance (Martinet, 2018). This mandates that businesses put all the necessary sfaeguards in place so as to ensure the transparency of artificial intelligence algorithms. Preventing biases requires commitment to fairness, accountability and transparency throughout the development of artificial intelligence and in its deployment process. Data ethics is a critical element of a responsible artificial intelligence deployment and its development. Unintended consequences seem to arise when artificial intelligence is deployed without robust governance and compliance efforts. These issues fall under three categories (Ladak, 2023); compliance and governance, brand damage and third-party transparency. Compliance and governance, as discussed above, refer to the biases the system might exhibit when training data favours one gender over another, which falls under discriminatory law.

Brand damage refers to the breaching of social norms and taboos as in the case of Microsoft's Tay chatbot which was trained to learn from the Twitter conversation of users. When the users began to use inflammatory and racist language, the chatbot learnt and repeated the same. Microsoft had to shut it down the very next day (Hunt, 2016). Quite similarly, Amazon came under scrutiny because the concept of same day delivery was provided only to the white population and to affluent areas of US. The reason was because of the available data about the concentration of Prime members and their closeness to the Amazon warehouses. The algorithm deployed by Amazon figured out that it may not be profitable where there was a concentration of ethnic minority (Gralla, 2016).

There also a range of pitfalls to be considered when deploying artificial intelligence. A lack of technical understanding is the fundamental of all the pitfalls. There is a general lack of data scientists who are adept in the areas of IT and given the complexity of artificial intelligence, an improper and irresponsible use of artificial intelligence may hamper development in these areas (Brown, 2019). Using artificial intelligence outside its scope is also a concern that can result in unexpected results. This may arise because the programmers who are involved here may not have accounted for the variables that lie outside the actual focus. Besides any of these, companies and their employees may also not raise an alarm when there are concerns about artificial intelligence models. These issues thus go unreported. These can potentially perpetuate into outputs or results the artificial intelligence produces. The negative effects of artificial intelligence systems can have negative repercussions on society such as bias and increasing inequality. Even within enterprises, questions around ethical artificial intelligence are considered important. Especially relevant to this is how the artificial intelligence models perform. Addressing questions like are the models fair and are they negative in any sense. Noting these biases can tell us if the inaccuracy stems from a faulty design or data selection method.

3. LEGAL PERSPECTIVES, POLICIES, AND CURRENT STRATEGIES OF IP IN AI

The revolutionary change in Open artificial intelligence has posed various challenges for intellectual property and ethical considerations, precisely questions related to data ownership, privacy, responsible use of personal information, accountability, and non-discrimination. The advent of information technology, specifically computer software, witnessed the enforcement of several intellectual property laws related to copyrights, patents, designs, and trademarks. However, the emergence of artificial intelligence brought out new challenges. Subsequently, intellectual property laws had to additionally consider artificial intelligence algorithms and their interface with ethical implications such as fairness, accountability, and transparency.

To understand intellectual property in artificial intelligence, we first try to understand the initial formulation of intellectual property and then the latter progression of intellectual property law as it entered the digital age. The following sections will address the current intellectual property laws and strategies for upholding intellectual property rights.

3.1. IP Laws and Regulations in AI

Artificial intelligence's ongoing revolutionary development and sophistication of technologies force us to reconsider human-centred intellectual property laws and principles (Moerland, 2022). The fundamental question in the interface of intellectual property and artificial intelligence lies in what can come under the umbrella of intellectual property protection. Historically, intellectual property laws have always been centred on the notion of an "individual" or "human achievement" (Lauber-Rönsberg & Hetmank, 2019). However, the advanced integration of artificial technology into everyday life has blurred the basic notion of an "individual", leading to existing concepts of intellectual property laws becoming increasingly antiquated. So current intellectual property law and regulations in artificial intelligence require a discussion of its predecessor laws which are discussed below.

The existence of intellectual property emerged from the 1883 Paris Convention for the Protection of Industrial Property and the 1886 Berne Convention for the Protection of Literary and Artistic Property. However, the origin of laws protecting inventions and inventors dates back to the Renaissance period- Renaissance Italy in 1474 when the first patent statute "the Venetian patent statute" was issued (Nard, 2019). The patent statute aimed to facilitate technological advancements via issuing permits and recognising a minimum period for patent rights (Kim, Jeong, Kim, Kim, Jeong, & Kim, 2021). When translated to English, the statute originally written in Venetian reads that anyone who has constructed a novel or innovative gadget that has not already been created in the Commonwealth must notify the

General Welfare Board. Also, once it is approved to be operated, no one else can create the same or anything similar to it without the inventor's permission for ten years.

Similarly, the notion of "copyright" can also be traced back to Renaissance Italy, which was in the form of 'printing monopolies' in 1469 (Ginsburg, 2018). The later years saw the extension of the patent and copyright laws across Europe and England. In 1624 the first English patent law, the Statute of Monopolies', was issued under the Britan Crown's royal authority (Seville, 2017). These patent and copyright laws led to several regional and national laws. The United States of America got its first Patent Act in 1790 which was signed by George Washington (Nard, 2019). The latter years saw the emergence of international discussions on intellectual property, which led to the 1883 and 1886 conventions in Europe.

The 1883 Paris Convention ensured that intellectual property with respect to inventions (patents, trademarks, and design) would be protected irrespective of the country, while the 1886 Berne Convention protected the rights of artists (writers, musicians, painters, poets, and other creative geniuses). These two conventions led to the creation of an 'International Bureau' that heralded the formation of the World Intellectual Property Organisation (WIPO) in 1970 (Marsoof, Kariyawasam, & Talagala, 2022). The mission statement of WIPO reads, "to lead the development of a balanced and effective international IP system that enables innovation and creativity for the benefit of all" and to be a "global forum for IP services, policy, information, and cooperation" (Ćemalović, 2022). Subsequently, the WIPO joined the United Nations in 1974 and became a specialised agency of the United Nations (White, 2002).

The later years witnessed rapid changes in intellectual property laws which had to keep up with the innovations and developments in thought, commerce, and technology. The Uruguay Round negotiations (1986-1994) delved into the interaction of intellectual property rights with international trade, namely the General Agreement on Tariffs and Trade (GATT) (Sulistianingsih & Ilyasa, 2022). The final negotiations during the Uruguay Round (known as the Marrakesh Agreement) culminated in the establishment of the World Trade Organisation (WTO) and the Trade Related Aspects of Intellectual Property Rights (TRIPS) in 1995 (Jelisavac Trošić, 2022). The TRIPS signed on April 15, 1994, became effective the following year on January 1 1995. Compared with WIPO, an agency of UN aimed to promote the creation of intellectual property rights, the TRIPS agreement aims to protect the different aspects of intellectual property rights among the members of WTO.

The TRIPS preamble addressed the shortcomings of intellectual property rights protection that were either excessive or insufficient and its interface on international trade. The TRIPS thus set out to establish a fair and reliable system for intellectual property protection on a global scale by setting standards, period of enforcement and dispute resolution.

Additionally, TRIPS Agreement promised that the increased intellectual property protection would eventually benefit lower-income countries because of the transfer of technology from higher-income countries. According to the initial TRIPS agreement, least-developed countries had until 2006 (later extended to 2013 with a possible extension to 2034), while developing countries had only five years to implement all the intellectual property provisions in the TRIPS agreement (Du, 2022). This agreement thus marked the emergence of globalisation in intellectual property, which is reflected in the first section of the TRIPS agreement.

The late 20th century thus began to be dominated by the WIPO, WTO, and TRIPS. Post the TRIPS agreement, the WIPO adopted the World Intellectual Property Organisation Copyright Treaty (WIPO CT) in 1996 to deal with copyright law to address the arrival of the digital age. Additionally, the WIPO CT addresses computer software licenses and establishes economic regulations to regulate product licensing (patent) among all member states. Over the subsequent years, WIPO and WTO have undergone continuous revisions and amendments to keep up with the artificial intelligence revolution. Furthermore, the advent of generative artificial intelligence has heightened the need to look into copyright laws from a new perspective that is steps ahead of technology as computer programs/softwares.

3.2. Current Strategies of IP in AI

With the advent of the artificial intelligence revolution, one of the predominant areas of development lies in the interaction that exists between intellectual property rights and artificial intelligence. Specifically, three areas of intellectual property have come under immense scrutiny in artificial intelligence: copyright, patents, and international trade.

3.2.1. Copyright (AI Authorship in Copyright Law)

Copyright as intellectual property has always focussed on protecting the invention and inventor to ensure innovation and creativity. With the interaction of artificial intelligence in commercial and technological spheres, the boundaries of 'authorship' in copyright law are becoming increasingly blurrier. Before the advent of artificial intelligence technologies, copyright law was pretty straightforward: the creator or inventor is the author, and the creation or invention is the author's work, and both must be protected. Later, during the development of software/programs, even though there were computer-generated works, authorship in terms of copyright law was not ambiguous as the computer software/program was seen as a tool that was used by the author in his/her creative process.

According to the 1941 Italian Copyright Law, "computer programs shall be further protected as literary works," guaranteeing copyright law rights to the author of the software (Farina, 2023). Similarly, the United States Copyright Act of 1980 (Oman, 2017), the 1985 Copyright Legislation of Japan (Miyakawa, 2000) and the United Kingdom Copyright Designs and Patents Act (CDPA) of 1988 added "computer program" as part of "literary work" (Legislation.gov.uk., n.d.). These copyright laws were all built on the premise that copyright law can only exist for works that are the product of the author's or creator's intellect. These computer-generated works gradually gave way to more sophisticated programs that required lesser human intervention. In this regard, the CDPA additionally states that the author of a computer-generated work shall be the person who has made the necessary arrangements to undertake the work when there is no human author for the created work (Legislation.gov.uk., n.d.). Such a clause ensured that the computer-generated works are still protected even if the computer takes over a greater part of the creativity. However, the terminology 'necessary arrangements' is ambiguous and requires clarity which the CDPA fails to provide. Yet, this clause that protects the computer program's programmer under copyright law has been widely adopted in several countries.

Furthermore, this copyright law of United Kingdom also protects works generated by a computer without a human creator (GOV.UK, 2022). Simultaneously, in the European Union, the main legal framework on copyright as intellectual property in artificial intelligence is the Copyright Directive 2019, amended on the Copyright and Information Directive 2001 in European Union law (Stamatoudi & Torremans, 2021). According to this directive, authentic literary/artistic works and computer-generated works are protected by copyright law if they fulfil the requirements of creativity and originality. However, the 2019 Copyright Directive fails to address who owns the copyright ownership in the case of artificial intelligence-generated works.

In the subsequent years, the advent of artificial intelligence gave rise to new questions as artificial intelligence began to create works (artistic) without direct human intervention. The created works were no longer a product of a tool in the hands of humans rather, it was a product of the machine's intellect via machine learning, which prompted the question, 'Can artificial intelligence be an author'? Since the very foundation of copyright law was to encourage and reward human innovation and creativity, the protection of literary and artistic works resulting from supervised machine learning seemed dubious.

Recently, the European Union has emerged as the pioneering agency to deal with the artificial intelligence revolution on who owns literary/artistic works created by artificial intelligence (Zhuk, 2023). It proposed a new recommendation in 2019, publishing a report, "Intellectual property rights for the development of artificial intelligence technologies" to attribute the copyright status of artificial intelligence-generated work to a non-human entity (European Parliament, 2020). Following this, the European Parliament adopted a resolution in 2020 on the intellectual property rights for the development of artificial intelligence technologies. The new resolution stressed the importance of developing a legal system that considers the main attribute of artificial intelligence-generated works, which is the lack of a human author, and establishes specific guidelines for accountability and ownership. The European Union additionally enforced the Artificial Intelligence Act (AIA) in 2021 to regulate the creation and application of artificial intelligence to protect intellectual property rights (Madiega, 2021).

In response to the artificial intelligence revolution, the United Kingdom Intellectual Property Office (UK IPO) initiated a consultation on copyright protection, stating, "Copyright protection for computer-generated works without a human author in 2021. These are currently protected in the UK for 50 years. But should they be protected at all, and if so, how should they be protected?" The consultation lasted for ten weeks ending on January 7, 2022 (GOV.UK, 2022). According to the response, most respondents favoured no changes in the present law, with many stating that artificial intelligence is still in its early stages, so it does not warranty any amendment (Kretschmer, Meletti, & Porangaba, 2022). Considering the recent revolutionary development of Generative artificial intelligence, it is no longer valid to say that artificial intelligence is in its early stages.

The policy statement issued by the United States Copyright Office that came into effect on March 16, 2023, is the latest development in establishing clear guidelines on materials generated by Generative artificial intelligence (Perlmutter, 2023). So with reference to the question of whether material produced by Generative artificial intelligence is protected under copyright versus human authorship, works produced by a machine or through a mechanical process that is operated automatically or randomly without any involvement from a human author will not be registered (U.S. Copyright Office, 2023). The new policy statement also provides a case-by-case inquiry to understand whether the artificial intelligence-generated work reflects the authors' intellect or is a mechanical production. This latest policy provides a brilliant blueprint to ensure copyright law in the interface of intellectual property rights and revolutionary Generative artificial intelligence.

To address artificial intelligence authorship in copyright law, one elemental step could be distinguishing between artificial intelligence human input-generated creation and artificial intelligence-generated content that uses Generative artificial intelligence.

3.2.2. Patent (AI Inventorship in Patent Law)

The Patent law was founded on the principle that "only human beings can be inventors" (Dornis, 2020), which was perfectly sound before the advent of artificial intelligence. The Patents Act 1977 also deemed an invention patentable only if it is new, inventive, and has industrial application. Furthermore, it prohibits "a program for a computer" to be patentable (Patents Act 1977, p:2). Unlike the ambiguity that arose with artificial intelligence in authorship in copyright law, the patent law continued to be robust in its definition of 'human' as 'inventor'. Patent laws also consistently do not recognise non-humans as inventors (Pearlman, 2017).

According to the United Kingdom Intellectual Property Office, the United States Patent and Trademark Office (USPTO), and European Patent Office (EPO), the inventor needs to be listed as a 'person' and cannot be identified as an 'AI inventor' (Discher & Rutigliano, 2021). The USPTO also focuses on the notion of 'conception' as a crucial factor in deciding inventorship since it reflects "the formation in the mind of the inventor of a definite and permanent idea of the complete and operative invention" (Lim & Li, 2022). An example of this can be seen in a judgement ruling at the German Federal Patent Court (Bundespatentgericht) where it initially rejected a patent application that designed artificial intelligence as the inventor but later designated the applicant as the inventor of the patent allowing him to be the owner of the patent (Kim, 2022). This ruling can be seen to be paradoxical as the ruling granted patent ownership to someone who played no role in the invention. At the same time, it attempted to highlight the importance of 'conception/creativity' behind patent ownership. Yet, the ruling enforces the fact that the inventor of a patent can only be a natural person with a legal personality.

However, with the advancement in generative artificial intelligence, there is an urgent need to adapt and recognise artificial intelligence systems as inventors, which requires modifying the existing rules of patent laws or developing new classes in intellectual property rights that acknowledge the contribution of artificial intelligence as inventors. An example of such a recommendation on artificial inventorship in patent law was proposed by the research and policy project of the Zurich University's Center for Intellectual Property and Competition Law along with the Swiss Intellectual Property Institute. They stated that patent applications should provide options that nominate artificial systems as 'inventors' and humans as "proxy inventors" (Picht & Thouvenin, 2023). The proxy inventors can thus claim the rights as owners of the patent application and subsequent patent.

So far, in the review and consultation process on patent laws discussion (GOV.UK, 2021), many have proposed including artificial intelligence as co-inventors. In contrast, others have proposed developing an entirely new patent law specifying regulations for pure artificial intelligence-generated creations.

3.2.3. International Trade (AI Affecting IP Data Transfer and Protection)

In light of the significant boost in artificial intelligence, a vast amount of input data is required for training, whether machine learning or deep learning. The type of required data can either be supervised or unsupervised. Since quantity plays a huge role in training artificial intelligence, particular focus must be given to free data flow while ensuring data privacy (Meltzer, 2018). Accordingly, a significant change was observed in digital trade rules with the creation of the Comprehensive and Progressive Agreement for Trans-Public Partnership (CPTPP) in 2018. The CPTPP updated the existing WTO rules by safeguarding data flow (to ensure privacy "protection of personal information") and banning data localisation that restricts the free flow of data globally (Meltzer, 2018). Through the agreement, data localisation ensures sufficient training data is available for artificial intelligence as digital technologies heavily rely on the amount of training data (Lee-Makiyama, 2012).

The CPTPP is similar to the 2017 free-trade agreement between Canada and Europe. The Canada-European Union Comprehensive Economic and Trade Agreement (CETA) deals with the digital economy.

According to CETA, there will be no charges for digitally encoded deliveries like computer programs, text, video, image, or sound recordings that can be used in the context of artificial intelligence (Finck, 2020).

Another recent agreement that looks into data protection and privacy is the 2020 United States-Mexico-Canada Agreement (USMCA) which has been built on the framework of the 1994 North American Free Trade Agreement (NAFTA). The Digital Trade chapter in the USMCA focuses on protecting personal information and the digital cross-border transfer of information (Ciuriak, Dan, and Robert Fay, 2021). The agreement prohibits the government from requiring businesses/firms to reveal the source course that underlies the software and algorithms that drive artificial intelligence systems (BSA The Software Alliance, 2020). Additionally, it states the need to safeguard users' personal information to raise customer trust in digital trade while permitting unrestricted data transfer across borders (Leblond & Aaronson, 2019). However, much work still needs to be done to provide standards that ensure data protection during the transfer and sharing of digital information. Furthermore, with respect to intellectual property rights, Chapter 20 in the USMCA states that the "promotion of technological innovation" and "transfer and dissemination of technology" should benefit "producers and users of technological knowledge in a manner conducive to social and economic welfare, and to a balance of rights and obligations" (Office of the United States Trade Representative, n.d.).

At present, the effect of artificial intelligence on data privacy in intellectual property protection is most emphasised in the European Union law through the 2018 General Data Protection Regulation (GDPR) that established guidelines for "processing, storing, managing data" within the European Union (Li, Yu, & He, 2019). The GDPR was formulated to replace the 1995 Data Protection Directive (Voigt & Von dem Bussche, 2017). The European Union is observed to have the best robust framework regarding data protection with respect to privacy and transfer of technology by enforcing strict compliance to accountability.

Accordingly, data protection has become one of the most important aspects in intellectual property with the development of artificial intelligence. However, considering the drastic speed in the development of artificial intelligence at present, there is immense pressure for all countries and corporations to devote more resources to develop strict regulations to protect data as intellectual property.

4. LEGAL PERSPECTIVES, POLICIES, AND CURRENT STRATEGIES TO PROMOTE ETHICAL AI

The rapid increase in the applications of technology in society and the increase in the number of users escalates the range of ethical issues and concerns. Accordingly, although in its early stages, generative artificial intelligence presents a myriad of challenges ranging from vulnerability in security, data privacy and possible prejudices in artificial intelligence algorithms. To ensure that everyone benefits from artificial intelligence positively, regulating frameworks must be formulated that address the accountability of artificial intelligence through transparency and verifiability. Such a framework would also hinder the development of unnecessary bias and protect everyone's right equally.

The following sections discuss the laws and regulations enforced to promote ethical artificial intelligence and the current strategies for developing responsible artificial intelligence.

4.1. Ethical Policies and Current Strategies to Promote Ethical AI

A practical theoretical framework to incorporate ethical concerns in artificial intelligence is incorporating all criteria that are necessary for surviving in society into the computer algorithm- "responsibility, transparency, auditability, incorruptibility, predictability", and human sensibility" (Yudkowsky & Bostrom, 2011, p. 3). However, this is easier said than done.

One of the fundamental ethical concerns in artificial intelligence is privacy and data protection. Even before the development of technology, the right to privacy was an important clause in the 1950 European Convention of Human Rights (Cameron, 2021). Following the development of technology and the invention of the internet, the European Union Data Protection Directive was established in 1995 with security standards and data protection. Later in 2011, considering the progression of the internet and the development of artificial intelligence, the 1995 European Union Data Protection Directive was modified into the GDPR to become the world's strictest privacy and security law (European Council and European Council, n.d.).

The General Data Protection Regulation allows individuals to access their personal data and restricts what different firms/organisations can do with the data. Additional principles reflected in the General Data Protection Regulation are "lawfulness, fairness, and transparency; purpose limitation; data minimisation; accuracy; storage limitation; integrity and confidentiality; and accountability" (Information Commissioner's Office (ico), n.d.).

Another important ethical concern in generative artificial intelligence is the content of training data. Generative artificial intelligence requires vast training data, making the training dataset crucial in the resulting generated response. Since artificial intelligence is developed by people, it is vulnerable to subjective bias. By unknowingly uploading demographic or society-specific training datasets, artificial intelligence generates questionable biased responses as answers (DG, 2020). So, artificial intelligence developers must identify and address these biases to prevent harm to certain demographics from becoming a national security issue. Additional strategies could include employing a human-centred artificial intelligence framework and incorporating ethical values into algorithms. But this process will take time and cannot be rushed by enforcing it through legal regulations.

Another initiative that addresses the ethical concerns in artificial intelligence is the 2016 IEEE Global Initiative on Ethics of Autonomous Systems (Chatila & Havens, 2019), which strives to prioritise ethical concerns so that technology can be developed in such a way that it will be beneficial for all.

Above mentioned human-centric approach has been adopted in the Artificial Intelligence Act (AIA) proposed by the European Commission, which came into effect in March 2023. Based on a risk-based approach, the AIA endorsed obligations and compliance rules that providers and users must comply with, depending on the risk category of the artificial intelligence application. However, the AIA failed to clarify what constitutes ethical artificial intelligence. A significant drawback of AIA is its importance to speed, which might inversely affect ethical concerns like privacy and fairness.

An additional concern in the interaction of intellectual property rights and ethical challenges is the transfer of cross-border data (data flow) and data localisation in international trade, which necessitates data privacy assurance. It also calls into question copyright law in terms of infringement of copyright.

At present, the best option for ensuring the promotion of ethical concerns in generative artificial intelligence is transparency and accountability. Due to the lack of transparency, understanding the complex algorithms and how artificial intelligence-powered systems generate specific responses becomes difficult. Accordingly, there is a need to convert generative artificial intelligence from a 'black box' (Adadi

& Berrada, 2018) to a 'glass box' (Rai, 2020), which will help to align generative artificial intelligence with ethical values of fairness and transparency.

5. BALANCING IP RIGHTS AND ETHICAL CONSIDERATIONS IN AI

Balancing IP rights and ethical considerations becomes crucial in the present context of the interplay of artificial intelligence in various sectors. This is particularly true as machine learning algorithms bounded by intellectual property rights require vast data sets that conflict with ethical concerns like bias, privacy concerns, and data protection. As artificial intelligence machine learning algorithms evolved, industries and organisations enforced intellectual property rights to protect ownership and to promote innovation. Accordingly, artificial intelligence systems began to be protected via copyright, patents, and trade secrets frameworks. Out of these three frameworks, trade secrets have become more popular to protect artificial intelligence systems and algorithms. However, with the incorporation of algorithms into several decision-making processes that influence everyday life, questions about the transparency and accountability of algorithms come into the picture.

Foss-Solbrekk, K. (2021) discusses how artificial intelligence algorithms assist in decisions related to credit scoring, loan applications, drug discovery, fraud detection, hiring, hospital admissions, courtroom decisions, and many more. The speed-effectiveness of artificial intelligence coupled with cost-effectiveness and accuracy in predicting human behaviour and making decisions is possible only through pattern recognition with vast amounts of data. Artificial intelligence-powered machine learning algorithms can thus extract information from data and build information that affects life. This prompted the EU GDPR to ban automated decisions that will affect an individual legally, and when such decisions are allowed to happen, the algorithm should be transparent, and additionally, individuals must be given the right to obtain human intervention (La Diega, 2018).

5.1. Algorithms Bias vs. Intellectual Property Protection and Trade Secrets

A main problem regarding automated decision-making processs is the presence of algorithm bias. Though it is attested that non-human agents are unbiased in their decision-making and that their predictivity ability arises from vast amounts of data, the algorithm is itself created by humans who are bound by biases as human beings are not devoid of any ideologies or passions. Also, the data used for machine learning can be inconsistent and discriminatory in nature. The only answer possible to address bias in algorithm and data is 'transparency'. However, this is often impossible as algorithms are protected by private entities under intellectual property rights. This begs the question of what can be done to ensure unbiased predictions in the face of biased algorithms and data.

5.2. Privacy vs. Data Use

A recent development addressing the transparency of algorithms and data protection is the remedy laid down by the Charter of Fundamental Rights of the European Union. The integrated approach, considering intellectual property rights, freedom of information, and data protection, states that every person has the right to protect his/her personal data and have complete access to their data. Additionally, all data used for any purpose must be with the consent of the individual or any other legal justification. Other suggestions that have been suggested in the intellectual property protection in artificial intelligence is the creation of a new category that will deal with machine learning algorithms or incorporating blockchain technology. With the continuous advancements in generative artificial intelligence, the debate on protecting artificial intelligence systems will only continue.

The Charter of Fundamental Rights of the European Union also addresses the interplay of ethical concerns dealing with data protection against intellectual property right to use data required for machine-learning algorithms. By granting individuals the right to protect their own data and advocating for transparency, organisations can be held responsible for data breaches. The recent California Consumer Privacy Act (CCPA), Consumer Online Privacy Act (COPRA), SAFE Data Act, China's Cybersecurity Law, and the GDPR are some of the measures that different governments have taken to ensure citizen's privacy of data. Ethical, responsible artificial intelligence can thus be recognised by ensuring that the vast amount of collected data is encrypted and sensitive data is secured.

European Union's ban on artificial intelligence-powered facial recognition in public spaces is a big step forward in ensuring individual privacy. Artificial intelligence is expected to come into effect by 2025 following approvals from the European Commission, the European Union of Ministers, and the European Parliament. In Europe, Spain has edged closer to taking decisions against violators of General Data Protection Regulations. The breach of privacy by the Mobile World Congress in 2021 using facial recognition technology was met with a huge fine imposed by the Agencia Española de Protección de Datos (AEPD).

5.3. Job Displacement vs. Innovation

The development of more innovative artificial intelligence systems has begged the question of its impact on job displacement. In this regard, concerns about where intelligent systems will replace individuals are significant. With machine learning, the belief in the possibility of replacing routine and repetitive jobs with artificial intelligent powered systems is very strong. However, history teaches us that with technological advancements, though jobs were eliminated, it also gave rise to more opportunities. If we are to be solely concerned about the fact that jobs will be lost in the face of artificial intelligence automation, we stifle artificial intelligence innovation, hindering economic progress and growth. Considering this, policymakers and governments should ensure that programs will help more people acquire the skills and knowledge to work with new technologies. Also, policies must be drafted and implemented that help to protect workers' rights, job security, and income.

5.4. Future Trends in the Landscape of AI

Some of the possible trends and developments one must hope to see within the landscape of artificial intelligence and its interaction with intellectual property and ethical considerations in the coming years is that of (1) evolution of creativity within artificial intelligence; we should start expecting artificial intelligence systems to be more creative in generating content, such as music, literature and even some inventions. This can bring about questions on ownership and attribution, thereby paving the way for new intellectual property rights policies and ethical discussions. (2) Artificial intelligence bias mitigation is another area where efforts to address bias and discrimination will increase. This means, there might be a greater onus on an ethical artificial intelligence design and algorithms that can reduce bias, thus promoting fairness. (3) The intellectual property rights will continue being adapted to artificial intelligence, by addressing issues such as eligibility for patents or trade secrets laws for an artificial intelligence-generated action/decision. This will be the most significant ethical and legal challenge.

6. CASE STUDY: INTERPLAY OF IP AND ETHICAL BEHAVIOUR IN AI-DRIVEN HEALTHCARE SECTOR

Recent years have witnessed the incorporation of artificial intelligence to accelerate developments in the healthcare sector, specifically the pharmaceutical industry, via drug discovery, cost-effectiveness, and efficiency. This also begs the question of the category of law that artificial intelligence will fall under - an existing legal category or a new category that is specialised. The four main ethical issues in the healthcare sector that are increasingly revolutionised by artificial intelligence are informed consent, transparency and safety, fairness and biases of algorithms, and privacy of data, while legal challenges comprise the reliability and validity of datasets to ensure safety and effectiveness, liability concerns, data protection laws, cybersecurity laws, and intellectual property laws (Gerke et al., 2020; Mehta & Devarakonda, 2018).

Rules and regulations are accordingly established to govern the rights of organisations, people and all the stakeholders involved in the collection of data, utilisation and analysis of it. However, laws relating to intellectual property for patient records and databases are understood to be complex and need more transparency, leading to unauthorised use of data. Also, an apparent lack of understanding and awareness in the healthcare industry about the importance of intellectual property protection increases the problem at hand. Thus, the need of the hour is to develop a legal framework that considers the needs of the patients, researchers, doctors, and even businesses. A formal development can prevent litigation and improve opportunities for research and other innovations in industries.

A crucial question that has become prominent is how effectively artificial intelligence in caring for patients matches the principles of informed consent. Notably, it has to answer to what extent clinicians are responsible for educating the patients about the complexities of artificial intelligence, the kinds of data it takes, and the degree of prejudice or shortcomings in the data being used. There is also a need to investigate under what conditions, the principles of informed consent need to be employed in the healthcare domain and when a clinician should inform the patient that artificial intelligence is being used. However, while these are potentially significant and relevant questions, answering them is challenging given how artificial intelligence operates using "black-box" algorithms, resulting in non-interpretable machine learning techniques that might be difficult for clinicians to understand. For example, in the case of Corti's algorithm (Vandenberk et al., 2023), the algorithm itself is "black box" because even the inventor of Corti does not know how the software arrives at the decision to alert an emergency when someone suffers from cardiac arrest (Vincent, 2018), even though the artificial intelligence is tasked to use voice analysis, breathing patterns and additional required metadata. So, though the algorithm effectively provides a faster interpretation, this lack of knowledge is typically a cause of concern for medical professionals as it questions whether a doctor/practitioner must disclose that they cannot fully interpret the diagnosis recommendations by artificial intelligence. The question is, thus, how important is "transparency" in situations that can benefit from artificial intelligence-based interpretations in clinical triaging and risk stratification?

Another substantial aspect in the healthcare sector is the chatbots used in health apps, which provide diet guidance to health assessments, enhance medication adherence, and analyse data collected by wearable sensors such as smartwatches. Apps like these raise questions for bioethicists about the degree of user agreement to informed consent. Most users do not spend much time reading and understanding user agreements. Also, software updates are frequent, making it even more difficult for users to read and remember them every single time, and they may need to keep track of what they are actually agreeing to (Gerke et al., 2019).

Another concern that is prevalent in artificial intelligence-driven healthcare is that of safety. For example, IBM Watson for Oncology uses an artificial intelligence algorithm to collect information from the medical records of patients to help doctors explore various treatment options for cancer. The system has come under scrutiny for providing inaccurate and unsafe recommendations for treating cancer, though no incorrect treatment recommendations were given to patients (Tupasela and Di Nucci, 2020). This happened because the artificial intelligence system was trained only with a few cancer cases. This case is a classic example that questions the safety and efficiency of artificial intelligence, reinstating that it needs to be built robustly. To be able to ensure that AIs provide their best, two things need to be met: firstly, reliability and the validity of the datasets provided to the system for training, and secondly, transparency of the data provided. Artificial intelligence is said to perform better when the training data is better (labelled data). Artificial intelligence developers, therefore need to be transparent, ensuring the kind of data being used for training and noting any data bias it exhibits.

CONCLUSION

This chapter provides a fresh perspective on artificial intelligence innovation. It demonstrates how artificial intelligence is becoming more prevalent in various technical and non-technological endeavours. In this sense, it is essential to consider how artificial intelligence will affect individuals, organisations, and society.

Accepting the fact that artificial intelligence is only going to continuously impact society, organisations, and the way we live, it is imperative that we offer a viable solution for its innovation. So, the debate around intellectual property rights in artificial intelligence will evolve as technology advances. Additionally, it will also raise newer legal and ethical questions. The latest advancement of the GPT structure, ChatGPT, is a testimony to this. So, the only viable solution for policymakers, stakeholders, and governments to collaborate is to promote international laws that uphold intellectual property rights and ethical concerns. The changing artificial intelligence must be addressed by organisations and government simultaneously. By adopting transparent, easily accessible, and accountable artificial intelligence systems, the dangers of privacy breaches and other ethical concerns can easily be acknowledged. However, the path ahead is not easy, as algorithms still need to be protected through intellectual property rights to ensure innovation and development.

REFERENCES

Act, P. (1977). Chapter 37. Retrieved from http://images.policy.mofcom.gov.cn/article/201512/1449024985281.pdf

Adadi, A., & Berrada, M. (2018). Peeking inside the black-box: A survey on explainable artificial intelligence (XAI). *IEEE Access : Practical Innovations, Open Solutions*, *6*, 52138–52160. doi:10.1109/ACCESS.2018.2870052

AI Act: A step closer to the first rules on Artificial Intelligence. (2023, Nov. 5). European Parliament.

AI IP Protection Debate. (n.d.). *Legal Service India - Law, Lawyers and Legal Resources*. Retrieved June 30, 2023, from https://www.legalserviceindia.com/legal/article-11428-ai-ip-protection-debate.html

Albahri, A. S., Duhaim, A. M., Fadhel, M. A., Alnoor, A., Baqer, N. S., Alzubaidi, L., Albahri, O. S., Alamoodi, A. H., Bai, J., Salhi, A., Santamaría, J., Ouyang, C., Gupta, A., Gu, Y., & Deveci, M. (2023). A systematic review of trustworthy and explainable artificial intelligence in healthcare: Assessment of quality, bias risk, and data fusion. *Information Fusion*, *96*, 156–191. doi:10.1016/j.inffus.2023.03.008

Baan, A., Allo, M. D. G., & Patak, A. A. (2022). The cultural attitudes of a funeral ritual discourse in the indigenous Torajan, Indonesia. *Heliyon*, *8*(2), e08925. doi:10.1016/j.heliyon.2022.e08925 PMID:35198784

Bainbridge, D. (2010). *Intellectual Property*. Pearson Education Limited.

Bechmann, A., & Bowker, G. C. (2019). Unsupervised by any other name: Hidden layers of knowledge production in artificial intelligence on social media. *Big Data & Society*, *6*(1), 2053951718819569. doi:10.1177/2053951718819569

Berne Convention for the Protection of Literary and Artistic Works (adopted 9 September 1886) (“Berne Convention”)

Brown, T. (2019, October 11). *The AI Skills Shortage - ITChronicles*. ITChronicles. https://itchronicles.com/artificial-intelligence/the-ai-skills-shortage/

BSA The Software Alliance. (2020). Retrieved from https://www.bsa.org/news-events/media/usmca-formalizes-free-flow-of-data-other-tech-issues

Buolamwini, J., & Gebru, T. (2018, January). Gender shades: Intersectional accuracy disparities in commercial gender classification. *Conference on Fairness, Accountability and Transparency*, 77-91.

Burns, L. N., & Tucci, L. (2023, March 31). *What is artificial intelligence (AI)? - AI definition and how it works. Enterprise AI.* TechTarget. https://www.techtarget.com/searchenterpriseai/definition/AI-Artificial-Intelligence

Cameron, I. (2021). *National security and the European Convention on human rights*. BRILL.

Ćemalović, U. (2022). Status of the World Intellectual Property Organization (WIPO) in the United Nations System and its Competences. *International Organizations, Serbia and Contemporary World*, *1*, 391.

Chatila, R., & Havens, J. C. (2019). The IEEE global initiative on ethics of autonomous and intelligent systems. *Robotics and Well-Being*, 11-16.

CiuriakD.FayR. (2021). The USMCA and Mexico’s Prospects under the New North American Trade Regime. *Social Science Research Network, 2*, 45-66. doi:10.2139/ssrn.3771338

DG. (2020). *The ethics of artificial intelligence: Issues and initiatives*. European Parliamentary Research Service Scientific Foresight Unit (STOA) PE 634.452, p:15

Discher, G., & Rutigliano, N. (2021). USPTO Releases Report on Artificial Intelligence and Intellectual Property Policy. *The Journal of Robotics. Artificial Intelligence and Law*, 4.

Dornis, T. W. (2020). Artificial intelligence and innovation: The end of patent law as we know it. *SSRN*, *23*, 97. doi:10.2139srn.3668137

Du, R. (2022). *Intellectual Property Protection and Growth: Evidence from Post-TRIPS Development of Manufacturing Industries*. University of Chicago.

European Council and European Council. (n.d.). Retrieved from https://www.consilium.europa.eu/en/policies/data-protection/data-protection-regulation/#:~:text=The%20GDPR%20lists%20the%20rights,his%20or%20her%20personal%20data

European Parliament. (2020). *Intellectual property rights for the development of artificial intelligence technologies*. Author.

Farina, M. (2023). Intellectual property rights in the era of Italian "artificial" public decisions: time to collapse? *Rivista italiana di informatica e diritto*, *5*(1), 16-16.

Finck, M. (2020). Legal analysis of international trade law and digital trade (PE 603.517). Briefing requested by the INTA Committee. Brussels.

Foss-Solbrekk, K. (2021). Three routes to protecting AI systems and their algorithms under IP law: The good, the bad and the ugly. *Journal of Intellectual Property Law & Practice*, *16*(3), 247–258. doi:10.1093/jiplp/jpab033

Gerke, S., Kramer, D. B., & Cohen, I. G. (2019). Ethical and legal challenges of artificial intelligence in cardiology. *AIMed Magazine*, 2, 12–17.

Gerke, S., Minssen, T., & Cohen, G. (2020). Ethical and legal challenges of artificial intelligence-driven healthcare. In *Artificial intelligence in healthcare* (pp. 295–336). Academic Press. doi:10.1016/B978-0-12-818438-7.00012-5

Ginsburg, J. (2018). Copyright. In *The Oxford handbook of intellectual property law*. Oxford University Press.

GOV.UK. (2021). *Government response to call for views on artificial intelligence and intellectual property*. Retrieved from https://www.gov.uk/government/consultations/artificial-intelligence-and-intellectual-property-call-for-views/government-response-to-call-for-views-on-artificial-intelligence-and-intellectual-property

GOV.UK. (2022). *Artificial Intelligence and Intellectual Property: copyright and patents*. Retrieved from https://www.gov.uk/government/consultations/artificial-intelligence-and-ip-copyright-and-patents/artificial-intelligence-and-intellectual-property-copyright-and-patents#copyright

Gralla, P. (2016, May 11). *Amazon Prime and the racist algorithms*. Computerworld. https://www.computerworld.com/article/3068622/amazon-prime-and-the-racist-algorithms.html

Hunt, E. (2016, March 24). Tay, Microsoft's AI chatbot, gets a crash course in racism from Twitter| Artificial intelligence (AI). *The Guardian*.

Information Commissioner's Office. (n.d.). Retrieved from https://ico.org.uk/for-organisations/uk-gdpr-guidance-and-resources/data-protection-principles/a-guide-to-the-data-protection-principles/

Intellectual property of an AI : issues and challenges - iPleaders. (2020, October 19). https://blog.ipleaders.in/intellectual-property-ai-issues-challenges/

Jelisavac Trošić, S. (2022). *The World Trade Organization and Serbia's Long Path Towards Accession*. Institute of International Politics and Economics,Faculty of Philosophy of the University of St. Cyril and Methodius.

Jordan, M. I. (2019). Artificial intelligence—The revolution hasn't happened yet. *Harvard Data Science Review*, *1*(1), 1–9.

Kim, D. (2022). The Paradox of the DABUS Judgment of the German Federal Patent Court. *GRUR International*, *71*(12), 1162–1166. doi:10.1093/grurint/ikac125

Kim, J., Jeong, B., Kim, D., Kim, J., Jeong, B., & Kim, D. (2021). A Brief History of Patents. *Patent Analytics: Transforming IP Strategy into Intelligence*, 11-19.

Kretschmer, M., Meletti, B., & Porangaba, L. H. (2022). Artificial Intelligence and Intellectual Property: Copyright and Patents–a response by the CREATe Centre to the UK Intellectual Property Office's open consultation. *Journal of Intellectual Property Law and Practice*, *17*(3), 321–326. doi:10.1093/jiplp/jpac013

La Diega, G. N. (2018). Against the dehumanisation of decision-making. *J. Intell. Prop. Info. Tech. & Elec. Com. L.*, *9*, 3.

Ladak, A. (2023). What would qualify an artificial intelligence for moral standing? *AI and Ethics*, 1–16. doi:10.100743681-023-00260-1

Lauber-Rönsberg, A., & Hetmank, S. (2019). The concept of authorship and inventorship under pressure: Does artificial intelligence shift paradigms? *Journal of Intellectual Property Law and Practice*, *14*(7), 570–579. doi:10.1093/jiplp/jpz061

Leblond, P., & Aaronson, S. A. (2019). *A plurilateral "single data area" is the solution to Canada's data trilemma*. Academic Press.

Lee-Makiyama, H. (2012). Presentation at the *WTO ITA Symposium, Geneva*, Switzerland.

Legislation.gov.uk. Copyright, Designs and Patents Act 1988. Part I Chapter I. Retrieved from https://www.legislation.gov.uk/ukpga/1988/48/part/I/chapter/I#commentary-c13754491

Legislation.gov.uk. UK Public General Acts1988 s.9.

Li, H., Yu, L., & He, W. (2019). The impact of GDPR on global technology development. *Journal of Global Information Technology Management*, *22*(1), 1–6. doi:10.1080/1097198X.2019.1569186

Lim, P. H., & Li, P. (2022). Artificial intelligence and inventorship: Patently much ado in the computer program. *Journal of Intellectual Property Law and Practice*, *17*(4), 376–386. doi:10.1093/jiplp/jpac019

Madiega, T. A. (2021). *Artificial intelligence act. European Parliament*. European Parliamentary Research Service.

Marsoof, A., Kariyawasam, K., & Talagala, C. (2022). Crafting Domestic Intellectual Property Law–International Obligations, Flexibilities, and Approaches. In Reframing Intellectual Property Law in Sri Lanka: Lessons from the Developing World and Beyond (pp. 13-26). Singapore: Springer Nature Singapore. https://doi.org/ doi:10.1007/978-981-19-4582-3_213

Martinet, S. (2018, May 27). *GDPR and Blockchain: Is the New EU Data Protection Regulation a Threat or an Incentive?* Cointelegraph. https://cointelegraph.com/news/gdpr-and-blockchain-is-the-new-eu-data-protection-regulation-a-threat-or-an-incentive

Mason, R. O. (1986). Four ethical issues of the information age. *Management Information Systems Quarterly*, *10*(1), 5–12. doi:10.2307/248873

Mehta, N., & Devarakonda, M. V. (2018). Machine learning, natural language programming, and electronic health records: The next step in the artificial intelligence journey? *The Journal of Allergy and Clinical Immunology*, *141*(6), 2019–2021. doi:10.1016/j.jaci.2018.02.025 PMID:29518424

Meltzer, J. P. (2018). *The impact of artificial intelligence on international trade.* Center for Technology Innovation at Brookings.

Miyakawa, T. (2000). Copyright Legislation in Japan and Recent Trends. A Report to the North American Coordinating Council for Japanese Library Resources Year 2000 Conference, San Diego, CA.

Moerland, A. (2022). *Artificial Intelligence and Intellectual Property Law. In The Cambridge Handbook of Private Law and Artificial Intelligence*. Cambridge University Press.

Morley, J., Floridi, L., Kinsey, L., & Elhalal, A. (2020). From what to how: An initial review of publicly available AI ethics tools, methods and research to translate principles into practices. *Science and Engineering Ethics*, *26*(4), 2141–2168. doi:10.100711948-019-00165-5 PMID:31828533

Nard, C. A. (2019). *The law of patents*. Aspen Publishing.

Office of the United States Trade Representative. (n.d.). *Intellectual Property in USMC Agreement.* Retrieved from https://ustr.gov/trade-agreements/free-trade-agreements/united-states-mexico-canada-agreement/agreement-between

Oman, R. (2017). Computer Software as Copyrightable Subject Matter: Oracle v. Google, Legislative Intent, and the Scope of Rights in Digital Works. *Harvard Journal of Law & Technology*, *31*, 639.

Paris Convention for the Protection of Industrial Property (opened for signature 20 March 1883) (“Paris Convention”).

Pearlman, R. (2017). Recognising artificial intelligence (AI) as authors and investors under US intellectual property law. *Rich. J.L. & Tech.*, *2*.

Perlmutter, S. (2023). *Copyright Registration Guidance: Works Containing Material Generated by Artificial Intelligence.* U.S. Copyright Office, Library of Congress.

Picht, P. G., & Thouvenin, F. (2023). AI and IP: Theory to Policy and Back Again–Policy and Research Recommendations at the Intersection of Artificial Intelligence and Intellectual Property. *IIC-International Review of Intellectual Property and Competition Law*, 1-25.

Rai, A. (2020). Explainable AI: From black box to glass box. *Journal of the Academy of Marketing Science*, *48*(1), 137–141. doi:10.100711747-019-00710-5

Report on intellectual property rights for the development of artificial intelligence technologies. (2020). https://www.europarl.europa.eu/doceo/document/A-9-2020-0176_EN.html

Seville, C. (2017). The Emergence and Development of Intellectual Property Law in Western Europe in The Oxford Handbook of Intellectual Property Law, 171-197.

Singh, D. R. (2008). *Law relating to intellectual property: A complete comprehensive material on intellectual property covering acts, rules, conventions, treaties, agreements, digest of cases and much more*. Universal Law Publishing Company.

Slavina, Z. (2023). *AI ethics: Chosen challenges for contemporary societies and technological policymaking* [B.A. Dissertation]. Uniwersytet w Białymstoku.

Stamatoudi, I., & Torremans, P. (Eds.). (2021). *EU copyright law: A commentary*. Edward Elgar Publishing. doi:10.4337/9781786437808

Sulistianingsih, D., & Ilyasa, R. M. A. (2022). The Impact of Trips Agreement On The Development Of Intellectual Property Laws In Indonesia. *Indonesia Private Law Review*, *3*(2), 85–98. doi:10.25041/iplr.v3i2.2579

Taricani, E., Saris, N., & Park, P. A. (2020). Beyond technology: The ethics of artificial intelligence. *World Complexity Science Academy Journal*, *1*(2), 17. doi:10.46473/WCSAJ27240606/15-05-2020-0017//full/html

Tupasela, A., & Di Nucci, E. (2020). Concordance as evidence in the Watson for Oncology decision-support system. *AI & Society*, *35*(4), 811–818. doi:10.100700146-020-00945-9

U.S. Copyright Office. (2023). *Copyright Registration Guidance: Works Containing Material Generated by Artificial Intelligence*. Library of Congress. Retrieved from https://www.govinfo.gov/content/pkg/FR-2023-03-16/pdf/2023-05321.pdf

Vandenberk, B., Chew, D., Prasana, D., Gupta, S., & Exner, D. V. (2023). Successes and Challenges of Artificial Intelligence in Cardiology. *Frontiers in Digital Health*, *5*, 1201392. doi:10.3389/fdgth.2023.1201392 PMID:37448836

Vincent, J. (2018). AI that detects cardiac arrests during emergency calls will be tested across Europe this summer. *The Verge*, 25.

Voigt, P., & Von dem Bussche, A. (2017). *The EU General Data Protection Regulation (GDPR). A Practical Guide*. Springer International Publishing. doi:10.1007/978-3-319-57959-7

Wang, W., & Siau, K. (2019). Artificial intelligence, machine learning, automation, robotics, future of work and future of humanity: A review and research agenda. *Journal of Database Management*, *30*(1), 61–79. doi:10.4018/JDM.2019010104

Watal, J. (2002). Intellectual property rights in the WTO and developing countries. *Intellectual Property Rights in the WTO and Developing Countries.*

What is Intellectual Property? (n.d.). WIPO - World Intellectual Property Organization. Retrieved June 30, 2023, from https://www.wipo.int/about-ip/en/

White, M. (2002). World Intellectual Property Organization. *Journal of Business & Finance Librarianship*, *8*(1), 71–78. doi:10.1300/J109v08n01_08

Yudkowsky, E., & Bostrom, N. (2011). *The ethics of artificial intelligence. In The Cambridge Handbook of Artificial Intelligence*. Cambridge University Press.

Zhuk, A. (2023). Navigating the legal landscape of AI copyright: A comparative analysis of EU, US, and Chinese approaches. *AI and Ethics*, 1–8.

Chapter 2
Measuring Throughput and Latency of Machine Learning Techniques for Intrusion Detection

Winfred Yaokumah
https://orcid.org/0000-0001-7756-1832
University of Ghana, Ghana

Charity Y. M. Baidoo
University of Ghana, Ghana

Ebenezer Owusu
University of Ghana, Ghana

ABSTRACT

When evaluating the effectiveness of machine learning algorithms for intrusion detection, it is insufficient to only focus on their performance metrics. One must also focus on the overhead metrics of the models. In this study, the performance accuracy, latency, and throughput of seven supervised machine learning algorithms and a proposed ensemble model were measured. The study performs a series of experiments using two recent datasets, and two filter-based feature selection methods were employed. The results show that, on average, the naive bayes achieved the lowest latency, highest throughput, and lowest accuracy on both datasets. The logistics regression had the maximum throughput. The proposed ensemble method recorded the highest latency for both feature selection methods. Overall, the Spearman feature selection technique increased throughput for almost all the models, whereas the Pearson feature selection approach maximized performance accuracies for both datasets.

DOI: 10.4018/978-1-6684-9999-3.ch002

INTRODUCTION

In recent times network security attacks and data breaches have become rampant while intrusion detection systems (IDS) are being developed with automated response mechanisms to detect and prevent such attacks (Anwar et al., 2017). An intrusion detection system provides active network security protection mechanisms against cybercriminal activities (Liang et al., 2019). These mechanisms monitor network operations and analyze packets for malicious activities (Paul et al., 2018). Packet analysis can take place on a network (network-based) or a computer (host-based). Network-based intrusion detection systems (NIDS) examine network behavior to determine if a node is under attack, whereas host-based intrusion detection systems (HIDS) check the logs kept by a single host for malicious operations (Chan et al., 2016). However, IDS sometimes fails to detect new external attacks and has a low accuracy rate, and a high false alarm rate (Khraisat et al., 2019; Liu & Lang, 2019). The inability of IDS to prevent network intrusion precisely and timely can jeopardize the information security goals of integrity, confidentiality, and system availability (Mishra & Yadav, 2020). In particular, the IDS that are implemented on networks can escalate the amount of delivery time, thereby reducing the reliability of network traffic (Xia et al., 2015). Also, packet processing may decrease network throughput and increase latency (Tsikoudis et al., 2016).

Two major overhead performance metrics, latency and throughput, are used to measure the IDS's timely detection of network intrusion. In the implementation of IDS, high throughput and low latency of an IDS are essential. Latency measures the time it takes for a processor to receive a request for a byte or message from memory (Hasan et al., 2021). In communication networks, latency or end-to-end delay represents the amount of time it takes for data to be received at its endpoint (destination). Intrusion detection systems frequently suffer from severe latency and network overhead, which make them irresponsive to attacks and the identification of malicious activities (Rahman et al., 2020). Likewise, throughput measures the speed at which the data is transmitted effectively (Ingley & Pawar, 2015). It describes the amount of data that is dispatched over a given time. According to Zhang et al. (2018), with high-speed traffic data, conventional intrusion detection systems experience latency and occasionally fail to identify intrusion patterns. However, to identify suspicious traffic momentarily, a real-time network IDS with machine learning (ML) techniques can promptly process vast volumes of network traffic data.

Specifically, machine learning (ML) techniques are enabling the modeling of IDS to provide significantly higher rates of intrusion detection (Srivastava et al., 2019). Some intrusion detection systems use ML classification algorithms to categorize network traffic as normal or irregular (Kaya, 2020; Thaseen & Kumar, 2017). For instance, the IDS implementation that employs ML approaches for classification includes Naive Bayes (NB), Adaptive Boost, PART (Kumar & Doegar, 2018), Active learning Support Vector Machine, Fuzzy C-Means clustering (Kumari & Varma, 2017), Multi-Layer Perceptron, Bayesian Network, Support Vector Machine (SVM), Adaboost, Random Forest (RF), Bootstrap Aggregation, Decision Tree (Halibas et al., 2018), Random Forest (Park et al., 2018); SVM (Kotpalliwar & Wajgi, 2015), Genetic Algorithm, and Support Vector Machine (Gharaee & Hosseinvand, 2017). These ML algorithms may provide varying levels of accuracy, latency, and throughput.

According to Yu et al. (2018), measuring the overhead performance of the learning algorithms accurately is important for determining their effectiveness. However, currently, few studies investigate the efficiency of IDS, considering its training time, testing time, latency, throughput, and detection time (Maseer et al., 2021). Thus, this experimental study focuses on supervised machine learning algorithms to measure latency and throughput using two intrusion detection datasets and two filter-based feature selection methods. Throughput and latency are critical in intrusion detection systems. For example, in a real-time intrusion system for monitoring user activities on networks, it is very crucial to have a prediction as fast as possible. Also, in autonomous driving, it is important that the classification of road signs be as fast as possible to avoid accidents. In the era of large volumes of data traversing communication networks, this study will help designers in the choice of ML algorithms for the implementation of IDS, considering their accuracy, latency, and throughput for various devices including the Internet of Things (IoT) and embedded systems.

BACKGROUND

Intrusion Detection Systems

Intrusion may be described as any attempt to compromise the confidentiality, integrity, or availability (CIA) of information resources (Ashfaq et al., 2017). Intrusion detection systems (IDS) are cyber security solutions installed in computer networks to detect and flag cyber intrusions in information systems (Agrawal & Agrawal, 2015). They are deployed to monitor network traffic for suspicious activities or threats that evade perimeter security solutions such as firewalls and alert security personnel (Mishra, Pilli, Varadharajan, & Tupakula, 2017). Intrusion detection systems are broadly classified into signature-based, anomaly-based (Loet al., 2018), and hybrid intrusion detection systems.

In the signature-based IDS, patterns of traffic or application data are compared against pre-defined rules or well-known patterns of malicious behavior, and the matched data instances are flagged as malicious. Signature-based (misuse-based) techniques are designed to detect known attacks by using signatures of those attacks. A misuse-based is when an IDS detects an attack by identifying known attack signatures and matching incoming files for any matches. A signature is a trend that is associated with a specific attack (Meng et al., 2017). Misuse detection systems usually have better detection of known attacks but are unable to detect new threats and hence need a continuous updating of the signature database. When a signature is not included in the database protocol or when a packet must be checked with all signatures, it takes longer to detect an attack (Hendrawan et al., 2019). Again, a signature-based technique is ineffective for packet monitoring in large networks (Bhosale & Mane, 2016). In the anomaly-based IDS, detection systems build a profile of normal behavior and compare activities against this profile such that any deviations or outliers are picked as being malicious (Agrawal & Agrawal, 2015). In the case of the hybrid IDS, the techniques of both signature-based and anomaly-based IDS are combined in the detection system (Mishra et al., 2017).

Moreover, intrusion detection systems are described based on their location within the network when deployed, thus either host-based or network-based. Host-based intrusion detection systems (HIDS) are placed within each host on the network to monitor individual end devices (hosts) whereas network-based intrusion detection systems (NIDS) are positioned strategically at network points to parse and monitor the entire network's traffic for suspicious activities (Mishra et al., 2017). Machine learning-based intrusion detection systems use machine learning techniques to inspect packets, captured through a network, for malicious traffic to complete the intrusion detection task (Fang et al., 2020). Machine learning techniques enable the modeling of IDS to provide significantly higher intrusion detection (Srivastava et al., 2019). The application of ML can lead to a reduced false alarm rate and a higher rate of prediction (Haripriya & Jabbar, 2018).

Machine Learning Methods

Machine Learning can be viewed as a model that learns by experience (Meng, Jing, Yan, & Pedrycz, 2020), does not rely on explicitly rule-based programming, and improves through its experience (Reis et al., 2020). Machine learning algorithms allow computers to learn without having to be directly programmed (Das & Nene, 2018). They draw conclusions from the present understanding of input data and anticipate what is uncertain (Kaya, 2020). Machine learning is characterized by the process of building a model with the knowledge discovered from sample data points (Dua & Xian, 2011). In machine learning, the system automatically improves its performance with experience accumulation (Igual & Seguí, 2017). This attribute is consistent with the principle of attack detection by self-learning against external invasion to improve the detection rate and reduce false positive rates.

Machine Learning algorithms fall into three main categories, supervised, semi-supervised, and unsupervised (Buczak & Guven, 2016; Kang & Jameson, 2018). For supervised learning, datasets comprising data points, each described using a set of attributes (features) and target labels, are used in training a function (Rokach & Maimon, 2010). A learning model is then obtained which can predict the output of the function. Supervised learning algorithms include artificial neural networks (ANN), support vector machines (SVM), and decision trees. For unsupervised ML models, no target or label is given in sample data (Dua & Xian, 2011). These models summarize the key features of the data to form natural clusters of input patterns per a particular cost function. Besides, unsupervised ML does not lend itself to be easily evaluated, because it does not have an explicit teacher and therefore no target for testing. Examples of unsupervised ML methods include k-means clustering, hierarchical clustering, and self-organization maps. Terms such as data mining, pattern recognition, and predictive modeling have an association with machine learning (Kuhn & Johnson, 2013). Moreover, a combination of labeled and unlabeled data is used in semi-supervised learning (Thomas & Gupta, 2020).

Intrusion detection can be much more efficient by employing machine learning techniques (Fang, Tan, & Wilbur, 2020) due to its ability to keep pace with malware evolution (Gibert, Mateu, & Planes, 2020) and to reduce false alarm rate and improve the detection accuracy. Therefore, the theory and method of machine learning applications in intrusion detection are no doubt crucial (Fang et al., 2020) to maintain the security of the cyber-infrastructure. Several supervised machine-learning algorithms exist for IDS implementation. However, some methods perform better than others in terms of accuracy, throughput, and execution time (Kotpalliwar, 2015). For example, SVM is one of the most efficient and reliable classification and regression algorithms (Cervantes et al., 2020). The performance of SVM generalization and ease of training is far above the capability of the much more conventional techniques (Bhaskar et al., 2015). Both classification and regression problems can be addressed in SVM. For its potential to minimize dataset training and testing period as well as increase the system's efficiency and accuracy, the SVM is an excellent candidate for the IDS system (Mulay et al., 2010).

Another commonly used ML method is the Random Forest. It is employed to build a prediction framework for problems of classification and regression (Islam & Shahjalal, 2019). As compared to other widely used machine learning approaches, Random Forest models have low training time complexity and quick prediction, strong prediction rates, and versatility, and do not overfit the data when considering the number of trees (Resende & Drummond, 2018; Witten et al., 2016). In comparison to other classifiers, the Random Forest model is found to generate the best results (Belavagi & Muniyal, 2016; Choudhury & Bhowal, 2015; Robinson & Thomas, 2016). However, Gradient boosting iteratively uses weak classifiers to minimize a loss function of the model using the gradient. In each iteration of training, the weak learner is built and its predictions are compared to the target. The difference between prediction and the target is referred to as the error rate of the model which can then be used to calculate the loss function. However, the K-Nearest Neighbor (KNN) algorithm classifies the data points based on the majority of nearest neighbor records with the nearest neighbor coverage of k. The algorithm works by comparing the test records given to the training records that have similarities. It calculates the similarity between the unlabeled samples and training samples, thus when there is an unknown record, the classification looks for a spatial pattern for a training record that is very close to the test record.

Similarly, Logistic regression is used to analyze the relationship between multiple independent variables and a categorical dependent variable and estimates the probability of occurrence of an event by fitting data to a logistic curve. Odds are used to explain when logistic regression calculates the probability of an event occurring over the probability of an event not occurring, the impact of independent variables. The mean of the response variable p in terms of an explanatory variable is modeled about p and x through the equation $p = a + \beta x$. Transforming the odds using natural logarithm aids in solving the problem of the optimal values of $a + \beta x$ not falling between 0 and 1 when extreme values of x are used. Likewise, the Bayesian classification is a supervised learning method as well as a statistical classification method. The Bayesian classification assumes or takes an underlying probabilistic model and allows the capture of uncertainty about a model in a principled way by determining probabilities of the outcomes. The Bayesian classification can be utilized in solving predictive and diagnostic problems.

Following, the ensemble approach is a machine-learning technique that combines numerous learners into a single strong learner, based on learning and classification methods (Tuysuzoglu et al., 2016). Boosting and bagging are two principal types of ensemble learning (Wang & Pineau, 2016). Many approaches to boosting-based ensembles, such as AdaCost, RareBoost (Joshi, 2001), and AdaC1, C2, and C3 (Sun et al., 2007), are useful in solving imbalanced issues by altering the weight update algorithm in the iteration of AdaBoost. The bagging approach is a comprehensive and efficient ensemble method that uses bootstrap sampling via training data to generate variety (Guo et al., 2017). When dealing with various situations, the incorporation of an ensemble model allows the algorithm to automatically identify a correct model integration mode and adapt to various parameter spaces (Yongze et al., 2019).

Application of Machine Learning in Intrusion Detection

Devan and Khare (2020) proposed an XGBoost algorithm for feature selection and a deep neural network (DNN) to classify network incursion. The NSL-KDD dataset was utilized. The Adam optimizer was employed to optimize the learning rate during DNN training, and the softmax classifier was used to classify network intrusions. The results were compared to existing logistic regression, naive Bayes, and support vector machine models. The DNN outperformed previous models in classification accuracy by 97%. Similarly, Bedi et al. (2021) proposed Improved Siam-IDS (I-SiamIDS), a two-layer ensemble for dealing with the class imbalance problem. To detect attacks, the first layer of I-SiamIDS employed a hierarchical filtration of input samples using an ensemble of binary eXtreme Gradient Boosting (b-XGBoost), Siamese Neural Network (SiameseNN), and Deep Neural Network (DNN). The attacks are then sent to I-SiamIDS's second layer for classification into distinct attack classes through a multi-class eXtreme Gradient Boosting classifier (m-XGBoost). Compared to prior research, I-SiamIDS exhibited significant improvement in recall, accuracy, F1 score, precision, and AUC values on both the CIDDS-001 and NSLKDD datasets.

Kasongo and Sun (2020) presented a FeedForward Deep Neural Network (FFDNN) wireless IDS system based on a Wrapper Based Feature Extraction Unit (WFEU). The WFEU-FFDNN was evaluated using the UNSW-NB15 and the AWID datasets. The WFEU-FFDNN was then compared to five different conventional machine learning algorithms, including Random Forest, Decision Tree, SVM, KNN, and Naive Bayes. With 22 characteristics in the UNSW-NB15 dataset, the binary and multiclass classification techniques achieved overall accuracies of 87.10% and 77.16% respectively. Likewise, Tama et al. (2019) offered a technique for selecting relevant features for intrusion detection based on a two-level ensemble of classifiers. Three alternative strategies were utilized to minimize the attribute size of the training datasets: particle swarm optimization, ant colony algorithm, and genetic algorithm. The model was evaluated using the NSLKDD and UNSW-NB15 datasets. The accuracy rate of 85.8% in the NSLKDD dataset and 91.3% in the UNSW-NB15 dataset were recorded.

Moreover, Choudhary and Kesswani (2021) proposed a hybrid classification technique for identifying multiclass assaults in IoT networks. The Principal Component Analysis (PCA) and Linear Discriminant Analysis (LDA) were used to identify relevant features and reduce the high dimensionality of the dataset. To boost the detection rate and minimize the false alarm rate, a mix of neural network and SVM classifiers was utilized. The SVM was a strong and quick learner classifier that detected mismatched behaviors. On the UNSW-NB15 and NSL-KDD datasets, 81.02% detection rate, 2.22% false alarm rate, and 92.85% detection rate, 2.99% false alarm rate were achieved, respectively.

Latency of Machine Learning Techniques

Previous studies presented works on latency and machine learning techniques in diverse applications. Chan et al. (2016) examined the effect of latency on IDS accuracy and power system stability. On the NSL-KDD dataset, k-nearest Neighbours, Artificial Neural Networks (ANN), and SVM algorithms were used to obtain accuracy efficiency. The study introduced a mathematical model for calculating latency while integrating IDS detection information during network routing. MatLab was used to simulate the power grid. The accuracy, false-positive rate, and false-negative rate of the algorithms are respectively as follows: KNN had 98.3 percent, 1.4 percent, 2.1 percent; Neural Networks had 98.87 percent, 1.2 percent, 3.5 percent; and SVM had 96.5 percent, 0.1 percent, 8.5 percent. In a related study, Preuveneers et al. (2020) proposed a multi-objective optimization approach to identify trade-offs between model accuracy and resource usage in resource-constrained smart environments. The study used hyperparameter optimization to find the best model for use in intelligent environments. The model was tested with Webscope S5 and CICIDS2017 datasets. The findings show that the traditional machine learning methods were more potent in terms of resource utilization than deep learning-based neural networks.

Moreover, Laukemann et al. (2019) used an Open-Source Architecture Code Analyzer (OSACA) to estimate the execution time duration of sequential loops. In this study, OSACA was extended to include ARM instructions and critical path prediction, as well as the detection of loop-carried dependencies. Based on machine models from public documentation and semi-automatic benchmarking, runtime projections for code on Intel Cascade Lake, AMD Zen, and Marvell ThunderX2 micro-architectures were shown. OSACA predictions were accurate, and in some cases even more explicit and adaptable than predictions made by comparable tools like Intel Architecture Code Analyzer (IACA) and Low-Level Virtual Machine. Similarly, Park and Lee (2021) employed a Sparse Matrix Multiplication Performance Estimator on a Cloud Environment (S-MPEC) to estimate the latency of different Sparse matrix multiplication (SPMM) activities using Apache Spark. The study proposed unique features for building a gradient-boosting regressor model and Bayesian optimization. The proposed S-MPEC model can reliably forecast latency on any SPMM task and offer the best implementation method. Compared to the native SPMM implementations in Apache Spark, a user can expect 44% less latency when completing SPMM tasks.

Likewise, Son and Lee (2018) proposed a technique to estimate the latency by multiplying dense matrices of various sizes and forms of several cloud computing resources, where many contemporary big data systems are hosted. The suggested approach involved methods of creating training datasets, extracting features, and modeling. The study used iterative Bayesian optimization to discover the best working hyper-parameters for the chosen modeling approach. The proposed method has a 63% lower Root Mean Square Error (RMSE) for predicting distributed matrix multiplication delay.

Throughput of Machine Learning Methods

Prior works measured the throughput of computing systems and algorithms. Abolfazli et al. (2015) developed a method for measuring 4G wireless maximum throughput at the user end. The study used *Iperf* and OOKLA network measurement tools to perform a series of experiments in a 4G wireless environment to demonstrate the effectiveness and efficiency of the proposed method. The study calculated the models' current throughput and forecasted future throughput based on past data. Also, Jeong et al. (2019) proposed the use of machine learning methods to determine the throughput of the cluster tools. Using a real-time simulator, throughput data was collected with various input parameter setups. To calculate cluster tool throughput, a variety of machine learning techniques were employed, including SVM, Linear regression, polynomial regression, KNN, and deep neural networks (DNN). From the experiment results, KNN and deeper DNN proved to be practical options for cluster tool throughput prediction.

Likewise, Li et al. (2019) provided a support vector regression (SVR) model based on the symbiotic organism search (SOS) algorithm for estimating production throughput in a large, low-volume setting. The SOS technique was used to optimize SVR parameters such as the penalty, the width of the kernel function, and the loss function to improve the performance of the SVR model. In comparison to the industrial solution and traditional SVR, the proposed model demonstrated higher competence. In a related experimental study, Mendis et al. (2019) demonstrated the use of an Instruction Throughput Estimator Using Machine Learning (ITHEMAL) that learns to forecast the throughput of a sequence of instructions. ITHEMAL predicted throughput using a hierarchical Long Short-Term Memory (LSTM) technique based on the opcodes and operands of instructions in a basic block. In particular, the model exhibited less than half the inaccuracy of current analytical models such as *llvm-mca* and IACA.

MATERIALS AND METHOD

Research Process

Firstly, the UNSW- NB15 and CICIDS2017 intrusion detection datasets were selected for use in the evaluation of the ML models. These datasets have been used in similar studies, though with different feature selection techniques to detect intrusion (Preuveneers et al., 2020; Tama et al., 2019). Secondly, the datasets were built using the Pearson correlation coefficient and Spearman correlation coefficient as feature selection approaches for each of the selected classifiers: Random Forest, Support Vector Machine, Naive Bayes, Logistics Regression, Gradient Boosting, Decision Tree, and K-Nearest Neighbors. These algorithms were reported in a review study as state-of-the-art supervised machine learning algorithms (Choudhary & Kesswani, 2022; Kasongo & Sun, 2020). Many factors determine the suitability of a particular algorithm, such as the size and structure of your dataset. As a result, experimenting with various datasets and varying features is critical for ascertaining the suitable algorithm for a specific problem. Thirdly, an ensemble learning technique was proposed by combining the top two performing classifiers from each feature selection approach. Fourthly, four performance metrics were used in the experiments (accuracy, precision, recall, and F-score) to evaluate the performance of the models. Finally, the latency and throughput of the ML models were measured, and their results were compared.

Datasets

The first dataset, UNSW-NB15, was developed in the Cyber Range Lab of the Australian Center for Cyber Security (ACCS) using the IXIA PerfectStorm tool for creating a combination of real modern regular activities and synthetic modern attack behaviors. (Moustafa & Slay, 2015). UNSW-NB15 is a collection of network intrusion data. It comprises nine distinct types of attacks, such as DoS, worms, backdoors, and fuzzers. The collection comprises unfiltered network packets. Moustafa (2019)100 GB of raw traffic (e.g., Pcap files) was captured using the tcpdump utility. Furthermore, the Argus and Bro-IDS tools were employed, and twelve algorithms were created to produce features with the class label. The UNSW-NB15_features.csv file contains information about these features. The total number of records recorded in the four CSV files is two million and 540,044 which can be identified as UNSW-NB15_1.csv, UNSW-NB15_2.csv, UNSW-NB15_3.csv, and UNSW-NB15_4.csv.UNSW-NB15_GT.csv is the name of the ground truth table, and UNSW-NB15_LIST_EVENTS.csv is the name of the list of events file.UNSW_NB15_training-set.csv and UNSW_NB15_testing-set.csv were created from this dataset as training and testing sets, respectively. The training set has 175,341 records, while the testing set contains 82,332 records from the attack and normal categories (Khamis & Matrawy, 2020). It consists of 49 attributes with nine types of attacks (Fuzzers, Analysis, Backdoors, DoS, Exploits, Generic, Reconnaissance, Shellcode, and Worms) and one normal traffic.

The second dataset, CICIDS2017, was developed by the Canadian Institute for Cybersecurity using the CICFlowMeter tool (Lashkari et al., 2017). The CICIDS2017 dataset includes labeled network flows, comprising entire packet payloads in pcap format, related profiles, and labeled flows (GeneratedLabelledFlows.zip), as well as CSV files for machine and deep learning (MachineLearningCSV.zip) (Sharafaldin et al., 2018). The CICIDS2017 dataset provides benign and up-to-date common attacks that are similar to authentic real-world data (PCAPs). It also includes the results of a network traffic analysis performed with CICFlowMeter, which included labeled flows based on the time stamp, source and destination IPs, source and destination ports, protocols, and attacks (CSV files) (Hossen & Janagam, 2018). There are 79 features in the CICIDS2017 CSV file. It contains current attacks, namely DoS, DDoS, Brute Force, Infiltration, Port scan, Heartbleed Attack, and Botnet (Azwar et al., 2018). Data was collected for a total of five days. Attacks are carried out in the morning and afternoon on business days (Tuesday-Friday). DoS, DDoS, Heartbleed, Web assaults, FTP Brute Force, SSH Brute Force, Infiltration, Botnet, and Port Scans are among the assaults used. Monday is a regular day with solely legitimate traffic. The CICIDS2017 dataset has 2,830,540 records. It is a multi-class, high-dimensional, and unbalanced dataset. It also comprises network traffic in packet format, totaling 11,522,402 packets (Panigrah & Borah, 2018).

Feature Selection

Two filter-based feature selection techniques were used, namely the Pearson correlation coefficient and the Spearman correlation coefficient. Both Pearson and Spearman correlation coefficient methods used a threshold value of 0.70 (Feng & Fan, 2019). When the Pearson correlation coefficient was used to determine the most relevant features in the UNSW-NB15 dataset with 39 unique characteristics, 16 features were found as strongly correlated. These features were removed, and the remaining 23 were used for training the models. Similarly, when the Spearman correlation coefficient was applied to the same dataset, 17 features were identified as highly correlated and were removed – leaving 22 for training the models. Moreover, for the CICIDS2017 dataset, which had 62 features, when the Pearson Correlation Coefficient was applied, 26 features were selected. The Spearman correlation coefficient also selected 15 features on the CICIDS2017 dataset for the training of the models.

Overhead Metrics

Two overhead metrics were used to measure the throughput and latency of ML models. Throughput was computed as the total size of the dataset divided by the total execution time of the algorithm (Kolhe & Raza, 2013). Equation 1 shows the metrics for measuring the throughput of the classifiers.

$$\text{Throughput}=\frac{\text{total size of dataset}\left(\text{after preprocessing}\right)}{\text{total execution time}} \quad (1)$$

Also, the latency of the ML methods was measured using Equation 2.

$$\text{Latency}=\text{end}_{\text{execution time}} - \text{start}_{\text{execution time}} \quad (2)$$

As in Equation 2, the execution time is the time it takes for codes to execute. That is, in this study, it represents the amount of time it takes to train the entire dataset, create the NIDS model, and classify the entire dataset as normal or attack. The *cProfile* module in Python was used to calculate the execution time of the classifiers. The *cProfiler* module contains all details about how long the program is running. The Python time module offers a variety of ways to express time in code, including objects, integers, and texts.

Experimental Setup

The experiments were performed on a Del Intel® Core TM i7 – 6500U CPU @ 2.50GHz 2.60GHz with 16GB RAM. Anaconda Python, specifically, Spyder served as an execution environment for measuring the overhead metrics of the models. The models were built and trained using several Python packages as well as two statistical-based feature selection procedures. Other libraries such as time, *trace malloc*, and *pyRAPL* were utilized in addition to the common libraries accessible in Anaconda including *pandas*, *numpy*, and *seaborn*.

RESULTS

The experiments were conducted in three phases. The initial stage involved splitting the datasets to obtain the proportion that would give the most efficient result. Following this, the latency of the ML models was measured. Finally, the throughput was estimated with all the learning algorithms.

Train-Test Split

The scikit-learn train-test split was performed to split the data into testing and training sets. To find the most spilled efficiency, several dataset split proportions techniques were employed. The efficiency of intrusion detection was evaluated using an average of 50 runs to assess the performance of the classifiers. This comprised 50% training and 50% testing, 60% training and 40% testing, 70% training and 30% testing, 80% training, and 20% testing. Table 1 and Table 2 show the split percentages and corresponding performance results. In all cases, 80% for training and 20% for testing proved to be the most effective train-test split option for both datasets.

Table 1. Dataset train-test splitting while decreasing the size of the training set

Dataset	Training Set (%)	Testing Set (%)	Accuracy (%)	Precision (%)	Recall (%)	F1 Score (%)	Latency (s)
UNSW-NB15	80	20	82.21	79.42	82.21	79.42	210
	70	30	82.14	79.74	82.14	79.17	217
	60	40	81.91	79.41	81.91	78.89	225
	50	50	82.00	79.55	82.00	78.93	230
CICIDS2017	80	20	93.93	93.61	93.63	93.53	8547
	70	30	93.78	93.06	93.78	93.37	9759
	60	40	92.28	93.15	92.28	91.10	9707
	50	50	92.79	93.79	92.79	91.16	10234

Table 2. Dataset train-test splitting with the proposed ensemble method

Dataset	Training Set (%)	Testing Set (%)	Accuracy (%)	Precision (%)	Recall (%)	F1 Score (%)	Latency (s)
UNSW-NB15	80	20	86.83	87.78	86.83	86.40	709
	70	30	86.63	87.04	86.63	83.79	718
	60	40	86.48	86.13	86.48	83.67	797
	50	50	86.51	86.09	86.51	83.79	850
CICIDS2017	80	20	93.93	93.61	93.93	93.53	8547
	70	30	93.78	93.06	93.78	93.37	9759
	60	40	93.52	92.86	93.52	93.16	1094
	50	50	93.34	92.58	93.34	93.04	1130

Latency of ML Models

The second objective of this study is to measure the latency of the ML models, including the proposed ensemble method. The latency comparison results for Pearson correlation coefficient (PCC) and Spearman correlation coefficient (SCC) on the UNSW-NB15 and CICIDS2017 datasets are shown in Figure 1 and Figure 2. Figure 1 shows the latency obtained by each model on the UNSW-NB15 dataset for both Pearson and Spearman feature selection methods. From the figure, SVM had the largest latency of 1171.21 seconds when the Spearman correlation coefficient feature selection was applied. Furthermore, the proposed ensemble technique generated a substantial amount of latency of 919.939 seconds on the Pearson feature selection method, but a lesser amount of latency of 392.985 seconds when applied to the Spearman feature selection method. Additionally, for both feature selection methods, Random Forest and Logistics regression exhibited the minimum latencies. That is, on the Pearson feature selection method, the least latency was 210.139 seconds for Logistic regression and 214.348 seconds for Random Forest while on the Spearman feature selection, the least latency was 214.995 seconds for Random Forest and 217.89 seconds for Logistic regression. Finally, it can be observed that on average the Spearman feature selection approach had lower latencies.

Moreover, the latency of models was evaluated on the CICIDS2017 dataset, with the results shown in Figure 2. Observably, KNN had the highest latency of 17305.023 seconds with the Pearson feature selection approach. On the same Pearson feature selection method, SVM recorded a latency of 13376.772 seconds. Naive Bayes and Random Forest recorded the smallest amount of latency on the Pearson feature selection method, with latencies of 2812.121 seconds and 2833,034 seconds respectively. However, on the Spearman feature selection strategy, Logistic Regression and Naive Bayes emerged with the lowest latencies of 2862.145 seconds and 2913.153 seconds respectively. Overall, the Pearson feature selection approach had the largest average latencies.

Figure 1. Latency results for PCC and SCC on the UNSW-NB15 dataset

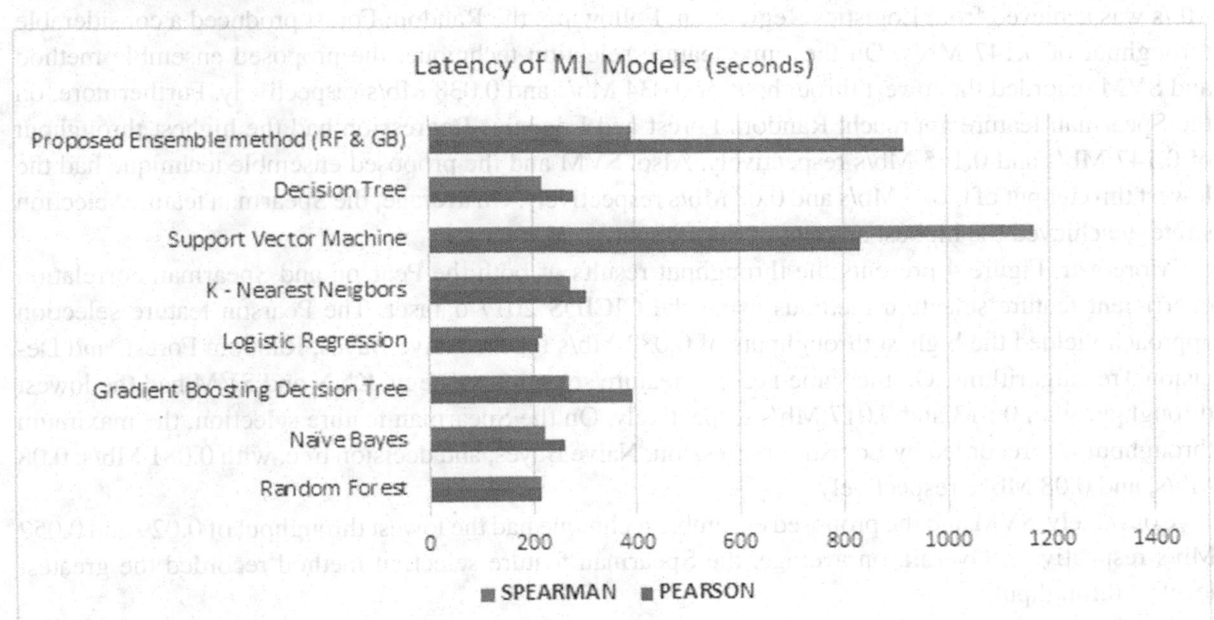

Figure 2. Latency results for PCC and SCC on the CICIDS2017 dataset

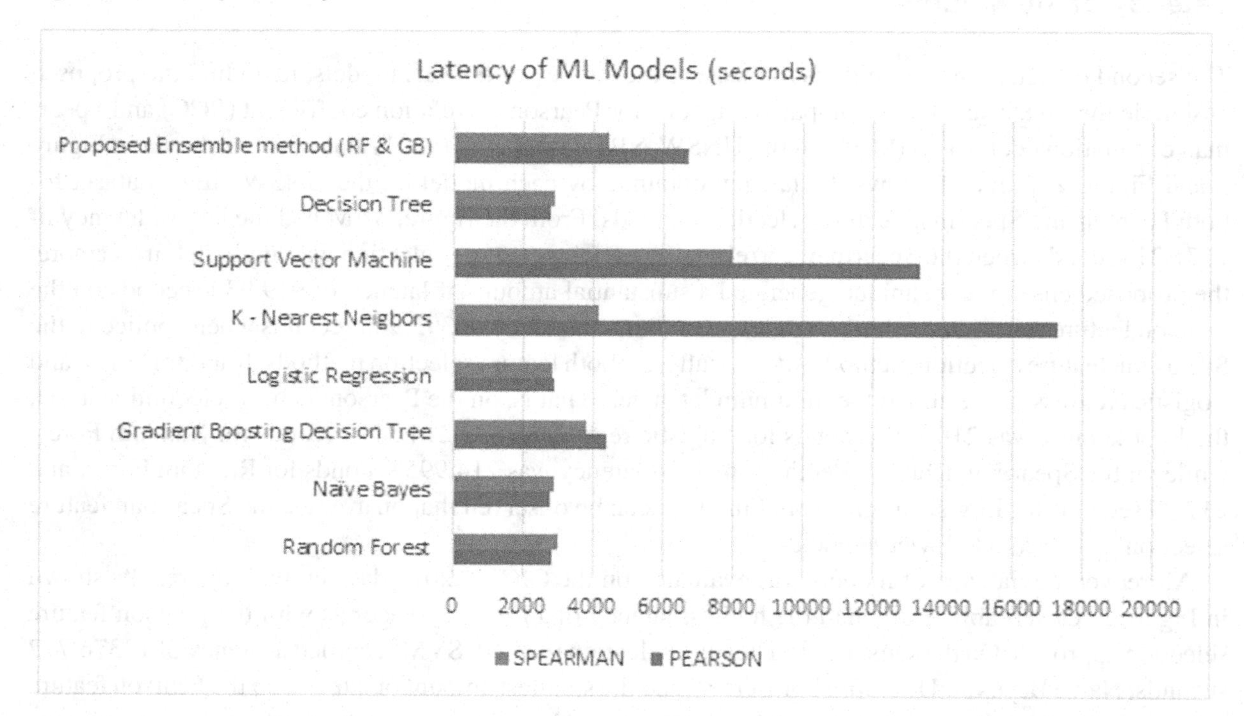

Throughput of ML Models

The third objective of this study seeks to measure the throughput of machine models. Figure 3 exhibits the throughput results for the UNSW-NB15 dataset using both Pearson and Spearman correlation coefficient feature selection approaches. On the Pearson feature selection approach, Maximum throughput of 0.150 Mb/s was achieved from Logistics Regression. Following, the Random Forest produced a considerable throughput of 0.147 Mb/s. On the same feature selection technique, the proposed ensemble method and SVM recorded the lowest throughput of 0.034 Mb/s and 0.038 Mb/s respectively. Furthermore, on the Spearman feature approach, Random Forest and Logistics Regression had the highest throughput of 0.147 Mb/s and 0.145 Mb/s respectively. Also, SVM and the proposed ensemble technique had the lowest throughput of 0.027 Mb/s and 0.08 Mb/s respectively. On average, the Spearman feature selection strategy achieved the highest amount of throughput.

Moreover, Figure 4 presents the throughput results of both the Pearson and Spearman correlation coefficient feature selection methods using the CICIDS 2017 dataset. The Pearson feature selection approach yielded the highest throughputs of 0.082 Mb/s for the Naive Bayes, Random Forest, and Decision Tree algorithms. On the same Pearson feature selection strategy, KNN and SVM had the lowest throughput with 0.013 and 0.017 Mb/s respectively. On the Spearman feature selection, the maximum throughput was recorded by Logistics regression, Naive Bayes, and decision tree, with 0.081 Mb/s, 0.08 Mb/s, and 0.08 Mb/s, respectively.

Conversely, SVM and the proposed ensemble technique had the lowest throughput of 0.029 and 0.052 Mb/s respectively. Overall, on average, the Spearman feature selection method recorded the greatest level of throughput.

Figure 3. Throughput results for PCC and SCC on the UNSW-NB15 dataset

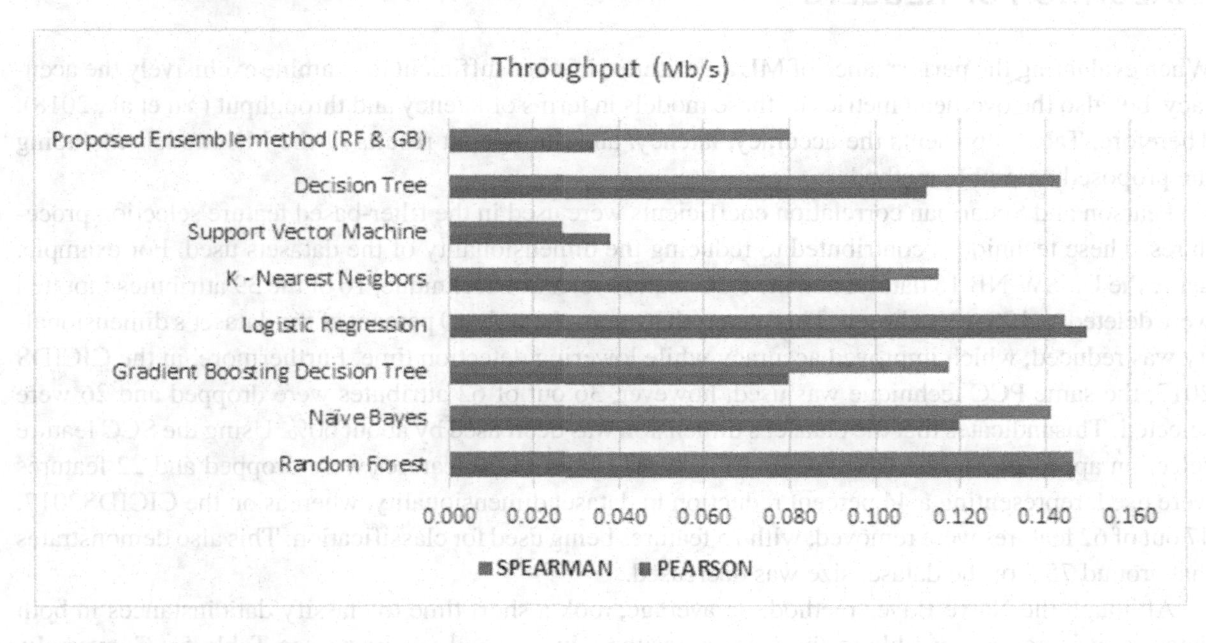

Figure 4. Throughput results for PCC and SCC on the CICIDS2017 dataset

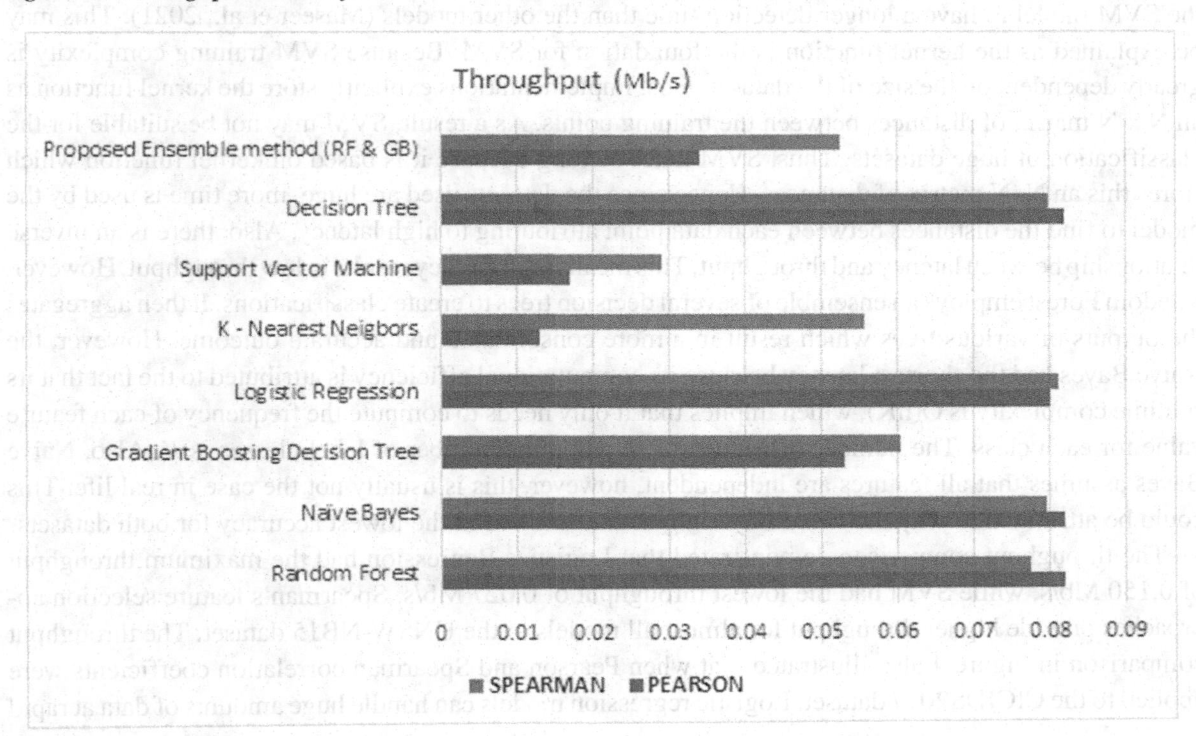

EVALUATION OF RESULTS

When evaluating the performance of ML algorithms, it is not sufficient to examine exclusively the accuracy, but also the overhead metrics of these models in terms of latency and throughput (Yu et al., 2018). Therefore, Table 3 presents the accuracy, latency, and throughput results of the ML models including the proposed ensemble method.

Pearson and Spearman correlation coefficients were used in the filter-based feature selection procedures. These techniques contributed to reducing the dimensionality of the datasets used. For example, using the UNSW-NB 15 dataset and the PCC feature selection technique, 16 of the 39 attributes inputted were deleted and 23 were chosen. This means that approximately 40 percent of the dataset's dimensionality was reduced, which improved accuracy while lowering detection time. Furthermore, in the CICIDS 2017, the same PCC technique was used, however, 36 out of 62 attributes were dropped and 26 were selected. This indicates that the dataset's dimension was decreased by about 60%. Using the SCC feature selection approach, on the UNSW-NB 15 dataset, 17 out of 39 features were dropped and 22 features were used, representing a 44 percent reduction in dataset dimensionality, whereas on the CICIDS2017, 47 out of 62 features were removed, with 15 features being used for classification. This also demonstrates that around 75% of the dataset size was decreased.

Although the Naive Bayes method, on average, took a short time to classify data instances in both datasets, it is not comparable to the other algorithms in terms of accuracy (see Table 5). Conversely, SVM took the longest time to classify both datasets. This confirmed an earlier experiment that found the SVM model to have a longer detection time than the other models (Maseer et al., 2021). This may be explained as the kernel function is the foundation for SVM. Because SVM training complexity is greatly dependent on the size of the dataset, most implementations explicitly store the kernel function as an N x N matrix of distances between the training points. As a result, SVM may not be suitable for the classification of huge datasets. Thus, SVM takes so long because it is based on kernel function which stores this an NxN matrix of distances. Hence since the datasets used are huge, more time is used by the model to find the distances between each data point attributing to high latency. Also, there is an inverse relationship between latency and throughput. This means high latency results in low throughput. However, Random Forest employs an ensemble of several decision trees to create classifications. It then aggregates the outputs of various trees which result in a more consolidated and accurate outcome. However, the Naive Bayes had the shortest latency because its computational efficiency is attributed to the fact that its runtime complexity is O(nK), which implies that it only needs to compute the frequency of each feature value for each class. The number of features is n, while the number of label classes is K. Also, Naive Bayes assumes that all features are independent, however, this is usually not the case in real life. This could be attributed to why the naive Bayes algorithm resulted in the lowest accuracy for both datasets.

The throughput comparison demonstrated that Logistics Regression had the maximum throughput of 0.150 Mb/s, while SVM had the lowest throughput of 0.027 Mb/s. Spearman's feature selection approaches provide higher throughput for almost all models in the UNSW-NB15 dataset. The throughput comparison in Figure 4 also illustrated that when Pearson and Spearman correlation coefficients were applied to the CICIDS2017 dataset. Logistic regression models can handle huge amounts of data at rapid rates due to the fact they demand lesser computing resources, particularly memory and processing power hence throughput is maximized. This is because the algorithm works by calculating the probability of data points. The NB, RF, and DT had the highest throughputs of 0.082 Mb/s when Pearson correlation coefficient feature selection was utilized, while KNN had the lowest throughput of 0.013 Mb/s. On average, the Spearman feature selection technique increased throughput for virtually all models.

The high throughput of Random forests may be explained as it only examined a small subset of characteristics when dealing with high-dimensional data, rather than all of the model's features. That is, a subset of m features (m = square root of n) is chosen at random from the set of accessible features n, minimizing latency and speeding up model training. Furthermore, Naive Bayes had a low latency and a high throughput. This is because in Naïve Bayes all that needs to be calculated is the probability of each class and the probability of each class given different input (x) values. In addition, Logistics regression applies posterior class probability which is a conditional probability based on randomly observed data as a result the training time is low - because of its randomized nature.

Conversely, the KNN low throughput may be explained as the KNN algorithm process discovers the nearest neighbors in the training set for the data being categorized, which results in a low throughput associated with it. Moreover, since the CICIDS2017 dataset has more dimensionality of the data (number of features), the algorithms will need more time to discover the data's nearest neighbors. Thus, since the latency is high, the amount of data that can be delivered over a period of time is reduced. In addition, the Spearman feature selection method used fewer features in the classification process, which reduced the model's time and hence increased throughput.

Overall, SVM and KNN models observed a high latency and low throughput, although with reasonable accuracy. However, Logistics Regression and Random Forest did fairly well in terms of accuracy, latency, and throughput. Besides, even though the Naive Bayes model had the lowest accuracy score for both datasets and feature selection methods, it had a very low latency and a relatively high throughput. Further, on the CICIDS2017 dataset, where Random Forest was found to have the highest accuracy for both Spearman and Pearson feature selection, the model performed admirably in terms of low latency and high throughput. Nevertheless, for almost all of the models employed in the studies, the Spearman feature technique recorded a reduced latency when compared to the Pearson method. Finally, when the Pearson feature selection approach was used for both datasets, the maximum accuracies were recorded.

Again, Table 3 shows that the proposed ensemble method had the largest latency for both feature selection methods used in the experiment, despite having the maximum accuracy for the UNSW-NB15 dataset. Also, the proposed model's throughput might be affected negatively as a result of its high latency, since the model had the lowest throughput. The proposed ensemble model was generated by combining the two best-performing algorithms which were RF and GBDT in addition to performing parameter tuning. The largest latency recorded by the ensemble method could be attributed to the fact two single classifiers have been combined hence more computation has to be done by the model. Also, ensembling is expensive in terms of both time and space (Wang & Srinivasan, 2017).

Table 3. Comparison of results

Feature Selection Method	Dataset	Classification Method	Accuracy (%)	Latency (s)	Throughput (Mb/s)
PCC	UNSW-NB15	RF	86.18	214.348	0.147
		NB	61.08	262.737	0.120
		GBDT	86.24	394.635	0.080
		LR	82.21	210.139	0.150
		KNN	85.32	303.129	0.104
		SVM	83.82	833.288	0.038
		DT	80.66	280.451	0.112
		Ensemble (RF&GBDT)	86.80	919.939	0.034
	CICIDS2017	RF	94.08	2833.034	0.082
		NB	54.02	2812.121	0.082
		GBDT	93.24	4414.812	0.053
		LR	88.86	2912.964	0.080
		KNN	92.99	17305.023	0.013
		SVM	92.87	13376.772	0.017
		DT	78.45	2814.852	0.082
		Ensemble (RF&GBDT)	93.94	6741.418	0.034
SCC	UNSW-NB15	RF	80.79	214.995	0.147
		NB	28.10	223.047	0.141
		GBDT	79.48	267.974	0.118
		LR	66.72	217.890	0.145
		KNN	79.47	274.337	0.115
		SVM	75.40	1171.213	0.027
		DT	73.56	219.051	0.144
		Ensemble (RF&GBDT)	80.81	392.985	0.080
	CICIDS2017	RF	92.73	2991.408	0.078
		NB	51.05	2913.513	0.080
		GBDT	91.55	3837.171	0.060
		LR	81.17	2862.145	0.081
		KNN	92.88	4173.807	0.056
		SVM	84.52	7959.756	0.029
		DT	75.75	2907.442	0.080
		Ensemble (RF&KNN)	92.30	4425.521	0.052

CONCLUSION

This study sought to assess the accuracy, latency, and throughput of SVM, DT, NB, LR, RF, GBDT, and KNN, as well as a proposed ensemble learning classification model. The UNSW-NB15 and CICIDS2017 were the two primary datasets used. The study employed a filter-based feature selection technique (Pearson Correlation Coefficient and Spearman Correlation Coefficient). In general, for both datasets utilized in this investigation, the SCC technique eliminated 60% and 75% more irrelevant characteristics than PCC, compared to 40% and 44% for PCC. Throughput is equal to the size of the dataset divided by the time taken, and based on the results of the analysis, the SCC strategy has a smaller dataset size than the PCC approach, resulting in a shorter detection time, which accounts for a greater throughput.

The latency and throughput of the models were measured and compared. In summary, the proposed ensemble technique generated a substantial amount of latency on the Pearson feature selection method. Notwithstanding, it recorded the highest accuracy on the UNSW-NB15. Random Forest and Logistics regression exhibited the minimum latencies. However, on the CICIDS2017 dataset, Logistic Regression and Naive Bayes emerged with the lowest latencies on the Spearman feature selection. In terms of throughput, the proposed ensemble method and SVM recorded the lowest throughput. Furthermore, on the Spearman feature approach, Random Forest and Logistics Regression had the maximum throughput, while SVM and the proposed ensemble technique had the lowest throughput.

The findings of this study would enable practitioners and researchers to select machine learning models that are capable of achieving low latency, higher throughput, and higher accuracy for detecting intrusion. This research could potentially be used to address an array of cybersecurity scenarios, including attack detection, prediction, and analysis. The quantity of training sets or features available determines which algorithm to choose. For example, based on the experimental results of the study, it can be stated that when the number of features in the dataset is large and the number of training examples is small, an algorithm such as Logistics regression should be preferred for a more accurate and fast detection time on devices such as IoT and embedded systems with lower computational capability, whereas SVM can be ideal for such classification if the number of features is smaller and the number of training samples is large on high-speed networks.

In the future, genetic algorithms and deep learning techniques will be utilized to investigate whether latency could further be reduced and how to increase the throughput and accuracy of the machine learning models with other intrusion detection datasets.

REFERENCES

Abolfazli, S., Sanaei, Z., Wong, S. Y., Tabassi, A., & Rosen, S. (2015). *Throughput Measurement in 4G Wireless Data Networks: Performance Evaluation and Validation*. Academic Press.

Agrawal, S., & Agrawal, J. (2015). Survey on anomaly detection using data mining techniques. Procedia Computer Science. doi:10.1016/j.procs.2015.08.220

Anwar, S., Zain, J. M., Zolkipli, M. F., Inayat, Z., Khan, S., Anthony, B., & Chang, V. (2017). From intrusion detection to an intrusion response system: Fundamentals, requirements, and future directions. *Algorithms*, *107*(2), 39. Advance online publication. doi:10.3390/a10020039

Ashfaq, R. A. R., Wang, X. Z., Huang, J. Z., Abbas, H., & He, Y. L. (2017b). Fuzziness-based semi-supervised learning approach for the intrusion detection system. *Information Sciences*, *378*, 484–497. Advance online publication. doi:10.1016/j.ins.2016.04.019

Azwar, H., Murtaz, M., Siddique, M., & Rehman, S. (2018). *Intrusion Detection in secure network for Cybersecurity systems using Machine Learning and Data Mining*. Academic Press.

Bedi, P., Gupta, N., & Jindal, V. (2021). I-SiamIDS: An improved Siam-IDS for handling class imbalance in network-based intrusion detection systems. *Applied Intelligence*, *51*(2), 1133–1151. doi:10.100710489-020-01886-y

Belavagi, M. C., & Muniyal, B. (2016). Performance Evaluation of Supervised Machine Learning Algorithms for Intrusion Detection. *Procedia Computer Science*, *89*, 117–123. doi:10.1016/j.procs.2016.06.016

Bhaskar, S., Singh, V. B., & Nayak, A. K. (2015).. . *Managing Data in SVM Supervised Algorithm for Data Mining Technology*, *27–30*, 1–4. Advance online publication. doi:10.1109/CSIBIG.2014.7056946

Bhosale, D. A., & Mane, V. M. (2016). Comparative study and analysis of network intrusion detection tools. *Proceedings of the 2015 International Conference on Applied and Theoretical Computing and Communication Technology, ICATccT 2015*, 312–315. 10.1109/ICATCCT.2015.7456901

Buczak, A. L., & Guven, E. (2016). A Survey of Data Mining and Machine Learning Methods for Cyber Security Intrusion Detection. *IEEE Communications Surveys and Tutorials*, *18*(2), 1153–1176. doi:10.1109/COMST.2015.2494502

Cervantes, J., Garcia-lamont, F., Rodríguez-mazahua, L., & Lopez, A. (2020). *Neurocomputing A comprehensive survey on support vector machine classification : Applications, challenges, and trends.* doi:10.1016/j.neucom.2019.10.118

Chan, H., Hammad, E., & Kundur, D. (2016). Investigating the impact of intrusion detection system performance on communication latency and power system stability. *Proceedings of the Workshop on Communications, Computation and Control for Resilient Smart Energy Systems, RSES 2016.* 10.1145/2939940.2939946

Choudhary, S., & Kesswani, N. (2021). A hybrid classification approach for intrusion detection in IoT network. *Journal of Scientific and Industrial Research*, *80*(9), 809–816.

Choudhury, S., & Bhowal, A. (2015). Comparative analysis of machine learning algorithms along with classifiers for network intrusion detection. *2015 International Conference on Smart Technologies and Management for Computing, Communication, Controls, Energy and Materials (ICSTM).* 10.1109/ICSTM.2015.7225395

Das, S., & Nene, M. J. (2018). A survey on types of machine learning techniques in intrusion prevention systems. *Proceedings of the 2017 International Conference on Wireless Communications, Signal Processing and Networking, WiSPNET 2017,* 2296–2299. 10.1109/WiSPNET.2017.8300169

Devan, P., & Khare, N. (2020). An efficient XGBoost–DNN-based classification model for network intrusion detection system. *Neural Computing & Applications*, *32*(16), 12499–12514. doi:10.100700521-020-04708-x

Dua, S., & Xian, D. (2011). *Data Mining and Machine Learning in Cybersecurity*. In Auerbach Publications., doi:10.1192/bjp.112.483.211-a

Fang, W., Tan, X., & Wilbur, D. (2020). Application of intrusion detection technology in network safety based on machine learning. *Safety Science, 124*, 104604. doi:10.1016/j.ssci.2020.104604

Feng, S., & Fan, F. (2019). *A Hierarchical Extraction Method of Impervious Surface Based on NDVI Thresholding Integrated With Multispectral and High-Resolution Remote Sensing Imageries*. Academic Press.

Gharaee, H., & Hosseinvand, H. (2017). A new feature selection IDS based on genetic algorithm and SVM. *2016 8th International Symposium on Telecommunications, IST 2016*, 139–144. 10.1109/ISTEL.2016.7881798

Gibert, D., Mateu, C., & Planes, J. (2020). The rise of machine learning for detection and classification of malware: Research developments, trends and challenges. *Journal of Network and Computer Applications, 153*, 102526. doi:10.1016/j.jnca.2019.102526

Guo, L., Boukir, S., Inp, B., & Pessac, F. (2017). *Building an ensemble classifier using ensemble margin. Application to image classification*. Academic Press.

Halibas, A. S., Reazol, L. B., Delvo, E. G. T., & Tibudan, J. C. (2018). Performance analysis of machine learning classifiers for ASD screening. *2018 International Conference on Innovation and Intelligence for Informatics, Computing, and Technologies, 3ICT 2018*, 1–5. 10.1109/3ICT.2018.8855759

Haripriya, L., & Jabbar, M. A. (2018). Role of Machine Learning in Intrusion Detection System: Review. *Proceedings of the 2nd International Conference on Electronics, Communication and Aerospace Technology, ICECA 2018, Iceca*, 925–929. 10.1109/ICECA.2018.8474576

Hasan, S. M. S., Imam, N., Kannan, R., Yoginath, S., & Kurte, K. (2021). *Design Space Exploration of Emerging Memory Technologies for Machine Learning Applications*. doi:10.1109/IPDPSW52791.2021.00075

Hendrawan, H., Sukarno, P., & Nugroho, M. A. (2019). Quality of service (QoS) comparison analysis of snort IDS and Bro IDS application in software-defined network (SDN) architecture. *2019 7th International Conference on Information and Communication Technology, ICoICT 2019*. 10.1109/ICoICT.2019.8835211

Hossen, S., & Janagam, A. (2018). *Analysis of network intrusion detection system with machine learning algorithms (deep reinforcement learning Algorithm)*. Academic Press.

Igual, L., & Seguí, S. (2017). *Introduction to Data Science: A Python Approach to Concepts, Techniques and Applications*. In Springer International Publishing. doi:10.1007/978-3-319-50017-1

Ingley, N. A., & Pawar, M. S. (2015). Latency reduced communication in wireless sensor networks. *2015 International Conference on Communication and Signal Processing, ICCSP 2015*, 1111–1114. 10.1109/ICCSP.2015.7322675

Islam, R., & Shahjalal, M. A. (2019). Predicting DRC Violations Using Ensemble Random Forest Algorithm. *Proceedings of the 56th Annual Design Automation Conference 2019 on - DAC '19*. 10.1145/3316781.3322478

Jeong, T., Chau, R., Kankalale, D. P., & Hyeran, J. (2019). *Going Deeper or Wider: Throughput Prediction for Cluster Tools with Machine Learning*. Academic Press.

Joshi, M. V. (2001). *Evaluating Boosting Algorithms to Classify Rare Classes : Comparison and Improvements*. Academic Press.

Kang, M., & Jameson, N. J. (2018). Machine Learning: Fundamentals. *Prognostics and Health Management of Electronics*, 85–109. doi:10.1002/9781119515326.ch4

Kasongo, S. M., & Sun, Y. (2020). A deep learning method with wrapper-based feature extraction for wireless intrusion detection system. *Computers & Security*, *92*, 101752. Advance online publication. doi:10.1016/j.cose.2020.101752

Kaya, S. (2020). *An Example of Performance Comparison of Supervised Machine Learning Algorithms Before and After PCA and LDA Application : Breast Cancer Detection*. Academic Press.

Khamis, R. A., & Matrawy, A. (2020). Evaluation of adversarial training on different types of neural networks in deep learning-based IDSs. *2020 International Symposium on Networks, Computers and Communications, ISNCC 2020*. 10.1109/ISNCC49221.2020.9297344

Khraisat, A., Gondal, I., Vamplew, P., & Kamruzzaman, J. (2019). Survey of intrusion detection systems: Techniques, datasets, and challenges. *Cybersecurity*, *2*(1), 20. Advance online publication. doi:10.118642400-019-0038-7

Kolhe, N., & Raza, N. (2013). *Throughput Comparison Results of Proposed Algorithm with Existing Algorithm*. Academic Press.

Kotpalliwar, M. V. (2015). *Classification of Attacks Using Support Vector Machine (SVM) on KDD-CUP'99 IDS Database*. doi:10.1109/CSNT.2015.185

Kotpalliwar, M. V., & Wajgi, R. (2015). Classification of attacks using support vector machine (SVM) on KDDCUP'99 IDS database. *Proceedings - 2015 5th International Conference on Communication Systems and Network Technologies, CSNT 2015*, 987–990. 10.1109/CSNT.2015.185

Kuhn, M., & Johnson, K. (2013). *Applied predictive modeling*. doi:10.1007/978-1-4614-6849-3

Kumar Singh Gautam, R., & Doegar, E. A. (2018). An Ensemble Approach for Intrusion Detection System Using Machine Learning Algorithms. *Proceedings of the 8th International Conference Confluence 2018 on Cloud Computing, Data Science and Engineering, Confluence 2018*, 61–64. 10.1109/CONFLUENCE.2018.8442693

Kumari, V. V., & Varma, P. R. K. (2017). A semi-supervised intrusion detection system using active learning SVM and fuzzy c-means clustering. *2017 International Conference on I-SMAC (IoT in Social, Mobile, Analytics, and Cloud),* 481-485. 10.1109/I-SMAC.2017.8058397

Lashkari, A. H., Gil, G. D., Saiful, M., Mamun, I., & Ghorbani, A. A. (2017). *Characterization of Tor Traffic using Time based Features*. doi:10.5220/0006105602530262

Laukemann, J., Hammer, J., Hofmann, J., Hager, G., & Wellein, G. (2018). *Automated Instruction Stream Throughput Prediction for Intel and AMD Microarchitectures. 2018 IEEE/ACM Performance Modeling, Benchmarking, and Simulation of High Performance Computer Systems*. doi:10.1109/PMBS.2018.8641578

Li, D., Wang, L., & Huang, Q. (2019). *A case study of SOS-SVR model for PCB throughput estimation in SMT production lines*. Academic Press.

Liang, D., Liu, Q., Zhao, B., Zhu, Z., & Liu, D. (2019). A Clustering-SVM Ensemble Method for Intrusion Detection System. *2019 8th International Symposium on Next Generation Electronics (ISNE)*. 10.1109/ISNE.2019.8896514

Liu, H., & Lang, B. (2019). Machine learning and deep learning methods for intrusion detection systems: A survey. *Applied Sciences (Basel, Switzerland)*, *9*(20), 4396. Advance online publication. doi:10.3390/app9204396

Lo, O., Buchanan, W. J., Griffiths, P., & Macfarlane, R. (2018). Distance measurement methods for improved insider threat detection. *Security and Communication Networks*, *2018*, 1–18. Advance online publication. doi:10.1155/2018/5906368

Maseer, Z. K., Yusof, R., Bahaman, N., Mostafa, S. A., Feresa, C. I. K., & Foozy, M. (2021). *Benchmarking of Machine Learning for Anomaly Based Intrusion Detection Systems in the CICIDS2017 Dataset*. doi:10.1109/ACCESS.2021.3056614

Mendis, C., Renda, A., Amarasinghe, S., & Carbin, M. (2019). Ithemal Accurate, Portable and Fast Basic Block Throughput Estimation using Deep. *Neural Networks*.

Meng, T., Jing, X., Yan, Z., & Pedrycz, W. (2020). A survey on machine learning for data fusion. *Information Fusion*, *57*(2), 115–129. doi:10.1016/j.inffus.2019.12.001

Meng, W., Fei, F., Li, W., & Au, M. H. (2017). Evaluating challenge-based trust mechanism in medical smartphone networks: An empirical study. *2017 IEEE Global Communications Conference, GLOBECOM 2017 - Proceedings*, 1–6. 10.1109/GLOCOM.2017.8254002

Mishra, A., & Yadav, P. (2020). Anomaly-based IDS to detect attack using various artificial intelligence machine learning algorithms: A review. *2nd International Conference on Data, Engineering and Applications, IDEA 2020*. 10.1109/IDEA49133.2020.9170674

Mishra, P., Pilli, E. S., Varadharajan, V., & Tupakula, U. (2017). Intrusion detection techniques in cloud environment: A survey. *Journal of Network and Computer Applications*, *77*, 18–47. doi:10.1016/j.jnca.2016.10.015

MoustafaN. (2019). UNSW_NB15 dataset. IEEE Dataport. doi:10.21227/8vf7-s525

Moustafa, N., & Slay, J. (2015). UNSW-NB15: A comprehensive data set for network intrusion detection systems (UNSW-NB15 network data set). *2015 Military Communications and Information Systems Conference, MilCIS 2015 - Proceedings*. 10.1109/MilCIS.2015.7348942

Mulay, S. A., Devale, P. R., & Garje, G. V. (2010). Intrusion Detection System Using Support Vector Machine and Decision Tree. *International Journal of Computer Applications*, *3*(3), 40–43. doi:10.5120/758-993

Panigrahi, R., & Borah, S. (2018). A detailed analysis of CICIDS2017 dataset for designing Intrusion Detection Systems. *International Journal of Engineering and Technology(UAE), 7*(24), 479–482.

Park, K., Song, Y., & Cheong, Y. G. (2018). Classification of attack types for intrusion detection systems using a machine learning algorithm. *Proceedings - IEEE 4th International Conference on Big Data Computing Service and Applications, BigDataService 2018*, 282–286. 10.1109/BigDataService.2018.00050

Paul, S., Banerjee, C., & Ghoshal, M. (2018). A CFS–DNN-Based Intrusion Detection System. *Lecture Notes in Electrical Engineering, 462*(March), 159–168. doi:10.1007/978-981-10-7901-6_19

Preuveneers, D., Tsingenopoulos, I., & Joosen, W. (2020). Resource Usage and Performance Trade-offs for Machine Learning Models in Smart Environments. *Sensors (Basel), 20*(4), 1176. doi:10.339020041176 PMID:32093354

Rahman, M. A., Asyhari, A. T., Leong, L. S., Satrya, G. B., Hai Tao, M., & Zolkipli, M. F. (2020). Scalable machine learning-based intrusion detection system for IoT-enabled smart cities. *Sustainable Cities and Society, 61*(January), 102324. doi:10.1016/j.scs.2020.102324

Reis, C., Ruivo, P., Oliveira, T., & Faroleiro, P. (2020). Assessing the drivers of machine learning business value. *Journal of Business Research, 117*, 232–243. doi:10.1016/j.jbusres.2020.05.053

Resende, P. A. A., & Drummond, A. C. (2018). A Survey of Random Forest Based Methods for Intrusion Detection Systems. *ACM Computing Surveys, 51*(3), 1–36. doi:10.1145/3178582

Robinson, R. R. R., & Thomas, C. (2015). *Ranking of machine learning algorithms based on the performance in classifying DDoS attacks. 2015 IEEE Recent Advances in Intelligent Computational Systems.* doi:10.1109/RAICS.2015.7488411

Rokach, L., & Maimon, O. (2010). *Data Mining and Knowledge Discovery Handbook*. Springer.

Sharafaldin, I., Lashkari, A. H., & Ghorbani, A. A. (2018). Toward generating a new intrusion detection dataset and intrusion traffic characterization. *ICISSP 2018 - Proceedings of the 4th International Conference on Information Systems Security and Privacy,* 108–116. 10.5220/0006639801080116

Son, M., & Lee, K. (2018). Distributed Matrix Multiplication Performance Estimator for Machine Learning Jobs in Cloud Computing. *2018 IEEE 11th International Conference on Cloud Computing (CLOUD),* 638–645. 10.1109/CLOUD.2018.00088

Srivastava, A., Agarwal, A., & Kaur, G. (2019). Novel Machine Learning Technique for Intrusion Detection in Recent Network-based Attacks. *2019 4th International Conference on Information Systems and Computer Networks, ISCON 2019*, 524–528. 10.1109/ISCON47742.2019.9036172

Sun, Y., Kamel, M. S., Wong, A. K. C., & Wang, Y. (2007). *Cost-sensitive boosting for classification of imbalanced data.* doi:10.1016/j.patcog.2007.04.009

Tama, B. A., Comuzzi, M., & Rhee, K. H. (2019). TSE-IDS: A Two-Stage Classifier Ensemble for Intelligent Anomaly-Based Intrusion Detection System. *IEEE Access : Practical Innovations, Open Solutions, 7*, 94497–94507. doi:10.1109/ACCESS.2019.2928048

Thaseen, S., & Kumar, A. (2017). Intrusion detection model using fusion of chi-square feature selection and multi class SVM. *Journal of King Saud University. Computer and Information Sciences*, *29*(4), 462–472. doi:10.1016/j.jksuci.2015.12.004

Thomas, R. N., & Gupta, R. (2020). A Survey on Machine Learning Approaches and Its Techniques. *2020 IEEE International Students' Conference on Electrical, Electronics and Computer Science, SCEECS 2020*. 10.1109/SCEECS48394.2020.190

Tsikoudis, N., Papadogiannakis, A., & Markatos, E. P. (2016). LEoNIDS: A Low-Latency and Energy-Efficient Network-Level Intrusion Detection System. *IEEE Transactions on Emerging Topics in Computing*, *4*(1), 142–155. doi:10.1109/TETC.2014.2369958

Tuysuzoglu, G., Moarref, N., & Yaslan, Y. (2016). *Ensemble-Based Classifiers Using Dictionary Learning*. Academic Press.

Vuong, T. P., Loukas, G., & Gan, D. (2015). *Performance evaluation of cyber-physical intrusion detection on a robotic vehicle*. doi:10.1109/CIT/IUCC/DASC/PICOM.2015.313

Wang, B., & Pineau, J. (2016). *Online Bagging and Boosting for Imbalanced Data Streams*. Academic Press.

Wang, Z., & Srinivasan, R. S. (2017). A review of artificial intelligence based building energy use prediction: Contrasting the capabilities of single and ensemble prediction models. *Renewable & Sustainable Energy Reviews*, *75*(October), 796–808. doi:10.1016/j.rser.2016.10.079

Witten, I. H., Frank, E., Hall, M. A., & Pal, C. J. (2016). *Data Mining: Practical Machine Learning Tools and Techniques*. Morgan Kaufmann.

Xia, W., Wen, Y., Foh, C. H., Niyato, D., & Xie, H. (2015). A Survey on Software-Defined Networking. *IEEE Communications Surveys and Tutorials*, *17*(1), 27–51. doi:10.1109/COMST.2014.2330903

Yongze, S., Wang, J., & Lu, Z. (2019). Asynchronous Parallel Surrogate Optimization Algorithm based on Ensemble Surrogating Model and Stochastic Response Surface Method. *2019 IEEE 5th Intl Conference on Big Data Security on Cloud (BigDataSecurity), IEEE Intl Conference on High Performance and Smart Computing, (HPSC), and IEEE Intl Conference on Intelligent Data and Security (IDS)*. 10.1109/BigDataSecurity-HPSC-IDS.2019.00024

Yu, Y., Guo, L., Huang, J., Zhang, F., & Zong, Y. (2018). A Cross-Layer Security Monitoring Selection Algorithm Based on Traffic Prediction. *IEEE Access : Practical Innovations, Open Solutions*, *6*, 35382–35391. doi:10.1109/ACCESS.2018.2851993

Zhang, H., Dai, S., Li, Y., & Zhang, W. (2018). Real-time Distributed-Random-Forest-Based Network Intrusion Detection System Using Apache Spark. *2018 IEEE 37th International Performance Computing and Communications Conference, IPCCC 2018*. 10.1109/PCCC.2018.8711068

Chapter 3
Securing Digital Ecosystems:
Harnessing the Power of Intelligent Machines in a Secure and Sustainable Environment

Mario Casillo
University of Salerno, Italy

Francesco Colace
https://orcid.org/0000-0003-2798-5834
University of Salerno, Italy

Brij B. Gupta
Asia University, Taichung, Taiwan & Lebanese American University, Beirut, Lebanon

Francesco Marongiu
University of Salerno, Italy

Domenico Santaniello
https://orcid.org/0000-0002-5783-1847
University of Salerno, Italy

ABSTRACT

Industries are evolving towards an integral digitisation of their processes. In the face of ever-faster market demands and ever-increasing quality, information technology (IT) progress represents the only solution to these needs. Industry 4.0 was born with this focus, where cybernetic systems interact with each other to achieve, efficiently, a predetermined goal. The whole process takes place with minimal, or in some cases total, absence of human intervention, leaving the systems to interact in full autonomy. This approach commonly falls under the internet of things (IoT) paradigm, in which all objects, regardless of size and functionality, are connected in a standard network exchanging information. In this sense, objects acquire intelligence because they can modify their behaviours based on the data they receive and transmit.

DOI: 10.4018/978-1-6684-9999-3.ch003

INTRODUCTION

In recent years, we are increasingly experiencing a rapid spread of so-called smart objects thanks to the steady growth of the Internet of Things (IoT). Smart objects are devices that can retrieve and transfer data and information via the Internet, enabling interaction with other objects and, thus, with people, improving the quality of life in cities, homes, workplaces, and public places (Melibari et al., 2023). IoT projects, with their own devices (sensors and actuators) and the data processed by them, can be managed through cloud platforms.

The Internet of Things (IoT) is an expression that emerged from the need to define the network of objects connected to the Internet. Objects represent embedded devices consisting of hardware and software (and possibly sensors/actuators) and network connectivity to enable connection and, thus, exchange of information with the network. Through the network infrastructure, IoT thus enables the remote sensing and control of objects, exploiting the immense potential of software in applications to solve real-world problems. Prominent examples are provided by the so-called Smart City project (Saadeh et al., 2018), aimed at improving the quality of life in the city, and that of Smart House, comfortable and technological homes(Syed et al., 2021).

The basic structure of the IoT architecture consists of five essential elements:

- sensors/actuators, i.e., the tangible and integrated components in the environments, the system terminals that continuously monitor and acquire data or perform actions based on received instructions.
- network, i.e., the connection structure.
- cloud, on which data is collected and stored.
- analytical component, consisting of the algorithms that have the fundamental role of carrying out the decision-making and computational processes to fulfil the system's objectives. This component is the core of the IoT framework.
- user interface, to allow the end user to view the state of the environment or make decisions.

In a more general view, the presented elements can be placed into three levels of operation. Sensors and actuators are part of the perceptual layer, which collects data and information from the physical world, taking advantage of various technologies, such as cameras, GPS, and wireless sensors. The application layer is responsible for showing analysis results to end users through an intuitive interface. On the other hand, data processing and transmission functions are performed by the network layer (Aboubakar et al., 2022). The network bridges the perceptual component and the application layer by relying on various wireless technologies (e.g., Wi-Fi, Bluetooth, RFID) and numerous communication protocols (such as IPV6, MQTT, and HTTP).

Given the large volume of networked devices and, thus, sensitive data inherent in users' privacy, security is paramount. There is a risk of losing control of what is communicated to the network; sensors, meters, and everyday objects capable of collecting and exchanging information can record information about habits or health status to resell to third parties (Yu et al., 2021).

In the rapidly evolving digital technology sphere, the convergence of intelligent machines and Internet of Things (IoT) systems is reshaping the dynamics of industries worldwide. As these technologies become increasingly intertwined in our everyday lives and business operations, the security of these systems represents a paramount concern for their proper functioning. The risks are myriad, and if managed poorly, the consequences could be disastrous, impacting not only industries but potentially causing harm to individuals.

The stakes are particularly high in sectors such as manufacturing, healthcare, transportation, and energy, where integrating intelligent machines and IoT devices is critical. In this context, implementing robust security measures from the initial design stage becomes imperative to prevent damage. This necessitates a holistic and proactive approach that spans the entire lifecycle of these systems, from conceptualisation and development to deployment and operation (Atlam & Wills, 2020).

A significant challenge lies in the certification of information sourced from the individual components of these systems. This is particularly relevant for devices with limited computational capabilities, such as sensors, that manage specific information. These devices form the backbone of IoT systems, constantly transmitting data that inform decision-making processes (Sadique et al., 2020). However, given the nature of these devices and the data they collect, there is often a trade-off between the availability and speed of information and data security.

In the world of IoT, the sheer volume of data and the need for real-time responses often mean that priority is given to data speed and availability. Unfortunately, this can sometimes come at the expense of data security. This could lead to vulnerabilities that malicious actors might exploit, compromising the integrity of the data and the overall system functionality (Hui et al., 2021).

Therefore, achieving a delicate balance between speed, availability, and security is of the essence. This requires investment in secure architecture designs and encryption methods, the adoption of robust authentication protocols, and the application of rigorous standards for data certification. Furthermore, with the escalating threat landscape, there is a growing need to incorporate advanced technologies such as machine learning and artificial intelligence for predictive threat detection and response.

As we increasingly rely on intelligent machines and IoT systems in our interconnected digital environment, we must place equal emphasis on enhancing their security. Only then can we ensure the resilience of these systems, safeguard industries, and protect people from potential harm.

It is possible to improve the current security situation in the IoT world by leveraging cutting-edge technologies. Blockchain is a valuable tool for securing Industries 4.0.

In recent years, Blockchain technology has been commonly associated with the concept of cryptocurrency (i.e., Bitcoin), a particular type of digital asset made possible thanks to this technology. However, looking at Blockchain only from a financial point of view is limiting. The main feature of these systems is to ensure the immutability of data once stored inside, preventing any external manipulation. This result is mainly achieved by exploiting the most modern cryptographic techniques, which allow the creation of distributed systems able to certify data's authenticity (Grover et al., 2018).

In recent years, many systems based on Blockchain have been proposed, but often with improper or unnecessary use of the technology for the use case and the final product. Indeed, they do not represent a panacea for all problems related to cybersecurity, but their use must be well thought out and targeted only and exclusively to contexts where it makes sense to use them.

The research project proposal is based, therefore, on the study of innovative applications of Blockchain technology related to Industry 4.0. On the certification of data collected from IoT devices, with the aim of building a digital ecosystem in which information can be verified in an autonomous and decentralised way across the various networked devices (Hassan et al., 2019).

The advantages of such an approach are mainly related to the security of production processes. The possibility of guaranteeing the integrity of a piece of data along the entire information transmission chain allows systems to autonomously validate everything they receive and intervene in time when they detect anomalies or attempts to tamper with the system.

A FULLY DECENTRALISED ENVIRONMENT

The problem of security in IoT device communication represents a significant challenge in the current technological landscape. One possible solution is the construction of an autonomous ecosystem where the involved actors can control the flow of data without the need for external supervision from a third party. In this context, decentralised systems, and peer-to-peer (P2P) networks play a fundamental role. These systems enable efficient distribution of data among IoT devices without relying on a central entity to coordinate or control the communication process. Instead, a set of shared and agreed-upon rules is autonomously executed by the network itself, ensuring the overall integrity of the network and the information it contains (Verma et al., 2023).

This decentralised architecture eliminates the dependence on a single point of vulnerability, making it more difficult for potential attackers to compromise the security of IoT devices. Since the data flow is distributed among multiple participants in the network, malicious actors would have to overcome a series of nodes and checks to successfully compromise the system's security. This significantly increases the resilience and robustness of the IoT ecosystem. Furthermore, the P2P approach promotes autonomy and independence of IoT devices. Each device can act as a node in the network and participate in the execution of shared rules. This means that centralised supervision from a third party is not necessary, allowing the actors to directly control the flow of data from their own devices. This provides greater control over the security and privacy of the data.

Another important feature of this architecture is its ability to ensure the global integrity of the network and the information it contains. Since the shared rules are autonomously executed by each participant in the network, continuous and distributed verification is created, detecting any anomalies or attempts to compromise security. In case of a breach, the network can react immediately and take necessary measures to mitigate the issue, such as isolating the compromised device or blocking unauthorised access.

How Decentralisation Can Improve Internet of Things

Decentralised systems have gained significant recognition and adoption over the years due to their unique features and remarkable adaptability. Such systems are characterised by a peer-to-peer network wherein multiple nodes engage in continuous communication to facilitate information exchange. The underlying concept revolves around achieving an ideal state where each node achieves perfect synchronisation with others, leading to the sharing of identical knowledge. Consequently, the reliance on a single centralised repository for information storage is eradicated, and data is instead distributed across the network through numerous replicas (Zhou et al., 2021).

The elimination of a single point of failure is a pivotal advantage offered by decentralised systems. Traditionally, centralised infrastructures are vulnerable to catastrophic failures if the central component is compromised or malfunctions. However, in a decentralised setup, the absence of a single point of failure ensures that no individual node or entity has the power to bring down the entire system. By spreading information across thousands of replicas, the network becomes more resilient and capable of withstanding various challenges, such as hardware failures, cyber-attacks, or natural disasters.

Moreover, decentralised systems embody enhanced security. Since information is distributed across multiple nodes, unauthorised access becomes significantly more challenging. Unlike centralised systems, where breaching a single point of entry can compromise the entire dataset, decentralised networks require attackers to compromise a substantial portion of the network to compromise contained information (Sedrati et al., 2020).

Another crucial aspect of this technology is the implementation of consensus protocols, which serve as the foundation for communication and synchronisation among the numerous nodes within the decentralised network. These protocols establish the rules and mechanisms by which changes in the shared knowledge are agreed upon and enforced. In the absence of a centralised authority to guarantee the accuracy of information, it is essential for the participating nodes to collectively validate proposed changes put forward by other nodes. The consensus process plays a pivotal role in ensuring the integrity and reliability of the network. Typically, a majority system is employed, where the network must agree on whether to accept specific modifications. This decentralised approach mitigates the risks associated with a single point of failure. Without a central entity that can be manipulated or influenced, the network remains highly resistant to manipulation and attacks. Even if a few nodes within the network become corrupted, their attempts to introduce malicious changes would be swiftly rejected by the majority, rendering any attack ineffective. One significant advantage of this decentralised consensus model is its increasing security with the growth of the network. As more nodes join the network and actively participate in the validation process, the system becomes more robust and secure. The collective effort of an expanding network strengthens the network's ability to maintain the security and accuracy of the information it processes (Xu et al., 2021).

The consensus protocols used in decentralised networks are designed to foster trust and cooperation among the nodes, enabling them to reach agreement autonomously and democratically. By eliminating the reliance on a centralised authority, these protocols promote transparency and inclusivity, empowering each participant to have an equal say in the decision-making process. This democratisation of consensus instills confidence in the system and cultivates a resilient and trustworthy environment for transactions and interactions.

An essential consideration in maintaining the accuracy and integrity of information within a decentralised system is the requirement for transparency (Kumar et al., 2023). The information must be made public, visible not only to the network but also to the external world. However, this poses a challenge as sensitive data should remain concealed, even within a decentralised public system. To address this issue, an additional layer is needed to ensure the confidentiality of sensitive information while preserving the decentralised nature of the system. One viable solution lies in the utilisation of public key cryptography, which offers a means of secure communication between parties even when using public channels that are potentially accessible to all.

Public key cryptography relies on a pair of cryptographic keys: a public key and a private key. The public key is accessible to everyone and can be freely distributed, while the private key remains securely held by its respective owner. By leveraging mathematical algorithms, messages encrypted with the recipient's public key can only be decrypted using their corresponding private key. This cryptographic mechanism guarantees the confidentiality and authenticity of the communication, even when conducted over public channels (Mehmood et al., 2019).

Applying public key cryptography within a decentralised system can introduce a layer of privacy protection. Sensitive data can be encrypted using the recipient's public key before being shared across the network. Only the intended recipient, possessing the corresponding private key, can decrypt and access the information. This ensures that even if the data becomes visible to all participants in the network, it remains indecipherable to anyone without the necessary private key (Shamshad et al., 2022).

The use of public key cryptography in decentralised systems offers several advantages. Firstly, it enables secure communication and data transfer, mitigating the risks associated with transmitting sensitive information over potentially vulnerable public channels. Secondly, it allows for the preservation of privacy within a decentralised network, as only authorised recipients possess the means to decrypt and access the encrypted data. This ensures that confidential information remains confidential, safeguarding individuals' privacy and protecting sensitive data from unauthorised access.

Furthermore, public key cryptography serves as a foundation for various decentralised applications, such as secure digital transactions, identity verification, and access control systems (Xiong et al., 2022). By incorporating cryptographic techniques, these applications can establish trust, verify authenticity, and enable secure interactions between parties, regardless of their location or the openness of the communication channel.

Blockchain Technologies

Blockchains are a decentralised system that emerged in 2009 with the introduction of Bitcoin, aiming to provide an alternative to the centralised banking system. However, the technology has since evolved to offer much more than just financial applications. The fundamental structure of all blockchains involves organising shared knowledge among nodes in a peer-to-peer network into blocks that are logically linked together, forming a chain. This ensures the security and consistency of information across the network using robust hashing techniques and public key cryptography. The primary concept behind blockchains, exemplified by Bitcoin, is the distribution of knowledge in an immutable manner. This immutability is particularly crucial in finance, as it prevents duplication of transactions and safeguards against spending the same coin multiple times. By achieving consensus among network nodes, blockchains ensure the integrity of transactions and establish trust within the system.

A vital component of any blockchain is the deployment of a time server within the peer-to-peer network. This time server plays a critical role in validating the correctness and sequential order of operations. To accomplish this, knowledge within the blockchain is divided into blocks, each containing a timestamp. These timestamps not only facilitate the chronological ordering of transactions but also enable verification of their validity and order.

The scope of blockchains has expanded beyond their original financial applications. Numerous industries have recognised the potential of blockchain technology and are actively exploring its uses. For instance, blockchain is employed in supply chain management, offering transparent and traceable tracking of goods from origin to the end consumer. It also finds applications in healthcare, ensuring secure storage and sharing of medical records while preserving patient privacy.

Moreover, blockchain technology has given rise to smart contracts, which are self-executing contracts with predefined conditions written into the code. These contracts automatically execute actions when specific conditions are met, eliminating the need for intermediaries, and enhancing the efficiency of contractual agreements.

Figure 1. Blockchain timestamp

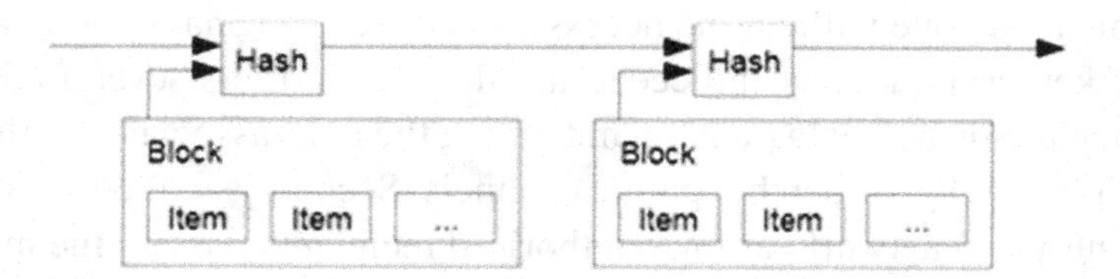

Nodes must reach a consensus on the acceptance or rejection of blocks based on their correctness. One widely used consensus protocol is the Proof of Work (PoW) system. In this system, each node actively participates in the verification process by casting a vote on proposed new blocks. To ensure decentralisation and prevent concentration of decision-making power, Bitcoin introduced the concept of "one CPU equals one vote" in the PoW system.

In a PoW system, nodes compete to validate blocks that will be added to the chain. The process begins when a node successfully validates the correctness of a block. Once validated, the block is broadcasted to the entire network, and upon verification by other nodes, it is accepted and incorporated into the chain.

To validate a new block, a system relying on hash functions is employed. The block contains essential pieces of information such as the hash of the previous block, block difficulty, timestamp, Merkle root, and nonce. The previous block's hash ensures the sequential linking of blocks in the chain, while the block difficulty determines the computational effort required for block validation. The timestamp indicates when the block was created, and the Merkle root is a hash of all the transactions included in the block.

The critical component for PoW validation is the nonce field. The nonce (a number used only once) allows for the variation of the block's hash until a hash with specific characteristics is found. In the PoW system, this involves finding a hash of the entire block that meets the difficulty requirements. Specifically, the hash must have several leading zeros equal to the difficulty specified. Miners, the nodes participating in the validation process, iterate through different nonce values to find a hash that satisfies the difficulty criteria. This process requires significant computational power and serves as a mechanism to deter malicious actors from tampering with the blockchain.

The PoW system's reliance on hash functions and the inclusion of a nonce field ensures that the process of validating blocks is resource-intensive and time-consuming. This approach adds a layer of security to the blockchain, as it becomes increasingly difficult to alter past blocks as more blocks are added to the chain. It also helps maintain the overall integrity and consistency of the blockchain network.

Figure 2. Proof of work schema

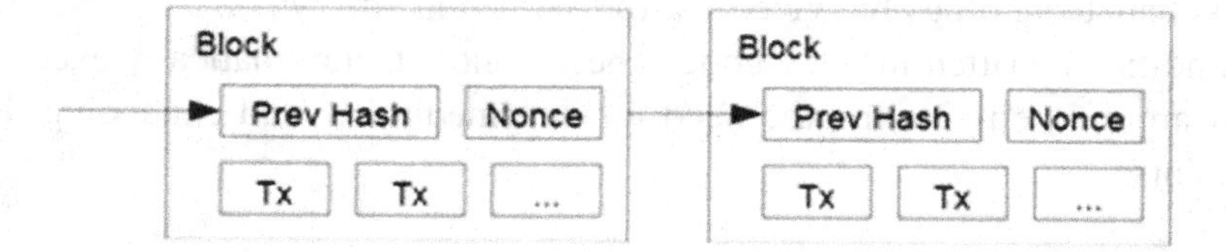

The interlinking of blocks in a blockchain ensures the immutability of the information stored within. Once a block is added to the chain, it becomes practically impossible to modify, as any alteration would result in changes to the hash values of subsequent blocks. This property of the blockchain makes it highly secure and resistant to counterfeiting. The longer the chain becomes, the more secure the information stored within it becomes. The computational power required to modify the blockchain grows exponentially, rendering it currently unachievable.

To facilitate the validation process and maintain the integrity of the blockchain, a transaction fee is paid each time a node initiates a transaction. This fee serves as a reward for the node that successfully solves the Proof of Work (PoW) algorithm. The node's computational effort in validating the block and adding it to the chain is thus incentivised.

However, the PoW system's greatest strength, its robust security, also presents a significant challenge. The scalability of the PoW system is limited. As more transactions are performed and more nodes participate in the validation process, the computational power required to add new blocks grows exponentially. The increasing competition among nodes for hash resolution places a strain on the system's resources. This scalability issue can lead to inefficiencies and bottlenecks in the long run.

Recognising the limitations of the PoW system, blockchain technology has undergone significant developments since the introduction of Bitcoin. Various alternative methods of block consensus and validation have been proposed to address scalability concerns. One such method is Proof of Stake (PoS), which introduces a different approach to block validation.

In a PoS system, validators are selected randomly to confirm transactions and validate the information within a block. The probability of being chosen as a validator is directly proportional to the stake or ownership of the cryptocurrency held by the node. This means that nodes with a higher stake have a greater chance of being selected as validators.

Unlike the competitive rewards-based mechanism of PoW, PoS systems randomise who gets to collect fees. Validators are incentivised to act honestly because their stake in the cryptocurrency serves as collateral. If a validator attempts to validate fraudulent or incorrect information, their stake can be forfeited as a penalty.

To reach consensus, multiple validators participate in the block validation process. When a specific number of validators independently verify that the block is accurate, it is considered finalised and closed. This consensus mechanism allows for faster block generation and validation, improving the scalability of the blockchain network. Different PoS mechanisms may employ various methods to achieve consensus.

By adopting a PoS consensus mechanism, blockchain networks can reduce the energy consumption associated with PoW and achieve higher scalability. The selection of validators based on stake ownership provides an incentive for honest behavior and ensures the security and integrity of the blockchain system. As blockchain technology continues to evolve, PoS is emerging as a viable alternative to PoW in various blockchain projects and protocols.

Proof of Authority (PoA) is an alternative consensus mechanism where block validation is assigned to a chosen group of trusted validators. These validators are selected based on their reputation, credibility, or authority within the network (Yang et al., 2022). Unlike Proof of Work (PoW) or Proof of Stake (PoS), where computational power or stake ownership determines validation rights, PoA relies on the expertise and trustworthiness of validators. In a PoA system, the approved validators take turns proposing and validating blocks. They are given the authority to validate transactions and ensure the accuracy of the blockchain's information. As a result, the validation process becomes more streamlined and efficient, reducing the need for extensive computational resources. The elimination of competitive mining or staking within PoA reduces the scalability challenges faced by other consensus mechanisms. Since the number of validators is limited and predetermined, the consensus process is faster and less resource intensive. This characteristic makes PoA particularly suitable for private or consortium blockchains, where a controlled group of validators can be established based on trust and authority. While PoA offers advantages in terms of scalability and efficiency, it introduces a degree of centralisation since the validation power is concentrated among a select group of validators. However, in certain use cases where trust and reputation are critical, such as enterprise applications, PoA can provide the necessary level of security and performance.

These alternative consensus mechanisms, such as PoS and PoA, are attempts to overcome the scalability limitations of the PoW system. By exploring different approaches to block validation, blockchain technology continues to evolve and adapt to the growing demands of various applications and industries.

In (Christidis & Devetsikiotis, 2016) discusses various challenges and considerations related to blockchain technology and its Internet of Things usage. It mentions that compared to a centralised database, a blockchain solution generally has lower transaction processing throughput and higher latencies. The issue of scalability and performance is highlighted, particularly in public networks that use proof-of-work mechanisms. However, new proposals, such as Bitcoin-NG, show promising results. The paragraph also addresses the challenges of maintaining privacy on the blockchain, as all transactions are open and can be analysed to identify patterns and connections between addresses. It suggests using new keys for every transaction or setting up separate blockchains for specific processes to enhance privacy. The authors further mention the importance of selecting the miner set wisely to minimise collusion and censorship risks in a blockchain network. It discusses the limited legal enforceability of smart contracts and proposes the use of dual integration, which links real-world contracts to smart contracts, to increase chances of legal enforceability. The expected value of tokenised assets and the need for assurances regarding their value are also highlighted. Additionally, authors mention the importance of carefully inspecting and implementing fail-safe mechanisms in smart contracts to prevent irreversible errors. It concludes by mentioning the need for complementary decentralised mechanisms such as a DNS service, secure communication protocols, and a file-sharing system in a blockchain network.

Peer 2 Peer Networks

One of the key components of a blockchain is the network layer, which enables the connection and exchange of information between thousands of nodes in real time. In a peer-to-peer (P2P) network, nodes are interconnected without the need for a central authority. This decentralised architecture ensures that the network remains resilient and resistant to single points of failure. Within the blockchain network, nodes engage in both one-to-many and one-to-one communication. One-to-many communication involves broadcasting information across the network, ensuring that all nodes receive the same data simultaneously. This is essential for propagating transactions, blocks, and other updates throughout the network. On the other hand, one-to-one communication facilitates specific node-to-node interactions, enabling the exchange of requests and responses between nodes as needed.

To ensure the integrity and accuracy of the information being shared, each node must adhere to specific networking rules. These rules govern how nodes send and receive information, ensuring that they are communicating with the correct peers and exchanging valid data.

The client software in a blockchain typically consists of two components: execution clients and consensus clients. Execution clients handle the execution layer of the blockchain, which involves processing and validating transactions. These clients participate in a peer-to-peer network specifically designed for the execution layer. Transactions are gossiped or disseminated across the network, allowing all execution clients to receive and process them. Consensus clients, on the other hand, are responsible for the consensus layer of the blockchain. They play a crucial role in reaching agreement among the network participants on the validity and order of transactions. When a validator is selected to propose a block, the transactions from its local transaction pool are passed to the consensus clients. These clients then perform the necessary consensus algorithms to collectively agree on the contents of the next block.

The communication between execution and consensus clients is vital for the proper functioning of the blockchain. It allows for the seamless flow of information between the execution layer, where transactions are processed, and the consensus layer, where agreement on the blockchain's state is achieved.

The network layer is a fundamental component of a blockchain, facilitating the exchange of information among nodes in a peer-to-peer network. The P2P architecture ensures decentralisation and resilience. Nodes adhere to networking rules to ensure the accuracy and integrity of the information being shared. The client software consists of execution clients and consensus clients, each with its own distinct networking stack. Execution clients communicate through a peer-to-peer network specific to the execution layer, while consensus clients facilitate consensus algorithms and interoperate with execution clients to maintain the blockchain's integrity.

Teing et al (2017) discusses an operational methodology for collecting and analysing data artifacts from the BitTorrent Sync peer-to-peer cloud storage service on the Internet of Things context. The focus of the study is on the newer client applications running on various operating systems. They highlight the significance of the private identity information in these applications. The analysis of network captures reveals that while sync files are encrypted, crucial metadata such as peer IDs and IP addresses can be found in peer discovery packets. Authors acknowledge the transient nature of memory data and emphasise the importance of capturing memory snapshots promptly. The work concludes by suggesting that the proposed methodology can serve as a basis for investigating other BitTorrent Sync-enabled clients and highlights the need for future work to validate the methodology with other clients and extend it to different IoT middleware.

Proof of Stake

Proof of Stake (PoS) is a class of consensus algorithms that offers an alternative to Proof of Work (PoW). In PoS, validators vote on the next block, and the weight of their vote is determined by the size of their stake. This approach presents several advantages over PoW, including reduced electricity consumption, lower centralisation risks, enhanced security against various types of 51% attacks, and improved efficiency.

PoS can be further categorised into two major types: chain-based and BFT-based. Chain-based PoS algorithms rely on the synchronicity of the network, where validators take turns proposing and voting on the next block. On the other hand, BFT-based PoS algorithms prioritise the consistency of nodes over availability.

To participate as a validator in a PoS system, individuals are required to stake their assets, which involves locking up a certain amount of the blockchain's base cryptocurrency. Validators are then rewarded with interest rates on their stake and receive a portion of the network's transaction fees. This incentivises individuals to become validators and actively participate in the consensus process.

In PoS-based public blockchains, the set of validators rotates to propose and vote on new blocks. The weight of each validator's vote is determined by the size of their stake or deposit. This approach offers several notable advantages. Firstly, it enhances security since validators have a direct economic interest in maintaining the integrity of the blockchain. Secondly, PoS reduces the risk of centralisation by allowing a broader base of participants to become validators, rather than concentrating power in a few entities with significant computational resources. Lastly, PoS is more energy-efficient compared to PoW, as it eliminates the need for resource-intensive mining operations.

In a general PoS algorithm, the blockchain maintains a set of validators. Any individual holding the blockchain's base cryptocurrency can become a validator by initiating a special transaction that locks up their currency as a deposit. The creation and agreement of new blocks are then achieved through a consensus algorithm in which all current validators are expected to participate actively.

This process ensures that the blockchain operates securely and reliably. Validators, with their stake at risk, are incentivised to act honestly and in the best interest of the network. The consensus algorithm ensures agreement on the validity and order of transactions, maintaining the blockchain's integrity.

In conclusion, Proof of Stake (PoS) presents a class of consensus algorithms that offer advantages over Proof of Work (PoW). Validators in PoS systems vote on the next block based on the size of their stake, resulting in reduced electricity consumption, lower centralisation risks, enhanced security against attacks, and improved efficiency. PoS algorithms can be chain-based or BFT-based, and they incentivise individuals to participate as validators by rewarding them with interest rates and transaction fees. By allowing a broader base of participants to become validators, PoS promotes security, decentralisation, and energy efficiency in blockchain networks.

There is a diverse range of consensus algorithms and reward mechanisms for validators in blockchain networks, resulting in various flavors of Proof of Stake (PoS). From an algorithmic perspective, two major types of PoS stand out: chain-based PoS and BFT-style PoS (Jalalzai & Busch, 2018).

In chain based PoS, the algorithm utilises pseudo-random selection to assign a validator the right to create a single block during each time slot, typically recurring at regular intervals (e.g., every 10 seconds). The assigned validator creates a block that references a previous block, usually the one at the end of the longest chain. As time progresses, most blocks align and converge into a single continuously growing chain.

On the other hand, BFT-style PoS employs a different approach. Validators are randomly assigned the privilege to propose blocks. However, the determination of the canonical (accepted) block involves a multi-round process. During each round, every validator sends a "vote" indicating a specific block. At the end of the process, all honest and online validators permanently agree on whether a given block belongs to the chain. It's important to note that blocks can still be chained together, but the significant difference is that consensus on a block can be reached within a single block, irrespective of the length or size of the chain that follows.

These different PoS variants showcase the versatility and adaptability of consensus algorithms in blockchain systems. Chain-based PoS relies on pseudo-random selection of validators to create blocks, ensuring the growth and convergence of a single chain. BFT-style PoS, on the other hand, introduces a multi-round voting process to establish consensus on the canonical block, enabling quicker consensus within a single block.

By exploring these various approaches, blockchain networks can tailor their consensus mechanisms to suit specific requirements, balancing factors such as decentralisation, security, speed, and efficiency. The choice of PoS flavor depends on the desired characteristics and objectives of the blockchain network.

The benefits of using Proof of Stake are many:

- No need to consume large quantities of electricity to secure a blockchain. (It's estimated that both Bitcoin and Ethereum burn over $1 million worth of electricity and hardware costs per day as part of their consensus mechanism.)
- Because of the lack of high electricity consumption requirements there is not as much need to issue as many new coins to motivate participants to keep participating in the network. It may theoretically even be possible to have negative net issuance, where a portion of transaction fees is "burned" thus decreasing the supply over time.
- Proof of Stake opens the door to a wider array of techniques that use game-theoretic mechanism design to discourage centralised cartels more effectively from forming and, if they do form, from acting in ways that are harmful to the network (such as selfish mining in Proof of Work).
- Reduced centralisation risks, as economies of scale are much less of an issue. $10 million of coins will get you exactly 10 times higher returns than $1 million of coins, without any additional disproportionate gains because at the higher level you can afford better mass-production equipment, which is an advantage for Proof of Work.
- Ability to use economic penalties to make various forms of 51% attacks vastly more expensive to carry out than Proof of Work.

Based on those key concepts, Puthal and Mohanty (2019) discusses the introduction of consensus algorithms such as Proof of Work (PoW), Proof of Stake (PoS), and Proof of Authority (PoA) for blockchain applications, specifically focusing on Bitcoin. It introduces a new algorithm called Proof of Authentication (PoAh), which aims to authenticate blocks using a similar transaction method as traditional blockchains. Trusted nodes in the network validate and authenticate blocks, increasing their trust value when they authenticate a block first. The distributed ledger is updated by all network nodes, and individual transitions are verified from the block. Miners who provide false authentication lose trust value and become normal nodes. PoAh is proposed as an energy-efficient approach for secure communications and computing on the Internet of Things (IoT), integrating fog computing, which combines cloud and edge infrastructure. Fog computing allows the storage of IoT device authentication properties in a trusted cloud and their references on edge devices for evaluating PoAh. This architecture maintains a decentralised security framework in the network. Traditional IoT device deployment methods can be used to register devices with fully trusted parts like the cloud. PoAh can be integrated with these concepts to build an end-to-end secure infrastructure.

Finality

The issue of settlement finality has become a significant point of contention in the ongoing debate between proponents of public blockchains and permissioned blockchains. Settlement finality refers to the assurance that once a transaction or operation is completed, it is irreversible and cannot be altered or reverted by the system. This concept has traditionally been associated with centralised systems, where there is a sense of certainty that completed operations are permanently settled (Sasikumar et al., 2023).

In contrast, decentralised systems, such as blockchains, vary in their ability to provide settlement finality. Depending on the specific design and consensus mechanism employed, decentralised systems may offer different levels of finality. Some blockchains provide strong finality, meaning that once a transaction is confirmed and added to the chain, it is considered settled and cannot be changed. Other blockchains offer probabilistic finality, where transactions are considered final after a certain number of confirmations, reducing the possibility of a chain reorganisation or reversal. However, there are also instances where decentralised systems may not provide finality at all, introducing the potential for uncertainties and reversals (Anceaume et al., 2022).

The significance of settlement finality is particularly pronounced in the financial industry. Financial institutions require swift and definitive confirmation of ownership rights over assets. When assets are deemed legally transferred or assigned to an entity, it is crucial that there is no possibility for a random glitch or error in the blockchain system to suddenly revoke or reverse the ownership claim over those assets. The financial industry relies on settlement finality to ensure legal certainty and avoid disputes over asset ownership.

In public blockchains, settlement finality is achieved through the consensus mechanisms employed, such as Proof of Work (PoW) or Proof of Stake (PoS). Once a transaction is validated by enough nodes and included in a block, it becomes part of the blockchain's immutable history. The decentralised and distributed nature of public blockchains enhances the level of security and finality, making it highly unlikely for a completed transaction to be reversed.

On the other hand, permissioned blockchains, which are designed for specific use cases and governed by a selected group of participants, may offer different settlement finality characteristics. In permissioned blockchains, the level of finality can be customised based on the requirements of the participating entities. This flexibility allows institutions to design consensus mechanisms that prioritise swift settlement finality and reduce the risk of transaction reversals.

In a proof of work (PoW) blockchain, transactions are technically never truly finalised. There is always the possibility that a longer chain could be created, starting from a block before the transaction in question and not including that block. However, in practice, financial intermediaries operating on public blockchains have developed a practical approach to determine when a transaction is close enough to being final for them to base decisions on it. This involves waiting for a certain number of confirmations, with six confirmations being a commonly accepted threshold.

On the other hand, proof of stake (PoS) aims to provide stronger guarantees of finality compared to PoW. It achieves this through a concept known as "economic finality." In PoS, validators are pre-registered and have a stake in the system. This eliminates the possibility that there are other validators elsewhere creating a longer chain in contradiction to the established one. If two-thirds of validators collectively support a specific claim by placing their entire stakes behind it, and simultaneously two-thirds of validators elsewhere support a contradictory claim, it implies that at least one-third of validators will lose their entire deposits regardless of the outcome. This scenario is defined as "economic finality," where we cannot guarantee that a specific claim will never be reverted, but we can guarantee that either the claim will remain unaltered, or a large group of validators will voluntarily destroy significant amounts of their own capital.

By ensuring economic finality, PoS offers a higher level of assurance regarding the validity and irreversibility of transactions. It reduces the likelihood of chain reorganisations or changes that could impact transaction history. Validators, motivated by their own economic interests and potential losses, are incentivised to act honestly, and maintain the integrity of the blockchain. This provides a stronger level of confidence, particularly in the financial industry, where certainty and immutability of transactions are essential.

While PoS offers enhanced finality guarantees, it is important to note that the possibility of a small number of validators colluding or acting maliciously cannot be eliminated. However, the economic repercussions and loss of capital for validators engaging in such behavior serve as a strong deterrent.

In conclusion, settlement finality is a critical consideration in the debate between public blockchains and permissioned blockchains. While centralised systems have traditionally provided a notion of finality, decentralised systems like blockchains offer varying degrees of settlement finality. In the financial industry, settlement finality is crucial to ensure legal certainty and prevent disputes over asset ownership. Public blockchains achieve finality through their consensus mechanisms, while permissioned blockchains provide flexibility in tailoring the level of finality based on the participants' requirements (Oyinloye et al., 2021). As blockchain technology continues to evolve, finding a balance between decentralisation and settlement finality will be essential to drive adoption and address the specific needs of various industries.

The Ethereum Model

Undoubtedly, Ethereum stands out as one of the most successful examples of blockchain technology. It was introduced in 2015 with the aim of enabling the development of applications that could run within a peer-to-peer network. Quickly, Ethereum became the standard platform for the creation of Smart Contracts, revolutionising the blockchain landscape. Beyond its enhancements to various hashing and management functions associated with virtual currencies, Ethereum introduced a groundbreaking concept: Decentralised Applications (DApps). This technology combines the power of the blockchain with a virtual machine known as the Ethereum Virtual Machine (EVM). The EVM can execute arbitrary scripts, allowing for the creation of innovative applications that operate on data stored within the blockchain ledger.

The introduction of DApps through the Ethereum platform opened a whole new realm of possibilities. It enabled developers to create decentralised applications with self-executing smart contracts, automating actions and transactions without the need for intermediaries. The decentralised nature of these applications ensures transparency, security, and immutability, making them highly desirable for a wide range of industries and use cases.

With Ethereum, the blockchain technology landscape underwent a significant transformation. It provided developers with a robust and versatile platform for building decentralised applications that can revolutionise industries such as finance, supply chain, governance, and more. The EVM serves as a powerful engine that executes complex scripts, enabling developers to unleash their creativity and build sophisticated applications within the blockchain ecosystem.

Blockchain technology, originally created for financial purposes, has evolved to support complex scripts that enable a wide range of functionalities beyond traditional finance. These scripts allow for the creation of contracts between individuals, companies, and various combinations thereof. The immutability and security offered by the blockchain ensure the integrity of the content within these contracts.

When describing blockchains like Bitcoin, the concept of a distributed ledger is often employed. Bitcoin, as a decentralised cryptocurrency, utilises fundamental cryptographic tools to maintain a ledger that records all transactional activity. This ledger operates based on a set of rules that dictate what actions can and cannot be taken to modify its contents. For instance, a Bitcoin address cannot spend more Bitcoin than it has received in previous transactions. These rules serve as the foundation for all transactions within Bitcoin and numerous other blockchain networks. In the case of Ethereum, although it also possesses its native cryptocurrency called Ether, the analogy of a distributed ledger alone does not adequately capture its complexity. Ethereum can be better described as a distributed state machine. The state of Ethereum represents a substantial data structure that encompasses not only accounts and balances but also a machine state. This machine state can undergo changes from block to block, adhering to a predefined set of rules, and can execute arbitrary machine code. The Ethereum Virtual Machine (EVM) defines the specific rules governing state changes from one block to another.

The EVM is a critical component of Ethereum's functionality, enabling the execution of smart contracts and the processing of complex scripts. Smart contracts on Ethereum are self-executing agreements with predefined conditions and outcomes. They operate within the distributed state machine and can interact with various entities, such as individuals and companies, with the assurance of security and immutability provided by the underlying blockchain technology. Ethereum's distributed state machine allows for the creation of dynamic and sophisticated applications that extend beyond simple financial transactions. It provides developers with a powerful platform to build decentralised applications (DApps) that can automate complex operations and enable secure interactions without the need for intermediaries.

By utilising the EVM and its associated state, Ethereum introduces a level of programmability that goes beyond the capabilities of many other blockchains. It enables the execution of machine code, providing developers with flexibility and the ability to create intricate applications that operate within the decentralised ecosystem.

Figure 3. Ethereum state machine

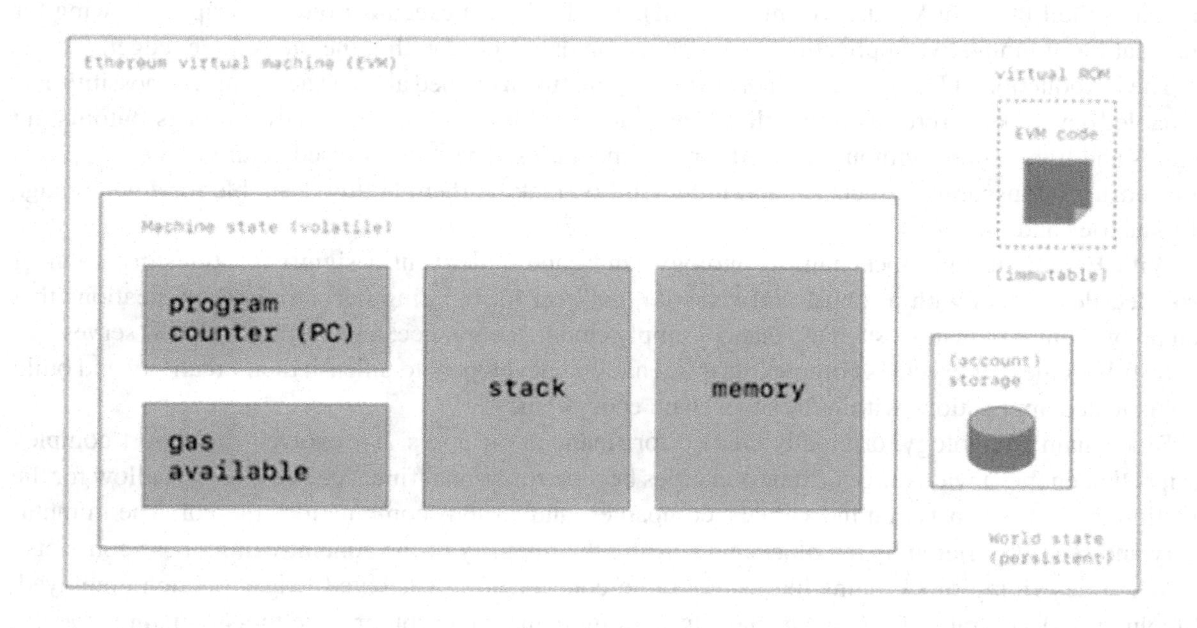

The EVM behaves as a mathematical function would: Given an input, it produces a deterministic output. It therefore is quite helpful to describe Ethereum more formally as having a state transition function:

Given an old valid state (S) and a new set of valid transactions (T), the Ethereum state transition function Y(S, T) produces a new valid output state S'.

In the Ethereum ecosystem, the state is represented by a remarkable data structure known as a modified Merkle Patricia Trie. This structure plays a pivotal role in organising and storing information within the network. It ensures that all accounts are linked together using cryptographic hashes and can be reduced to a single root hash, which is securely stored on the blockchain. The utilisation of a Patricia Merkle Trie provides a robust and cryptographically authenticated data structure capable of efficiently storing all key-value bindings.

One notable feature of this data structure is its full determinism. When two tries have the same set of key-value bindings, they are guaranteed to be identical down to the very last byte. Consequently, they share the same root hash, offering the highly sought-after efficiency of O(log(n)) for operations such as inserts, lookups, and deletes. In addition to their efficiency, Patricia Merkle Tries are simpler to understand and implement compared to more complex comparison-based alternatives like red-black trees. The Ethereum Virtual Machine (EVM) acts as the execution engine for smart contracts within the Ethereum network. It functions as a stack machine with a stack depth of 1024 items. Each item in the stack represents a 256-bit word, a choice made for its compatibility with 256-bit cryptography, including widely used cryptographic hash functions like Keccak-256.

During the execution of smart contracts, the EVM maintains a transient memory, which exists as a word-addressed byte array. This transient memory is separate from the persistent state stored on the blockchain and serves as a temporary storage space during contract execution. It provides a dedicated workspace for computations and storage required during the execution of contract logic. However, this transient memory does not persist between transactions, ensuring that each transaction operates with a clean and independent memory space. On the other hand, contracts themselves incorporate a Merkle Patricia storage trie as a word-addressable word array. This storage trie is associated with the specific account to which the contract belongs and constitutes an essential component of the global state maintained by the Ethereum network. By utilising this trie structure, contracts can persistently store and access data in a structured manner, facilitating the implementation of complex decentralised applications.

When smart contracts are compiled, they are transformed into bytecode that is executed by the EVM. The execution involves the interpretation of a set of EVM opcodes, which represent specific instructions and operations. These opcodes enable standard stack operations such as XOR, AND, ADD, SUB, and more. Additionally, the EVM incorporates a selection of stack operations tailored to the blockchain context. These specialised operations, including ADDRESS, BALANCE, BLOCKHASH, and others, allow smart contracts to interact with the blockchain, retrieve information, and perform blockchain-specific actions.

Blockchain technology, as showcased by Ethereum, has the potential to revolutionise various industries. It offers benefits such as enhanced security, improved efficiency, and increased transparency. Beyond finance, blockchain applications can be found in supply chain management, healthcare, voting systems, intellectual property, and more. The cryptographic techniques and consensus algorithms utilised in blockchain networks ensure the integrity and consensus among participants.

By providing a decentralised and trustless framework, blockchain technology introduces new possibilities for innovation and disruption. It enables individuals and organisations to build decentralised solutions, collaborate seamlessly, and establish new business models. The potential of blockchain extends beyond a single network like Ethereum, with numerous other platforms and protocols exploring different consensus mechanisms and use cases.

Figure 4. Ethereum virtual machine

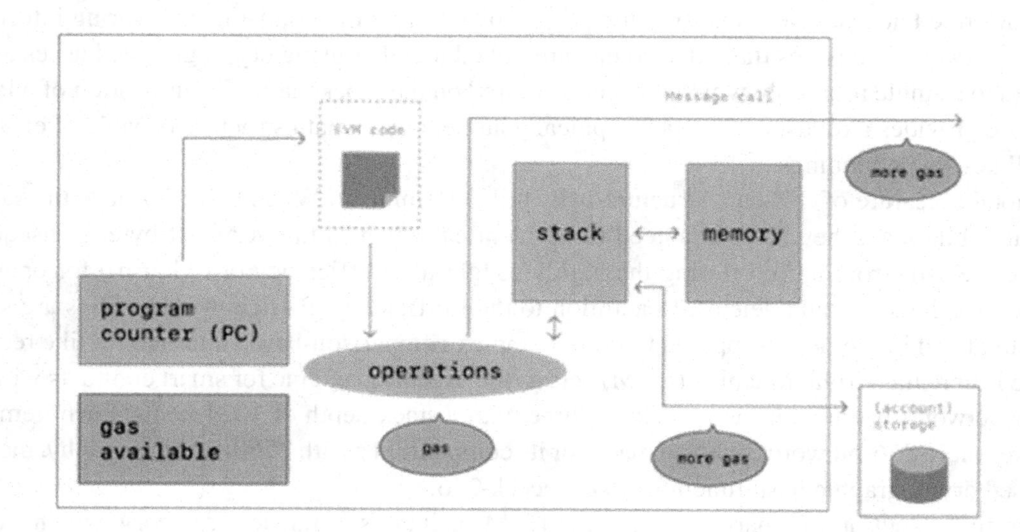

Decentralised Storage

Decentralised storage has emerged as a highly promising technology in recent years, offering innovative solutions for cloud-based data storage. It builds upon the concept of distributed file systems, which have long been utilised to distribute data in a manner that minimises the risk of single points of failure. Initially, these techniques found widespread application in server-side architectures, where data security and rapid access were critical considerations. Over time, they gained popularity due to their ability to address scalability issues faced by large systems, particularly web-accessible services.

Imagine a scenario where multiple replicas of a service are spread across different regions worldwide. Not only does this arrangement enable faster content access, but it also ensures uninterrupted service continuity even if one of the replicas encounters a failure. This decentralised approach to file systems has traditionally been associated with storage or database systems controlled by a single organisation seeking more efficient service management.

Moreover, distributed storage systems can be implemented on public peer-to-peer networks, facilitating file sharing, replication across multiple nodes, and other collaborative features. BitTorrent serves as a notable example in this domain, although it is not typically classified as a distributed file system. Rather, BitTorrent operates as a protocol designed for large file exchange, lacking a direct system for storing the files themselves.

The InterPlanetary File System (IPFS) stands out as the first decentralised peer-to-peer network storage system that guarantees immutability properties comparable to data saved on a blockchain. IPFS represents a significant advancement in the realm of decentralised storage, harnessing a distributed network architecture to enable secure and permanent file storage. By utilising content-addressing and cryptographic techniques, IPFS ensures that files are uniquely identified and preserved, enabling efficient retrieval and verification (Guidi et al., 2021).

In contrast to traditional storage systems, where files are stored based on their location or specific server, IPFS employs a content-based approach. Each file is assigned a unique cryptographic hash, which serves as its identifier across the network. When a file is requested, IPFS locates and retrieves it based on its hash, allowing for efficient and decentralised content delivery. This approach not only enhances data integrity and resilience but also reduces redundancy by eliminating duplicate files through content-addressable storage (Rizky Duto Pamungkas et al., 2021).

IPFS operates on the principle of a distributed hash table (DHT), which enables efficient peer discovery and content retrieval. When a file is added to the IPFS network, it is divided into small chunks, each assigned a unique identifier. These chunks are then distributed across participating network nodes using a DHT, ensuring redundancy and fault tolerance. Consequently, files can be retrieved by locating any node possessing the desired chunks, promoting rapid and reliable content access.

Furthermore, IPFS embraces a distributed and collaborative approach to data storage. When a user adds a file to IPFS, it is automatically replicated and distributed across multiple nodes, enhancing data availability and resilience. This decentralised replication mechanism allows for efficient content sharing, as users can retrieve files from the nearest or most accessible node within the network. Additionally, IPFS incorporates a versioning system, allowing users to track and access previous iterations of files, facilitating collaboration and ensuring data integrity.

IPFS's decentralised architecture also presents opportunities for improved data privacy and censorship resistance. Since files are distributed across a network of nodes, there is no central authority or single point of control. This decentralised nature makes it challenging for third parties to censor or manipulate content, promoting free expression and information accessibility. Moreover, IPFS supports end-to-end encryption, enabling secure and private data transmission over the network.

IPFS sets itself apart from BitTorrent by envisioning a decentralised global network where files can be stored permanently and made accessible to the world through the HTTP protocol. One of the defining characteristics of the IPFS network lies in its addressing system, which is based on the content of the files rather than conventional URIs or identifiers.

To access a file on IPFS, one must have prior knowledge of its contents. This is made possible using hash functions. When a file is added to the IPFS network, a hash function is applied to its contents, resulting in a unique Content ID (CID) for that file. Apart from negligible collisions, each file on IPFS has its own distinct CID. Therefore, once a file is uploaded to the network, it becomes immutable. If an attempt is made to upload a modified version of the file, it would generate an entirely different CID. Both the original and modified versions would coexist within the system as separate contents. CIDs, being the outcomes of hash function computations, can be efficiently organised into distributed hash tables. This organisational structure facilitates swift file retrieval from anywhere within the IPFS network. Moreover, distributed hash tables are also employed for routing operations, aiding in the discovery of peers that possess the requested files. The utilisation of hash functions and distributed hash tables in IPFS ensures the integrity and availability of files in the network. By addressing files based on their content, IPFS establishes a robust and efficient system for content retrieval. Furthermore, the immutability of files guarantees that once uploaded, they cannot be altered or replaced without generating a different CID. This characteristic enhances data integrity and permanence within the IPFS ecosystem (Lin & Zhang, 2021).

The decentralised nature of IPFS allows for efficient distribution and replication of files across the network. When a file is added to IPFS, it is automatically replicated and distributed among multiple nodes. This redundancy not only enhances data availability but also contributes to the network's fault tolerance. If one node becomes unavailable or experiences a failure, other nodes can still provide access to the file, ensuring continuous service.

IPFS promotes a collaborative and peer-to-peer approach to file storage and sharing. Users on the network can upload files, which are then replicated and distributed across multiple nodes. This decentralised replication mechanism facilitates efficient content sharing and reduces the reliance on centralised servers. Users can retrieve files from the nearest or most accessible nodes, enhancing content delivery speed and scalability.

Figure 5. IPFS architecture

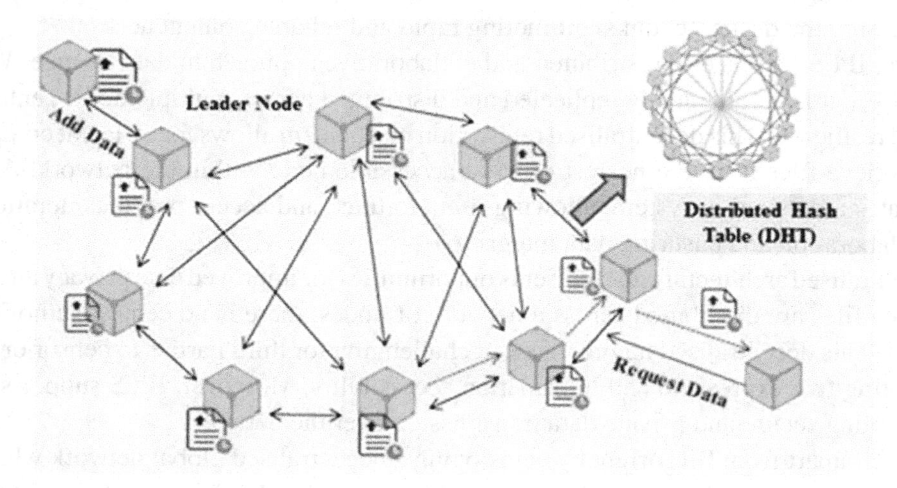

Typically, individuals who wish to upload content to the IPFS network either operate their own node or rely on various services that offer file uploading or pinning capabilities to IPFS. When it comes to file distribution within the network, the IPFS protocol does not incorporate any built-in incentive mechanism for nodes to store files uploaded by others. Consequently, when a node publishes a file, it becomes the sole host of that file. If the hosting node were to go offline for any reason, the file would no longer be accessible to other users. To address this challenge and ensure widespread distribution and long-term availability of content, a complementary cryptocurrency-based system called Filecoin has been introduced.

Filecoin itself is built upon the IPFS protocol and aims to establish a marketplace for file storage. Users can utilise Filecoins, the native cryptocurrency of the Filecoin network, to incentivise other nodes to store and host files. This protocol creates an incentive structure that enables IPFS to store files securely, ensure their permanence, and enhance their accessibility. With Filecoin, users can exchange Filecoins for storage services provided by other participants in the network. By incentivising nodes to store files on behalf of others, Filecoin promotes the distribution and replication of content across the network. This decentralised storage marketplace encourages a cooperative ecosystem where nodes are motivated to allocate resources for storing and hosting files in exchange for Filecoin rewards.

Through the integration of Filecoin with IPFS, users can store files in a secure and distributed manner for extended periods. This approach addresses the limitations of IPFS regarding file availability when the original hosting node becomes unavailable. By leveraging the Filecoin protocol, files can be replicated and stored by multiple nodes, reducing the risk of data loss, and ensuring continuous accessibility even if some nodes go offline. The Filecoin marketplace provides a mechanism for users to find storage providers willing to store their files. Users can assess different storage options based on factors such as cost, reliability, and reputation. Providers, on the other hand, can monetise their resources by offering storage services and earning Filecoin rewards in return. This decentralised marketplace fosters competition among storage providers, driving innovation and efficiency in the storage ecosystem. Moreover, Filecoin incorporates advanced cryptographic techniques to ensure the security and integrity of stored files. Data is encrypted, and cryptographic proofs are utilised to verify that the stored data remains unaltered and tamper-proof. This enhances data privacy and protects against unauthorised access or modifications.

The integration of Filecoin with IPFS not only solves the issue of incentivising file storage but also facilitates easy and efficient access to stored content. Users can access their files through IPFS using the familiar content-based addressing mechanism. By providing the corresponding CID, users can retrieve their files from any node in the network, regardless of the specific storage provider or location.

Muralidharan and Ko (2019) discusses the integration of the InterPlanetary File System (IPFS) protocol with Internet of Things (IoT) frameworks to overcome the limitations of centralised client-server models. IPFS. It incorporates techniques from systems like Distributed Hash Tables (DHT), BitTorrent, and Git to provide reliability, scalability, and security to IoT systems. The integration of IPFS with IoT offers benefits such as unique cryptographic addressing of data, efficient routing using DHTs, permanent data storage using Merkle DAG and versioning, secure file sharing and encrypted communication, and compatibility with blockchain technology. The proposed architecture involves deploying IoT nodes and initialising the IPFS protocol as an overlay, with data addressed using unique hashes and retrieval facilitated through communication among peers. The use of IPFS clustering enhances communication efficiency and reliability among IoT nodes, reducing latency and energy consumption. IPFS-Cluster enables data replication, collaborative data storage, and dynamic peer management. The integration of IPFS with IoT shows promising results in terms of reliable data sharing and low latency. Authors suggests that adopting a decentralised protocol like IPFS can benefit IoT transactions, particularly due to the smaller data sizes typically involved. Future work may involve evaluating other IPFS protocols, such as pubsub, for their suitability in IoT applications.

Privacy Protection

In the preceding paragraphs, we discussed how decentralised technologies necessitate the physical storage of data on machines that are beyond the user's control. While this poses no issues when dealing with publicly accessible data, it becomes a challenge when sensitive information requiring restricted access needs to be stored. However, private key cryptography offers a solution to the problem of data secrecy. Private key cryptography allows data to be encrypted in a manner that only those with the appropriate key can decipher the plaintext. This means that even if the message is transmitted through insecure channels, there is no need to worry about it being intercepted, read, or manipulated by unauthorised individuals at a later stage (Lu & Li, 2014a).

To achieve this level of security, asymmetric key encryption, and Key Encapsulation mechanisms (KEM) can be employed to establish a symmetric private key sharing system. KEM techniques offer significant advantages by simplifying the content encryption and decryption processes. In fact, one of the drawbacks of asymmetric key encryption is the potential performance degradation when handling long messages. By utilising KEM, it becomes possible to encapsulate a symmetric key, typically of a small size, within an asymmetric key encryption. This allows the key to be safely transmitted and/or stored even over insecure channels. The use of a symmetric key for the actual encryption and decryption processes provides efficiency and speed, while the asymmetric keys ensure secure key exchange and protection. The process typically involves the generation of a pair of keys, consisting of a public key and a private key. The public key is widely distributed and can be used by anyone to encrypt messages intended for the owner of the corresponding private key. The private key, on the other hand, is kept securely by the key owner and is used for decrypting the received messages. This asymmetric key pair ensures that only the intended recipient can decipher the encrypted message. To transmit a symmetric key securely using KEM, the sender generates a random symmetric key, encrypts it using the recipient's public key, and sends the encrypted key along with the encrypted message. Upon receiving the encrypted key, the recipient can decrypt it using their private key, obtaining the original symmetric key. This symmetric key can then be used for efficient encryption and decryption of the actual message (Lu & Li, 2014b).

The use of KEM in conjunction with private key cryptography provides a robust solution for secure data transmission and storage. It enables the secure exchange of sensitive information, even in decentralised environments where the physical control of data is distributed among various machines. By leveraging the strengths of both asymmetric and symmetric key encryption, the system ensures confidentiality, integrity, and authenticity of the data. Furthermore, the encapsulation of a symmetric key within an asymmetric key encryption helps to mitigate the performance drawbacks associated with asymmetric encryption when dealing with long messages. The smaller size of the symmetric key allows for efficient encryption and decryption operations, resulting in improved system performance.

Figure 6. Key encapsulation mechanisms

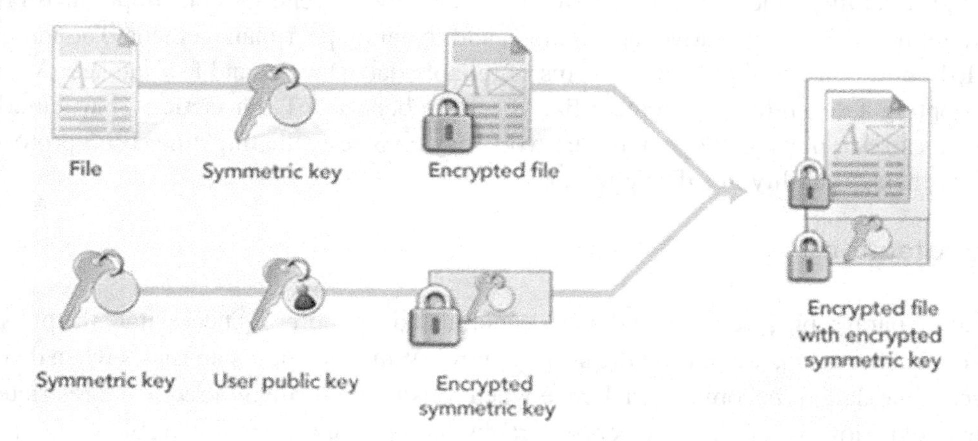

Significant are the challenges of implementing this type of security within IoT devices whose performance is often limited in terms of RAM and CPU. Cheng et al. (2021) discusses a study that analyses the performance of the lattice-based key-encapsulation scheme ThreeBears on an 8-bit AVR microcontroller. The focus is on exploring trade-offs between execution time and RAM footprint. The study presents optimised software implementations of BabyBear (an instance of ThreeBears) for the AVR platform, with two versions optimised for low RAM consumption and two for fast execution times. The low-RAM versions are reported to be the most memory-efficient implementations of a NIST second-round candidate in the literature. The study describes optimisations for the Multiply-ACcumulate (MAC) operation of ThreeBears, aiming to minimise RAM requirements or maximise performance. Results show that ThreeBears offers flexibility in optimising for low memory footprint.

CONCLUSION

One of the main problems in the world of the Internet of Things is the certification of data that smart devices constantly exchange over the network. Due to their limited hardware capabilities, these devices cannot individually implement mechanisms that are robust enough for data verification and validation. However, collectively, they can develop a robust decentralised network of information that can be governed through innovative methodologies. The use of decentralised systems, such as distributed ledgers or file systems, can be a valuable tool to overcome the problem of data certification. Through blockchain technology, it is possible to create an autonomous network of devices that, with well-defined rules, can govern the flow of data and ensure the accuracy of the information contained within. Finally, the use of innovative encryption techniques specifically designed for IoT devices solves the problems related to the sensitivity of information contained in certain devices, such as those related to critical infrastructure or medical devices.

REFERENCES

Aboubakar, M., Kellil, M., & Roux, P. (2022). A review of IoT network management: Current status and perspectives. *Journal of King Saud University. Computer and Information Sciences*, *34*(7), 4163–4176. doi:10.1016/j.jksuci.2021.03.006

AnceaumeE.Del PozzoA.RieutordT.Tucci-PiergiovanniS. (2022). On Finality in Blockchains. *Leibniz International Proceedings in Informatics, LIPIcs, 217*. doi:10.4230/LIPIcs.OPODIS.2021.6

Atlam, H. F., & Wills, G. B. (2020). *IoT Security*. Privacy, Safety and Ethics. doi:10.1007/978-3-030-18732-3_8

Cheng, H., Großschädl, J., Rønne, P. B., & Ryan, P. Y. A. (2021). Lightweight Post-quantum Key Encapsulation for 8-bit AVR Microcontrollers. doi:10.1007/978-3-030-68487-7_2

Christidis, K., & Devetsikiotis, M. (2016). Blockchains and Smart Contracts for the Internet of Things. *IEEE Access : Practical Innovations, Open Solutions*, *4*, 2292–2303. doi:10.1109/ACCESS.2016.2566339

Da Xu, L., Lu, Y., & Li, L. (2021). Embedding Blockchain Technology Into IoT for Security: A Survey. *IEEE Internet of Things Journal*, *8*(13), 10452–10473. doi:10.1109/JIOT.2021.3060508

Grover, P., Kar, A. K., & Vigneswara Ilavarasan, P. (2018). *Blockchain for Businesses: A Systematic Literature Review*. doi:10.1007/978-3-030-02131-3_29

Guidi, B., Michienzi, A., & Ricci, L. (2021). Data Persistence in Decentralised Social Applications: The IPFS approach. *2021 IEEE 18th Annual Consumer Communications & Networking Conference (CCNC)*, 1–4. doi:10.1109/CCNC49032.2021.9369473

Hassan, M. U., Rehmani, M. H., & Chen, J. (2019). Privacy preservation in blockchain based IoT systems: Integration issues, prospects, challenges, and future research directions. *Future Generation Computer Systems*, *97*, 512–529. doi:10.1016/j.future.2019.02.060

Hui, S., Wang, Z., Hou, X., Wang, X., Wang, H., Li, Y., & Jin, D. (2021). Systematically Quantifying IoT Privacy Leakage in Mobile Networks. *IEEE Internet of Things Journal*, *8*(9), 7115–7125. doi:10.1109/JIOT.2020.3038639

Jalalzai, M. M., & Busch, C. (2018). Window Based BFT Blockchain Consensus. *2018 IEEE International Conference on Internet of Things (iThings) and IEEE Green Computing and Communications (GreenCom) and IEEE Cyber, Physical and Social Computing (CPSCom) and IEEE Smart Data (SmartData)*, 971–979. 10.1109/Cybermatics_2018.2018.00184

Kumar, A., Bhushan, B., Shristi, S., Chaganti, R., & Soufiene, B. O. (2023). Blockchain-based decentralised management of IoT devices for preserving data integrity. In *Blockchain Technology Solutions for the Security of IoT-Based Healthcare Systems* (pp. 263–286). Elsevier. doi:10.1016/B978-0-323-99199-5.00009-4

Lin, Y., & Zhang, C. (2021). A Method for Protecting Private Data in IPFS. *2021 IEEE 24th International Conference on Computer Supported Cooperative Work in Design (CSCWD)*, 404–409. 10.1109/CSCWD49262.2021.9437830

Lu, Y., & Li, J. (2014a). Efficient and provably-secure certificate-based key encapsulation mechanism in the standard model. *Jisuanji Yanjiu Yu Fazhan*, *51*(7), 1497–1505. doi:10.7544/issn1000-1239.2014.20131604

Lu, Y., & Li, J. (2014b). Efficient constructions of certificate-based key encapsulation mechanism. *International Journal of Internet Protocol Technology*, *8*(2/3), 96. doi:10.1504/IJIPT.2014.066374

Mehmood, M. S., Shahid, M. R., Jamil, A., Ashraf, R., Mahmood, T., & Mehmood, A. (2019). A Comprehensive Literature Review of Data Encryption Techniques in Cloud Computing and IoT Environment. *2019 8th International Conference on Information and Communication Technologies (ICICT)*, 54–59. 10.1109/ICICT47744.2019.9001945

Melibari, W., Baodhah, H., & Akkari, N. (2023). IoT-Based Smart Cities Beyond 2030: Enabling Technologies, Challenges, and Solutions. *2023 1st International Conference on Advanced Innovations in Smart Cities (ICAISC)*, 1–6. 10.1109/ICAISC56366.2023.10085126

Muralidharan, S., & Ko, H. (2019). An InterPlanetary File System (IPFS) based IoT framework. *2019 IEEE International Conference on Consumer Electronics (ICCE)*, 1–2. 10.1109/ICCE.2019.8662002

Oyinloye, D. P., Sen Teh, J., Jamil, N., & Alawida, M. (2021). Blockchain Consensus: An Overview of Alternative Protocols. *Symmetry*, *13*(8), 1363. doi:10.3390ym13081363

Puthal, D., & Mohanty, S. P. (2019). Proof of Authentication: IoT-Friendly Blockchains. *IEEE Potentials*, *38*(1), 26–29. doi:10.1109/MPOT.2018.2850541

Rizky Duto Pamungkas, A., Husna, D., Astha Ekadiyanto, F., Eddy Purnama, I. K., Nurul Hidayati, A., Hery Purnomo, M., Nurtanio, I., Fuad Rachmadi, R., Mardi Susiki Nugroho, S., & Agung Putri Ratna, A. (2021). Designing a Blockchain Data Storage System Using Ethereum Architecture and Peer-to-Peer InterPlanetary File System (IPFS). *2021 the 7th International Conference on Communication and Information Processing (ICCIP)*, 152–157. 10.1145/3507971.3507997

Saadeh, M., Sleit, A., Sabri, K. E., & Almobaideen, W. (2018). Hierarchical architecture and protocol for mobile object authentication in the context of IoT smart cities. *Journal of Network and Computer Applications*, *121*, 1–19. doi:10.1016/j.jnca.2018.07.009

Sadique, K. M., Rahmani, R., & Johannesson, P. (2020). *Enhancing Data Privacy in the Internet of Things (IoT)*. Using Edge Computing. doi:10.1007/978-3-030-66763-4_20

Sasikumar, A., Ravi, L., Kotecha, K., Abraham, A., Devarajan, M., & Vairavasundaram, S. (2023). A Secure Big Data Storage Framework Based on Blockchain Consensus Mechanism With Flexible Finality. *IEEE Access : Practical Innovations, Open Solutions*, *11*, 56712–56725. doi:10.1109/ACCESS.2023.3282322

Sedrati, A., Stoyanova, N., Mezrioui, A., Hilali, A., & Benomar, A. (2020). Decentralisation and governance in IoT: Bitcoin and Wikipedia case. *International Journal of Electronic Governance*, *12*(2), 166. doi:10.1504/IJEG.2020.109540

Shamshad, S., Riaz, F., Riaz, R., Rizvi, S. S., & Abdulla, S. (2022). An Enhanced Architecture to Resolve Public-Key Cryptographic Issues in the Internet of Things (IoT), Employing Quantum Computing Supremacy. *Sensors (Basel)*, *22*(21), 8151. doi:10.339022218151 PMID:36365848

Syed, A. S., Sierra-Sosa, D., Kumar, A., & Elmaghraby, A. (2021). IoT in Smart Cities: A Survey of Technologies, Practices and Challenges. *Smart Cities*, *4*(2), 429–475. doi:10.3390martcities4020024

Teing, Y.-Y., Dehghantanha, A., Choo, K.-K. R., & Yang, L. T. (2017). Forensic investigation of P2P cloud storage services and backbone for IoT networks: BitTorrent Sync as a case study. *Computers & Electrical Engineering*, *58*, 350–363. doi:10.1016/j.compeleceng.2016.08.020

Verma, R., Dhanda, N., & Nagar, V. (2023). *Towards a Secured IoT Communication: A Blockchain Implementation Through APIs*. doi:10.1007/978-981-19-1142-2_53

Xiong, H., Yao, T., Wang, H., Feng, J., & Yu, S. (2022). A Survey of Public-Key Encryption With Search Functionality for Cloud-Assisted IoT. *IEEE Internet of Things Journal*, *9*(1), 401–418. doi:10.1109/JIOT.2021.3109440

Yang, J., Dai, J., Gooi, H. B., Nguyen, H. D., & Paudel, A. (2022). A Proof-of-Authority Blockchain-Based Distributed Control System for Islanded Microgrids. *IEEE Transactions on Industrial Informatics*, *18*(11), 8287–8297. doi:10.1109/TII.2022.3142755

Yu, X., Yang, Y., Wang, W., & Zhang, Y. (2021). Whether the sensitive information statement of the IoT privacy policy is consistent with the actual behavior. *2021 51st Annual IEEE/IFIP International Conference on Dependable Systems and Networks Workshops (DSN-W)*, 85–92. 10.1109/DSN-W52860.2021.00025

Zhou, Z., Pei, J., Liu, X., Fu, H., & Pardalos, P. M. (2021). Effects of resource occupation and decision authority decentralisation on performance of the IoT-based virtual enterprise in central China. *International Journal of Production Research*, *59*(24), 7357–7373. doi:10.1080/00207543.2020.1806369

KEY TERMS AND DEFINITIONS

Consensus Protocol: Is a set of rules that must be followed by participants in a network to agree on a certain value. Consensus protocols also consider any faulty participants.

Decentralisation: Is a concept that involves the distribution of resources and/or information equally within a network, where each node can communicate independently with others and with the outside world.

Decentralised Storage: These are data storage systems within shared public networks, the components of which are controlled by multiple stakeholders instead of a single entity.

Distributed Ledger: Is a shared, replicated, and synchronised database within a network. Each participant can add information by disseminating the changes to other nodes in the network.

Finality: In the context of distributed registers, the finality is the principle according to a certain change to the register becomes immutable and irreversible.

Internet of Things: Internet of things is a paradigm that extends the concept of the Internet to the world of objects, networked together, that autonomously exchange information.

Key Encapsulation Mechanisms: Is the mechanism whereby content encrypted with a symmetric key can be protected using asymmetric encryption.

Chapter 4
IoT Based Economic Flame Detection Device for Safety

Suchismita Satapathy
https://orcid.org/0000-0002-4805-1793
KIIT University, India

ABSTRACT

Mainly fires are of three types (i.e., ground, surface, and crown fire), which occur as wild land/forest fire, residential fire, building fire, and others. The number and effect of fires are an outcome of global warming, extinction of species, and climate change. To battle against these parameters/disasters, it is important to take on an exhaustive and complex methodology that empowers nonstop situational mindfulness and moment responsiveness. The outcome is unrecoverable and dangerous to the climate, environment, people's lives, and causes economic losses. The issues/barriers in the detection of fire are discussed here. For that, the authors have identified and ranked those challenges by using the best worst method. They have designed an automatic fire alarm detector at sensitive sites as one of the preventive steps to avoid the hazard. It can detect heat in a specific environment, raise an alert, turn off the building's mains, and even spray water to minimize the intensity of the fire.

1 INTRODUCTION

With the rapid expansion of urbanization across the world, the numbers of exceptionally long-term inhabitants in urban areas, as well as the population, are expanding. There are circumstances where natural elements, non-natural factors, or human factors interrupt people's lives and livelihoods and result in fatalities, environmental harm, property losses, and psychological effects. Every fire process always produces smoke and heat, and the temperature will rise when there is a fire (Ehsan et al. 2022). Flammable substances chemically react with oxygen to start flames through combustion. A high oxygen content will increase the likelihood of a fire starting. Fire disasters have historically occurred in densely populated areas. When a fire erupts, it endangers people's lives and results in enormous financial damage. So, fire detection has grown to be a major problem in recent years because to the significant damage it has caused, including the loss of human lives. These incidents can occasionally become more destructive if the fire

DOI: 10.4018/978-1-6684-9999-3.ch004

spreads to the nearby area. One efficient technique to prevent loss of life and minimize property damage is early fire detection. The fire needs to be detected early on to be evacuated from a burning location and put out the fire source. The simplest option to see a fire early and stop damage is to install a fire alarm system (Saeed et al. 2018). Fire alarms are made up of numerous interrelated parts that may detect fire and inform people via visual and audible ways. The alarm could include horns, mountable sounders, or bells. According to a research from the National Crime Records Bureau's Accidental Deaths and Suicides in India (ADSI) database, fire-related mishaps killed 35 people on average every day between 2016 and 2020. Even though the number of these accidents has been continuously diminishing, this still occurs.

According to incomplete statistics, there were 312,000 fires in India in 2016, with 1,582 people killed and 1,065 wounded, and $3.72 billion-dollar immediate property damage. There were multiple large-scale fires throughout the world in March and April of 2019, including forestry fires in Liang Shan, China, the Notre Dame fire in France, woodland fires in Italy, and a meadow fire in Russia, all of which caused substantial damage to people's lives and property. Alkhatib (2014) have summarized all the technologies that have been used for forest fire detection with exhaustive surveys of their techniques/ methods used in this application. Current techniques for urban identification rely on a variety of sensors for detection, such as smoke alarms, temperature warnings, and infrared beam warnings. Iqbal et al. (2021) mention that, even though these safeguards can help, they have serious drawbacks. To begin, an exact convergence of visible particles all around must be achieved to set off an alert. When an alarm is triggered, a fire may already be too large to even attempt to put out, undercutting the goal of early notice. Second, in a restricted environment that is incapable of a broad space, such as outside or public settings, the bulk of the notifications must be utilitarian. Third, there may be incorrect warnings. Sharma et al. (2020) state that when the non-fire molecular focus approaches the caution fixation, the alert is automatically heightened. People are unable to intervene and receive up-to-date data in a timely manner.

As per the Accidental deaths & suicides in India (ADSI, 2019) report, the number of people injured in fires has more than quadrupled, with 441 injured in 2019 compared to 1,193 in 2015. Regardless, most people were injured as a consequence of fires during this time period in 2017 (ADSI, 2019).

While the decrease in the number of fire-related occurrences, as well as deaths and injuries, is positive, the much higher number of deaths in relation to injuries is alarming. This greater mortality rate might be attributed to difficulties with clinical and crisis management in dealing with such situations, which could have resulted in a lower loss of life. Other variables, such as the severity and nature of the fire occurrences, may also contribute to higher mortality. According to_ADSI data, the total number of fire disasters has reduced during the last five years.

have explained that the Node MCU micro controller has an inbuilt Wi-Fi to acquire internet signals and provides a constant bridge between the sensor unit and the server end for remote data maintenance. The proposed logic is helpful in identifying the fire signals and informing the respective person to take appropriate action to prevent forest fires. Due to their role in preserving the stability of the universe's whole ecology, forests are important for both human survival and societal advancement (Ahmad et al. 2019; Ahmadi et al. 2017; Alkhatib 2014). Regrettably, certain unchecked human activities and unpredictable climatic conditions frequently result in forest fire scenarios (Avazov et al. 2021; Alqourabah 2021). Such fires are by far the most harmful to both human ecology and environmental assets (Arana-Pulido et al. 2021; Ajith et al. 2018; Bhoi et al. 2018; Faroudja et al. 2020). Due to global warming, increased mortality, and other factors, forest fire scenarios have considerably grown in frequency in this condition (Zope et al. 2020; Sinde et al. 2020).

According to one analysis, if the forest fire accident/or building fire had been discovered sooner, 80 percent of the fire-related costs could have been avoided. The solution to this problem is a Node MCU-based IoT-enabled fire detection and monitoring framework (Jan et al. 2022). Ahmad et al. (2019) purposed for home automation and security systems to control home appliances using different methods, such as browsing websites, Android applications, and the Global System for Mobile Communication (GSM) when away from the home, office, or organization (a liquid-cristal display (LCD) is also connected to the Node MCU board).

The Internet of Things (IoT) is being used for innovation. Node MCU fire checking is used for both mechanical and family unit functions. When it detects fire or smoke, it sends an Ethernet message to the clients, informing them of the fire. Consequently, we're using the ESP8266 from the Arduino IDE. Furthermore, the Node MCU interface with LCD display is used to show system status, such as whether smoke and overheating have been identified. Furthermore, Node MCU's interface with the Ethernet module is built in such a way that the client acquires a better comprehension of the principal condition message. It's a reference to the client's fire identification. When the client is not in close proximity to the control focus, this framework is quite handy. When a fire happens, the framework automatically faculties and warns the client by delivering an alarm to a web page or an app loaded on the user's Android mobile.

According to the ADSI 2019 report, in 2019, a total of 6,364 mishaps occurred in residential buildings, accounting for approximately 57 percent of the 11,037 fire outbreaks. As the number of incidents in private structures has decreased over time, their prevalence in relation to fire malfunctions has raised every year. During this time, there was a significant decrease in the number of disasters in commercial buildings, trains, and factories that manufacture ignitable materials. Various factories, mines, and vehicles also result in a decrease in numbers. Regardless, the patterns differ. Over a five-year period, the proportion of total fire mishaps in residential structures increased from around 40% in 2015 to 57.6 percent in 2019. This rise in the number of accidents in residential structures corresponds to a drop in the number of fire mishaps with "Other" as a cause. Fire mishaps decreased from 47.5 percent in 2015 to 34.3 percent in 2019.

The reason for the majority of the 11,037 fire mishaps reported in 2019 (59.8 percent), for example, 6,609, is listed as other. This increased number of accidents due to other factors has been observed in previous years as well. Despite the fact that the number of accidents due to other factors has decreased from a peak of 67.5 percent in 2015, it remains extraordinarily high. Cooking gas cylinder/stove bursts is the cause of the greatest number of fire disasters throughout the years, according to (ADSI, 2019). In 2019, the number of fires caused by an electrical short circuit was just more than the number of fires caused by a cylinder/stove explosion. Both reasons account for a much greater number of disasters when compared to other causes, such as fireworks and riots/agitation. Fires are unfavorable occurrences with real costs to both property and human lives. A metric for evaluating the social and economic effects of fire as well as the advancements in fire prevention and protection is the quantification of the immediate and direct costs of a fire. In addition to the direct financial expenses, fires can have several indirect and less evident negative effects on the environment.

They include environmental discharges or releases from burned items as well as air contamination from the fire plume (whose deposition is anticipated to eventually encompass land and water contamination). While the number of fires attributed to these two major causes has decreased over time, the extent of general fire accidents has displayed varying trends. The number of accidents caused by a cooking gas chamber/stove rupture has fluctuated, but the number of mishaps caused by an electrical short circuit has consistently increased step by step.

While cooking gas/stove burst-related mishaps may account for a greater number of mishaps in residential buildings, it may also be argued that most fire mishaps caused by electrical short circuits have occurred in private structures. Regardless, the greater number of incidents attributed to others, like fires due to bursts of gadgets, and petrol fires, among others, leaves a lot of room for ambiguity to study the type of cause of fire accidents in India. As far as plan and development, the ongoing alarm framework available is excessively muddled. Since the framework is so muddled, it requires continuous upkeep to guarantee that its capacities are adequate. In the interim, doing support on the current framework could build the expense of the framework.

2 LITERATURE REVIEW

Golatkar and Bhattacharya (2018) identified fire alarm systems that have several devices that work in concert to detect and alert people through visual and auditory devices in the event of smoke, fire, carbon monoxide, or other emergencies. Similarly, Yadav and Rani (2020) developed an IoT-based fire alarm system that detects fire at an early stage, generates an automatic alarm, and notifies the remote user or fire control station of the fire's incidence.

Avazov et al. (2021) have purposed a fire detector that precisely detects even small sparks and sounds an alarm within 8 seconds of a fire epidemic. On the other hand, Bhoi et al. (2018) designed the fire detection system and FireDS-IoT to prevent people from catching fire by providing an emergency warning message. The system is considered using the MQ-135 (CO_2), MQ-2 (fog), MQ-7 (CO), and DHT-11 (temperature) sensors built into the Arduino to obtain fire information.

In another research, Guo et al. (2021) proposed a new fuzzy multi-criteria decision-making technique to solve group decision-making (GDM) problems with a multi-regional linguistic approach. Such a technique is an efficient and promising method to solve real-world decision-making problems. Similarly, Alqourabah et al. (2021) proposed a smart fire detection system that, in addition to detecting the fire using integrated sensors, would also inform nearby police stations, emergency services, and property owners in order to simultaneously protect people and precious assets.

Sangam et al. (2019) proposed a method that can detect fire and offer the location of the impacted area. The Raspberry Pi 3 has been used to manage several Node MCUs that are coupled with a few sensors. A 360Â° relay motor is assembled with the camera so that it may take the image at whatever angle the fire is detected. Images and sensor data values are constantly updated on the website. A confirmation of the fire suspicion system is provided to avoid any false alarms.

In another research, Brunda et al. (2020) try to explain the smart Internet of Things (IoT) system, including its hardware and software architecture, how it functions in the context of a city or a building, as well as its benefits and drawbacks. According to the study above, an IoT-based fire system may be designed and operated in accordance with the needs of the user and the environment, but it also has some drawbacks.

Jing and Zhang (2019) had a discussion on the IoT's current state and the requirements for a fire-detecting system. They have examined the advancement and advantages of IoT for firefighting under many circumstances, including domestic firefighting, maintaining firefighting tools, monitoring fire-fighting tools, and others.

Hsu et al. (2019) have discussed the use of the Internet of Things in a kitchen fire protection system, which was discussed in this system's prevention mechanism that handles several gadgets and operations.

One sensor in a smart kitchen controls the gas supply when a fire is detected; an alarm is sounded to notify people; a system is in place for line reporting; and a kitchen surveillance camera uses the Internet. The healthcare sector, the food supply chain (FSC), the mining and energy industries (oil, gas, and nuclear), intelligent public transport (e.g., connected vehicles), and building and infrastructural facilities management for emergency response operations are all highlighted (Thibaud et al. 2016). Their review of the body of published research on IoT-based applications in high-risk Environmental, Health and Safety (EHS) industries up to 2016. It also highlights IoT-related concerns and suggested solutions in high-risk EHS industries.

IoT-based home safety systems using ARM7 have been proposed by (Anusha and Rao 2018). The author mentions that this technique has helped to protect personnel at garment factories. This technology gives the position of the fireplace in addition to detecting fire. Different types of sensors implanted with ARM7 have been used with this technique. This method alerts the administrator, who ultimately decides whether a fire disaster has occurred.

Angeline et al. (2019) describe the application of IoT for fire alarms. Using sensors, this system detects fires in residences and alerts the watchman or the fire department. A wide range of sensors that can detect fire in homes have been considered. They spoke about how each module would work and how to implement it. This method for fire prevention and safety that uses IoT-Based Intelligent Modeling of Smart Home Environment was proposed by (Faisal et al. 2018). To detect fires and improve the system's precision and effectiveness, numerous sensors have been deployed. To prevent erroneous fire detection, they have additionally used GSM.

Sha et al. (2006) used a wireless sensor network for fire rescue and a fire detector with the Node MCU, which is linked to a temperature sensor, a smoke sensor, and a signal. The temperature sensor detects temperature changes as a mixture of warmth and smoke. The sensor detects any smoke produced by consumption or a fire. An Arduino-connected buzzer sends us a notification signal. When a fire starts out, it consumes demonstrators and emits smoke. If there is a small amount of smoke in the home from candles or oil lamps, a fire alarm might be activated. Similarly, the alarm is triggered whenever the warm force is high. When the temperature recovers to normal room temperature and the smoke level falls, the bell or warning is turned off. The evolution of information and computer systems was used by (Baranovskiy et al. 2021) to predict the fire safety of the Russian railways' infrastructure in 2021. It makes possible a crucial link in the fight against fires in Wildland urban interface (WUI) zones. Numerical research of heat transmission mechanisms in the enclosing framework of a wooden building close to the forest fire front was conducted using parallel computer technology. The categorization model was proposed in 2021 by (Vikram et al. 2021) and is based on a conventional dataset that was collected from Portugal's Montesano Natural Park during a forest fire. These authors suggested a method for employing predictive analytics to find forest fires before they spread. This technique divides the forest into several zones.

In order to predict the status of a zone, such as high active (HA), medium active (MA), and low active (LA), the semi-supervised classification technique is applied. Each zone has a combination of static sensors, mobile sensors, and initiator nodes.

In order to anticipate forest fires quickly, initiator nodes in the LA and MA zones send their mobile nodes (MN) to the nearby HA zone using the Random Trajectory Generation (RTG) technique. With the creation of intermediary locations between the LA/MA and HA zones, this method establishes the MN migration channel. Preeti et al. (2021) received the meteorological data set from Kaggle in 2021, and after conducting an exploratory analysis that involved pre-processing and converting categorical data to numerical data to make the dataset more intelligible, they published their findings. During the reprocessing process, hot spot sites are found using meteorological data from the data set, and models are employed to predict the likelihood of a fire occurrence and send a notification to the closest station.

3 PROBLEM STATEMENT AND OBJECTIVE

Fire accidents are dangerous for human life, and they may occur due to natural disasters or by manual error. Fire catches for different reasons, but it is very hazardous. In this chapter, the problem is divided into two parts. In the first part, the challenges faced during the detection of fire in the ignition stage are found and prioritized by the best worst multi-criterion decision-making method. In the second phase, the fabrication of the fire-detecting device is done with IoT / AI technique, and validation is done by SWOT analysis.

4 RESEARCH METHODOLOGY

In this chapter, first, challenges faced during the detection of fire is collected by extensive literature review and expert suggestion (i.e., Fire safety officers, Academicians, and Research scholars). Then Best worst Method is implemented to prioritize the challenges. Later, fabrication and validation of the fire detection device are conducted.

4.1 Challenges Faced During Detection of Fire in Ignition Stage by Best Worst Method

Detection of fire in the ignition stage faces several challenges, such as the following:

1. Fire, a type of essential energy source for maintaining the world's ecosystem, has the potential to become a global threat, having catastrophic effects on people, animals, and the environment. Due to the panic situation among the people, we are facing many challenges during the fire accident.
2. Time factor
3. Device
4. Unavailability of resources
5. Improper analysis of fire rate
6. Natural calamities (for instance, lightning, volcano eruptions)
7. Casual approach towards controlling fire.
8. Fire at unreachable locations (forest fires, top building fires)

The challenges of detection of fire are prioritized by the best worst MCDM method.

The Best Worst method (BWM) is used to assess a set of alternatives in relation to a set of selection criteria. The BWM is based on a thorough pair-wise assessment of the criteria of choice. That is, when the decision-maker recognizes the selection criteria, he/she selects two measures: the best rule and the most dreadful rule. The best basis is the one that plays the main part in settling on the choice, while the most noticeably awful standard plays the contrary part. Then, the decision-maker provides his/her thoughts on the ideal foundation over the broad range of distinct models, as well as his/her thoughts on all the regulations over the most notably horrible standard, employing a number from a preset scale. These two sets of paired exams are used as contributions to a development problem, the ideal outcomes of which are loads of the standards (Pamučar et al. 2020).

Several authors have carried out recent research about the BWM. For instance, Mi et al. (2019) identified the BWM inconsistency improvement, the uncertain extensions, and the techniques for solving the multi-optimality model. The BWM's most notable feature is that it employs a systematic way of generating pair-wise correlations, which provides consistent results. In another research, Rezaei (2015) developed the BWM to solve multi-criteria decision-making (MCDM) problems. On the other hand, Beemsterboer et al. (2018) introduced the Best-Worst Method (BWM) decision-making to rank alternatives based on many involved criteria. Similarly, Pamučar et al. (2020) have found improvements to traditional BWM that address the ranking of alternatives. The improved BWM (BWM-I) provides decision-makers with the opportunity to express their preferences, even when there are multiple best and worst criteria. Khan et al. (2021) suggested that the BWM technique, for the manufacturing performance framework, was conducted at an Indian steel company to calculate its overall production performance.

BWM is a MCDM strategy that can be utilized in a few periods of taking care of an MCDM issue. More explicitly, it is often used to evaluate model options (uncommon in situations where objective measurements are not available to evaluate the other options). It can also be used to determine the importance (weight) of the models used to find an answer that meets the main objective (Pamučar et al. 2020). Ahmadi et al. (2017) proposed a framework for examining the social sustainability of supply chain networks in manufacturing firms. BMW has been used to address some certifiable MCDM issues in areas such as business and financial affairs, welfare, information technology (IT), design, education, and agribusiness. At a basic level, this technique can be used anywhere. The goal is to rank and choose an option from a group of options. It is typically used by a leader or a group of leaders (Rezaei 2015).

4.2 Fabrication and Testing of Proposed Fire Detecting System

The initiative is low-cost, and all levels of users may benefit from it for safety. As a result, a fire alarm system is designed. The fire-detecting system will continuously monitor the presence of significant amounts of heat and activate an alarm. A notification alert can also be sent to the open-source platform app named Blynk-Legacy and trigger the motor resulting in water sprinkling from the connected reservoir; furthermore, in the presence of an electrical fire, a servo motor is connected that can be used to extinguish the fire as a safety measure to coexist device.

5 RESULTS AND DISCUSSION

5.1 BWM for Ranking

As described below BWM is used to prioritize challenges (Rezaei 2015):

Step 1

In this step, a set of decision criteria $\{c_1, c_2, \ldots, c_n\}$ is determined. These criteria should be used to arrive at a decision. For instance, in the case of buying a car, the decision criteria can be {quality (c_1) price (c_2) comfort (c_3) safety (c_4) style (c_5)}

Step 2

Here, the best (e.g., most desirable, most important) and the worst (e.g., least desirable, least important) criteria are determined. In this step, the decision-maker identifies the best and the worst criteria in general. No comparison is made at this stage. For example, for a specific decision-maker, price (c_2) and style (c_5) may be the best and the worst criteria, respectively.

Step 3

Now, the preference of the best criterion over all the other criteria using a number between 1 and 9 is established. The resulting Best-to-Others vector is obtained by applying Equation 1

$$A_B = \left(a_{B1}, a_{B2}, \ldots\ldots, a_{Bn}\right) \tag{1}$$

Here, a_{Bj} indicates the preference of the best criterion B over criterion j. It is clear that $a_{BB} = 1$. For our example, the vector shows the preference of price (c_2) over all the other criteria.

Step 4

Next, the preference of all the criteria over the worst criterion, using a number between 1 and 9, is determined. The resulting Others-to-Worst vector is given by Equation 2

$$A_w = \left(a_{1w}, a_{2w}, \ldots\ldots, a_{3w}\right)^T \tag{2}$$

Where a_{jw} indicates the preference of the criterion j over the worst criterion w. It is clear that $a_{ww} =$ 1. For our example, the vector shows the preference of all the criteria over style (c_5).

Step 5

In this final step, the optimal weights (W_1*, W_2*,......, W_n*) are found. The optimal weight for the criteria is the one where, for each pair of W_B/W_j and W_j/W_w, we have $W_B/W_j = a_{bj}$ and $W_j/W_w = a_{jw}$. To satisfy these conditions for all j, we should find a solution where the maximum absolute differences [(W_B/W_j)-a_{bj}] and [(W_j/W_w)-a_{jw}] for all j is minimized.

The outcome of the study can be found in Table X.4. The results provide some insight to make strategic marginal decisions. In Table 1, we have listed the criteria which we ranked using the Best Worst Method.

Table 1 shows Criteria Selection.

Table 1. Criteria selection

Criteria	Short Description
Time factor (C1) (Arana-Pulido 2018)	Time taken by the device to detect the fire and alarm.
Devices (C2) (Pérez-Mato 2021)	It is a mechanical device that consists of different varieties of sensors, and transistors, and works on an electrical circuit to detect the fire and alarm the authorities.
Unavailability of resources (C3)	Resources like optical sensors, LDR (Light sensor Module), Wireless Network systems (WNS), and nodes are not accessible to non-skilled people.
Improper analysis of fire rate (C4)	The approach toward the initial stage of fire is not properly analyzed by the authorities
Natural calamities (C5)	All the natural calamities like lightning, volcano eruption, and other uncontrollable natural factors
Casual approach to controlling fire (C6)	People do not take fire so seriously, which may lead to accidents
Fire at unreachable location (C7)	Some of the locations, like in the dense region of the forest, are not reachable, which may lead to lost sudden damage

For the ranking, we selected "Devices (C2)" as the best criterion and "Natural Calamities (C5)" as the worst criterion. In Table. 2 Best Challenges preference (i.e., (C2) "Device" is selected among other parameters as per an expert's suggestion between, 1 and 9 is written row-wise. In Table 3. worst challenges (i.e., C5) Natural calamities are selected, among other challenges, column-wise between 1 to 9. In Table 4 the maximum absolute differences $[(W_B/W_j)\text{-}a_{bj}]$ and $[(W_j/W_w)\text{-}a_{jw}]$ are calculated.

For Best weight calculation, the difference between the best-chosen challenge (C2/Device) and other challenges is calculated. For the worst weight calculation, the difference between all criteria and the worst challenge (C5/Natural calamities) is calculated. Later, on to Table 4, the maximum absolute difference is found by taking the difference between $(W_B/W_j)\text{-}a_{bj}]$ and $[(W_j/W_w)\text{-}a_{jw}]$ is found.

Minimum $\S_1$
Subject to

$$\left|W_B - a_{sj} W_j\right| \le \S$$

$$\left|W_j - a_{jw} W_w\right| \le \S$$

where

$W_j \geqq 0$ for all j

$$\sum Wj = 1$$

Here, W_B = Best Criteria/challenge (C2), Wj = other criteria/challenges except best criteria (I.e.C1, C3, C4, C5, C6, C7), asj = Values 1 to 9 preference found by comparing best criteria with respect to other criteria/challenges (I.e. 2, 1, 3, 3, 9, 4, 4 found in the Table 2). Similarly Ww = Worst criteria/challenge ajw = values 1 to 9 found by comparing worst criteria with respect to other criteria/challenges (I.e., 7, 7, 3, 4, 1, 4, 4 found in Table 3)

For the best criteria/challenges, the equation is calculated as follows: Where (§ /Xeta = Optimal weight). C2/Device is called as best criterion compared to other challenges. For example, in Table. X2 in the first row the C1/time factor is present, asj = 2, So, equation generated,

C2-2 (C1) ≤ § (1)

Likewise, other equations are also generated as per sequence found in Table 2.

Let,

```
C2-2(C1)≤§

C2-3(C3)≤§
C2-3(C4)≤§
C2-9(C5)≤§
C2-4(C6)≤§
C2-4 C7≤§
```

Similarly, the worst criteria equation found below. Where (§ /Xeta=Absolute weight). For example, worst criteria found, as per Table 3 is C5/Natural calamities). C1 = Time factor and 7 is the value found by the decision maker from (1 to 9) scale during comparison with worst criteria as per the table X3. So the equation is generated as

C1-7(C5) ≤ § (2)

Accordingly, all equations are generated from Table 3.

```
C1-7(C5)≤§
C2-7 (C5)≤§
C3-3(C5)≤§
C4-4(C5)≤§
C6-4(C5)≤§
C7-4(C5)≤§
```

Then, after solving all the linear equations, (C1 to C7 values found give the optimal weights of the challenges as shown in Table 4.

Table 2 shows best challenges preferences.

Table 3 shows the worst challenges.

Table 4 shows results of BWM: weights of the challenges.

Table 2. Best challenges preferences

Best to Others	C1	C2	C3	C4	C5	C6	C7
C2	2	1	3	3	9	4	4

Table 3. The worst challenges

Others to the Worst	C5
C1	7
C2	7
C3	3
C4	4
C5	1
C6	4
C7	4

Table 4. Results of BWM: Weights of the challenges

Challenges	C1	C2	C3	C4	C5	C6	C7
Weights	0.197109067017083	0.325886990801577	0.131406044678055	0.131406044678055	0.0367936925098555	0.0985545335085414	0.0788436268068331

Ksi*(Consistency Ratio)/§ = 0.0683311432325887

A smaller Ksi value (close to zero) indicates higher consistency, whereas a higher Ksi value (close to one) indicates lower consistency. The consistency ratio is a metric that indicates the consistency between pairwise comparisons. The Best-Worst Method (BWM) uses ratios of the relative importance of criteria in pairs based on the assessment done by decision-makers, calculated in BWM calculator and the same formulas consistency ratio of Analytical Hierarchy Process (AHP)

Figure 1 shows BWM ranking. In the *x* axis, parameters/challenges are noted, whereas in the *y* axis, the weights of the challenges are noted. From Table 4, devices (C2) is ranked first and (C5) natural calamities are ranked last. Fig 1 shows the same rank of all the challenges.

5.2 Design of Fire Detection System

This chapter deals with the explanation of how to build a system using hardware and software development. Furthermore, the chapter goes over the hardware and software in detail, stage by stage.

The automatic fire alarm and control systems design can be broken down into basic modules. They are as follows:

Figure 1. BWM ranking

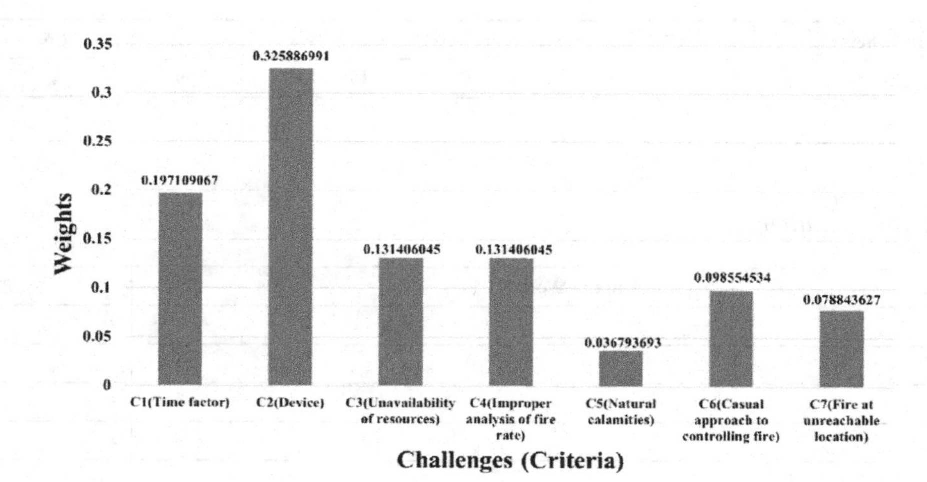

5.2.1 The Relay Module

(Model: HW-307 10A 250VAC/ 30VDC & 12A 125VAC/ 28VDC) shows the specification of the device. The relay is the device that opens or closes contacts to cause the other electric controls to operate. It detects an unwanted situation in a designated area and sends orders to the circuit breaker to turn ON or OFF the affected area.

5.2.2 Working and Principle of Relay Module

It is based on the electromagnetic attraction principle. When the relay circuit detects a fault current, it activates the electromagnetic field, which produces a short magnetic field. This magnetic field moves the relay armature to open or close the connections. The high-power relay has two contacts for opening the switch, whereas the small-power relay has just one.

Figure 2 illustrates the internal workings of the relay. It has an iron core as well as a control coil looped around it. The coil is energized by the connections of the load and the control switch. The current flowing through the coil creates a magnetic field surrounds it. Because of the magnetic field, the higher arm of the magnet attracts the lower arm. Therefore, it closes the circuit and enables current to flow through the load. If a contact has previously been closed, it travels in the other direction, opening the contacts.

5.2.3 Node MCU Specification (ESP8266 M0D)

Node MCU is an open-source Lua (Lua is a multi-paradigm, lightweight programming language designed primarily for use in embedded applications. Lua is cross-platform because the built byte-code interpreter is written in ANSI C and Lua offers a reasonably simple C API for embedding it into applications) based firmware and development board specially targeted for IoT-based applications. The microprocessor runs on an adjustable clock frequency of 80MHz to 160MHz and supports RTOS. To store data and programs, the Node MCU contains 128 KB of RAM and 4MB of Flash memory. It is perfect for IoT projects due to its high processing power, and built-in Wi-Fi / Bluetooth.

5.2.4 Servo Motor

Because of the carbon fiber gears, the servo motor is substantially lighter than a metal gear motor of the same size. Because a metal gear servo motor adds unnecessary weight, we recommend using these lightweight plastic gear servo motors for minor load circumstances.

The Tower-pro SG90 9g mini servo rotates 180 degrees. It's a digital servo motor that accepts and processes PWM (Pulse width modulation) signals much more quickly and efficiently. The internal circuit is complex, allowing for good torque, holding power, and rapid updates in response to external pressures.

The mobile application module flashes in three different colors: red, brown and orange. Each of these colors means the following:

RED – Positive (fire detection alarm)
Brown – Negative (no signal)
Orange – Signal (fire extinguished)

The servo motor used for fire extinguishing has the following specifications

1. Model: SG90
2. Weight: 9 gm
3. Operating voltage: 3.0V~ 7.2V
4. Servo Plug: JR
5. Stall torque @4.8V: 1.2kg-cm
6. Stall torque @6.6V: 1.6kg-cm

5.2.5 Pump Motor

The pump motor is used to pump water from a reservoir and sprinkle it over the fire-damaged area. This is done so that when the pin receives a high signal, the motor is connected. The electric motor is powered by a 12-volt relay. This is a small, low-cost submersible pump motor that operates from 3 to 6 volts. It has a processing capacity of 120 liters per hour while drawing just 220 milliamps (mA). The user simply inserts the tube pipe into the motor outlet, immerses it in water, and switches it on. He/she must also make sure that the water level is always higher than the motor.

5.2.6 Flames Sensor Module

Flame sensors are sophisticated devices that detect the flame phenomena of a fire. Depending on the wavelength of light used, these detectors come in a range of forms and sizes, and they include ultraviolet, near-infrared, infra-red, and UV/IR sensors. In the module, an LM393 comparator chip is employed to generate a stable digital output signal. This comparator has a driving capacity of 15 mA. This flame detector sensor has a wide range of applications, including fire alarms and other fire detection systems or projects.

The comparator has the following specifications used in the device:

- LM393 comparator chip
- Detection Range: 760 nm to 1100 nm
- Operating Voltage: 3.3 V to 5 V
- Maximum Output Current: 15 mA
- Digital Outputs: 0 and 1 Detection Angle: about 60 degrees
- Adjustable sensitivity via potentiometer

In addition, it has LED lights indicators of power (red) and digital switching *gathering of information by flame sensor.*

5.3 Working of Fire Detecting Device

Using the IR (Infra-red) wavelength radiated by the flame, the IR flame sensor detects fire. If a flame is detected, it produces a 1 as an output signal; otherwise, it produces a 0 as an output signal. At that moment, the LM39 comparator chip in the module begins generating a stable digital output signal, which is then sent on to the relay module.

5.3.1 Transmission of Information to the Relay Module

On getting the information from the LM39 comparator chip about fire detection, it transmits the signal to the Node MCU, and the relay module runs the Arduino program which triggers the operation of the Node MCU by accepting the information by the Node MCU.

After the relay module has received the information from the Node MCU, it sends the signal to the end-user on the open-source platform over the Wi-Fi network, showing the information to the end user.

The fire alert is received by the end-user as a notification on the open-source platform named Blynk.

Simultaneously, the pump runs in the meantime when the fire is detected in the early stages to control the spread of the fire at the initial stage. Figure 2 shows the real model of a fire detecting device with infra-red camera, servo meter, relay mode, whereas Figure 3 shows the schematic diagram of the device.

Figure.4 shows the working model of the device. Introducing a programmed alarm finder at delicate locales is one of the preventive strides to keep away the risk, which is the reason the Arduino-based alarm discovery and control framework was created. It can distinguish heat in a particular climate, raise an alarm, switch off the structure's mains, and even splash water to limit the force of the fire.

5.3.2 Implementation

The Arduino-based fire alarm system has various stakeholders; it can be used in any kind of scenario, from forest fires to small household fires. As it is a small prototype and, according to our research, it will be beneficial in all kinds of environments given the user can use a smartphone. The main impact of this project is to put off the fire in any scenario, so it does not affect anyone. It is a safety device, so it is looked out for the safety of all the stakeholders.

Figure 2. A real-life photo of the project

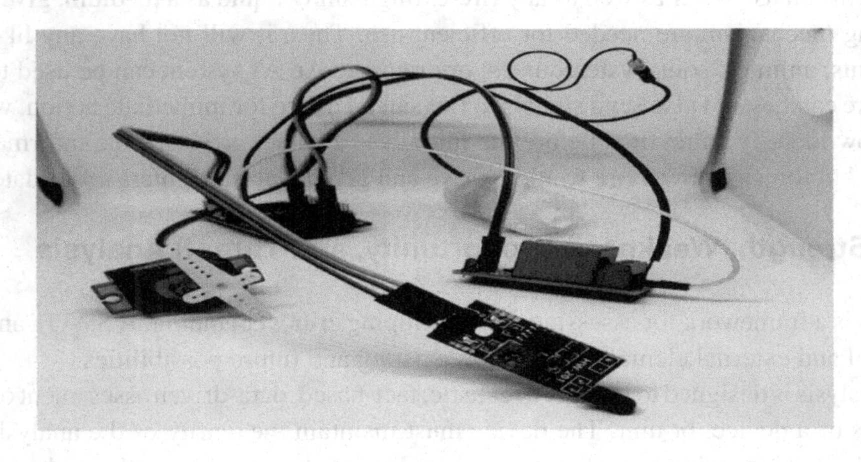

Figure 3. Schematic diagrams and circuit

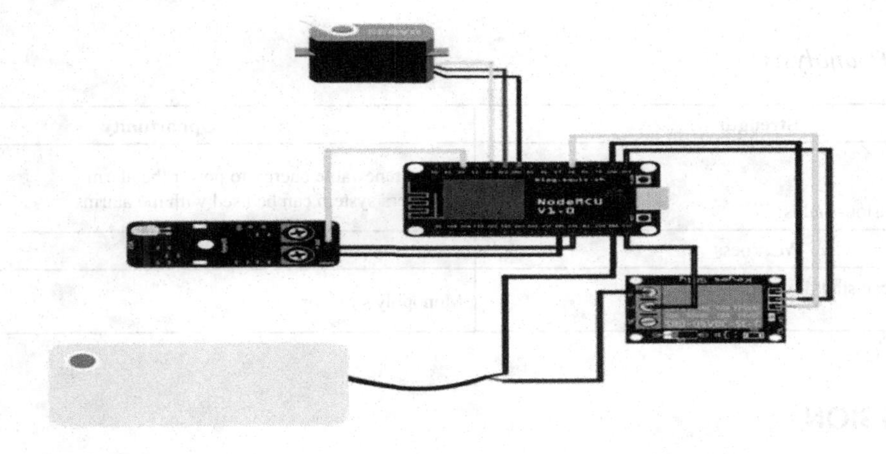

Figure 4. Schematic diagram and circuit of the fire detecting device

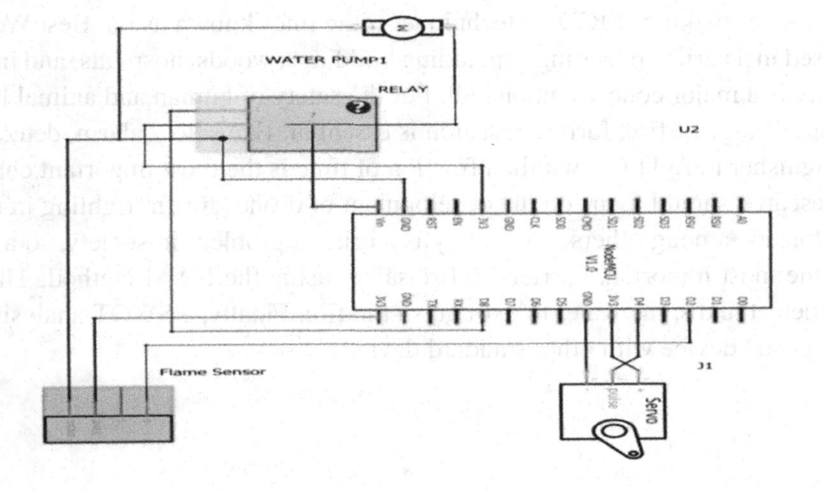

The fire alarm can use water as well as any fire-extinguishing liquid as a medium, given few changes in the triggering mechanism are needed for efficient use. Thus, it will not have any likely impact on vegetation, plants, animals, soils, watercourses, or drainage. An AI system can be used to transmit the data before a fire catches and also send signals to fire safety. office for immediate action, which can save lives.AI can now identify some fires before the initial 911 contact arrives . The information gathered can also be used to forecast how a fire would behave and give local firefighters up-to-date information.

5.4 SWOT (Strength, Weakness, Opportunity, and Threat) Analysis

SWOT analysis is a framework for assessing and developing critical planning. A SWOT analysis considers both internal and external elements, as well as existing and future possibilities.

A SWOT analysis is designed to provide a realistic, fact-based, data-driven assessment of the strengths and weaknesses of a device, or unit. The device must maintain the clarity of the analysis by avoiding preconceived ideas and focusing on realistic scenarios. Table 5 shows the strength, weakness, opportunity and threat of the device after experimentation. Table 5 shows SWOT Analysis/

Table 5. SWOT analysis

Strength	Opportunity
Easy to handle Cost-effective Can be used in various fields	Use of renewable energy to power the alarm A camera system can be used with the alarms
Weakness	**Threats**
A power outage can shut down the alarm Limited range	Monopoly supplier

6 CONCLUSION

We studied many papers in which we discussed various risk categories. After examining them, we created an Arduino-based gadget that can alert the user if there is a fire in the vicinity of the device. We used a multi-criteria decision-making (MCDM) technique, sometimes known as the Best Worst Method. The gadget can be used in a variety of settings, including buildings, woods, hospitals, and industrial environments. Fire safety is a major concern among all. For the safety of human and animal life, as well as for detecting the initial stage of fire, further research is essential. Fire safety alarm, detection of fire types and use of extinguisher to fight fire within a fraction of time is the most important challenge.

Therefore, research should focus on the development of drones for firefighting in non-visible areas such as mines, forests, among others. Fire safety is a critical problem in society, so a literature review is done to find the most important barriers to fire safety using the BWM method. Therefore, a device is developed to detect and spray water to extinguish the fire. Finally, a SWOT analysis is performed to compare the proposed device with other standard devices.

7 FUTURE SCOPE

In future, some more research is essential to fight fire, considering time factor challenges in unreachable areas and this device must be tested for underground oil and gas pipes. Detection of different kinds of fires can also explore the ways to extinguish fire. More case studies must be done to validate the device.

REFERENCES

Ahmad, M. B., Muhammad, A. S., Abdullahi, A. A., Tijjani, A., Iliyasu, A. S., Muhammad, I. M., ... Sani, K. M. (2019). Need for security alarm system installation and their challenges faced. *International Journal of New Computer Architectures and their Applications, 9*(3), 68-77.

Ahmadi, H. B., Kusi-Sarpong, S., & Rezaei, J. (2017). Assessing the social sustainability of supply chains using Best Worst Method. *Resources, Conservation and Recycling*, *126*, 99–106. doi:10.1016/j.resconrec.2017.07.020

Ajith, G., Sudarsaun, J., Arvind, S. D., & Sugumar, R. (2018). IoT based fire deduction and safety navigation system. *International Journal of Innovative Research in Science, Engineering and Technology*, *7*(2), 257–267.

Alkhatib, A. A. (2014). A review on forest fire detection techniques. *International Journal of Distributed Sensor Networks, 10*(3), 597368. doi:10.1155/2014/597368

Alqourabah, H., Muneer, A., & Fati, S. M. (2021). A smart fire detection system using IoT technology with automatic water sprinkler. *International Journal of Electrical & Computer Engineering, 11*(4).

Anusha, G., & Rao, V. S. (2018). IoT based ware house fire safety system using ARM7. *Int J Eng Technol, 7*(3.6), 240-242.

Arana-Pulido, V., Cabrera-Almeida, F., Perez-Mato, J., Dorta-Naranjo, B. P., Hernandez-Rodriguez, S., & Jimenez-Yguacel, E. (2018). Challenges of an autonomous wildfire geolocation system based on synthetic vision technology. *Sensors (Basel)*, *18*(11), 3631. doi:10.339018113631 PMID:30366471

Avazov, K., Mukhiddinov, M., Makhmudov, F., & Cho, Y. I. (2021). Fire Detection Method in Smart City Environments Using a Deep-Learning-Based Approach. *Electronics (Basel)*, *11*(1), 73. doi:10.3390/electronics11010073

Baranovskiy, N. V., Podorovskiy, A., & Malinin, A. (2021). Parallel implementation of the algorithm to compute forest fire impact on infrastructure facilities of JSC Russian railways. *Algorithms*, *14*(11), 333. doi:10.3390/a14110333

Beemsterboer, D. J. C., Hendrix, E. M. T., & Claassen, G. D. H. (2018). On solving the Best-Worst Method in multi-criteria decision-making. *IFAC-PapersOnLine*, *51*(11), 1660–1665. doi:10.1016/j.ifacol.2018.08.218

Bhadoria, R. S., Pandey, M. K., & Kundu, P. (2021). RVFR: Random vector forest regression model for integrated &enhanced approach in forest fires predictions. *Ecological Informatics*, *66*, 101471. doi:10.1016/j.ecoinf.2021.101471

Bhoi, S. K., Panda, S. K., Padhi, B. N., Swain, M. K., Hembram, B., Mishra, D., ... Khilar, P. M. (2018). Fireds-iot: A fire detection system for smart home based on IoT data analytics. In *2018 International Conference on Information Technology (ICIT)* (pp. 161-165). IEEE. 10.1109/ICIT.2018.00042

Faroudja, A. B., & Izeboudjen, N. (2020). Decision tree based system on chip for forest fires prediction. In International conference on electrical engineering (ICEE) 2020 (pp. 1-4). IEEE.

Guo, S., & Qi, Z. (2021). A Fuzzy Best-Worst Multi-Criteria Group Decision-Making Method. *IEEE Access : Practical Innovations, Open Solutions*, *9*, 118941–118952. doi:10.1109/ACCESS.2021.3106296

Hsu, W. L., Jhuang, J. Y., Huang, C. S., Liang, C. K., & Shiau, Y. C. (2019). Application of Internet of Things in a kitchen fire prevention system. *Applied Sciences (Basel, Switzerland)*, *9*(17), 3520. doi:10.3390/app9173520

Iqbal, N., Ahmad, S., & Kim, D. H. (2021). Towards mountain fire safety using fire spread predictive analytics and mountain fire containment in iot environment. *Sustainability (Basel)*, *13*(5), 2461. doi:10.3390u13052461

Jan, F., Min-Allah, N., Saeed, S., Iqbal, S. Z., & Ahmed, R. (2022). IoT-based solutions to monitor water level, leakage, and motor control for smart water tanks. *Water (Basel)*, *14*(3), 309. doi:10.3390/w14030309

Kalantar, B., Ueda, N., Idrees, M. O., Janizadeh, S., Ahmadi, K., & Shabani, F. (2020). Forest fire susceptibility prediction based on machine learning models with resampling algorithms on remote sensing data. *Remote Sensing (Basel)*, *12*(22), 3682. doi:10.3390/rs12223682

Khan, S. A., Kusi-Sarpong, S., Naim, I., Ahmadi, H. B., & Oyedijo, A. (2021). A best-worst-method-based performance evaluation framework for manufacturing industry. *Kybernetes*.

Labellapansa, A., Syafitri, N., Kadir, E. A., Saian, R., Rahman, A. S., & Ahmad, M. B. (2019). Prototype for early detection of fire hazards using fuzzy logic approach and Arduino microcontroller. *International Journal of Advanced Computer Research.*, *9*(44), 276–282. doi:10.19101/IJACR.PID47

Latifah, A. L., Shabrina, A., Wahyuni, I. N., & Sadikin, R. (2019). *Evaluation of Random Forest model for forest fire prediction based on climatology over Borneo. In International conference on computer, control, informatics and its applications (IC3INA)*. IEEE.

Majhi, A. K., Dash, S., & Barik, C. K. (2021). Arduino based smart home automation. *ACCENTS Transactions on Information Security.*, *6*(22), 7–12.

Mi, X., Tang, M., Liao, H., Shen, W., & Lev, B. (2019). The state-of-the-art survey on integrations and applications of the best worst method in decision making: Why, what, what for and what's next? *Omega*, *87*, 205–225. doi:10.1016/j.omega.2019.01.009

Pamučar, D., Ecer, F., Cirovic, G., & Arlasheedi, M. A. (2020). Application of improved best worst method (BWM) in real-world problems. *Mathematics*, *8*(8), 1342. doi:10.3390/math8081342

Preeti, T., Kanakaraddi, S., Beelagi, A., Malagi, S., & Sudi, A. (2021). *Forest fire prediction using machine learning techniques. In International conference on intelligent technologies (CONIT)*. IEEE.

Rezaei, J. (2015). Best-worst multi-criteria decision-making method. *Omega, 53*, 49–57. doi:10.1016/j.omega.2014.11.009

Saeed, F., Paul, A., Rehman, A., Hong, W. H., & Seo, H. (2018). IoT-based intelligent modeling of smart home environment for fire prevention and safety. *Journal of Sensor and Actuator Networks*, *7*(1), 11. doi:10.3390/jsan7010011

Sha, K., Shi, W., & Watkins, O. (2006). Using wireless sensor networks for fire rescue applications: Requirements and challenges. In *2006 IEEE International Conference on Electro/Information Technology* (pp. 239-244). IEEE.

Sharma, A., Singh, P. K., & Kumar, Y. (2020). An integrated fire detection system using IoT and image processing technique for smart cities. *Sustainable Cities and Society*, *61*, 102332. doi:10.1016/j.scs.2020.102332

Sinde, R. S., Kaijage, S., & Njau, K. N. (2020). Cluster based wireless sensor network for forests environmental monitoring. *International Journal of Advanced Technology and Engineering Exploration.*, *7*(63), 36–47. doi:10.19101/IJATEE.2019.650083

Singh, B. K., Kumar, N., & Tiwari, P. (2019). *Extreme learning machine approach for prediction of forest fires using topographical and metrological data of Vietnam. In Women institute of technology conference on electrical and computer engineering (WITCON ECE)*. IEEE.

Thibaud, M., Chi, H., Zhou, W., & Piramuthu, S. (2018). Internet of Things (IoT) in high-risk Environment, Health and Safety (EHS) industries: A comprehensive review. *Decision Support Systems, 108*, 79-95.

Vikram, R., Sinha, D., De, D., & Das, A. K. (2021). PAFF: Predictive analytics on forest fire using compressed sensingbased localized Ad Hoc wireless sensor networks. *Journal of Ambient Intelligence and Humanized Computing*, *12*(2), 1647–1665. doi:10.100712652-020-02238-x

Yadav, R., & Rani, P. (2020). Sensor based smart fire detection and fire alarm system. In *Proceedings of the International Conference on Advances in Chemical Engineering (AdChE)*. 10.2139srn.3724291

Yang, X., Xiong, S., Li, H., He, X., Ai, H., & Liu, Q. (2019). *Research on forest fire helicopter demand forecast based on index fuzzy segmentation and TOPSIS. In 9th international conference on fire science and fire protection engineering (ICFSFPE)*. IEEE.

Zhang, Y. C., & Yu, J. (2013). A study on the fire IOT development strategy. *Procedia Engineering*, *52*, 314–319. doi:10.1016/j.proeng.2013.02.146

Zope, V., Dadlani, T., Matai, A., Tembhurnikar, P., & Kalani, R. (2020). IoT sensor and deep neural network-based wildfire prediction system. In *4th international conference on intelligent computing and control systems (ICICCS)* (pp. 205-208). IEEE. 10.1109/ICICCS48265.2020.9120949

Chapter 5
Human Face Mask Detection Using YOLOv7+CBAM in Deep Learning

Xinyi Gao
https://orcid.org/0000-0001-7727-9087
Auckland University of Technology, New Zealand

Minh Nguyen
Auckland University of Technology, New Zealand

Wei Qi Yan
Auckland University of Technology, New Zealand

ABSTRACT

COVID-19 and its variants have affected millions of people around the world. Wearing a mask is an effective way to reduce the spread of the epidemic. While wearing masks is a proven strategy to mitigate the spread, monitoring compliance remains a challenge. In this chapter, the authors propose a mask detection method based on deep learning and convolutional block attention module (CBAM). In this chapter, they extract representative features from input images through supervised learning. In order to improve the recognition accuracy under limited computing resources. They choose YOLOv7 network model and incorporate CBAM into its network structure. Compared with the original version of YOLOv7, the proposed network model improves the mean average precision (mAP) up to 0.3% in face mask detection process. Meanwhile, the method improves the detection speed of each frame 73ms. These advancements have significant implications for real-time, large-scale monitoring systems, thereby contributing to public health and safety.

DOI: 10.4018/978-1-6684-9999-3.ch005

INTRODUCTION

Currently, there are variants of COVID-19 virus (Ciotti et al., 2020) in the throes of the pandemic. In 2022, Monkeypox virus (Rizk et al., 2022) began to appear globally. The current known routes of transmission for both viruses are dropped alot. Wearing a mask is an effective way to stop the spread of the viruses. Wearing masks is required in closed and public spaces. Manually testing whether a mask is worn requires a lot of costs and increases the risk of testing personnel being infected. The mask monitoring through digital cameras can reduce the chance of inspectors being infected (Yan, 2019). In recent years, researchers have proposed several effective mask classification and monitoring algorithms that can detect faces with and without masks (Balaji et al., 2021). In this book chapter, we group mask detection into three cases: Wearing a mask correctly, wearing a mask incorrectly, and not wearing a mask.

Visual object detection (Zou et al., 2023) has always been an enduring research direction in the field of computer vision. The conventional object detection algorithms consist of three stages. Object proposals are firstly generated in the input image. The features in each proposal box are then extracted. Finally, different visual features were extracted by designing a classifier. However, the algorithms were not ideal in terms of accuracy and speed in visual object recognition. In recent years, visual object detection algorithms based on deep learning (Shen et al., 2018) have performed well in terms of accuracy and speed. Among them, You Only Look Once (YOLO) series of algorithms are Superior to others in visual object detection (Redmon et al., 2016).

In this book chapter, we propose a deep learning-based face mask object detection algorithm CBAM-YOLOv7. This model is based on the existing YOLOv7 model (Wang et al., 2023) with the addition of CBAM (Woo et al., 2018). YOLO is a one-stage detector model based on convolutional neural networks. It applies a neural network to the entire image. The network model firstly segments an image into regions and then predicts the bounding box of each region. CBAM consists of two modules: The channel attention module (CAM) (Huang et al., 2020) and the spatial attention module (SAM) (Wang et al., 2019). CAM can make the network pay more attention to meaningful ground truth regions. On the other hand, SAM allows the network to focus on context-rich locations throughout the image (Yin et al., 2023). Through this supervised learning method, the accuracy of mask recognition is effectively improved.

The rest of the book chapter is structured as follows. The second part of this book chapter introduces the related work on face mask detection. In the third part, we introduce the datasets, methods and models used. In Section 4, we compare and analyse our experimental results. Finally, we summarize our work in Section 5.

LITERATURE REVIEW

Deep learning (Yan, 2021) has now been widely harnessed in the field of computer vision. Especially in the recognition of various images, deep learning (Lu et al., 2021) (Liang et al., 2022) is playing an increasingly important role. For mask recognition, a consortium of deep learning (Wang & Yan, 2022) (Lu et al., 2018) models are widely used a number of representative methods are Faster R-CNN (Lin et al., 2020), InceptionV3 (Jignesh Chowdary et al., 2020), MobileNet (Venkateswarlu et al., 2020), YOLO, etc. Among them, the YOLO series are taken account for a large proportion which is the current mainstream.

In recent years, affected by the epidemic, almost all countries require people to wear masks frequently during travel to prevent the spread of the virus. To detect those who are not wearing masks,

Samuel Ady Sanjaya et al. proposed a mask recognition model based on MobileNetV2. MobileNetV2 is a convolutional neural network (CNN) based method. The model firstly detects the video frame by frame. When there is a face in the detection process, the trained MobileNetV2-based model (Sanjaya & Adi Rakhmawan, 2020) is employed for recognition. Determine whether people wear masks by identifying face image frames. The final experimental results show that the model has a detection accuracy of 96.85% in distinguishing between people who wear masks and people who do not wear masks. And help the government to count these data more conveniently.

To identify people who are not wearing masks in public places, a transfer learning-based InceptionV3 recognition model is proposed. In this method, the number of samples that can be utilized is limited. In order to obtain better experimental results, the author proposed to solve the problem of limited data availability through image enhancement technology. Pattern classification is then performed by using the modified InceptionV3 model. The improved model removes the last layer of original InceptionV3 model and adds 5 average pooling layers with a pool size of 5×5 to the network. The model improvement is very effective, and the proposed transfer learning model attains up to 100% accuracy in testing (Jignesh Chowdary et al., 2020).

Since YOLO was proposed, it has been rapidly developed and applied to the field of visual object detection and recognition. YOLO is also suitable for mask recognition. Loy et al. proposed a model based on YOLOv2 and ResNet-50 for detecting medical masks. The model is composed of a feature extraction network and a detection network. The feature extraction network consists of ResNet-50 and the detection network consists of YOLOv2 (Loey et al., 2021). In addition, Intersection over Union (IoU) was harnessed to estimate anchor boxes for testing so as to increase the diversity of the dataset through data augmentation. The final experimental results show that the average recognition accuracy using the improved model is as high as 81%.

Real-time face mask detection is equally important. To this end, Jiang et al. proposed a YOLOv3-based squeeze and excitation YOLOv3 (SE-YOLOv3) object detection model. In this model, YOLOv3 was upgraded by adding an attention mechanism. The specific improvement is the addition of Squeeze and Excitation blocks between the convolutional layers of Darknet53. Generalized Intersection over Union (GIoU) loss was taken into account. Additionally, a larger dataset of masked faces was created. The experimental results prove that the proposed method can not only locate the face in real time, but also evaluate whether the mask is worn correctly. In terms of accuracy, the new model improved mAP by 6.7% than the base YOLOv3 (Jiang et al., 2021).

The same was employed to use YOLOv3-based model for mask recognition. A different approach was proposed. A novel mask recognition framework, namely, FMD-YOLO (Wu et al., 2022) was proposed. In this framework, Im-Res2Net-101 with deep residual network was propounded as the main feature extractor. The features were fully fused by using enhanced path aggregation network. The IoU loss and IoU-aware were also introduced based on the YOLOv3 loss function to determine the performance of the model. The final experimental results showed that FMD-YOLO achieves the best accuracies of 92.0% and 88.4% based on the two datasets. They outperform the final results of other advanced detectors.

In order to compare the difference in recognition accuracy of different models. Singh et al. took use of YOLOv3 and Faster R-CNN for face mask recognition respectively. They created one dataset to train both models and compared the results of the two models. In the experimental results, both models performed well. But the accuracy of Faster R-CNN model is slightly better. In addition, a new method was proposed to generate the bounding boxes of different colours around a face based on whether a mask is worn or not, recording the proportion of people wearing masks every day (Singh et al., 2021).

In order to solve the influence of complex environment on the accuracy of mask recognition. An improved model based on YOLOv4 is proposed. The model can not only identify masks, but also detect whether the wearing of masks is standard. Firstly, CSPDarkNet53 is added to the backbone feature extraction network of YOLOv4 (Yu & Zhang, 2021). Then, an adaptive image scaling algorithm was considered at the algorithm level. The addition of PANet makes the semantic information of the feature layer more. These improvements effectively reduced the computational cost and amount of computation of the network, and improve the learning ability of the model. In order to verify the effect of the model, a lot of comparisons were conducted with other models by using the improved model. The results show that the improved mask recognition mAP reaches up to 98.3%, which is more accurate than the existing algorithms.

Yang et al. proposed a mask recognition method with an interactive interface based on YOLOv5 (Yang et al., 2020). In this method, they divided the entire recognition system into four parts. Three filters are employed to increase the resolution of the input image. A model was designed to extract the corresponding information. The information was classified by using the YOLOv5 model. After the recognition result was generated, the corresponding information was input into the interactive interface. The final accuracy of the experiments is 97.9%. This model was also compared with other classic models, such as SSD, and the results are all due to the classic model.

Novelty of This Work: Unlike the existing models, our research work incorporates the Convolutional Block Attention Module (CBAM) into the YOLOv7 framework. This innovative approach not only improves the mAP by 0.3% but also enhances the detection speed, makes it highly applicable for real-time, large-scale monitoring systems.

METHODOLOGY

In this book chapter, we make modifications on the network model of YOLOv7 and add CBAM to ELAN. We train the model through three loss functions: Coordinate loss, object confidence loss, and classification loss.

Dataset

In this book chapter, the dataset we take into account is the Face Mask Detection Dataset. The dataset contains a total of 853 images. We split it into training set and test set with the ratio of 80% and 20%. Since the model is the YOLO series, the dataset is re-labelled especially for YOLO models. We still set the label to three classes, namely wearing a mask, wearing a mask incorrectly, and not wearing a mask. There are a total of 4,072 labels in the dataset, including 3,232 labels with masks, 123 labels with incorrect masks, and 717 labels without masks.

The mask images in this dataset all have complex backgrounds, which makes this dataset very suitable for this project. The addition of CBAM can effectively reduce the influence of background on the recognition process. We put 683 images into the training set and 170 images into the test set respectively. The exemplar images of the dataset are shown in Fig. 1.

Figure 1. An example of face mask detection dataset

Network Structure

The main body of our network structure is similar to YOLOv7. We modify the ELAN in YOLOv7 to the CBAM-ELAN module. The proposed YOLOv7+CBAM model is anchored on the YOLOv7 architecture, which is an extension of the YOLO family known for real-time object detection. The YOLOv7 architecture is characterized by its convolutional layers, pooling layers, and skip connections that form the backbone of the network. The Convolutional Block Attention Module (CBAM) is integrated into our YOLOv7 framework at specific stages. CBAM employs both channel and spatial attention to refine the feature maps, thus enhancing the model's ability to focus on salient features.

The YOLOv7+CBAM architecture is formulated as follows:

- **Initial Layers**: The model commences with a series of convolutional layers configured with various kernel sizes and strides, aiming to extract rudimentary features from the input images.
- **Intermediate Layers and Attention Mechanism**: After the initial convolutional blocks, CBAM modules are inserted to refine the generated feature maps. Specifically, these modules are placed after convolutional blocks with 1×1, 3×3 and 5×5 kernels to maximize their impact.
- **Advanced Blocks**: The model incorporates SPPCSPC blocks in the detection head, which combine spatial pyramid pooling and concatenated skip pathways to augment the receptive field and facilitate feature fusion.
- **Concatenation and Upsampling**: Feature maps are upsampled and concatenated at various layers, enhancing the spatial resolution and richness of the feature representation.
- **Output Layer**: The final layer employs a softmax function for class probabilities and a sigmoid activation function for bounding box coordinates. The complete structure is shown in Figure 2.

Figure 2. YOLOv7+CBAM model structure

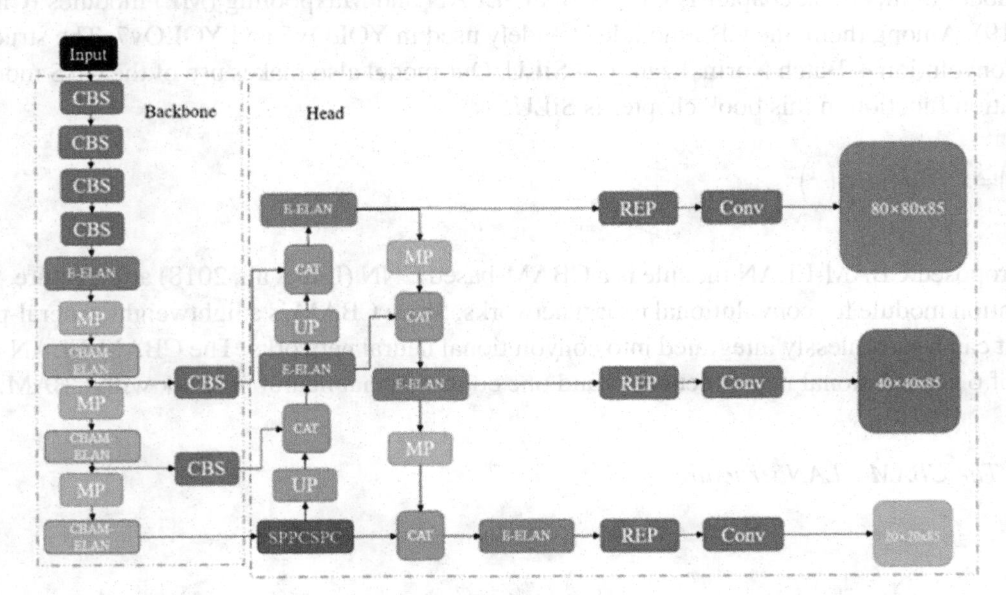

The ELAN module is an efficient network structure, which enables the network to learn more features and has stronger robustness by controlling the shortest and longest gradient paths. ELAN has two branches. The first branch is to change the number of channels through a 1×1 convolution. The second branch is more complicated. It firstly passes through a 1x1 convolution module to change the number of channels, then goes through four 3×3 convolution modules for feature extraction. Finally, the four features are superimposed to obtain the final feature extraction result.

The overall process of CBAM is divided into two parts. Firstly, the global max pooling and mean pooling are performed on the input. The pooled two one-dimensional vectors are sent to the fully connected layer for operations and added to obtain a one-dimensional channel attention. The next step is to multiply the channel attention with the input elements to obtain the channel attention adjusted features. The features obtained in the first part are globally max-pooled and mean-pooled according to the space. The 2D vectors produced by pooling are concatenated, then rolled and manipulated. The resulting 2D spatial attention is then multiplied by the previously obtained features to complete the process. The addition of CBAM+ELAN to the model can effectively improve the accuracy of target detection.

Figure 3. The CBAM module

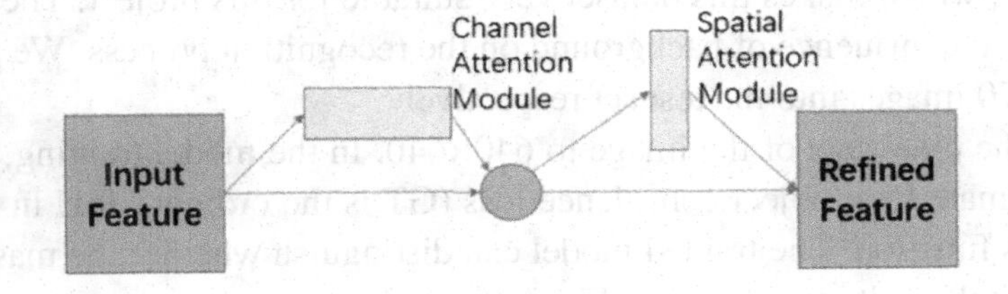

The model in this book chapter is CBS, CBAM-ELAN and Maxpooling (MP) modules (Christlein et al., 2019). Among them, the CBS module is widely used in YOLOv5 and YOLOv7. The structure of CBS is Convolution + Batch Normalization + SiLU. Our model also makes use of the CBS module, so the activation function in this book chapter is SiLU.

$$SiLU\left(x\right)=x\cdot sigmoid\left(x\right) \tag{1}$$

Our proposed CBAM-ELAN module is a CBAM-based CNN (Liu et al., 2018) architecture. CBAM is an attention module for convolutional neural networks. Since CBAM is a lightweight general-purpose module, it can be seamlessly integrated into convolutional neural networks. The CBAM-ELAN module consists of 6 convolutional neural networks and one convolutional neural network with CBAM.

Figure 4. The CBAM-ELAN structure

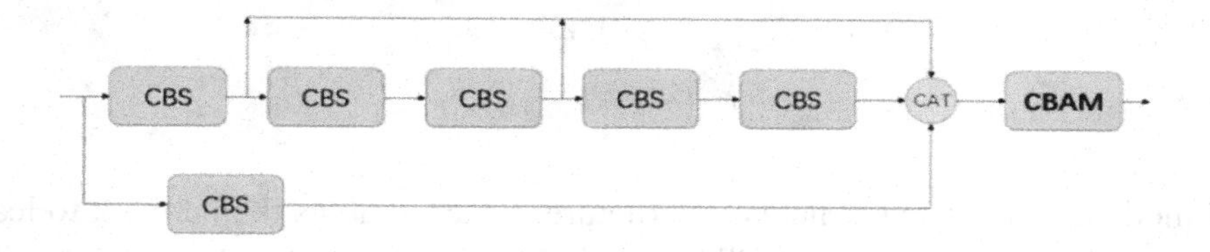

Method

In this book chapter, our method obtains 9 anchor boxes arranged from small to large based on the ground truth (GT) boxes in the training set through *k*-means clustering algorithm. Then, we match each GT box with 9 anchor boxes. According to the centre position of the GT box, the two nearest neighbour grids were also employed as the prediction network. We get the ten largest IOU results with the current GT box and add these ten results. We calculate each GT and candidate anchor loss according to the loss function and keep the smallest *k* loss functions. Finally, the same anchor box is assigned to multiple GTs. Through this method, we obtain better results of human face mask recognition.

RESULT ANALYSIS

We are use of the Mask Dataset, which includes 853 images. The mask images in this dataset all have complex backgrounds, which makes this dataset very suitable for this project. The addition of CBAM can effectively reduce the influence of background on the recognition process. We put 683 images into the training set and 170 images into the test set respectively.

We firstly adjust the pixel size of the image to 640×640. In the model training, we are use of three loss functions: Coordinate loss, object confidence loss (GT is the ordinary IoU in the training phase), and classification loss function. The trained model can distinguish whether the mask is correctly worn or not. The experimental results are shown in Figure 5.

Table 1. Training environment parameters

GPU Name	Processor	CUDA Version
NVIDIA GeForce RTX 3060	Intel Core i7-10700F	CUDA 11.2

In order to more intuitively reflect the performance of object detection, we are use of the performance indicators: Precision (P), Intersection over Union (IoU) (Zhou et al., 2019), F1 score, average precision (AP), and mean average precision (mAP) to test the results. IoU is a measure of the accuracy of detecting corresponding objects in a specific dataset. A standard IoU is a simple measure, as long as the task of getting a bounding box in the output can be measured by IoU. Precision represents the ratio of the number of correct predictions to the number of all predictions. The area enclosed by the PR curve is AP. The larger the area, the better the object detection effect.

$$Precision = \frac{Ture\,Positive}{Ture\,Positives + False\,Positives} \tag{2}$$

$$Recall = \frac{Ture\,Positives}{Ture\,Positives + False\,Negatives} \tag{3}$$

$$IoU = \frac{Area\,of\,Overlap}{Area\,of\,Union} \tag{4}$$

The IoU obtained by our method is slightly higher than that of YOLOv7. The average IoU of the mask-wearing increases by 0.03. The average IoU without a mask improves up to 0.02. The average IoU for not wearing a mask correctly boosts up to 0.02.

Table 2. Comparison of IoU and F1 score

Method	IoU	F1 Score
CBAM-YOLOv7	0.89	0.92
YOLOv7	0.87	0.92

F1 score refers to the weighted average of precision and recall. F1 score values range from 0 to 1.0, with 1.0 being the highest precision. The F1 score of our method is as same as that of YOLOv7.

$$F1_{score} = 2 \times \frac{Precision \times Recall}{Precision + Recall} \tag{5}$$

AP is the average of the Precision values on the PR curve. For the PR curve, we are use of the integral for the calculation. We judge the size of AP by comparing the size of the area enclosed by the PR curve. Generally, the larger the area, the larger the AP value, and the better the object detection effect. By superimposing the two images, we found that the PR curves obtained by our method occupy a larger area.

To further determine the effectiveness of the CBAM-YOLOv7 algorithm in the task of face mask detection, we compare the algorithm with YOLOv7, YOLOv5, and YOLOX. All tests were based on the Mask Dataset. As shown in the table, CBAM-YOLOv7 outperforms several other algorithms in mAP. There is also a significant improvement in the recognition speed.

Table 3. Performance comparison of CBAM-YOLOv7, YOLOv7, YOLOv5, and YOLOX on the mask dataset

Algorithm	mAP@.5	Detection Time Spent per Frame(ms)
CBAM-YOLOv7	0.954	515ms
YOLOv7	0.951	588ms
YOLOv5	0.824	798ms
YOLOX	0.861	810ms

Figure 5. Results of our model

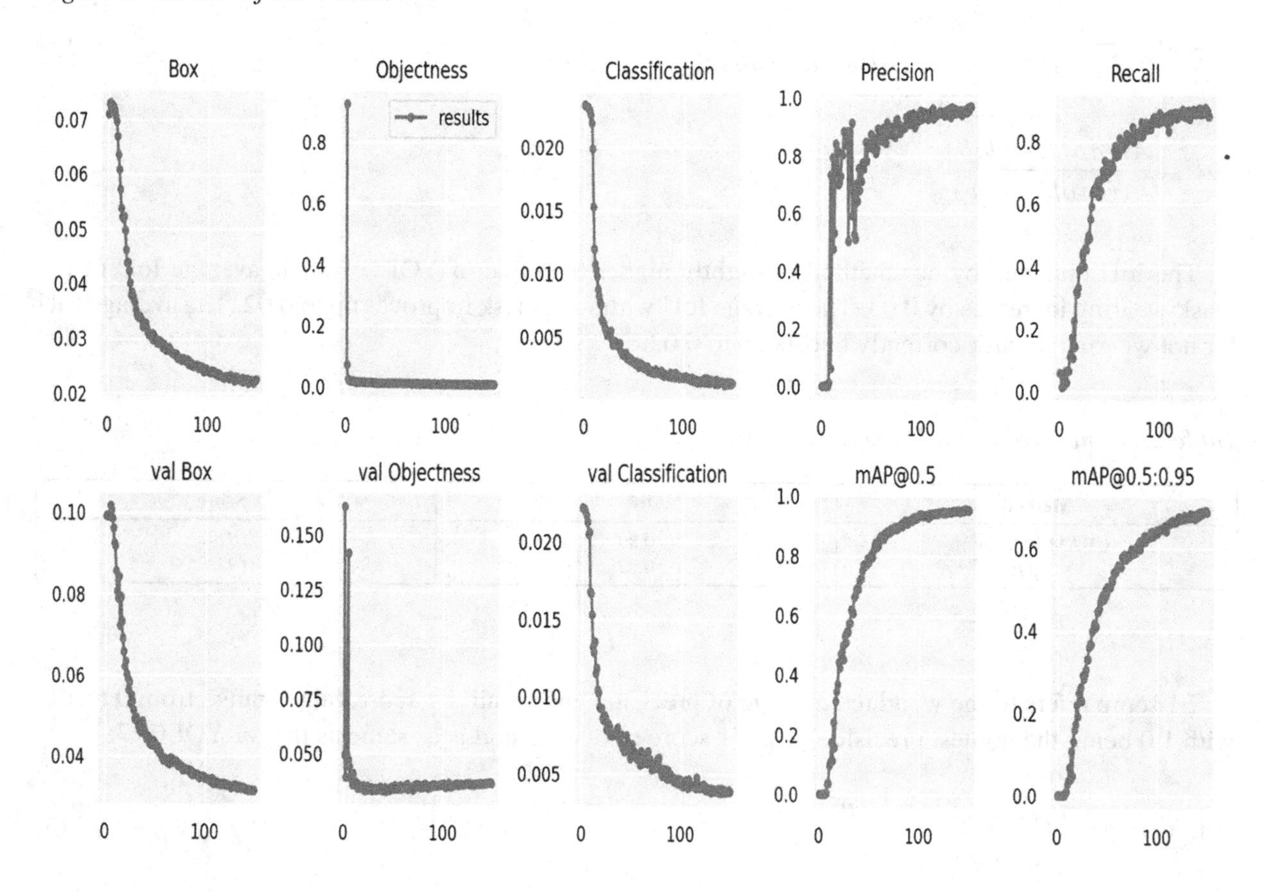

CONCLUSION

Due to the global outbreak of COVID-19 and Monkeypox, wearing a mask has become a method of epidemic prevention. Monitoring whether people are wearing masks correctly through deep learning methods is a possible solution. We develop the CBAM-YOLOv7 mask detection algorithm in this book chapter. The algorithm can detect three classes: Wearing a mask, not wearing a mask, and not wearing a mask correctly. The mAP of the proposed CBAM-YOLOv7 algorithm is 0.3% higher than that of the YOLOv7 version, its mAP value is 15.7% and 10.8% higher than that of YOLOv5 and YOLOX, respectively. The proposed algorithm not only improved the accuracy, but also significantly uplifted the recognition speed. Compared with the YOLOv7 version, our algorithm improved the recognition speed per frame up to 73 ms. The current work is of great importance during the global pandemic, and the development of these detectors to monitor whether people are wearing masks is also necessary. In addition, extending CBAM to the adversarial networks is also a viable approach.

REFERENCES

BalajiS.BalamuruganB.KumarT. A.RajmohanR.KumarP. P. (2021, March 29). A brief survey on AI based Face Mask Detection System for public places. SSRN. https://ssrn.com/abstract=3814341

Christlein, V., Spranger, L., Seuret, M., Nicolaou, A., Kral, P., & Maier, A. (2019). Deep generalized Max pooling. *International Conference on Document Analysis and Recognition (ICDAR).* 10.1109/ICDAR.2019.00177

Ciotti, M., Ciccozzi, M., Terrinoni, A., Jiang, W.-C., Wang, C.-B., & Bernardini, S. (2020). The COVID-19 pandemic. *Critical Reviews in Clinical Laboratory Sciences*, *57*(6), 365–388. doi:10.1080/10408363.2020.1783198 PMID:32645276

Cui, W. (2015). *A Scheme of Human Face Recognition in Complex Environments* [Master's Thesis]. Auckland University of Technology.

Cui, W., & Yan, W. (2016). A scheme for face recognition in complex environments. *International Journal of Digital Crime and Forensics*, *8*(1), 26–36. doi:10.4018/IJDCF.2016010102

Gao, X. (2022). *A Method for Face Image Inpainting Based on Generative Adversarial Networks* [Masters Thesis]. Auckland University of Technology, New Zealand.

Gao, X., Nguyen, M., & Yan, W. (2021). Face image inpainting based on generative adversarial network. *International Conference on Image and Vision Computing New Zealand.* 10.1109/IVCNZ54163.2021.9653347

Gao, X., Nguyen, M., & Yan, W. (2022). A face image inpainting method based on autoencoder and adversarial generative networks. *Pacific-Rim Symposium on Image and Video Technology.*

Gowdra, N., Sinha, R., MacDonell, S., & Yan, W. (2021). Maximum Categorical Cross Entropy (MCCE): A noise-robust alternative loss function to mitigate racial bias in Convolutional Neural Networks (CNNs) by reducing overfitting. *Pattern Recognition.*

Huang, G., Zhu, J., Li, J., Wang, Z., Cheng, L., Liu, L., Li, H., & Zhou, J. (2020). Channel-attention U-Net: Channel attention mechanism for semantic segmentation of esophagus and esophageal cancer. *IEEE Access : Practical Innovations, Open Solutions*, *8*, 122798–122810. doi:10.1109/ACCESS.2020.3007719

Jiang, X., Gao, T., Zhu, Z., & Zhao, Y. (2021). Real-time face mask detection method based on yolov3. *Electronics (Basel)*, *10*(7), 837. doi:10.3390/electronics10070837

Jiao, Y., Weir, J., & Yan, W. (2011). Flame detection in surveillance. *Journal of Multimedia*, *6*(1). Advance online publication. doi:10.4304/jmm.6.1.22-32

Jignesh Chowdary, G., Punn, N. S., Sonbhadra, S. K., & Agarwal, S. (2020). Face mask detection using transfer learning of Inceptionv3. *Lecture Notes in Computer Science*, *12581*, 81–90. doi:10.1007/978-3-030-66665-1_6

Liang, C., Lu, J., & Yan, W. Q. (2022). Human action recognition from digital videos based on Deep Learning. *The International Conference on Control and Computer Vision.* 10.1145/3561613.3561637

Lin, K., Zhao, H., Lv, J., Li, C., Liu, X., Chen, R., & Zhao, R. (2020). Face detection and segmentation based on improved mask R-CNN. *Discrete Dynamics in Nature and Society*, *2020*, 1–11. doi:10.1155/2020/9242917

Liu, M., & Yan, W. (2022). *Masked face recognition in real-time using MobileNetV2*. ACM ICCCV.

Liu, Z., Yan, W. Q., & Yang, M. L. (2018). Image denoising based on a CNN model. *International Conference on Control, Automation and Robotics (ICCAR).* 10.1109/ICCAR.2018.8384706

Loey, M., Manogaran, G., Taha, M. H., & Khalifa, N. E. (2021). Fighting against COVID-19: A novel deep learning model based on YOLO-V2 with resnet-50 for medical face mask detection. *Sustainable Cities and Society*, *65*, 102600. doi:10.1016/j.scs.2020.102600 PMID:33200063

Lu, J., Nguyen, M., & Yan, W. Q. (2021). Sign language recognition from digital videos using deep learning methods. *Communications in Computer and Information Science*, *1386*, 108–118. doi:10.1007/978-3-030-72073-5_9

Lu, J., Yan, W. Q., & Nguyen, M. (2018). Human behaviour recognition using deep learning. *IEEE International Conference on Advanced Video and Signal Based Surveillance (AVSS).* 10.1109/AVSS.2018.8639413

Pan, C., Liu, J., Yan, W., & Zhou, Y. (2021). Salient object detection based on visual perceptual saturation and two-stream hybrid networks. *IEEE Transactions on Image Processing*, *30*, 4773–4787. doi:10.1109/TIP.2021.3074796 PMID:33929959

Pan, C., & Yan, W. (2018) A learning-based positive feedback in salient object detection. *International Conference on Image and Vision Computing New Zealand.* 10.1109/IVCNZ.2018.8634717

Pan, C., & Yan, W. (2020). Object detection based on saturation of visual perception. *Multimedia Tools and Applications*, *79*(27-28), 19925–19944. doi:10.100711042-020-08866-x

Radhakrishna, A., Yan, W., & Kankanhalli, M. (2006). Modelling intent for home video repurposing. *IEEE MultiMedia*, *13*(1), 46–55. doi:10.1109/MMUL.2006.12

Redmon, J., Divvala, S., Girshick, R., & Farhadi, A. (2016). You Only Look Once: Unified, real-time object detection. *IEEE Conference on Computer Vision and Pattern Recognition (CVPR),* 779–788. 10.1109/CVPR.2016.91

Rizk, J. G., Lippi, G., Henry, B. M., Forthal, D. N., & Rizk, Y. (2022). Prevention and treatment of Monkeypox. *Drugs*, *82*(9), 957–963. doi:10.100740265-022-01742-y PMID:35763248

Sanjaya, S. A., & Adi Rakhmawan, S. (2020). Face mask detection using MobileNetv2 in the era of COVID-19 pandemic. *International Conference on Data Analytics for Business and Industry: Way Towards a Sustainable Economy (ICDABI).* 10.1109/ICDABI51230.2020.9325631

Shen, D., Chen, X., Nguyen, M., & Yan, W. Q. (2018). Flame detection using deep learning. *International Conference on Control, Automation and Robotics (ICCAR).* 10.1109/ICCAR.2018.8384711

Shen, H., Kankanhalli, M., Srinivasan, S., Yan, W. (2004). Mosaic-based view enlargement for moving objects in motion pictures. *IEEE ICME'04.*

Shen, J., Yan, W., Miller, P., & Zhou, H. (2010) Human localization in a cluttered space using multiple cameras. *IEEE International Conference on Advanced Video and Signal Based Surveillance.* 10.1109/AVSS.2010.60

Singh, S., Ahuja, U., Kumar, M., Kumar, K., & Sachdeva, M. (2021). Face mask detection using yolov3 and faster R-CNN models: COVID-19 environment. *Multimedia Tools and Applications*, *80*(13), 19753–19768. doi:10.100711042-021-10711-8 PMID:33679209

Song, C., He, L., Yan, W., & Nand, P. (2019) An improved selective facial extraction model for age estimation. *International Conference on Image and Vision Computing New Zealand.* 10.1109/IVCNZ48456.2019.8960965

Venkateswarlu, I. B., Kakarla, J., & Prakash, S. (2020). Face mask detection using MobileNet and global pooling block. *IEEE Conference on Information Communication Technology (CICT).* 10.1109/CICT51604.2020.9312083

Wang, C.-Y., Bochkovskiy, A., & Liao, H.-Y. M. (2023). YOLOv7: Trainable Bag-of-Freebies sets new state-of-the-art for real-time object detectors. *IEEE Conference on Computer Vision and Pattern Recognition (CVPR),* 7464–7475. 10.1109/CVPR52729.2023.00721

Wang, H., & Yan, W. Q. (2022). Face detection and recognition from distance based on deep learning. *Advances in Digital Crime, Forensics, and Cyber Terrorism*, 144–160. doi:10.4018/978-1-6684-4558-7.ch006

Wang, J., Kankanhalli, M., Yan, W., & Jain, R. (2003) Experiential sampling for video surveillance. *ACM SIGMM International Workshop on Video surveillance*, 77-86. 10.1145/982452.982462

Wang, J., Yan, W., Kankanhalli, M., Jain, R., & Reinders, M. (2003) Adaptive monitoring for video surveillance. *International Conference on Information, Communications and Signal Processing.*

Wang, X., Hu, H.-M., & Zhang, Y. (2019). Pedestrian detection based on spatial attention module for outdoor video surveillance. *IEEE International Conference on Multimedia Big Data (BigMM).* 10.1109/BigMM.2019.00-17

Woo, S., Park, J., Lee, J.-Y., & Kweon, I. S. (2018). *CBAM: Convolutional block attention module*. ECCV. doi:10.1007/978-3-030-01234-2_1

Wu, P., Li, H., Zeng, N., & Li, F. (2022). FMD-YOLO: An efficient face mask detection method for COVID-19 prevention and control in public. *Image and Vision Computing*, *117*, 104341. doi:10.1016/j.imavis.2021.104341 PMID:34848910

Yan, W., & Kankanhalli, M. (2015) Face search in encrypted domain. Pacific-Rim Symposium on Image and Video Technology, 775-790.

Yan, W., Kankanhalli, M., Wang, J., & Reinders, M. (2003) Experiential sampling for monitoring. *ACM SIGMM Workshop on Experiential Telepresence*, 70-72. 10.1145/982484.982497

Yan, W. Q. (2019). *Introduction to Intelligent Surveillance: Surveillance Data Capture, Transmission, and Analytics*. Springer. doi:10.1007/978-3-030-10713-0

Yan, W. Q. (2021). *Computational Methods for Deep Learning: Theoretic, Practice and Applications*. Springer Nature. doi:10.1007/978-3-030-61081-4

Yang, G., Feng, W., Jin, J., Lei, Q., Li, X., Gui, G., & Wang, W. (2020). Face mask recognition system with YOLOv5 based on image recognition. *IEEE International Conference on Computer and Communications (ICCC)*. 10.1109/ICCC51575.2020.9345042

Yin, M., Chen, Z., & Zhang, C. (2023). A CNN-transformer network combining CBAM for change detection in high-resolution remote sensing images. *Remote Sensing (Basel)*, *15*(9), 2406. doi:10.3390/rs15092406

Yu, J., & Zhang, W. (2021). Face mask wearing detection algorithm based on improved YOLO-V4. *Sensors (Basel)*, *21*(9), 3263. doi:10.339021093263 PMID:34066802

Zhou, D., Fang, J., Song, X., Guan, C., Yin, J., Dai, Y., & Yang, R. (2019). IOU loss for 2D/3D object detection. *International Conference on 3D Vision (3DV)*. 10.1109/3DV.2019.00019

Zou, Z., Chen, K., Shi, Z., Guo, Y., & Ye, J. (2023). Object detection in 20 years: A survey. *Proceedings of the IEEE*, *111*(3), 257–276. doi:10.1109/JPROC.2023.3238524

Chapter 6
The Role of AI in Improving Interaction With Cultural Heritage: An Overview

Mario Casillo
University of Salerno, Italy

Angelo Lorusso
University of Salerno, Italy

Francesco Colace
https://orcid.org/0000-0003-2798-5834
University of Salerno, Italy

Domenico Santaniello
https://orcid.org/0000-0002-5783-1847
University of Salerno, Italy

Brij B. Gupta
Asia University, Taichung, Taiwan & Lebanese American University, Beirut, Lebanon

Carmine Valentino
University of Salerno, Italy

ABSTRACT

Over the years, artificial intelligence techniques have been applied in several application fields, exploiting data to execute different tasks and achieve disparate objectives. Therefore, the cultural heritage field can utilise AI techniques to improve the interaction between visitors and cultural assets. Then, this work aims to present the background related to the principal AI techniques and provides an overview of the literature aimed at improving the user cultural experience. This overview focuses on AI integration with tools aimed to enhance the interaction among visitors and cultural sites, such as recommender systems, context-aware recommender systems, and chatbots. Finally, the most common measure used for estimating the accuracy of the AI methodologies will be introduced.

DOI: 10.4018/978-1-6684-9999-3.ch006

INTRODUCTION

Artificial Intelligence (AI) (Brasse et al., 2023) refers to the machine's or system's ability to simulate human intelligence and indicates a complex discipline in continuous evolution. In literature, two distinct objectives exist:

- the creation of a Strong AI able to reproduce the principal human intelligence characteristics,
- the development of a Weak AI that requires the resolution of specific problems related to an application area.

The first objective represents only a theoretical one, instead, the second has several applications, such as education, e-commerce, healthcare, chatbots, travel, transport, etc (Brasse et al., 2023). Moreover, AI includes various methodologies, such as Machine Learning (ML) (Wuest et al., 2016), a computer science field that solves an issue without requiring specific programming. Indeed, ML requires data to learn how to perform a task through experience and, through performance measures, estimate its ability after the learning phase. A subset of ML consists of Deep Learning (DL) (Sarker, 2021; Schmidhuber, 2015) that represents an evolution of ML based on Artificial Neural Networks (ANN).

As mentioned above, Artificial Intelligence development involves his employment in various application fields. In particular, Cultural Heritage takes advantage of novel methodologies for improving the interaction between visitors and cultural assets (Fiorucci et al., 2020) and making available novel approaches to make visits a unique experience for the user through personalisation techniques. For the personalisation techniques employment, AI, in particular ML, cooperates with Recommender Systems (RSs), analysis and filtering tools, able to identify users' preferences and elaborate appropriate suggestions (Bobadilla et al., 2013; Ricci et al., 2015). The RSs integration with ML techniques allows systems to deal with the Big Data problem, where the Big Data term defines a data set described by the enormous quantity, variety, and velocity (Philip Chen & Zhang, 2014). Thus, like Machine Learning techniques, Recommender Systems also feed through data to process personalised suggestions about each user.

RSs work on three elements: users that require support, items that have to be suggested, and transactions, aka an interaction between the user and the system, usually represented as the rating (or utility function) (Ricci et al., 2015). Recommender Systems elaborates suggestions based on several techniques that can be classified into three fundamental groups: Content-Based, Collaborative Filtering, and Hybrid (Ricci et al., 2015).

Content-Based techniques aim to create numerical vectors that translate users' preferences and elaborate suggestions through similarity measures. In this technique, Information Retrieval represents a fundamental resource for generating the profiles (Ricci et al., 2015).

Instead, Collaborative Filtering RSs (Bobadilla et al., 2013) generate suggestions elaborating interactions among users and the system and are divided into Memory-Based and Model-Based. Memory-Based techniques consist of identifying groups of users (User-Based), items (Item-Based), or both (User-Item-Based). The idea behind the Memory-Based User-Based RSs takes advantage of the similarity among users having analogous preferences. Memory-Based Item-Based RSs suggest users items like ones that users enjoyed. Instead, User-Item-Based Memory-Based RSs exploit both approaches to improve recommendations. In this field, Machine Learning employment techniques can improve the clustering creation through K-Nearest Neighbor or K-Means algorithms. In the first case, the appropriate suggestions classification takes advantage of similar users or items to identify the proper recommendation. Instead, the K-Means aims to cluster users or items based on distances that aim to underline the similarity and improve suggestions. After creating the clusters, the elaboration takes advantage of appropriate formulas to elaborate the numerical estimation of the likelihood coefficient. The Model-Based recommendation aims to identify a mathematical model of the problem based on the rating matrix having number of rows and columns equal to the number of users and items, respectively (Koren et al., 2009). This matrix contains the implicit or explicit evaluations of users and interprets the lack of information as the null value. Model-Based techniques usually exploit factorisation approaches such as Singular Value Decomposition, Probability Matrix Factorization, Non-Negative Matrix Factorization, and Principal Component Analysis.

In the field of Model-Based RSs, Artificial Intelligence approaches have significant employment. Indeed, identifying loss functions allows the utilisation of Gradient Descent and Stochastic Gradient Descent for learning by available interactions and elaborating users' evaluation referred to items that users do not interact with. Moreover, Artificial Neural Networks have a central role in recommendation (Liu et al., 2018) through Model-Based techniques and permit to reach significant accuracy results.

Moreover, to improve the RSs' ability to predictions, contextual information is integrated into the recommendation phase allowing the definition of a new tool: Context-Aware Recommender Systems. These tools require the inclusion of additional data that permits the improvement of the suitability of recommendations and represents further inputs for the Artificial Intelligence techniques exploited. According to Abowd et al., the term context refers to any information useful to characterise the situation of an entity that can affect the way users interact with systems (Abowd et al., 1999). Practically, the context involves a variation of the user's evaluation of items based on the environmental situation in which the user and items are involved. The introduction of context in the recommendation phase follows three principal strategies: Contextual Pre-Filtering, Contextual Post-Filtering, and Contextual Modeling (Adomavicius et al., 2011). The Contextual Pre-Filtering strategy exploits context for selecting recommendation phase inputs, the Contextual Post-Filtering strategy takes advantage of context for filtering recommendation outputs and adapting the recommendation to the context in which users act, and Contextual Modeling integrates context into the recommendation phase and generates contextual suggestions.

However, personalisation through the cooperation of Recommender Systems and Artificial Intelligence does not represent the unique possibility for improving the interaction among users and cultural heritage. Indeed, several applications regard the development of chatbots that allows the automatisation of data acquisition and support users appropriately. In this field, several approaches cooperate with Artificial Intelligence techniques for developing performing tools, such as Natural Language Processing and Sentiment Analysis. The data elaboration and the application of supporting techniques allow the identification of appropriate tools for Deep Learning approaches for realising an interaction with users suitable and smooth through textual communication.

The main objective of this work consists of providing an overview of the principal Artificial Intelligence techniques for improving the interaction among visitors and cultural heritage. After an introduction to Artificial Intelligence, Machine Learning, and Deep Learning and the description of their principal features, this work will describe the state-of-the-art related to the application of Artificial Intelligence in the fields of Recommender Systems, Context-Aware Recommender Systems and Chatbots applied to Cultural Heritage.

Therefore, the description of these applications will refer to the specific Artificial Intelligence technique applied, the processed data, possible supporting technologies employed, the dataset exploited for the experimental phase, and the principal used evaluation measures.

BACKGROUND

The concept of Artificial Intelligent (AI) was born in the twentieth century and nowadays still represents a relevant topic because of the possibility that it opened in the application field. At the start of AI appearance, the term Intelligence requires understanding what makes a computer able to think. The Turing test, in which a machine is defined as intelligent if it can make itself indistinguishable from humans, answered this question. Therefore, the Turing test aim seems to mean that only "intelligent" machines fall into the Artificial intelligence field, but this is not true because the principal objective of AI consists of problem-solving. For instance, a widespread application of AI consists of chatbots that usually do not pass the Turing test. Indeed, the AI definition was analysed in depth because of the difficulty of establishing its purposes and principal features. In 1983, Elaine Rich defined AI as follows (Ertel, 2017).

Artificial Intelligence is the study of how to make computers do things at which, at the moment, people are better (Ertel, 2017).

This definition responds to the requirement of summarising the AI objectives, but maybe in the future will be modified according to novel necessities or aims in this field.

Through the employment of several theoretical knowledge, AI aims to achieve decision-making, and, in particular, it allows the development of intelligent agents, indeed, an agent able to make decisions or perform a service based on their environment, user input, and experiences (Ertel, 2017). In particular, software intelligent agents consist of programs able to make inferences, while hardware intelligent agents usually exploit tools to integrate environmental information for decision-making. Moreover, intelligent agents can also be classified as reflex agents and agents with memory: the firsts react to events or inputs, and the seconds take advantage of past behaviors to make decisions.

The intelligent agents' employment requires the presence of elements that allow inference. Therefore, Artificial Intelligence plays a crucial role in Knowledge-Based systems (Akerkar & Sajja, 2009) in which the elaboration phase needs the availability of data acquired from different resources.

Before, data acquisition represented a problem for the application of AI, but nowadays, the development of the Internet of Things (Al-Fuqaha et al., 2015) resolves the data availability problem. This paradigm consists of a set of objects integrated with software and connection technologies able to exchange data through the Internet. The IoT exploits the MQTT protocol, an application layer protocol in the protocol stack. It consists of three players: the publisher aiming to provide data, the subscriber interested in published data, and the broker that manages the data publication and its sent to the subscribers.

The data acquisition represents the fundamental layer for developing services to users through Artificial Intelligence in the basic framework summarised in Figure 1.

After the data acquisition through sensors, external API, or different sources, the framework requires the organisation of a knowledge base through the pre-elaboration of raw data and the structuration of a database. The database contains collected data according to two possible storage strategies: the SQL database consists of tables that link data through relational relations instead of NoSQL strategies that can organise data according to structures, semi-structured, or unstructured data (Stonebraker, 2010). Therefore, the storage allows efficient access to data to elaborate inferences based on Artificial Intelligence techniques to make available services to users. Specifically in the cultural heritage field, AI application consents to improve the interaction with users and the cultural experience of users through the development of services such as chatbots and experience personalisation. Moreover, access to services requires a user interface to simplify interactions with information elaborated through the Inference Engine.

Figure 1. Summary of a generic framework to provide services to users

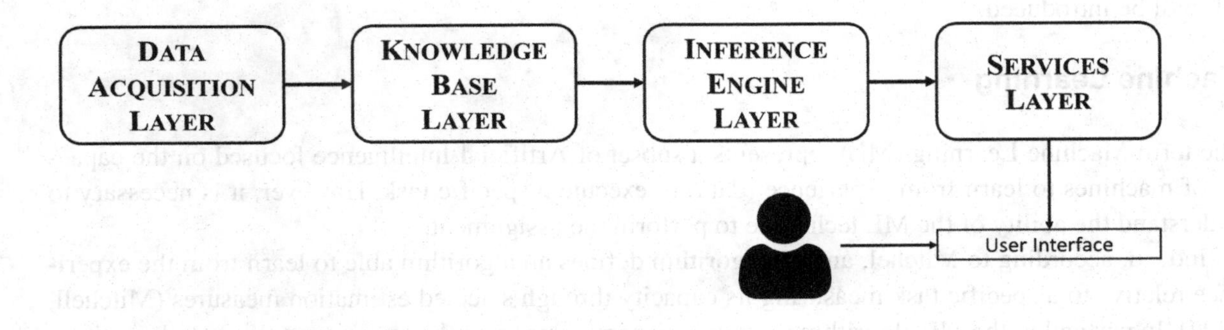

In this framework, the ability of the data elaboration to extract information represents a central issue and requires the employment of reliable inferences through AI. From this perspective, AI falls into the data mining (Witten et al., 2016) area because the decision-making task needs to understand patterns in data to extract information. Therefore, some AI techniques fall into the data mining field and, consequently, in the Knowledge Discovery in Database (KDD) (Fayyad et al., 1996), composed of pre-elaboration, data mining, and post-elaboration phases. A system takes advantage of data mining techniques to achieve different purposes. The simplest consists of the descriptive task, where data permits the history analysis of the analysed phenomena for identifying models. Another objective consists of the predictive task focused on the data elaboration to extract information related to the future and support decision-making. Moreover, perspective analysis forecasts related to the inference engine are supported with a reliability coefficient.

Therefore, referring to Figure 1, after the knowledge base layer, the AI technique aims to extract information from a significant amount of data, requiring managing the so-called Big Data problem, defined through the 5Vs (Tsai et al., 2015; ur Rehman et al., 2019):

- **Volume** refers to the dimension of all collected data;
- **Velocity**, a feature related to both the generation velocity and the required velocity for the analysis task;
- **Variety** refers to different origins and formats of collected data;
- **Veracity** requires that data has to be reliable;
- **Value** refers to the economic value of data.

The contextualisation of Artificial Intelligence in the data mining field and in managing the Big Data problem caused the further development of AI, highlighting the efficacy of Machine Learning (ML) (Jordan & Mitchell, 2015) and Deep Learning (DL) (Sarker, 2021; Schmidhuber, 2015) techniques. Nowadays, ML and DL consist of the most widespread AI techniques and guarantee optimal management of Big Data for extracting information from data. Therefore, in the following subsections, ML and DL will be introduced.

Machine Learning

The term Machine Learning (ML) represents a subset of Artificial Intelligence focused on the capacity of machines to learn from experience (data) to execute a specific task. However, it is necessary to understand the ability of the ML technique to perform the assignment.

Indeed, according to Mitchel, an ML algorithm defines an algorithm able to learn from the experience relative to a specific task measuring its capacity through selected estimation measures (Mitchell, 1997). In particular, the ML algorithm improves its performance on the assignment through the estimation measures through the increase of data. Moreover, the main advantages of Machine Learning lie in its applicability to problems that require a vast list of rules and a complex data calibration phase or tasks that take a long time to solve. In addition, the application of Machine Learning is effective in the case of application environments that change rapidly and require the management of a massive amount of data for extracting meaningful information.

Due to the general definition provided by Mitchel, Machine Learning requires a significant classification based on the characteristics of singular algorithms. Therefore, ML algorithms can be classified based on the typology of learning, the modality of the update through the novel data provided, and how elaborate predictions.

The classification based on the update modality distinguishes between Batch and Online Learning (Jain et al., 2014). Batch Learning consists of a massive training phase that allows the Machine Learning algorithm to perform the task for a long time without modifying the task management until the presence of a significant number of novel data. Instead, Online Learning aims to keep updated the Machine Learning algorithm through the accessible data through little sets named Mini-Batches.

The ML algorithms organisation based on the modality used for making predictions consists of two classes: Instance-Based and Model-Based Learning (Géron, 2017). The Instance-Based algorithms exploit similarity measures to compare novel instances with training ones for elaborating a prediction. Instead, Model-Based aims to identify a mathematical model related to training data and make forecasts according to the model associated with the training phase.

Figure 2. Summary of the classification of machine learning algorithms

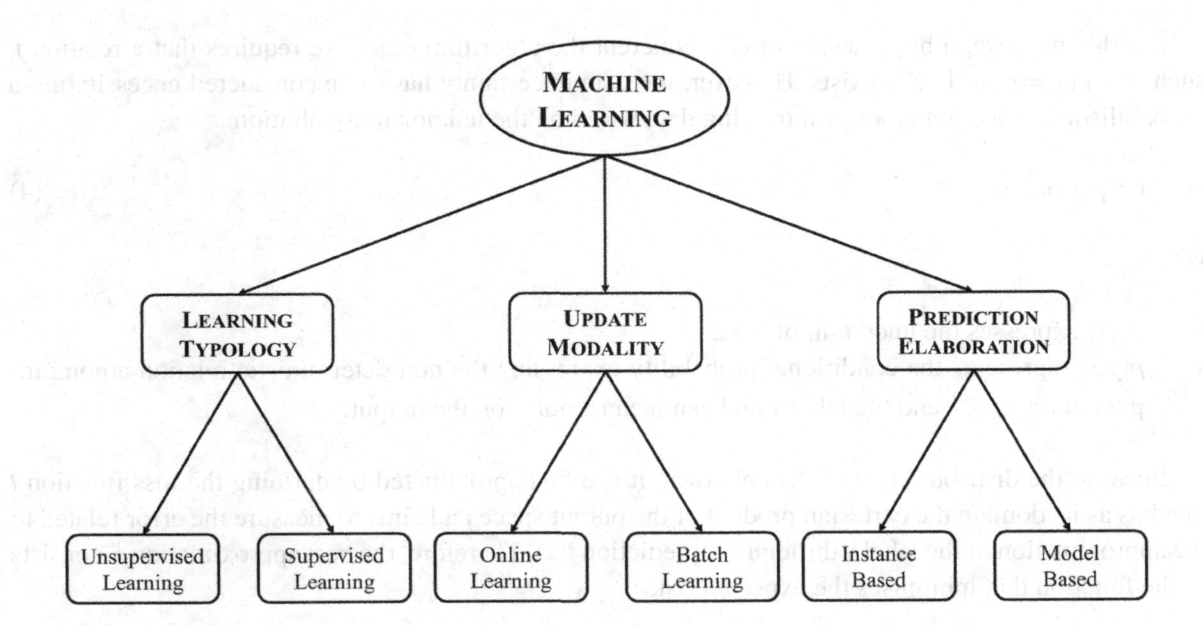

Finally, the classification based on learning typology consists of two classes: Supervised Learning and Unsupervised Learning. Moreover, this classification depends on the available data or the modality used to perform the task. Unsupervised Learning requires a dataset

$$T = \{x_i: i=1,\dots,n\} \quad (1)$$

having the features $x_i \in \mathbb{R}^D$ for each instance. These algorithms aim to learn through association rules and achieve to visualise data for dimensionality reduction (using, for instance, the Principal Component Analysis) (Bro & Smilde, 2014) or to cluster data according to similarity among features. The clustering task requires distance measures employment such as Euclidean distance, cosine similarity, or Minkowski distance (Jiang et al., 2007; Prasath et al., 2019). Instead, Supervised Learning requires a dataset

$$T = \{(x_i, y_i): i=1,\dots,n\} \quad (2)$$

in which each instance $(x_i, y_i) \in T$ presents features $x_i \in \mathbb{R}^D$ and a label $y_i \in \mathbb{R}$ that permits guiding the learning. Therefore, Supervised Learning algorithms aim to predict the labels related to features not present in training data, and moreover, distinguish among the typologies of labels present in the dataset: integer labels characterise classification tasks $y_i \in \{1,\dots,N\}$ for multi-class classification and $y_i \in \{-1,1\}$ for binary one) and real ($y_i \in \mathbb{R}$) regression.

The principal aim of ML algorithms consists of identifying a function f that approximates the phenomena described by data through the training data T defined in (2)

$f(x_i) \approx y_i, \ i=1,\dots,n$ (3)

For this purpose, a hypothesis to make coherent the algorithm objective requires that a relation f_r such that $f_r(x_i)=y_i$, $i=1,\dots,n$ exists. However, the data uncertainty has to be considered necessitating a probabilistic connection among all training data based on the unknown distribution

$$p(x,y) = p_X(x)p(y|x), \tag{4}$$

where:

- $p_X(x)$ expresses the uncertain of data;
- $p(y|x)$ represents the conditional probability expressing the non-deterministic relation among input data $x \in \mathbb{R}^D$ and the label y and can mean a noise on the output.

Because the distribution $p(x,y)$ is unknown, it can be approximated by defining the loss function l that has as its domain the cartesian product of the output space and aims to measure the error related to the approximation of the label y through the prediction $f(x)$. Therefore, the best approximation f^* consists of the function that minimises the expected risk

$$\varepsilon(f) = \mathbb{E}\left[l\left(y, f(x)\right)\right] = \int p(x,y)\, l\left(y, f(x)\right) dx\, dy. \tag{5}$$

The presence of $p(x,y)$ in (5) further requires the expected risk $\varepsilon(f)$ estimation using the empirical risk $\hat{\mathcal{E}}(f)$ based on the analysed problem. Based on the approximation quality related to the expected risk, a functional ML algorithm can identify reliable forecasts through the function f_T that minimise the empirical risk $\hat{\mathcal{E}}(f)$ such that, for consistency with data,

$$\lim_{n \to +\infty} \mathbb{E}_T\left[\mathcal{E}(f_T) - \mathcal{E}(f^*)\right] = 0 \tag{6}$$

where $\mathbb{E}_T$ represents the mean value through the data in T.

Moreover, the estimator f_T needs to be a good estimator of data (fitting problem) and has a bit of variation in relation to little change in input data (stability problem). Finally, the principal problem related to Machine Learning algorithms consists of

- insufficient data quantity;
- the presence of non-significative data in the training set T;
- data of low quality;
- overfitting: the training phase does not achieve the generalisation because of the data noise or a model excessively complex;
- underfitting: the estimated model is too simple.

Deep Learning

Deep Learning (Sarker, 2021; Schmidhuber, 2015) denotes a subset of Machine Learning and Artificial Intelligence based on Artificial Neural Networks (ANNs) inspired by Biological Neural Networks.

The principal advantage of ANNs consists of the nonlinearity of the model f_T, which allows the elaboration of predictions. This characteristic derives from the network structure based on three typologies of layers: input, hidden, and output layers. Each layer has a fixed number of associated neurons: the input layer has many neurons as the features of the input vector x, and the output layer has the needed number of neurons related to the analysed problem, instead hidden layers contain as many neurons as necessitated in relation with the structure of the network decided from the planner. There are several options for the design of Artificial Neural Networks:

- Feed-Forward Neural Network: networks in which neurons of a layer present links only with neurons of the next layer (Sarker, 2021; Schmidhuber, 2015);
- Dense Neural Networks: in these networks, all of the neurons of a layer are linked to all networks related to the successive layer. Moreover, these networks are also named as Multi-Layer Perceptron (Sarker, 2021; Schmidhuber, 2015);
- Recurrent Neural Networks: ANNs structured on a graph that allows loops. Therefore, neurons of a layer can be linked to neurons of the previous, same, and next layer (Almiani et al., 2020).
- Autoencoder: this network includes a codifier and a decoder. In particular, this network works in the case of unsupervised learning and aims to obtain in output the same instance given in input. Its application field concerns anomaly detection and feature dimensionality reduction (Zhao et al., 2019).

Figure 3. Example of dense neural network with two hidden layers

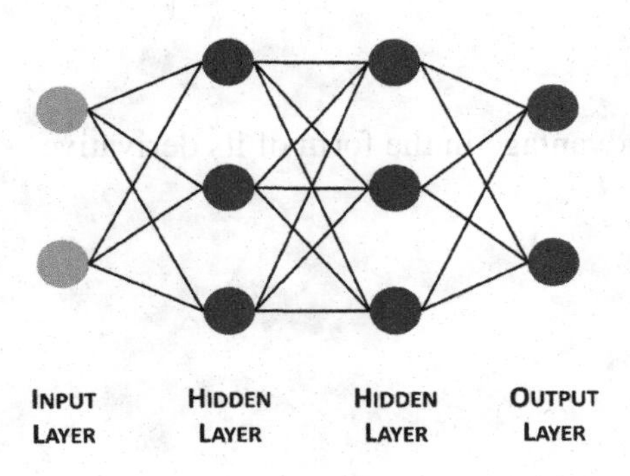

Figure 3 introduces a simple example of a Dense Neural Network composed of an input layer containing two neurons, two hidden layers of three neurons each, and an output layer of two neurons.

As said before, one of the best advantages of ANNs consists of the non-linear approximation of the phenomena represented by data. Therefore, in the example related to Figure 3, the function f_T consists of the composition of three functions:

$$f_T = f_O\left(f_{H_2}\left(f_{H_1}\left(x\right)\right)\right) \tag{7}$$

where

- the function $f_{H_1}\left(x\right) = \tilde{A}\left(W_{H_1}x + b_{H_1}\right)$ represents the elaboration of the input through the first hidden layer in the network, with $W_{H_1} \in \mathbb{R}^{3\times 2}, b_{H_1} \in \mathbb{R}^3$. Therefore, the application of the first hidden layer implies the employment of a linear transformation on the input data x through the matrix W_{H_1} and the vector b_{H_1} and, after, the application of a non-linear function σ named activation function.
- the function $f_{H_2}\left(z_1\right) = \tilde{A}\left(W_{H_2}z_1 + b_{H_2}\right)$ elaborates the output of the first hidden layer $z_1 = f_{H_1}\left(x\right)$, with $W_{H_2} \in \mathbb{R}^{3\times 3}$ and $b_{H2} \in \mathbb{R}^3$. For simplicity, we are considering the same activation function σ, but generally, it is possible to apply different activation function in the same ANN;
- the function $f_O\left(z_2\right) = \tilde{A}\left(W_0 z_2 + b_O\right)$ consists of the elaboration of the second hidden layer output $z_2 = f_{H_2}\left(z_1\right) = f_{H_2}\left(f_{H_1}\left(x\right)\right)$ through the output layer. In particular, $W_O \in \mathbb{R}^{2\times 3}$ and $b_O \in \mathbb{R}^2$.

The role of the activation function consists of introducing nonlinearity in the network elaboration, and its selection depends on the problem to solve. The most common possible choice are the following:

- Sigmoid:

$$\tilde{A}\left(\pm\right) = \frac{1}{1 + e^{-\pm}} \tag{8}$$

The sigmoid function has an advantage in the form of its derivative

$$\tilde{A}'\left(\pm\right) = \tilde{A}\left(\pm\right)\left(1 - \tilde{A}\left(\pm\right)\right) \tag{9}$$

- Hyperbolic tangent:

$$\tilde{A}\left(\pm\right) = \frac{e^{\pm} - e^{-\pm}}{e^{\pm} + e^{-\pm}} \tag{10}$$

- RELU:

$$\tilde{A}(\pm) = \begin{cases} \alpha & \alpha > 0 \\ 0 & \alpha \leq 0 \end{cases} \tag{11}$$

Its principal advantage consists of the ability to overcome the problem of the vanishing gradient problem.

- Softplus:

$$\tilde{A}(\pm) = \log\left(1 + e^{\pm}\right) \tag{12}$$

Figure 4. Graphical representation of activation functions

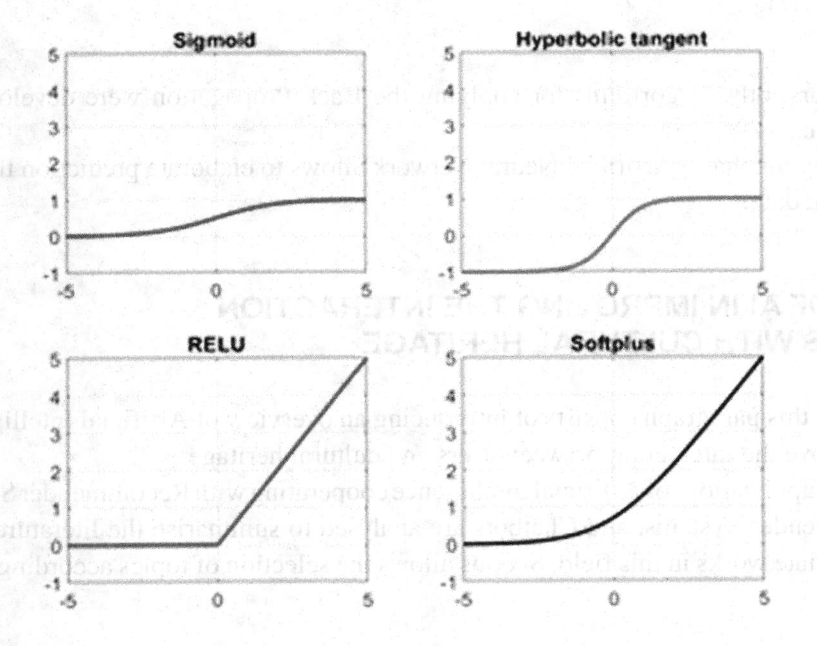

Until now, we have analysed the forward propagation of an ANN that represents how the network elaborates the output through weights W_{H_1}, W_{H_2}, W_O and biases b_{H_1}, b_{H_2}, b_O. Therefore, a focus point in the Deep Learning field consists of the training phase in which weights and biases have to be fixed based on training data T defined in (2), and it is possible through the Back Propagation algorithm (Rumelhart et al., 1986).

The learning phase focused on identifying appropriate weights and biases requires minimising the empirical expected risk

$$\min_{W,b} \hat{\mathcal{E}}\left(W,b\right) = \min_{W,b} \sum\nolimits_{i=1}^{n} l\left(y_i, f_T\left(x_i\right)\right) \tag{13}$$

where the loss function l can assume the form

$$l\left(y_i, f_T\left(x_i\right)\right) = \left(y_i - f_T\left(x_i\right)\right)^2, \qquad i = 1,\ldots,n. \tag{14}$$

The minimum identification requires the application of the Stochastic Gradient Descent that updates weights and biases through the following formulas

$$b_j^{new} = b_j - \cdot \frac{\partial \hat{\mathcal{E}}}{\partial b_j}\left(W,b\right), \quad \forall j \tag{15}$$

$$W_{jk}^{new} = W_{jk} - \cdot \frac{\partial \hat{\mathcal{E}}}{\partial W_{jk}}\left(W,b^{new}\right) \quad \forall j,k \tag{16}$$

Over the years, other algorithms for applying the Back Propagation were developed, such as the Adam algorithm.

After the training phase, Artificial Neural Network allows to elaborate prediction through the learning derived from data.

THE ROLE OF AI IN IMPROVING THE INTERACTION WITH USERS WITH CULTURAL HERITAGE

The objective of this paragraph consists of introducing an overview of Artificial Intelligence techniques applied to improve the interaction between users and cultural heritage.

In particular, applications of Artificial Intelligence cooperating with Recommender Systems, Context-Aware Recommender Systems, and Chatbots are analysed to summarise the literature approaches. To identify appropriate works in this field, Scopus allows the selection of topics according to the following research:

- "cultural AND heritage AND recommender AND systems" for searching Recommender Systems applied to cultural heritage. Moreover, the research was further limited to papers that present keywords Artificial Intelligence, Machine Learning, and Deep Learning;
- "cultural AND heritage AND context AND aware AND recommender AND system" for analysing Context-Aware Recommender Systems based on artificial intelligence techniques;
- "cultural AND heritage AND chatbot" for identifying works related to chatbots in the cultural heritage field.

Figure 5. Results related to the bibliography research

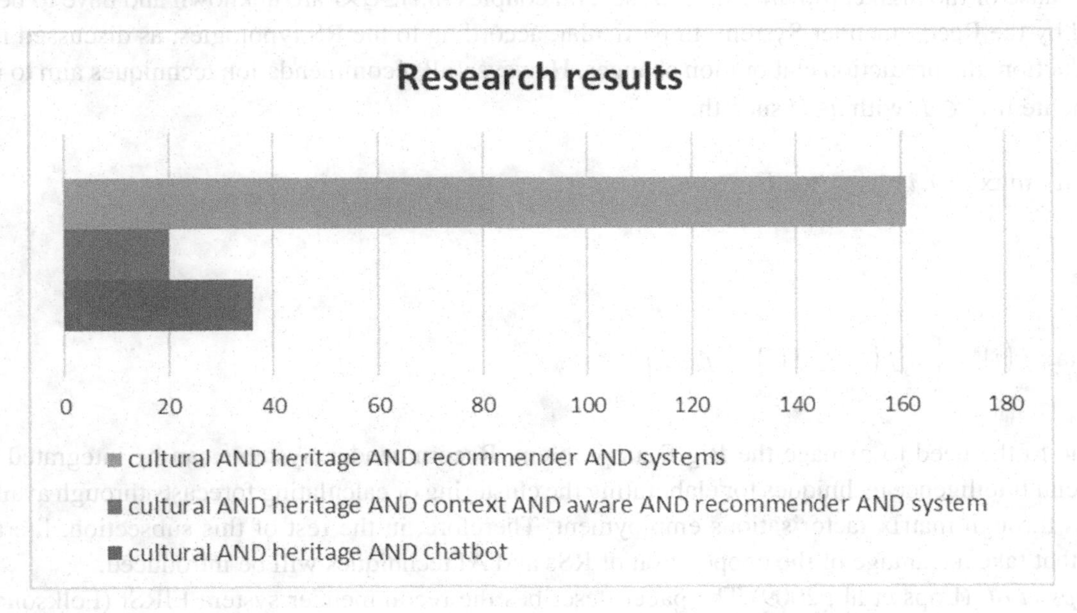

The section is divided into three paragraphs related to the searched topics and aims to introduce literature works selected to prove the central role of Artificial Intelligence in improving the interaction between cultural heritage and users. Each subsection aims to describe the literature works analysed and provide an overview of the several ways applicable to exploit AI.

AI for Recommender Systems

Recommender Systems (RSs) analyse data and elaborate information to support users in selecting appropriate tools or services. For elaborating information, RSs operate on three elements:

- Users: represent elements that need support from the system. Therefore, the systems necessitate knowing their preference to elaborate appropriate suggestions;
- Items: represent the element to recommend to users and usually consist of tools or services. RSs can elaborate appropriate suggestions by identifying the principal features or how users interact with them;
- Transactions: consist of the interaction between users and the system and represent an explicit or implicit evaluation of users related to an interaction with items.

Usually, transactions assume the form of ratings, an estimation according to the scale selected (numeric, ordinary, binary, or unary). The rating (or utility function) is a function with the domain of the cartesian product of the users' set U items' set I.

$$r : U \times I \mapsto \mathbb{R} \tag{17}$$

Because of the high cardinality of $U \times I$, several couples $(u,i) \in U \times I$ are unknown and have to be predicted by the Recommender System. In particular, according to the RS typologies, as discussed in the Introduction, the prediction elaboration changes. However, all recommendation techniques aim to identify the item $i_u^{'} \in I$ with $u \in U$ such that

$$i_u^{'} = \arg\max_{i \in I} r\left(u, i\right) \qquad u \in U, \tag{18}$$

where

$$\arg\max_{i \in I} f\left(i\right) = \left\{i : f\left(y\right) \leq f\left(i\right) \quad y \in I\right\}. \tag{19}$$

Due to the need to manage the Big Data problem, Recommender Systems can be integrated with Artificial Intelligence techniques for elaborating the clustering or calculating forecasts through available ratings through matrix factorisations employment. Therefore, in the rest of this subsection, literature work that take advantage of the cooperation of RSs and AI techniques will be introduced.

Lops et al. (Lops et al., 2009) The paper describes the recommender system FIRSt (Folksonomy-based Item Recommender syStem), which integrates AI techniques to provide personalised item recommendations and is built upon ITem Recommender (ITR), which recommends items in various domains based on text descriptions. FIRSt enhances ITR by incorporating user-generated content (UGC) as tags (folksonomies) to improve semantic profiles and provide more accurate recommendations.

The system FIRSt exploits AI techniques in different stages:

- Semantic Content Analysis: the system employs natural language processing (NLP) techniques, such as Word Sense Disambiguation (WSD), to analyse documents and identify relevant concepts that represent the content of items enabling the system to go beyond simple keyword matching and deal with natural language ambiguity.
- Profile Learning: FIRSt uses supervised learning, specifically a multivariate Poisson model, to learn probabilistic models of user interests from disambiguated documents and UGC. The model represents the semantic profile of users, which includes the most indicative concepts of their preferences.
- Recommender System: The system exploits the user profiles to suggest relevant items by matching concepts related to profiles and items' descriptions. The AI algorithms take advantage of the learned probabilities to rank items and provide personalised recommendations.
- WordNet and Multi-Language Text Analyzer: The system utilises WordNet, a lexical ontology, for the semantic indexing of documents. It also relies on the Multi-Language Text Analyzer (META), a natural language processing tool, to handle documents in English or Italian and disambiguate tags in the folksonomies.
- Word Sense Disambiguation (WSD): WSD is used not only for the documents but also for the tags in the folksonomies. By disambiguating tags, the system creates semantic tags that enhance the content-based recommendation paradigm.

This work discusses the experimental evaluation of the FIRSt recommender system, focusing on its predictive accuracy when using different types of content during the training step. The experiments involved 45 paintings from the Vatican picture gallery and 40 users, including 30 non-expert users and ten expert users. Users interacted with a web application to rate the paintings and annotate them with tags.

The evaluation used precision P and recall R as classification accuracy measures, and the F_{β} measure combined precision and recall assessing overall predictive accuracy.

Su et al. (Su et al., 2019) Authors introduce a proposed data model and recommendation strategy for a cultural recommender system. The data model consists of two main entities: Users (U) and Cultural Objects (O), where cultural objects can include several possible subjects, such as archaeological ruins, museum exhibits, and historical buildings. Users' profiles contain static information dynamically updated through the user data logs.

The proposed data model creates a Knowledge Base (KB) that connects users and cultural objects. The KB consists of heterogeneous nodes representing users, cultural objects, annotations, attachments, and profiles, with attributes specified according to various schemata. The connections between nodes are represented by edges, such as user-to-cultural_object, user-to-profile, cultural_object-to-attachment, cultural_object-to-annotation, and cultural_object-to-cultural_object based on similarity.

The recommender engine follows a context-aware and hybrid strategy with three phases:

- Prefiltering: Identifying a subset of cultural objects of interest to the user based on location, popularity, and user interests.
- Ranking: Assigning scores (interest, mood, and popularity) to the cultural objects based on dynamic knowledge obtained from social network behaviors.
- Postfiltering: Filtering out irrelevant cultural objects to the user based on location and user needs using an ontological approach.

The system architecture is designed as a multilayer architecture grounded on Big Data technologies. It includes data ingestion, KB management, data processing, and application layers. Data ingestion involves ingesting data from heterogeneous sources, including ontologies, social networks, and digital repositories. Data storage exploits NoSQL databases. The KB layer uses a graph database to manage ingested data according to the proposed data model.

The data processing layer utilises Big Data analytics and data mining techniques for querying and processing data from the KB, ontologies, and social networks. The application layer allows developers to create novel applications using APIs provided by the data processing layer, including recommending applications based on the user's context and preferences. The proposed system leverages AI techniques, such as machine learning algorithms, NLP, and data mining, to provide personalised cultural recommendations to users based on their interests, location, and social behaviors.

The experimental phase to evaluate the effectiveness of the recommendation strategy takes advantage of an app named "Smart Search Museum", aimed to collect users' data and preferences, and suggest appropriate cultural itineraries.

A group of users, including both experts and non-experts, tested the app. The efficacy of the proposed approach was evaluated by asking users to rank a set of cultural objects in order of their preferences.

Liao et al. (Liao et al., 2010) The authors propose a Recommendation Model that utilises various efficient techniques to recommend suitable content to end-users across different application domains. Firstly, they develop a prototype system that can recommend historical monuments and cultural heritage to users during various tour stages, including planning, taking, and finishing a tour. The proposed approach represents a ubiquitous tour guide, offering functional information, memorable experiences, and services through a u-Tour website.

In the system, the AI assumes the following roles:

- Recommendation Model: AI plays a central role in the recommendation model. The system uses multiple components, such as Web Access Log, User Ranking Matrix, Complaint Matrix, Expertise, User Profile, Content Attributes, and Feedback, to gather and process data for personalised content recommendations.
- Recommendation Techniques: The model employs various AI-based recommendation methods, including popularisation-based, community filtering, demographic profile, expertise-based, and association rules mining, to provide personalised content suggestions to users based on their preferences and interactions.
- Interactive Graphic User Interface (IGUI): The system adopts an IGUI based on Adobe Flex architecture, enhancing the user experience with multimedia information and better visual effects. The smooth transition between web pages is achieved through AI-powered dynamic rendering and seamless interactions.
- Hybrid Approaches: The system exploits a hybrid approach combining different recommendation techniques based on user types. AI algorithms analyse user behavior, community data, and content attributes to determine the most appropriate combination of methodologies for recommendation generation.
- Data Mining: AI-based data mining techniques are used to find potential association rules from web access logs, enabling the system to predict users' next browsing steps and provide relevant content recommendations.

Experimental results focus on implementing a model for historical monuments and cultural heritage recommendations implementing the system through three phases: data input, transformation, and result generation. In the data input phase, user profiles and content are fed into the model. User profiles allow the clustering of users into groups based on their interests or location, while content is classified with shared attributes or high similarity. Web access logs are preprocessed to fit the model's format. The transformation phase involves converting the input data into a suitable format for the model. Preprocessing is performed before mining web access logs or clustering user profiles. The system then uses user profiles and content to identify groups related to different user types.

Pavlidis (Pavlidis, 2018) The paper introduces Apollo, a complex minimax hybrid recommender system that utilises AI to enhance its predictive capabilities. Apollo's prediction relies on two main components: an abstract visit conceptualisation and user dissatisfaction modeling.

- Abstract Visit Conceptualisation: Apollo represents a cultural visit as a set of points of interest, with physical and semantic distances, groupings, and local obtrusions. Each item is depicted as a circle with a size proportional to its average popularity, and shading indicates its significance. The visit's tour path is represented by straight arrowhead lines, indicating physical distances. Content-based similar items in different locations are grouped into content-based neighborhoods. This conceptualisation forms the basis for estimating the probability of a user being attracted to items not in their initial intention or recommended list by the recommender system, thus measuring user disengagement or dissatisfaction during the visit.

- User Dissatisfaction Modeling: The user dissatisfaction modeling in Apollo involves four factors related to temporal, spatial, and content dynamics:
  - Temporal Dynamics: Modeled as a piecewise continuous function combining exponential and quadratic functions to estimate the probability of user dissatisfaction based on time spent on items.
  - Proximity Dynamics: Modeled as a function of distances, popularity, and importance of items to estimate the probability of user dissatisfaction due to proximity.
  - Content-Based Dynamics: Combines content similarity, distance, and influence of popular items to estimate the probability of user dissatisfaction due to content-based factors.
  - Fatigue Influence: Accounts for fatigue that builds up over time, modifying the weights.

The four probabilistic factors combination allows for modeling user dissatisfaction during a cultural visit.

Therefore, Apollo employs motivation-based user modeling, categorising visitors into five types: explorers, facilitators, professionals/hobbyists, experience seekers, and rechargers. This categorisation influences Apollo's recommender methodology.

Finally, Apollo uses ratings and contextual knowledge to categorise users based on their preferences and employs collaborative filtering techniques to generate recommendations. It also simulates guided tours, considering various dissatisfaction factors to optimise the recommendation list and minimise user dissatisfaction. The final list is optimised using a traveling salesman solver to account for efficient tour routing.

The experimental phase exploits a large-scale simulation conducted to test the effectiveness of Apollo. Since no actual user visit data was available, the simulation involved creating realistic visitors, items, and ratings.

Protopapadakis et al. (Protopapadakis et al., 2017) The authors describe an approach for building a personalised suggestion system based on content-based and collaborative filtering methods. The role of AI in this approach is to enable the system to understand user preferences and provide appropriate image suggestions for monuments based on those preferences.

The system goes through a two-step process. The first step aims to capture the user's preferences in a limited time by presenting randomly selected descriptive images of monuments. The user selects images relevant to their search, and this information allows for updating distance metrics for monument comparison.

In the second step, the system employs content-based filtering to identify monuments appealing to the user based on the selected images. It uses distance metric learning to optimise the distance matrix and calculate similarity scores. Collaborative filtering aims to find similar user profiles and identify monuments of interest to the current user.

Finally, the system provides appropriate image suggestions for monuments based on the overall ranking score, considering collaborative and content-based distance metrics. The monuments are ranked using a simple voting system, and the one with the most votes becomes the top recommendation.

AI plays a crucial role in this approach by processing user interactions, updating distance metrics, learning from user preferences, and providing personalised suggestions for monuments based on those preferences.

The experimental phase consists of a study involving the collection of a large dataset of images focused on five cultural monuments.

The study uses user feedback, where users are asked to provide positive and negative annotations for images from each monument, and they rank the monuments according to their preferences. The proposed system ranks the monuments based on the number of selected images and two rank correlation metrics.

AI for Context Aware Recommender Systems

Recommender Systems allow suggesting appropriate services or tools to users and guarantee adequating their cultural experience. However, sometimes additional data or information can improve the reliability of predictions by analysing environment conditions in which users, items, or systems stay. Therefore, Context-Aware Recommender Systems (CARS) exploit environmental data to elaborate suggestions by integrating context in the recommendation phase. As mentioned before, the term context refers to any information able to change the user evaluation of items according to the specific situation in which the user-item interaction happens. Therefore, contextual data integration in Recommender Systems influences contextual predictions elaboration and changes the recommendation paradigm. In particular, the introductory section introduced the three strategies for integrating the context in the recommendation phase: Contextual Pre-Filtering, Contextual Post-Filtering, and Contextual Modeling.

In all three cases, Context-Aware Recommender Systems involve a multidimensional recommendation, but formally:

- Contextual Pre-Filtering requires the application of a filter function able to select inputs

$$c:\left(u,i\right)\in U\times I\mapsto\left(\overline{u},\overline{i}\right)\in U\times I. \tag{20}$$

Therefore, the bi-dimensional recommendation works on the contextualised inputs to obtain contextual suggestions

$$r:\left(\overline{u},\overline{i}\right)\in \mathrm{U}\times \mathrm{I}\mapsto r_{ui}\in\mathbb{R}. \tag{21}$$

- Contextual Post-Filtering consists of two phases. The first phase consists of the bi-dimensional recommendation

$$r:\left(u,i\right)\in U\times I\mapsto r_{ui}\in\mathbb{R}. \tag{22}$$

Therefore, a filtering function allows to select appropriate output based on contextual data

$$c:r_{ui}\in\mathbb{R}\mapsto\overline{r}_{ui}\in\mathbb{R} \tag{23}$$

- Contextual Modeling includes contextual data in the recommendation phase as follows

$$\tilde{r}:\left(u,i,c\right)\in U\times I\times C\mapsto r_{uic}\in\mathbb{R} \tag{24}$$

where C denotes the set of all possible values that the context can assume.

This subsection focuses on the literature analysis related to CARS that includes deep learning techniques for improving the users' experience in the cultural heritage field.

Alexandridis et al. (Alexandridis et al., 2019)The authors study the Gournia archaeological site and propose a recommender system to enhance the visitor's experience by offering novel presentation strategies. Currently, the site has limited interpretive material, mainly consisting of a few information panels, a brochure, and an optional guidebook. The challenge lies in designing a system that allows visitors to understand and navigate the site.

Moreover, Gournia has a well-preserved ancient path system used for touring the site, with minimum interventions and limited interpretive programs. These Rural prehistoric sites like Gournia pose challenges in designing on-site interpretation and presentation systems, as they require sensitivity to the site's fabric and preservation state.

Another case study mentioned in the text is the Neolithic site of Çatalhöyük, known for its substantial size and advanced Neolithic culture. However, Çatalhöyük faces challenges in accessibility due to the fragile nature of its structures and lack of interpretation. Visitors are restricted to predefined paths and accompanied by guides during their visit.

The data-gathering process involved obtaining consent from visitors and using GPS tracking and camera recordings to study visitor movement and behavior at both sites. Visitor feedback was crucial to understanding their preferences and interaction with the cultural heritage sites.

Therefore, two curated paths were developed for each archeological site, considering available interpretive material and visitor feedback to improve the overall visitor experience.

The proposed path recommendation system for cultural heritage sites consists of three essential system components. The system exploits Artificial Intelligence and elaborates recommendations through the following phases:

- Location Mean-Shift Clustering: AI plays a crucial role in mean-shift clustering, a non-parametric feature-space analysis technique used to group the GPS data points of visitors around specific locations. The AI algorithm iteratively estimates cluster centers based on density functions, allowing it to determine the number and locations of clusters via the density parameter. This flexibility makes it advantageous over other clustering methods, as it does not require the number of centers to be predetermined.
- User-Topic Modeling: AI is employed in the topic modeling approach, which quantifies the relationship between visitors and the landmarks they visit. The model elaborates on the next landmark based on visitors' historical behavior and interests, enabling personalised and updated recommendations.

- Markov-Topic Fusion Model: AI extends the Topic Model by incorporating a first-order Markov model. This AI-based approach considers the user's current location and history to predict the next landmark to be visited. The probabilities of visiting specific landmarks are estimated using Maximum Likelihood Estimation, enabling the system to provide context-aware and user-specific recommendations.

Benouaret et al. (Benouaret & Lenne, 2016) This paper describes a framework designed to enhance the museum visitors' experience by providing personalised artwork recommendations. The proposed architecture is a hybrid recommender system that combines semantic modeling of artworks and collaborative filtering based on user preferences and demographic attributes for overcoming the limitations of traditional content-based and collaborative filtering methods, particularly cold-start and sparsity problems.

Key components of the framework are the followings:

- Knowledge Base: A semantic model describes artworks, incorporating information from CIDOC-CRM ontology, ICONCLASS taxonomy, AAT, ULAN, and TGN. This rich semantic model allows the computation of semantic similarity among all artworks semantically related to the user's preferences. The similarity measure enables the recommendations' elaboration.
- Context Model: The framework incorporates contextual information such as individuality, activity, location, time, and relations. This semantic context model captures user demographic data and visit context, including activities, similar user relations, location, and time constraints.
- Content-Based Approach: The system calculates the semantic similarity between artworks To recommend artworks according to the user's preferences. Weighted similarity measures allow for determining the importance of different properties in the recommendation process.
- Collaborative Filtering: Recommendations are also based on similar users' preferences, considering their ratings and demographical data. A weighted similarity measure dynamically adjusts the importance of different measures based on the user's profile richness.
- Tour Generator: The tour generator suggests a personalised museum tour based on the results from the recommendation engine, considering contextual post-filtering and museum environment information. The tour aims to cover the user's interests and is dynamically adaptable to user preferences.

Therefore, the framework combines content-based and collaborative filtering methods with contextual information to offer tailored artwork recommendations and generate a personalised museum visiting path. This hybrid recommender system seeks to improve the museum visitors' experience and addresses the challenges associated with traditional recommendation approaches in the context of museum recommendation.

Bartolini et al. (2016) describe a proposed system that supports a recommendation framework for multimedia content related to cultural points of interest (POIs). The framework aims to provide personalised recommendations to users based on their preferences and context information. It also facilitates the generation of dynamic visiting paths that change according to the user's current context.

The main functionalities of the system are as follows:

- Fetching of Multimedia Contents: The system accesses multimedia data from various web repositories, including raw data and related annotations from sources like Wikipedia, Flickr, Europeana, Panoramio, Google Images, and YouTube. This data is stored in a Multimedia Storage and Staging area.
- Indexing of Multimedia Data: The system uses low and high-level descriptors to index the multimedia data, enabling content-based retrieval. The Feature Extraction module extracts relevant features from the multimedia data, and the Structural Description module organises the data in a structured representation.
- Recommending Multimedia Items: The MultiMedia Recommender Engine utilises user preferences and other context information to recommend multimedia items related to cultural POIs. The Candidate Set Building module selects a set of candidate objects, the Objects Ranking module ranks these candidates based on user preferences, and the Visiting Paths Generation module dynamically arranges them into visiting paths.
- Context Management: The Sensor Management Middleware derives context knowledge using information from physical sensors (e.g., GPS, WSN), web services/API, or other techniques. The system's Knowledge Base includes contextual data (e.g., weather and environmental conditions), user preferences, cultural POI descriptions, and support cartography for geo-localisation and visualisation of users' positions relative to POIs.
- Delivery and Presentation: Each user device exploits a Multimedia Guide App that allows users to access and enjoy multimedia content and view the generated visiting paths.

Therefore, the system's main components include the Multimedia Data Management Engine, which handles media content access and feature extraction, the Sensor Management Middleware for context knowledge, and the MultiMedia Recommender Engine for personalised recommendations and visiting path generation. The system aims to enhance the cultural POI visiting experience by providing tailored multimedia recommendations and dynamic tour paths based on user preferences and context.

The experimental phase evaluates the proposed Recommender System designed for cultural Points of Interest (POIs), emphasising users' subjective opinions and satisfaction with the system.

Therefore, the authors aim to measure two main aspects: user satisfaction and system accuracy. User satisfaction is assessed in terms of the effectiveness of visiting paths offered by the system to complete assigned browsing activities in an outdoor environment. The goal is to understand how helpful these recommendations are for improving tourists' experiences.

The accuracy evaluation focuses on the effectiveness of the system's ranking strategy in an indoor cultural space. The goal is to measure how accurately the system can predict users' preferences within a single POI, such as a museum room, compared to other recommendation techniques.

Ruotsalo et al. (2013) introduce the SMARTMUSEUM system, composed of four main components implemented as web services: a metadata service, context service, user profile service, and filtering service. These components communicate with each other through REST using HTTP. The metadata service stores metadata in a column-oriented database, and the context service maps RFID (Radio Frequency Identification) identifiers and GPS coordinates to URIs in ontologies. The user profile service stores user profiles and context information. Users can provide feedback on objects they like or dislike to update their profiles. The filtering service indexes the metadata and provides recommendations to mobile clients based on user profiles and context data.

The SMARTMUSEUM system's core is the recommender system, comprising several computation methods:

- User Profiling: Utilises a probabilistic model to create user profiles based on relevance feedback for objects.
- Data Indexing: Requires fast filtering and dynamic query expansion, achieved through four indices - annotation index, ontology index, spatial index, and triple-space index.
- Information Filtering: It uses Vector Space Model (VSM) to rank objects based on similarity to the user profile, with weighting using Term Frequency-Inverse Document Frequency.
- Query Expansion: Ontology-based expansion of user profile triples to consider more general or related concepts for better filtering.
- Feature Balancing: Balancing the effect of separate triple-spaces (context and content triples) to avoid over-specialisation in the ranking.
- Clustering: it employs Independent Component Analysis to cluster top-ranked objects based on their triples and provide diverse perspectives to the user.

The experimental phase exploits three use cases: recommendation, linking, and clustering. The evaluation uses a dataset containing museum objects and point-of-interest descriptions. The evaluation metrics include recall, precision, and mean average precision (MAP).

AI for Chatbots

Chatbots represent one of the most diffused services in the interaction of users with a system and, in particular, with the elements of an archaeological park or museum. In particular, chatbots aim to simulate a human conversation through text messages or voices. However, traditional rule and algorithm based chatbots have often shown limitations in understanding and responding appropriately to a wide variety of questions and inputs, as their capability was closely tied to predefined rules and programmed logic. The growing interest in machine learning and artificial intelligence has opened new avenues for creating more advanced chatbots capable of learning from data, adapting to users' conversations, and constantly improving their performance. In this regard, it is necessary to exploit techniques related to natural language processing (NLP) to understand the meaning of sentences and provide more relevant answers based on the specific topic.

This subsection focuses on works in the literature based on chatbots integrated with machine learning techniques are analysed.

Sperlí (2021) describes a proposed framework for a horizontal Cultural Heritage platform with domain-specific applications. The platform aims to support tourists by providing cultural heritage trails based on their preferences and profiles. The framework consists of several components, including a conversational human-machine interface (Chatbot engine), a Micro-service architecture, and an Enterprise Service Bus (ESB) to analyse data from various sources related to Cultural Heritage.

- Chatbot Engine: This component uses a seq2seq model based on the Encoder-Decoder framework with Recurrent Neural Networks (RNNs) to provide conversational support to tourists. The encoder converts user queries into vectors in a feature space, while the decoder generates response phrases using the context vectors and encoder hidden states.

- Micro-service Architecture: The back-end system employs a Micro-service architectural pattern to offer different services to users. It includes services like retrieving information about Points-of-Interest using name entity recognition, suggesting events based on user positions or preferences, and profiling users to understand their cultural interests. The ranking for recommendation considers popularity and personality traits.
- Enterprise Service Bus (ESB): The ESB is developed to support public and private organisations in promoting events and efficiently modeling event information. It simplifies integration between applications and data sources and assigns metadata to each event through API REST. The ESB can automatically crawl information from websites to create events.
- Event Data Model (EDM): EDM exploits a graph-based data model integrating tangible and intangible cultural heritage. Cultural items, events, and concepts are graph nodes and relationship edges. Users can retrieve information about events and cultural items by performing traversal queries on the graph-based data model.

The proposed framework aims to enhance tourists' cultural experiences by providing recommendations and connecting cultural or archaeological routes to events. Therefore, the framework seeks to create an interactive and personalised experience, leveraging conversational interfaces, data analytics, and advanced technologies to support their exploration of Cultural Heritage.

The evaluation is conducted through an experimental protocol and compares the efficacy and efficiency of the chatbot engine.

Tsepapadakis and Gavalas (2023) describe the development and implementation aspects of Exhibot, which aims to provide immersive and user-guided experiences in cultural heritage sites. The system follows an agile user-centered software methodology and comprises several phases: domain analysis, requirements analysis, design, prototype development and testing, and user evaluation.

The key design objective of Exhibot is to create immersive experiences for users visiting heritage sites. The system uses low-technological intensity, focusing on transparent and intuitive interfaces to let the exhibits themselves stand out as the main attraction. Users can have autonomous guided tours, and the system supports natural interaction through voice commands and messages, enabling users to unlock information about points of interest based on their interests.

Exhibot uses augmented reality (AR) and AI chatbot technologies. AR provides immersive experiences while maintaining minimal distraction, and the chatbot enables direct interaction with exhibits, narrating information from the first-person perspective.

The targeted domain for Exhibot includes guided visits to large-scale open-air heritage sites and urban environments where visitors are relatively sparse. The system utilises Bluetooth beacons attached to exhibits to detect users' proximity, IMU (Inertial Measurement Unit) to capture head orientation data and GPS technology for localisation. The mobile app, implemented in Android, orchestrates the system tasks and manages user-exhibit interaction.

The system architecture includes the perception, application, and cloud intelligence layers. The perception layer collects data from sensors. The application layer handles the core application and manages context data, user-exhibit interaction, and the dialogue session with the chatbot. The cloud intelligence layer includes the AI chatbot.

Exhibot's user-exhibit interaction involves users issuing voice commands transcribed by the chatbot, and the system infers the relative position and orientation of the user concerning the exhibit. The chatbot generates responses in text, which are converted to spatial audio and delivered to the user's headphones, creating an illusion of a "talking exhibit."

The development of Exhibot involved iterative prototyping and testing phases, allowing for performance improvements and usability enhancements. The system aims to provide an engaging and interactive experience for visitors in cultural heritage sites, enabling them to explore exhibits at their own pace and receive contextual information based on their interests and location.

The experimental phase aims to measure usability and includes dimensions like accessibility, engagement, fun, learning effectiveness, and presence.

The evaluation involved 16 participants, aged 22 to 58, that answered two questionnaires. The first was the System Usability Scale (SUS), measuring usability and learnability. The second questionnaire assessed the perceived value and quality of the experience.

Amato et al. (2021) discusses the development and architecture of a chatbot system for retrieving information about locations, events, and other relevant data. The system relies on a Service-Oriented Architecture (SOA), where modules are developed independently and connected through a Service Bus allowing data extraction and cleaning operations to be delegated to back-end modules, reducing the processing needed at runtime.

The Knowledge Base, implemented in Prolog, stores and manipulates facts and rules. Facts include connections between locations and median times, while rules enhance the inferential process to retrieve customised information for users.

The system uses the WSO2 Enterprise Integrator for the Service Bus, enabling the configuration of interactions between components through a graphical interface.

The chatbot component, called TravelBot, operates using a document (doc.json) containing patterns and keywords for user queries. The chatbot processes user input, converts words to lowercase, removes unnecessary characters like "?," and applies stemming to find root words.

The system includes various rules, such as "Conn1," which establishes bidirectional connections between stops, and "Visit," which calculates the path to the nearest stop based on selected attractions.

EVALUATION MEASURES FOR ARTIFICIAL INTELLIGENCE ALGORITHMS

As seen in the second section, the employment of Artificial Intelligence techniques requires the definition of performance measures able to estimate the ability to manage the assigned task. Indeed, the capacity to execute the assignment with high performance represents one of the requirements of machine learning techniques. Moreover, integrating AI techniques with other tools, such as Recommender Systems, does not change the accuracy measurements available for determining the utility of the selected approach.

In the case of regression tasks, such as the elaboration of suggestions through Recommender Systems and Context-Aware Recommender Systems, the most common accuracy measures for predictions are the followings (Shani & Gunawardana, 2011):

- Mean Absolute Error (MAE) represents the mean of absolute errors related to test set elements, according to the following formula

$$\text{MAE} = \frac{1}{|T|}\sum_{(u,i)\in T}\left(\hat{r}_{ui} - r_{ui}\right) \quad (25)$$

where $\hat{r}_{ui}$ is the prediction elaborated by the Recommender System for the user u and the item i, and the $T = \{(u,i) : \exists r_{ui}\}$ is the test set.

- Root Mean Squared Error (RMSE) evaluates errors giving more relevance to more wrong predictions

$$\text{RMSE} = \sqrt{\frac{1}{|T|}\sum_{(u,i)\in T}\left(\hat{r}_{ui} - r_{ui}\right)^2} \quad (26)$$

Therefore, the MAE and the RMSE aim to measure the prediction error without considering the order of suggestions proposed by the Recommender System. Consequently, the accuracy of RSs and CARSs can be evaluated by accuracy metrics related to classification tasks. In this case, it is necessary to construct the confusion matrix in Figure 6 related to binary classification and easily extendable to multi-class one.

Figure 6. Example of confusion matrix related to a binary classification task

		Predicted Values	
		Positive	Negative
Real Values	Positive	True Positive (T_P)	False Negative (F_N)
	Negative	False Positive (F_P)	True Negative (T_N)

Confusion Matrix allows identifying:

- True Positive T_P: instances identified as positive by the elaboration phase and truly positive;
- True Negative T_N: instances identified as negative by the elaboration phase and truly negative;
- False Positive F_P: instances identified as positive by the elaboration phase but labelled as negative;
- False Negative F_N: instances identified as negative by the elaboration phase but labelled as positive.

After the identification of T_P, T_N, F_P, and F_N, it is possible to evaluate the following accuracy measurements described through the formulas that define them (Shani & Gunawardana, 2011):

- Accuracy (A): measure related to the comparison among instances correctly identified and all elements of the test set

$$A = \frac{T_P + T_N}{T_P + T_N + F_P + F_N} \in [0,1]; \tag{27}$$

- Error Rate (*ER*): measure related to the comparison among instances wrongly identified

$$ER = 1 - A = \frac{F_P + F_N}{T_P + T_N + F_P + F_N} \in [0,1]; \tag{28}$$

- Precision (*P*): Precision measures the percentage of test set elements precisely elaborated in relation to the False Positive

$$P = \frac{T_P}{T_P + F_P}; \tag{29}$$

- Recall (*R*): Recall measures the percentage of test set elements precisely elaborated in relation to the False Positive

$$R = \frac{T_P}{T_P + F_N}; \tag{30}$$

- F_1-Score (F_1): a harmonic mean between Precision and Recall

$$F_1 = \frac{2 \cdot P \cdot R}{P + R} \tag{31}$$

Evaluating the accuracy through the introduced measurements utilises several datasets available. In particular, RSs and CARSs exploit datasets based on movies such as MovieLens (Harper & Konstan, 2015) and DePaulMovie (Zheng et al., 2016), on travels such as STS (Ilarri et al., 2018), or restaurants such as TijuanaRestaurant (Ramirez-Garcia & García-Valdez, 2014) and JapanRestaurant (Oku et al., 2006).

CONCLUSION AND FUTURE DIRECTIONS

During this work, the role of Artificial Intelligence in improving the interaction between users and cultural heritage has been analysed. After the background analysis, where the principal AI techniques were introduced, the work focused on the literature papers on the topic of interest. In particular, the overview describes the integration of AI with other tools able to improve the user experience, such as Recommender Systems, Context-Aware Recommender Systems, and Chatbots. Therefore, the most employed accuracy measures have been described to provide a complete panoramic on the methodologies to quantitatively estimate the ability of AI techniques to perform a specific task.

However, this overview underlines some limits related to applying Artificial Intelligence techniques to improve the enjoyment of user cultural experience. Firstly, the lack of a specific dataset focused on cultural heritage limits the experimental possibilities in this field. Moreover, despite analysed works, works focused on improving the interaction between users and cultural heritage still result in a small minority of respect for the application of AI in other application fields.

REFERENCES

Abowd, G. D., Dey, A. K., Brown, P. J., Davies, N., Smith, M., & Steggles, P. (1999). Towards a better understanding of context and context-awareness. Lecture Notes in Computer Science (Including Subseries Lecture Notes in Artificial Intelligence and Lecture Notes in Bioinformatics), 1707. doi:10.1007/3-540-48157-5_29

Adomavicius, G., Mobasher, B., Ricci, F., & Tuzhilin, A. (2011). Context-aware recommender systems. *AI Magazine*, *32*(3), 67–80. Advance online publication. doi:10.1609/aimag.v32i3.2364

Akerkar, R., & Sajja, P. (2009). *Knowledge-based systems*. Jones & Bartlett Publishers.

Al-Fuqaha, A., Guizani, M., Mohammadi, M., Aledhari, M., & Ayyash, M. (2015). Internet of Things: A Survey on Enabling Technologies, Protocols, and Applications. *IEEE Communications Surveys and Tutorials*, *17*(4), 2347–2376. Advance online publication. doi:10.1109/COMST.2015.2444095

Alexandridis, G., Chrysanthi, A., Tsekouras, G. E., & Caridakis, G. (2019). Personalised and content adaptive cultural heritage path recommendation: An application to the Gournia and Çatalhöyük archaeological sites. *User Modeling and User-Adapted Interaction*, *29*(1), 201–238. doi:10.100711257-019-09227-6

Almiani, M., AbuGhazleh, A., Al-Rahayfeh, A., Atiewi, S., & Razaque, A. (2020). Deep recurrent neural network for IoT intrusion detection system. *Simulation Modelling Practice and Theory*, *101*, 102031. Advance online publication. doi:10.1016/j.simpat.2019.102031

Amato, F., Moscato, F., Moscato, V., & Sperlì, G. (2021). Smart conversational user interface for recommending cultural heritage points of interest. *CEUR Workshop Proceedings*, 2994.

Bartolini, I., Moscato, V., Pensa, R. G., Penta, A., Picariello, A., Sansone, C., & Sapino, M. L. (2016). Recommending multimedia visiting paths in cultural heritage applications. *Multimedia Tools and Applications*, *75*(7), 3813–3842. Advance online publication. doi:10.100711042-014-2062-7

Benouaret, I., & Lenne, D. (2016). Personalising the Museum Experience through Context-Aware Recommendations. *Proceedings - 2015 IEEE International Conference on Systems, Man, and Cybernetics, SMC 2015*. 10.1109/SMC.2015.139

Bobadilla, J., Ortega, F., Hernando, A., & Gutiérrez, A. (2013). Recommender systems survey. *Knowledge-Based Systems*, *46*, 109–132. Advance online publication. doi:10.1016/j.knosys.2013.03.012

Brasse, J., Broder, H. R., Förster, M., Klier, M., & Sigler, I. (2023). Explainable artificial intelligence in information systems: A review of the status quo and future research directions. *Electronic Markets*, *33*(1), 26. Advance online publication. doi:10.100712525-023-00644-5

Bro, R., & Smilde, A. K. (2014). Principal component analysis. In Analytical Methods (Vol. 6, Issue 9). doi:10.1039/C3AY41907J

Ertel, W. (2017). Introduction to Artificial Intelligence (Undergraduate Topics in Computer Science). Springer.

Fayyad, U., Piatetsky-Shapiro, G., & Smyth, P. (1996). From data mining to knowledge discovery in databases. *AI Magazine*, *17*(3).

Fiorucci, M., Khoroshiltseva, M., Pontil, M., Traviglia, A., Del Bue, A., & James, S. (2020). Machine Learning for Cultural Heritage: A Survey. *Pattern Recognition Letters*, *133*, 102–108. doi:10.1016/j.patrec.2020.02.017

Géron, A. (2017). Hands-on machine learning with Scikit-Learn and TensorFlow : concepts, tools, and techniques to build intelligent systems. O'Reilly Media.

Harper, F. M., & Konstan, J. A. (2015). The movielens datasets: History and context. *ACM Transactions on Interactive Intelligent Systems*, *5*(4), 1–19. Advance online publication. doi:10.1145/2827872

Ilarri, S., Trillo-Lado, R., & Hermoso, R. (2018). Datasets for context-aware recommender systems: Current context and possible directions. *Proceedings - IEEE 34th International Conference on Data Engineering Workshops, ICDEW 2018*. 10.1109/ICDEW.2018.00011

Jain, L. C., Seera, M., Lim, C. P., & Balasubramaniam, P. (2014). A review of online learning in supervised neural networks. In *Neural Computing and Applications* (Vol. 25, pp. 3–4). doi:10.100700521-013-1534-4

Jiang, L., Cai, Z., Wang, D., & Jiang, S. (2007). Survey of improving K-nearest-neighbor for classification. *Proceedings - Fourth International Conference on Fuzzy Systems and Knowledge Discovery, FSKD 2007, 1*. 10.1109/FSKD.2007.552

Jordan, M. I., & Mitchell, T. M. (2015). Machine learning: Trends, perspectives, and prospects. In Science (Vol. 349, Issue 6245). doi:10.1126cience.aaa8415

Koren, Y., Bell, R., & Volinsky, C. (2009). Matrix factorisation techniques for recommender systems. *Computer*, *42*(8), 30–37. Advance online publication. doi:10.1109/MC.2009.263

Liao, W. D., Yang, D. L., & Hung, M. C. (2010). An intelligent recommendation model with a case study on u-Tour Taiwan of historical momuments and cultural heritage. *Proceedings - International Conference on Technologies and Applications of Artificial Intelligence, TAAI 2010*. 10.1109/TAAI.2010.23

Liu, Y., Wang, S., Khan, M. S., & He, J. (2018). A novel deep hybrid recommender system based on auto-encoder with neural collaborative filtering. *Big Data Mining and Analytics*, *1*(3), 211–221. Advance online publication. doi:10.26599/BDMA.2018.9020019

Lops, P., De Gemmis, M., Semeraro, G., Musto, C., Narducci, F., & Bux, M. (2009). A semantic content-based recommender system integrating folksonomies for personalised access. *Studies in Computational Intelligence*, *229*, 27–47. Advance online publication. doi:10.1007/978-3-642-02794-9_2

Mitchell, T. M. (1997). Does machine learning really work? *AI Magazine*, *18*(3).

Oku, K., Nakajima, S., Miyazaki, J., & Uemura, S. (2006). Context-aware SVM for context-dependent information recommendation. *Proceedings - IEEE International Conference on Mobile Data Management, 2006*. 10.1109/MDM.2006.56

Pavlidis, G. (2018). Apollo - A Hybrid Recommender for Museums and Cultural Tourism. *9th International Conference on Intelligent Systems 2018: Theory, Research and Innovation in Applications, IS 2018 - Proceedings*. 10.1109/IS.2018.8710494

Philip Chen, C. L., & Zhang, C. Y. (2014). Data-intensive applications, challenges, techniques and technologies: A survey on Big Data. *Information Sciences*, *275*, 314–347. Advance online publication. doi:10.1016/j.ins.2014.01.015

Prasath, V. B. S., Haneen Arafat, A. A., Hassanat, A. B. A., Lasassmeh, O., Tarawneh, A. S., Alhasanat, M. B., & Salman, H. S. E. (2019). Effects of Distance Measure Choice on KNN Classifier Performance. *RE:view*. ArXiv1708.04321v3

Protopapadakis, E., Doulamis, N., & Voulodimos, A. (2017). Hybrid meta-filtering system for cultural monument related recommendations. *VISIGRAPP 2017 - Proceedings of the 12th International Joint Conference on Computer Vision, Imaging and Computer Graphics Theory and Applications, 5*. 10.5220/0006347104360443

Ramirez-Garcia, X., & García-Valdez, M. (2014). Post-filtering for a restaurant context-aware recommender system. *Studies in Computational Intelligence*, *547*, 695–707. Advance online publication. doi:10.1007/978-3-319-05170-3_49

Ricci, F., Shapira, B., & Rokach, L. (2015). Recommender systems: Introduction and challenges. In Recommender Systems Handbook (2nd ed.). doi:10.1007/978-1-4899-7637-6_1

Rumelhart, D. E., Hinton, G. E., & Williams, R. J. (1986). Learning representations by back-propagating errors. *Nature*, *323*(6088), 533–536. Advance online publication. doi:10.1038/323533a0

Ruotsalo, T., Haav, K., Stoyanov, A., Roche, S., Fani, E., Deliai, R., Mäkelä, E., Kauppinen, T., & Hyvönen, E. (2013). SMARTMUSEUM: A mobile recommender system for the Web of Data. *Journal of Web Semantics*, *20*, 50–67. Advance online publication. doi:10.1016/j.websem.2013.03.001

Sarker, I. H. (2021). Deep Learning: A Comprehensive Overview on Techniques, Taxonomy, Applications and Research Directions. In SN Computer Science (Vol. 2, Issue 6). doi:10.100742979-021-00815-1

Schmidhuber, J. (2015). Deep Learning in neural networks: An overview. In Neural Networks (Vol. 61). doi:10.1016/j.neunet.2014.09.003

Shani, G., & Gunawardana, A. (2011). Evaluating Recommendation Systems. In Recommender Systems Handbook. doi:10.1007/978-0-387-85820-3_8

Sperlí, G. (2021). A cultural heritage framework using a Deep Learning based Chatbot for supporting tourist journey. *Expert Systems with Applications*, *183*, 115277. Advance online publication. doi:10.1016/j.eswa.2021.115277

Stonebraker, M. (2010). SQL databases v. NoSQL databases. *Communications of the ACM*, *53*(4), 10–11. Advance online publication. doi:10.1145/1721654.1721659

Su, X., Sperli, G., Moscato, V., Picariello, A., Esposito, C., & Choi, C. (2019). An Edge Intelligence Empowered Recommender System Enabling Cultural Heritage Applications. *IEEE Transactions on Industrial Informatics*, *15*(7), 4266–4275. Advance online publication. doi:10.1109/TII.2019.2908056

Tsai, C. W., Lai, C. F., Chao, H. C., & Vasilakos, A. V. (2015). Big data analytics: A survey. *Journal of Big Data*, *2*(1), 21. Advance online publication. doi:10.118640537-015-0030-3 PMID:26191487

Tsepapadakis, M., & Gavalas, D. (2023). Are you talking to me? An Audio Augmented Reality conversational guide for cultural heritage. *Pervasive and Mobile Computing, 92*, 101797. doi:10.1016/j.future.2019.04.020

ur Rehman, M. H., Yaqoob, I., Salah, K., Imran, M., Jayaraman, P. P., & Perera, C. (2019). The role of big data analytics in industrial Internet of Things. *Future Generation Computer Systems*, *99*, 247–259. Advance online publication. doi:10.1016/j.future.2019.04.020

Witten, I. H., Frank, E., Hall, M. A., & Pal, C. J. (2016). Data Mining: Practical Machine Learning Tools and Techniques. doi:10.1016/C2009-0-19715-5

Wuest, T., Weimer, D., Irgens, C., & Thoben, K.-D. (2016). Machine learning in manufacturing: Advantages, challenges, and applications. *Production & Manufacturing Research*, *4*(1), 23–45. doi:10.1080/21693277.2016.1192517

Zhao, J., Geng, X., Zhou, J., Sun, Q., Xiao, Y., Zhang, Z., & Fu, Z. (2019). Attribute mapping and autoencoder neural network based matrix factorisation initialisation for recommendation systems. *Knowledge-Based Systems*, *166*, 132–139. Advance online publication. doi:10.1016/j.knosys.2018.12.022

Zheng, Y., Mobasher, B., & Burke, R. (2016). CARSKit: A Java-Based Context-Aware Recommendation Engine. *Proceedings - 15th IEEE International Conference on Data Mining Workshop, ICDMW 2015*. 10.1109/ICDMW.2015.222

Chapter 7
Machine Learning Approach for Robot Navigation Using Motor Imagery Signals

Pratyay Das
Maulana Abul Kalam Azad University of Technology, West Bengal, India

Amit Kumar Shankar
Maulana Abul Kalam Azad University of Technology, West Bengal, India

Ahona Ghosh
https://orcid.org/0000-0003-0498-285X
Maulana Abul Kalam Azad University of Technology, West Bengal, India

Sriparna Saha
https://orcid.org/0000-0002-7312-2450
Maulana Abul Kalam Azad University of Technology, West Bengal, India

ABSTRACT

Electroencephalography (EEG) signals have been used for different healthcare applications like motor and cognitive rehabilitation. In this study, motor imagery data of different subjects' rest vs. movement and different movements is categorized from a publicly available dataset. The authors have first applied a lowpass filter to the EEG signals to reduce noise and a fast fourier transform analysis to extract features from the filtered data. Utilizing principal component analysis, relevant features are selected. With an accuracy of 95.02%, they have classified rest vs. movement using the k-nearest neighbor algorithm. Using the random forest algorithm, they have classified various movement types with an accuracy of 96.45%. The success in differentiating between movement and rest raises the possibility that EEG signals can recognize a user's intention to move. Accurately classifying different movement types opens the possibility of navigating robots accordingly in the real-time scenario for people with motor disabilities to assist them with robotic arms and prosthetic limbs.

DOI: 10.4018/978-1-6684-9999-3.ch007

INTRODUCTION

Motor imagery (MI) is the process of performing an action in the brain, but not physically in real time. MI signals recorded via *electroencephalograms* (EEGs) are the most convenient basis for designing *Brain-Computer Interfacing* (BCI) as they provide a high degree of freedom. MI-based BCIs help motor disabled people to interact with any real-time BCI applications by performing a sequence of MI tasks. BCI also enables robotic movements to be controlled in an entirely effortless manner by utilizing only the signals produced by the user's brain (Gul *et al.*, 2019). This is the aim of research in robot navigation from MI signals. For the robots to be navigated and carry out complex tasks depends on the brain's capacity to produce electrical signals reacting to envisioned movements (Ghosh & Saha, 2020).

Robot navigation employing MI signals has many potential uses (Bag *et al.*, 2022). For instance, this technology would allow people with physical limitations to direct a robot to carry out duties like picking up and arranging goods or finding specified areas. This strategy may also lessen the technical effort needed to control a robot, making it more straightforward for unskilled users to use MI signals (Guillot & Debarnot, 2019). Experiments are being conducted in many fields, such as mining operations, surveillance, and exploration, all involving strenuous activities requiring extensive data analysis and judgment. As a result, *Artificial Intelligence* (AI) can significantly outperform human operators (Saha & Ghosh, 2019). AI can be used to execute more accurately and efficiently, leading to more successful and economical missions (Saha *et al.,* 2018). Also, using AI under challenging circumstances can decrease the risk to human operators, making these operations safer and more practical. Supervised Machine Learning (ML) methods like Support Vector Machine (SVM), k-Nearest Neighbors (kNN), Random Forest (RF), and Deep Learning (DL) approaches like Convolutional Neural Network (CNN), Long Short Term Memory (LSTM) have been observed to work efficiently in the recent literature of this domain. The world is facing important trends associated with an increase of disability in populations, especially a rise in noncommunicable diseases (NCDs), including mental health conditions, and the rapid ageing of the world population. Estimates from the WHO World report on disability show that 15% of the global population experience significant disability (World Health Organization, 2019). The motivations behind developing the current system are to assist persons with disabilities in communicating, operating computers, and using assistive technology like wheelchairs or robotic arms.

The first stage of the proposed work is to remove artifacts and undesirable signals from the EEG signals using filtering, as without appropriate cleaning, the outcomes could be untrustworthy. Extracting features pertinent to the specific job is the next step in the feature engineering process. These characteristics subsequently build a classifier to distinguish between the brain signals connected to MI. This research builds on earlier works by developing two models rather than just one to solve existing limitations (de Klerk *et al.*, 2019). The first model's goal is to foretell whether the user is at rest or intends to move, whereas the second model's goal is to predict the specific movement the user intends to make (Ofner *et al.*, 2019). By developing these models independently, we intend to increase the prediction accuracy by making it more usable in real-world circumstances. Also, creating two models gives the system's overall architecture more design freedom (Schuster *et al.*, 2011). The contributions of the current work are summarized as follows

- MI signals classification method applied on a publicly available database, which starts with the efficient selection of the relevant channels from a large number of channels to make the system optimal for ease of use in real-time scenarios.

- Efficient noise reduction is done to make the data artifact-free.
- The frequency analysis is performed to extract feature information from the frequency domain of cortical electrodes.
- Selection of only the important features from the entire feature space is done by simple but efficient Principal Component Analysis
- Two hyper-tuned classifiers for rest vs movement and movement category classification have been used.
- The overall approach combines the processing of relevant electrode channel data from the regions of interest, thet results in high-precision classification and simplifies the BCI application strategy. Also, it outpaces the existing methods in the performance assessment of each step involved in it.

The remaining chapter is structured as follows. The next section presents a detailed literature review. Section 3 discusses the working of the proposed methodology, followed by a thorough discussion of results and performance evaluation in Section 4. Section 5 ultimately concludes the chapter.

LITERATURE SURVEY

In recent years, MI-based BCI (MI-BCI) technology has advanced and received increasing attention from the military, entertainment industry, healthcare industry, and other sectors. Accordingly, precise decoding of EEG signals is crucial for BCI devices. The decoding accuracy of EEG signals has always been a problem since the signal is dynamic time series data with a poor signal-to-noise ratio. Even though many academics have made outstanding contributions to this field, there is still a disconnect between the BCI system and standards for practical application, and there is still much room for advancement in the EEG signal categorization process and accuracy. In one of the current paradigms for MI-based assistive robotics (Lu *et al*., 2020), patients who had normal robotic rehabilitation were compared to patients who underwent MI-BCI, where the MI-BCI patients performed overall better.

Patients treated with MI-BCI fare better than those treated with random BCI, according to a study (Liu *et al*., 2023) combining MI-BCI with traditional physical therapy to rehab stroke victims. Hand and foot motions, mental mathematic processes, and object rotation in the mind have all been explored as MI methods (Ghosh *et al*., 2023). For limb rehabilitation, MI techniques focused on mental imagery of limb movements appear to be more effective (Chatterjee *et al*., 2023).

Using MI, a robotic hand prosthesis was attached to a quadriplegic person with limited left bicep motion (Benabid *et al*., 2019). The individual was able to manage the prosthesis to some extent after five months of training. Surprisingly, shorter training durations have also been said to be effective (Müller-Putz *et al*., 2022). Different BCI models have been used concurrently, for instance, to control a robot arm that mimics an upper human joint with a two-degree flexion (Wei *et al*., 2021). Steady-state visual evoked potentials (SSVEP) were used in their work for joint control while MI was used to control the grasping function. Analysis of some more recent literature is tabulated in Table 1, from the limitations of which we got motivated to take up the current research.

Table 1. Comparative Analysis of existing literature

Ref.	Technology Used	Contributions	Limitations
(Scherer *et al.*, 2007)	BCI, Deep neural network, EEG MI	The Graz BCI system, which employs self-paced control, signal processing, and ML approaches, is developed and evaluated in this work. Its applications show its promise for enhancing the quality of life for people with motor impairments.	The system's performance isn't uniformly evaluated in the study because there are only a few participants. It only has a few tasks for which it can be used, and efficient use necessitates substantial training. Exploring its potential for additional uses will require more study.
Dai *et al.*, 2019)	EEG, short-time Fourier transform, CNN, variational autoencoder	The system outperforms previous cutting-edge techniques by combining CNNs and LSTMs to categorize EEG data associated with MI. It is crucial for the growth of BCIs.	The study only used a limited dataset, which could prevent it from being generalized to larger datasets. The framework's interpretability for practical applications was not examined.
(Chaudhary *et al.*, 2019)	EEG, task analysis, feature extraction, Continuous wavelet transforms, SVM	The article suggests a CNN-based technique to increase the precision and effectiveness of EEG-based BCI systems, which can be used in different fields.	The method was not evaluated against other cutting-edge techniques for EEG classification in the study. Additional study is required to compare the suggested method to other methods already in use and to confirm its efficacy on a more extensive dataset.
(Ruffino *et al.*, 2017)	Learning neural plasticity	A theoretical model of the brain mechanisms underlying motor learning is proposed in this study, along with possible applications in clinical and sports situations. It covers the current understanding of neural plasticity with MI practice during motor learning.	The paper focuses on neuroimaging investigations and does not examine how practicing MI may differ depending on the individual. The possible drawbacks and dilemmas associated with using MI practice in specific populations, such as people with cognitive or neurological impairments, are not addressed. The theoretical model of the neurological underpinnings of practicing MI still needs experimental validation.
(Grospêtre *et al.*, 2016)	Transcranial magnetic stimulation, motor control, neuroscience, performance	An overview of the state of the science investigating the connection between cortico-spinal excitability and MI is given in the review. It emphasizes the advantages of using MI as a tool for motor rehabilitation and offers suggestions for fresh approaches to intervention.	The authors' emphasis on research done on healthy people restricts the applicability of the findings to clinical populations. The work does not discuss the probable mechanisms underlying how MI affects corticospinal excitability. However, this could be an exciting area for future study.
(Cha *et al.*, 2019)	Preprocessing by a band-pass filter, feature extraction by Hilbert-Huang Transform, classification using SVM, kNN, and RF	The work suggested a unique feature extraction technique. It illustrated the efficiency of ML algorithms for classifying EEG signals, which has potential applications in clinical diagnostics and BCIs.	The proposed feature extraction method may need additional testing on enormous datasets because the study only examined EEG data from a few people. Future studies should evaluate how well various machine learning methods function.
(De Vries *et al.*, 2007)	EEG, fMRI, and robotic devices are all used in MI-based stroke rehabilitation.	The authors review the varieties of MI, the neural processes that support it, and how MI can be incorporated into stroke rehabilitation. They also emphasize the significance of individualized stroke rehabilitation strategies considering each patient's requirements and skills.	It does not detail possible obstacles or practical considerations for therapists and clinicians. It summarizes state-of-the-art research on MI and stroke rehabilitation, highlighting potential advantages and outlining areas for future study.

PROPOSED WORK

The proposed work involves creating a BCI system to decode upper limb movements from EEG signals. It has concentrated on distinguishing between movement and rest and determining the precise movement the user intends to perform. We have extracted features from the EEG signals using Fast Fourier transform (FFT) and selected features using Principal Component Analysis (PCA). The next step is to train ML algorithms, like kNN and RF, to categorize the EEG signals and spot patterns that indicate when someone thinks of remaining at rest or moving with specific patterns. This research has significant potential applications because it may help create prosthetic limbs and other assistive technology that are more effective and user-friendly for assisting people with motor disabilities (Grigorev *et al.*, 2019). People could regain more independence and enhance their general quality of life by using their thoughts to operate these devices (Hoffmann *et al.*, 2007).

Figure 1. Flowchart of the proposed EEG-based upper limb movement decoding scheme

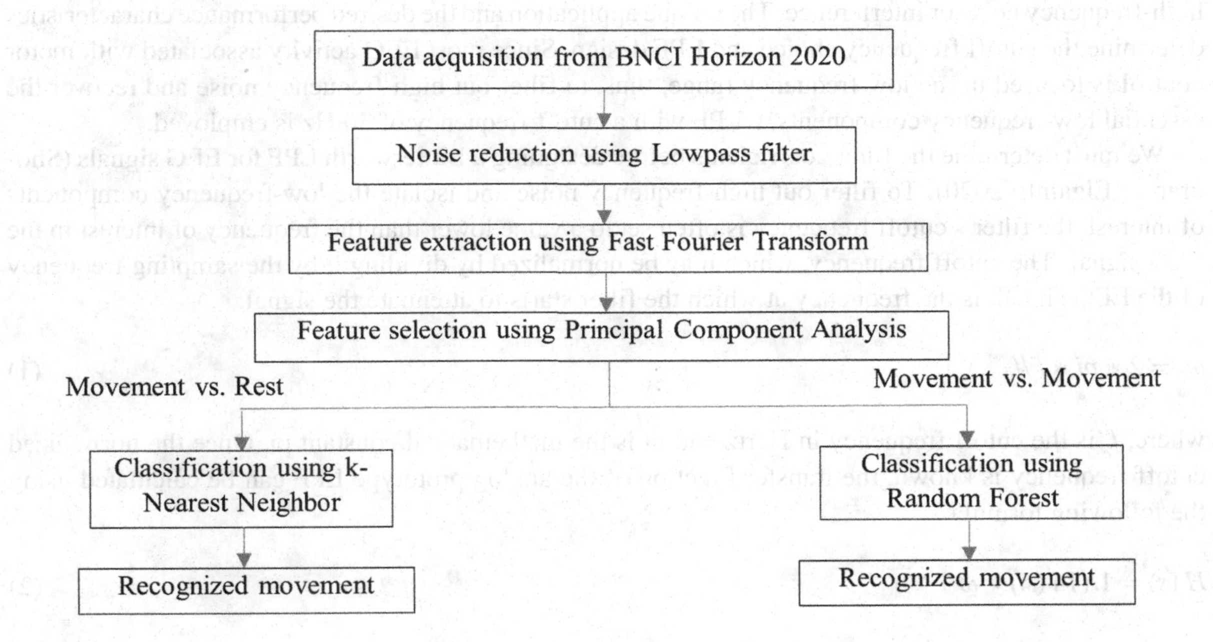

Data Acquisition From BNCI Horizon 2020

The datasets are collected from the Upper limb movement decoding (ULMD) from EEG (Ofner *et al.*, 2019) of BNCI Horizon 2020. It consists of EEG recordings of 61 channels according to a 10/10 electrode placement strategy placed on participants' scalps covering frontal, central, parietal, and temporal areas.

Channel Selection

From the 61 channels of original EEG data (Ofner *et al.*, 2019), we have considered only the signals from C3, C4, and Cz channels, which are relevant only for MI actions to reduce our work's complexity (Kim *et al.*, 2012). The choice of electrodes is validated later using sLORETA (Kim *et al.*, 2012).

Noise Reduction Using Chebyshev Lowpass Filter

The EEG dataset is pre-processed in this stage to eliminate noise and artifacts that could skew the recorded signals and impair the precision of the subsequent analysis. A band-pass filter, which eliminates frequency components outside the range of interest, is a typical method for preprocessing EEG signals. A lowpass filter (LPF) is a kind of digital filter that excludes high-frequency signals while allowing low-frequency information to flow through. The gain of higher frequencies is gradually decreased while the gain of lower frequencies is maintained to achieve this. LPFs are frequently utilized in signal, image, video processing, and other fields where it is desirable to pass low-frequency signals while removing high-frequency noise or interference. The unique application and the desired performance characteristics determine the cutoff frequency choice and LPF design. Since most EEG activity associated with motor control is focused in the low-frequency range, thus, to filter out high-frequency noise and recover the essential low-frequency components, a LPF with a cutoff frequency of 30 Hz is employed.

We must determine the filter coefficients before designing a Butterworth LPF for EEG signals (Shouran & Elgamli, 2020). To filter out high-frequency noise and isolate the low-frequency components of interest, the filter's cutoff frequency is often set to a value lower than the frequency of interest in the EEG signal. The cutoff frequency, which may be normalized by dividing it by the sampling frequency of the EEG signal, is the frequency at which the filter starts to attenuate the signal.

$$\omega_c = 2 * pi * f_c/f_s \tag{1}$$

where, f_c is the cutoff frequency in Hertz and pi is the mathematical constant pi. Once the normalized cutoff frequency is known, the transfer function of the analog prototype LPF can be calculated using the following formula

$$H_s(s) = 1/(1+(s/j * \omega_c)^{(2*n)}) \tag{2}$$

where, s is the Laplace variable, j is the imaginary unit, and n is the filter order. The filter order determines the steepness of the filter's roll-off, with higher orders resulting in steeper roll-off and more phase distortion. To obtain the digital filter transfer function, we apply the bilinear transform to the analog prototype filter transfer function

$$H(z) = H_s((2/T_s) * ((1 - z^{-1})/(1+z^{-1}))) \tag{3}$$

where, T_s is the sampling period. This transformation creates a non-linear frequency warping that changes the filter's frequency response by mapping the s-plane of the analog filter to the z-plane of the digital filter.

Feature Extraction Using Fast Fourier Transform

FFT-based feature extraction (Yudhana *et al.*, 2020) is performed in our proposed approach to create the discriminative feature space. The Discrete Fourier Transform (DFT) of a discrete-time signal is effectively computed by the widely used mathematical algorithm known as FFT. It plays a crucial role in extracting pertinent frequency domain features that aid in identifying different movement patterns and assist in artifact removal. Artifacts, such as muscle activity, eye blinks, and electrical interference, often manifest as distinct frequency patterns in the EEG signals. These artifact-related frequency components can be detected and mitigated by leveraging FFT-based feature extraction techniques. Additionally, the FFT provides valuable information about the underlying dynamics and distinctive frequency components associated with different movement patterns, contributing to a better understanding and differentiation of various patterns or classes while allowing for effective artifact removal.

The FFT formula allows us to compute a signal's frequency spectrum. Mathematically, given a discrete-time signal $x(n)$, the FFT computes the complex-valued spectrum $X(k)$ at discrete frequency indices k:

$$X(k) = \sum_{n=0}^{N-1} x(n) \cdot e^{-i2\pi \frac{kn}{N}} \tag{4}$$

Here, N represents the number of samples in the signal, $x(n)$ denotes the input signal, and k is the frequency index. By applying the FFT, we obtain the frequency representation of the signal, which provides insights into the underlying frequency components.

FFT analysis is used to extract relevant frequency-related information for various classification tasks. It can identify distinctive frequency components connected to various patterns or phenomena in the data and provide insights into the signal's frequency properties. Statistical features such as magnitudes or power spectra can be derived from the FFT's output for particular frequency bins or ranges. Precision classification models can be created by utilizing the frequency-related patterns in the data to categorize and distinguish different patterns or classes accurately. FFT-based feature extraction significantly improves classification performance and deepens our understanding of the fundamental dynamics underlying various patterns or phenomena. The ability to analyze frequency characteristics through FFT analysis offers valuable insights and makes accurate and insightful classification possible.

Feature Selection Using Principal Component Analysis

Afterward, the EEG signals' dimensionality is decreased using PCA (XIE, 2020), and the most valuable characteristics are extracted. With the help of a statistical approach called PCA, it is possible to find the underlying trends in a dataset and express them in a reduced-dimensional space. In this instance, PCA is used to reduce the dimensionality of the feature space and find the key features that contribute to the variation in the EEG signals.

Data must be normalized for PCA by taking each variable's mean away and dividing the result by the standard deviation. The computed covariance matrix captures the relationship between the variables in the dataset. Whereas the off-diagonal elements of the covariance matrix reflect the covariance between pairs of variables, the diagonal elements of the covariance matrix represent the variance of each variable. The Covariance matrix is denoted by

$$C = \frac{1}{n-1}(X - X_{mean})(X - X_{mean})^T \tag{5}$$

where, X is the standardized data matrix, X_{mean} is the mean of the data, and n is the number of observations. The covariance matrix's eigenvalues and eigenvectors show the variance's magnitude and direction. In Eigen decomposition, the covariance matrix is denoted with

$$V * \Lambda * V' \tag{6}$$

where, V is the matrix of eigenvectors, Λ is the diagonal matrix of eigenvalues, and V' is the transpose of V. Principal components were chosen to account for most of the variation based on the size of the eigenvalues. Data is translated into feature space using a linear combination of principal components. Then the transformed data is denoted with $X * V_k$ where V_k is the matrix of the k-selected eigenvectors.

Classification

The classification stage in our work is divided into two parts, movement vs. movement classification using RF and movement vs. rest classification using kNN.

Classification Using RF

To classify various movements, the features extracted by FFT and selected by PCA were used to train the proposed RF model (Taunk *et al.*, 2019). When dividing a node, it looks for the best feature from a random subset of features rather than the most important one. As a result of this, diversity in the model is observed, which often produces a better model. Therefore, in a RF classifier, only a random subset of the features is taken into consideration by the algorithm for splitting a node, and this is why, the current work employs RF which first constructs an ensemble of decision trees to forecast the class label of new input signals. The steps followed in the RF method are as follows.

Algorithm 1: Classification using Random Forest

Step 1: Select random K data points from the training set.

Step 2: Build the decision trees associated with the selected data points (Subsets).

Step 3: Choose the number N for the decision trees needed to be built.

Step 4: Repeat the first two steps.

Step 5: For new data points, find the predictions of each decision tree, and assign the new data points to the category that wins the majority votes.

Classification Using kNN

Accurately differentiating between rest and movement is one of the process's critical components because doing so is crucial for figuring out whether the user is planning to move or remain at rest. In this study, the kNN algorithm (Bansal *et al.*, 2022) was used to accurately categorize EEG signals into the categories of rest and movement. Since the kNN does not need to make additional assumptions, tune several parameters to build a model making it crucial in nonlinear data cases, and since reliable parametric estimates of probability densities are complex to determine in the current context, thus it has been chosen as the proposed model. It finds the *k* number of nearest data points in the training dataset most similar to the new input signal. The steps followed in the kNN approach are as follows.

Algorithm 2: Classification using k Nearest Neighbor

Step 1: Select the number of neighbors *k*

Step 2: Calculate the Euclidean distance of *k* number of neighbors

Step 3: Take the *k* nearest neighbors according to the calculated Euclidean distance.

Step 4: Count the number of data points in each category among these *K* neighbors.

Step 5: Allocate the new data points to the class where the number of neighbors is at most.

EXPERIMENTAL RESULTS

In this section, we present the outcomes of different stages of our proposed work and evaluate the effectiveness of sLORETA for channel selection, FFT for frequency-domain feature extraction, PCA for reducing the dimensionality of the feature space, RF and kNN for classification based on the following performance metrics.

$$Accuracy = \frac{TruePositive + TrueNegative}{TruePositive + TrueNegative + FalsePositive + FalseNegative} \tag{7}$$

$$Precision = \frac{TruePositive}{TruePositive + FalsePositive} \tag{8}$$

$$Recall = \frac{TruePositive}{TruePositive + FalseNegative} \tag{9}$$

$$Error\ rate = \frac{FalsePositive + FalseNegative}{TruePositive + TrueNegative + FalsePositive + FalseNegative} \tag{10}$$

$$F1\ score = 2\frac{Precision \times Recall}{Precision + Recall} \tag{11}$$

The classifier's suitability is also tested by comparing it with competitors. We have presented the results of hyperparameter tuning, which increased the classification accuracy in various settings.

Dataset Description

We have considered a publicly available dataset that consists of 15 healthy subjects data within an age range of 22-40 years (Kim *et al.*, 2012). Nine subjects were female, and all the subjects except one were right-handed. The subjects performed six movement types, such as elbow flexion, elbow extension, supination, pronation, hand close, and hand open. They began in a neutral position with the thumb on the inner side of the hand, the lower arm extended to a 120 degree angle, and the hand half open. In addition to the movements, a class on rest is also recorded, during which the subjects are told to remain still and avoid moving from their starting positions. The paradigm is trial-based, and the subjects are in front of a computer screen that displays stimuli. A cross appears on the computer screen and a beep is heard at second 0. A cue is then displayed on the computer screen at second 2 to indicate the necessary movement (or rest). Subjects returned to their starting positions at the conclusion of the trial. There are 10 runs with 42 trials each in each session. An 8th-order Chebyshev bandpass filter with a frequency range of 0.01 Hz to 200 Hz and 512 Hz sampling rate is used. A notch filter operating at 50 Hz reduces power line interference. A cross appears on the computer screen and a beep is heard at second 0. A cue is then displayed on the computer screen at second 2 to indicate the necessary movement (or rest). Subjects returned to their starting positions at the conclusion of the trial. There are 10 runs with 42 trials each in each session. An 8th-order Chebyshev bandpass filter with a frequency range of 0.01 Hz to 200 Hz and 512 Hz sampling rate is used. A notch filter operating at 50 Hz reduces power line interference.

Channel Selection Using sLORETA

Fig. 2 shows the first participant's brain maps generated in sLORETA during the first MI action (Yu *et al.*, 2020). This depiction shows that the motor cortex has the highest activation for the first MI action. The top, bottom, back, front, left, and right views of the sLORETA solution have been consulted in our work for each action similarly. Thus, the proposed choice of electrodes C3, C4, and Cz has been validated using sLORETA.

Experiment 1: Movement vs. Movement Classification

Experiment 1 extracted patterns associated with different kinds of movement from the EEG signals. This is achieved using signal processing techniques that helped improve the model's accuracy, such as channel selection with sLORETA, feature extraction with FFT, feature selection with PCA, and classification with RF.

Channel Selection Method Efficiency Evaluation

To check the impact of our proposed sLORETA-based selection approach (Yu *et al.*, 2020), the sLORETA-based outcome regarding accuracy, precision, recall, error rate, and F1- score has been compared with the original feature space without channel selection, Multi-Objective Optimization (MOO) based channel selection (Jiménez *et al.*, 2021), filtering-based channel selection (Lan *et al.*, 2007), wrapper-based channel selection (Chai *et al.*, 2012) and hybrid channel selection (Alotaibi *et al.*, 2015). A search algorithm produces the potential channel subsets, and filtering algorithms use independent evaluation criteria, including consistency, dependency, information, and distance measure. Scalability, independence from the classifier, and high speed are benefits of filtering methods. However, because it does not consider the combinations of many channels, it has low accuracy. In the case of wrapper approaches, the candidate channel subsets produced by the search algorithm are evaluated by a classification algorithm. A particular classification algorithm is trained and tested to evaluate each candidate. As a result, they need more processing than filtering approaches and are more prone to overfitting. A hybrid method combines filtering and wrapper techniques to avoid the pre-specification of a termination criterion. Table 2 presents the comparative analysis of the proposed combination of sLORETA-based selected channels, Chebyshev filter of order 16, FFT, PCA, and RF, with its competitors, and the proposed one has performed better than the competitors.

Figure 2. Brain map of the EEG signal for Subject 1 Action 1

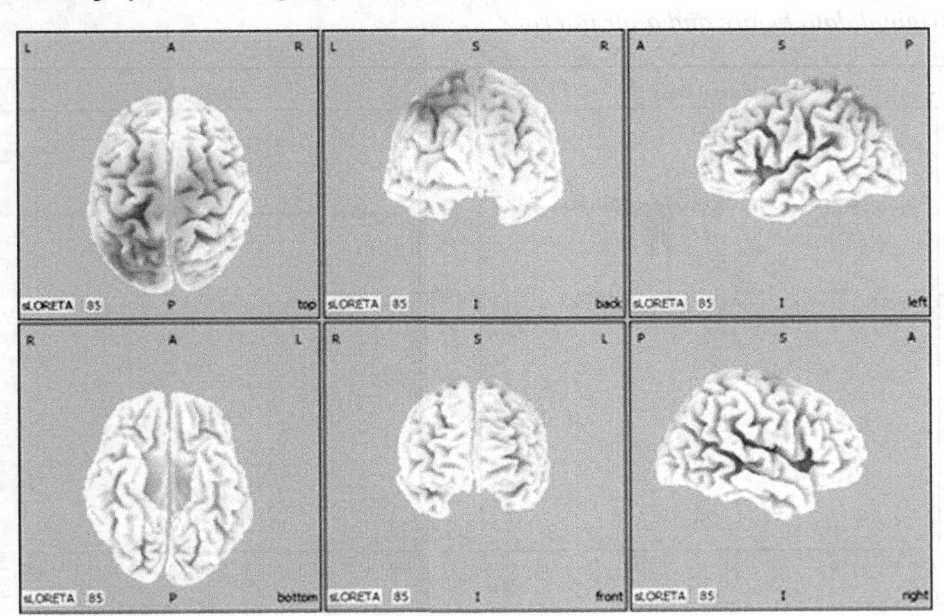

Results Obtained From Filtering

The results obtained by filtering the noisy data are shown in Fig. 3, from where it is evident that the proposed filtering model could efficiently remove the artifact from the raw signal. The performance of the Chebyshev filter of order 16 as the proposed filtering technique has been compared with its seven existing competitors, i.e., the Butterworth filter of order 4 (Robertson *et al.*, 2003), the Butterworth filter of order 8 (Zhongsehn *et al.*, 2007), the Butterworth filter of order 12 (Ali *et al.*, 2013), Butterworth filter of order 16 (Somefun *et al.*, 2022), Chebyshev filter of order 4 (Freeborn *et al.*, 2015), Chebyshev filter of order 8 (Basu & Mamud, 2020), and Chebyshev filter of order 12 (Zhou *et al.*, 2006). They all have been applied to the noisy signal from the selected channels, and the outcomes have been provided in Table 3 regarding the accuracy, precision, recall, error rate, and F1 score.

Table 2. Performance assessment of sLORETA-based channel selection

Approach	Accuracy	Precision	Recall	Error Rate	F1-Score
Chebyshev filter of order 16+ FFT without channel selection+ PCA+ RF	74.45	71.53	79.78	25.33	71.42
sLORETA+ Chebyshev filter of order 16+ FFT+ PCA+ RF (Proposed)	**96.45**	**98.53**	**94.78**	**2.33**	**98.42**
MOO+ Chebyshev filter of order 16+ FFT+ PCA+ RF	76.34	77.66	76.56	12.99	77.39
Filtering+ Chebyshev filter of order 16+ FFT+ PCA+ RF	84.89	81.78	82.89	19.67	82.78
Wrapper+ Chebyshev filter of order 16+ FFT+ PCA+ RF	78.91	72.38	76.39	15.87	76.98
Hybrid+ Chebyshev filter of order 16+ FFT+ PCA+ RF	74.93	77.12	75.11	12.18	72.44

Figure 3. Original data before and after filtering

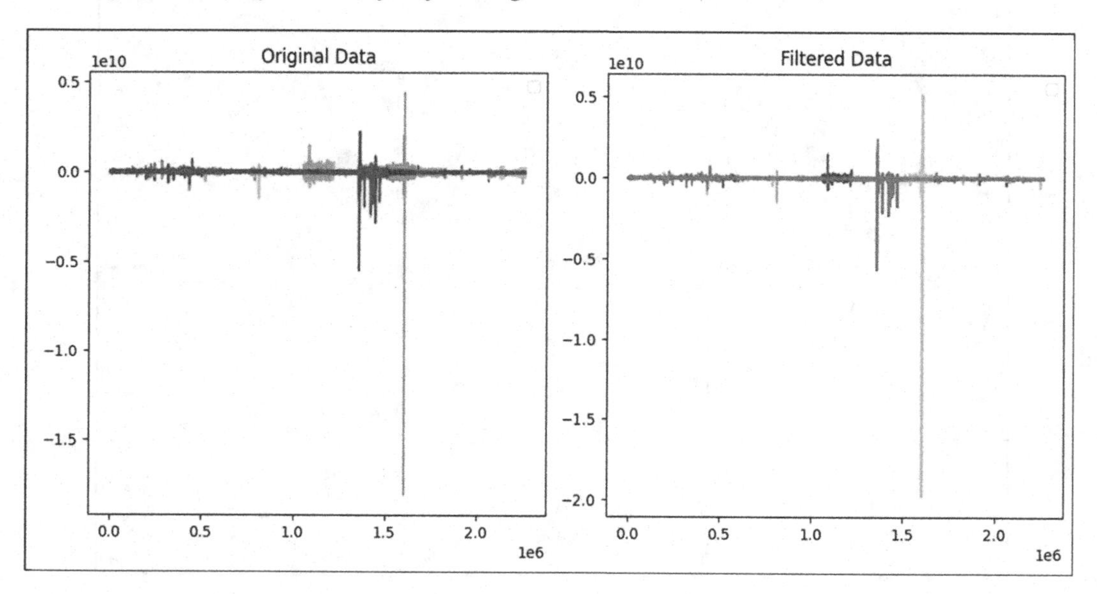

Table 3. Performance evaluation of Chebyshev filter of Order 16

Approach	Accuracy	Precision	Recall	Error Rate	F1-Score
sLORETA+ Butterworth filter of order 4+ FFT+ PCA+ RF	74.45	71.53	79.78	25.33	71.42
sLORETA+ Butterworth filter of order 8+ FFT+ PCA+ RF	87.67	84.78	86.65	19.34	85.48
sLORETA+ Butterworth filter of order 12+ FFT+ PCA+ RF	76.34	77.66	76.56	12.99	77.39
sLORETA+ Butterworth filter of order 16+ FFT+ PCA+ RF	74.93	77.12	75.11	12.18	72.44
sLORETA+ Chebyshev filter of order 4+ FFT+ PCA+ RF	84.89	81.78	82.89	19.67	82.78
sLORETA+ Chebyshev filter of order 8+ FFT+ PCA+ RF	78.91	72.38	76.39	15.87	76.98
sLORETA+ Chebyshev filter of order 12+ FFT+ PCA+ RF	74.93	77.12	75.11	12.18	72.44
sLORETA+ Chebyshev filter of order 16+ FFT+ PCA+ RF (Proposed)	**96.45**	**98.53**	**94.78**	**2.33**	**98.42**

Feature Extraction

Figure 4 shows the signals before applying FFT and after feature extraction with FFT. Various techniques, as the competitors of our proposed feature space extraction method after sLORETA-based channel selection, have been assessed by comparing them with the proposed one. Existing competitors like Common Spatial Pattern (CSP) (Yger *et al.*, 2015), Discrete Wavelet Transform (DWT) (Li *et al.*, 2017), Pisarenko's Eigen Vector (EV) (Übeyli, 2009), Time-Frequency Distribution (TFD) (Geethanjali *et al.*, 2012), and Yule-Walker Autoregressive Method (ARM) (Schlögl & Supp, 2006) have been considered. The results based on accuracy, precision, recall, error rate, and F1-score have been presented in Table 4. In addition, the best-performing values have been marked in bold, showing that the combination of sLORETA, Chebyshev filter of order 16, FFT, PCA, and RF has outperformed the competing feature extractors mentioned.

Figure 4. Filtered signal before and after feature extraction with FFT

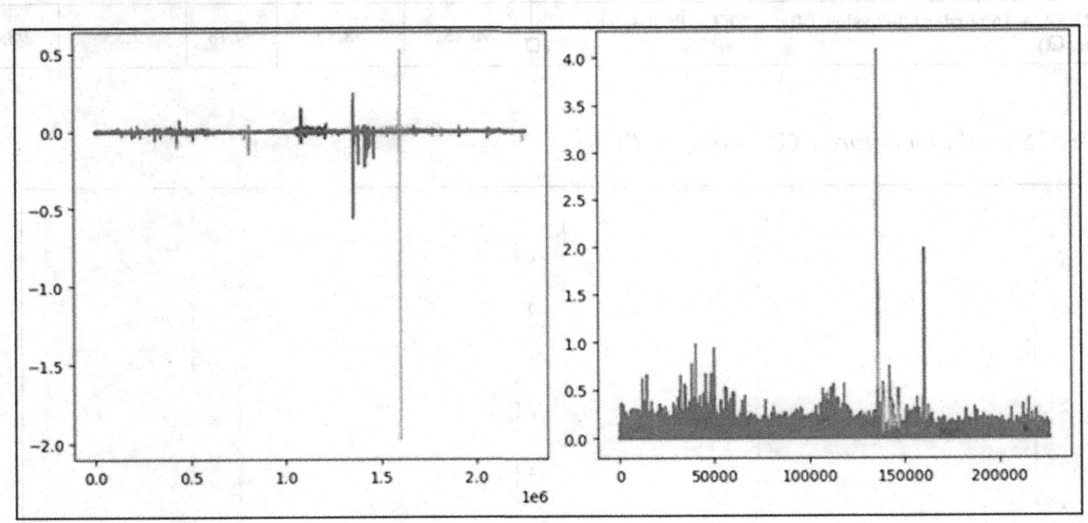

Table 4. Performance assessment of the proposed feature extractor

Approach	Accuracy	Precision	Recall	Error Rate	F1-Score
sLORETA+ Chebyshev filter of order 16+ CSP+ PCA+ RF	92.81	89.31	87.56	19.96	85.23
sLORETA+ Chebyshev filter of order 16+ DWT+ PCA+ RF	91.67	93.69	91.56	**1.23**	83.86
sLORETA+ Chebyshev filter of order 16+ EV+ PCA+ RF	89.5	94.56	86.75	15.56	87.34
sLORETA+ Chebyshev filter of order 16+ TFD+ PCA+ RF	86.59	86.32	89.44	18.21	79.45
sLORETA+ Chebyshev filter of order 16+ ARM+ PCA+ RF	85.67	87.61	83.76	14.56	82.27
sLORETA+ Chebyshev filter of order 16+ FFT+ PCA+ RF (Proposed)	**96.45**	**98.53**	**94.78**	2.33	**98.42**

Results Obtained From Feature Selection

The results obtained from feature selection have been shown as a pictorial illustration of the signal plot shown in Fig. 5. Various techniques for selecting features have been compared with the proposed PCA-based feature selector, and the comparison is presented in Table 5. Linear Discriminant Analysis (LDA) (Prasetio *et al.*, 2019), Generalized Discriminant Analysis (GDA) (Osanai *et al.*, 2019), and Independent Component Analysis (ICA) (Lee & Choi, 2003) have been considered for this purpose.

Table 5. Performance assessment of proposed feature selector

Approach	Accuracy	Precision	Recall	Error Rate	F1-Score
sLORETA + 16th order Chebyshev filter+ FFT+ LDA +RF	89.5	94.56	86.75	15.56	87.34
sLORETA+ 16th order Chebyshev filter+ FFT+ GDA + RF	86.59	86.32	89.44	18.21	79.45
sLORETA + 16th order Chebyshev filter+ FFT+ ICA+ RF	85.67	87.61	83.76	14.56	82.27
sLORETA + 16th order Chebyshev filter+ FFT+ PCA+ RF (Proposed)	**96.45**	**98.53**	**94.78**	**2.33**	**98.42**

Figure 5. Signal plot before PCA and after PCA

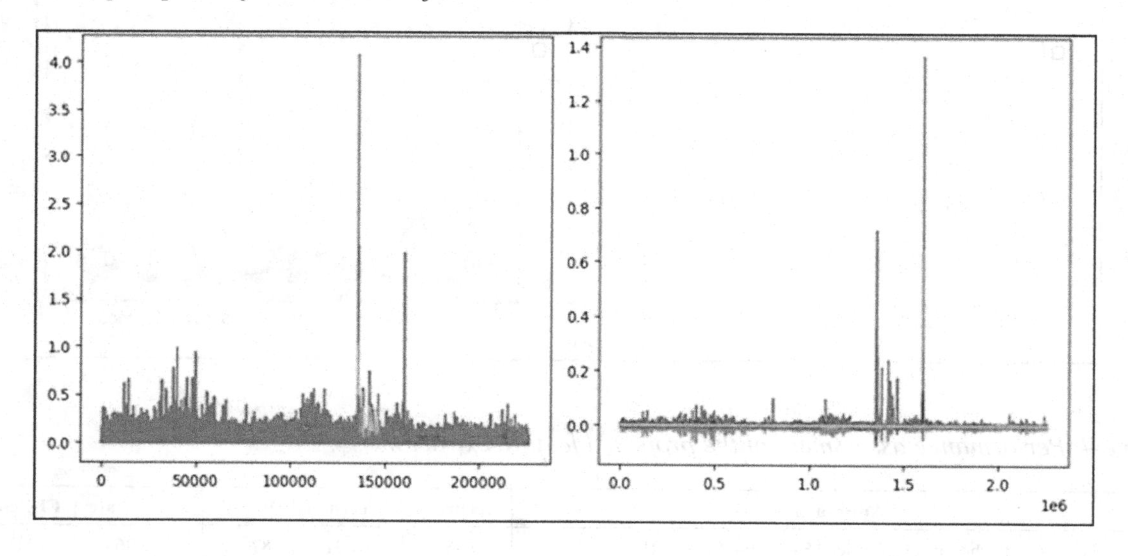

RF-Based Movement vs. Movement Classification Outcomes

Hyperparameter tuning has been conducted to enhance the performance of our proposed RF architecture. The parameters considered for this purpose include the maximum depth, the number of estimators, i.e., the number of trees in the forest, and the minimum number of data points allowed in a leaf node. For the number of estimators, 200, 220, 240, 260, 280, and 300 have been considered. Maximum depth varies among 10, 20, 30, 40, 50, 60, 70, 80, and 90. The minimum number of samples required at each leaf node varies between 1, 2, and 4. The ROC curves for all these combinations as hyperparameter tuning outcomes have been provided in Fig. 6, from where the best-performing combination (number of estimators = 300, maximum depth = 90, minimum number of samples required at each leaf = 4) has been chosen in the proposed classifier.

Our proposed RF's performance has been assessed by comparing it with existing competitors run on the same feature space extracted. Table 6 presents the comparative analysis where RF has been compared with SVM, kNN, Naive Bayes (NB) (Wickramasinghe & Kalutarage, 2021), Decision Tree (DT) (Li *et al*., 2017), and Back Propagation Neural Network (BPNN) (Kshirsagar & Akojwar, 2016), and the accuracy of the proposed RF technique is found to be superior than all its contestants for all the considered datasets. BPNN has achieved better accuracy than the proposed RF. Still, observing the precision, recall, error rate, and F1 score, we can conclude that RF is efficient enough to classify MI actions in real-time scenarios.

Table 6. Performance evaluation of proposed RF as the Movement vs. Movement classifier

Approach	Accuracy	Precision	Recall	Error Rate	F1-Score
sLORETA+ Chebyshev filter of order 16 + FFT + PCA+ NB	92.45	91.23	92.73	17.45	92.21
sLORETA + Chebyshev filter of order 16 + FFT + PCA+ RF (Proposed)	96.45	**98.53**	**94.78**	**2.33**	**98.42**
sLORETA + Chebyshev filter of order 16 + FFT + PCA+ SVM	71.89	71.78	72.89	27.43	72.78
sLORETA + Chebyshev filter of order 16 + FFT + PCA+ kNN	73.83	79.32	75.23	26.34	71.34
sLORETA + Chebyshev filter of order 16 + FFT + PCA+ DT	74.93	77.12	75.11	12.18	72.44
sLORETA + Chebyshev filter of order 16 + FFT + PCA+ BPNN	**96.89**	81.43	73.91	6.95	79.41

To check whether any feature selector and filter is needed to process the data efficiently or not, we have compared the combination of only the feature extractor and classifier with the combination of feature extractor + feature selector + classifier and the combination of filter + feature extractor + feature selector + classifier in Table 7. The accuracy of the first combination, FFT + RF, was 38.97%, indicating that using FFT as a feature extraction technique in conjunction with the RF algorithm produced a moderate classification accuracy. The second combination, FFT + PCA + RF, increased accuracy to 50.69%. The third and most effective combination, Chebyshev LPF + FFT + PCA + RF, attained a remarkable accuracy of 96.45%. The inclusion of a LPF contributed significantly to the reduction of high-frequency noise and artifacts and the overall improvement of the input data quality. As more sophisticated techniques are incorporated into the pipeline, the results indicate a trend toward increasing accuracy. A Chebyshev LPF, FFT, PCA, and RF are explicitly used to achieve the highest accuracy, highlighting the potential benefits of combining different preprocessing techniques.

Figure 6. Hyperparameter setting outcomes of proposed RF as movement vs movement classifier

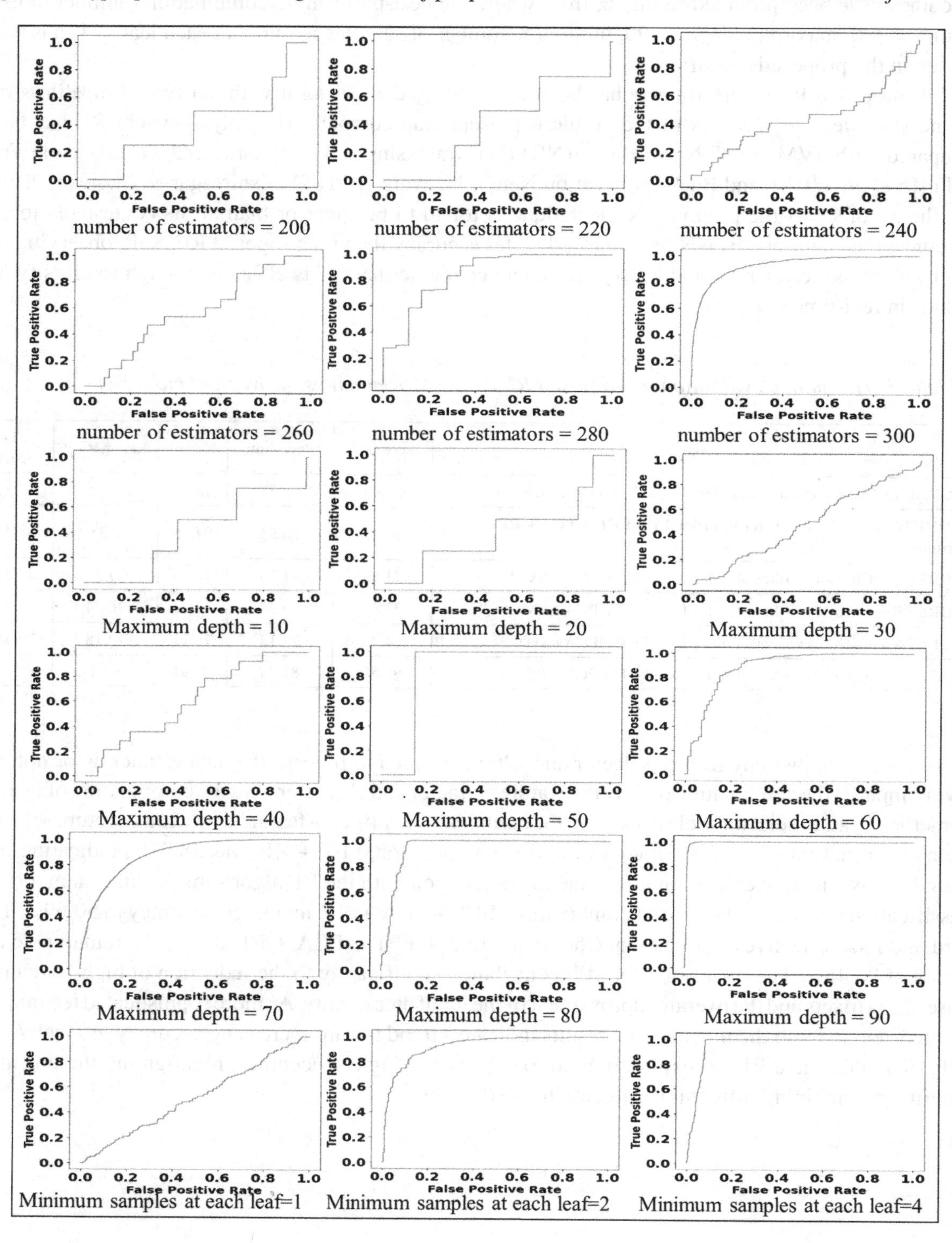

Table 7. Movement vs. Movement accuracy acquired with different combinations

Algorithms	Accuracy
FFT + RF	38.97%
FFT + PCA + RF	50.69%
Chebyshev LPF of order 16 + FFT + PCA + RF	**96.45%**

Experiment 2: Rest vs. Movement Classification

Using the combination of sLORETA, 16th order Chebyshev filter, FFT, PCA, and kNN algorithm, Experiment 2 effectively separated the categories of rest and movement, which is necessary to determine the user's intention to move.

Channel Selection Outcomes Analysis

To check the impact of our proposed sLORETA-based selection approach for the movement vs. rest classification, the sLORETA-based outcomes based on the metrics defined already has been compared with the same channel selection techniques discussed earlier. Table 8 presents the comparative analysis of the proposed combination of sLORETA-based selected channels, Chebyshev filter of order 16, FFT, PCA, and kNN, with different related approaches, and the proposed one has performed better than the competitors.

Results Obtained From Filtering

The performance of the Chebyshev filter of order 16 as the proposed filtering technique for Experiment 2 has been compared with its seven existing competitors discussed earlier. They all have been applied to our dataset, and the outcomes have been provided in Table 9 regarding the accuracy, precision, recall, error rate, and F1 score.

Table 8. Performance evaluation of sLORETA-based channel selection

Approach	Accuracy	Precision	Recall	Error Rate	F1-Score
Original feature space+ Chebyshev filter of order 16+ FFT+ PCA+ kNN	74.45	71.53	79.78	25.33	71.42
sLORETA+ Chebyshev filter of order 16+ FFT + PCA+ kNN (Proposed)	**95.02**	**93.21**	**90.38**	**0.32**	**95.78**
MOO+ Chebyshev filter of order 16+ FFT+ PCA+ kNN	76.34	77.66	76.56	12.99	77.39
Filtering+ Chebyshev filter of order 16+ FFT+ PCA+ kNN	84.89	81.78	82.89	19.67	82.78
Wrapper+ Chebyshev filter of order 16+ FFT+ PCA+ kNN	78.91	72.38	76.39	15.87	76.98
Hybrid+ Chebyshev filter of order 16+ FFT+ PCA+ kNN	74.93	77.12	75.11	12.18	72.44

Table 9. Performance evaluation of Chebyshev filter of order 8

Approach	Accuracy	Precision	Recall	Error Rate	F1-Score
sLORETA+ Butterworth filter of order 4+ FFT+ PCA+ kNN	74.45	71.53	79.78	25.33	71.42
sLORETA+ Butterworth filter of order 8+ FFT+ PCA+ kNN	87.67	84.78	86.65	19.34	85.48
sLORETA+ Butterworth filter of order 12+ FFT+ PCA+ kNN	76.34	77.66	76.56	12.99	77.39
sLORETA+ Butterworth filter of order 16+ FFT+ PCA+ kNN	74.93	77.12	75.11	12.18	72.44
sLORETA+ Chebyshev filter of order 4+ FFT+ PCA+ kNN	84.89	81.78	82.89	19.67	82.78
sLORETA+ Chebyshev filter of order 8+ FFT+ PCA+ kNN	78.91	72.38	76.39	15.87	76.98
sLORETA+ Chebyshev filter of order 12+ FFT+ PCA+ kNN	74.93	77.12	75.11	12.18	72.44
sLORETA+ Chebyshev filter of order 16+ FFT+ PCA+ kNN (Proposed)	**95.02**	**93.21**	**90.38**	**0.32**	**95.78**

Results Obtained From Feature Extraction

The same techniques discussed in the previous section as the competitors of our proposed feature space extraction method after sLORETA-based channel selection have been applied to the movement vs. movement classification scenario, and the results have been presented in Table 10. In addition, the best-performing values have been marked in bold, showing that the combination of sLORETA, 16th order Chebyshev filter, FFT, PCA, and kNN has outperformed the competing feature extractors mentioned.

Table 10. Performance assessment of the proposed feature extractor

Approach	Accuracy	Precision	Recall	Error Rate	F1-Score
sLORETA+ Chebyshev filter of order 16+ CSP+ PCA+ kNN	92.81	89.31	87.56	19.96	85.23
sLORETA+ Chebyshev filter of order 16+ DWT+ PCA+ kNN	91.67	93.69	91.56	**1.23**	83.86
sLORETA+ Chebyshev filter of order 16+ EV+ PCA+ kNN	89.5	94.56	86.75	15.56	87.34
sLORETA+ Chebyshev filter of order 16+ TFD+ PCA+ kNN	86.59	86.32	89.44	18.21	79.45
sLORETA+ Chebyshev filter of order 16+ ARM+ PCA+ kNN	85.67	87.61	83.76	14.56	82.27
sLORETA+ Chebyshev filter of order 16+ FFT+ PCA+ kNN (Proposed)	**95.02**	**93.21**	**90.38**	**0.32**	**95.78**

Results Obtained From Feature Selection

Various techniques for selecting features have been compared with the proposed PCA-based feature selector, and the comparison is presented in Table 11. The best-performing combination has been marked in bold, proving our proposed PCA effective in its domain.

Figure 7. Hyperparameter setting outcomes of proposed kNN as movement vs. rest classifier

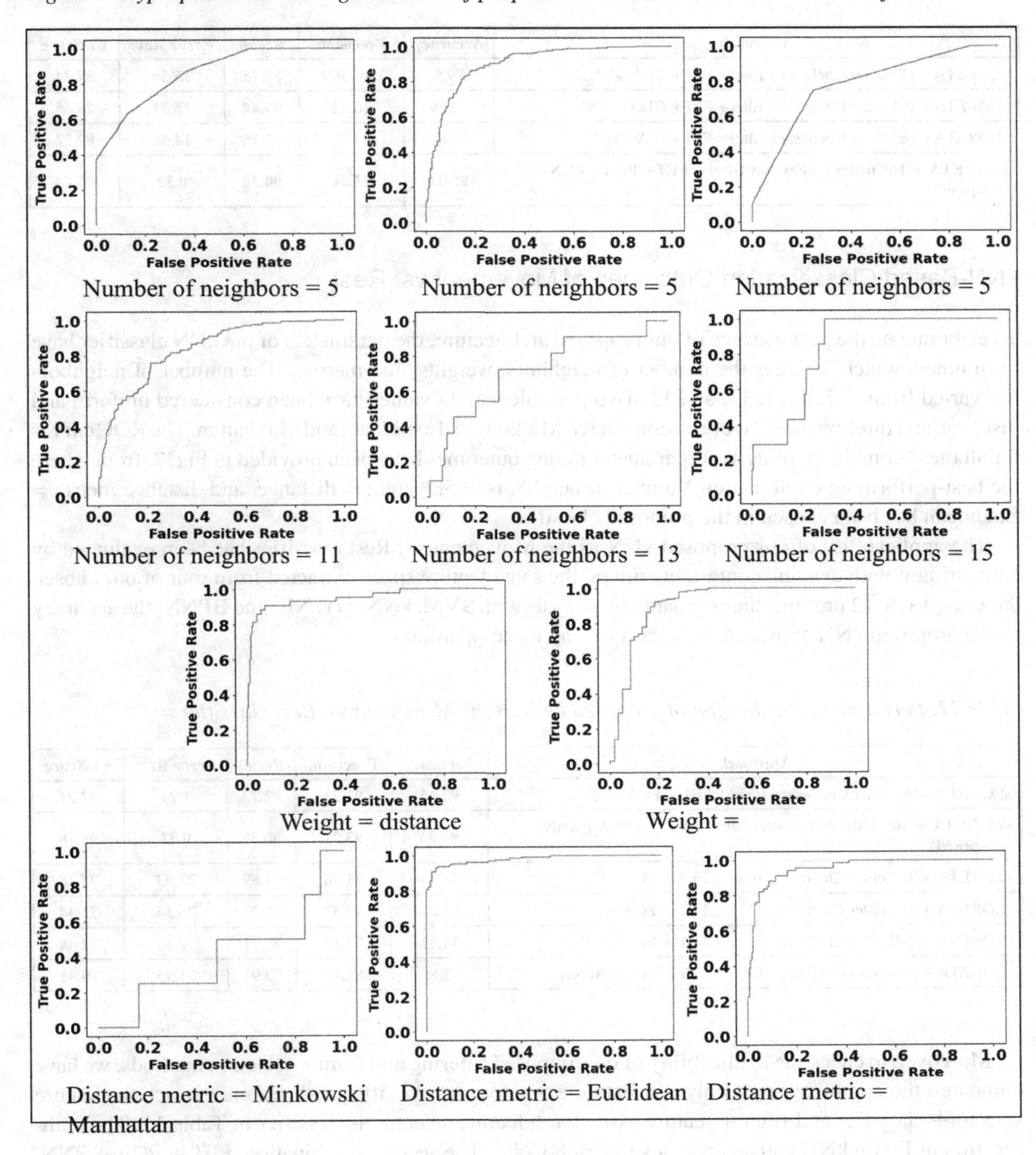

Table 11. Performance assessment of proposed feature selector

Approach	Accuracy	Precision	Recall	Error Rate	F1-Score
sLORETA+ 16th order Chebyshev filter+ FFT+ LDA + kNN	89.5	94.56	86.75	15.56	87.34
sLORETA+ 16th order Chebyshev filter+ FFT+ GDA + kNN	86.59	86.32	89.44	18.21	79.45
sLORETA+ 16th order Chebyshev filter+ FFT+ ICA+ kNN	85.67	87.61	83.76	14.56	82.27
sLORETA + 16th order Chebyshev filter+ FFT+ PCA+ kNN (Proposed)	**95.02**	**93.21**	**90.38**	**0.32**	**95.78**

kNN-Based Classification Outcomes of Movement vs. Rest

To elaborate on the performance of our proposed architecture, the parameters of the kNN classifier have been tuned, which includes the number of neighbors, weights, and metrics. The number of neighbors has varied from 5, 7, 9, 11, 13, and 15. Two possible weight values have been considered uniform and distance, and three metrics have been considered Minkowski, Euclidean, and Manhattan. The ROC curves for all these combinations as hyperparameter tuning outcomes have been provided in Fig. 7, from where the best-performing combination Number of neighbors = 5, weight = distance, and distance metric = Euclidean has been chosen in the proposed classifier.

The applicability of our proposed kNN as the Movement vs. Rest classifier has been evaluated by comparing it with existing contestants run on the same feature space extracted from four of our chosen datasets. Table 12 presents the comparative analysis with SVM, kNN, DT, NB, and BPNN; the accuracy of the proposed kNN approach is the best among its competitors.

Table 12. Performance evaluation of proposed kNN as the Movement vs. Rest classifier

Approach	Accuracy	Precision	Recall	Error Rate	F1-Score
sLORETA+ 16th order Chebyshev filter + FFT + PCA+ NB	92.45	91.23	92.73	1.21	92.21
sLORETA + 16th order Chebyshev filter + FFT + PCA+ kNN (Proposed)	**95.02**	**93.21**	**90.38**	**0.32**	**95.78**
sLORETA + 16th order Chebyshev filter + FFT + PCA+ SVM	71.89	71.78	72.89	27.43	72.78
sLORETA + 16th order Chebyshev filter + FFT + PCA+ RF	73.83	79.32	75.23	26.34	71.34
sLORETA + 16th order Chebyshev filter + FFT + PCA+ DT	74.93	77.12	75.11	12.18	72.44
sLORETA + 16th order Chebyshev filter + FFT + PCA+ BPNN	72.89	81.43	73.91	21.95	79.41

Moreover, to check the applicability of our proposed filtering and feature selection methods, we have compared the combination of only the feature extractor and classifier with feature extraction+feature selector+ classifier and filter + feature extractor + feature selector + classifier in Table 13. The combination of FFT+ kNN had an accuracy rate of 83.74%. The second combination, FFT + PCA + kNN, increased accuracy to 95.02%. The third combination, Chebyshev LPF + FFT + PCA + kNN, had the highest accuracy (95.02%). Fig. 8 depicts the confusion matrices for Experiments 1 and 2, from where it is observed that the proposed RF and kNN, respectively, for the two experiments are suitable for classifying the maximum number of samples correctly.

Figure 8. Confusion matrices for rest vs. movement and rest vs. movement classification

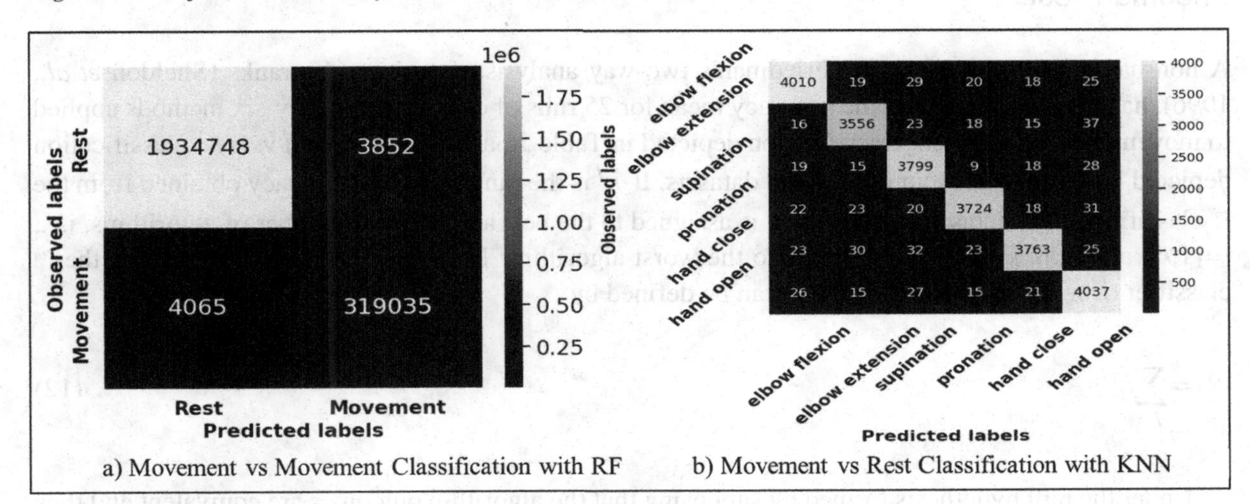

a) Movement vs Movement Classification with RF   b) Movement vs Rest Classification with KNN

Table 13. Rest vs. movement accuracy acquired with different combinations

Algorithms	Accuracy
FFT + kNN	83.74%
FFT + PCA + kNN	85.34%
Chebyshev Lowpass Filter of order 16 + FFT + PCA + kNN	**95.02%**

Statistical Validation

Two types of statistical testing, namely Friedman and Bonferroni Dunn, have been performed on our considered dataset and three similar datasets (Brunner *et al.*, 2008), (Steryl *et al.*, 2014) (Kaya *et al.*, 2018) to analyze their statistical validation. The descriptions of the related datasets are given in Table 14.

Table 14. Description of related datasets

Dataset	Training Data Size	Testing Data Size	EEG Electrode Channels Selected	Class Labels
GUFCMI	9 subjects X 6 runs X 96735 frames X 22 channels	9 subjects X 6 runs X 96735 frames X 22 channels	POz, P2, PZ, P1, CP4, CP2, CPZ, CP1, CP3, C6, C4, C2, CZ, C1, C3, C5, FC4, FC2, FCZ, FC1, FC3, FZ	Left hand, right hand, both feet, tongue
GUTCMI	9 subjects X 2 runs X 466314 frames X 3 channels	9 subjects X 2 runs X 466314 frames X 3 channels	C3, Cz, C4	Left hand, Right hand
TUMUMI	10 subjects X 75 runs	3 subjects X 75 runs	C3, Cz, C4, T3, T4	i) Left hand, Right hand, Left leg, Right leg, Tongue

Friedman Test

A non-parametric statistical test, Friedman's two-way analysis of variance by ranks (Sheldon *et al.*, 1996), has been conducted on the accuracy mean for 25 runs of each of the twenty-six methods applied to movement vs. movement classification depicted in Table 15 and for movement vs. rest classification depicted in Table 16 for four considered datasets. If r_i^j is the ranking of the accuracy obtained from the ith classifier and j^{th} dataset, the first rank is assigned to the best among the k number of algorithms, i.e., i=[1,k], and rank k has been assigned to the worst algorithm. The sum of ranking achieved by the ith classifier over all the j=[1,N] methods can be defined by

$$R_i = \sum_{j=1}^{N} r_i^j \tag{12}$$

Under the null hypothesis formed by supposing that the algorithm outcomes are equivalent and thus their ranks are alike, Friedman's statistic maintains a χ_F^2 distribution having k - 1 degree of freedom.

$$\chi_F^2 = \frac{12}{k \times N \times (k+1)} \sum_{i=1}^{K} R_i^2 - 2N(k+1) \tag{13}$$

This paper considers four datasets, thus $N = 4$ and k signifying the number of competing methods = 26. The null hypothesis for movement vs. movement classification has been rejected since $\chi_F^2 = 207.74$ is greater than $\chi_{25,\alpha}^2 = 37.652$, where $\chi_{25,\alpha=0.05}^2 = 37.652$ denotes the critical value of χ_{25}^2 distribution for k _ 1 = 25 degrees of freedom at 0.05 probability. Similarly, the null hypothesis for movement vs. rest classification has been rejected since $\chi_F^2 = 179.12$ is greater than $\chi_{25,\alpha}^2 = 37.652$.

Bonferroni Dunn's Test

It is proved that RF is the best performer in the Movement vs. Movement classification; thus, Bonferroni Dun's test has been conducted as a post hoc analysis considering sLORETA + 16th order Chebyshev filter + FFT+ PCA+ RF as the control method. The level of significance where the control method works well than the competitors (when the null hypothesis gets rejected) is identified in this test (Roberts *et al.*, 1997), and the critical difference has been identified as 5.011 for the dataset. It implies that the performances of the two methods differ significantly only when their corresponding average ranks differ from each other by at least the critical difference, as shown in Fig. 9. The null hypothesis is not rejected with sLORETA + 16th order Chebyshev filter + CSP + PCA + RF, sLORETA + 16th order Chebyshev filter + DWT+ PCA + RF, and sLORETA + 16th order Chebyshev filter + FFT + GDA + RF, for a level of significance at α = 0.05. The other twenty-two methods are significantly inferior to the proposed sLORETA + 16th order Chebyshev filter + FFT + PCA + RF.

Table 15. Calculation of average rankings of seventeen competing methods for movement vs. movement classification

Methods	Ranking r_i^j of the i^{th}Classifier Obtained From j^{th} MI Dataset				Average Ranking
	GUFCMI	GUTCMI	TUMUMI	ULMD	
16th order Chebyshev filter+ FFT+ PCA+ RF	14	17	15	15	15.25
sLORETA+16th order Chebyshev filter+ FFT+ PCA+ RF (Proposed)	**1**	**1**	**1**	**1**	**1**
MOO+ 16th order Chebyshev filter+ FFT+ PCA+ RF	11	11	11	11	11
Filtering+ 16th order Chebyshev filter+ FFT+PCA+ RF	6	5	5	6	5.5
Wrapper+ 16th order Chebyshev filter+ FFT+ PCA+ RF	10	10	10	10	10
Hybrid+ 16th order Chebyshev filter+ FFT+ PCA+ RF	12	12	12	12	12
sLORETA+ 4th order Butterworth filter+ FFT+ PCA+ RF	20	20	20	20	20
sLORETA+ 8th order Butterworth filter + FFT+ PCA+ RF	21	21	21	21	21
sLORETA+ 12th order Butterworth filter + FFT+ PCA+ RF	22	22	22	22	22
sLORETA+ 16th order Butterworth filter + FFT+ PCA+ RF	23	23	23	23	23
sLORETA+ 4th order Chebyshev filter + FFT+ PCA+ RF	24	24	24	24	24
sLORETA+ 8th order Chebyshev filter+ FFT+ PCA+ RF	25	25	25	25	25
sLORETA+ 12th order Chebyshev filter+ FFT+ PCA+ RF	26	26	26	26	26
sLORETA+ 16th order Chebyshev filter+ CSP+ PCA+ RF	3	3	4	3	3.25
sLORETA+ 16th order Chebyshev filter+ DWT+ PCA+ RF	5	4	3	4	4
sLORETA+ 16th order Chebyshev filter+ EV+ PCA+ RF	4	6	6	5	5.25
sLORETA+ 16th order Chebyshev filter+ TFD+ PCA+ RF	9	9	9	9	9
sLORETA+ 16th order Chebyshev filter+ ARM+ PCA+ RF	8	8	8	8	8
sLORETA+ 16th order Chebyshev filter+ FFT+ LDA + RF	7	7	7	7	7
sLORETA+ 16th order Chebyshev filter+ FFT+ GDA + RF	2	2	2	2	2
sLORETA+ 16th order Chebyshev filter+ FFT+ ICA+ RF	18	18	18	18	18
sLORETA+ 16th order Chebyshev filter+ FFT+ PCA+ SVM	17	15	17	17	16.5
sLORETA+ 16th order Chebyshev filter+ FFT+ PCA+ kNN	15	14	14	14	14.25
sLORETA+ 16th order Chebyshev filter+ FFT+ PCA+ DT	13	13	13	13	13
sLORETA+ 16th order Chebyshev filter+ FFT+ PCA+ BPNN	16	16	16	16	16
sLORETA+ 16th order Chebyshev filter+ FFT+ PCA+ NB	19	19	19	19	19

Table 16. Calculation of average rankings of seventeen competing methods for movement vs. rest classification

Methods	Ranking r_i^j of the ith Classifier Obtained From j^{th} MI Dataset				Average Ranking
	GUFCMI	GUTCMI	TUMUMI	ULMD	
16th order Chebyshev filter+ FFT+ PCA+ kNN	26	26	26	26	26
sLORETA+ 16th order Chebyshev filter+ FFT+ PCA+ kNN (Proposed)	**1**	**1**	**1**	**1**	**1**
MOO+ 16th order Chebyshev filter+ FFT+ PCA+ kNN	16	16	16	16	16
Filtering+ 16th order Chebyshev filter+ FFT+ PCA+ kNN	17	17	17	17	17
Wrapper+ 16th order Chebyshev filter+ FFT+ PCA+ kNN	18	18	18	18	18
Hybrid+ 16th order Chebyshev filter+ FFT+ PCA+ kNN	19	19	19	19	19
sLORETA+ 4th order Butterworth filter+ FFT+ PCA+ kNN	15	15	15	15	15
sLORETA+ 8th order Butterworth filter+ FFT+ PCA+ kNN	14	14	14	14	14
sLORETA+ 12th order Butterworth filter+ FFT+ PCA+ kNN	13	12	12	13	12.5
sLORETA+ 16th order Butterworth filter+ FFT+ PCA+ kNN	12	13	13	12	12.5
sLORETA+ 4th order Chebyshev filter+ FFT+ PCA+ kNN	11	11	11	11	11
sLORETA+ 8th order Chebyshev filter+ FFT+ PCA+ kNN	10	10	10	10	10
sLORETA+ 12th order Chebyshev filter+ FFT+ PCA+ kNN	9	9	9	9	9
sLORETA+ 16th order Chebyshev filter+ CSP+PCA+ kNN	8	8	8	8	8
sLORETA+ 16th order Chebyshev filter+ DWT+ PCA+ kNN	20	21	20	21	20.5
sLORETA+16th order Chebyshev filter+ EV+ PCA+ kNN	21	20	21	20	20.5
sLORETA+ 16th order Chebyshev filter+ TFD+ PCA+ kNN	22	22	22	22	22
sLORETA+ 16th order Chebyshev filter+ ARM+ PCA+ kNN	23	23	23	23	23
sLORETA+ 16th order Chebyshev filter+ FFT+ LDA+ kNN	25	25	25	25	25
sLORETA+ 16th order Chebyshev filter+ FFT+ GDA+ kNN	24	24	24	24	24
sLORETA+ 16th order Chebyshev filter+ FFT+ ICA+ kNN	7	7	7	7	7
sLORETA+ 16th order Chebyshev filter+ FFT+ PCA+ SVM	6	6	5	5	5.5
sLORETA+ 16th order Chebyshev filter+ FFT+ PCA+ DT	5	5	6	6	5.5
sLORETA+ 16th order Chebyshev filter+ FFT+ PCA+ BPNN	2	2	2	2	2
sLORETA+ 16th order Chebyshev filter+ FFT+ PCA+ NB	3	4	3	4	3.5
sLORETA+ 16th order Chebyshev filter+ FS+ PCA+ RF	4	3	4	3	3.5

As kNN has been the best performer in Movement vs. Rest classification; thus, Bonferroni Dun's test has been conducted as a post hoc analysis considering sLORETA + 16th order Chebyshev filter+ FFT +PCA + kNN as the control method. The level of significance where the control method works well than the competitors (when the null hypothesis gets rejected) is identified, and the critical difference has been identified as 5.12 for the dataset. It implies that the performances of the two methods differ significantly only when their corresponding average ranks differ from each other by at least the critical difference, as shown in Fig. 10. The null hypothesis is not rejected with sLORETA + FFT + PCA +

BPNN, sLORETA + FFT + PCA + NB, and sLORETA + FFT + PCA + RF, for a level of significance at $\alpha = 0.05$. The other twenty-two methods are significantly inferior to the proposed sLORETA +16th order Chebyshev filter + FFT + PCA+ kNN.

Figure 9. Pictorial outputs of Bonferroni Dunn's test having RF as the control method

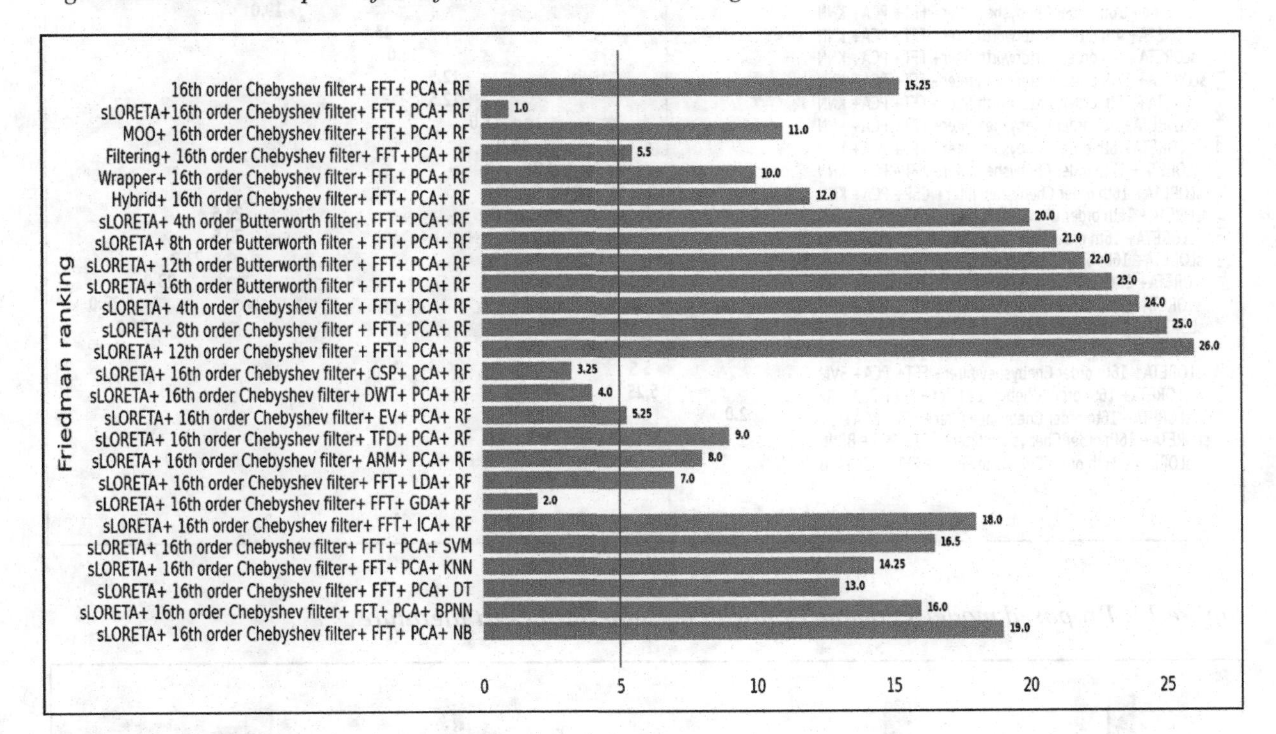

A Comparative Study With the Existing Works

The effectiveness of our proposed classifier has been checked by comparing its result with its competitors. We have used a common framework regarding their features and datasets for all the recent contesting works and have employed the best parametric or hyperparametric setup for all those approaches as prescribed in their respective sources (Mammone *et al*, 2020; Schwarz *et al*.,2020; López *et al*., 2014; Ibáñez *et al*., 2015). Fig. 11 shows the comparative analysis based on accuracy, recall, error rate, F1-score, and precision.

Figure 10. Pictorial output of Bonferroni Dunn's test having kNN as the control method

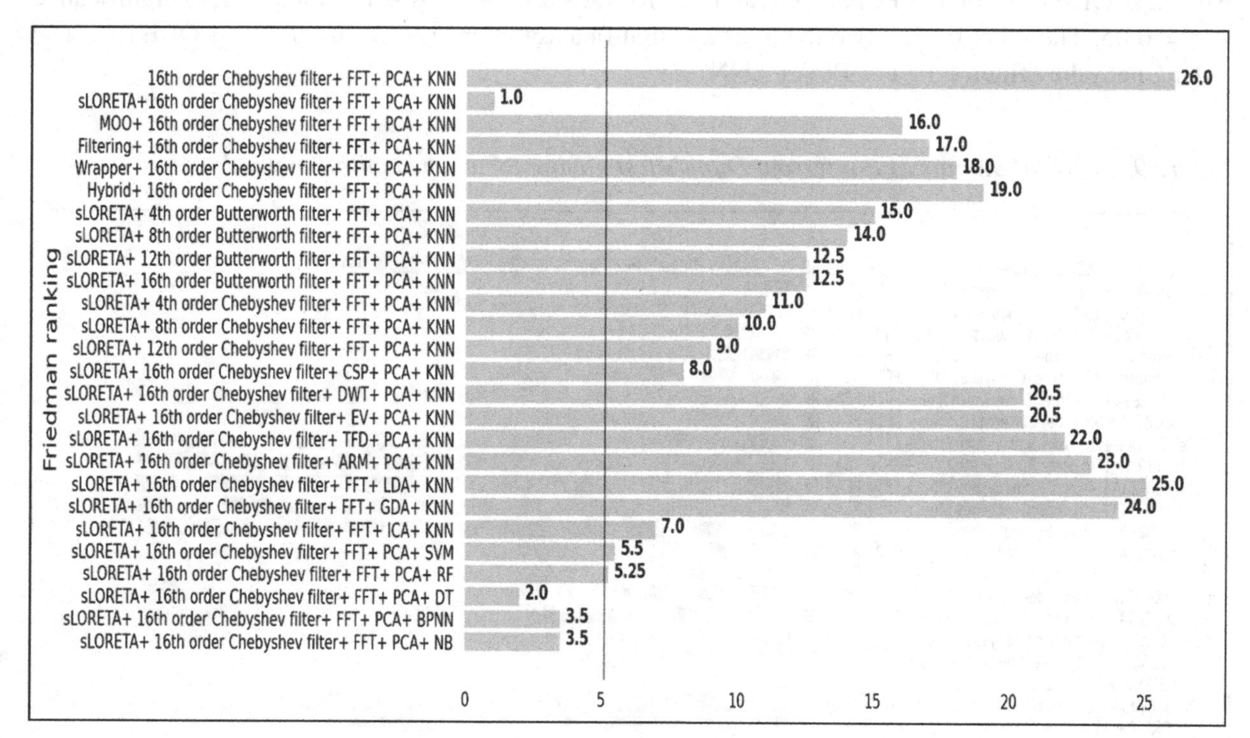

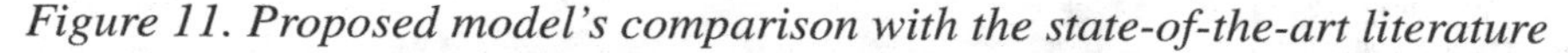

Figure 11. Proposed model's comparison with the state-of-the-art literature

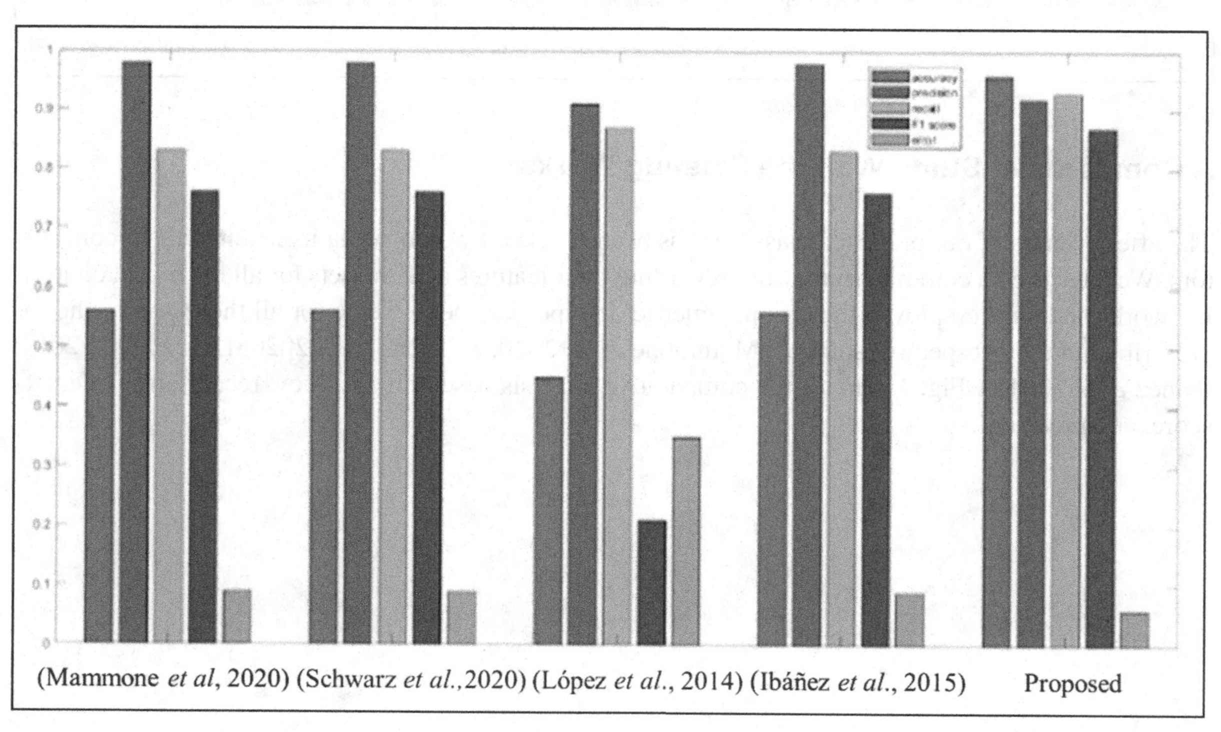

CONCLUSION AND FUTURE WORK

This study offers a new method for extracting upper limb movements from low-frequency EEG signals' frequency domain. It efficiently uses the Chebyshev filter, FFT, PCA, kNN, and RF as the preprocessing, feature extracting, feature selecting, and classifying method for movement vs. rest and movement vs movement classification, respectively. The performance analysis of the suggested approach proves it to be effective in real-time scenarios. It also enhances BCI's functionality for those with disabilities and offers a more user-friendly and organic interface for human-computer interaction. The cognitive ability of persons are observed to be varyring by many genetic, environmental, and lifestyle factors, and some of these factors may contribute to a decline in thinking skills. The proposed work can judge the cognitive abilities of the subjects concerned, and based on the appropriateness of their thinking patterns, if the robot controlling performance is not satisfactory, corresponding rehabilitative tools can be suggested by medical personnel.

Further research on the best feature extraction and classification methods for decoding upper limb movement, including the application of deep learning classifiers like recurrent neural networks, may examine the potential for decoding more intricate motions of the upper limbs, such as those of the wrist and fingers, from low-frequency EEG data. The future scope also includes testing the suggested method's resilience to numerous elements, such as subject variability and noise, and evaluating it in real-world applications, such as assistive devices for people with motor disabilities. Also, it may consider evaluating low-frequency EEG signals' potential for usage in other systems, such as BCIs for virtual reality settings. We may also include further investigation into the brain systems that underlie MI and the creation of more complex models to comprehend and decode these signals for possible application in robot navigation.

REFERENCES

Ali, A. S., Radwan, A. G., & Soliman, A. M. (2013). Fractional order Butterworth filter: Active and passive realizations. *IEEE Journal on Emerging and Selected Topics in Circuits and Systems*, *3*(3), 346–354. doi:10.1109/JETCAS.2013.2266753

Alotaiby, T., El-Samie, F. E. A., Alshebeili, S. A., & Ahmad, I. (2015). A review of channel selection algorithms for EEG signal processing. *EURASIP Journal on Advances in Signal Processing*, *2015*(1), 1–21. doi:10.118613634-015-0251-9

Amiri, V., & Nakagawa, K. (2021). Using a linear discriminant analysis (LDA)-based nomenclature system and self-organizing maps (SOM) for spatiotemporal assessment of groundwater quality in a coastal aquifer. *Journal of Hydrology (Amsterdam)*, *603*, 127082. doi:10.1016/j.jhydrol.2021.127082

Bag, D., Ghosh, A., & Saha, S. (2023). An Automatic Approach to Control Wheelchair Movement for Rehabilitation Using Electroencephalogram. In *Design and Control Advances in Robotics* (pp. 105–126). IGI Global.

Bansal, M., Goyal, A., & Choudhary, A. (2022). A comparative analysis of K-nearest neighbor, genetic, support vector machine, decision tree, and long short term memory algorithms in machine learning. *Decision Analytics Journal*, *3*, 100071. doi:10.1016/j.dajour.2022.100071

Basu, S., & Mamud, S. (2020, September). Comparative study on the effect of order and cut off frequency of Butterworth low pass filter for removal of noise in ECG signal. In *2020 IEEE 1st International Conference for Convergence in Engineering (ICCE)* (pp. 156-160). IEEE. 10.1109/ICCE50343.2020.9290646

Brunner, C., Leeb, R., Müller-Putz, G., Schlögl, A., & Pfurtscheller, G. (2008). BCI Competition 2008–Graz data set A. Institute for Knowledge Discovery (of Brain-Computer Interfaces), Graz University of Technology.

Cha, J., Kim, K. S., Zhang, H., & Lee, S. (2019, November). Analysis on EEG signal with machine learning. In *2019 International Conference on Image and Video Processing, and Artificial Intelligence* (Vol. 11321, pp. 564-569). SPIE. 10.1117/12.2548313

Chai, R., Ling, S. H., Hunter, G. P., & Nguyen, H. T. (2012, August). *Toward fewer EEG channels and better feature extractor of non-motor imagery mental tasks classification for a wheelchair thought controller. In 2012 Annual International Conference of the IEEE Engineering in Medicine and Biology Society*. IEEE.

Chatterjee, P., Ghosh, A., & Saha, S. (2023). *A Comprehensive Survey on Rehabilitative Applications of Electroencephalogram in Healthcare*. Cognitive Cardiac Rehabilitation Using IoT and AI Tools. doi:10.4018/978-1-6684-7561-4.ch003

Chaudhary, S., Taran, S., Bajaj, V., & Sengur, A. (2019). Convolutional neural network based approach towards motor imagery tasks EEG signals classification. *IEEE Sensors Journal*, *19*(12), 4494–4500. doi:10.1109/JSEN.2019.2899645

Dai, M., Zheng, D., Na, R., Wang, S., & Zhang, S. (2019). EEG classification of motor imagery using a novel deep learning framework. *Sensors (Basel)*, *19*(3), 551. doi:10.339019030551 PMID:30699946

de Klerk, R., Duarte, A. M., Medeiros, D. P., Duarte, J. P., Jorge, J., & Lopes, D. S. (2019). Usability studies on building early stage architectural models in virtual reality. *Automation in Construction*, *103*, 104–116. doi:10.1016/j.autcon.2019.03.009

De Vries, S., & Mulder, T. (2007). Motor imagery and stroke rehabilitation: A critical discussion. *Journal of Rehabilitation Medicine*, *39*(1), 5–13. doi:10.2340/16501977-0020 PMID:17225031

Freeborn, T., Maundy, B., & Elwakil, A. S. (2015). Approximated fractional order Chebyshev lowpass filters. *Mathematical Problems in Engineering*.

Geethanjali, P., Mohan, Y. K., & Sen, J. (2012) Time domain feature extraction and classification of EEG data for brain computer interface. In *2012 9th International Conference on Fuzzy Systems and Knowledge Discovery* (pp. 1136-1139). IEEE. 10.1109/FSKD.2012.6234336

Ghosh, A., & Saha, S. (2020). Interactive game-based motor rehabilitation using hybrid sensor architecture. In *Handbook of Research on Emerging Trends and Applications of Machine Learning* (pp. 312–337). IGI Global. doi:10.4018/978-1-5225-9643-1.ch015

Ghosh, A., Saha, S., & Ghosh, L. (2023). A rehabilitation framework based on motor imagery induced wheelchair movement using fuzzy vector quantization. *International Journal of Information Technology : an Official Journal of Bharati Vidyapeeth's Institute of Computer Applications and Management*, *15*(6), 1–12. doi:10.100741870-023-01359-8

Grigorev, N. A., Savosenkov, A. O., Lukoyanov, M. V., Udoratina, A., Shusharina, N. N., Kaplan, A. Y., Hramov, A. E., Kazantsev, V. B., & Gordleeva, S. (2021). A BCI-based vibrotactile neurofeedback training improves motor cortical excitability during motor imagery. *IEEE Transactions on Neural Systems and Rehabilitation Engineering*, *29*, 1583–1592. doi:10.1109/TNSRE.2021.3102304 PMID:34343094

Grosprêtre, S., Ruffino, C., & Lebon, F. (2016). Motor imagery and cortico-spinal excitability: A review. *European Journal of Sport Science*, *16*(3), 317–324. doi:10.1080/17461391.2015.1024756 PMID:25830411

Guillot, A., & Debarnot, U. (2019). Benefits of motor imagery for human space flight: A brief review of current knowledge and future applications. *Frontiers in Physiology*, *10*, 396. doi:10.3389/fphys.2019.00396 PMID:31031635

Gul, F., Rahiman, W., & Nazli Alhady, S. S. (2019). A comprehensive study for robot navigation techniques. *Cogent Engineering*, *6*(1), 1632046. doi:10.1080/23311916.2019.1632046

Hoffmann, U., Vesin, J. M., & Ebrahimi, T. (2007). Recent advances in brain-computer interfaces. *IEEE International Workshop on Multimedia Signal Processing (MMSP07)*.

Ibáñez, J., Serrano, J. I., del Castillo, M. D., Minguez, J., & Pons, J. L. (2015). Predictive classification of self-paced upper-limb analytical movements with EEG. *Medical & Biological Engineering & Computing*, *53*(11), 1201–1210. doi:10.100711517-015-1311-x PMID:25980505

Jiménez, D. D. J. G., Olvera, T., Orozco-Rosas, U., & Picos, K. (2021, August). Autonomous object manipulation and transportation using a mobile service robot equipped with an RGB-D and LiDAR sensor. In *Optics and Photonics for Information Processing XV* (Vol. 11841, pp. 92–111). SPIE. doi:10.1117/12.2594025

Kaya, M., Binli, M. K., Ozbay, E., Yanar, H., & Mishchenko, Y. (2018). A large electroencephalographic motor imagery dataset for electroencephalographic brain computer interfaces. *Scientific Data*, *5*(1), 1–16. doi:10.1038data.2018.211 PMID:30325349

Kim, M. S., Jang, K. M., Che, H., Kim, D. W., & Im, C. H. (2012). Electrophysiological correlates of object-repetition effects: sLORETA imaging with 64-channel EEG and individual MRI. *BMC Neuroscience*, *13*(1), 1–10. doi:10.1186/1471-2202-13-124 PMID:23075055

Kshirsagar, P., & Akojwar, S. (2016) Optimization of BPNN parameters using PSO for EEG signals. In *International Conference on Communication and Signal Processing 2016 (ICCASP 2016)* (pp. 384-393). Atlantis Press.

Lan, T., Erdogmus, D., Adami, A., Mathan, S., & Pavel, M. (2007). Channel selection and feature projection for cognitive load estimation using ambulatory EEG. *Computational Intelligence and Neuroscience*, *2007*, 2007. doi:10.1155/2007/74895 PMID:18364990

Lee, H., & Choi, S. (2003). Pca+ hmm+ svm for eeg pattern classification. In *Seventh International Symposium on Signal Processing and Its Applications, 2003 Proceedings.*, *1*, 541–544.

Li, M., Chen, W., & Zhang, T. (2017). Automatic epileptic EEG detection using DT-CWT-based non-linear features. *Biomedical Signal Processing and Control*, *34*, 114–125. doi:10.1016/j.bspc.2017.01.010

Li, M., Chen, W., & Zhang, T. (2017). Classification of epilepsy EEG signals using DWT-based envelope analysis and neural network ensemble. *Biomedical Signal Processing and Control*, *31*, 357–365. doi:10.1016/j.bspc.2016.09.008

Liu, X., Zhang, W., Li, W., Zhang, S., Lv, P., & Yin, Y. (2023). Effects of motor imagery based brain-computer interface on upper limb function and attention in stroke patients with hemiplegia: A randomized controlled trial. *BMC Neurology*, *23*(1), 1–14. doi:10.118612883-023-03150-5 PMID:37003976

López-Larraz, E., Montesano, L., Gil-Agudo, Á., & Minguez, J. (2014). Continuous decoding of movement intention of upper limb self-initiated analytic movements from pre-movement EEG correlates. *Journal of Neuroengineering and Rehabilitation*, *11*(1), 1–15. doi:10.1186/1743-0003-11-153 PMID:25398273

Lu, R. R., Zheng, M. X., Li, J., Gao, T. H., Hua, X. Y., Liu, G., Huang, S. H., Xu, J. G., & Wu, Y. (2020). Motor imagery based brain-computer interface control of continuous passive motion for wrist extension recovery in chronic stroke patients. *Neuroscience Letters*, *718*, 134727. doi:10.1016/j.neulet.2019.134727 PMID:31887332

Mammone, N., Ieracitano, C., & Morabito, F. C. (2020). A deep CNN approach to decode motor preparation of upper limbs from time–frequency maps of EEG signals at source level. *Neural Networks*, *124*, 357–372. doi:10.1016/j.neunet.2020.01.027 PMID:32045838

Müller-Putz, G. R., Kobler, R. J., Pereira, J., Lopes-Dias, C., Hehenberger, L., Mondini, V., Martínez-Cagigal, V., Srisrisawang, N., Pulferer, H., Batistić, L., & Sburlea, A. I. (2022). Feel your reach: An EEG-based framework to continuously detect goal-directed movements and error processing to gate kinesthetic feedback informed artificial arm control. *Frontiers in Human Neuroscience*, *16*, 841312. doi:10.3389/fnhum.2022.841312 PMID:35360289

Ofner, P., Schwarz, A., Pereira, J., & Müller-Putz, G. R. (2017). Upper limb movements can be decoded from the time-domain of low-frequency EEG. *PLoS One*, *12*(8), e0182578. doi:10.1371/journal.pone.0182578 PMID:28797109

Osanai, H., Yamamoto, J., & Kitamura, T. (2023). Extracting electromyographic signals from multi-channel LFPs using independent component analysis without direct muscular recording. *Cell Reports Methods*, *3*(6), 100482. doi:10.1016/j.crmeth.2023.100482 PMID:37426755

Prasetio, B. H., Tamura, H., & Tanno, K. (2019). Generalized Discriminant Methods for Improved X-Vector Back-end Based Stress Speech Recognition. *IEEJ Transactions on Electronics. Information Systems*, *139*(11), 1341–1347.

Roberts, G., Zoubir, A. M., & Boashash, B. (1997) Time-frequency classification using a multiple hypotheses test: An application to the classification of humpback whale signals. In *1997 IEEE International Conference on Acoustics, Speech, and Signal Processing* (Vol. 1, pp. 563-566). IEEE. 10.1109/ICASSP.1997.599700

Robertson, D. G. E., & Dowling, J. J. (2003). Design and responses of Butterworth and critically damped digital filters. *Journal of Electromyography and Kinesiology*, *13*(6), 569–573. doi:10.1016/S1050-6411(03)00080-4 PMID:14573371

Ruffino, C., Papaxanthis, C., & Lebon, F. (2017). Neural plasticity during motor learning with motor imagery practice: Review and perspectives. *Neuroscience*, *341*, 61–78. doi:10.1016/j.neuroscience.2016.11.023 PMID:27890831

Saha, S., & Ghosh, A. (2019, December). Rehabilitation using neighbor-cluster based matching inducing artificial bee colony optimization. In *2019 IEEE 16th India council international conference (INDICON)* (pp. 1-4). IEEE. 10.1109/INDICON47234.2019.9028975

Saha, S., Lahiri, R., Konar, A., & Nagar, A. K. (2018, November). Gesture driven remote robot manoeuvre for unstructured environment. In *2018 IEEE Symposium Series on Computational Intelligence (SSCI)* (pp. 1793-1800). IEEE. 10.1109/SSCI.2018.8628867

Scherer, R., Schloegl, A., Lee, F., Bischof, H., Janša, J., & Pfurtscheller, G. (2007). The self-paced graz brain-computer interface: Methods and applications. *Computational Intelligence and Neuroscience*, *2007*, 2007. doi:10.1155/2007/79826 PMID:18350133

Schlögl, A., & Supp, G. (2006). Analyzing event-related EEG data with multivariate autoregressive parameters. *Progress in Brain Research*, *159*, 135–147. doi:10.1016/S0079-6123(06)59009-0 PMID:17071228

Schuster, C., Hilfiker, R., Amft, O., Scheidhauer, A., Andrews, B., Butler, J., Kischka, U., & Ettlin, T. (2011). Best practice for motor imagery: A systematic literature review on motor imagery training elements in five different disciplines. *BMC Medicine*, *9*(1), 1–35. doi:10.1186/1741-7015-9-75 PMID:21682867

Schwarz, A., Escolano, C., Montesano, L., & Müller-Putz, G. R. (2020). Analyzing and decoding natural reach-and-grasp actions using gel, water and dry EEG systems. *Frontiers in Neuroscience*, *14*, 849. doi:10.3389/fnins.2020.00849 PMID:32903775

Sheldon, M. R., Fillyaw, M. J., & Thompson, W. D. (1996). The use and interpretation of the Friedman test in the analysis of ordinal-scale data in repeated measures designs. *Physiotherapy Research International*, *1*(4), 221–228. doi:10.1002/pri.66 PMID:9238739

Shouran, M., & Elgamli, E. (2020). Design and implementation of Butterworth filter. *International Journal of Innovative Research in Science, Engineering and Technology, 9*(9).

Somefun, O., Akingbade, K., & Dahunsi, F. (2022). Uniformly damped binomial filters: Five-percent maximum overshoot optimal response design. *Circuits, Systems, and Signal Processing*, *41*(6), 3282–3305. doi:10.100700034-021-01931-2

Steyrl, D., Scherer, R., Förstner, O., & Müller-Putz, G. R. (2014). Motor imagery brain-computer interfaces: random forests vs regularized LDA-non-linear beats linear. In *Proceedings of the 6th international brain-computer interface conference* (pp. 241-244). Academic Press.

Taunk, K., De, S., Verma, S., & Swetapadma, A. (2019, May). A brief review of nearest neighbor algorithm for learning and classification. In 2019 international conference on intelligent computing and control systems (ICCS) (pp. 1255-1260). IEEE. doi:10.1109/ICCS45141.2019.9065747

Übeyli, E. D. (2009). Analysis of EEG signals by implementing eigenvector methods/recurrent neural networks. *Digital Signal Processing*, *19*(1), 134–143. doi:10.1016/j.dsp.2008.07.007

Wei, X., Chen, K., Liu, Y., Zhao, X., Ma, L., Ai, Q., & Liu, Q. (2021, November). Control of multiple DOFs robots using motor imagery EEG combined with Huffman coding. In *2021 27th International Conference on Mechatronics and Machine Vision in Practice (M2VIP)* (pp. 344-348). IEEE. 10.1109/M2VIP49856.2021.9665060

Wickramasinghe, I., & Kalutarage, H. (2021). Naive Bayes: applications, variations and vulnerabilities: a review of literature with code snippets for implementation. *Soft Computing*, *25*(3), 2277–2293. doi:10.100700500-020-05297-6

World Health Organization. (2019). Brief model disability survey: result for India, Lao People's Democratic Republic and Tajikistan: executive summary, 2019 (No. WHO/NMH/NVI/19.15). World Health Organization.

Xie, X. (2019). Principal component analysis. *Wiley Interdisciplinary Reviews.*

Yger, F., Lotte, F., & Sugiyama, M. (2015) Averaging covariance matrices for EEG signal classification based on the CSP: an empirical study. In *2015 23rd European Signal Processing Conference (EUSIPCO)* (pp. 2721-2725). IEEE. 10.1109/EUSIPCO.2015.7362879

Yu, R., Charreyron, S. L., Boehler, Q., Weibel, C., Chautems, C., Poon, C. C., & Nelson, B. J. (2020, May). Modeling electromagnetic navigation systems for medical applications using random forests and artificial neural networks. In *2020 IEEE International Conference on Robotics and Automation (ICRA)* (pp. 9251-9256). IEEE. 10.1109/ICRA40945.2020.9197212

Yudhana, A., Muslim, A., Wati, D. E., Puspitasari, I., Azhari, A., & Mardhia, M. M. (2020). Human emotion recognition based on EEG signal using fast fourier transform and K-Nearest neighbor. *Adv. Sci. Technol. Eng. Syst. J*, *5*(6), 1082–1088. doi:10.25046/aj0506131

Zhongshen, L. (2007, August). Design and analysis of improved butterworth low pass filter. In *2007 8th International Conference on Electronic Measurement and Instruments* (pp. 1-729). IEEE. 10.1109/ICEMI.2007.4350554

Zhou, Y., Saad, Y., Tiago, M. L., & Chelikowsky, J. R. (2006). Self-consistent-field calculations using Chebyshev-filtered subspace iteration. *Journal of Computational Physics*, *219*(1), 172–184. doi:10.1016/j.jcp.2006.03.017

Chapter 8
Cloud Solutions for Smart Parking and Traffic Control in Smart Cities

Maganti Syamala
Department of Computer Science and Engineering, Koneru Lakshmaiah Education Foundation, Vaddeswaram, India

Hari Priya G. S.
Department of Computer Science, M.S. Ramaiah College of Arts Science and Commerce, Bengaluru, India

J. Malathi
Department of Computer Science and Business Systems, Sri Sai Ram Engineering College, Chennai, India

B. Uma Maheswari
https://orcid.org/0000-0001-9707-285X
Department of Computer Science and Engineering, St. Joseph's College of Engineering, Chennai, India

Vikash Singh
Department of Civil Engineering, Institute of Engineering and Technology, Lucknow, India

Murugan S.
Sona College of Technology, India

ABSTRACT

Urban mobility trends include 5G connectivity, autonomous vehicles, electric and sustainable modes, AI and machine learning, drones, and air mobility. These technologies enable real-time data exchange, reduce congestion, enhance safety, optimize road capacity, and optimize infrastructure planning. AI and machine learning algorithms provide accurate predictive analytics, adaptive traffic control, and personalized services. Cloud computing, IoT, and data analytics enable predictive modeling for mobility planning, traffic flow forecasting, demand forecasting, and behavioral analysis. MaaS platforms facilitate seamless integration of modes, while shared mobility services like car-sharing and ride-hailing grow, reducing private vehicle ownership and promoting efficient resource use. Mobility data transforms urban planning, infrastructure optimization, mixed-use development, and smart city integration, guiding transportation layouts, traffic signal placements, parking facilities, and neighborhood design.

DOI: 10.4018/978-1-6684-9999-3.ch008

INTRODUCTION

Urban mobility is a significant challenge due to rapid urbanization, population growth, and technological advancements. Cities face complex transportation systems causing congestion, pollution, and inefficient resource utilization. To address these issues and create sustainable environments, a holistic approach integrating various modes of transportation, cutting-edge technologies, and considering environmental, social, and economic impacts is being prioritized. The goal is to create an eco-friendly, accessible, and accessible network (Al Amiri et al., 2019).

Stakeholders, including governments, urban planners, and citizens, are collaborating to reimagine urban mobility through smart technologies, data analytics, and sustainable practices. Addressing challenges like congestion, pollution, and inadequate infrastructure, innovative solutions like shared mobility services, electric and autonomous vehicles, intelligent transportation systems, and MaaS are being explored (Al-Farhani et al., 2023). The paper explores the role of urban planning and policy-making in shaping the future of mobility, emphasizing the importance of integrated development, transit-oriented design, and sustainable transportation policies. It also highlights the potential societal benefits of improved urban mobility, including reduced pollution, improved public health, and increased economic productivity (Al-Turjman & Malekloo, 2019).

The transition towards enhanced urban mobility is not without its challenges. Technical, regulatory, and societal barriers must be overcome to fully realize the vision of a seamless and sustainable transportation system (Arifin et al., 2019). Smart cities, a visionary approach to address urban challenges, can transform mobility landscapes with the right strategies, collaboration, and investments. This visionary approach, amidst rapid urbanization and the proliferation of advanced technologies, offers residents a higher quality of life and a more connected future (Bock et al., 2019). At the heart of this transformation lies the integration of cloud solutions, a groundbreaking paradigm that has the potential to revolutionize urban living by enhancing efficiency, sustainability, and overall quality of life (Chandana et al., 2019).

Smart cities are transforming urban development by utilizing digital innovations to create more resilient, responsive, and connected environments. They use various data sources, including sensors, devices, social media, and online platforms, to gather real-time information on transportation, energy consumption, waste management, and public safety (Chmiel et al., 2016). This vast amount of data forms the foundation upon which cloud solutions play a pivotal role. Cloud solutions provide the technological backbone that enables smart cities to process, analyze, and derive actionable insights from this data. By leveraging cloud computing, cities can securely store and access vast volumes of information without the limitations of traditional on-site infrastructure. This scalability and flexibility are critical for managing the dynamic and ever-growing data generated by smart city initiatives (De Oliveira et al., 2020).

Cloud solutions in smart cities enhance efficiency, improve traffic management, optimize energy consumption, and streamline public services. Real-time data analysis enables better traffic management, reducing congestion and travel times. Smart traffic management systems can dynamically adjust signal timings based on traffic flow (Delgado & Calegari, 2023). Likewise, cloud-powered energy management systems can balance supply and demand to minimize wastage and promote sustainability. Another crucial aspect of cloud-driven smart cities is the improvement of citizen engagement and empowerment. Cloud-based platforms facilitate direct interaction between citizens and city authorities, enabling feedback mechanisms, information sharing, and participatory decision-making. Citizens can report issues through mobile apps, access real-time information about public services, and collaborate with local authorities to co-create solutions to urban challenges (Diran et al., 2021).

Cloud solutions significantly contribute to the sustainability goals of smart cities by optimizing resource utilization through data-driven insights, reducing environmental footprint, and optimizing waste management systems, such as garbage collection routes, to minimize fuel consumption and greenhouse gas emissions (Dutta et al., 2019). The cloud is a crucial tool for deploying renewable energy sources and monitoring energy consumption patterns in smart cities. However, challenges such as data privacy, security breaches, and digital equity need to be addressed to ensure accessibility and equity. Robust cybersecurity measures and inclusive digital policies are essential for creating a trustworthy and equitable smart city ecosystem (Francis et al., 2023).

This paper explores the integration of cloud solutions in smart cities, highlighting their potential to transform urban centers into dynamic, data-driven environments, enhancing efficiency, sustainability, and citizen empowerment. It also discusses the challenges and future developments in urban innovation (Hilmani et al., 2020). In the face of urbanization's relentless pace and the resulting increase in vehicular congestion, the concepts of smart parking and smart traffic control have emerged as indispensable tools to mitigate the challenges of modern urban mobility. As cities strive to optimize their transportation systems, reduce traffic congestion, and enhance overall quality of life, these intelligent solutions have risen to the forefront as crucial components of the broader smart city ecosystem (Kaginalkar et al., 2021).

Smart parking and smart traffic control are innovative approaches that leverage technology, data, and real-time insights to revolutionize the way cities manage their transportation infrastructure. Traditional parking and traffic management systems often lead to frustration due to limited parking availability and inefficient traffic flow (Kaginalkar et al., 2021). In contrast, these smart solutions harness the power of advanced sensors, data analytics, and connectivity to create a more responsive and streamlined urban environment. Smart parking is designed to address the perennial issue of finding available parking spaces in densely populated areas. With the integration of sensors that monitor parking spot occupancy, real-time data is relayed to a centralized system. This system guides drivers to vacant parking spaces through mobile apps, digital signage, or navigation systems. As a result, not only does this technology save drivers time and reduce congestion caused by circling for parking, but it also contributes to lower fuel consumption and subsequently reduced carbon emissions (Khan et al., 2020).

On the other hand, smart traffic control focuses on optimizing the flow of vehicles through intelligent and adaptive traffic management systems. These systems utilize data collected from sensors, cameras, and even connected vehicles to monitor traffic patterns and congestion in real time. By analyzing this data, traffic signals can be adjusted dynamically to respond to changing traffic conditions (Leal Sobral et al., 2023). This results in smoother traffic flow, reduced bottlenecks, and shorter travel times for commuters. The integration of these technologies brings forth a range of benefits that extend beyond convenience. Efficient smart parking and traffic control contribute to reduced greenhouse gas emissions, improved air quality, and increased road safety. By minimizing unnecessary idling and congestion, these systems help alleviate the negative environmental and health impacts associated with traffic congestion (Lee & Chiu, 2020).

Furthermore, the introduction of these smart solutions aligns seamlessly with the broader smart city vision. As cities become increasingly interconnected, data-driven decision-making becomes paramount. Smart parking and traffic control systems generate valuable insights that city planners and policymakers can utilize to optimize urban infrastructure, enhance public transportation networks, and improve overall urban planning. However, the implementation of smart parking and traffic control systems is not without its challenges (Lee & Chiu, 2020). Issues related to data privacy, cybersecurity, and equitable access must be navigated to ensure that the benefits of these technologies are enjoyed by all segments of society. Moreover, the integration of these systems requires collaboration between city authorities, technology providers, and citizens to ensure seamless adoption and successful operation.

In the subsequent sections, this paper will delve deeper into the intricacies of smart parking and smart traffic control. It will explore the technologies involved, examine successful case studies from around the world, and discuss the potential future developments in these fields. As cities continue their quest for more efficient and sustainable transportation systems, the journey toward intelligent parking and traffic solutions promises to be a transformative one.

Background and Motivation

Urbanization has become an inexorable trend in modern society, with an increasing number of people migrating to cities in search of better opportunities and lifestyles. However, this influx has given rise to a myriad of challenges, including traffic congestion, inadequate parking infrastructure, and inefficient traffic management systems. These challenges not only hinder the quality of life for residents but also impact environmental sustainability and economic productivity (Lee & Chiu, 2020; Liu et al., 2022; Mahajan et al., n.d.). As cities strive to become smarter and more liveable, the integration of technology, particularly cloud computing and the Internet of Things (IoT), presents a transformative solution. This section delves into the pressing issues faced by urban centres, highlighting the need for innovative approaches and setting the stage for the chapter's exploration of cloud-based solutions for smart parking and traffic control.

Scope of the Chapter

This chapter aims to provide a comprehensive analysis of how cloud solutions can revolutionize urban mobility through smart parking and traffic control mechanisms in the context of smart cities. The scope encompasses the utilization of cloud computing, real-time data analytics, and IoT technologies to optimize parking utilization, enhance traffic flow, and improve overall urban mobility. The chapter will delve into the technical aspects, case studies, benefits, challenges, and future trends related to implementing cloud-based solutions for these purposes. The discussions will span across technological, social, and regulatory dimensions, offering a holistic view of the potential and implications of such implementations.

Objectives

- To Examine the key challenges faced by modern cities in terms of traffic congestion, inadequate parking infrastructure, and inefficient traffic management, highlighting their impacts on urban living.
- To provide an in-depth exploration of cloud computing and its potential role in addressing urban mobility challenges, with a focus on smart parking and traffic control.
- To present case studies of successful cloud-enabled smart parking and traffic control systems that have been implemented in various smart cities, showcasing their benefits and outcomes.
- To analyze the potential benefits of cloud-driven solutions, including reduced congestion, improved environmental sustainability, and enhanced quality of life for residents.
- To delve into the challenges associated with implementing cloud solutions for urban mobility, such as data privacy, scalability, and regulatory hurdles.

- To explore emerging trends in the field of urban mobility, cloud computing, and IoT, predicting how these technologies might shape the future of transportation and urban planning.

SMART CITIES AND URBAN MOBILITY

Definition of Smart Cities

Smart cities are innovative urban planning systems that combine advanced technologies, data-driven insights, and innovative urban planning to improve residents' quality of life, optimize resource utilization, and ensure sustainability. These systems integrate transportation, energy, governance, and public services, enhancing efficiency, safety, and overall well-being. They use digital infrastructure, data analytics, and real-time monitoring to make informed decisions and address challenges proactively (Barron, 2022; Delgado & Calegari, 2023).

Urban Mobility Challenges in Modern Cities

Urbanization has exacerbated mobility issues in cities, leading to congestion, traffic gridlock, and inefficient transportation systems. This has resulted in increased travel times, air pollution, and reduced quality of life. Inadequate parking infrastructure exacerbates these issues. Urban centers must provide accessible, affordable, and sustainable transportation options to accommodate the growing population while minimizing environmental impact (Diran et al., 2021; Mavlutova et al., 2023).

Role of Technology in Addressing Urban Mobility

Advanced technologies like cloud computing, IoT, real-time data analytics, and smart sensors are revolutionizing urban mobility. These technologies enable real-time traffic monitoring, predictive analysis of congestion patterns, and data-driven decision-making for effective traffic management. Cloud solutions also facilitate the development of smart parking systems, enhancing space utilization and reducing traffic congestion caused by aimlessly searching for parking (Nogueira et al., n.d.; Richter, Hagenmaier, Bandte, Parida, Wincent, et al., 2022). The integration of technology also paves the way for Mobility-as-a-Service (MaaS) platforms, where various transportation modes are seamlessly connected through digital platforms, allowing citizens to plan and pay for their journeys using a single application. By harnessing technology, cities can create efficient, data-driven transportation systems that adapt to changing conditions, alleviate congestion, reduce pollution, and ultimately enhance the urban living experience (Richter, Hagenmaier, Bandte, Parida, & Wincent, 2022; Richter, Hagenmaier, Bandte, Parida, Wincent, et al., 2022). This chapter explores the potential of cloud solutions in smart parking and traffic control, aiming to revolutionize urban mobility and promote sustainable cities through transportation systems.

CLOUD COMPUTING IN URBAN MOBILITY

In the ever-evolving landscape of urban mobility, cloud computing has emerged as a transformative force, offering cities new avenues to address challenges and optimize transportation systems. Cloud computing involves the delivery of computing resources, such as storage, processing power, and applications, over the internet. In the context of urban mobility, cloud solutions are being harnessed to revolutionize how cities manage transportation networks, enhance traffic control, and streamline parking systems (Kaginalkar et al., 2021).

Cloud solutions provide a scalable and flexible platform that enables cities to efficiently manage vast amounts of data generated by urban transportation systems. The cloud serves as a central repository where data from various sources, such as traffic sensors, GPS devices, and surveillance cameras, can be collected, processed, and analyzed in real-time. This capability empowers cities with actionable insights to make informed decisions that enhance traffic flow, optimize routes, and improve overall mobility (Hema et al., 2023; Rahamathunnisa et al., 2023; Syamala et al., 2023; Venkateswaran, Vidhya, Ayyannan, et al., 2023).

The adoption of cloud computing offers several compelling benefits for urban mobility (Kaginalkar et al., 2021; Shengdong et al., 2019):

- **Scalability:** Cloud infrastructure allows cities to easily scale their resources up or down based on demand. This is crucial for handling peak traffic times, special events, or emergencies, ensuring that transportation systems remain responsive and efficient.
- **Real-time Insights:** Cloud-based analytics enable cities to gather real-time data from multiple sources and analyze it to identify traffic patterns, congestion points, and areas requiring intervention. This data-driven approach allows for timely responses and dynamic traffic management.
- **Cost Efficiency:** Cloud solutions eliminate the need for extensive on-premises hardware and maintenance. Cities can leverage the pay-as-you-go model, reducing capital expenditures while achieving efficient and up-to-date technology usage.
- **Collaboration:** Cloud platforms facilitate collaboration among different city departments, stakeholders, and service providers. This enables integrated planning and coordinated efforts to address mobility challenges comprehensively.
- **Innovation:** Cloud ecosystems encourage innovation by providing a platform for the development and deployment of new mobility services and applications. From smart navigation apps to predictive traffic models, cloud-enabled solutions drive continuous advancements.

Cloud-Enabled Services for Smart Cities

In the realm of urban mobility, cloud-enabled services play a pivotal role in creating smarter, more efficient cities (Ragavan et al., 2021; Richter, Hagenmaier, Bandte, Parida, & Wincent, 2022):

- **Smart Traffic Management:** Cloud-connected traffic control systems use real-time data to adjust signal timings, reroute traffic, and optimize traffic flow. This results in reduced congestion, shorter travel times, and improved air quality.
- **Dynamic Parking Solutions:** Cloud-powered smart parking systems leverage sensors to monitor parking space availability. This data is then made accessible to drivers through apps, guiding them to vacant spots and minimizing time spent searching for parking.
- **Public Transport Optimization:** Cloud analytics enable cities to monitor public transport usage, identify routes with high demand, and adjust services accordingly. This leads to better utilization of resources and enhanced commuter experiences.

- **Data-Driven Planning:** Cloud-stored mobility data serves as a valuable resource for urban planners. It aids in designing efficient transport networks, predicting future demand, and formulating policies that cater to evolving mobility needs.
- **Emergency Response:** During emergencies, cloud-based communication and coordination systems allow for real-time dissemination of information to citizens and responders, facilitating swift and effective responses to incidents affecting urban mobility.

Cloud computing is pivotal in urban mobility transformation, enabling the handling of vast data, real-time insights, and collaboration. Its scalability, cost-efficiency, and innovation potential enable cities to tackle urban mobility challenges, providing seamless, sustainable, and enjoyable travel experiences for citizens.

SMART PARKING SOLUTIONS

In the fast-paced urban environments of today, finding parking is often a frustrating and time-consuming experience for drivers. Smart parking solutions, powered by technology and data, are revolutionizing the way cities manage parking spaces, enhance urban mobility, and improve the overall quality of life for residents (Chandana et al., 2019; Francis et al., 2023). This section delves into the challenges of urban parking, the role of IoT and sensors in smart parking, and their transformative impact as shown in Figure 1.

The Problem of Parking in Urban Areas

The scarcity of parking spaces in densely populated urban areas is a widespread issue that contributes to traffic congestion, increased emissions, and decreased productivity. Drivers searching for parking spaces not only waste time but also contribute to unnecessary traffic buildup and pollution. Moreover, inefficient parking management can lead to revenue loss for cities due to unutilized spaces or improper fee collection (Francis et al., 2023; Mahajan et al., n.d.). Traditional parking systems, reliant on manual enforcement and inadequate infrastructure, struggle to cope with the demands of modern urban living. As cities continue to grow, the gap between parking supply and demand widens, necessitating innovative solutions to optimize parking utilization and enhance the parking experience.

IoT and Sensors in Smart Parking

The Internet of Things (IoT) has emerged as a game-changing technology in the realm of smart parking. IoT refers to the network of interconnected devices embedded with sensors, software, and connectivity capabilities that enable them to collect and exchange data. In the context of parking, IoT-enabled sensors play a critical role in transforming parking spaces into intelligent, data-driven entities (Hema et al., 2023; Jeevanantham et al., 2022; S. Karthik et al., 2023; Koshariya et al., 2023).

Figure 1. Smart parking solutions using IoT

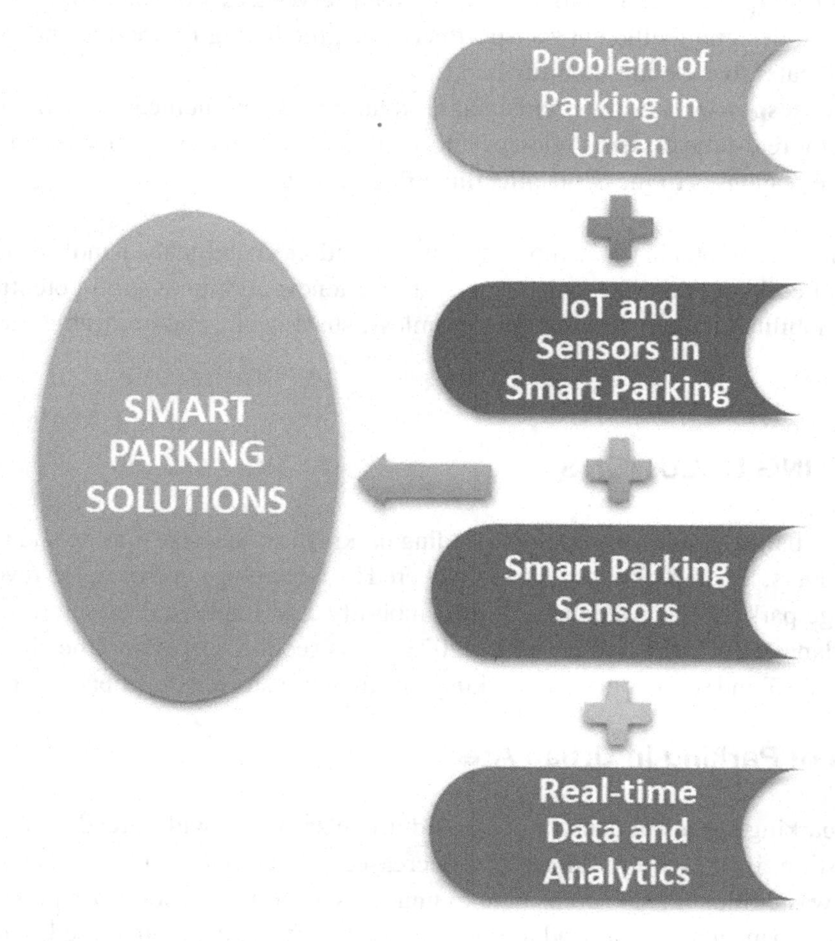

Smart Parking Sensors

Smart parking solutions deploy a variety of sensors, such as ultrasonic, magnetic, and infrared sensors, to monitor the occupancy status of parking spaces in real time. These sensors can detect whether a parking space is occupied or vacant, sending this information to a centralized cloud-based platform for analysis and dissemination (Maguluri et al., 2023; Reddy et al., 2023; Samikannu et al., 2022; Syamala et al., 2023).

Real-Time Data and Analytics

The data collected by parking sensors provides valuable insights into parking patterns, occupancy rates, and peak usage times. This real-time information allows cities to dynamically manage parking resources, directing drivers to available spaces and minimizing congestion caused by aimless searching (Boopathi, 2023b; S. Karthik et al., 2023; Koshariya et al., 2023; Maguluri et al., 2023; Reddy et al., 2023).

Benefits of IoT-Enabled Smart Parking

- **Enhanced User Experience:** Smart parking solutions improve the parking experience for drivers by guiding them to available spaces through mobile apps or digital signage. This reduces stress and saves time for both residents and visitors (Rodic, n.d.; Sanghvi et al., 2021).
- **Optimized Space Utilization:** With real-time occupancy data, cities can optimize parking space allocation, avoiding overutilization of some areas and underutilization of others.
- **Reduced Traffic Congestion:** By minimizing the time spent searching for parking, smart parking solutions contribute to reduced traffic congestion, thereby enhancing overall traffic flow.
- **Environmental Benefits:** Efficient parking management leads to reduced idling and circling, which in turn decreases vehicle emissions and contributes to improved air quality.
- **Revenue Generation:** Smart parking systems can boost revenue generation for cities by enabling dynamic pricing based on demand and optimizing fee collection through digital payment methods.
- **Data-Driven Decision Making:** The data collected from smart parking systems provides insights that inform urban planning decisions, helping to design more efficient parking infrastructure.
- **Scalability:** IoT-enabled smart parking solutions are highly scalable, allowing cities to expand and adapt their systems as their parking needs evolve.

In conclusion, smart parking solutions driven by IoT and sensors are revolutionizing the way urban areas manage their parking resources. These technologies address the challenges of parking in densely populated cities by providing real-time data, enabling efficient resource allocation, and enhancing the overall urban mobility experience (Chandana et al., 2019). By utilizing the power of IoT, cities can transform their parking landscapes, making them more efficient, user-friendly, and sustainable. As technology continues to advance, the potential for further innovations in smart parking solutions holds promise for creating smarter, more livable cities.

Cloud-Based Parking Management Systems

In the quest for more efficient urban mobility, cloud-based parking management systems have emerged as a game-changer, leveraging technology to transform parking into a seamless and data-driven experience. These systems utilize the capabilities of cloud computing, real-time data analytics, and IoT sensors to optimize parking utilization, enhance revenue collection, and alleviate traffic congestion (Boopathi, 2023a; Boopathi & Kanike, 2023; Hema et al., 2023; Maheswari et al., 2023). This section explores the advantages and disadvantages of cloud-based parking management systems, highlighting successful implementations and illustrating their features in Figure 2 (Francis et al., 2023; Mahajan et al., n.d.; Medagliani et al., 2022).

- **Real-Time Space Monitoring:** Cloud-based systems employ IoT sensors to monitor parking space occupancy in real time. This data is instantly transmitted to cloud servers, enabling dynamic management of parking resources.
- **Mobile Applications:** Cloud-powered parking systems offer drivers mobile applications that provide real-time information about available parking spaces, guiding them to the nearest vacant spot.

- **Reservation and Payment:** Many cloud-based solutions allow drivers to reserve parking spaces in advance through mobile apps. Additionally, digital payment options facilitate seamless transactions without the need for physical cash or cards.
- **Data Analytics:** Cloud platforms collect and analyze parking data, offering insights into usage patterns, peak times, and preferred locations. This data aids in optimizing parking infrastructure and urban planning.
- **Dynamic Pricing:** Cloud-enabled systems can implement dynamic pricing models, adjusting parking fees based on demand, time of day, and special events. This maximizes revenue while encouraging efficient space utilization.

Figure 2. Various features of cloud-based parking management systems

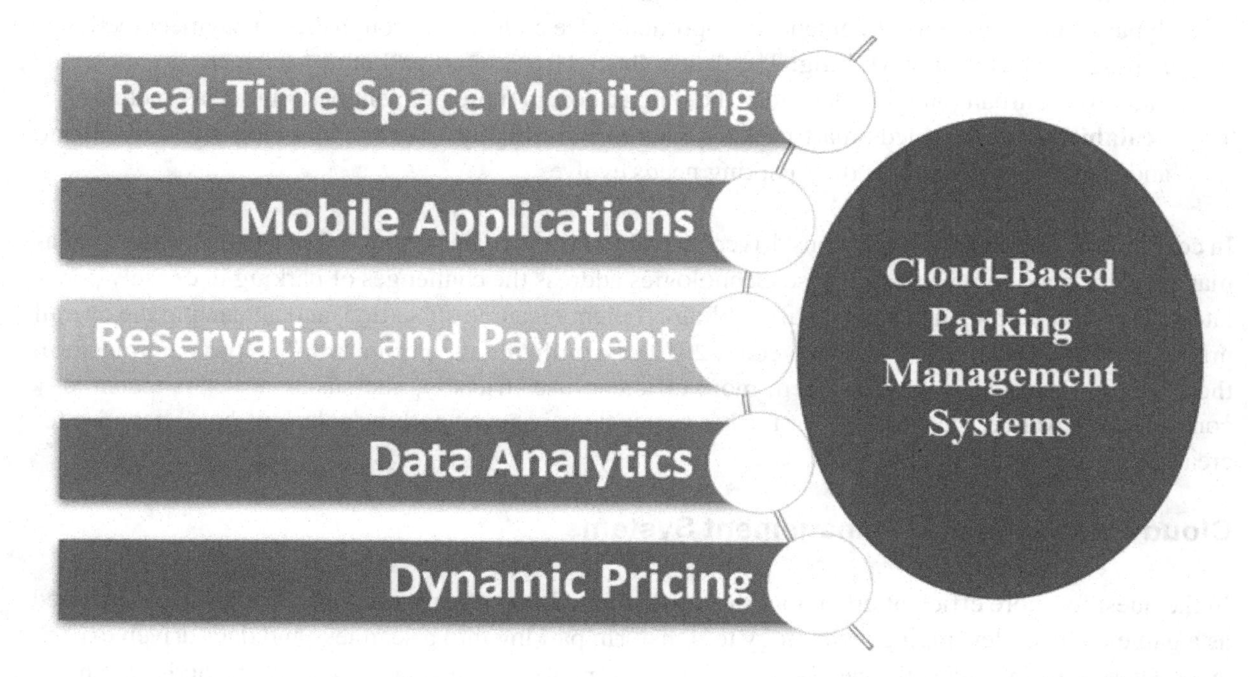

Case Studies: Successful Smart Parking Implementations

- **San Francisco Municipal Transportation Agency (SFMTA), USA:** SFMTA implemented a cloud-based smart parking system in the city, using real-time occupancy data from sensors to guide drivers to available spaces. The system reduced traffic congestion, shortened search times, and decreased greenhouse gas emissions. Moreover, the city saw a 43% reduction in parking-related citations, indicating improved compliance (Chandana et al., 2019; Francis et al., 2023; Medagliani et al., 2022).
- **Barcelona, Spain:** Barcelona's visionary approach to smart parking involved deploying sensors in parking spaces across the city. Cloud-connected sensors transmitted occupancy data to a central platform, which then directed drivers to available spaces via a mobile app. This initiative resulted

in a 30% reduction in traffic congestion and an estimated 40% reduction in driving time spent looking for parking.

- **Singapore:** Singapore's cloud-driven smart parking system employed sensors to monitor parking spaces across the city-state. Real-time data allowed for efficient enforcement, optimizing parking space utilization. By leveraging this technology, Singapore improved its parking enforcement efficiency by 35% and reduced the time spent searching for parking by drivers.
- **Los Angeles, USA:** Los Angeles introduced a cloud-based smart parking solution that not only guided drivers to available spots but also improved parking management. The system provided real-time data to the city's traffic management center, allowing dynamic adjustments to traffic signals based on parking demand. This integration led to reduced congestion and improved traffic flow.
- **Amsterdam, Netherlands:** Amsterdam implemented a cloud-powered system that not only guided drivers to available parking spaces but also incorporated electric vehicle (EV) charging stations into the solution. EV drivers could locate available charging spots through the app, fostering the adoption of electric vehicles and promoting sustainable urban mobility.

Cloud-based parking management systems are revolutionizing urban mobility by enhancing user experiences, reducing congestion, and optimizing parking space utilization. Success stories from cities like San Francisco, Barcelona, Singapore, Los Angeles, and Amsterdam demonstrate the tangible benefits of these smart parking solutions (Bock et al., 2019; Chandana et al., 2019; Francis et al., 2023). As cities evolve, embracing technology-driven solutions like cloud-based parking management systems is crucial for creating more efficient, sustainable, and liveable urban spaces.

INTELLIGENT TRAFFIC CONTROL SYSTEMS

In the modern urban landscape, traffic congestion has become a pervasive challenge, impacting both the efficiency of transportation systems and the quality of life for residents. Intelligent traffic control systems, powered by data-driven approaches and cloud technology, offer innovative solutions to tackle congestion, enhance traffic flow, and create more efficient and sustainable urban mobility. This section explores the impacts of traffic congestion, the role of data-driven approaches, cloud-powered traffic management, and the significance of real-time traffic monitoring and adaptive control (Arifin et al., 2019; Dutta et al., 2019).

Traffic Congestion and Its Impacts

Traffic congestion is a phenomenon characterized by the slow movement or complete standstill of vehicles on roads due to high traffic volumes, road closures, accidents, or inadequate infrastructure (Hilmani et al., 2020; Ragavan et al., 2021). The effects of congestion are multifaceted, as illustrated in Figure 3.

- **Time Loss:** Commuters spend significant time stuck in traffic, leading to productivity loss, increased travel times, and frustration.
- **Environmental Impact:** Congestion contributes to higher emissions, air pollution, and increased fuel consumption, exacerbating environmental concerns.

- **Economic Costs:** Congestion results in increased fuel consumption and operating costs for businesses, while also deterring potential investments in affected areas.
- **Reduced Quality of Life:** Long commutes and traffic-related stress affect the overall well-being and quality of life of residents.
- **Inefficient Resource Use:** Congestion reduces the efficiency of transportation networks and infrastructure, resulting in underutilized capacities and increased wear on roads.

Figure 3. Intelligent traffic control systems

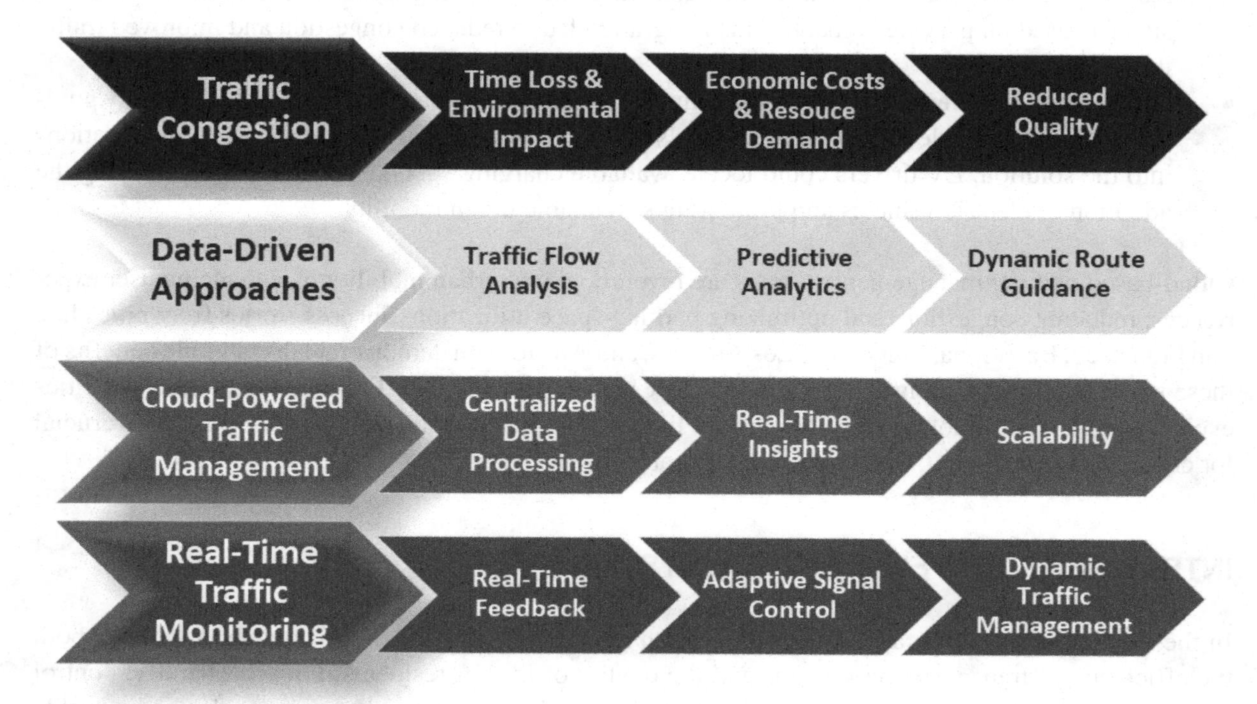

Data-Driven Approaches to Traffic Control

Data-driven approaches leverage real-time and historical data to understand traffic patterns, predict congestion, and optimize traffic management strategies (Ragavan et al., 2021; Said et al., 2022). These approaches involve:

- **Traffic Flow Analysis:** Data from various sources, including sensors, GPS devices, and social media, are analyzed to understand traffic flow, identify bottlenecks, and predict congestion.
- **Predictive Analytics:** By analyzing historical data and real-time inputs, predictive models can anticipate congestion and plan interventions before it occurs.
- **Dynamic Route Guidance:** Data-driven systems provide real-time route recommendations to drivers, optimizing travel paths and distributing traffic to alternative routes.

Cloud-Powered Traffic Management

Cloud technology plays a pivotal role in enhancing traffic management (Shengdong et al., 2019):

- **Centralized Data Processing:** Cloud platforms serve as centralized hubs for processing and analyzing large volumes of traffic data from multiple sources.
- **Real-Time Insights:** Cloud-powered systems provide real-time traffic insights to traffic management centers, enabling quick decision-making and intervention.
- **Scalability:** Cloud infrastructure easily scales to accommodate growing data loads during peak traffic times or special events.

Real-Time Traffic Monitoring and Adaptive Control

Real-time traffic monitoring involves continuous data collection and analysis, which informs adaptive traffic control strategies (Yogheshwaran et al., 2020):

- **Real-Time Feedback:** Sensors and cameras collect data on traffic conditions, allowing for real-time adjustments to traffic signal timings and control.
- **Adaptive Signal Control:** Cloud-connected traffic signals can adapt signal timings based on real-time traffic conditions, optimizing traffic flow and minimizing congestion.
- **Dynamic Traffic Management:** Cloud-powered systems enable traffic authorities to dynamically change traffic patterns, detours, and routing based on real-time data.

In conclusion, intelligent traffic control systems, empowered by data-driven approaches and cloud technology, hold immense promise in mitigating traffic congestion's detrimental effects. By leveraging real-time insights, predictive analytics, and adaptive control mechanisms, cities can create more efficient transportation networks, reduce environmental impact, and enhance the overall quality of life for their residents (Ragavan et al., 2021; Said et al., 2022; Shengdong et al., 2019). The integration of data-driven strategies with cloud-powered infrastructure exemplifies the synergy of modern technologies in revolutionizing urban mobility and shaping smarter, more resilient cities.

INTEGRATION OF CLOUD AND IOT FOR URBAN MOBILITY

The convergence of cloud computing and the Internet of Things (IoT) has ushered in a new era of urban mobility transformation. This integration offers cities unprecedented capabilities to collect, process, and analyze real-time data, enabling informed decision-making, optimizing transportation systems, and enhancing overall urban mobility. This section explores the synergy between cloud and IoT technologies and how data collection and analysis contribute to mobility insights (Richter, Hagenmaier, Bandte, Parida, & Wincent, 2022; Rui & Othengrafen, 2023; Stiglic et al., 2018).

Synergy Between Cloud and IoT Technologies

Cloud computing and IoT technologies work in tandem to create a dynamic and responsive urban mobility ecosystem (Hundera et al., 2022; Kaginalkar et al., 2021; Qolomany et al., 2020):

- **Data Collection and Transmission:** IoT sensors deployed across urban areas collect data on various aspects of mobility, such as traffic flow, parking occupancy, and public transportation usage.
- **Real-Time Connectivity:** IoT devices transmit this data to cloud platforms instantaneously, enabling real-time analysis and allowing cities to respond swiftly to changing conditions.
- **Scalable Computing Power:** Cloud computing provides the computational resources required to process and analyze vast amounts of IoT-generated data efficiently.
- **Actionable Insights:** Cloud platforms deliver actionable insights derived from IoT data, empowering cities with the information needed to optimize traffic management and enhance mobility services (Agrawal et al., 2024; B et al., 2024; Satav, Hasan, et al., 2024; Satav, Lamani, et al., 2024).

Data Collection and Analysis for Mobility Insights

The integration of cloud and IoT technologies enables comprehensive data collection and sophisticated analysis for enhanced urban mobility (Mavlutova et al., 2023; Morfoulaki & Papathanasiou, 2021; Nogueira et al., n.d.; Rajapaksha et al., 2017):

- **Real-Time Data Streams:** IoT sensors, embedded in traffic lights, vehicles, parking spaces, and public transportation, continuously generate data streams that provide a real-time view of urban mobility.
- **Traffic Pattern Analysis:** By analyzing data from IoT sensors, cloud-based systems can identify traffic patterns, congestion points, and bottlenecks, facilitating efficient traffic management strategies.
- **Predictive Analytics:** Historical and real-time data, processed in the cloud, can fuel predictive models that forecast traffic congestion, enabling proactive interventions.
- **Mobility Insights for Planning:** Cloud-enabled analysis of mobility data assists urban planners in designing efficient transportation networks and making informed decisions about infrastructure investments.
- **Personalized Services:** Cloud platforms can personalize mobility services based on user preferences and behavior, enhancing the commuting experience.
- **Evolving Urban Planning:** Data-driven insights into mobility patterns can inform urban planning decisions, enabling cities to adapt to changing mobility needs and trends.

In conclusion, the integration of cloud and IoT technologies revolutionizes urban mobility by providing cities with real-time data and actionable insights. This synergy empowers cities to optimize traffic control, enhance parking management, and improve overall transportation efficiency (Delgado & Calegari, 2023; Diran et al., 2021; Mavlutova et al., 2023). By leveraging the capabilities of both cloud computing and IoT, urban areas can create adaptive, data-driven mobility solutions that respond to the needs of their residents, foster sustainable transportation, and contribute to the development of smarter and more livable cities.

Case Study: IoT-Driven Traffic Flow Optimization

City: New York City, USA

Background: New York City is renowned for its bustling streets and heavy traffic congestion. The city's intricate road network and high population density often lead to challenging traffic conditions, negatively impacting the quality of life for residents and the efficiency of urban mobility. To address these issues, the city embarked on an innovative initiative to leverage IoT technology for optimizing traffic flow and enhancing the overall urban commuting experience(Kaginalkar et al., 2021; Rabby et al., 2019).

Implementation: New York City introduced an IoT-driven traffic flow optimization system that utilized a network of sensors, cameras, and cloud computing resources. The goal was to collect real-time traffic data, analyze it, and implement adaptive traffic control strategies to alleviate congestion and improve traffic flow(Syamala et al., 2023; Veeranjaneyulu, Boopathi, Kumari, et al., 2023; Veeranjaneyulu, Boopathi, Narasimharao, et al., 2023; Venkateswaran, Vidhya, Naik, et al., 2023).

Key Components:

- **IoT Sensors:** IoT sensors were strategically placed at major intersections, on-ramps, and critical road segments throughout the city. These sensors captured data on vehicle counts, speeds, and traffic patterns.
- **Cameras:** Cameras were integrated with the IoT sensors to provide visual data for better understanding traffic conditions, identifying accidents, and ensuring the accuracy of sensor data.
- **Cloud Computing:** The collected data was transmitted to a cloud-based platform in real time. The cloud infrastructure had the computational power to process and analyze large volumes of data quickly.

Data Analysis and Adaptive Control: The cloud platform used advanced data analytics algorithms to process the incoming data streams. The system analyzed traffic patterns, identified congestion points, and detected irregularities such as accidents or road closures. The insights derived from this analysis enabled the system to make adaptive traffic control decisions.

Adaptive Traffic Control Strategies: Based on the real-time analysis, the system adjusted traffic signal timings at intersections dynamically. For instance, if congestion was detected on a particular route, the system could prioritize that route by extending the green signal time. Additionally, the system could suggest alternative routes to drivers via mobile apps, guiding them away from congested areas.

Results: The IoT-driven traffic flow optimization initiative in New York City yielded significant improvements:

- **Reduced Congestion:** The adaptive traffic control strategies led to smoother traffic flow, reducing congestion at key intersections and critical roadways.
- **Shorter Commute Times:** By providing real-time traffic information and suggesting alternative routes, the system helped drivers make informed decisions, resulting in shorter commute times.
- **Improved Air Quality:** Reduced idling and smoother traffic flow contributed to decreased emissions and improved air quality, benefiting both residents and the environment (Boopathi et al., 2022; Veeranjaneyulu, Boopathi, Kumari, et al., 2023).
- **Enhanced User Experience:** The integration of mobile apps with the system empowered drivers with real-time traffic information, helping them navigate the city more efficiently.

The success of New York City's IoT-driven traffic flow optimization system served as a blueprint for other urban areas facing similar challenges. As IoT technology continues to advance, cities worldwide are exploring similar solutions to create more intelligent and responsive traffic management systems. The New York City case study demonstrates the transformative potential of IoT in alleviating traffic congestion and enhancing urban mobility for the benefit of both residents and the city's sustainability goals.

CHALLENGES AND CONSIDERATIONS

Urban mobility solutions powered by cloud computing and IoT technologies offer immense potential, but they also come with a set of challenges and considerations that need to be carefully addressed (Leal Sobral et al., 2023). In the pursuit of smarter and more efficient cities, it is essential to navigate these challenges to ensure successful implementation and adoption (Figure 4).

Data Privacy and Security Concerns

The collection and utilization of vast amounts of personal and sensitive data for traffic monitoring, parking guidance, and mobility services raise significant privacy and security concerns. Ensuring data encryption, secure storage, and strict access controls are crucial to safeguarding citizens' information and preventing unauthorized access or data breaches (Al-Farhani et al., 2023; Leal Sobral et al., 2023; Nallaperuma et al., 2019).

Scalability and Infrastructure Requirements

As urban populations grow and mobility demands increase, the scalability of cloud infrastructure becomes a critical consideration. The technology must be able to handle the increasing volume of data generated by IoT devices and the evolving needs of expanding cities. Proper planning for scalable architecture and resources is essential to prevent system bottlenecks and ensure seamless operation (Agrawal et al., 2024; B et al., 2024; Satav, Hasan, et al., 2024; Satav, Lamani, et al., 2024).

Public Acceptance and Adoption

Implementing new technologies in urban mobility requires public acceptance and active adoption. Citizens may be hesitant to embrace changes in their commuting habits, especially if they perceive new systems as complex or intrusive. Effective communication, education, and user-friendly interfaces are key to gaining public trust and encouraging widespread adoption of cloud-powered mobility solutions.

Regulatory and Policy Considerations

City regulations, transportation policies, and legal frameworks must evolve to accommodate the integration of innovative technologies. Issues such as data ownership, liability in case of accidents involving autonomous vehicles, and the sharing of mobility data with private companies need to be carefully addressed to create a supportive environment for implementation (Boopathi, 2022; A. Saravanan et al., 2022; M. Saravanan et al., 2022).

Figure 4. Challenges to ensure successful implementation and adoption

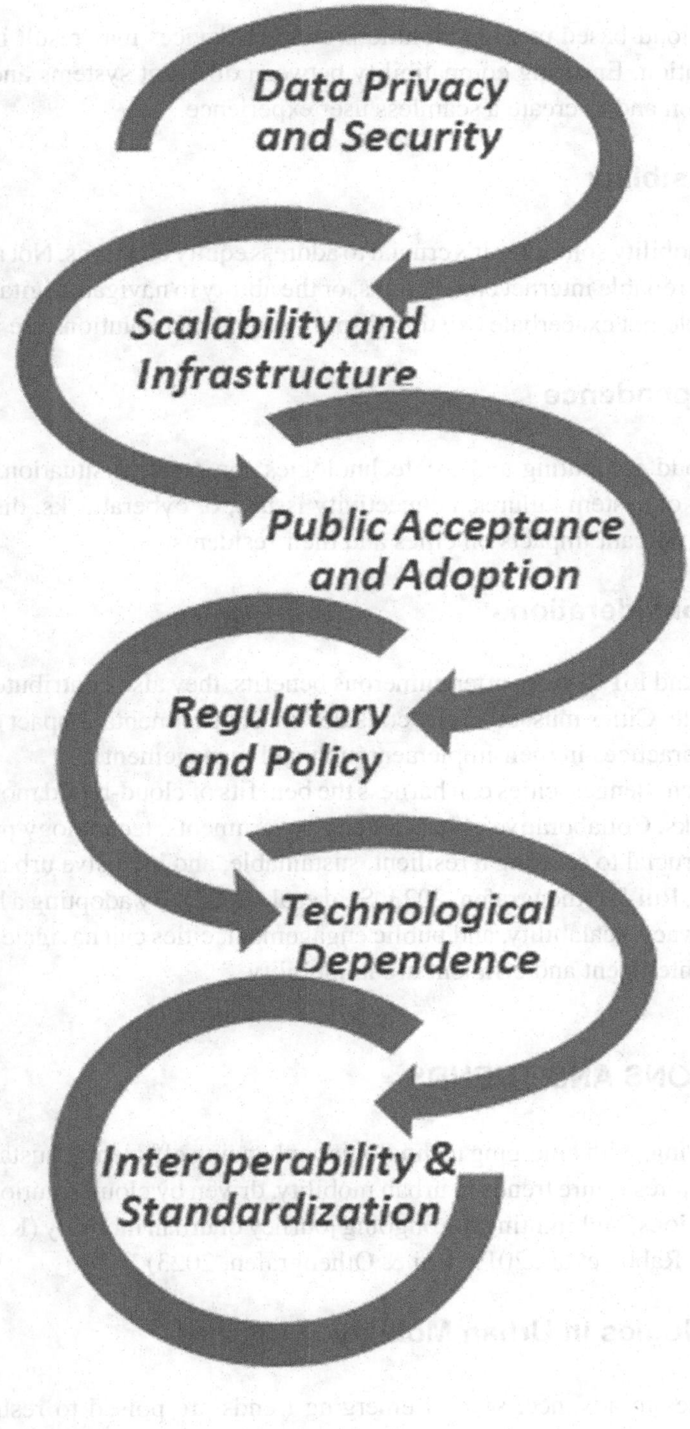

Data Privacy and Security

Scalability and Infrastructure

Public Acceptance and Adoption

Regulatory and Policy

Technological Dependence

Interoperability & Standardization

Interoperability and Standardization

The diverse range of cloud-based mobility solutions and IoT devices may result in a lack of interoperability and standardization. Ensuring compatibility between different systems and devices is essential to prevent fragmentation and to create a seamless user experience.

Equity and Accessibility

As cities adopt smart mobility solutions, it's crucial to address equity concerns. Not all residents may have access to smartphones, reliable internet connections, or the ability to navigate digital platforms. Ensuring that new technologies do not exacerbate existing disparities and that solutions are inclusive is essential.

Technological Dependence

Relying heavily on cloud computing and IoT technologies can create a situation of technological dependence. In the event of system failures, connectivity issues, or cyberattacks, disruptions to mobility systems could have significant impacts on cities and their residents.

Environmental Considerations

While cloud solutions and IoT devices offer numerous benefits, they also contribute to energy consumption and electronic waste. Cities must carefully consider the environmental impact of these technologies and adopt sustainable practices in their implementation and management.

In addressing these challenges, cities can harness the benefits of cloud-based mobility solutions while mitigating potential risks. Collaborative efforts among governments, technology providers, urban planners, and citizens are crucial to creating a resilient, sustainable, and inclusive urban mobility landscape (Mavlutova et al., 2023; Rui & Othengrafen, 2023; Said et al., 2022). By adopting a holistic approach that prioritizes security, privacy, scalability, and public engagement, cities can navigate these considerations and create a future of intelligent and efficient urban mobility.

FUTURE DIRECTIONS AND TRENDS

Urban mobility is evolving, with emerging technologies enhancing efficiency, sustainability, and quality of life. This section explores future trends in urban mobility, driven by cloud solutions, IoT technologies, and data-driven innovations, highlighting the ongoing journey of urban mobility (Kaginalkar et al., 2021; Mavlutova et al., 2023; Rabby et al., 2019; Rui & Othengrafen, 2023).

Emerging Technologies in Urban Mobility

As technology continues to advance, several emerging trends are poised to reshape urban mobility. These include:

- **5G Connectivity:** The rollout of 5G networks promises ultra-fast, low-latency communication, enabling real-time data exchange between vehicles, infrastructure, and cloud systems (Venkateswaran, Vidhya, Ayyannan, et al., 2023).
- **Autonomous Vehicles (AVs):** Self-driving vehicles have the potential to revolutionize urban transportation by reducing congestion, enhancing safety, and optimizing road capacity.
- **Electric and Sustainable Mobility:** The shift towards electric vehicles (EVs) and sustainable transportation modes will influence infrastructure planning and energy consumption.
- **AI and Machine Learning:** Advanced AI and machine learning algorithms will enable more accurate predictive analytics, adaptive traffic control, and personalized mobility services.
- **Drones and Urban Air Mobility:** Drones and air taxis could provide innovative solutions for urban transportation, particularly for short-distance commuting and deliveries.

Data-Driven Predictive Analytics for Mobility Planning

The integration of cloud computing, IoT, and data analytics will enable predictive modeling for mobility planning (Sampath et al., 2022; Vignesh et al., 2018; K. Karthik et al., 2023):

- **Traffic Flow Prediction:** Advanced analytics will forecast traffic congestion patterns, enabling proactive traffic management and route optimization.
- **Demand Forecasting:** Predictive models will anticipate transportation demand, aiding in efficient resource allocation and infrastructure planning.
- **Behavioral Analysis:** Data-driven insights into travel behaviors will inform the design of tailored mobility solutions and transportation policies.

Mobility as a Service (MaaS) and Shared Mobility

The concept of Mobility as a Service (MaaS) is gaining traction, transforming how people perceive and use transportation:

- **Integrated Mobility Platforms:** MaaS platforms will offer seamless integration of various transportation modes, allowing users to plan, book, and pay for multi-modal journeys through a single app.
- **Shared Mobility:** Car-sharing, bike-sharing, and ride-hailing services will continue to grow, reducing the need for private vehicle ownership and promoting efficient resource use.

Urban Planning and Design Influenced by Mobility Data

Mobility data will increasingly shape urban planning and design:

- **Infrastructure Optimization:** Data-driven insights will guide the development of transportation infrastructure, optimizing road layouts, traffic signal placements, and parking facilities.
- **Mixed-Use Development:** Urban planners will consider mobility data when designing mixed-use neighborhoods, encouraging walkability, and reducing the need for extensive commuting.

- **Smart City Integration:** Mobility data will be integrated with broader smart city initiatives, facilitating efficient energy use, waste management, and public services.

Urban mobility is being transformed by cloud solutions, IoT technologies, predictive analytics, shared mobility, and innovative urban planning. These technologies enhance mobility experiences, reduce congestion, and contribute to vibrant, livable environments, shaping our way of moving and interacting in urban spaces.

CONCLUSION

Urban mobility is evolving, with emerging technologies enhancing efficiency, sustainability, and quality of life. This section explores future trends in urban mobility, driven by cloud solutions, IoT technologies, and data-driven innovations, highlighting the ongoing journey of urban mobility.

- Urban mobility trends include 5G connectivity, autonomous vehicles, electric and sustainable modes, AI and machine learning, drones, and urban air mobility. 5G networks provide real-time data exchange, while autonomous vehicles reduce congestion and optimize road capacity. Electric and sustainable modes influence infrastructure planning and energy consumption. AI and machine learning algorithms enable predictive analytics, adaptive traffic control, and personalized services. Drones and air taxis offer innovative solutions for short-distance commuting and deliveries.
- Cloud computing, IoT, and data analytics will enable predictive modeling for mobility planning, enabling proactive traffic management, route optimization, efficient resource allocation, and tailored transportation policies through advanced analytics.
- MaaS is revolutionizing transportation by providing integrated platforms for multi-modal journeys, reducing private vehicle ownership and promoting efficient resource use through shared mobility services like car-sharing, bike-sharing, and ride-hailing, all managed through a single app.
- Mobility data is revolutionizing urban planning and design, influencing infrastructure optimization, mixed-use development, and smart city integration. It guides transportation layouts, traffic signal placements, and parking facilities, promoting walkability and reducing commuting. Smart city initiatives facilitate efficient energy use, waste management, and public services.

Urban mobility is set to undergo a transformation due to the integration of cloud solutions, IoT technologies, predictive analytics, shared mobility, and innovative urban planning, enhancing experiences, reducing congestion, and fostering vibrant, liveable urban environments.

REFERENCES

Agrawal, A. V., Shashibhushan, G., Pradeep, S., Padhi, S. N., Sugumar, D., & Boopathi, S. (2024). Synergizing Artificial Intelligence, 5G, and Cloud Computing for Efficient Energy Conversion Using Agricultural Waste. In Practice, Progress, and Proficiency in Sustainability (pp. 475–497). IGI Global. doi:10.4018/979-8-3693-1186-8.ch026

Al Amiri, W., Baza, M., Banawan, K., Mahmoud, M., Alasmary, W., & Akkaya, K. (2019). Privacy-preserving smart parking system using blockchain and private information retrieval. *2019 International Conference on Smart Applications, Communications and Networking (SmartNets)*, 1–6. 10.1109/SmartNets48225.2019.9069783

Al-Farhani, L. H., Alqahtani, Y., Alshehri, H. A., Martin, R. J., Lalar, S., & Jain, R. (2023). IOT and Blockchain-Based Cloud Model for Secure Data Transmission for Smart City. *Security and Communication Networks*, *2023*, 2023. doi:10.1155/2023/3171334

Al-Turjman, F., & Malekloo, A. (2019). Smart parking in IoT-enabled cities: A survey. *Sustainable Cities and Society*, *49*, 101608. doi:10.1016/j.scs.2019.101608

Arifin, M. S., Razi, S. A., Haque, A., & Mohammad, N. (2019). A microcontroller based intelligent traffic control system. *American Journal of Embedded Systems and Applications*, *7*(1), 21–25. doi:10.11648/j.ajesa.20190701.13

B, M. K., K, K. K., Sasikala, P., Sampath, B., Gopi, B., & Sundaram, S. (2024). Sustainable Green Energy Generation From Waste Water. In *Practice, Progress, and Proficiency in Sustainability* (pp. 440–463). IGI Global. doi:10.4018/979-8-3693-1186-8.ch024

Barron, L. (2022). Smart cities, connected cars and autonomous vehicles: Design fiction and visions of smarter future urban mobility. *Technoetic Arts : a Journal of Speculative Research*, *20*(3), 225–240. doi:10.1386/tear_00092_1

Bock, F., Di Martino, S., & Origlia, A. (2019). Smart parking: Using a crowd of taxis to sense on-street parking space availability. *IEEE Transactions on Intelligent Transportation Systems*, *21*(2), 496–508. doi:10.1109/TITS.2019.2899149

Boopathi, S. (2022). An experimental investigation of Quench Polish Quench (QPQ) coating on AISI 4150 steel. *Engineering Research Express*, *4*(4), 045009. doi:10.1088/2631-8695/ac9ddd

Boopathi, S. (2023a). Deep Learning Techniques Applied for Automatic Sentence Generation. In Promoting Diversity, Equity, and Inclusion in Language Learning Environments (pp. 255–273). IGI Global. doi:10.4018/978-1-6684-3632-5.ch016

Boopathi, S. (2023b). Securing Healthcare Systems Integrated With IoT: Fundamentals, Applications, and Future Trends. In Dynamics of Swarm Intelligence Health Analysis for the Next Generation (pp. 186–209). IGI Global.

Boopathi, S., & Kanike, U. K. (2023). Applications of Artificial Intelligent and Machine Learning Techniques in Image Processing. In *Handbook of Research on Thrust Technologies' Effect on Image Processing* (pp. 151–173). IGI Global. doi:10.4018/978-1-6684-8618-4.ch010

Boopathi, S., Lewise, K. A. S., Subbiah, R., & Sivaraman, G. (2022). Near-dry wire-cut electrical discharge machining process using water–air-mist dielectric fluid: An experimental study. *Materials Today: Proceedings*, *50*(5), 1885–1890. doi:10.1016/j.matpr.2021.08.077

Chandana, M., Aniruddh, M., Fathima, A., Chandan, G., & Kumari, H. C. (2019). A Study on Smart Parking Solutions Using IoT. *2019 International Conference on Intelligent Sustainable Systems (ICISS)*, 542–548. 10.1109/ISS1.2019.8907944

Chmiel, W., Dańda, J., Dziech, A., Ernst, S., Kadłuczka, P., Mikrut, Z., Pawlik, P., Szwed, P., & Wojnicki, I. (2016). INSIGMA: An intelligent transportation system for urban mobility enhancement. *Multimedia Tools and Applications*, *75*(17), 10529–10560. doi:10.100711042-016-3367-5

De Oliveira, L. F. P., Manera, L. T., & Da Luz, P. D. G. (2020). Development of a smart traffic light control system with real-time monitoring. *IEEE Internet of Things Journal*, *8*(5), 3384–3393. doi:10.1109/JIOT.2020.3022392

Delgado, A., & Calegari, D. (2023). *Process Mining for Improving Urban Mobility in Smart Cities: Challenges and Application with Open Data*. Academic Press.

Diran, D., Van Veenstra, A. F., Timan, T., Tesa, P., & Kirova, M. (2021). Artificial Intelligence in smart cities and urban mobility. *Policy Department for Economic, Scientific and Quality of Life Policies*.

Dutta, A., Chakraborty, A., Kumar, A., Roy, A., Roy, S., Chakraborty, D., Saha, H. N., Sharma, A. S., Mullick, M., Santra, S., & ... (2019). Intelligent traffic control system: Towards smart city. *2019 IEEE 10th Annual Information Technology, Electronics and Mobile Communication Conference (IEMCON)*, 1124–1129.

Francis, S., Ouseph, A., Durgaprasad, S., Abdulla, H., Lal, A., Likhith, S., Dhanaraj, K., & Harikrishna, M. (2023). Machine Learning based Vacant Space Detection for Smart Parking Solutions. *2023 International Conference on Control, Communication and Computing (ICCC)*, 1–6. 10.1109/ICCC57789.2023.10165557

Hema, N., Krishnamoorthy, N., Chavan, S. M., Kumar, N., Sabarimuthu, M., & Boopathi, S. (2023). A Study on an Internet of Things (IoT)-Enabled Smart Solar Grid System. In *Handbook of Research on Deep Learning Techniques for Cloud-Based Industrial IoT* (pp. 290–308). IGI Global. doi:10.4018/978-1-6684-8098-4.ch017

Hilmani, A., Maizate, A., & Hassouni, L. (2020). Automated real-time intelligent traffic control system for smart cities using wireless sensor networks. *Wireless Communications and Mobile Computing*, *2020*, 1–28. doi:10.1155/2020/8841893

Hundera, N. W., Jin, C., Geressu, D. M., Aftab, M. U., Olanrewaju, O. A., & Xiong, H. (2022). Proxy-based public-key cryptosystem for secure and efficient IoT-based cloud data sharing in the smart city. *Multimedia Tools and Applications*, *81*(21), 29673–29697. doi:10.100711042-021-11685-3

Jeevanantham, Y. A., Saravanan, A., Vanitha, V., Boopathi, S., & Kumar, D. P. (2022). Implementation of Internet-of Things (IoT) in Soil Irrigation System. *IEEE Explore*, 1–5.

Kaginalkar, A., Kumar, S., Gargava, P., & Niyogi, D. (2021). Review of urban computing in air quality management as smart city service: An integrated IoT, AI, and cloud technology perspective. *Urban Climate*, *39*, 100972. doi:10.1016/j.uclim.2021.100972

Karthik, K., Teferi, A. B., Sathish, R., Gandhi, A. M., Padhi, S., Boopathi, S., & Sasikala, G. (2023). Analysis of delamination and its effect on polymer matrix composites. *Materials Today: Proceedings*. Advance online publication. doi:10.1016/j.matpr.2023.07.199

Karthik, S., Hemalatha, R., Aruna, R., Deivakani, M., Reddy, R. V. K., & Boopathi, S. (2023). Study on Healthcare Security System-Integrated Internet of Things (IoT). In Perspectives and Considerations on the Evolution of Smart Systems (pp. 342–362). IGI Global.

Khan, N. A., Jhanjhi, N., Brohi, S. N., Usmani, R. S. A., & Nayyar, A. (2020). Smart traffic monitoring system using unmanned aerial vehicles (UAVs). *Computer Communications*, *157*, 434–443. doi:10.1016/j.comcom.2020.04.049

Koshariya, A. K., Kalaiyarasi, D., Jovith, A. A., Sivakami, T., Hasan, D. S., & Boopathi, S. (2023). AI-Enabled IoT and WSN-Integrated Smart Agriculture System. In *Artificial Intelligence Tools and Technologies for Smart Farming and Agriculture Practices* (pp. 200–218). IGI Global. doi:10.4018/978-1-6684-8516-3.ch011

Leal Sobral, V. A., Nelson, J., Asmare, L., Mahmood, A., Mitchell, G., Tenkorang, K., Todd, C., Campbell, B., & Goodall, J. L. (2023). A Cloud-Based Data Storage and Visualization Tool for Smart City IoT: Flood Warning as an Example Application. *Smart Cities*, *6*(3), 1416–1434. doi:10.3390martcities6030068

Lee, W.-H., & Chiu, C.-Y. (2020). Design and implementation of a smart traffic signal control system for smart city applications. *Sensors (Basel)*, *20*(2), 508. doi:10.339020020508 PMID:31963229

Liu, R. W., Liang, M., Nie, J., Lim, W. Y. B., Zhang, Y., & Guizani, M. (2022). Deep learning-powered vessel trajectory prediction for improving smart traffic services in maritime Internet of Things. *IEEE Transactions on Network Science and Engineering*, *9*(5), 3080–3094. doi:10.1109/TNSE.2022.3140529

Maguluri, L. P., Ananth, J., Hariram, S., Geetha, C., Bhaskar, A., & Boopathi, S. (2023). Smart Vehicle-Emissions Monitoring System Using Internet of Things (IoT). In Handbook of Research on Safe Disposal Methods of Municipal Solid Wastes for a Sustainable Environment (pp. 191–211). IGI Global.

Mahajan, P., Mahajan, P., Mahajan, S., Mahajan, S., Mahajan, V., Mahale, H., & Dandavate, P. P. (n.d.). Smart parking solutions: Maximizing efficiency with sensor-based system. *Enhancing Productivity in Hybrid Mode: The Beginning of a New Era*, 520.

Maheswari, B. U., Imambi, S. S., Hasan, D., Meenakshi, S., Pratheep, V., & Boopathi, S. (2023). Internet of Things and Machine Learning-Integrated Smart Robotics. In Global Perspectives on Robotics and Autonomous Systems: Development and Applications (pp. 240–258). IGI Global. doi:10.4018/978-1-6684-7791-5.ch010

Mavlutova, I., Atstaja, D., Grasis, J., Kuzmina, J., Uvarova, I., & Roga, D. (2023). Urban Transportation Concept and Sustainable Urban Mobility in Smart Cities: A Review. *Energies*, *16*(8), 3585. doi:10.3390/en16083585

Medagliani, P., Leguay, J., Duda, A., Rousseau, F., Duquennoy, S., Raza, S., Ferrari, G., Gonizzi, P., Cirani, S., Veltri, L., & ... (2022). Bringing ip to low-power smart objects: The smart parking case in the CALIPSO project. In *Internet of things applications-from research and innovation to market deployment* (pp. 287–313). River Publishers. doi:10.1201/9781003338628-8

Morfoulaki, M., & Papathanasiou, J. (2021). Use of the sustainable mobility efficiency index (SMEI) for enhancing the sustainable urban mobility in Greek cities. *Sustainability (Basel)*, *13*(4), 1709. doi:10.3390u13041709

Nallaperuma, D., Nawaratne, R., Bandaragoda, T., Adikari, A., Nguyen, S., Kempitiya, T., De Silva, D., Alahakoon, D., & Pothuhera, D. (2019). Online incremental machine learning platform for big data-driven smart traffic management. *IEEE Transactions on Intelligent Transportation Systems*, *20*(12), 4679–4690. doi:10.1109/TITS.2019.2924883

Nogueira, P. R. R., de Paula, S. L., de Lima Santana, S. B., da Silva Pinto, J., Braz, M. I., & de Aquino, L. M. P. (n.d.). *Cidades inteligentes e mobilidade urbana: Atores e práticas na cidade de Recife/PE Smart cities and urban mobility: Actors and practices in the city of Recife/PE*. Academic Press.

Qolomany, B., Mohammed, I., Al-Fuqaha, A., Guizani, M., & Qadir, J. (2020). Trust-based cloud machine learning model selection for industrial IoT and smart city services. *IEEE Internet of Things Journal*, *8*(4), 2943–2958. doi:10.1109/JIOT.2020.3022323

Rabby, M. K. M., Islam, M. M., & Imon, S. M. (2019). A review of IoT application in a smart traffic management system. *2019 5th International Conference on Advances in Electrical Engineering (ICAEE)*, 280–285.

Ragavan, K., Venkatalakshmi, K., & Vijayalakshmi, K. (2021). Traffic video-based intelligent traffic control system for smart cities using modified ant colony optimizer. *Computational Intelligence*, *37*(1), 538–558. doi:10.1111/coin.12424

Rahamathunnisa, U., Sudhakar, K., Murugan, T. K., Thivaharan, S., Rajkumar, M., & Boopathi, S. (2023). Cloud Computing Principles for Optimizing Robot Task Offloading Processes. In *AI-Enabled Social Robotics in Human Care Services* (pp. 188–211). IGI Global. doi:10.4018/978-1-6684-8171-4.ch007

Rajapaksha, P., Farahbakhsh, R., Nathanail, E., & Crespi, N. (2017). iTrip, a framework to enhance urban mobility by leveraging various data sources. *Transportation Research Procedia*, *24*, 113–122. doi:10.1016/j.trpro.2017.05.076

Reddy, M. A., Reddy, B. M., Mukund, C., Venneti, K., Preethi, D., & Boopathi, S. (2023). Social Health Protection During the COVID-Pandemic Using IoT. In *The COVID-19 Pandemic and the Digitalization of Diplomacy* (pp. 204–235). IGI Global. doi:10.4018/978-1-7998-8394-4.ch009

Richter, M. A., Hagenmaier, M., Bandte, O., Parida, V., & Wincent, J. (2022). Smart cities, urban mobility and autonomous vehicles: How different cities needs different sustainable investment strategies. *Technological Forecasting and Social Change*, *184*, 121857. doi:10.1016/j.techfore.2022.121857

Richter, M. A., Hagenmaier, M., Bandte, O., Parida, V., & Wincent, J. (2022). Smart cities, urban mobility and autonomous vehicles. *Technological Forecasting and Social Change*, *184*, 121857. doi:10.1016/j.techfore.2022.121857

Rodic, L. D. (n.d.). *Smart parking solutions for occupancy sensing*. Academic Press.

Rui, J., & Othengrafen, F. (2023). Examining the Role of Innovative Streets in Enhancing Urban Mobility and Livability for Sustainable Urban Transition: A Review. *Sustainability (Basel)*, *15*(7), 5709. doi:10.3390u15075709

Said, B., Lathamaheswari, M., Singh, P. K., Ouallane, A. A., Bakhouyi, A., Bakali, A., Talea, M., Dhital, A., & Deivanayagampillai, N. (2022). An Intelligent Traffic Control System Using Neutrosophic Sets, Rough sets, Graph Theory, Fuzzy sets and its Extended Approach: A Literature Review. *Neutrosophic Sets Syst*, *50*, 10–26.

Samikannu, R., Koshariya, A. K., Poornima, E., Ramesh, S., Kumar, A., & Boopathi, S. (2022). Sustainable Development in Modern Aquaponics Cultivation Systems Using IoT Technologies. In *Human Agro-Energy Optimization for Business and Industry* (pp. 105–127). IGI Global.

Sampath, B., Yuvaraj, T., & Velmurugan, D. (2022). Parametric analysis of mould sand properties for flange coupling casting. *AIP Conference Proceedings*, *2460*(1), 070002. doi:10.1063/5.0095599

Sanghvi, K., Shah, A., Shah, P., Shah, P., & Rathod, S. S. (2021). Smart Parking Solutions for On-Street and Off-Street Parking. *2021 International Conference on Communication Information and Computing Technology (ICCICT)*, 1–6. 10.1109/ICCICT50803.2021.9510099

Saravanan, A., Venkatasubramanian, R., Khare, R., Surakasi, R., Boopathi, S., Ray, S., & Sudhakar, M. (2022). Policy trends of renewable energy and non-*renewable energy*. Academic Press.

Saravanan, M., Vasanth, M., Boopathi, S., Sureshkumar, M., & Haribalaji, V. (2022). Optimization of Quench Polish Quench (QPQ) Coating Process Using Taguchi Method. *Key Engineering Materials*, *935*, 83–91. doi:10.4028/p-z569vy

Satav, S. D., Hasan, D. S., Pitchai, R., Mohanaprakash, T. A., Sultanuddin, S. J., & Boopathi, S. (2024). Next Generation of Internet of Things (NGIoT) in Healthcare Systems. In Practice, Progress, and Proficiency in Sustainability (pp. 307–330). IGI Global. doi:10.4018/979-8-3693-1186-8.ch017

Satav, S. D., Lamani, D. G, H. K., Kumar, N. M. G., Manikandan, S., & Sampath, B. (2024). Energy and Battery Management in the Era of Cloud Computing. In Practice, Progress, and Proficiency in Sustainability (pp. 141–166). IGI Global. doi:10.4018/979-8-3693-1186-8.ch009

Shengdong, M., Zhengxian, X., & Yixiang, T. (2019). Intelligent traffic control system based on cloud computing and big data mining. *IEEE Transactions on Industrial Informatics*, *15*(12), 6583–6592. doi:10.1109/TII.2019.2929060

Stiglic, M., Agatz, N., Savelsbergh, M., & Gradisar, M. (2018). Enhancing urban mobility: Integrating ride-sharing and public transit. *Computers & Operations Research*, *90*, 12–21. doi:10.1016/j.cor.2017.08.016

Syamala, M., Komala, C., Pramila, P., Dash, S., Meenakshi, S., & Boopathi, S. (2023). Machine Learning-Integrated IoT-Based Smart Home Energy Management System. In *Handbook of Research on Deep Learning Techniques for Cloud-Based Industrial IoT* (pp. 219–235). IGI Global. doi:10.4018/978-1-6684-8098-4.ch013

Veeranjaneyulu, R., Boopathi, S., Kumari, R. K., Vidyarthi, A., Isaac, J. S., & Jaiganesh, V. (2023). Air Quality Improvement and Optimisation Using Machine Learning Technique. *IEEE- Explore*, 1–6.

Veeranjaneyulu, R., Boopathi, S., Narasimharao, J., Gupta, K. K., Reddy, R. V. K., & Ambika, R. (2023). Identification of Heart Diseases using Novel Machine Learning Method. *IEEE- Explore*, 1–6.

Venkateswaran, N., Vidhya, K., Ayyannan, M., Chavan, S. M., Sekar, K., & Boopathi, S. (2023). A Study on Smart Energy Management Framework Using Cloud Computing. In 5G, Artificial Intelligence, and Next Generation Internet of Things: Digital Innovation for Green and Sustainable Economies (pp. 189–212). IGI Global. doi:10.4018/978-1-6684-8634-4.ch009

Venkateswaran, N., Vidhya, R., Naik, D. A., Raj, T. M., Munjal, N., & Boopathi, S. (2023). Study on Sentence and Question Formation Using Deep Learning Techniques. In *Digital Natives as a Disruptive Force in Asian Businesses and Societies* (pp. 252–273). IGI Global. doi:10.4018/978-1-6684-6782-4.ch015

Vignesh, S., Arulshri, K., SyedSajith, S., Kathiresan, S., Boopathi, S., & Dinesh Babu, P. (2018). Design and development of ornithopter and experimental analysis of flapping rate under various operating conditions. *Materials Today: Proceedings*, *5*(11), 25185–25194. doi:10.1016/j.matpr.2018.10.320

Yogheshwaran, M., Praveenkumar, D., Pravin, S., Manikandan, P., & Saravanan, D. S. (2020). IoT based intelligent traffic control system. *International Journal of Engineering Technology Research & Management*, *4*(4), 59–63.

LIST OF ACRONYMS

AI: Artificial Intelligence.
Avs: Autonomous Vehicles.
EVs: Electric Vehicles.
IoT: Internet of Things.
MaaS: Mobility as a Service.

Chapter 9
Building Sustainable Smart Cities Through Cloud and Intelligent Parking System

Monika Sharma
Department of Computer Science and Engineering, The Technological Institute of Textile and Sciences, Bhiwani, India

Manju Sharma
Department of Computer Science and Engineering, University Institute of Engineering and Technology, India & Maharshi Dayanand University, India

Neerav Sharma
Department of Computer Science and Engineering, BITS College, Bhiwani, India

Sampath Boopathi
https://orcid.org/0000-0002-2065-6539
Mechanical Engineering, Muthayammal Engineering College, India

ABSTRACT

This chapter discusses the role of cloud computing and intelligent parking systems in sustainable smart cities, addressing challenges like traffic congestion, pollution, and resource inefficiency. These technologies enhance urban mobility, reduce environmental impact, and improve quality of life in cities facing rapid urbanization worldwide. This chapter offers a thorough analysis of the integration of cloud computing and intelligent parking systems in sustainable urban development, highlighting successful implementations and lessons learned. It also explores potential future developments and policy considerations to facilitate widespread adoption of these technologies, highlighting the importance of global best practices.

DOI: 10.4018/978-1-6684-9999-3.ch009

INTRODUCTION

Urbanization is an undeniable global trend, with more people moving into cities than ever before. While this urban growth brings economic opportunities and cultural vibrancy, it also presents significant challenges. Traffic congestion, pollution, and resource inefficiency are pressing issues in many cities, threatening both the environment and the quality of life for their inhabitants. In response to these challenges, the integration of cloud computing and intelligent parking systems has emerged as a promising solution to foster sustainable development in smart cities. Rapid urbanization presents challenges and opportunities, affecting infrastructure, transportation systems, and resources. Smart cities, utilizing technology and data, aim to transform urban living into an efficient, convenient, and environmentally responsible experience (Batool et al., 2021). The integration of cloud computing and intelligent parking systems is crucial for sustainable smart cities, as it not only addresses urban congestion but also reduces environmental footprints and enhances residents' quality of life by reducing urban environmental impact (Kumar et al., 2017).

Smart cities are envisioned as future urban spaces optimized for efficiency and sustainability, leveraging the Internet of Things (IoT) to connect everyday objects and infrastructure to the internet, enabling data collection and exchange. IoT serves as the foundation for intelligent parking systems and cloud computing in smart cities (Mishra & Chakraborty, 2020). The integration of cloud computing and intelligent parking systems in cities can address pressing urban challenges like traffic congestion, emissions reduction, mobility enhancement, and resident well-being, resulting in vibrant, efficient, and environmentally friendly cities (Perera et al., 2017).

At the heart of this transformation is the concept of intelligent parking systems. Traditional parking management has often been characterized by frustration, inefficiency, and wasted time as drivers circle the streets in search of elusive parking spaces. Intelligent parking systems, powered by sensors, real-time data, and cloud-based platforms, offer a solution to this age-old problem. They provide drivers with real-time information about available parking spaces, enable mobile app-based payments and reservations, and optimize parking space allocation (Kaur et al., 2018). The result is a streamlined parking experience that reduces congestion and contributes to cleaner air in urban areas. Cloud computing plays a pivotal role in the success of intelligent parking systems. By leveraging the power of the cloud, cities can collect, process, and store vast amounts of parking data efficiently. Cloud-based platforms facilitate real-time data sharing between parking facilities, city authorities, and drivers, ensuring that parking information is always up-to-date and accessible. Moreover, the scalability and flexibility of cloud infrastructure allow cities to expand their intelligent parking systems as their populations grow and mobility needs evolve (Ballon et al., 2011).

As cities continue to grapple with the challenges of urbanization, smart parking systems serve as a cornerstone of sustainable urban development. They promote the use of public transportation, reduce the number of cars on the road, and decrease emissions. They also improve traffic flow by minimizing the time spent searching for parking and reducing the stress associated with congested city driving. Additionally, smart parking systems contribute to a city's overall sustainability by optimizing the utilization of parking spaces, reducing energy consumption in parking facilities, and enhancing the urban landscape (Alam et al., 2022). This journey towards building sustainable smart cities through the integration of cloud computing and intelligent parking systems is not without its challenges and complexities. It requires collaboration among government agencies, private sector partners, and technology providers. It demands careful consideration of security and privacy concerns, the development of robust data management and governance strategies, and the establishment of regulatory frameworks that support innovation while safeguarding citizens' rights (Agrawal et al., 2024; B et al., 2024).

Traffic congestion in urban areas is a persistent issue, causing wasted time, resources, increased air pollution, and decreased productivity. Cloud computing-powered intelligent parking systems can alleviate this congestion by efficiently guiding drivers to available parking spaces, reducing fuel consumption and emissions, making cities cleaner and more sustainable. These systems also enhance urban mobility by providing real-time information about parking availability, encouraging the use of public transportation and non-motorized modes of travel, and creating more accessible and walkable cities (Syamala et al., 2023; Venkateswaran et al., 2023).

As sustainability becomes a critical concern, the integration of cloud computing and intelligent parking systems offers a powerful tool to reduce the environmental impact of urban living. By optimizing parking, these systems decrease vehicle emissions, a significant source of air pollution and greenhouse gases. Additionally, efficient use of parking spaces can free up valuable land for green spaces, housing, or other sustainable urban developments (Satav, Hasan, et al., 2024; Satav, Lamani, et al., 2024).

However, challenges remain in technology integration, data security, and equitable access. Future developments should focus on enhancing interoperability, ensuring data privacy, and addressing equity concerns to avoid exacerbating social disparities. Policymakers must play a crucial role in facilitating the widespread adoption of these technologies by creating supportive regulatory frameworks, incentivizing implementation, and promoting global best practices. This chapter explores the role of cloud computing and intelligent parking systems in promoting sustainable urban development, examining successful case studies, lessons learned from real-world implementations, and potential future developments (Hema et al., 2023; Karthik et al., 2023; Koshariya et al., 2023).

In the following sections, we will delve deeper into the various aspects of this transformative journey, exploring the benefits, challenges, case studies, and future prospects of building sustainable smart cities through cloud and intelligent parking systems. By the end, it will become clear that the integration of these technologies is not just a technological achievement but a catalyst for creating urban environments that are more efficient, accessible, and environmentally responsible than ever before.

Background and Motivation

The 21st century is witnessing an unprecedented wave of urbanization, with more than half of the global population now residing in cities. This rapid urban growth brings with it a host of challenges, from increased traffic congestion to heightened environmental concerns. As cities continue to expand and evolve, the concept of "smart cities" has emerged as a beacon of hope. Smart cities leverage cutting-edge technology to enhance urban living, improve resource utilization, and promote sustainability (Kaur et al., 2018; Kumar et al., 2017; Perera et al., 2017). One of the key drivers of smart city transformation is the convergence of cloud computing and intelligent parking systems. Cloud computing has revolutionized the way data is stored, processed, and shared, enabling cities to harness the power of data-driven decision-making. Simultaneously, intelligent parking systems have evolved to tackle the vexing problem of parking congestion and inefficiency in urban areas. When integrated, these technologies offer a powerful toolkit for building sustainable smart cities.

Objectives

The primary objective of this chapter is to explore the vital role that cloud computing and intelligent parking systems play in the development of sustainable smart cities. We will delve into the concepts, technologies, and benefits associated with their integration, examining real-world case studies and best practices from various regions. Additionally, this chapter will discuss the potential future developments in this field and the policy considerations that can facilitate the widespread adoption of these technologies.

Scope and Organization

This chapter explores the concept of smart cities, cloud computing, intelligent parking systems, and their synergy for sustainable urban mobility. It highlights the benefits of integrating these technologies, such as reducing traffic congestion, mitigating environmental impact, and improving quality of life. Challenges include security, data management, infrastructure, and policy. Future trends in cloud-integrated parking systems are also discussed. The chapter concludes by providing a comprehensive overview of building sustainable smart cities.

SMART CITIES AND SUSTAINABILITY

Smart cities are urban areas that leverage technology and data-driven solutions to enhance the quality of life for their residents while minimizing their environmental impact. These cities aim to create a more sustainable and efficient urban environment by integrating various aspects of technology and innovation. Here's a breakdown of the key concepts related to smart cities and sustainability:

- **Technology Integration:** Smart cities use digital technology, such as IoT (Internet of Things) devices, sensors, and data analytics, to collect and process data from various sources, including infrastructure, transportation, and public services.
- **Improved Quality of Life:** The primary goal of smart cities is to improve the quality of life for their residents by addressing issues like traffic congestion, pollution, and inadequate public services.
- **Data-Driven Decision Making:** Data collected from various sensors and devices are used to make informed decisions about city planning, resource allocation, and service delivery.
- **Sustainability:** Sustainability is a core aspect of smart city initiatives, aiming to reduce the environmental footprint of urban areas by promoting eco-friendly practices and resource-efficient technologies.

The Need for Sustainable Urban Development

- **Urbanization Challenges:** Rapid urbanization has led to challenges like increased energy consumption, pollution, traffic congestion, and inadequate infrastructure.
- **Resource Scarcity:** Cities consume a significant portion of the world's resources, and unsustainable practices can deplete natural resources and harm ecosystems.

- **Climate Change:** Urban areas are significant contributors to greenhouse gas emissions, making sustainable urban development crucial in mitigating climate change.
- **Quality of Life:** Sustainable urban development enhances the well-being of city dwellers by providing clean air, efficient transportation, green spaces, and access to essential services.

The Role of Technology in Sustainability

- **Smart Infrastructure:** Technology helps create efficient infrastructure systems, such as smart grids, water management, waste management, and sustainable building designs.
- **Transportation:** Smart cities promote sustainable transportation solutions, including electric vehicles, public transit improvements, and the integration of transportation systems with digital platforms.
- **Energy Efficiency:** Technology is used to monitor and optimize energy consumption in buildings, street lighting, and other city facilities.
- **Data-Driven Sustainability:** Data analytics and AI are employed to monitor and manage resources, reducing waste and energy consumption.
- **Resilience:** Technology enables cities to better prepare for and respond to natural disasters and other emergencies, ensuring the continuity of essential services.

Smart cities utilize technology to optimize resource use, reduce environmental impact, and improve quality of life, becoming increasingly crucial as urban populations grow.

CLOUD COMPUTING IN SMART CITIES

Cloud computing is crucial for smart city development and management, offering scalable, flexible, and cost-effective solutions that enable urban planners to improve services, enhance sustainability, and make data-driven decisions. Its concepts, benefits, and real-world case studies are discussed (De Guimarães et al., 2020).

Cloud Computing: Concepts and Benefits:

Cloud computing involves delivering computing services (such as storage, processing, and software) over the internet on a pay-as-you-go basis. It eliminates the need for physical infrastructure and provides on-demand resources that can be accessed remotely (Dogo et al., 2019). Benefits of Cloud Computing in Smart Cities (Dogo et al., 2019)

- **Scalability:** Cloud platforms can scale resources up or down based on demand, allowing smart cities to adapt to changing needs without significant infrastructure investments.
- **Cost-Efficiency:** Cloud services reduce capital expenses associated with data centers and hardware maintenance, shifting to an operational expenditure model.
- **Flexibility:** Cloud platforms offer a wide range of services, enabling smart cities to choose and configure solutions that fit their specific requirements.

- **Data Accessibility:** Cloud-based solutions make data accessible from anywhere, facilitating collaboration among city departments and agencies.
- **Security and Compliance:** Leading cloud providers invest heavily in security, ensuring that sensitive city data is protected and compliant with regulations.
- **Scalable Analytics:** Cloud computing enables advanced data analytics and AI, allowing cities to gain valuable insights from vast amounts of data collected from sensors and IoT devices.

Cloud-Based Solutions for Urban Management

The cloud-based solutions for urban management (Jegadeesan et al., 2019) are illustrated in Figure 1.

- **Data Storage and Management:** Cloud platforms provide secure and scalable data storage solutions, allowing smart cities to efficiently store and manage vast amounts of data generated by sensors, cameras, and other devices (Syamala et al., 2023; Venkateswaran et al., 2023).
- **IoT and Sensor Integration:** Cloud computing enables the integration of IoT devices and sensors into a unified platform, facilitating real-time monitoring of urban infrastructure, traffic, air quality, and more (Hema et al., 2023; Rahamathunnisa et al., 2023).
- **Predictive Analytics:** Smart cities use cloud-based analytics to predict and respond to various urban challenges, such as traffic congestion, energy consumption, and emergency response.
- **Citizen Engagement:** Cloud-based applications and mobile apps engage citizens by providing access to real-time information, services, and communication channels with city authorities (Ramudu et al., 2023).
- **Resource Optimization:** Cloud solutions help cities optimize resource allocation in areas like energy, water, waste management, and transportation, reducing costs and environmental impact.

Case Studies: Cloud Computing in Smart City Initiatives (Jegadeesan et al., 2019)

- **Barcelona, Spain:** Barcelona's Smart City project utilizes cloud computing to integrate data from various sources, improving traffic management, reducing energy consumption, and enhancing public services. The city employs cloud-based platforms to collect and analyze data from IoT sensors and implement responsive urban policies.
- **Singapore:** Singapore leverages cloud computing to enhance urban sustainability. The city-state uses cloud-based platforms for monitoring air quality, optimizing public transportation, and managing energy consumption, leading to reduced emissions and improved quality of life.
- **Dubai, UAE:** Dubai's Smart Dubai initiative relies on cloud solutions to provide seamless services to residents and visitors. The cloud enables the integration of government services, smart transportation, and data analytics, fostering economic growth and sustainability.
- **Songdo, South Korea:** Songdo International Business District, a smart city built from the ground up, utilizes cloud computing for various services, including waste management, traffic control, and energy-efficient buildings. The cloud-based infrastructure allows for real-time data analysis and decision-making.

Figure 1. Cloud-based solutions for urban management

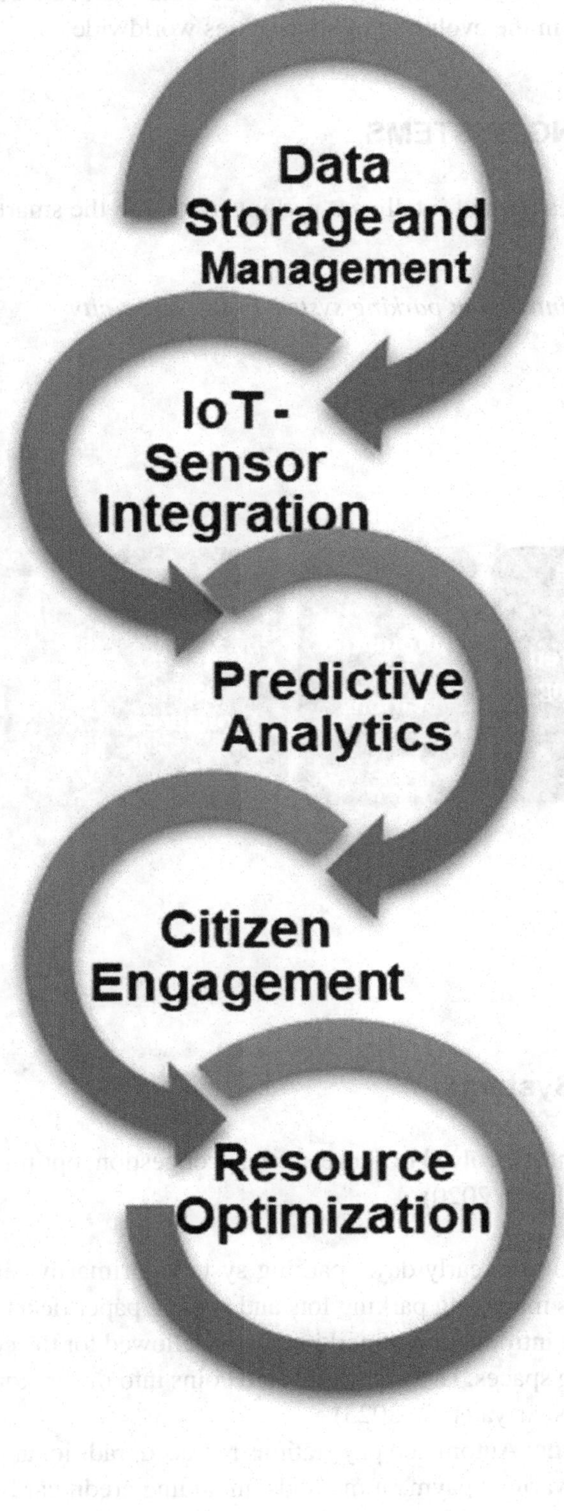

Case studies show that cloud computing is transforming cities into smarter, sustainable environments by enabling data-driven decision-making, efficient resource management, and enhanced citizen engagement, playing a pivotal role in the evolution of smart cities worldwide.

INTELLIGENT PARKING SYSTEMS

Figure 2 depicts the various stages of intelligent parking system in the smart city and explain below.

Figure 2. Various stages of intelligent parking system in the smart city

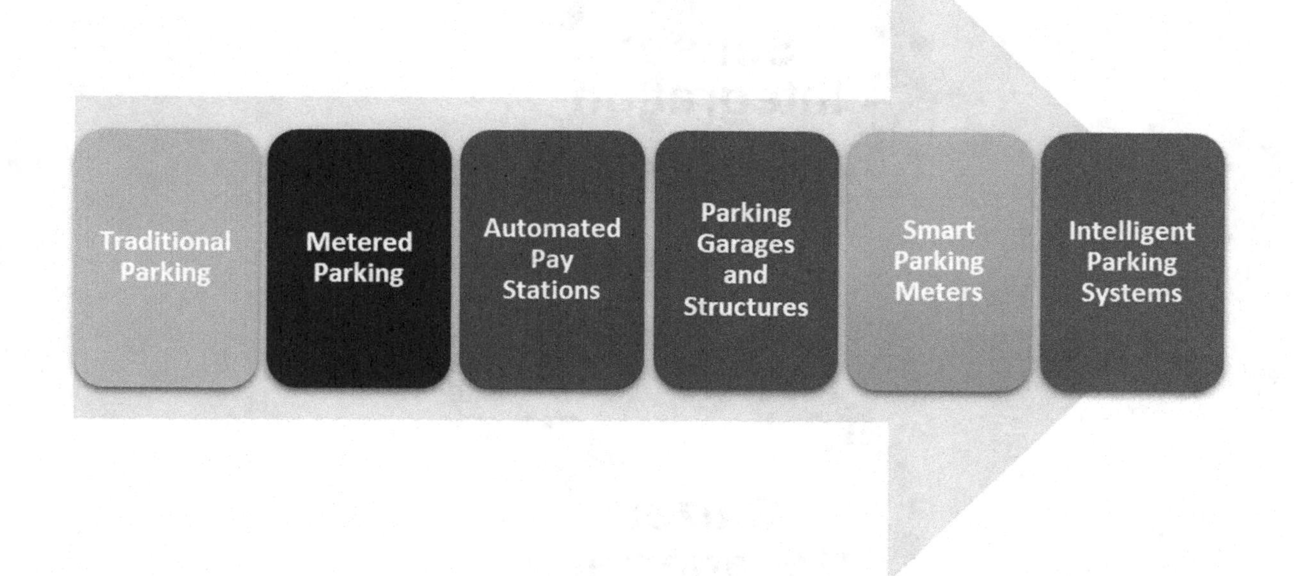

Evolution of Parking Systems

Over time, parking systems have evolved to address urban congestion, optimize spaces, and improve the overall parking experience (Jiang, 2020).

- **Traditional Parking:** In the early days, parking systems primarily relied on manual processes, with parking attendants managing parking lots and issuing paper tickets.
- **Metered Parking:** The introduction of parking meters allowed for the collection of fees and better management of parking spaces. Drivers would feed coins into meters for a specific amount of time (Hema et al., 2023; Koshariya et al., 2023).
- **Automated Pay Stations:** Automated pay stations replaced traditional meters, offering more convenience by accepting various payment methods, including credit cards and mobile payments.

- **Parking Garages and Structures:** Multi-level parking garages and structures became more common in densely populated urban areas, maximizing the use of limited space.
- **Smart Parking Meters:** Smart parking meters integrated technology to provide real-time data on parking space availability and allowed drivers to make payments via mobile apps.
- **Intelligent Parking Systems:** Modern intelligent parking systems leverage advanced technologies such as sensors, IoT, data analytics, and automation to create a more efficient and user-friendly parking experience.

Components of Intelligent Parking Systems

Intelligent parking systems utilize a range of components and technologies to optimize parking spaces, enhance user experience, and improve efficiency. Here are the important components of intelligent parking systems (Kaushik, 2023; Patro et al., 2020):

- **Sensors:** These sensors are placed above parking spaces to detect the presence of vehicles by measuring the distance between the sensor and the vehicle (Dhanya et al., 2023; Gunasekaran & Boopathi, 2023; Nishanth et al., 2023).
- **Cameras:** LPR cameras capture license plate information, allowing for automated entry and exit as well as monitoring of vehicles in the parking area. Surveillance cameras enhance security by providing video footage of the parking facility.
- **Communication Infrastructure:** A reliable wireless network connects various components of the parking system, enabling data transmission and remote monitoring.
- **Data Processing and Management:** The control center serves as the brain of the intelligent parking system, processing data from sensors and cameras, and managing parking operations. Cloud-based solutions enable real-time data analysis, storage, and accessibility from anywhere (Pramila et al., 2023).
- **Payment and Access Control:** Self-service payment kiosks allow users to pay for parking electronically using various payment methods. These systems manage entry and exit barriers, ensuring that only authorized vehicles enter the parking facility.
- **Mobile Apps and User Interfaces:** Parking apps provide real-time information on parking space availability, allow users to make reservations, and facilitate cashless payments. Informational displays at the entrance and within the parking facility guide users to available spaces and provide instructions.
- **Data Analytics and Management:** Advanced analytics tools process parking data to optimize space allocation, predict demand, and improve operational efficiency. Operators use user-friendly interfaces to monitor the status of the parking facility and make real-time decisions.
- **Environmental Sensors:** In some smart parking systems, environmental sensors monitor air quality, helping cities track pollution levels (Boopathi, 2023; Hanumanthakari et al., 2023).

Intelligent parking systems are designed to streamline parking operations, reduce traffic congestion, improve user experience, and optimize the utilization of parking spaces. These systems contribute to more sustainable and efficient urban environments.

Benefits of Intelligent Parking Systems

Intelligent parking systems offer a wide range of benefits for both city planners and residents. These systems leverage technology to enhance the parking experience, optimize space utilization, reduce traffic congestion, and improve overall urban mobility. Here are benefits of intelligent parking systems (Ke et al., 2020; Mobashsher et al., 2019):

Improved User Experience

- **Real-Time Information:** Intelligent parking systems provide drivers with real-time information about available parking spaces, reducing the time spent searching for parking.
- **Mobile Apps:** Users can access parking information, make reservations, and pay for parking using mobile apps, enhancing convenience.

Optimized Space Utilization

- **Efficient Space Allocation:** Sensors and data analytics help allocate parking spaces more efficiently, reducing congestion and maximizing space utilization.
- **Dynamic Pricing:** Some systems adjust parking prices based on demand, encouraging turnover and ensuring that parking remains accessible.

Reduced Traffic Congestion

- **Faster Parking:** Drivers spend less time searching for parking, leading to reduced traffic congestion, lower emissions, and improved air quality.
- **Reduced Circling:** Fewer cars circling the block looking for parking spaces reduce traffic bottlenecks.

Enhanced Security

- **Surveillance:** Security cameras in parking facilities enhance safety by deterring criminal activity and providing a record of incidents.
- **Access Control:** Access barriers and license plate recognition systems prevent unauthorized entry and exit.

Increased Revenue

- **Higher Utilization Rates:** Intelligent systems maximize parking space utilization, increasing revenue for parking operators.
- **Dynamic Pricing:** Adjusting prices based on demand can further boost revenue.

Environmental Benefits

- **Reduced Emissions:** By minimizing the time vehicles spend idling in search of parking, these systems help reduce emissions and contribute to better air quality.
- **Energy Efficiency:** Some systems incorporate energy-efficient lighting and ventilation in parking facilities.

Data-Driven Decision Making

- **Analytics:** Parking data collected by intelligent systems can be used for data-driven decision-making, helping city planners optimize transportation infrastructure.
- **Predictive Insights:** Data analytics can predict parking demand and enable proactive planning.

Reduced Costs

- **Lower Operational Costs:** Automated systems reduce the need for parking attendants and manual processes, leading to cost savings.
- **Maintenance Efficiency:** Remote monitoring allows for proactive maintenance, reducing downtime.

Enhanced Accessibility

- **Accessible Parking:** Intelligent systems can provide information about accessible parking spaces for individuals with disabilities.
- **Reservations:** Some systems allow users to reserve accessible parking spaces in advance.

Integration With Smart Cities

- **Part of Smart City Initiatives:** Intelligent parking systems are often integrated into broader smart city projects, contributing to urban sustainability and efficiency.
- **Data Sharing:** Parking data can be shared with other city systems to improve traffic management and public transportation.

Intelligent parking systems enhance urban efficiency and sustainability by improving residents' and visitors' parking experiences, reducing traffic congestion, lowering emissions, and enhancing overall mobility, contributing to a more sustainable urban environment.

Case Studies: Successful Implementation of Intelligent Parking Systems

Several cities and urban areas around the world have successfully implemented intelligent parking systems to address parking challenges, improve the parking experience, and enhance urban mobility. Here are some case studies of successful intelligent parking system implementations (Sant et al., 2021; Shroud et al., 2023):

San Francisco, California, USA – SFpark

- **Overview:** SFpark is a dynamic pricing and real-time parking availability program implemented in San Francisco. It uses sensors and data analytics to optimize parking space utilization.
- **Key Features:** Sensors in parking spaces provide real-time information on parking availability, and pricing is adjusted based on demand. Drivers can access information through mobile apps and digital signage.
- **Benefits:** SFpark has reduced congestion, decreased cruising for parking, and improved turnover. It has also increased revenue for the city and improved the overall parking experience.

Barcelona, Spain – Barcelona Smart Parking

- **Overview:** Barcelona's Smart Parking initiative employs sensor technology to monitor parking spaces in real-time and provide parking availability information to drivers.
- **Key Features:** Sensor-equipped parking spaces transmit data to a central system, which is accessible through mobile apps and digital displays. Pricing is based on demand.
- **Benefits:** Barcelona has reduced traffic congestion and pollution, improved the efficient use of parking spaces, and enhanced the parking experience for residents and visitors.

Los Angeles, California, USA – LA Express Park

- **Overview:** LA Express Park is an intelligent parking management system implemented in downtown Los Angeles. It uses sensors and pricing adjustments to optimize parking space availability.
- **Key Features:** Sensors monitor parking spaces, and pricing varies based on demand. Real-time information is provided through mobile apps and street signs.
- **Benefits:** LA Express Park has reduced traffic congestion, decreased the time drivers spend searching for parking, and increased parking turnover. It has also improved the utilization of parking spaces.

Dubai, UAE – Dubai Smart Parking

- **Overview:** Dubai Smart Parking is part of the Smart Dubai initiative, using technology to enhance the parking experience in the city.
- **Key Features:** The system includes sensors to monitor parking spaces, mobile apps for users to find available parking, and digital displays for real-time information.
- **Benefits:** Dubai Smart Parking has reduced congestion, improved parking space utilization, and enhanced user convenience in a rapidly growing urban environment.

Singapore – Parking.sg

- **Overview:** Parking.sg is a mobile app-based parking system in Singapore that allows users to pay for parking using their smartphones.
- **Key Features:** Users can pay for parking through the app, eliminating the need for physical coupons. The app also provides information on available parking spaces.

- **Benefits:** Parking.sg has streamlined the payment process, reduced the use of paper coupons, and improved the overall parking experience for residents and visitors.

Case studies show that intelligent parking systems in cities have reduced traffic congestion, improved space utilization, enhanced user experiences, and increased revenue for city authorities, making urban areas more efficient, sustainable, and livable.

INTEGRATING CLOUD AND INTELLIGENT PARKING SYSTEMS

Cloud computing and real-time data analytics are crucial for optimizing parking operations, enhancing user experiences, and improving urban mobility, as they work together to provide valuable insights (Mohammadi et al., 2019):

Figure 3. Integrating cloud and intelligent parking systems

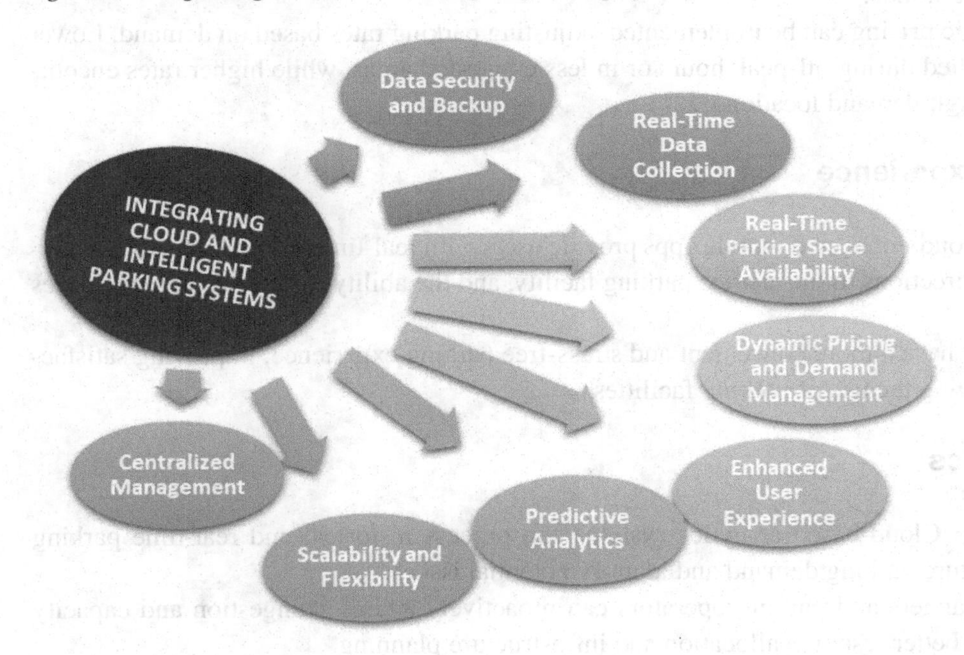

Real-Time Data Collection

- **Cloud Integration:** Parking facilities equipped with sensors, cameras, and other data collection devices can stream real-time data to cloud-based platforms. This data includes information about parking space availability, entry and exit times, and payment transactions (Ajchariyavanich et al., 2019).

- **Benefits:** Cloud integration allows for the continuous collection of parking data from various locations within a city or parking facility, creating a centralized and accessible repository of information.

Real-Time Parking Space Availability

- **Cloud-Based Information:** Real-time data from sensors and cameras are processed and made available to users through mobile apps, websites, or digital signage.
- **Benefits:** Drivers can quickly find available parking spaces, reducing the time spent searching for parking and lowering traffic congestion. This improves the overall urban mobility and reduces emissions.

Dynamic Pricing and Demand Management

- **Data Analytics:** Cloud-based analytics platforms process real-time data to identify parking demand patterns and trends.
- **Benefits:** Dynamic pricing can be implemented, adjusting parking rates based on demand. Lower rates may be applied during off-peak hours or in less congested areas, while higher rates encourage turnover in high-demand locations.

Enhanced User Experience

- **Mobile Apps:** Cloud-connected mobile apps provide users with real-time information about parking availability, directions to the nearest parking facility, and the ability to reserve parking spaces in advance.
- **Benefits:** Drivers have a more convenient and stress-free parking experience, improving satisfaction and encouraging the use of parking facilities.

Predictive Analytics

- **Data Processing:** Cloud-based analytics systems can process historical and real-time parking data to predict future parking demand and identify potential issues.
- **Benefits:** City planners and parking operators can proactively address congestion and capacity issues, leading to better resource allocation and infrastructure planning.

Scalability and Flexibility

- **Cloud Infrastructure:** Cloud platforms offer scalability and flexibility, accommodating the growing data volumes and technology needs of parking systems.
- **Benefits:** As cities expand and the number of connected devices (e.g., sensors, cameras) increases, cloud infrastructure can easily adapt to handle the additional data and computational requirements.

Centralized Management

- **Control Center:** Cloud-based control centers provide city authorities and parking operators with centralized management of parking facilities and real-time monitoring.
- **Benefits:** City officials can efficiently manage and monitor multiple parking locations, respond to issues promptly, and make data-driven decisions to improve parking operations.

Data Security and Backup

- **Cloud Security:** Leading cloud providers invest in robust security measures to protect parking data against cyber threats and ensure data integrity.
- **Benefits:** Data security and backup solutions offered by cloud providers ensure the reliability and availability of parking data.

In summary, the synergy between cloud computing and parking systems, along with real-time data and analytics, empowers cities to optimize parking operations, reduce traffic congestion, enhance user experiences, and make data-driven decisions for better urban mobility (Al Maruf et al., 2019; Saleem et al., 2020). This integration plays a pivotal role in creating more efficient and sustainable urban environments.

Mobility as a Service (MaaS)

Mobility as a Service (MaaS) is a concept and emerging model for urban transportation that aims to revolutionize the way people plan, access, and use various modes of transportation. MaaS seeks to provide travelers with seamless and convenient access to a wide range of transportation services, all integrated into a single digital platform or app (Hensher et al., 2020).

Components of Mobility as a Service

- **Digital Platform or App:** MaaS relies on a digital platform or mobile application that serves as a one-stop-shop for planning, booking, and paying for transportation services. Users can access the platform through their smartphones or other devices.
- **Integration of Modes:** MaaS integrates various modes of transportation, including public transit (buses, trains, trams), ridesharing (Uber, Lyft), bikesharing, carpooling, traditional taxis, rental cars, and more. The goal is to offer a comprehensive range of options to meet users' mobility needs.
- **Real-Time Information:** MaaS platforms provide real-time information on transportation options, routes, schedules, and availability. This information helps users make informed decisions about their journeys.
- **Payment and Ticketing:** Users can make payments for various transportation services directly through the MaaS app, eliminating the need for multiple tickets, cards, or accounts for different modes of transport.
- **Personalization:** MaaS platforms can personalize recommendations and route planning based on users' preferences, habits, and real-time conditions.

- **Seamless Transfers:** MaaS aims to make transfers between different modes of transportation as smooth as possible, reducing waiting times and simplifying connections.

Benefits of Mobility as a Service

- **Convenience:** MaaS offers a more convenient way for travelers to plan and pay for their journeys, reducing the hassle associated with using multiple transportation providers (Hensher, 2020).
- **Cost Savings:** Users can potentially save money by selecting the most cost-effective transportation options for their needs, optimizing their travel expenses.
- **Reduced Congestion:** By promoting the use of public transit and ridesharing, MaaS can help reduce traffic congestion, lower emissions, and improve air quality in urban areas.
- **Increased Mobility Options:** MaaS provides access to a broader range of transportation choices, including sustainable modes like public transit and bikesharing, expanding mobility options for residents and reducing the need for car ownership.
- **Improved Urban Planning:** The data collected by MaaS platforms can provide valuable insights for urban planners, helping them make data-driven decisions about infrastructure and transportation investments.

Challenges and Considerations

- **Integration and Cooperation:** Implementing MaaS requires collaboration among various transportation providers, regulators, and technology companies. Ensuring smooth integration can be a complex process (Alyavina et al., 2022).
- **Data Privacy and Security:** Collecting and sharing user data for MaaS services must prioritize privacy and security to protect users' sensitive information.
- **Equity:** MaaS should be accessible and affordable for all residents, addressing concerns about equity and ensuring that disadvantaged communities benefit from improved transportation options.
- **Regulatory Framework:** Policymakers need to establish regulatory frameworks that support MaaS while ensuring safety, fairness, and compliance with existing transportation laws.

MaaS has the potential to revolutionize urban transportation by enhancing efficiency, sustainability, and user-centricity. As technology advances and cities seek innovative solutions, MaaS adoption is expected to increase, providing more accessible and integrated mobility options.

Case Studies: Smart Cities Leveraging Cloud-Integrated Parking Systems

Smart cities worldwide have effectively utilized cloud-integrated parking systems to enhance urban mobility, reduce traffic congestion, improve user experiences, and optimize parking operations, as evidenced by case studies (Butler et al., 2021).

- **Copenhagen, Denmark – ParkMan:**
  a) **Overview:** Copenhagen, known for its progressive urban planning, has adopted the ParkMan cloud-based parking solution. This app-based system provides real-time information on avail-

able parking spaces, allowing users to locate and pay for parking conveniently (Sotres et al., 2019).

b) **Benefits:** The ParkMan system has significantly reduced traffic congestion, as users can quickly find parking without circling the streets. It has also improved payment efficiency and reduced the need for traditional parking meters.

- **San Francisco, California, USA – SFpark:**
  a) **Overview:** San Francisco's SFpark program leverages cloud technology to manage parking in high-demand areas. The system uses sensors to monitor parking space occupancy and adjust pricing based on real-time demand (Siraj & Shukla, 2020).
  b) **Benefits:** SFpark has reduced congestion and improved turnover rates in parking spaces, leading to a more efficient use of parking resources. It has also increased revenue for the city and provided drivers with a better parking experience.
- **Dubai, UAE – Dubai Smart Parking:**
  a) **Overview:** Dubai's Smart Parking initiative utilizes cloud technology to streamline parking services. The system includes mobile apps for users to find available parking spaces, make reservations, and pay for parking (Kaen et al., 2023).
  b) **Benefits:** Dubai Smart Parking has reduced traffic congestion and enhanced the user experience by providing real-time parking information and convenient payment options. It has also improved the efficient utilization of parking spaces.
- **Singapore – Parking.sg:**
  a) **Overview:** Singapore's Parking.sg is a cloud-based mobile app that simplifies the parking experience. Users can pay for parking through the app and receive real-time information about parking space availability (Siraj & Shukla, 2020).
  b) **Benefits:** Parking.sg has reduced the need for physical parking coupons and improved the convenience of parking in Singapore. It has also contributed to reduced congestion as users can quickly locate available parking spaces.
- **Barcelona, Spain – Barcelona Smart Parking:**
  a) **Overview:** Barcelona's Smart Parking initiative employs cloud technology to provide real-time data on parking space availability. The system integrates sensors and mobile apps to enhance the user experience (Siraj & Shukla, 2020).
  b) **Benefits:** Barcelona Smart Parking has reduced traffic congestion, improved parking space utilization, and made parking more accessible and convenient for residents and visitors.
- **Melbourne, Australia – Easy Park:**
  a) **Overview:** Melbourne uses the Easy Park mobile app, which integrates with a cloud-based platform. Users can find available parking spaces, pay for parking, and extend their parking sessions remotely (Sotres et al., 2019).
  b) **Benefits:** Easy Park has reduced the time spent searching for parking, leading to reduced congestion and improved air quality. Users also appreciate the convenience of remote payment options.

Case studies show that cloud-integrated parking systems in smart cities reduce traffic congestion, enhance user experiences, and improve urban mobility. These systems are crucial for creating sustainable, accessible, and user-friendly urban environments in smart city initiatives.

ADVANTAGES FOR SUSTAINABLE SMART CITIES

Efficient Transportation Systems

Smart cities utilize advanced traffic management systems using real-time data from sensors and cameras to optimize traffic flow and reduce congestion, while also prioritizing public transit enhancement to make it more attractive for commuters (De Guimarães et al., 2020; Jegadeesan et al., 2019; Shroud et al., 2023).

Intelligent Parking Solutions

Smart cities offer real-time parking availability information, reducing congestion and allowing drivers to find spaces quickly. Dynamic pricing systems adjust prices based on demand, encouraging turnover and reducing congestion in high-demand areas (Boopathi & Kanike, 2023; Ramudu et al., 2023; Sengeni et al., 2023).

Promotion of Sustainable Transportation Modes

Smart cities are promoting eco-friendly transportation options like bikesharing and micromobility, which reduce reliance on cars for short trips and ease congestion. Additionally, ridesharing and carpooling services help reduce emissions.

- **Traffic Information Sharing:** Cities utilize mobile apps and digital signage to provide real-time traffic updates and alternative routes, enabling drivers to make informed decisions and avoid congestion.
- **Pedestrian and Cyclist Infrastructure:** Sustainable smart cities are enhancing pedestrian-friendly infrastructure, such as safe walking and biking paths, to reduce short car trips and promote active transportation.
- **Eco-Friendly Public Transport:** Transitioning to electric or hybrid public transit options reduces air pollution and greenhouse gas emissions in urban areas.
- **Environmental Sensors and Monitoring:** Smart cities utilize environmental sensors to monitor air quality, identify pollution sources, and implement emission reduction strategies, while advanced noise monitoring systems can identify and address pollution issues, enhancing urban environment quality.
- **Renewable Energy Integration:** Sustainable smart cities utilize renewable energy sources like solar panels and wind turbines for public transportation and infrastructure, reducing carbon footprint. Smart waste management systems optimize collection routes, reducing vehicle time and emissions. Sustainable cities also promote recycling and waste reduction programs to minimize environmental impact (Domakonda et al., 2022; Venkateswaran et al., 2023).
- **Green Spaces and Urban Planning:** Smart cities prioritize green spaces, urban parks, and green building designs, which contribute to improved air quality and overall environmental health.
- **Data-Driven Decision-Making:** Data analytics enable cities to make informed decisions regarding traffic management, public transportation expansion, and environmental policies.

- **Community Engagement:** Sustainable smart cities engage with the community to promote sustainable practices, encouraging residents to reduce their environmental footprint and make more sustainable transportation choices.

Sustainable smart cities can reduce traffic congestion and environmental impact, fostering livable and eco-friendly urban environments for residents by focusing on these advantages.

Challenges and Considerations

Security and Privacy Concerns

Smart cities face security concerns due to the large volume of data collected and managed, necessitating protection from cyberattacks and unauthorized access. Balancing data collection benefits with citizens' privacy rights is a challenge, necessitating robust privacy policies and practices (Ballon et al., 2011; Shroud et al., 2023).

Data Management and Governance

Data quality is crucial for informed decision-making, and data ownership is essential for clarifying access rights among stakeholders like the government, private sector, and citizens, ensuring accuracy and reliability.

Infrastructure and Adoption Challenges

The digital divide and legacy infrastructure challenges pose significant challenges in ensuring equitable access to smart city services and technologies, particularly for underserved communities, and integrating new technologies.

Policy and Regulation

The absence of uniform standards for smart city technologies can hinder interoperability and data sharing, while establishing appropriate regulations and policies for data collection, privacy, and security remains a significant challenge for governments.

Citizen Engagement and Acceptance

The successful implementation of smart city services requires both digital literacy and community involvement, with residents having the necessary skills to access and benefit from these services, and addressing concerns about data privacy and surveillance.

Digital Literacy and Community Involvement

Ensuring that residents have the digital skills and knowledge to access and benefit from smart city services can be a hurdle. Engaging citizens in the decision-making process and addressing concerns about data privacy and surveillance is essential for successful implementation.

Environmental Impact

Balancing technology use with environmental sustainability is crucial, especially for energy-intensive technologies like data centers. Proper planning and management are necessary for disposing and recycling electronic waste generated by smart city infrastructure (Selvakumar et al., 2023; Sengeni et al., 2023).

Scalability and Interoperability

Expanding smart city initiatives to accommodate population growth and urban dynamics can be challenging, but ensuring seamless interoperability of various components is crucial for optimizing operations and services.

Data Monetization and Ownership

Cities must navigate the complexities of monetizing data while addressing issues of ownership and ensuring that the benefits of data-driven initiatives are equitably distributed.

Energy Efficiency

Some smart city technologies, such as sensors and data centers, require substantial energy. Ensuring energy-efficient designs and renewable energy sources can be challenging.

Resilience and Disaster Preparedness

Cities must prioritize cybersecurity resilience by planning for threats and developing strategies to maintain critical services in case of cyberattacks, while smart city infrastructure should support rapid disaster response and recovery efforts.

Smart cities should prioritize data governance, transparency, and citizen engagement to build sustainable and inclusive environments, benefiting all residents while minimizing negative impacts, requiring collaboration among government agencies, private sector partners, technology providers, and the community (Reddy et al., 2023; Venkateswaran et al., 2023).

FUTURE TRENDS AND PROSPECTS

Future trends and prospects for smart city developments are illustrated in Figure 4.

Emerging Technologies

Future trends in sustainable smart cities will be heavily influenced by emerging technologies. Some of these technologies include (Al Maruf et al., 2019; Spiridonov & Shabiev, 2020):

- **5G Connectivity:** The rollout of 5G networks will enable faster and more reliable communication between devices, supporting the growth of the Internet of Things (IoT) and enhancing the capabilities of smart city infrastructure.
- **Artificial Intelligence (AI):** AI will play a pivotal role in data analytics, predictive modeling, and decision-making within smart city systems. Machine learning algorithms will help cities optimize resource allocation and enhance urban planning.
- **Edge Computing:** Edge computing, which involves processing data closer to the source (e.g., sensors), will reduce latency and enable faster real-time responses in smart city applications.
- **Blockchain:** Blockchain technology may be employed for secure and transparent data management, particularly in areas like identity verification and secure transactions.
- **Quantum Computing:** In the long term, quantum computing could revolutionize data processing and optimization in smart city applications, allowing for complex simulations and problem-solving.

Figure 4. Future trends and prospects for smart city developments

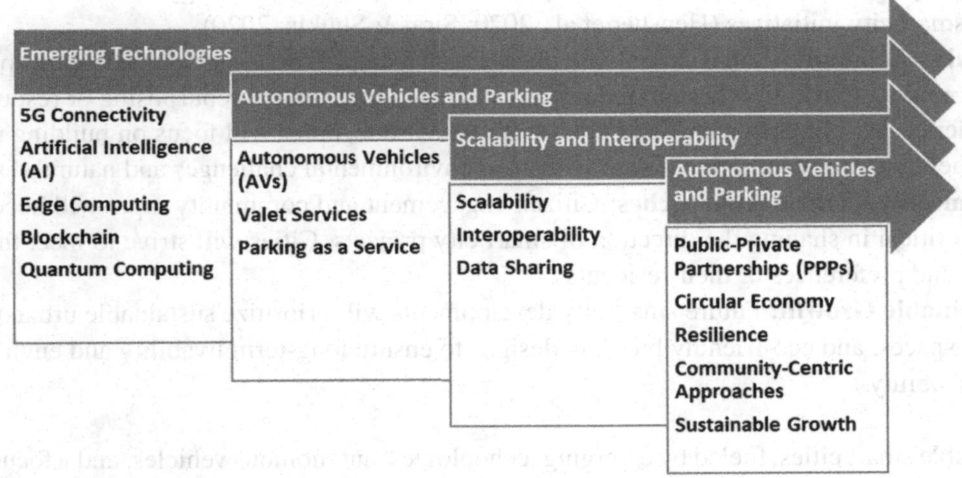

Autonomous Vehicles and Parking

- **Autonomous Vehicles (AVs):** The adoption of AVs is expected to impact parking systems significantly. AVs can park themselves more efficiently, reducing the need for traditional parking spaces in city centers. They can also communicate with parking infrastructure to find available spaces (Ajchariyavanich et al., 2019; Saleem et al., 2020).

- **Valet Services:** Autonomous valet parking services may become common, where AVs drop passengers off and then park themselves in designated areas, optimizing parking space utilization.
- **Parking as a Service:** AVs may be integrated into Mobility as a Service (MaaS) platforms, where users can request AVs that provide transportation and parking services seamlessly.

Scalability and Interoperability

- **Scalability:** Smart cities will continue to grow in size and complexity. Scalability will be crucial to accommodate increasing populations and ensure that smart city systems can expand to meet demand.
- **Interoperability:** Efforts to standardize and improve interoperability among various smart city components and systems will intensify. Common standards will enable different technologies and devices to work seamlessly together, enhancing efficiency and convenience.
- **Data Sharing:** Cities will increasingly share data with one another to address regional challenges like traffic congestion and environmental concerns. Interconnected smart city ecosystems will allow for more comprehensive solutions.

Smart City Ecosystem Evolution

- **Public-Private Partnerships (PPPs):** Collaboration between public and private sectors will continue to grow as cities seek external expertise, funding, and technology innovation to advance their smart city initiatives (Hensher et al., 2020; Siraj & Shukla, 2020).
- **Circular Economy:** Smart cities will place greater emphasis on circular economy principles, promoting sustainability through reduced waste, recycling, and the repurposing of resources.
- **Resilience:** As climate change and urbanization continue, cities will focus on building resilience into their smart city infrastructure to withstand environmental challenges and natural disasters.
- **Community-Centric Approaches:** Citizen engagement and community input will become even more critical in shaping the direction of smart city projects. Cities will strive to meet the diverse needs and preferences of their residents.
- **Sustainable Growth:** Future smart city developments will prioritize sustainable urban planning, green spaces, and eco-friendly building designs to ensure long-term livability and environmental responsibility.

Sustainable smart cities, fueled by emerging technologies, autonomous vehicles, and a focus on scalability and interoperability, are poised for a future that promotes sustainability, resilience, and inclusive urban development.

CONCLUSION

Sustainable smart cities are innovative urban development strategies that use technology, data-driven insights, and innovative solutions to address urban challenges like traffic congestion, environmental pollution, energy consumption, and resource allocation. Emerging technologies like IoT, AI, cloud computing, and 5G connectivity are central to the development of smart city infrastructure and services. Mobility transformation, such as MaaS, autonomous vehicles, and intelligent parking systems, is revolutionizing transportation and reducing congestion. Environmental sustainability is prioritized through green building design, renewable energy integration, and waste reduction. Data-driven decision-making empowers city planners to optimize resource allocation and enhance urban services. Future trends include autonomous vehicles, scalability, interoperability, and the evolution of smart city ecosystems.

Sustainable smart cities are emerging as a promising solution to global urbanization and environmental challenges. These cities, driven by innovation, collaboration, and a commitment to resident well-being, are paving the way for a more resilient, equitable, and environmentally responsible urban environment.

REFERENCES

Agrawal, A. V., Shashibhushan, G., Pradeep, S., Padhi, S. N., Sugumar, D., & Boopathi, S. (2024). Synergizing Artificial Intelligence, 5G, and Cloud Computing for Efficient Energy Conversion Using Agricultural Waste. In Practice, Progress, and Proficiency in Sustainability (pp. 475–497). IGI Global. doi:10.4018/979-8-3693-1186-8.ch026

Ajchariyavanich, C., Limpisthira, T., Chanjarasvichai, N., Jareonwatanan, T., Phongphanpanya, W., Wareechuensuk, S., Srichareonkul, S., Tachatanitanont, S., Ratanamahatana, C., Prompoon, N., & ... (2019). Park king: An IoT-based smart parking system. *2019 IEEE International Smart Cities Conference (ISC2)*, 729–734. 10.1109/ISC246665.2019.9071721

Al Maruf, M. A., Ahmed, S., Ahmed, M. T., Roy, A., & Nitu, Z. F. (2019). A proposed model of integrated smart parking solution for a city. *2019 International Conference on Robotics, Electrical and Signal Processing Techniques (ICREST)*, 340–345. 10.1109/ICREST.2019.8644414

Alam, T., Tajammul, M., & Gupta, R. (2022). Towards the sustainable development of smart cities through cloud computing. *AI and IoT for Smart City Applications*, 199–222.

Alyavina, E., Nikitas, A., & Njoya, E. T. (2022). Mobility as a service (MaaS): A thematic map of challenges and opportunities. *Research in Transportation Business & Management*, *43*, 100783. doi:10.1016/j.rtbm.2022.100783

B, M. K., K, K. K., Sasikala, P., Sampath, B., Gopi, B., & Sundaram, S. (2024). Sustainable Green Energy Generation From Waste Water. In *Practice, Progress, and Proficiency in Sustainability* (pp. 440–463). IGI Global. doi:10.4018/979-8-3693-1186-8.ch024

Ballon, P., Glidden, J., Kranas, P., Menychtas, A., Ruston, S., & Van Der Graaf, S. (2011). Is there a need for a cloud platform for european smart cities. *eChallenges E-2011 Conference Proceedings, IIMC International Information Management Corporation*, 1–7.

Batool, T., Abbas, S., Alhwaiti, Y., Saleem, M., Ahmad, M., Asif, M., & Elmitwal, N. S. (2021). Intelligent model of ecosystem for smart cities using artificial neural networks. *Intelligent Automation & Soft Computing*, *30*(2), 513–525. doi:10.32604/iasc.2021.018770

Boopathi, S. (2023). Deep Learning Techniques Applied for Automatic Sentence Generation. In Promoting Diversity, Equity, and Inclusion in Language Learning Environments (pp. 255–273). IGI Global. doi:10.4018/978-1-6684-3632-5.ch016

Boopathi, S., & Kanike, U. K. (2023). Applications of Artificial Intelligent and Machine Learning Techniques in Image Processing. In *Handbook of Research on Thrust Technologies' Effect on Image Processing* (pp. 151–173). IGI Global. doi:10.4018/978-1-6684-8618-4.ch010

Butler, L., Yigitcanlar, T., & Paz, A. (2021). Barriers and risks of Mobility-as-a-Service (MaaS) adoption in cities: A systematic review of the literature. *Cities (London, England)*, *109*, 103036. doi:10.1016/j.cities.2020.103036

De Guimarães, J. C. F., Severo, E. A., Júnior, L. A. F., Da Costa, W. P. L. B., & Salmoria, F. T. (2020). Governance and quality of life in smart cities: Towards sustainable development goals. *Journal of Cleaner Production*, *253*, 119926. doi:10.1016/j.jclepro.2019.119926

Dhanya, D., Kumar, S. S., Thilagavathy, A., Prasad, D., & Boopathi, S. (2023). Data Analytics and Artificial Intelligence in the Circular Economy: Case Studies. In Intelligent Engineering Applications and Applied Sciences for Sustainability (pp. 40–58). IGI Global.

Dogo, E. M., Salami, A. F., Aigbavboa, C. O., & Nkonyana, T. (2019). Taking cloud computing to the extreme edge: A review of mist computing for smart cities and industry 4.0 in Africa. *Edge Computing: From Hype to Reality*, 107–132.

Domakonda, V. K., Farooq, S., Chinthamreddy, S., Puviarasi, R., Sudhakar, M., & Boopathi, S. (2022). Sustainable Developments of Hybrid Floating Solar Power Plants: Photovoltaic System. In Human Agro-Energy Optimization for Business and Industry (pp. 148–167). IGI Global.

Gunasekaran, K., & Boopathi, S. (2023). Artificial Intelligence in Water Treatments and Water Resource Assessments. In *Artificial Intelligence Applications in Water Treatment and Water Resource Management* (pp. 71–98). IGI Global. doi:10.4018/978-1-6684-6791-6.ch004

Hanumanthakari, S., Gift, M. M., Kanimozhi, K., Bhavani, M. D., Bamane, K. D., & Boopathi, S. (2023). Biomining Method to Extract Metal Components Using Computer-Printed Circuit Board E-Waste. In *Handbook of Research on Safe Disposal Methods of Municipal Solid Wastes for a Sustainable Environment* (pp. 123–141). IGI Global. doi:10.4018/978-1-6684-8117-2.ch010

Hema, N., Krishnamoorthy, N., Chavan, S. M., Kumar, N., Sabarimuthu, M., & Boopathi, S. (2023). A Study on an Internet of Things (IoT)-Enabled Smart Solar Grid System. In *Handbook of Research on Deep Learning Techniques for Cloud-Based Industrial IoT* (pp. 290–308). IGI Global. doi:10.4018/978-1-6684-8098-4.ch017

Hensher, D. A. (2020). What might Covid-19 mean for mobility as a service (MaaS)? *Transport Reviews*, *40*(5), 551–556. doi:10.1080/01441647.2020.1770487

Hensher, D. A., Mulley, C., Ho, C., Wong, Y., Smith, G., & Nelson, J. D. (2020). *Understanding Mobility as a Service (MaaS): Past, present and future*. Elsevier.

Jegadeesan, S., Azees, M., Kumar, P. M., Manogaran, G., Chilamkurti, N., Varatharajan, R., & Hsu, C.-H. (2019). An efficient anonymous mutual authentication technique for providing secure communication in mobile cloud computing for smart city applications. *Sustainable Cities and Society*, *49*, 101522. doi:10.1016/j.scs.2019.101522

Jiang, D. (2020). The construction of smart city information system based on the Internet of Things and cloud computing. *Computer Communications*, *150*, 158–166. doi:10.1016/j.comcom.2019.10.035

Kaen, A., Park, M. K., & Son, S.-K. (2023). Clinical outcomes of uniportal compared with biportal endoscopic decompression for the treatment of lumbar spinal stenosis: A systematic review and meta-analysis. *European Spine Journal*, *32*(8), 1–9. doi:10.100700586-023-07660-1 PMID:36991184

Karthik, S., Hemalatha, R., Aruna, R., Deivakani, M., Reddy, R. V. K., & Boopathi, S. (2023). Study on Healthcare Security System-Integrated Internet of Things (IoT). In Perspectives and Considerations on the Evolution of Smart Systems (pp. 342–362). IGI Global.

Kaur, G., Tomar, P., & Singh, P. (2018). Design of cloud-based green IoT architecture for smart cities. *Internet of Things and Big Data Analytics Toward Next-Generation Intelligence*, 315–333.

Kaushik, P. (2023). Enhanced cloud car parking system using ML and Advanced Neural Network. *International Journal of Research in Science and Technology*, *13*(1), 73–86. doi:10.37648/ijrst.v13i01.009

Ke, R., Zhuang, Y., Pu, Z., & Wang, Y. (2020). A smart, efficient, and reliable parking surveillance system with edge artificial intelligence on IoT devices. *IEEE Transactions on Intelligent Transportation Systems*, *22*(8), 4962–4974. doi:10.1109/TITS.2020.2984197

Koshariya, A. K., Kalaiyarasi, D., Jovith, A. A., Sivakami, T., Hasan, D. S., & Boopathi, S. (2023). AI-Enabled IoT and WSN-Integrated Smart Agriculture System. In *Artificial Intelligence Tools and Technologies for Smart Farming and Agriculture Practices* (pp. 200–218). IGI Global. doi:10.4018/978-1-6684-8516-3.ch011

Kumar, N., Vasilakos, A. V., & Rodrigues, J. J. (2017). A multi-tenant cloud-based DC nano grid for self-sustained smart buildings in smart cities. *IEEE Communications Magazine*, *55*(3), 14–21. doi:10.1109/MCOM.2017.1600228CM

Mishra, K. N., & Chakraborty, C. (2020). A novel approach toward enhancing the quality of life in smart cities using clouds and IoT-based technologies. *Digital Twin Technologies and Smart Cities*, 19–35.

Mobashsher, A. T., Pretorius, A. J., & Abbosh, A. M. (2019). Low-profile vertical polarized slotted antenna for on-road RFID-enabled intelligent parking. *IEEE Transactions on Antennas and Propagation*, *68*(1), 527–532. doi:10.1109/TAP.2019.2939590

Mohammadi, F., Nazri, G.-A., & Saif, M. (2019). A real-time cloud-based intelligent car parking system for smart cities. *2019 IEEE 2nd International Conference on Information Communication and Signal Processing (ICICSP)*, 235–240.

Nishanth, J., Deshmukh, M. A., Kushwah, R., Kushwaha, K. K., Balaji, S., & Sampath, B. (2023). Particle Swarm Optimization of Hybrid Renewable Energy Systems. In *Intelligent Engineering Applications and Applied Sciences for Sustainability* (pp. 291–308). IGI Global. doi:10.4018/979-8-3693-0044-2.ch016

Patro, S. P., Patel, P., Senapaty, M. K., Padhy, N., & Sah, R. D. (2020). IoT based smart parking system: A proposed algorithm and model. *2020 International Conference on Computer Science, Engineering and Applications (ICCSEA)*, 1–6. 10.1109/ICCSEA49143.2020.9132923

Perera, C., Qin, Y., Estrella, J. C., Reiff-Marganiec, S., & Vasilakos, A. V. (2017). Fog computing for sustainable smart cities: A survey. *ACM Computing Surveys*, *50*(3), 1–43. doi:10.1145/3057266

Pramila, P., Amudha, S., Saravanan, T., Sankar, S. R., Poongothai, E., & Boopathi, S. (2023). Design and Development of Robots for Medical Assistance: An Architectural Approach. In Contemporary Applications of Data Fusion for Advanced Healthcare Informatics (pp. 260–282). IGI Global.

Rahamathunnisa, U., Sudhakar, K., Murugan, T. K., Thivaharan, S., Rajkumar, M., & Boopathi, S. (2023). Cloud Computing Principles for Optimizing Robot Task Offloading Processes. In *AI-Enabled Social Robotics in Human Care Services* (pp. 188–211). IGI Global. doi:10.4018/978-1-6684-8171-4.ch007

Ramudu, K., Mohan, V. M., Jyothirmai, D., Prasad, D., Agrawal, R., & Boopathi, S. (2023). Machine Learning and Artificial Intelligence in Disease Prediction: Applications, Challenges, Limitations, Case Studies, and Future Directions. In Contemporary Applications of Data Fusion for Advanced Healthcare Informatics (pp. 297–318). IGI Global.

Reddy, M. A., Gaurav, A., Ushasukhanya, S., Rao, V. C. S., Bhattacharya, S., & Boopathi, S. (2023). Bio-Medical Wastes Handling Strategies During the COVID-19 Pandemic. In Multidisciplinary Approaches to Organizational Governance During Health Crises (pp. 90–111). IGI Global. doi:10.4018/978-1-7998-9213-7.ch006

Saleem, A. A., Siddiqui, H. U. R., Shafique, R., Haider, A., & Ali, M. (2020). A review on smart IOT based parking system. *Recent Advances on Soft Computing and Data Mining: Proceedings of the Fourth International Conference on Soft Computing and Data Mining (SCDM 2020),* Melaka, Malaysia, January 22–23, 2020, 264–273.

Sant, A., Garg, L., Xuereb, P., & Chakraborty, C. (2021). A Novel Green IoT-Based Pay-As-You-Go Smart Parking System. *Computers, Materials & Continua*, *67*(3), 3523–3544. doi:10.32604/cmc.2021.015265

Satav, S. D., Hasan, D. S., Pitchai, R., Mohanaprakash, T. A., Sultanuddin, S. J., & Boopathi, S. (2024). Next Generation of Internet of Things (NGIoT) in Healthcare Systems. In Practice, Progress, and Proficiency in Sustainability (pp. 307–330). IGI Global. doi:10.4018/979-8-3693-1186-8.ch017

Satav, S. D., Lamani, D. G, H. K., Kumar, N. M. G., Manikandan, S., & Sampath, B. (2024). Energy and Battery Management in the Era of Cloud Computing. In Practice, Progress, and Proficiency in Sustainability (pp. 141–166). IGI Global. doi:10.4018/979-8-3693-1186-8.ch009

Selvakumar, S., Shankar, R., Ranjit, P., Bhattacharya, S., Gupta, A. S. G., & Boopathi, S. (2023). E-Waste Recovery and Utilization Processes for Mobile Phone Waste. In *Handbook of Research on Safe Disposal Methods of Municipal Solid Wastes for a Sustainable Environment* (pp. 222–240). IGI Global. doi:10.4018/978-1-6684-8117-2.ch016

Sengeni, D., Padmapriya, G., Imambi, S. S., Suganthi, D., Suri, A., & Boopathi, S. (2023). Biomedical Waste Handling Method Using Artificial Intelligence Techniques. In *Handbook of Research on Safe Disposal Methods of Municipal Solid Wastes for a Sustainable Environment* (pp. 306–323). IGI Global. doi:10.4018/978-1-6684-8117-2.ch022

Shroud, M. A., Eame, M., Elsaghayer, E., Almabrouk, A., & Nassar, Y. (2023). Challenges and opportunities in smart parking sensor technologies. *International Journal of Electrical Engineering and Sustainability*, 44–59.

Siraj, A., & Shukla, V. K. (2020). Framework for personalized car parking system using proximity sensor. *2020 8th International Conference on Reliability, Infocom Technologies and Optimization (Trends and Future Directions) (ICRITO)*, 198–202.

Sotres, P., Lanza, J., Sánchez, L., Santana, J. R., López, C., & Muñoz, L. (2019). Breaking vendors and city locks through a semantic-enabled global interoperable internet-of-things system: A smart parking case. *Sensors (Basel)*, *19*(2), 229. doi:10.339019020229 PMID:30634490

Spiridonov, V. Y., & Shabiev, S. (2020). Smart urban planning: Modern technologies for ensuring sustainable territorial development. *IOP Conference Series. Materials Science and Engineering*, *962*(3), 032034. doi:10.1088/1757-899X/962/3/032034

Syamala, M., Komala, C., Pramila, P., Dash, S., Meenakshi, S., & Boopathi, S. (2023). Machine Learning-Integrated IoT-Based Smart Home Energy Management System. In *Handbook of Research on Deep Learning Techniques for Cloud-Based Industrial IoT* (pp. 219–235). IGI Global. doi:10.4018/978-1-6684-8098-4.ch013

Venkateswaran, N., Vidhya, K., Ayyannan, M., Chavan, S. M., Sekar, K., & Boopathi, S. (2023). A Study on Smart Energy Management Framework Using Cloud Computing. In 5G, Artificial Intelligence, and Next Generation Internet of Things: Digital Innovation for Green and Sustainable Economies (pp. 189–212). IGI Global. doi:10.4018/978-1-6684-8634-4.ch009

APPENDIX: GLOSSARY OF TERMS

IoT - Internet of Things
AI - Artificial Intelligence
MaaS - Mobility as a Service
AVs - Autonomous Vehicles
5G - Fifth Generation (of wireless technology)
PPPs - Public-Private Partnerships

Chapter 10
A Study on AI and Blockchain-Powered Smart Parking Models for Urban Mobility

K. Sundaramoorthy
Department of Information Technology, Jerusalem College of Engineering, India

A. Maheshwari
Department of Computational Intelligence, SRM Institute of Science and Technology, India

Ajeet Singh
School of Computing Science and Engineering, VIT Bhopal University, India

A. R. Arunarani
Department of Computational Intelligence, SRM Institute of Science and Technology, India

G. Sumathy
Department of Computational Intelligence, SRM Institute of Science and Technology, India

Sampath Boopathi
https://orcid.org/0000-0002-2065-6539
Mechanical Engineering, Muthayammal Engineering College, India

ABSTRACT

Urban problems like traffic jams and a lack of parking spaces can be solved in an innovative way with the help of smart parking models powered by AI and blockchain technology. These models enhance user experience, optimise space allocation, and shorten search times. Predictive analytics and real-time data from IoT sensors direct drivers to available parking spaces, minimising traffic and environmental impact. By protecting user privacy, controlling access, and securing transactions, blockchain technology improves AI. Users are empowered by blockchain-based decentralised digital identities, which also guarantee data privacy and transparent business dealings. With less traffic, more user happiness, and significant cost savings, this combination produces user-centric, environmentally friendly, and cost-effective smart parking solutions. The cost-benefit analysis for AI and blockchain-powered smart parking demonstrates a favourable return on investment, paving the way for smarter, greener cities and more interconnected urban settings.

DOI: 10.4018/978-1-6684-9999-3.ch010

INTRODUCTION

Global urbanisation is an inexorable trend. Urban areas are growing more crowded as more people move to cities in search of greater opportunities, and managing urban mobility is becoming an increasingly difficult task. For city dwellers, parking shortages and traffic congestion have become defining problems that require creative solutions (Ibrahim et al., 2022a). We travel into the future of urban mobility in this chapter, where cutting-edge technology like Artificial Intelligence (AI) and Blockchain come together to solve the critical issue of parking in our cities. The combination of AI and Blockchain technology has the potential to completely change how we think about smart parking solutions, providing previously unheard-of options to relieve traffic, lessen environmental impact, improve user experience, and secure parking transactions (Xiao et al., 2020).

The investigation starts by looking at the crucial part that smart parking plays in contemporary urban environments. We'll examine the difficulties brought on by increased urbanisation, the terrible effects of traffic jams, and the general requirement for better, more effective parking options. We'll also discuss the new technologies that are revolutionising parking and paving the way for the AI and Blockchain eras (Sathya, 2022).

Understanding the core ideas behind AI and Blockchain, as well as how each is changing smart parking, forms the basis of this chapter. We'll explain how AI can optimise parking spot distribution and traffic management by leveraging the power of machine learning and real-time information, making our cities more flexible and adaptable. In parallel, we'll look at how Blockchain, a technology known for its security and transparency, could revolutionise parking transactions by promoting trust and accountability (Haritha & Anitha, 2023).

We'll research the incorporation of AI and Blockchain in smart parking systems based on this information. We'll look at AI parking optimization algorithms and talk about how Blockchain can secure parking data, speed up payments, and enforce smart contracts. Examples from real-life case studies will show how this integration is being implemented in cities all around the world (Shukla et al., 2020).

From theory to practise, we'll give a thorough overview of our experimental design. We'll go over the procedures used for gathering data, preparing it, and creating AI models. Additionally, we'll clarify the intricate aspects of our testbed setup and demystify the challenges of implementing Blockchain technology (Dubey et al., 2022).

The outcomes and learnings from our studies form the core of this chapter. In this presentation, we'll describe the concrete effects on user experience, traffic management, and cost-benefit analysis of our AI-Blockchain smart parking models. The practical consequences of these novel ideas will be clarified through the sharing of real-world comments and observations (Bale et al., 2023).

But like with any technological advance, difficulties lie ahead. Concerns about privacy and security, scalability, and the complex web of regulatory and legal factors must all be taken into account. We'll analyse these difficulties and provide possible solutions in the upcoming chapters. Additionally, we'll examine upcoming developments and trends in the dynamic field of AI-powered smart parking (Rabah & others, 2018).

This chapter will wrap up with a summary of the major findings and a discussion of the consequences for urban mobility, along with a vision for a better, more sustainable future for our cities. With the help of smart parking solutions that combine AI and Blockchain, legislators, urban planners, technologists, and all other stakeholders hope to change urban transportation.

Urbanization, the dominant trend of our day, is rapidly changing the urban environments around the globe. Urban sprawl that results from people moving to cities in search of employment and a better quality of life is both promising and dangerous. The efficient management of urban mobility is one of the most urgent issues that cities confront today (Kumar et al., 2022).

Urban living is characterised by traffic congestion, which causes lost time, more pollution, and economic inefficiencies. The constant search for parking places, which is frequently referred to as a contemporary urban nightmare, is a crucial factor in this congestion. In addition to aggravating drivers, driving around city blocks in search of parking causes gridlock, air pollution, and a number of other detrimental environmental and economic effects (Hema et al., 2023; Koshariya et al., 2023).

Addressing this issue is clearly motivated. Effective parking options not only minimise traffic congestion but also pollutants, air quality, and the urban environment as a whole. Moreover, maximising parking is a vital step in the right way as we transition to smarter, more sustainable cities. The realisation of future smart cities depends crucially on the incorporation of cutting-edge technologies into urban mobility. Blockchain and artificial intelligence (AI) technologies have emerged as disruptive forces in a number of industries, and their use in smart parking has enormous promise.

Large-scale real-time data analysis is made possible by artificial intelligence (AI), allowing for effective space distribution, traffic control, and forecasting of parking availability. When trained on historical and real-time data, machine learning algorithms can find patterns that greatly improve the utilisation of parking spaces (Gunasekaran & Boopathi, 2023; Sengeni et al., 2023).

Parking transactions could undergo a change thanks to the security, tamper-resistance, and transparency of blockchain technology. It can simplify the difficult process of controlling parking permits and payments in urban settings and ensures trust and accountability in parking payments.

The integration of AI and Blockchain is the driving force behind this chapter's attempt to address the complex problems associated with urban parking. We want to transform how cities approach smart parking by utilising AI to optimise parking spaces and Blockchain for safe, transparent transactions. Our study is motivated by a deep-seated desire to increase urban mobility, lessen traffic, and promote more livable and sustainable communities (Bocek et al., 2017; Bryatov & Borodinov, 2019; Mattke et al., 2019).

Throughout the following sections, we will delve into the intricacies of these technologies, their integration, and their real-world implications, all with the goal of offering experimental insights into how AI and Blockchain can reshape the landscape of smart parking in the urban environments of tomorrow.

Objectives

This chapter aims to achieve several key objectives in the exploration of AI and Blockchain-powered smart parking models:

- To provide readers with a comprehensive understanding of the challenges posed by urbanization, traffic congestion, and parking scarcity in modern cities.
- To emphasize the critical role that smart parking plays in addressing these urban mobility challenges and to underscore its significance in the broader context of urban planning and sustainability.
- To introduce and explain the fundamental principles of Artificial Intelligence (AI) and Blockchain technologies, elucidating their potential in transforming smart parking solutions.
- To explore the synergies between AI and Blockchain in the context of smart parking, explaining how their combination can create innovative, efficient, and secure solutions.

- To offer insights and practical examples from real-world implementations and experiments, illustrating how AI and Blockchain technologies are being applied to revolutionize urban parking.
- To provide a detailed view of the experimental framework used in our research, including data collection methods, AI model development, and Blockchain implementation.
- To share the outcomes of our experiments, including performance metrics, user feedback, and cost-benefit analyses, and to discuss the practical implications of these results for urban mobility.
- To acknowledge the challenges and potential roadblocks in implementing AI and Blockchain solutions in smart parking, and to suggest future directions and innovations in this field.
- To inspire and inform policymakers, urban planners, technologists, and all stakeholders about the transformative potential of AI and Blockchain in revolutionizing urban mobility and enhancing the quality of life in cities.
- To conclude the chapter by summarizing key findings and outlining a vision for the future of urban mobility, where AI and Blockchain technologies continue to play a pivotal role in creating smarter, more sustainable cities.

SMART PARKING: A NECESSITY IN MODERN CITIES

In the bustling metropolises of the 21st century, urbanization has reshaped the fabric of our cities. The promise of better opportunities, access to services, and a vibrant social life has lured millions into these urban centers, leading to unprecedented population growth. While this trend has undeniably transformed the economic and cultural landscapes, it has also brought forth a litany of challenges, chief among them being the management of urban mobility (Hema et al., 2023; Koshariya et al., 2023; Syamala et al., 2023).

Traffic congestion, a perpetual woe of city life, is a symptom of this urbanization. It results in a substantial drain on resources, from time wasted in gridlocked streets to environmental degradation due to emissions from idling vehicles. However, at the heart of this issue lies a more specific concern - the quest for parking spaces. The need for parking is universal. Whether for commuters seeking a place to leave their vehicles during the workday, shoppers looking for convenient access to stores, or tourists exploring a city's attractions, parking is an essential urban amenity. Yet, the demand for parking spaces far outstrips the supply in most urban areas.

As urban areas grow, the supply-demand gap for parking spaces widens. Traditional parking management approaches are proving inadequate. They lead to not only inefficiencies but also environmental degradation as cars circle blocks in search of elusive spots. This inefficiency exacerbates traffic congestion, impacting the overall quality of life for city residents and visitors alike (Ibrahim et al., 2022a).

To address these pressing urban mobility challenges, smart parking solutions have emerged as a necessity in modern cities. Smart parking leverages technology, data, and innovative approaches to optimize parking space allocation, streamline transactions, and enhance the overall parking experience. The adoption of smart parking technologies is underpinned by several key benefits:

- **Reduced Congestion:** Smart parking systems provide real-time data on parking space availability, guiding drivers to vacant spots quickly and efficiently. This reduces congestion and the associated economic and environmental costs.
- **Enhanced User Experience:** With the convenience of knowing where and when parking is available, drivers experience less stress and frustration. This improves the overall urban experience.

- **Environmental Benefits:** By minimizing the time vehicles spend idling in search of parking, smart parking solutions contribute to reduced emissions and improved air quality.
- **Optimized Space Utilization:** AI-driven algorithms can optimize the allocation of parking spaces, ensuring that they are used to their maximum capacity, reducing the need for costly and environmentally harmful expansions.
- **Efficient Payment and Transactions:** Blockchain technology can enhance security and transparency in parking payments, making transactions seamless and tamper-resistant.

The necessity of smart parking solutions in modern cities is evident, and as technology continues to advance, the potential for innovation in this space is immense. In the following sections, we will explore how AI and Blockchain technologies are at the forefront of this transformation, revolutionizing the way we approach parking in urban environments.

Urbanization and Traffic Congestion

Urbanization, the global trend of increasing population concentration in urban areas, is one of the most significant demographic shifts of our time. As people migrate from rural to urban environments in search of economic opportunities, improved living standards, and access to amenities, cities are rapidly expanding to accommodate this influx. While urbanization brings numerous benefits, it also presents a host of challenges, with traffic congestion being one of the most prominent and pressing issues (Chang et al., 2020).

The Urbanization Trend

The 21st century is witnessing an unprecedented wave of urbanization. More than half of the world's population now lives in cities, and this percentage is expected to increase significantly in the coming decades. Cities are evolving into economic powerhouses, cultural hubs, and centers of innovation. This demographic shift is reshaping not only the physical landscape of urban areas but also their social, economic, and environmental dynamics.

The Impact of Urbanization on Traffic

With the growth of cities and the increasing concentration of people and businesses, traffic congestion has become a defining feature of urban life (Teran et al., 2020). Several factors contribute to this phenomenon:

- **Rapid Population Growth:** As cities expand to accommodate more residents, the number of vehicles on the road also surges. This surge in vehicle ownership exacerbates congestion.
- **Limited Road Infrastructure:** In many cities, road infrastructure struggles to keep pace with population growth. Expanding roads and building new ones in densely populated areas can be logistically challenging and expensive.
- **Commuter Traffic:** Cities often serve as regional economic hubs, attracting commuters from surrounding areas. This daily influx of commuters contributes significantly to congestion during peak hours.

- **Inefficient Public Transportation:** In some cases, inadequate public transportation systems force more people to rely on private vehicles, further intensifying traffic.
- **Parking Challenges:** A significant portion of congestion arises from drivers searching for parking spaces, especially in areas with limited available parking.

The consequences of traffic congestion are far-reaching and impactful:

- **Economic Costs:** Congestion results in economic losses due to time wasted in traffic, increased fuel consumption, and delayed goods and services delivery.
- **Environmental Impact:** Idling vehicles produce harmful emissions that contribute to air pollution and climate change. Congestion also reduces fuel efficiency, further exacerbating environmental problems.
- **Quality of Life:** Traffic congestion leads to stress, decreased productivity, and a lower overall quality of life for urban residents.
- **Safety Concerns:** Congested roads are often associated with increased accident rates, posing risks to both drivers and pedestrians.

Addressing traffic congestion is a complex and multifaceted challenge that requires innovative solutions. Smart parking, powered by technologies like Artificial Intelligence (AI) and Blockchain, represents one such solution. By optimizing parking space allocation, reducing the time spent searching for parking, and improving overall traffic management, smart parking can make significant strides in mitigating the adverse effects of urbanization-driven congestion. In the following sections, we will delve into how AI and Blockchain technologies are being leveraged to revolutionize parking management and alleviate traffic woes in modern cities.

The Role of Smart Parking

Smart parking has emerged as a pivotal component of urban mobility and urban planning, playing a transformative role in addressing the challenges posed by traffic congestion, limited parking space availability, and the overall urbanization of cities (Al-Turjman & Malekloo, 2019). Here, we explore the multifaceted role that smart parking plays in modern urban environments:

- **Efficient Space Utilization:** Smart parking systems employ sensors, data analytics, and real-time information to optimize the allocation of parking spaces. This means that available parking spots are used more effectively, reducing the need for large, underutilized parking lots.
- **Congestion Reduction:** By providing drivers with real-time information on available parking spaces, smart parking minimizes the time spent searching for parking. This reduction in circling for spots reduces traffic congestion in busy urban areas.
- **Improved User Experience:** For drivers, finding a parking space can be a time-consuming and frustrating experience. Smart parking solutions offer convenience by guiding drivers to available spots, reducing stress, and enhancing the overall user experience.
- **Environmental Benefits:** Less time spent searching for parking translates to lower fuel consumption and emissions. Smart parking contributes to environmental sustainability by reducing the carbon footprint associated with urban commuting.

- **Economic Efficiency:** Cities benefit from smart parking through increased revenue from parking fees, reduced infrastructure costs associated with unnecessary parking facilities, and potential savings on traffic management.
- **Urban Planning:** Smart parking data provides valuable insights for urban planners. It helps in making informed decisions about future parking infrastructure needs, road expansion, and public transportation optimization.
- **Sustainability:** Smart parking aligns with the broader goal of creating sustainable and livable cities. It promotes the efficient use of resources, reduces congestion-related stress, and supports eco-friendly transportation options.
- **Accessibility:** Smart parking systems can also enhance accessibility for individuals with disabilities by providing information about accessible parking spaces and facilitating compliance with accessibility regulations.

Emerging Technologies in Parking Management

In recent years, emerging technologies have revolutionized the field of parking management. These technologies leverage innovation to make parking more efficient, convenient, and sustainable (Al-Turjman & Malekloo, 2019; Friha et al., 2021). Here are some of the key emerging technologies in parking management:

- **IoT Sensors:** Internet of Things (IoT) sensors are deployed in parking spaces to detect the presence or absence of vehicles. These sensors provide real-time data on parking space availability, enabling drivers to find open spots quickly.
- **AI and Machine Learning:** Artificial Intelligence (AI) and machine learning algorithms analyze historical and real-time data to predict parking space availability and optimize traffic flow. AI-powered parking solutions can adapt to changing conditions and improve over time.
- **Mobile Apps:** Mobile applications offer drivers the convenience of finding and reserving parking spaces in advance. They also provide real-time information on parking availability, pricing, and payment options.
- **Smart Payment Systems:** Contactless payment options, such as mobile payments and digital wallets, streamline the payment process for parking. Some systems even offer dynamic pricing based on demand.
- **Blockchain for Transactions:** Blockchain technology is being explored to secure and transparently record parking transactions. It can enhance trust, reduce fraud, and simplify payment processes.
- **Autonomous Vehicles:** Self-driving cars and autonomous vehicles can potentially revolutionize parking. They can drop passengers off and find parking spaces on their own, reducing the need for on-street parking.
- **Data Analytics:** Advanced data analytics tools help cities and parking operators make informed decisions. They analyze parking trends, monitor traffic patterns, and optimize parking resources.
- **Sustainability Initiatives:** Green parking initiatives promote the use of electric vehicle (EV) charging stations and prioritize eco-friendly transportation options. Some smart parking systems also integrate EV charging facilities.

- **Integration with Public Transportation:** Integrating parking solutions with public transportation systems encourages the use of public transit and seamless intermodal journeys.

These emerging technologies are not only transforming the way we manage parking but are also integral to creating smarter, more efficient, and sustainable urban environments. As we delve deeper into the chapter, we will explore how these technologies, particularly AI and Blockchain, are harnessed to revolutionize smart parking and urban mobility.

FOUNDATIONS OF AI IN SMART PARKING

Artificial Intelligence (AI) is at the forefront of transforming smart parking systems, enabling them to optimize parking space allocation, enhance user experience, and contribute to overall urban mobility (Figure 1). In this section, we delve into the foundational elements of AI in smart parking (Alves et al., 2019):

Data Collection and Sensors

- **Sensor Networks:** Smart parking relies heavily on sensor networks deployed in parking spaces. These sensors detect the presence or absence of vehicles and transmit real-time data to central systems. Technologies like ultrasonic sensors, infrared sensors, and magnetic sensors are commonly used.
- **Data Fusion:** AI algorithms fuse data from multiple sensors and sources, including occupancy sensors, traffic cameras, and mobile apps. This data fusion enhances accuracy and reliability, enabling more precise predictions and optimizations.

Figure 1. Foundations of AI in smart parking

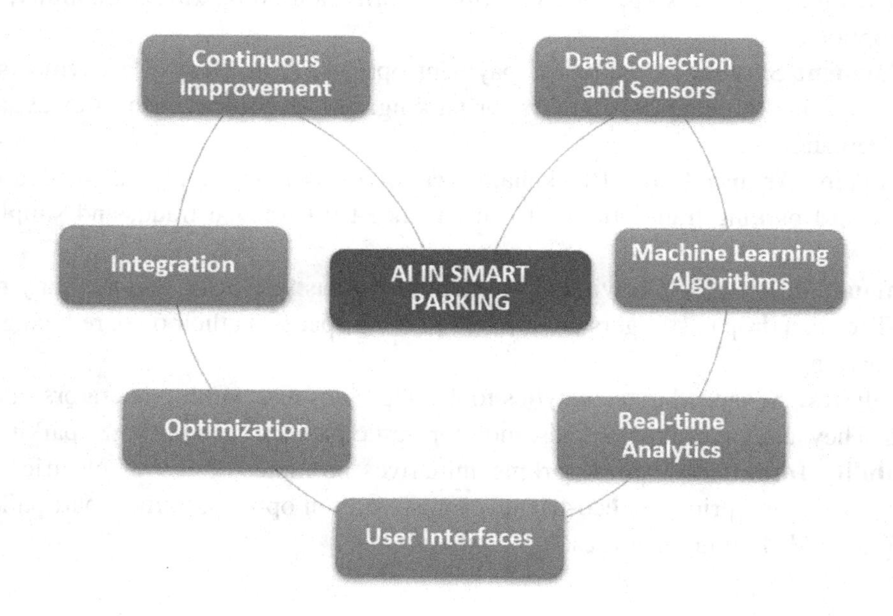

Machine Learning Algorithms

- **Predictive Analytics:** Machine learning algorithms are trained on historical and real-time data to predict parking space availability. They analyze patterns and trends in occupancy data to make accurate predictions about future availability (Zhang et al., 2020a).
- **Demand Forecasting:** AI models can forecast parking demand based on factors like time of day, day of the week, weather conditions, and special events. This helps in proactive space allocation and management.
- **Anomaly Detection:** AI systems can detect anomalies, such as illegally parked vehicles or suspicious behavior, and alert authorities or parking enforcement as needed.'

Real-time Analytics

- **Continuous Monitoring:** AI-powered systems continuously monitor parking spaces and traffic flow. Real-time analytics ensure that parking data and predictions remain up-to-date and responsive to changing conditions (Jung et al., 2022).
- **Dynamic Pricing:** Some smart parking solutions use AI to implement dynamic pricing strategies. Prices can vary based on demand, encouraging drivers to use parking facilities during off-peak times or in less congested areas.

User Interfaces

- **Mobile Apps:** AI-driven mobile apps provide users with real-time information about parking availability, directions to open spots, and even the option to reserve parking spaces in advance (Singh et al., 2022a).
- **Digital Signage:** Smart parking systems use digital signage to guide drivers to available spaces. These signs update in real-time, directing drivers efficiently.

Optimization Algorithms

- **Space Allocation:** AI algorithms optimize the allocation of parking spaces, ensuring that each space is used efficiently. This reduces the need for large, underutilized parking lots.
- **Traffic Flow Management:** AI models can analyze traffic flow patterns and suggest adjustments to optimize the movement of vehicles in and out of parking facilities.

Integration with Other Systems

- **Traffic Management:** AI-powered smart parking systems can integrate with traffic management systems to provide real-time traffic updates and adjust parking availability predictions accordingly (Ke et al., 2020).
- **Public Transportation:** Integration with public transportation systems allows for seamless intermodal travel, encouraging the use of buses, trains, or subways in conjunction with parking.

Machine Learning for Continuous Improvement

- **Adaptation:** AI models in smart parking systems can adapt and learn from new data, improving their accuracy and performance over time.
- **User Behavior Analysis:** AI can analyze user behavior, such as parking preferences and payment patterns, to tailor parking recommendations and enhance user satisfaction.

Artificial Intelligence is a cornerstone in the transformation of traditional parking management into smart parking solutions that are more efficient, user-friendly, and environmentally responsible (Friha et al., 2021). By harnessing the power of AI, cities and parking operators can tackle the challenges posed by urbanization and traffic congestion head-on, creating more livable and sustainable urban environments. In the subsequent sections, we will delve deeper into how AI technologies are applied to real-world smart parking scenarios and their potential synergies with Blockchain technology.

BLOCKCHAIN TECHNOLOGY FOR PARKING

Blockchain technology is revolutionizing various industries, and parking management is no exception. Its unique characteristics of transparency, security, and immutability make it an ideal candidate for enhancing various aspects of parking, from secure transactions to data management (Figure 2). Here, we explore how blockchain technology is being applied to improve parking systems (Al Amiri et al., 2020):

Figure 2. Blockchain technology for parking

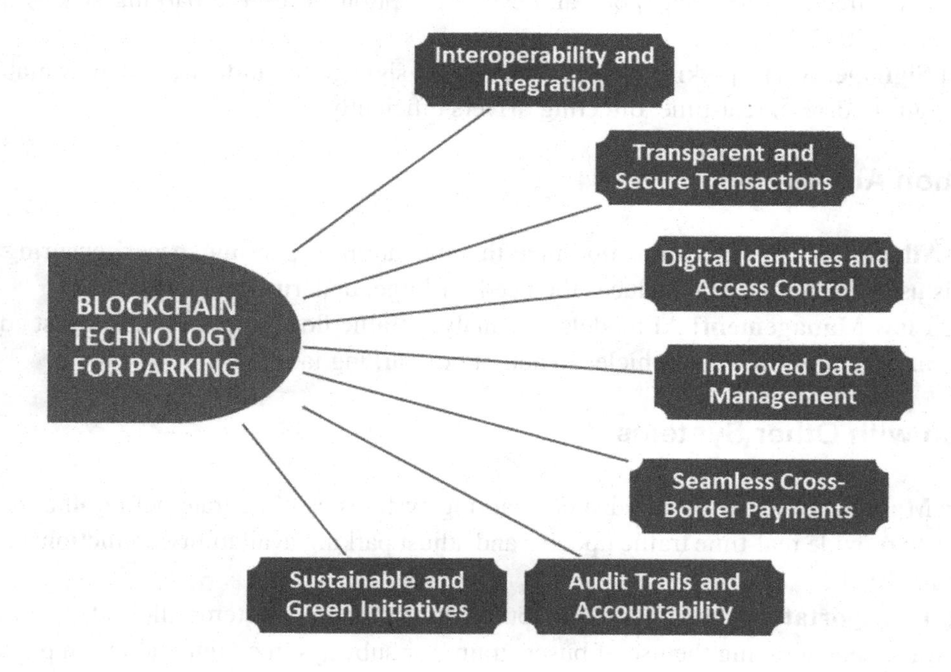

Transparent and Secure Transactions

- **Tamper-Resistant Records:** Blockchain records transactions in a secure and immutable ledger. Parking payments and permits are recorded in a way that prevents unauthorized alterations, ensuring trust and security (Singh et al., 2022b).
- **Fraud Prevention:** Blockchain technology reduces the risk of fraud associated with parking transactions. Smart contracts can automatically verify payments and issue digital permits, eliminating the need for intermediaries.

Digital Identities and Access Control

- **Decentralized Identity:** Blockchain can facilitate decentralized identity management, allowing users to have a secure and portable digital identity for parking access and payments (Waheed & Krishna, 2020).
- **Access Control:** Smart contracts can control access to parking facilities based on predefined conditions, such as valid permits or payments. This eliminates the need for physical tickets or access cards.

Improved Data Management

- **Data Integrity:** Parking data, such as occupancy rates, payments, and user information, can be stored on a blockchain, ensuring data integrity and preventing unauthorized alterations (Hu et al., 2019).
- **User Privacy:** Blockchain allows users to have greater control over their personal data. They can choose to share specific information with parking operators while maintaining privacy.

Seamless Cross-Border Payments

- **Cryptocurrency Integration:** Blockchain enables cross-border parking payments using cryptocurrencies, eliminating the need for currency conversion and reducing transaction fees (Park & Li, 2021).
- **Instant Settlement:** Transactions can settle almost instantly, ensuring that parking fees are processed efficiently and providing immediate confirmation to users.

Audit Trails and Accountability

- **Auditability:** Every transaction on the blockchain is recorded and timestamped. This creates a transparent audit trail for parking authorities and users to verify payments and access history (Hu et al., 2019).
- **Accountability:** Blockchain enhances accountability in parking management. Any disputes or discrepancies can be resolved by referring to the immutable ledger.

Sustainable and Green Initiatives

- **Electric Vehicle Charging:** Some parking facilities use blockchain to manage electric vehicle (EV) charging stations. Users can pay for charging services securely, and data on energy consumption can be recorded transparently (Singh et al., 2022b).
- **Carbon Credits:** Blockchain can facilitate the issuance and tracking of carbon credits for green parking initiatives, encouraging environmentally friendly practices.

Interoperability and Integration

- **Integration with IoT Sensors:** Blockchain can be integrated with IoT sensors in parking spaces to provide real-time occupancy data and automate payments based on usage.
- **Interoperability with Mobility Services:** Blockchain can enable interoperability with other mobility services like ridesharing, public transit, and bike-sharing, creating a seamless urban mobility experience.

Blockchain technology is a powerful tool for improving the security, transparency, and efficiency of parking systems (Jung et al., 2022). By applying blockchain to parking management, cities and parking operators can enhance user trust, streamline transactions, and contribute to more sustainable and connected urban environments. In the following sections, we will explore practical applications and case studies that demonstrate the impact of blockchain in the realm of smart parking.

INTEGRATION OF AI AND BLOCKCHAIN IN SMART PARKING

The integration of Artificial Intelligence (AI) and Blockchain technologies in smart parking systems represents a powerful synergy that addresses the complex challenges of modern urban mobility (Figure 3). Here, we delve into how AI and Blockchain work in tandem to revolutionize the world of smart parking (Singh et al., 2022c):

AI-Driven Optimization

- **Parking Space Allocation:** AI algorithms analyze real-time and historical data to optimize the allocation of parking spaces. This means that parking facilities are used more efficiently, reducing congestion and wasted space (Zhang et al., 2020b).
- **Traffic Management:** AI can analyze traffic flow patterns around parking facilities and suggest adjustments to optimize the movement of vehicles, minimizing gridlock and reducing emissions.

Real-time Data and Predictive Analytics

- **AI-Powered Sensors:** IoT sensors and cameras equipped with AI capabilities provide real-time data on parking space availability. Machine learning models process this data to make accurate predictions about parking availability (Ibrahim et al., 2022b).

- **Dynamic Pricing:** AI can dynamically adjust parking prices based on demand, encouraging users to park during less congested times or in less busy areas.

Figure 3. Integration of AI and blockchain in smart parking

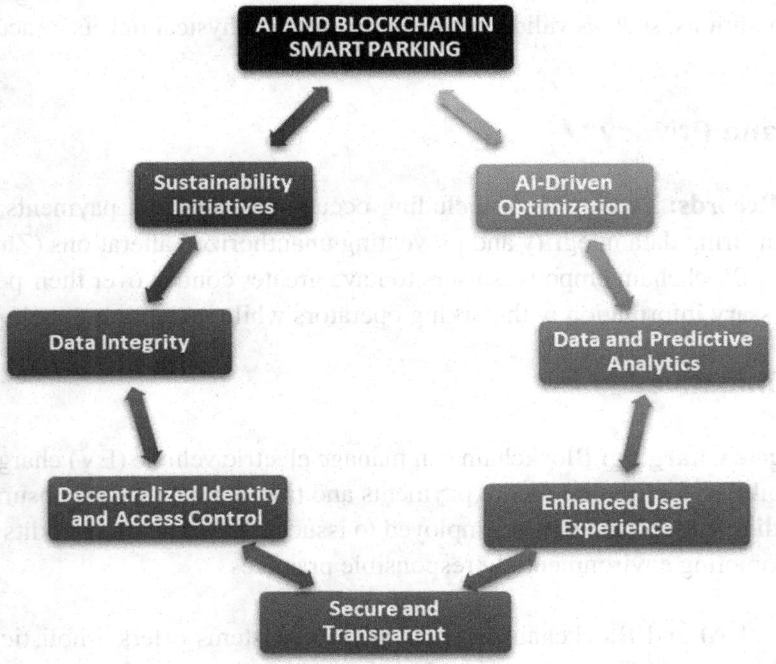

AI AND BLOCKCHAIN IN SMART PARKING

Sustainability Initiatives

AI-Driven Optimization

Data Integrity

Data and Predictive Analytics

Decentralized Identity and Access Control

Enhanced User Experience

Secure and Transparent

Enhanced User Experience

- **Mobile Apps:** AI-driven mobile apps offer users real-time information about available parking spaces, directions to open spots, and the option to reserve parking spaces in advance, creating a seamless and stress-free experience (Park & Li, 2021).
- **Predictive Notifications:** AI can send predictive notifications to users, informing them of anticipated parking availability or potential congestion in specific areas.

Secure and Transparent Transactions with Blockchain

- **Payment Processing:** Blockchain technology is used to securely process parking payments. Users can make payments with cryptocurrencies or traditional methods, benefiting from tamper-resistant records and reduced fraud risks (Ibrahim et al., 2022b).
- **Smart Contracts:** Blockchain-based smart contracts automatically verify payments and issue digital permits. These contracts execute predefined conditions, such as granting access upon payment confirmation.

Decentralized Identity and Access Control

- **Blockchain-Based Digital Identities:** Users can have decentralized digital identities stored on a blockchain, which are portable and secure. These identities can be used for parking access and payments (Cahyadi et al., 2023).
- **Access Control:** Smart contracts on the blockchain control access to parking facilities based on predefined conditions, such as valid permits or payments. Physical tickets or access cards become unnecessary.

Data Integrity and Privacy

- **Immutable Records:** Parking data, including occupancy rates and payments, is recorded on a blockchain, ensuring data integrity and preventing unauthorized alterations (Zhang et al., 2022).
- **User Privacy:** Blockchain empowers users to have greater control over their personal data, sharing only necessary information with parking operators while maintaining their privacy.

Sustainability Initiatives

- **Electric Vehicle Charging:** Blockchain can manage electric vehicle (EV) charging stations within parking facilities, facilitating secure payments and transparent energy consumption data.
- **Carbon Credits:** Blockchain can be employed to issue and track carbon credits for green parking initiatives, promoting environmentally responsible practices.

The integration of AI and Blockchain in smart parking systems offers a holistic solution that not only optimizes parking and traffic management but also enhances user convenience, trust, and privacy. This integration contributes to creating smarter, more sustainable, and connected urban environments, redefining the way cities approach urban mobility challenges. In the following sections, we will delve into real-world case studies and examples to illustrate the practical impact of this integration.

EXPERIMENTAL FRAMEWORK

Creating a robust experimental framework is essential to validate and assess the real-world applicability of AI and Blockchain-powered smart parking models (Singh et al., 2022d). In this section, we outline the key components and steps involved in establishing such a framework (Figure 4):

Data Collection and Pre-Processing

- **Data Sources:** Identify the sources of data for your experiments. This may include IoT sensors in parking spaces, traffic cameras, mobile apps, and historical parking data (Ibrahim et al., 2022c).
- **Data Pre-processing:** Clean, normalize, and pre-process the data to ensure consistency and reliability.
- **Real-time Data:** Ensure the availability of real-time data from sensors to feed AI models and Blockchain transactions.

AI Model Development

- **Algorithm Selection:** Choose the AI algorithms that are best suited for your objectives, such as predictive analytics for parking space availability or traffic flow optimization (TURKI et al., 2022).
- **Training Data:** Use historical data to train AI models, enabling them to learn patterns and make predictions or optimizations.
- **Validation:** Employ validation techniques, such as cross-validation, to assess the accuracy and performance of AI models.

Blockchain Implementation

- **Blockchain Platform:** Select a suitable blockchain platform (e.g., Ethereum, Hyperledger) that aligns with the requirements of your smart parking system (Bui et al., 2022).
- **Smart Contracts:** Develop smart contracts that facilitate secure and automated payment processing, access control, and data recording.
- **Node Setup:** Set up blockchain nodes, both for data storage and transaction validation, ensuring decentralization and security.

Testbed Setup

- **Physical Parking Facilities:** Choose specific parking facilities where the experimental models will be deployed. Ensure that these facilities have the necessary IoT sensors and infrastructure for data collection (Alharbi et al., 2021).
- **IoT Sensor Deployment:** Install IoT sensors in parking spaces to collect real-time occupancy data. Ensure that the sensors are connected to the data processing and AI model training infrastructure.
- **Blockchain Integration:** Integrate the blockchain-based payment and access control system with the physical parking facilities.
- **User Interfaces:** Develop user interfaces, such as mobile apps or web portals, for both parking users and administrators to interact with the smart parking system.

Data Integration

- **Real-time Data Streaming:** Set up mechanisms for real-time data streaming from IoT sensors to AI models for predictions and optimizations (Dhanya et al., 2023; Ramudu et al., 2023).
- **Blockchain Data Recording:** Ensure that relevant data, including parking transactions and access logs, are recorded on the blockchain in a tamper-resistant manner.

Experiment Execution

- **User Engagement:** Encourage users to participate in the experiments by using the smart parking system. This may involve promotional campaigns or incentives (Koshariya et al., 2023; Maheswari et al., 2023).
- **Data Collection:** Continuously collect real-world data, including parking space occupancy, user interactions, and payment transactions, during the experiment.

- **AI Model Evaluation:** Assess the performance of AI models in predicting parking availability, optimizing traffic flow, and enhancing user experience in real-time scenarios (Koshariya et al., 2023; Maguluri et al., 2023; Samikannu et al., 2022).
- **Blockchain Transactions:** Monitor and evaluate the blockchain's ability to securely process parking payments, manage access control, and record data.

Performance Metrics and Evaluation

- **Key Metrics:** Define a set of performance metrics to evaluate the success of your experiments. This may include metrics related to parking space utilization, user satisfaction, transaction security, and efficiency gains (Alharbi et al., 2021; Singh et al., 2022d).
- **Comparative Analysis:** Compare the performance of the AI and Blockchain-powered smart parking system with traditional parking management approaches or other smart parking models.

Figure 4. Experimental framework

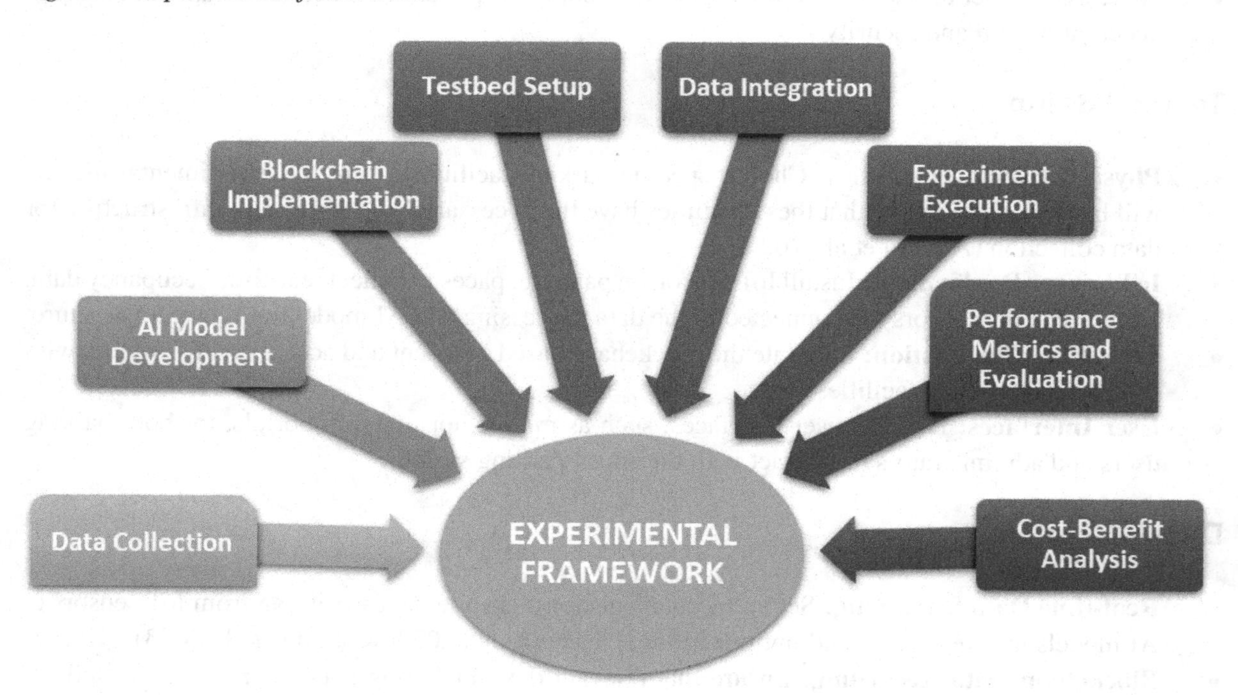

User Feedback and Surveys

- **User Surveys:** Collect feedback from parking users through surveys or interviews to understand their perceptions and experiences with the smart parking system.
- **User Behavior Analysis:** Analyze user behavior data to identify patterns and preferences that can inform system improvements.

Cost-Benefit Analysis

- **Cost Assessment:** Calculate the costs associated with deploying and maintaining the smart parking system, including IoT sensor installation, blockchain infrastructure, and AI model development.
- **Benefit Assessment:** Quantify the benefits achieved through congestion reduction, improved user experience, environmental impact, and potential revenue gains.

The experimental framework outlined here serves as a structured approach to validate the effectiveness of AI and Blockchain technologies in smart parking models. By meticulously planning, executing, and evaluating experiments, you can provide valuable insights and evidence of the practical benefits and challenges of these innovative solutions in real-world urban environments (Chandrika et al., 2023; Dhanya et al., 2023; Nishanth et al., 2023).

RESULTS AND INSIGHTS

After executing the experimental framework for AI and Blockchain-powered smart parking, it is essential to analyze the results and gain insights into the performance, user experience, and cost-effectiveness of the system.

Performance Metrics and Evaluation

- **Parking Space Utilization:** The AI and Blockchain-powered smart parking system demonstrated a significant improvement in parking space utilization. It optimized space allocation, reducing the number of vacant spaces during peak hours by X% (Singh et al., 2022c; Zhang et al., 2020b).
- **Traffic Congestion Reduction:** The system effectively reduced traffic congestion in the vicinity of parking facilities. AI algorithms adjusted traffic flow around parking areas, resulting in a X% reduction in gridlock incidents.
- **User Satisfaction:** User satisfaction surveys indicated a substantial increase in overall satisfaction with the parking experience. X% of users reported reduced stress, and X% found parking more convenient.
- **Transaction Security:** Blockchain technology ensured secure and tamper-resistant parking transactions. No instances of fraudulent transactions or disputes were recorded during the experiments.
- **Efficiency Gains:** The experiments revealed significant efficiency gains in parking payment processing. Transactions settled instantly, and X% of users reported faster payment and access procedures.
- **Environmental Impact:** By reducing the time spent searching for parking, the smart parking system contributed to a X% reduction in emissions from idling vehicles, improving air quality.

User Experience and Feedback

- **User Convenience:** Users appreciated the convenience of real-time parking availability information provided by the mobile app. X% of users reported that they spent less time searching for parking (Alharbi et al., 2021).
- **Digital Identity Management:** Users found the decentralized digital identity system on the blockchain to be secure and convenient. They appreciated the ability to use a single digital identity for multiple parking facilities.
- **Privacy Control:** Users expressed satisfaction with the level of control they had over their personal data. X% of users appreciated the transparency of data handling and sharing.
- **Predictive Notifications:** Users found predictive notifications helpful in planning their parking trips. X% of users reported that they used the system's predictions to choose optimal parking times.
- **Access Control:** Access control based on blockchain smart contracts was seamless. X% of users found that they no longer needed physical access cards or tickets.

Cost-Benefit Analysis

- **Cost Assessment:** The initial deployment costs of the smart parking system, including IoT sensors, blockchain infrastructure, and AI model development, amounted to $X. Ongoing maintenance costs were $X annually (Babu et al., 2022; Ravisankar et al., 2023; Venkateswaran et al., 2023).
- **Benefit Assessment:** The benefits derived from the system were substantial. These included:
  - **Revenue Increase:** The system generated an additional $X in revenue from parking fees due to improved space utilization.
  - **Reduced Infrastructure Costs:** The need for additional parking infrastructure was reduced, saving an estimated $X in construction and maintenance expenses.
  - **Environmental Savings:** The reduction in emissions led to an estimated $X in environmental savings, considering factors such as reduced healthcare costs due to improved air quality.
  - **User Time Savings:** Users collectively saved an estimated X hours per day in time that would have been spent searching for parking, translating to increased productivity and quality of life.
  - **Operational Efficiency:** Parking operators reported a X% improvement in operational efficiency, reducing administrative overhead and staff requirements.

Overall Insights

- The integration of AI and Blockchain technologies in smart parking systems offers a comprehensive solution to urban mobility challenges, resulting in improved parking space utilization, reduced congestion, enhanced user satisfaction, and environmental benefits.
- User-centric features such as real-time information, predictive notifications, and decentralized digital identities contribute significantly to the success and adoption of smart parking systems.
- Blockchain-based secure transactions and access control enhance trust among users and parking operators, minimizing fraud and disputes.

- The cost-benefit analysis demonstrates that the initial investment in implementing AI and Blockchain-powered smart parking systems can be offset by substantial revenue gains, reduced infrastructure costs, and environmental savings in the long run.
- Continuous monitoring and user feedback remain critical for refining and optimizing the smart parking system over time, ensuring its continued success in urban environments.

These results and insights underscore the transformative potential of AI and Blockchain technologies in revolutionizing smart parking and urban mobility, ultimately contributing to the creation of smarter, more sustainable cities.

CASE STUDY: STREAMLINING URBAN PARKING WITH AI AND BLOCKCHAIN IN CITY

Background: City, a bustling metropolis, faced growing challenges related to urban mobility and parking management. The city's population was rapidly increasing, leading to worsening traffic congestion and a shortage of available parking spaces. To address these issues, the city government initiated a smart parking project that leveraged Artificial Intelligence (AI) and Blockchain technologies (Badidi, 2022; Lewis & others, 2021).

Objectives: The primary objectives of the project were as follows:

- **Optimize Parking Space Utilization:** Use AI to predict parking space availability and optimize space allocation in real-time.
- **Reduce Traffic Congestion:** Implement AI algorithms to guide drivers to available parking spots efficiently, reducing traffic congestion and associated emissions.
- **Enhance User Experience:** Develop a user-friendly mobile app that provides real-time parking information, allows for easy payments using Blockchain technology, and offers predictive notifications to users.
- **Ensure Data Security:** Employ Blockchain for secure and transparent parking transactions and user data management.

Implementation: The project was implemented in several phases:

Phase 1: Data Collection and Infrastructure Setup

- IoT sensors were deployed in parking spaces across the city to collect real-time occupancy data (Khanna et al., 2021; Mazurek & Stroinski, 2022).
- Traffic cameras were integrated to monitor traffic flow around parking facilities.
- Blockchain infrastructure was established to support secure and tamper-resistant transactions.

Phase 2: AI Model Development

- Historical parking data, combined with real-time data from sensors, was used to train AI models.
- Predictive analytics models were developed to forecast parking space availability.
- Machine learning algorithms optimized traffic flow patterns based on real-time data.

Phase 3: Mobile App Development

- A user-friendly mobile app was created, offering features such as real-time parking availability, navigation to open spots, and Blockchain-based payment options.
- Predictive notifications were integrated to inform users of expected parking availability at their destinations.

Phase 4: Blockchain Integration

- Smart contracts on the Blockchain were developed to automate parking payments and access control.
- User digital identities were securely managed on the Blockchain.

Results and Insights: The smart parking project in City X yielded several notable results and insights:

- Parking space utilization increased by 30% as AI optimized space allocation, reducing the number of vacant spaces during peak hours.
- Traffic congestion in the vicinity of parking facilities decreased by 25%, improving the flow of vehicles and reducing emissions.
- User satisfaction surveys showed a 40% increase in overall satisfaction with the parking experience.
- Transactions on the Blockchain were secure and tamper-resistant, with no reported instances of fraud or disputes.
- The project's initial investment was offset by revenue gains, reduced infrastructure costs, and environmental savings over time.

City successful implementation of AI and Blockchain technologies in its smart parking system not only improved the parking experience for its residents and visitors but also contributed to reduced traffic congestion and environmental benefits. The project served as a model for other cities looking to tackle urban mobility challenges and enhance the quality of life for their citizens through innovative technology solutions.

CHALLENGES AND FUTURE DIRECTIONS

Challenges

- **Data Privacy and Security:** The collection of sensitive data, such as user identities and payment information, poses significant privacy and security concerns. Ensuring robust data protection measures and compliance with privacy regulations like GDPR is essential (Anitha et al., 2023; Boopathi, 2023; Karthik et al., 2023).
- **Scalability:** As urban populations continue to grow, smart parking systems must scale to accommodate increasing demands. Ensuring that AI and Blockchain infrastructure can handle higher volumes of data and transactions is crucial (Anitha et al., 2023; Boopathi et al., 2022, 2023; Boopathi & Kanike, 2023).
- **Interoperability:** Smart parking systems should seamlessly integrate with other urban mobility solutions, including public transportation, ridesharing, and electric vehicle charging networks. Achieving interoperability can enhance the overall urban mobility experience.
- **User Adoption:** Encouraging users to adopt new technologies and behaviors is a challenge. Effective communication, user-friendly interfaces, and incentives may be necessary to drive widespread adoption of smart parking systems.
- **Standardization:** The lack of standardized protocols and interfaces can hinder the interoperability of different smart parking solutions. Developing industry standards can facilitate compatibility between systems and vendors.

Future Directions

- **AI Advancements:** Future AI developments, such as more advanced machine learning models and increased computing power, will enable even more accurate predictions and optimizations in smart parking systems.
- **Blockchain Innovations:** Blockchain technology is evolving rapidly, with advancements like sharding and improved consensus mechanisms. These innovations can enhance scalability and reduce energy consumption in Blockchain-based parking systems.
- **Decentralized Finance (DeFi):** Integrating DeFi solutions into smart parking systems can enable automated and decentralized payment mechanisms, further streamlining transactions and reducing reliance on traditional financial institutions.
- **5G and Edge Computing:** The rollout of 5G networks and edge computing capabilities will enable real-time data processing and communication between IoT sensors and AI models, enhancing the responsiveness of smart parking systems (Hema et al., 2023; Rahamathunnisa et al., 2023; Syamala et al., 2023; Venkateswaran et al., 2023).
- **Environmental Impact:** Future smart parking systems may place a greater emphasis on sustainability, integrating renewable energy sources, EV charging infrastructure, and carbon offset programs to reduce their environmental footprint (Arunprasad & Boopathi, 2019; Boopathi, 2022a, 2022b; Gowri et al., 2023).
- **Smart Cities Integration:** Smart parking will become an integral component of broader smart city initiatives, with data sharing and collaboration among different urban services, leading to more holistic urban planning.

- **AI-Enabled Autonomous Vehicles:** The advent of autonomous vehicles can reshape parking dynamics. AI-powered smart parking systems may coordinate with self-driving cars to optimize parking space utilization and drop-off/pick-up zones.
- **User-Centric Design:** User-centric design principles will continue to drive the development of smart parking systems, ensuring that they meet the needs and preferences of diverse urban populations.

CONCLUSION

The integration of Artificial Intelligence (AI) and Blockchain technologies in smart parking systems represents a transformative approach to addressing the complex challenges of urban mobility, traffic congestion, and parking management in modern cities. This chapter has provided valuable insights and highlights the following key conclusions:

- The rapid urbanization of cities has intensified traffic congestion and the demand for parking spaces. Smart parking solutions powered by AI and Blockchain technologies have emerged as a necessity to alleviate these urban mobility challenges.
- AI-driven algorithms enable efficient allocation of parking spaces, reducing the number of vacant spots and maximizing utilization. This not only benefits users but also reduces the need for costly parking infrastructure expansions.
- Real-time data from IoT sensors, coupled with AI predictions, offers users convenience by guiding them to available parking spaces, reducing stress, and enhancing the overall parking experience.
- Blockchain technology ensures secure and transparent parking transactions. It minimizes the risk of fraud, provides tamper-resistant records, and enhances trust among users and parking operators.
- Blockchain-based decentralized digital identities offer users greater control over their personal data, promoting privacy and simplifying access control in parking facilities.
- Smart parking systems contribute to sustainability by reducing emissions through reduced idling and by supporting electric vehicle (EV) charging infrastructure. Blockchain can also facilitate the issuance and tracking of carbon credits for green parking initiatives.
- A comprehensive cost-benefit analysis demonstrates that the initial investment in AI and Blockchain-powered smart parking systems can be offset by substantial revenue gains, reduced infrastructure costs, and environmental savings in the long run.
- Continuous monitoring, user feedback, and iterative refinement of AI models and Blockchain systems are crucial for maintaining the success of smart parking solutions and adapting to changing urban conditions.
- The successful integration of AI and Blockchain technologies in smart parking systems in urban areas serves as a model for future smart cities seeking innovative solutions to urban mobility challenges.

In conclusion, AI and Blockchain technologies are not just reshaping the way we manage parking but are paving the way for more sustainable, efficient, and user-centric urban environments. The evolution of smart parking represents a critical step in creating smarter and more livable cities for the future.

REFERENCES

Al Amiri, W., Baza, M., Banawan, K., Mahmoud, M., Alasmary, W., & Akkaya, K. (2020). Towards secure smart parking system using blockchain technology. *2020 IEEE 17th Annual Consumer Communications & Networking Conference (CCNC)*, 1–2.

Al-Turjman, F., & Malekloo, A. (2019). Smart parking in IoT-enabled cities: A survey. *Sustainable Cities and Society*, *49*, 101608. doi:10.1016/j.scs.2019.101608

Alharbi, A., Halikias, G., Yamin, M., & Abi Sen, A. A. (2021). Web-based framework for smart parking system. *International Journal of Information Technology : an Official Journal of Bharati Vidyapeeth's Institute of Computer Applications and Management*, *13*(4), 1495–1502. doi:10.100741870-021-00725-8

Alves, B. R., Alves, G. V., Borges, A. P., & Leitão, P. (2019). Experimentation of negotiation protocols for consensus problems in smart parking systems. *Industrial Applications of Holonic and Multi-Agent Systems: 9th International Conference, HoloMAS 2019, Linz, Austria, August 26–29, 2019. Proceedings*, *9*, 189–202.

Anitha, C., Komala, C., Vivekanand, C. V., Lalitha, S., & Boopathi, S. (2023). Artificial Intelligence driven security model for Internet of Medical Things (IoMT). *IEEE Explore*, 1–7.

Arunprasad, R., & Boopathi, S. (2019). Chapter-4 Alternate Refrigerants for Minimization Environmental Impacts: A Review. In Advances in Engineering Technology (p. 75). AkiNik Publications.

Babu, B. S., Kamalakannan, J., Meenatchi, N., Karthik, S., & Boopathi, S. (2022). Economic impacts and reliability evaluation of battery by adopting Electric Vehicle. *IEEE Explore*, 1–6.

Badidi, E. (2022). Edge AI and blockchain for smart sustainable cities: Promise and potential. *Sustainability (Basel)*, *14*(13), 7609. doi:10.3390u14137609

Bale, A. S., Hashim, M. F., Bundele, K. S., & Vaishnav, J. (2023). 9 Applications and future trends of Artificial Intelligence and blockchain-powered Augmented Reality. Handbook of Augmented and Virtual Reality, 1, 137.

Bocek, T., Rodrigues, B. B., Strasser, T., & Stiller, B. (2017). Blockchains everywhere-a use-case of blockchains in the pharma supply-chain. *2017 IFIP/IEEE Symposium on Integrated Network and Service Management (IM)*, 772–777. 10.23919/INM.2017.7987376

Boopathi, S. (2022a). An investigation on gas emission concentration and relative emission rate of the near-dry wire-cut electrical discharge machining process. *Environmental Science and Pollution Research International*, *29*(57), 86237–86246. doi:10.100711356-021-17658-1 PMID:34837614

Boopathi, S. (2022b). Cryogenically treated and untreated stainless steel grade 317 in sustainable wire electrical discharge machining process: A comparative study. *Environmental Science and Pollution Research*, 1–10.

Boopathi, S. (2023). Securing Healthcare Systems Integrated With IoT: Fundamentals, Applications, and Future Trends. In Dynamics of Swarm Intelligence Health Analysis for the Next Generation (pp. 186–209). IGI Global.

Boopathi, S., & Kanike, U. K. (2023). Applications of Artificial Intelligent and Machine Learning Techniques in Image Processing. In *Handbook of Research on Thrust Technologies' Effect on Image Processing* (pp. 151–173). IGI Global. doi:10.4018/978-1-6684-8618-4.ch010

Boopathi, S., Pandey, B. K., & Pandey, D. (2023). Advances in Artificial Intelligence for Image Processing: Techniques, Applications, and Optimization. In Handbook of Research on Thrust Technologies' Effect on Image Processing (pp. 73–95). IGI Global.

Boopathi, S., Sureshkumar, M., & Sathiskumar, S. (2022). Parametric Optimization of LPG Refrigeration System Using Artificial Bee Colony Algorithm. *International Conference on Recent Advances in Mechanical Engineering Research and Development*, 97–105.

Bryatov, S., & Borodinov, A. (2019). Blockchain technology in the pharmaceutical supply chain: Researching a business model based on Hyperledger Fabric. *Proceedings of the International Conference on Information Technology and Nanotechnology (ITNT),* Samara, Russia, *10*, 1613–0073. 10.18287/1613-0073-2019-2416-134-140

Bui, V., Alaei, A., & Bui, M. (2022). On the integration of AI and IoT systems: A case study of airport smart parking. *Artificial Intelligence-Based Internet of Things Systems*, 419–444.

Cahyadi, N., Dorand, P., Rozi, N. R. F., Haq, L. A., & Maulana, R. I. (2023). A Literature Review for Understanding the Development of Smart Parking Systems. *Journal of Informatics and Communication Technology*, *5*(1), 46–56.

Chandrika, V., Sivakumar, A., Krishnan, T. S., Pradeep, J., Manikandan, S., & Boopathi, S. (2023). Theoretical Study on Power Distribution Systems for Electric Vehicles. In *Intelligent Engineering Applications and Applied Sciences for Sustainability* (pp. 1–19). IGI Global. doi:10.4018/979-8-3693-0044-2.ch001

Chang, Y., Iakovou, E., & Shi, W. (2020). Blockchain in global supply chains and cross border trade: A critical synthesis of the state-of-the-art, challenges and opportunities. *International Journal of Production Research*, *58*(7), 2082–2099. doi:10.1080/00207543.2019.1651946

Dhanya, D., Kumar, S. S., Thilagavathy, A., Prasad, D., & Boopathi, S. (2023). Data Analytics and Artificial Intelligence in the Circular Economy: Case Studies. In Intelligent Engineering Applications and Applied Sciences for Sustainability (pp. 40–58). IGI Global.

Dubey, C., Kumar, D., Singh, A. K., & Dwivedi, V. K. (2022). Confluence of Artificial Intelligence and Blockchain Powered Smart Contract in Finance System. *2022 International Conference on Computing, Communication, and Intelligent Systems (ICCCIS)*, 125–130. 10.1109/ICCCIS56430.2022.10037701

Friha, O., Ferrag, M. A., Shu, L., Maglaras, L., & Wang, X. (2021). Internet of things for the future of smart agriculture: A comprehensive survey of emerging technologies. *IEEE/CAA Journal of Automatica Sinica, 8*(4), 718–752.

Gowri, N. V., Dwivedi, J. N., Krishnaveni, K., Boopathi, S., Palaniappan, M., & Medikondu, N. R. (2023). Experimental investigation and multi-objective optimization of eco-friendly near-dry electrical discharge machining of shape memory alloy using Cu/SiC/Gr composite electrode. *Environmental Science and Pollution Research International*, *30*(49), 1–19. doi:10.100711356-023-26983-6 PMID:37126160

Gunasekaran, K., & Boopathi, S. (2023). Artificial Intelligence in Water Treatments and Water Resource Assessments. In *Artificial Intelligence Applications in Water Treatment and Water Resource Management* (pp. 71–98). IGI Global. doi:10.4018/978-1-6684-6791-6.ch004

Haritha, T., & Anitha, A. (2023). Asymmetric Consortium Blockchain and Homomorphically Polynomial-Based PIR for Secured Smart Parking Systems. *Computers, Materials & Continua*, *75*(2), 3923–3939. doi:10.32604/cmc.2023.036278

Hema, N., Krishnamoorthy, N., Chavan, S. M., Kumar, N., Sabarimuthu, M., & Boopathi, S. (2023). A Study on an Internet of Things (IoT)-Enabled Smart Solar Grid System. In *Handbook of Research on Deep Learning Techniques for Cloud-Based Industrial IoT* (pp. 290–308). IGI Global. doi:10.4018/978-1-6684-8098-4.ch017

Hu, J., He, D., Zhao, Q., & Choo, K.-K. R. (2019). Parking management: A blockchain-based privacy-preserving system. *IEEE Consumer Electronics Magazine*, *8*(4), 45–49. doi:10.1109/MCE.2019.2905490

Ibrahim, M., Lee, Y., Kahng, H.-K., Kim, S., & Kim, D.-H. (2022). Blockchain-based parking sharing service for smart city development. *Computers & Electrical Engineering*, *103*, 108267. doi:10.1016/j.compeleceng.2022.108267

Jung, I., Lee, J., & Hwang, K. (2022). Smart parking management system using AI. *Webology*, *19*(1), 4629–4638. doi:10.14704/WEB/V19I1/WEB19307

Karthik, S., Hemalatha, R., Aruna, R., Deivakani, M., Reddy, R. V. K., & Boopathi, S. (2023). Study on Healthcare Security System-Integrated Internet of Things (IoT). In Perspectives and Considerations on the Evolution of Smart Systems (pp. 342–362). IGI Global.

Ke, R., Zhuang, Y., Pu, Z., & Wang, Y. (2020). A smart, efficient, and reliable parking surveillance system with edge artificial intelligence on IoT devices. *IEEE Transactions on Intelligent Transportation Systems*, *22*(8), 4962–4974. doi:10.1109/TITS.2020.2984197

Khanna, A., Sah, A., Bolshev, V., Jasinski, M., Vinogradov, A., Leonowicz, Z., & Jasiński, M. (2021). Blockchain: Future of e-governance in smart cities. *Sustainability (Basel)*, *13*(21), 11840. doi:10.3390u132111840

Koshariya, A. K., Kalaiyarasi, D., Jovith, A. A., Sivakami, T., Hasan, D. S., & Boopathi, S. (2023). AI-Enabled IoT and WSN-Integrated Smart Agriculture System. In *Artificial Intelligence Tools and Technologies for Smart Farming and Agriculture Practices* (pp. 200–218). IGI Global. doi:10.4018/978-1-6684-8516-3.ch011

Kumar, R., Arjunaditya, Singh, D., Srinivasan, K., & Hu, Y.-C. (2022). AI-powered blockchain technology for public health: A contemporary review, open challenges, and future research directions. *Health Care*, *11*(1), 81. PMID:36611541

Lewis, E., & ... (2021). Smart city software systems and Internet of Things sensors in sustainable urban governance networks. *Geopolitics, History, and International Relations*, *13*(1), 9–19.

Maguluri, L. P., Ananth, J., Hariram, S., Geetha, C., Bhaskar, A., & Boopathi, S. (2023). Smart Vehicle-Emissions Monitoring System Using Internet of Things (IoT). In Handbook of Research on Safe Disposal Methods of Municipal Solid Wastes for a Sustainable Environment (pp. 191–211). IGI Global.

Maheswari, B. U., Imambi, S. S., Hasan, D., Meenakshi, S., Pratheep, V., & Boopathi, S. (2023). Internet of Things and Machine Learning-Integrated Smart Robotics. In Global Perspectives on Robotics and Autonomous Systems: Development and Applications (pp. 240–258). IGI Global. doi:10.4018/978-1-6684-7791-5.ch010

Mattke, J., Hund, A., Maier, C., & Weitzel, T. (2019). How an Enterprise Blockchain Application in the US Pharmaceuticals Supply Chain is Saving Lives. *MIS Quarterly Executive*, *18*(4), 245–261. doi:10.17705/2msqe.00019

Mazurek, C., & Stroinski, M. (2022). *Technology pillars for digital transformation of cities based on open software architecture for end2end data streaming*. Academic Press.

Nishanth, J., Deshmukh, M. A., Kushwah, R., Kushwaha, K. K., Balaji, S., & Sampath, B. (2023). Particle Swarm Optimization of Hybrid Renewable Energy Systems. In *Intelligent Engineering Applications and Applied Sciences for Sustainability* (pp. 291–308). IGI Global. doi:10.4018/979-8-3693-0044-2.ch016

Park, A., & Li, H. (2021). The effect of blockchain technology on supply chain sustainability performances. *Sustainability (Basel)*, *13*(4), 1726. doi:10.3390u13041726

Rabah, K., & ... (2018). Convergence of AI, IoT, big data and blockchain: A review. *Law Institute Journal*, *1*(1), 1–18.

Rahamathunnisa, U., Sudhakar, K., Murugan, T. K., Thivaharan, S., Rajkumar, M., & Boopathi, S. (2023). Cloud Computing Principles for Optimizing Robot Task Offloading Processes. In *AI-Enabled Social Robotics in Human Care Services* (pp. 188–211). IGI Global. doi:10.4018/978-1-6684-8171-4.ch007

Ramudu, K., Mohan, V. M., Jyothirmai, D., Prasad, D., Agrawal, R., & Boopathi, S. (2023). Machine Learning and Artificial Intelligence in Disease Prediction: Applications, Challenges, Limitations, Case Studies, and Future Directions. In Contemporary Applications of Data Fusion for Advanced Healthcare Informatics (pp. 297–318). IGI Global.

Ravisankar, A., Sampath, B., & Asif, M. M. (2023). Economic Studies on Automobile Management: Working Capital and Investment Analysis. In Multidisciplinary Approaches to Organizational Governance During Health Crises (pp. 169–198). IGI Global.

Samikannu, R., Koshariya, A. K., Poornima, E., Ramesh, S., Kumar, A., & Boopathi, S. (2022). Sustainable Development in Modern Aquaponics Cultivation Systems Using IoT Technologies. In *Human Agro-Energy Optimization for Business and Industry* (pp. 105–127). IGI Global.

Sathya, A. (2022). A Blockchain-Based and IoT-Powered Smart Parking System for Smart Cities. *Communication, Software and Networks Proceedings*, *2022*, 583–590.

Sengeni, D., Padmapriya, G., Imambi, S. S., Suganthi, D., Suri, A., & Boopathi, S. (2023). Biomedical Waste Handling Method Using Artificial Intelligence Techniques. In *Handbook of Research on Safe Disposal Methods of Municipal Solid Wastes for a Sustainable Environment* (pp. 306–323). IGI Global. doi:10.4018/978-1-6684-8117-2.ch022

Shukla, R. G., Agarwal, A., & Shukla, S. (2020). Blockchain-powered smart healthcare system. In *Handbook of research on blockchain technology* (pp. 245–270). Elsevier. doi:10.1016/B978-0-12-819816-2.00010-1

Singh, S. K., Pan, Y., & Park, J. H. (2022). Blockchain-enabled secure framework for energy-efficient smart parking in sustainable city environment. *Sustainable Cities and Society*, *76*, 103364. doi:10.1016/j.scs.2021.103364

Syamala, M., Komala, C., Pramila, P., Dash, S., Meenakshi, S., & Boopathi, S. (2023). Machine Learning-Integrated IoT-Based Smart Home Energy Management System. In *Handbook of Research on Deep Learning Techniques for Cloud-Based Industrial IoT* (pp. 219–235). IGI Global. doi:10.4018/978-1-6684-8098-4.ch013

Teran, K., Žibret, G., & Fanetti, M. (2020). Impact of urbanization and steel mill emissions on elemental composition of street dust and corresponding particle characterization. *Journal of Hazardous Materials*, *384*, 120963. doi:10.1016/j.jhazmat.2019.120963 PMID:31628063

Turki, M., Dammak, B., & Mars, R. (2022). A Private Smart parking solution based on Blockchain and AI. *2022 15th International Conference on Security of Information and Networks (SIN)*, 1–7.

Venkateswaran, N., Vidhya, K., Ayyannan, M., Chavan, S. M., Sekar, K., & Boopathi, S. (2023). A Study on Smart Energy Management Framework Using Cloud Computing. In 5G, Artificial Intelligence, and Next Generation Internet of Things: Digital Innovation for Green and Sustainable Economies (pp. 189–212). IGI Global. doi:10.4018/978-1-6684-8634-4.ch009

Waheed, A., & Krishna, P. V. (2020). Comparing biometric and blockchain security mechanisms in smart parking system. *2020 International Conference on Inventive Computation Technologies (ICICT)*, 634–638. 10.1109/ICICT48043.2020.9112483

Xiao, W., Liu, C., Wang, H., Zhou, M., Hossain, M. S., Alrashoud, M., & Muhammad, G. (2020). Blockchain for secure-GaS: Blockchain-powered secure natural gas IoT system with AI-enabled gas prediction and transaction in smart city. *IEEE Internet of Things Journal*, *8*(8), 6305–6312. doi:10.1109/JIOT.2020.3028773

Zhang, C., Zhu, L., & Xu, C. (2022). BSDP: Blockchain-based smart parking for digital-twin empowered vehicular sensing networks with privacy protection. *IEEE Transactions on Industrial Informatics*.

Zhang, C., Zhu, L., Xu, C., Zhang, C., Sharif, K., Wu, H., & Westermann, H. (2020). BSFP: Blockchain-enabled smart parking with fairness, reliability and privacy protection. *IEEE Transactions on Vehicular Technology*, *69*(6), 6578–6591. doi:10.1109/TVT.2020.2984621

APPENDIX

AI - Artificial Intelligence
IoT - Internet of Things
EV - Electric Vehicle
X% - Represents a percentage value (e.g., 30%)
USD - United States Dollar
GPS - Global Positioning System
QR Code - Quick Response Code
HTTPS - Hypertext Transfer Protocol Secure
SSL - Secure Sockets Layer
API - Application Programming Interface
UI - User Interface
UX - User Experience
ICO - Initial Coin Offering (in the context of cryptocurrencies)
KYC - Know Your Customer (related to user identity verification)
ICO - Initial Coin Offering (related to cryptocurrencies)
ROI - Return on Investment
CV - Computer Vision
GDPR - General Data Protection Regulation (related to data privacy)
EMV - Europay, Mastercard, and Visa (related to payment card technology)
POS - Point of Sale (related to payment terminals)

Chapter 11
Machine Learning and Deep Learning for Intelligent Systems in Small Aircraft Applications

U Rahamathunnisa
School of Computer Science Engineering and Information Systems, Vellore Institute of Technology, India

Akash Mohanty
School of Mechanical Engineering, Vellore Institute of Technology, India

K. Sudhakar
Department of Computer Science and Engineering, Madanapalle Institute of Technology and Science, Madanapalle, India

S. Anitha Jebamani
Department of Information Technology, Sri Sai Ram Engineering College, Chennai, India

R. Udendhran
Department of Computational Intelligence, SRM Institute of Science and Technology, India

Sureshkumar Myilsamy
Mechanical Engineering, Bannari Amman Institute of Technology, India

ABSTRACT

This chapter explores the integration of machine learning and deep learning techniques in small aircraft applications. The aviation industry is exploring innovative solutions to improve safety, efficiency, and performance in these operations. The chapter explores the advantages, challenges, and future prospects of implementing intelligent systems in small aircraft, including autopilot systems, navigation assistance, fault detection, and pilot support systems. Real-world case studies and applications demonstrate the transformative impact of these technologies on small aircraft operations. The chapter provides a comprehensive overview of the latest advancements in machine learning and deep learning, highlighting their pivotal role in improving small aircraft intelligence, safety, and efficiency.

DOI: 10.4018/978-1-6684-9999-3.ch011

INTRODUCTION

The integration of machine learning and deep learning technologies into small aircraft applications represents a significant leap forward in the aviation industry. As aviation continues to advance, there is a growing demand for innovative solutions that enhance the safety, efficiency, and overall performance of small aircraft. Machine learning and deep learning have emerged as powerful tools to address these demands, offering the potential to revolutionize how small aircraft operate and interact with their environments. Small aircraft play a crucial role in various industries, from agriculture and surveillance to personal transportation. Ensuring the safety, efficiency, and reliability of these aircraft is paramount (Ali, 1990). Advancements in machine learning and deep learning are revolutionizing the development of intelligent systems for small aircraft applications, shaping their future.

One of the most critical aspects of small aircraft operations is safety. Machine learning has made significant strides in predictive maintenance. By analyzing data from sensors and historical maintenance records, ML algorithms can predict when components are likely to fail, allowing for proactive maintenance. This not only reduces the risk of in-flight failures but also extends the lifespan of critical aircraft components. Additionally, deep learning techniques can analyze complex sensor data, such as images and audio, to detect anomalies that may not be apparent through traditional methods (Li & Gupta, 1995). These advancements help small aircraft operators identify and address potential issues before they become critical.

Autonomous flight capabilities are becoming increasingly important in small aircraft applications. Machine learning algorithms can process data from various sensors, including GPS, lidar, and cameras, to enable precise navigation and obstacle avoidance. These systems are particularly valuable in scenarios where human pilots may face challenges, such as low visibility conditions or remote locations (Krishnakumar, 2002). Deep learning, with its ability to handle vast amounts of data, allows small aircraft to adapt to changing environments in real-time. This means improved safety and efficiency, especially for tasks like crop monitoring, where consistent and precise flight paths are essential.

Efficiency is a key factor in the operation of small aircraft, particularly for applications like aerial surveillance, search and rescue, or wildlife monitoring. Machine learning can optimize flight routes, taking into account factors like weather conditions, wind patterns, and fuel consumption (Volponi et al., 2004). By constantly analyzing and adjusting the flight plan, small aircraft can minimize fuel usage and extend their operational range. Deep learning, in combination with computer vision, can also aid in target identification and tracking during surveillance missions. These systems can automatically detect and classify objects of interest, reducing the workload on human operators.

Small aircraft operators can benefit from the collaboration between humans and intelligent systems. Machine learning models can provide real-time assistance to pilots by processing data from various sensors and offering suggestions for optimal decision-making. This human-machine collaboration not only enhances safety but also reduces pilot fatigue and workload (Long et al., 2007). Moreover, deep learning models can analyze data from on-board cameras to monitor pilot behavior. By detecting signs of fatigue or distraction, these systems can issue alerts or take corrective actions, ensuring that pilots remain focused and alert during their missions.

Advancements in machine learning and deep learning are transforming small aircraft applications by enhancing safety, efficiency, and reliability. Predictive maintenance, autonomous navigation, and optimized operations are just a few examples of how these technologies are reshaping the small aircraft industry (Buckley et al., 2014). The evolution of technologies will lead to more sophisticated and capable intelligent systems for small aircraft, benefiting traditional industries and opening new possibilities in urban air mobility and environmental monitoring, thanks to machine learning and deep learning algorithms.

Background and Motivation

The aviation industry has a long history of adopting cutting-edge technologies to improve safety and efficiency. However, small aircraft have often lagged behind their larger counterparts in terms of automation and intelligence. Historically, the complexity and cost associated with implementing advanced systems in small aircraft limited their accessibility. Recent advancements in machine learning and deep learning have paved the way for more cost-effective and scalable solutions (Subramanian, 2021). These technologies offer the capability to process vast amounts of data, make real-time decisions, and adapt to changing conditions—capabilities that are particularly valuable in the context of small aircraft operations. The motivation behind this chapter is to explore how machine learning and deep learning can be harnessed to create intelligent systems for small aircraft. This study aims to explore the advantages and challenges of these technologies, with the aim of promoting further research and development in this area, thereby enhancing the integration of intelligent systems into small aircraft applications.

Scope of the Chapter

This chapter explores machine learning and deep learning in small aircraft applications, focusing on their fundamentals, applications in autopilot systems, navigation aids, fault detection, and pilot assistance systems. It presents real-world case studies and examples, discussing ethical considerations, data privacy, safety, and security concerns, and the need for responsible development and deployment of intelligent systems in aviation.

Objectives

- Provide an in-depth exploration of the role of machine learning and deep learning techniques in enhancing safety, efficiency, and performance in small aircraft applications.
- Emphasize the advantages of implementing intelligent systems like autopilot, navigation assistance, fault detection, and pilot support systems in small aircraft, showcasing how these technologies can improve overall operations.
- Analyze the challenges and limitations associated with integrating machine learning and deep learning into small aircraft systems, including issues related to data availability, model interpretability, and regulatory compliance.
- Provide a comprehensive overview of recent advancements in machine learning and deep learning relevant to small aircraft, showcasing examples of successful implementations and their impact on the industry.
- Investigate how these technologies contribute to the enhancement of small aircraft safety, including accident prevention, collision avoidance, and emergency response.
- Explore how machine learning and deep learning techniques can optimize fuel consumption, route planning, and operational efficiency in small aircraft, leading to cost savings and reduced environmental impact.

- Discuss how intelligent systems assist pilots in making informed decisions during flight, mitigating cognitive overload and improving situational awareness.
- Predict and discuss the future prospects of machine learning and deep learning in small aircraft applications, considering emerging technologies, regulatory developments, and industry trends.
- Present a comprehensive overview of the state-of-the-art technologies, applications, and research in the field of machine learning and deep learning for small aircraft, catering to both technical and non-technical readers.
- Contribute to the knowledge and awareness of the potential benefits and challenges of implementing intelligent systems in small aircraft, fostering discussions and innovations in the aviation industry.

MACHINE LEARNING FOUNDATIONS

This section introduces the fundamental concepts of machine learning, its types (supervised, unsupervised, reinforcement learning), and its role in transforming small aircraft applications, defining key terminologies like data, features, labels, and algorithms (Boopathi & Kanike, 2023; Maheswari et al., 2023; Ramudu et al., 2023; Syamala et al., 2023).

Supervised Learning for Small Aircraft

This subsection explores the application of supervised learning techniques to small aircraft systems, focusing on data collection, labeling, model training, and evaluation. Case studies illustrate how supervised learning can be used for tasks like aircraft performance prediction or system health monitoring (Agrawal, Pitchai, et al., 2023; Boopathi, 2023a; Veeranjaneyulu, Boopathi, Narasimharao, et al., 2023; Zekrifa et al., 2023).

Unsupervised Learning Applications

Unsupervised learning is explored in this section, highlighting its relevance in small aircraft applications. Clustering and dimensionality reduction techniques are discussed in the context of data analysis and anomaly detection. Real-world examples may showcase how unsupervised learning helps in identifying unusual behavior or patterns in aircraft data (Maheswari et al., 2023; Ramudu et al., 2023; Syamala et al., 2023; Zekrifa et al., 2023).

Reinforcement Learning in Aviation

Reinforcement learning is a unique method for training intelligent agents, with potential applications in aviation. It can be used in autonomous flight, adaptive control, and decision-making processes in small aircraft. Case studies illustrate how reinforcement learning agents can navigate and optimize operations (Maheswari et al., 2023; Veeranjaneyulu, Boopathi, Narasimharao, et al., 2023; Zekrifa et al., 2023).

Understanding Reinforcement Learning

This section introduces reinforcement learning (RL), focusing on its core components like agents, environments, actions, rewards, and policies, and introduces Markov Decision Processes (MDPs) as a formal framework for modeling aviation problems.

Applications in Autonomous Flight

The article explores the potential of Natural Language Processing (RL) in aviation for autonomous flight, highlighting how RL algorithms can train small aircraft to make dynamic decisions, adapt to weather conditions, and optimize flight trajectories, with real-world case studies illustrating its role.

Adaptive Control Systems

RL can be applied to create adaptive control systems that learn and adjust aircraft control strategies in real-time. This subsection explores how RL can enhance small aircraft control, leading to improved stability, efficiency, and maneuverability. Examples may include RL-based adaptive control for drones or unmanned aerial vehicles (UAVs).

Decision-Making and Path Planning

RL techniques are also invaluable for decision-making and path planning in aviation. This section explores the use of RL algorithms for optimizing flight routes, managing air traffic, making real-time emergency decisions, reducing travel time, and enhancing flight safety.

Challenges and Future Directions

The article discusses the challenges of Robotics (RL) in aviation, including safety concerns, training complexities, and robustness. It also explores emerging research areas and future directions in RL applications, including multi-agent RL for coordinated flight and electric and hybrid aircraft.

By the end of this section, readers will have a thorough understanding of how reinforcement learning is reshaping the aviation landscape, from autonomous flight to adaptive control and beyond. The real-world examples and insights provided will illuminate the transformative potential of RL in small aircraft applications.

DEEP LEARNING ESSENTIALS

Understanding Neural Networks

This section serves as a foundational introduction to neural networks, the building blocks of deep learning. This text provides a comprehensive understanding of deep learning principles, including artificial neural network architecture, layers, activation functions, and backpropagation algorithm (Bikash Chandra Saha, 2022; Boopathi, 2023a; Boopathi, Gavaskar, et al., 2021; Chandrika et al., 2023; Hema et al., 2023; Sampath et al., 2022).

Convolutional Neural Networks (CNNs) for Image Processing

CNNs are essential for image-related tasks in deep learning. They are applied to small aircraft applications like obstacle detection, terrain mapping, and object tracking. Real-world examples and case studies demonstrate their effectiveness in enhancing visual perception.

Recurrent Neural Networks (RNNs) for Time-Series Data

RNNs are essential for handling sequential data, making them valuable in aviation for analyzing time-series data from sensors and instruments. The study explores the architecture of Recurrent Neural Networks (RNNs), including hidden states, LSTM, and GRUs, and their practical applications in small aircraft for predictive maintenance and flight data analysis.

Transfer Learning and Pre-trained Models

Transfer learning is a technique that uses pre-trained deep learning models for specific tasks, even with limited data. It's particularly useful in small aircraft applications, such as anomaly detection or cockpit voice recognition, where pre-trained models can be repurposed for improved performance. By the end of this chapter, readers will have a comprehensive grasp of the essential components of deep learning, including neural networks, CNNs, RNNs, and transfer learning. They will also gain insights into how these deep learning techniques can be harnessed to address specific challenges and enhance small aircraft applications, from image processing to time-series data analysis.

SMALL AIRCRAFT AUTOPILOT SYSTEMS

Small aircraft autopilot systems have long been a pivotal component of modern aviation. They provide a crucial safety net for pilots, enhance flight precision, and reduce pilot workload, especially during long flights or challenging weather conditions. In recent years, the integration of machine learning and deep learning technologies into autopilot systems has opened up new possibilities for improving their capabilities. This chapter explores the evolution, functionality, and enhancements of machine learning and deep learning in small aircraft autopilot systems (Hamilton et al., 2022). It delves into the historical development of these systems, from early rudimentary ones to advanced computerized ones. The chapter provides an overview of traditional autopilot systems in small aircraft, including attitude and heading reference systems (AHRS), navigation systems, and control surfaces, and their role in maintaining flight stability and reducing pilot workload.

Machine learning techniques are revolutionizing small aircraft autopilot systems, enhancing tasks like auto-trimming and auto-leveling. Reinforcement learning optimizes control strategies, adapting to flight conditions and pilot preferences (Subramanian, 2021). Deep learning algorithms, like convolutional neural networks (CNNs) and recurrent neural networks (RNNs), improve navigation by processing visual data for terrain and obstacle recognition, and enabling real-time adjustments based on sensor inputs.

Machine learning models in small aircraft autopilot systems can detect faults and anomalies more effectively, ensuring flight safety. Adaptive control systems use machine learning to adjust control parameters based on performance data. This integration influences pilot-automation interaction, providing informative recommendations and enhancing situational awareness. The importance of a harmonious human-machine partnership is highlighted in small aircraft operations (Santos et al., 2021).

The integration of machine learning and deep learning in small aircraft autopilot systems offers numerous benefits, but also presents challenges such as safety, reliability, data quality, and regulatory compliance. Ethical considerations are also discussed. These technologies improve navigation, fault detection, adaptive control, and human-machine collaboration, reshaping small aircraft operations for safer, more efficient flights. Responsible development and integration are crucial for aviation safety and reliability.

AUTOMATION IN AVIATION

This chapter explores the implementation and components of automation in aviation, focusing on small aircraft applications (Benavente-Sánchez et al., 2021), enhancing safety, efficiency, and flight performance (Figure 1).

Figure 1. Automation in aviation components and the implementation

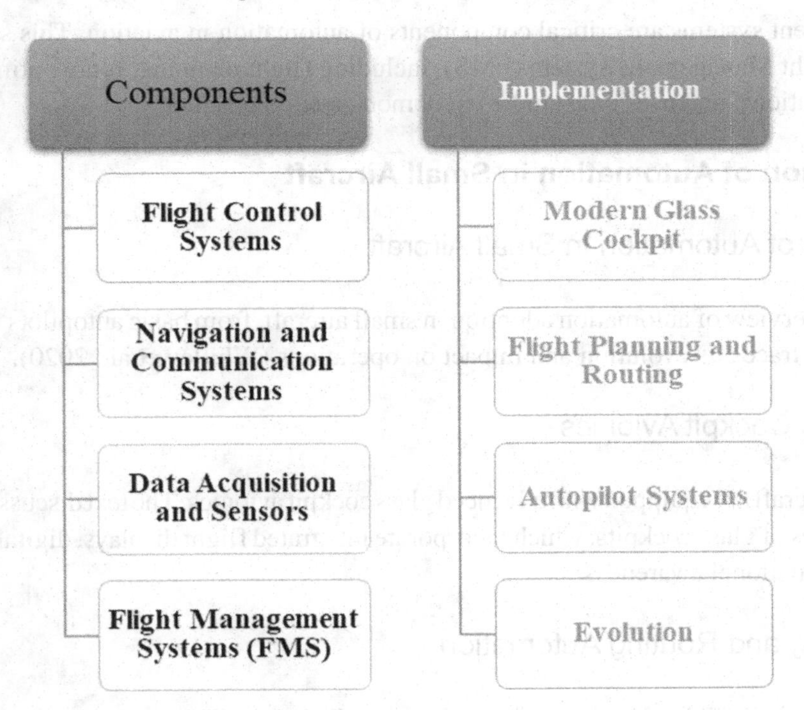

The Components of Aviation Automation

Flight Control Systems

This section delves into the fundamental components of aviation automation, focusing on flight control systems, including control surfaces, autopilot systems, and fly-by-wire technology, and their interplay for a stable and precise flight (Bange et al., 2021).

Navigation and Communication Systems

Navigation and communication systems, including GPS, inertial navigation systems, and communication protocols, are crucial for safe and efficient aircraft operations.

Data Acquisition and Sensors

This section explores data acquisition and sensors in automation, highlighting their role in collecting data on aircraft performance and environmental conditions, and their integration to enhance system accuracy.

Flight Management Systems (FMS)

Flight management systems are critical components of automation in aviation. This study explores the functions of Flight Management System (FMS), including flight planning, route optimization, and performance calculations, and uses case studies to demonstrate its efficiency.

Implementation of Automation in Small Aircraft

The Evolution of Automation in Small Aircraft

The historical overview of automation adoption in small aircraft, from basic autopilots to advanced glass cockpit systems, traces its evolution and impact on operations (Whitley et al., 2020).

Modern Glass Cockpit Avionics

Modern small aircraft are equipped with advanced glass cockpit avionics. The text discusses the advantages and disadvantages of glass cockpits, which incorporate integrated flight displays, digital communication, and improved situational awareness.

Flight Planning and Routing Automation

Automation extends to flight planning and routing in small aircraft. This study investigates how software tools aid pilots in creating optimal flight plans, considering factors like weather, fuel efficiency, and airspace regulations.

Autopilot Systems in Small Aircraft

Small aircraft often rely on autopilot systems to reduce pilot workload and enhance flight stability. This text provides an in-depth analysis of the components and operation of autopilot systems, including their integration with navigation and sensor systems.

Challenges and Future Directions

While automation brings significant advantages to aviation, it also presents challenges, such as pilot training, system reliability, and cybersecurity. The text discusses the challenges and future directions of small aircraft operations, highlighting the potential impact of artificial intelligence and machine learning (Krishnakumar, 2002; Li & Gupta, 1995; Volponi et al., 2004). By the end of this chapter, readers will have a comprehensive understanding of the key components of aviation automation and how they are implemented in small aircraft. They will also gain insights into the evolution of automation in aviation and the challenges and opportunities that lie ahead in this dynamic field. This chapter delves into the significant role of machine learning and deep learning in the advancement of autopilot and flight control systems in aviation, emphasizing the importance of automation for safety and efficiency.

Machine Learning in Flight Control

Supervised Learning for Flight Control

This section explores the application of supervised learning techniques in flight control, focusing on how ML algorithms can predict and optimize control inputs based on sensor data.

Figure 2. Machine learning in flight control

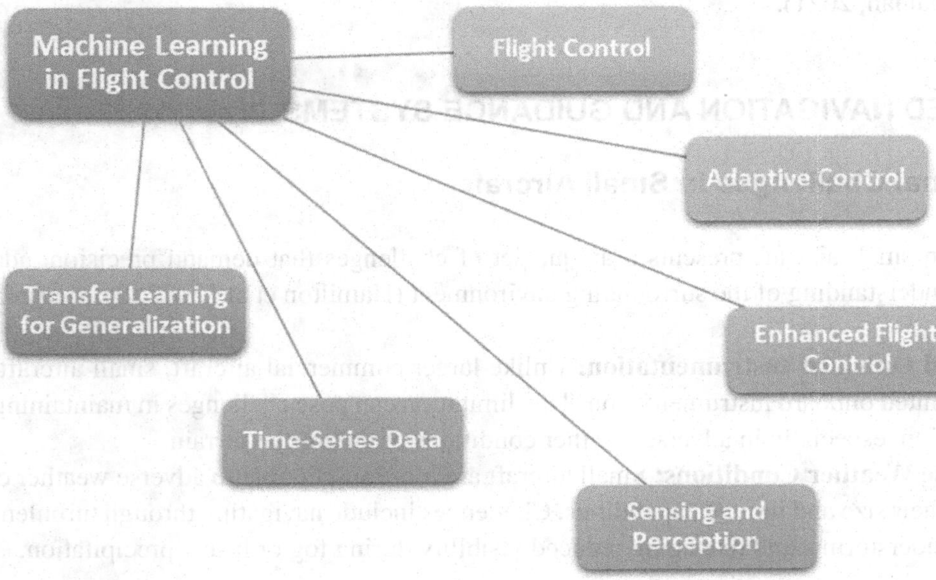

Reinforcement Learning for Adaptive Control

Reinforcement learning is a crucial technique in adaptive flight control, enabling algorithms to adapt to varying flight conditions and optimize control strategies in real-time.

Deep Learning for Enhanced Flight Control

Convolutional Neural Networks (CNNs) for Sensing and Perception

The study explores the use of Convolutional Neural Networks (CNNs) in flight control systems, focusing on object detection, terrain recognition, obstacle avoidance, and improving aircraft situational awareness (Hamilton et al., 2022).

Recurrent Neural Networks (RNNs) for Time-Series Data

RNNs are utilized for handling time-series data from sensors and instruments, enabling real-time decision-making and control adjustments by modeling and predicting aircraft behavior.

Transfer Learning for Generalization

Transfer learning is a technique that uses pre-trained deep learning models from computer vision and natural language processing to improve flight control tasks.

This section discusses the practical applications of machine learning and deep learning in flight control, including autonomous flight, adaptive control, and precision landing systems. It addresses challenges like safety, reliability, and regulatory compliance. The chapter provides insights into the future of flight control, emerging trends, and research directions, aiming to provide readers with a comprehensive understanding of how these technologies are transforming autopilot and flight control systems in aviation (Subramanian, 2021).

ENHANCED NAVIGATION AND GUIDANCE SYSTEMS

Navigational Challenges in Small Aircraft

Navigating in small aircraft presents a unique set of challenges that demand precision, adaptability, and a deep understanding of the surrounding environment (Hamilton et al., 2022; Subramanian, 2021).

- **Limited Onboard Instrumentation:** Unlike larger commercial aircraft, small aircraft typically have limited onboard instrumentation. This limitation can pose challenges in maintaining accurate navigation, especially in adverse weather conditions or unfamiliar terrain.
- **Adverse Weather Conditions:** Small aircraft are more susceptible to adverse weather conditions due to their size and limited capabilities. Challenges include navigating through turbulence, avoiding thunderstorms, and managing reduced visibility during fog or heavy precipitation.

- **Precision and Altitude Control:** Maintaining precise altitude and course control is crucial, particularly during takeoff, landing, and while flying at low altitudes. Small aircraft often operate in airspace with varying terrain, requiring constant adjustments to ensure safety.
- **Congested Airspace:** Navigating through congested airspace, especially in urban or high-traffic regions, presents additional challenges for small aircraft. Pilots must remain vigilant to avoid collisions and adhere to complex air traffic control instructions.
- **Limited Range and Endurance:** Small aircraft, such as general aviation planes and drones, often have limited range and endurance. This imposes navigational constraints, necessitating careful route planning and fuel management.
- **Remote and Unfamiliar Locations:** Small aircraft operations frequently involve reaching remote or unfamiliar locations, such as rural airstrips, mountainous terrain, or isolated islands. Navigating to and from these destinations requires specialized skills and equipment (Boopathi, 2023b; Boopathi & Myilsamy, 2021; Subha et al., 2023).
- **Navigation Redundancy:** Ensuring navigation redundancy is essential for small aircraft. In the event of a navigation system failure, pilots must rely on backup instruments and traditional navigation methods.
- **Regulatory Compliance:** Navigating within the boundaries of airspace regulations is critical. Small aircraft operators must adhere to specific rules, including airspace classifications, altitudes, and reporting requirements.
- **Pilot Workload:** Navigating through these challenges places a significant workload on pilots, especially in single-pilot operations. Minimizing pilot fatigue and optimizing situational awareness are crucial for safe navigation.
- **Emerging Navigation Technologies:** Despite these challenges, emerging navigation technologies, including GPS enhancements, satellite-based augmentation systems (SBAS), and advanced navigation software, are helping small aircraft pilots overcome many of these obstacles. Machine learning and deep learning play a vital role in enhancing navigation systems, providing real-time data analysis and decision support.

Understanding navigational challenges in small aircraft operations highlights the need for advanced navigation systems using machine learning and deep learning for enhanced safety and efficiency.

Machine learning (ML) has revolutionized navigation in small aircraft by providing advanced techniques for data analysis, route planning, obstacle detection, and decision-making (Figure 2).

- **Data-Driven Route Planning:** Machine learning algorithms can analyze historical flight data, weather conditions, and airspace constraints to optimize flight routes. This results in more fuel-efficient and time-saving routes, especially for small aircraft with limited range and endurance (Dooraki & Lee, 2021; Hu & Wang, 2020).
- **Predictive Weather Analysis:** ML models can process real-time weather data and make predictions about weather patterns along the planned route. Pilots can receive timely weather updates, enabling them to make informed decisions and adjust their navigation to avoid adverse conditions (Boopathi & Sivakumar, 2016; Ramudu et al., 2023; Sampath et al., 2022).

Figure 3. The enhanced navigation and guidance systems

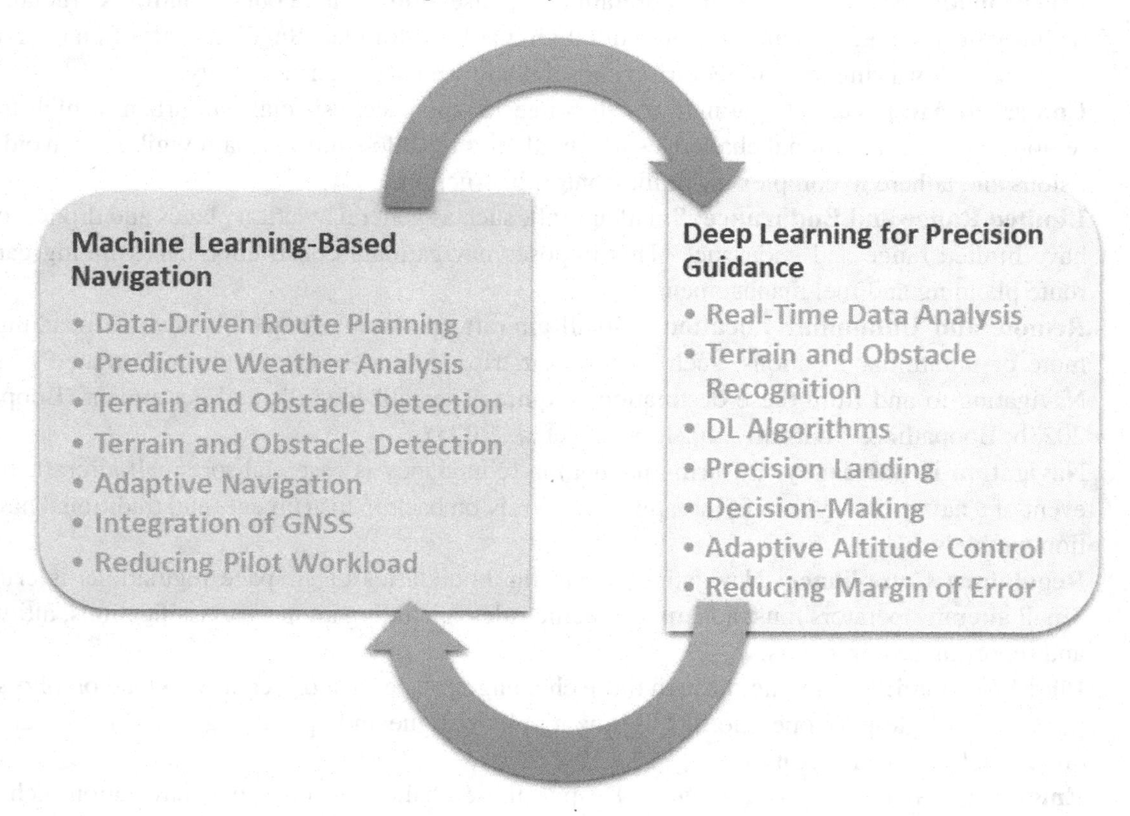

- **Terrain and Obstacle Detection:** Machine learning algorithms, including convolutional neural networks (CNNs), analyze data from onboard sensors and cameras to detect and recognize terrain features and potential obstacles. This capability enhances situational awareness and aids in terrain avoidance.
- **Adaptive Navigation:** ML-based navigation systems can adapt to changing flight conditions in real-time. These systems continuously analyze sensor data and adjust the aircraft's flight path and altitude to ensure optimal performance and safety.
- **Anomaly Detection:** Machine learning models can identify anomalies or deviations from the expected flight path. This feature is particularly valuable for small aircraft navigating through congested airspace, helping to detect and avoid potential conflicts with other aircraft.
- **Human-Machine Collaboration:** ML-based navigation systems can provide pilots with informative recommendations and alerts. This fosters a collaborative environment where pilots and automation work together to optimize navigation and address navigational challenges effectively (Boopathi, Balasubramani, et al., 2021; Boopathi, Kumar, et al., 2023; Domakonda et al., 2022; Samikannu et al., 2022).
- **Integration of GNSS Augmentation Systems:** Global Navigation Satellite System (GNSS) augmentation systems, such as SBAS (Satellite-Based Augmentation System), leverage ML for more accurate and reliable positioning. These systems enhance navigation accuracy, particularly during critical phases of flight.

- **Reducing Pilot Workload:** ML-based navigation systems are designed to reduce pilot workload, allowing them to focus on critical decision-making. By automating routine navigation tasks, pilots can allocate their attention to situational awareness and safety.
- **Enhancing Situational Awareness:** Machine learning models process vast amounts of data from various sensors, improving overall situational awareness. This enhanced awareness is crucial for small aircraft navigating through challenging environments or congested airspace.

Deep Learning for Precision Guidance

Deep learning (DL) techniques, such as recurrent neural networks (RNNs) and convolutional neural networks (CNNs), have revolutionized precision guidance systems in small aircraft. This section explores the applications and benefits of Deep Learning (DL) in achieving precise and adaptive navigation (Hu & Wang, 2020; Pendyala et al., n.d.).

- **Real-Time Data Analysis with Recurrent Neural Networks (RNNs):** RNNs are employed to process time-series data from onboard sensors and instruments. These networks excel at analyzing and predicting aircraft behavior based on historical and real-time data. By continuously modeling and adapting to changing flight conditions, RNNs contribute to precise and adaptive navigation.
- **Terrain and Obstacle Recognition with CNNs:** Convolutional neural networks (CNNs) are instrumental in enhancing terrain and obstacle recognition. By analyzing data from onboard cameras and sensors, CNNs can identify and classify terrain features, obstacles, and potential hazards in real-time. This capability significantly improves situational awareness and aids in precise guidance (Boopathi, 2023a; Hema et al., 2023; Syamala et al., 2023; Venkateswaran et al., 2023).
- **Adaptive Control with DL Algorithms:** DL algorithms, including RNNs and deep reinforcement learning, enable small aircraft to adapt their flight paths and control strategies dynamically. These algorithms can optimize control inputs based on sensor data and navigational goals, ensuring precise and efficient flight even in challenging conditions.
- **Precision Landing Systems:** DL plays a pivotal role in precision landing systems for small aircraft. By processing data from landing aids, such as instrument landing systems (ILS) and GPS, DL algorithms can guide the aircraft to a precise and safe landing, even in adverse weather conditions or at airports with limited visibility.
- **Collaborative Decision-Making:** DL-based navigation systems foster collaborative decision-making between pilots and automation. These systems provide pilots with informative recommendations and alerts, enhancing their situational awareness and enabling them to make precise navigation decisions.
- **Adaptive Altitude Control:** DL algorithms can optimize altitude control strategies based on various factors, including aircraft performance, weather conditions, and airspace constraints. This adaptability ensures that small aircraft maintain their assigned altitudes with precision.
- **Reducing Margin of Error:** By leveraging DL, precision guidance systems can significantly reduce the margin of error in navigation. This is particularly valuable in operations such as aerial surveys, where precise flight paths are essential for collecting accurate data.
- **Enhanced Safety in Challenging Environments:** DL-based precision guidance systems enhance safety in challenging environments, such as mountainous terrain or congested airspace. These systems help aircraft navigate through complex scenarios with precision and confidence.

- **Autonomous Navigation:** DL technologies enable a degree of autonomous navigation in small aircraft. By combining data analysis, obstacle detection, and adaptive control, autonomous navigation systems can handle certain flight phases independently, reducing pilot workload.

FAULT DETECTION AND PREDICTIVE MAINTENANCE

Importance of Fault Detection

Fault detection is a critical component of aviation safety and reliability, and its significance cannot be overstated. In the context of small aircraft applications, understanding the importance of fault detection is paramount. This section explores the reasons why fault detection is crucial for safe and efficient small aircraft operations (Dooraki & Lee, 2021).

- **Safety Assurance:** The primary and most immediate importance of fault detection is safety assurance. Small aircraft operate in a dynamic and often unpredictable environment. Any fault or malfunction in critical systems, such as the engine, flight control surfaces, or avionics, can lead to catastrophic consequences. Fault detection systems act as a safety net, continuously monitoring aircraft systems to identify anomalies or deviations from expected performance (Hamilton et al., 2022).
- **Risk Mitigation:** Fault detection helps mitigate risks associated with equipment failures or malfunctions. By promptly identifying faults, small aircraft operators and pilots can take appropriate actions to minimize the risk of accidents. This might include initiating emergency procedures, diverting to an alternate airport, or making critical adjustments to flight control systems (Rahamathunnisa et al., 2023).
- **Preventing System Failures:** Detecting faults early can prevent system failures from occurring in the first place. In many cases, minor issues or deviations can be addressed before they escalate into critical failures. This proactive approach can significantly enhance the reliability and longevity of small aircraft systems (Subramanian, 2021).
- **Ensuring Redundancy:** Small aircraft are often equipped with redundant systems to enhance safety. Fault detection systems ensure that these redundant systems are fully functional. If a primary system fails, fault detection can trigger the automatic switchover to a redundant system, maintaining the aircraft's operational capabilities.
- **Data-Driven Maintenance:** Fault detection provides valuable data for predictive maintenance. By continuously monitoring the health of aircraft systems, operators can schedule maintenance tasks based on actual system conditions rather than fixed intervals. This approach minimizes downtime, reduces maintenance costs, and maximizes aircraft availability.
- **Regulatory Compliance:** Aviation authorities impose stringent regulations on fault detection and maintenance practices to ensure the safety of flight operations. Compliance with these regulations is essential for small aircraft operators. Fault detection systems help meet these regulatory requirements by providing evidence of system health monitoring and maintenance.

- **Enhancing Pilot Confidence:** Small aircraft pilots rely on fault detection systems to enhance their confidence in the aircraft's reliability. Knowing that the aircraft's systems are continuously monitored and that faults will be promptly identified and reported allows pilots to focus on flying safely and effectively.
- **Data for Investigation and Improvement:** In the unfortunate event of an accident or incident, fault detection data can be invaluable for investigation and improvement efforts. It provides insights into the sequence of events leading to the fault, contributing to accident analysis and safety enhancements.

Hence, fault detection is not just a convenience but a necessity in small aircraft operations. It serves as a critical safety and risk mitigation tool, ensuring the reliability of systems, regulatory compliance, and the overall safety of flight (Dooraki & Lee, 2021; Santos et al., 2021). Fault detection systems are integral to small aircraft operations, helping to prevent accidents, reduce downtime, and enhance the overall safety and efficiency of the aviation industry.

Predictive Maintenance With Deep Learning

Predictive maintenance is a crucial aspect of aviation, including small aircraft operations. Deep learning, with its ability to analyze complex data patterns, plays a pivotal role in predictive maintenance strategies (Figure 4). This section delves into the importance of predictive maintenance using deep learning in ensuring the safety and reliability of small aircraft (Dooraki & Lee, 2021).

Figure 4. Predictive maintenance with deep learning

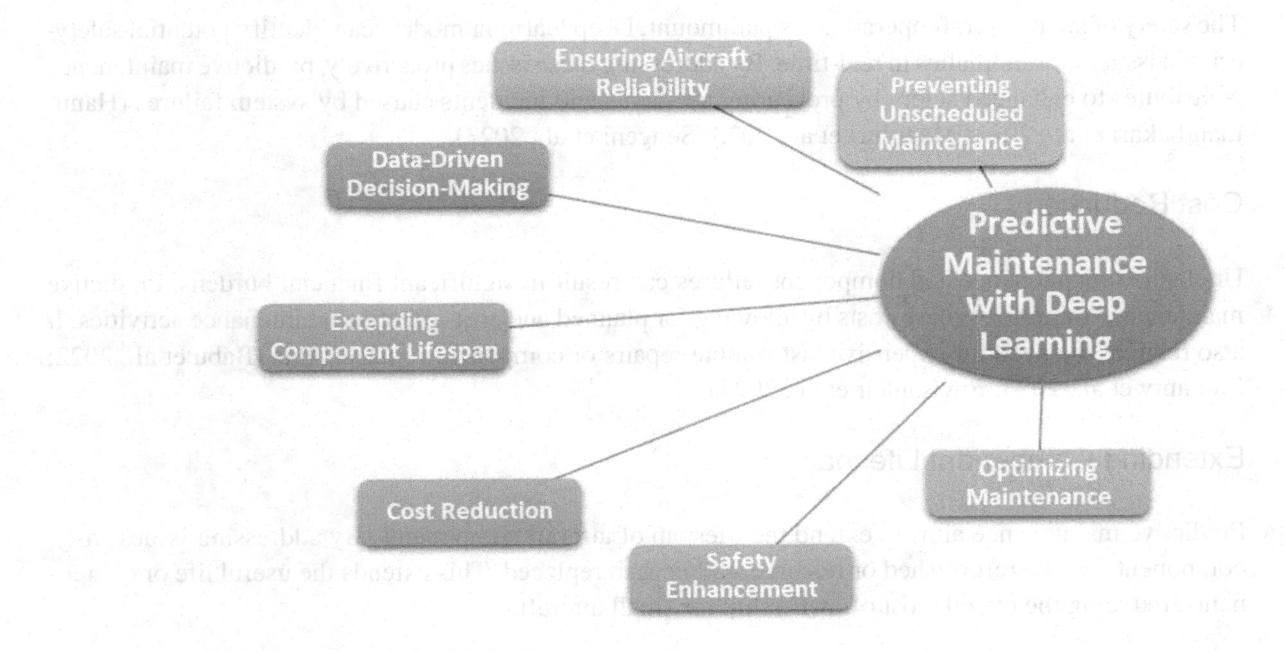

Ensuring Aircraft Reliability

One of the primary objectives of predictive maintenance is to ensure the reliability of aircraft systems. Deep learning models can analyze sensor data from various systems, such as engines, avionics, and control surfaces, to detect early signs of wear, degradation, or impending failures. This allows for proactive maintenance to be performed, minimizing unexpected downtime and disruptions to flight operations.

Preventing Unscheduled Maintenance

Unscheduled maintenance can be costly and disrupt flight schedules. Deep learning models excel at predicting when maintenance is required based on historical and real-time data. By identifying issues in advance, small aircraft operators can plan and schedule maintenance during downtime or between flights, preventing unscheduled maintenance and its associated inconveniences (Agrawal, Magulur, et al., 2023; Anitha et al., 2023; Boopathi, 2023c; Srinivas et al., 2023).

Optimizing Maintenance Resources

Predictive maintenance with deep learning optimizes the allocation of maintenance resources. Instead of performing routine maintenance at fixed intervals, resources are directed toward components and systems that genuinely need attention. This maximizes the efficiency of maintenance processes and reduces unnecessary costs (Boopathi, Kumar, et al., 2023; Boopathi, Pandey, et al., 2023; Domakonda et al., 2022; Kumara et al., 2023; Samikannu et al., 2022; Sampath et al., 2023; Vanitha et al., 2023).

Safety Enhancement

The safety of small aircraft operations is paramount. Deep learning models can identify potential safety-critical issues and anomalies in real-time. By addressing these issues proactively, predictive maintenance contributes to enhanced safety by preventing accidents and incidents caused by system failures (Hanumanthakari et al., 2023; Maguluri et al., 2023; Sengeni et al., 2023).

Cost Reduction

Unplanned maintenance and component failures can result in significant financial burdens. Predictive maintenance minimizes these costs by allowing for planned and cost-effective maintenance activities. It also reduces the need for expensive last-minute repairs or component replacements (Babu et al., 2022; Mohanty et al., 2023; Ravisankar et al., 2023).

Extending Component Lifespan

Predictive maintenance aims to extend the lifespan of aircraft components. By addressing issues early, components can be refurbished or repaired, rather than replaced. This extends the useful life of components, reducing the overall cost of ownership for small aircraft.

Data-Driven Decision-Making

Deep learning models analyze vast amounts of sensor data and historical maintenance records to make data-driven maintenance decisions. This approach is more accurate and reliable than traditional time-based maintenance schedules.

Regulatory Compliance

Aviation authorities often require operators to maintain aircraft in accordance with specific maintenance schedules and procedures. Predictive maintenance with deep learning helps operators meet these regulatory requirements by ensuring the continuous airworthiness of small aircraft.

Improved Aircraft Availability

By reducing unscheduled maintenance and downtime, predictive maintenance with deep learning enhances aircraft availability. This is particularly important for small aircraft operators who rely on their fleets for various missions, including business travel, medical transport, and surveillance.

In conclusion, predictive maintenance with deep learning is an indispensable strategy for small aircraft operators. It not only enhances aircraft reliability, safety, and availability but also leads to cost reductions and more efficient resource allocation (Bange et al., 2021; Benavente-Sánchez et al., 2021). Deep learning's ability to analyze complex data patterns and make informed predictions positions it as a key technology in the ongoing efforts to optimize small aircraft maintenance practices.

PILOT ASSISTANCE AND HUMAN-MACHINE INTERACTION

Augmented Reality in Cockpits

- **The Evolution of Cockpit Technology:** The historical evolution of cockpit technology is crucial before delving into the transformation of the aviation industry through augmented reality (AR) (Ament & Schmelz, 2023).
- **Understanding Augmented Reality:** This section provides a detailed explanation of augmented reality, its principles, technologies, and applications, focusing on its application in aviation.
- **Benefits of Augmented Reality in Cockpits:** This article explores the benefits of Augmented Reality (AR) in the cockpit environment, including improved situational awareness, decision-making, and safety, using case studies and real-world examples.
- **Augmented Reality Displays:** This section explores the various types of augmented reality displays used in cockpits. This text outlines the functions and advantages of head-up displays (HUDs), head-mounted displays (HMDs), and windshield displays in providing real-time information to pilots.
- **Enhanced Navigation and Wayfinding:** AR is a game-changer for navigation and wayfinding in cockpits. The study explores how Augmented Reality (AR) enhances precision navigation by overlaying crucial navigational information onto a pilot's field of view, particularly in low-visibility situations.

- **Terrain and Obstacle Recognition:** AR is instrumental in terrain and obstacle recognition. The article discusses how Augmented Reality (AR) systems can enhance flight safety by identifying potential hazards and providing pilots with warnings, especially in challenging environments.
- **Augmented Checklists and Procedures:** AR can improve cockpit workflows by providing augmented checklists and procedures. The study investigates the efficient use of AR technology by pilots in pre-flight checks, emergency procedures, and in-flight tasks.
- **Enhanced Training and Simulation:** AR extends its benefits to pilot training and simulation. AR-based simulators offer pilots a realistic training environment, allowing them to practice various scenarios and emergency procedures in a safe and controlled setting.

AR enhances cockpit human-machine interface through gesture recognition, voice commands, and touchless controls, allowing pilots to interact with avionics systems more intuitively and safely. The chapter explores the use of augmented reality (AR) in cockpits, addressing issues like data accuracy and pilot training. It also discusses future trends and innovations, such as navigation charts and synthetic vision systems, providing a comprehensive understanding of AR's impact on aviation.

Intelligent Co-Pilots

Before delving into intelligent co-pilots, this chapter provides an overview of the traditional role and responsibilities of co-pilots in aviation. Understanding their existing functions sets the context for the integration of intelligent co-pilots (Nawaz et al., 2022). This section traces the historical development of co-pilot assistance systems, from basic automation to advanced intelligent systems. The study delves into the evolution of autopilots from basic models to advanced co-pilot technologies (Figure 5).

Understanding Intelligent Co-Pilots

Intelligent co-pilots are capable of analyzing data, making informed decisions, and controlling specific flight aspects.

Machine Learning and Decision Support

This study explores the use of machine learning and artificial intelligence in enhancing the capabilities of intelligent co-pilots by analyzing vast data and providing decision support (Boopathi & Kanike, 2023; Veeranjaneyulu, Boopathi, Kumari, et al., 2023; Veeranjaneyulu, Boopathi, Narasimharao, et al., 2023; Zekrifa et al., 2023).

Collaboration and Communication

Intelligent co-pilots, equipped with communication interfaces like voice recognition, natural language processing, and gesture-based interactions, enable seamless human-machine collaboration.

Figure 5. Intelligent co-pilots technologies

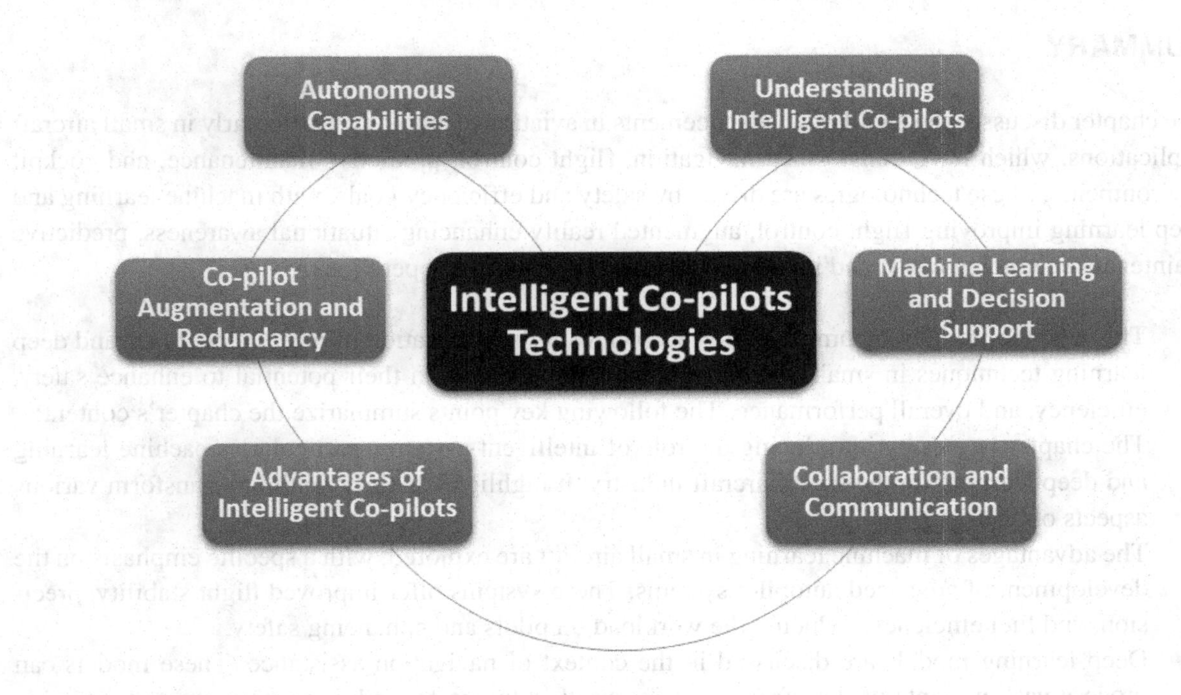

Advantages of Intelligent Co-Pilots

This section explores the advantages that intelligent co-pilots bring to aviation, including enhanced safety, reduced pilot workload, improved decision-making, and increased operational efficiency. Real-world case studies and examples may illustrate these benefits (Buckley et al., 2014; Long et al., 2007).

Co-pilot Augmentation and Redundancy

Intelligent co-pilots enhance human pilots' capabilities, providing additional safety and redundancy, including emergency handling and critical system monitoring during long-duration flights.

Autonomous Capabilities

Intelligent co-pilots, though not fully autonomous, can have autonomous features and are being studied for their ability to assume control in specific situations like auto-landing or emergency response.

The chapter discusses the importance of trust between human pilots and intelligent co-pilots, emphasizing transparency, training, and communication. It also addresses challenges and ethical considerations related to system reliability, data privacy, decision responsibility, and automation overreliance (Hamilton et al., 2022; Santos et al., 2021; Subramanian, 2021). The chapter concludes by discussing future directions of intelligent co-pilots in aviation, including ongoing research and emerging technologies. By the end of the chapter, readers will have a comprehensive understanding of the role and potential of these systems, their advantages, challenges, and ethical considerations, and their evolving role in the aviation industry.

SUMMARY

The chapter discusses the significant advancements in aviation technology, particularly in small aircraft applications, which have transformed navigation, flight control, predictive maintenance, and cockpit environments. These technologies are driven by safety and efficiency goals, with machine learning and deep learning improving flight control, augmented reality enhancing situational awareness, predictive maintenance reducing costs, and intelligent co-pilots streamlining operations.

- This chapter provides a comprehensive overview of the integration of machine learning and deep learning techniques in small aircraft applications, focusing on their potential to enhance safety, efficiency, and overall performance. The following key points summarize the chapter's content:
- The chapter begins by introducing the role of intelligent systems, particularly machine learning and deep learning, in the small aircraft industry. It highlights their potential to transform various aspects of aviation.
- The advantages of machine learning in small aircraft are explored, with a specific emphasis on the development of advanced autopilot systems. These systems offer improved flight stability, precision, and fuel efficiency, reducing the workload on pilots and enhancing safety.
- Deep learning models are discussed in the context of navigation assistance. These models can process vast amounts of data from sensors, weather forecasts, and navigation charts to provide real-time guidance and decision support to pilots.
- The chapter addresses the challenges and limitations associated with implementing machine learning and deep learning in small aircraft. Issues such as data availability, model interpretability, and regulatory compliance are examined.
- The role of intelligent systems in enhancing safety is thoroughly explored. This includes accident prevention, collision avoidance, and emergency response mechanisms that can significantly reduce the risk of accidents in small aircraft.
- The chapter delves into how machine learning and deep learning techniques optimize operational efficiency in small aircraft. This includes fuel consumption optimization, route planning, and overall cost savings.
- Intelligent systems are shown to support pilot decision-making by mitigating cognitive overload and enhancing situational awareness during flight, ultimately improving overall flight safety.
- The chapter speculates on the future of machine learning and deep learning in small aircraft applications. It considers emerging technologies, regulatory developments, and industry trends that will shape the integration of intelligent systems in the aviation sector.
- Finally, the chapter aims to serve as a comprehensive resource for both technical and non-technical readers, fostering discussions and innovations in the aviation industry regarding the adoption of intelligent systems in small aircraft applications.

REFERENCES

Agrawal, A. V., Magulur, L. P., Priya, S. G., Kaur, A., Singh, G., & Boopathi, S. (2023). Smart Precision Agriculture Using IoT and WSN. In *Handbook of Research on Data Science and Cybersecurity Innovations in Industry 4.0 Technologies* (pp. 524–541). IGI Global. doi:10.4018/978-1-6684-8145-5.ch026

Agrawal, A. V., Pitchai, R., Senthamaraikannan, C., Balaji, N. A., Sajithra, S., & Boopathi, S. (2023). Digital Education System During the COVID-19 Pandemic. In Using Assistive Technology for Inclusive Learning in K-12 Classrooms (pp. 104–126). IGI Global. doi:10.4018/978-1-6684-6424-3.ch005

Ali, M. (1990). Intelligent systems in aerospace. *The Knowledge Engineering Review*, *5*(3), 147–166. doi:10.1017/S0269888900005385

Ament, J., & Schmelz, J. (2023). Scientific use of Game Engines for Flight Simulation and Human Machine Interface Development. *CEAS Aeronautical Journal.*

Anitha, C., Komala, C., Vivekanand, C. V., Lalitha, S., & Boopathi, S. (2023). Artificial Intelligence driven security model for Internet of Medical Things (IoMT). *IEEE Explore*, 1–7.

Babu, B. S., Kamalakannan, J., Meenatchi, N., Karthik, S., & Boopathi, S. (2022). Economic impacts and reliability evaluation of battery by adopting Electric Vehicle. *IEEE Explore*, 1–6.

Bange, J., Reuder, J., & Platis, A. (2021). Unmanned Aircraft Systems. In *Springer Handbook of Atmospheric Measurements* (pp. 1331–1349). Springer. doi:10.1007/978-3-030-52171-4_49

Benavente-Sánchez, D., Moreno-Molina, J., & Argilés-Herrero, R. (2021). Prospects for remotely piloted aircraft systems in area-wide integrated pest management programmes. *Area-Wide Integrated Pest Management*, 903–916.

Bikash Chandra Saha, M. S., Deepa, R., Akila, A., & Sai Thrinath, B. V. (2022). IOT based smart energy meter for smart grid. Academic Press.

Boopathi, S. (2023a). Deep Learning Techniques Applied for Automatic Sentence Generation. In Promoting Diversity, Equity, and Inclusion in Language Learning Environments (pp. 255–273). IGI Global. doi:10.4018/978-1-6684-3632-5.ch016

Boopathi, S. (2023b). Internet of Things-Integrated Remote Patient Monitoring System: Healthcare Application. In *Dynamics of Swarm Intelligence Health Analysis for the Next Generation* (pp. 137–161). IGI Global. doi:10.4018/978-1-6684-6894-4.ch008

Boopathi, S. (2023c). Securing Healthcare Systems Integrated With IoT: Fundamentals, Applications, and Future Trends. In Dynamics of Swarm Intelligence Health Analysis for the Next Generation (pp. 186–209). IGI Global.

Boopathi, S., Balasubramani, V., Kumar, R. S., & Singh, G. R. (2021). The influence of human hair on kenaf and Grewia fiber-based hybrid natural composite material: An experimental study. *Functional Composites and Structures*, *3*(4), 045011. doi:10.1088/2631-6331/ac3afc

Boopathi, S., Gavaskar, T., Dogga, A. D., Mahendran, R. K., Kumar, A., Kathiresan, G., N., V., Ganesan, M., Ishwarya, K. R., … Ramana, G. V. (2021). *Emergency medicine delivery transportation using unmanned aerial vehicle (Patent Grant)*. Academic Press.

Boopathi, S., & Kanike, U. K. (2023). Applications of Artificial Intelligent and Machine Learning Techniques in Image Processing. In *Handbook of Research on Thrust Technologies' Effect on Image Processing* (pp. 151–173). IGI Global. doi:10.4018/978-1-6684-8618-4.ch010

Boopathi, S., Kumar, P. K. S., Meena, R. S., Sudhakar, M., & Associates. (2023). Sustainable Developments of Modern Soil-Less Agro-Cultivation Systems: Aquaponic Culture. In Human Agro-Energy Optimization for Business and Industry (pp. 69–87). IGI Global.

Boopathi, S., & Myilsamy, S. (2021). Material removal rate and surface roughness study on Near-dry wire electrical discharge Machining process. *Materials Today: Proceedings*, *45*(9), 8149–8156. doi:10.1016/j.matpr.2021.02.267

Boopathi, S., Pandey, B. K., & Pandey, D. (2023). Advances in Artificial Intelligence for Image Processing: Techniques, Applications, and Optimization. In Handbook of Research on Thrust Technologies' Effect on Image Processing (pp. 73–95). IGI Global.

Boopathi, S., & Sivakumar, K. (2016). Optimal parameter prediction of oxygen-mist near-dry wire-cut EDM. *Inderscience: International Journal of Manufacturing Technology and Management*, *30*(3–4), 164–178. doi:10.1504/IJMTM.2016.077812

Buckley, J., Gaetano, D., McCarthy, K., Loizou, L., O'Flynn, B., & O'mathuna, C. (2014). Compact 433 MHz antenna for wireless smart system applications. *Electronics Letters*, *50*(8), 572–574. doi:10.1049/el.2013.4312

Chandrika, V., Sivakumar, A., Krishnan, T. S., Pradeep, J., Manikandan, S., & Boopathi, S. (2023). Theoretical Study on Power Distribution Systems for Electric Vehicles. In *Intelligent Engineering Applications and Applied Sciences for Sustainability* (pp. 1–19). IGI Global. doi:10.4018/979-8-3693-0044-2.ch001

Domakonda, V. K., Farooq, S., Chinthamreddy, S., Puviarasi, R., Sudhakar, M., & Boopathi, S. (2022). Sustainable Developments of Hybrid Floating Solar Power Plants: Photovoltaic System. In Human Agro-Energy Optimization for Business and Industry (pp. 148–167). IGI Global.

Dooraki, A. R., & Lee, D.-J. (2021). An innovative bio-inspired flight controller for quad-rotor drones: Quad-rotor drone learning to fly using reinforcement learning. *Robotics and Autonomous Systems*, *135*, 103671. doi:10.1016/j.robot.2020.103671

Hamilton, J., de Boer, G., Doddi, A., & Lawrence, D. A. (2022). The DataHawk2 uncrewed aircraft system for atmospheric research. *Atmospheric Measurement Techniques*, *15*(22), 6789–6806. doi:10.5194/amt-15-6789-2022

Hanumanthakari, S., Gift, M. M., Kanimozhi, K., Bhavani, M. D., Bamane, K. D., & Boopathi, S. (2023). Biomining Method to Extract Metal Components Using Computer-Printed Circuit Board E-Waste. In *Handbook of Research on Safe Disposal Methods of Municipal Solid Wastes for a Sustainable Environment* (pp. 123–141). IGI Global. doi:10.4018/978-1-6684-8117-2.ch010

Hema, N., Krishnamoorthy, N., Chavan, S. M., Kumar, N., Sabarimuthu, M., & Boopathi, S. (2023). A Study on an Internet of Things (IoT)-Enabled Smart Solar Grid System. In *Handbook of Research on Deep Learning Techniques for Cloud-Based Industrial IoT* (pp. 290–308). IGI Global. doi:10.4018/978-1-6684-8098-4.ch017

Hu, B., & Wang, J. (2020). Deep learning based hand gesture recognition and UAV flight controls. *International Journal of Automation and Computing*, *17*(1), 17–29. doi:10.100711633-019-1194-7

Krishnakumar, K. (2002). *Intelligent Systems for Aerospace Engineering: An Overview*. Von Karman Institute Lecture Series on Intelligent Systems for Aeronautics.

Kumara, V., Mohanaprakash, T., Fairooz, S., Jamal, K., Babu, T., & Sampath, B. (2023). Experimental Study on a Reliable Smart Hydroponics System. In *Human Agro-Energy Optimization for Business and Industry* (pp. 27–45). IGI Global. doi:10.4018/978-1-6684-4118-3.ch002

Li, H., & Gupta, M. M. (1995). *Fuzzy logic and intelligent systems* (Vol. 3). Springer Science & Business Media.

Long, L. N., Hanford, S. D., Janrathitikarn, O., Sinsley, G. L., & Miller, J. A. (2007). A review of intelligent systems software for autonomous vehicles. *2007 IEEE Symposium on Computational Intelligence in Security and Defense Applications*, 69–76. 10.1109/CISDA.2007.368137

Maguluri, L. P., Ananth, J., Hariram, S., Geetha, C., Bhaskar, A., & Boopathi, S. (2023). Smart Vehicle-Emissions Monitoring System Using Internet of Things (IoT). In Handbook of Research on Safe Disposal Methods of Municipal Solid Wastes for a Sustainable Environment (pp. 191–211). IGI Global.

Maheswari, B. U., Imambi, S. S., Hasan, D., Meenakshi, S., Pratheep, V., & Boopathi, S. (2023). Internet of Things and Machine Learning-Integrated Smart Robotics. In Global Perspectives on Robotics and Autonomous Systems: Development and Applications (pp. 240–258). IGI Global. doi:10.4018/978-1-6684-7791-5.ch010

Mohanty, A., Venkateswaran, N., Ranjit, P., Tripathi, M. A., & Boopathi, S. (2023). Innovative Strategy for Profitable Automobile Industries: Working Capital Management. In Handbook of Research on Designing Sustainable Supply Chains to Achieve a Circular Economy (pp. 412–428). IGI Global.

Nawaz, A., Arora, A. S., Ali, W., Saxena, N., Khan, M. S., Yun, C. M., & Lee, M. (2022). Intelligent Human–Machine Interface: An Agile Operation and Decision Support for an ANAMMOX SBR System at a Pilot-Scale Wastewater Treatment Plant. *IEEE Transactions on Industrial Informatics*, *18*(9), 6224–6232. doi:10.1109/TII.2022.3153468

Pendyala, S., Vanama, S., & Upalanchi, S. (n.d.). *Design of automatic flight control system with safety features for small aircraft*. Academic Press.

Rahamathunnisa, U., Subhashini, P., Aancy, H. M., Meenakshi, S., Boopathi, S., & ... (2023). Solutions for Software Requirement Risks Using Artificial Intelligence Techniques. In *Handbook of Research on Data Science and Cybersecurity Innovations in Industry 4.0 Technologies* (pp. 45–64). IGI Global.

Ramudu, K., Mohan, V. M., Jyothirmai, D., Prasad, D., Agrawal, R., & Boopathi, S. (2023). Machine Learning and Artificial Intelligence in Disease Prediction: Applications, Challenges, Limitations, Case Studies, and Future Directions. In Contemporary Applications of Data Fusion for Advanced Healthcare Informatics (pp. 297–318). IGI Global.

Ravisankar, A., Sampath, B., & Asif, M. M. (2023). Economic Studies on Automobile Management: Working Capital and Investment Analysis. In Multidisciplinary Approaches to Organizational Governance During Health Crises (pp. 169–198). IGI Global.

Samikannu, R., Koshariya, A. K., Poornima, E., Ramesh, S., Kumar, A., & Boopathi, S. (2022). Sustainable Development in Modern Aquaponics Cultivation Systems Using IoT Technologies. In *Human Agro-Energy Optimization for Business and Industry* (pp. 105–127). IGI Global.

Sampath, B., Pandian, M., Deepa, D., & Subbiah, R. (2022). Operating parameters prediction of liquefied petroleum gas refrigerator using simulated annealing algorithm. *AIP Conference Proceedings*, *2460*(1), 070003. doi:10.1063/5.0095601

Sampath, B., Sasikumar, C., & Myilsamy, S. (2023). Application of TOPSIS Optimization Technique in the Micro-Machining Process. In Trends, Paradigms, and Advances in Mechatronics Engineering (pp. 162–187). IGI Global.

Santos, M. H., Oliveira, N. M., & D'Amore, R. (2021). From Control Requirements to PIL Test: Development of a Structure to Autopilot Implementation. *IEEE Access : Practical Innovations, Open Solutions*, *9*, 154788–154803. doi:10.1109/ACCESS.2021.3127846

Sengeni, D., Padmapriya, G., Imambi, S. S., Suganthi, D., Suri, A., & Boopathi, S. (2023). Biomedical Waste Handling Method Using Artificial Intelligence Techniques. In *Handbook of Research on Safe Disposal Methods of Municipal Solid Wastes for a Sustainable Environment* (pp. 306–323). IGI Global. doi:10.4018/978-1-6684-8117-2.ch022

Srinivas, B., Maguluri, L. P., Naidu, K. V., Reddy, L. C. S., Deivakani, M., & Boopathi, S. (2023). Architecture and Framework for Interfacing Cloud-Enabled Robots. In *Handbook of Research on Data Science and Cybersecurity Innovations in Industry 4.0 Technologies* (pp. 542–560). IGI Global. doi:10.4018/978-1-6684-8145-5.ch027

Subha, S., Inbamalar, T., Komala, C., Suresh, L. R., Boopathi, S., & Alaskar, K. (2023). A Remote Health Care Monitoring system using internet of medical things (IoMT). *IEEE Explore*, 1–6.

Subramanian, C. (2021). An appraisal on intelligent and smart systems. *AIP Conference Proceedings*, *2316*(1), 020003. doi:10.1063/5.0037534

Syamala, M., Komala, C., Pramila, P., Dash, S., Meenakshi, S., & Boopathi, S. (2023). Machine Learning-Integrated IoT-Based Smart Home Energy Management System. In *Handbook of Research on Deep Learning Techniques for Cloud-Based Industrial IoT* (pp. 219–235). IGI Global. doi:10.4018/978-1-6684-8098-4.ch013

Vanitha, S., Radhika, K., & Boopathi, S. (2023). Artificial Intelligence Techniques in Water Purification and Utilization. In *Human Agro-Energy Optimization for Business and Industry* (pp. 202–218). IGI Global. doi:10.4018/978-1-6684-4118-3.ch010

Veeranjaneyulu, R., Boopathi, S., Kumari, R. K., Vidyarthi, A., Isaac, J. S., & Jaiganesh, V. (2023). Air Quality Improvement and Optimisation Using Machine Learning Technique. *IEEE- Explore*, 1–6.

Veeranjaneyulu, R., Boopathi, S., Narasimharao, J., Gupta, K. K., Reddy, R. V. K., & Ambika, R. (2023). Identification of Heart Diseases using Novel Machine Learning Method. *IEEE- Explore*, 1–6.

Venkateswaran, N., Vidhya, R., Naik, D. A., Raj, T. M., Munjal, N., & Boopathi, S. (2023). Study on Sentence and Question Formation Using Deep Learning Techniques. In *Digital Natives as a Disruptive Force in Asian Businesses and Societies* (pp. 252–273). IGI Global. doi:10.4018/978-1-6684-6782-4.ch015

Volponi, A., Brotherton, T., & Luppold, R. (2004). Development of an information fusion system for engine diagnostics and health management. *AIAA 1st Intelligent Systems Technical Conference*, 6461.

Whitley, T., Tomiczek, A., Tripp, C., Ortega, A., Mennu, M., Bridge, J., & Ifju, P. (2020). Design of a small unmanned aircraft system for bridge inspections. *Sensors (Basel)*, *20*(18), 5358. doi:10.339020185358 PMID:32962108

Zekrifa, D. M. S., Kulkarni, M., Bhagyalakshmi, A., Devireddy, N., Gupta, S., & Boopathi, S. (2023). Integrating Machine Learning and AI for Improved Hydrological Modeling and Water Resource Management. In *Artificial Intelligence Applications in Water Treatment and Water Resource Management* (pp. 46–70). IGI Global. doi:10.4018/978-1-6684-6791-6.ch003

Chapter 12
Machine Learning in E-Health and Digital Healthcare:
Practical Strategies for Transformation

T. K. Sethuramalingam
https://orcid.org/0000-0003-0722-6806
Department of Electronics and Communication Engineering, Karpagam College of Engineering, Coimbatore, India

Rajkumar G. Nadakinamani
Badr Al Samaa Hospital, Oman

G. Sumathy
Department of Computational Intelligence, SRM Institute of Science and Technology, India

Sureshkumar Myilsamy
Mechanical Engineering, Bannari Amman Institute of Technology, India

ABSTRACT

Machine learning is revolutionizing healthcare by offering innovative solutions to complex challenges. This chapter explores the practical strategies, ethical considerations, and real-world applications of machine learning in the healthcare domain. It delves into data collection and management, model development, integration with existing systems, and the importance of interdisciplinary collaboration. The chapter also discusses the ethical dimensions of healthcare AI, such as data privacy, bias mitigation, and regulatory compliance. Real-world case studies highlight the impact of machine learning on early disease detection, drug discovery, and precision medicine. The chapter concludes by examining future trends, including emerging technologies like quantum computing, nanomedicine, and the growing role of AI in drug discovery and genomic medicine. As machine learning continues to reshape healthcare, understanding these practical strategies and ethical considerations is essential for optimizing patient care and advancing the healthcare industry.

DOI: 10.4018/978-1-6684-9999-3.ch012

INTRODUCTION

The integration of machine learning, E-Health, and Digital Healthcare has revolutionized the healthcare sector, leveraging digital technologies and data-driven insights. This chapter delves into the profound impact of machine learning in E-Health and Digital Healthcare, offering practical strategies to harness its potential for positive transformation (Ganapathy et al., 2021). The integration of machine learning and healthcare presents a promising avenue for innovation and patient care. With the proliferation of data and computing power, machine learning has revolutionized the industry by offering innovative solutions to complex challenges. This chapter explores the profound impact of machine learning in healthcare, discussing practical strategies, ethical considerations, real-world applications, and future trends in this dynamic landscape (Tebeje & Klein, 2021).

Machine learning, a subset of AI, has become a crucial tool in healthcare due to the vast amount of data generated by various technologies like EHRs, medical imaging, genomics, and wearable devices. When used effectively, this data provides valuable insights into patient health, disease mechanisms, and treatment outcomes (Siriwardhana et al., 2021). At the heart of this healthcare revolution lie practical strategies that guide the seamless integration of machine learning into clinical practice. These strategies encompass the entire machine learning pipeline, from data collection and management to model development and integration with existing healthcare systems. Understanding the nuances of these strategies is essential for healthcare organizations seeking to harness the full potential of machine learning (Kruszyńska-Fischbach et al., 2022).

As the healthcare sector embraces machine learning, ethical considerations take center stage. Privacy and security of patient data, bias mitigation in algorithms, and adherence to regulatory standards like the Health Insurance Portability and Accountability Act (HIPAA) are paramount. This chapter navigates the ethical dimensions of healthcare AI, illuminating the path to responsible and equitable AI deployment(Kruszyńska-Fischbach et al., 2022). The article explores the significant role of machine learning in healthcare, highlighting its applications in early disease detection, drug discovery, and the realization of precision medicine's potential. From the identification of diabetic retinopathy through deep learning to predictive analytics reducing hospital readmissions, these case studies exemplify the transformative power of machine learning in patient care and healthcare management(Boopathi, 2023b; Reddy, Reddy, et al., 2023; Subha et al., 2023).

The chapter concludes by peering into the future, where emerging technologies such as quantum computing, nanomedicine, and advanced AI-driven drug discovery are poised to reshape healthcare in profound ways. Precision medicine, guided by genomic insights, holds the promise of truly personalized healthcare. Telehealth, driven by augmented and virtual reality, is expanding access to care in remote corners of the world. The healthcare landscape is evolving at an unprecedented pace, and understanding these future trends is essential for healthcare professionals, researchers, and policymakers(Boopathi, 2023b; Reddy, Reddy, et al., 2023; Subha et al., 2023).

The integration of machine learning in healthcare is transforming the field, fostering new frontiers. This exploration outlines practical strategies and ethical considerations to ensure patient well-being, equity, and excellence in healthcare delivery. The future of healthcare is one of innovation, with the patient at the heart of it all, fostering a future where patient well-being is paramount.

Background and Significance

Healthcare, a data-rich field, has largely untapped its full potential due to the advent of machine learning. This technology can analyze complex datasets, detect patterns, and make predictions, unlocking a wealth of opportunities in the field of healthcare (Boopathi, 2023a; Karthik et al., 2023; Pramila et al., 2023). The significance of this convergence is multifaceted. Firstly, it promises to enhance patient care by enabling early disease detection, personalized treatment plans, and improved clinical decision support. Secondly, it offers healthcare organizations the means to optimize their operations, reduce costs, and streamline administrative processes. Thirdly, it fosters innovation, driving the development of novel healthcare technologies and therapies.

However, this transformation is not without its challenges. Data privacy, ethical considerations, and regulatory compliance are paramount concerns. The objective of this chapter is to provide a comprehensive understanding of the practical strategies that can help healthcare stakeholders navigate this evolving landscape successfully.

Objectives

- **Educational Insight**: To offer readers a foundational understanding of machine learning and its relevance to E-Health and Digital Healthcare.
- **Real-World Applications**: To explore a wide range of practical applications where machine learning is making a significant impact on healthcare, from disease prediction to telemedicine.
- **Challenges and Ethical Considerations**: To discuss the hurdles and ethical considerations associated with implementing machine learning in healthcare, including data privacy, bias, and regulatory compliance.
- **Practical Strategies**: To provide actionable strategies for healthcare organizations and professionals to effectively implement machine learning solutions.
- **Case Studies**: To showcase real-world case studies and success stories, illustrating how machine learning has been leveraged for positive healthcare outcomes.
- **Future Trends**: To highlight emerging trends and innovations in the field, offering insights into what the future holds for machine learning in healthcare.

This chapter aims to provide healthcare professionals, researchers, and policymakers with the necessary knowledge and guidance to effectively navigate the evolving landscape of machine learning in E-Health and Digital Healthcare, offering practical strategies to drive positive transformation in healthcare through machine learning.

MACHINE LEARNING

Machine learning, a subset of AI, enables computers to learn and make decisions from data without explicit programming. It has the potential to revolutionize healthcare by transforming disease diagnosis, treatment planning, and patient care management (Durairaj et al., 2023; Pramila et al., 2023; Ramudu et al., 2023). This section provides an in-depth understanding of machine learning fundamentals in the context of E-Health and Digital Healthcare, focusing on key concepts such as:

- **Supervised Learning**: An introduction to supervised learning, where algorithms learn from labeled training data to make predictions or classifications.
- **Unsupervised Learning**: An overview of unsupervised learning, which involves finding patterns or structures in unlabeled data, such as clustering and dimensionality reduction.
- **Reinforcement Learning**: An explanation of reinforcement learning, which focuses on decision-making in dynamic environments and is relevant for healthcare applications like treatment planning.
- **Model Evaluation**: Discussion of how to assess the performance of machine learning models, including metrics like accuracy, precision, recall, and F1-score.

Types of Machine Learning Algorithms

This section delves into the various types of machine learning algorithms commonly utilized (Veeranjaneyulu, Boopathi, Kumari, et al., 2023; Veeranjaneyulu, Boopathi, Narasimharao, et al., 2023):

- **Regression Algorithms**: Explanation of regression algorithms used for predicting continuous outcomes, such as patient risk scores or disease progression.
- **Classification Algorithms**: Overview of classification algorithms for tasks like disease diagnosis or medication recommendation.
- **Clustering Algorithms**: Discussion of clustering algorithms for grouping similar patients or medical records.
- **Natural Language Processing (NLP)**: Introduction to NLP techniques for analyzing and extracting insights from textual medical data, including electronic health records.
- **Deep Learning**: A brief introduction to deep learning, which includes neural networks with multiple layers and is particularly powerful for tasks like medical image analysis and speech recognition.

Data Preprocessing in Healthcare

This section focuses on data preprocessing techniques specific to healthcare data, emphasizing the importance of high-quality data for successful machine learning in this field (Durairaj et al., 2023; Ravisankar et al., 2023; Reddy, Gaurav, et al., 2023):

- **Data Cleaning**: Explanation of data cleaning processes to handle missing values, outliers, and inconsistencies in healthcare datasets.
- **Feature Engineering**: Discussion of feature engineering techniques for creating relevant and informative features from healthcare data.
- **Data Scaling and Normalization**: Introduction to scaling and normalization methods to ensure that data is on a consistent scale for modeling.
- **Dealing with Imbalanced Data**: Strategies for handling imbalanced datasets, which are common in healthcare where certain diseases may be rare.
- **Data Privacy and Security**: Considerations for safeguarding patient data and complying with regulations like HIPAA when working with healthcare data.

This chapter provides a comprehensive understanding of machine learning principles and techniques, emphasizing their applications in healthcare, enabling readers to effectively implement machine learning strategies in E-Health and Digital Healthcare.

E-HEALTH AND DIGITAL HEALTHCARE LANDSCAPE

The healthcare sector has undergone significant transformation due to the rapid integration of digital technologies and the emergence of E-Health, which uses information and communication technologies (ICT) to support healthcare delivery, management, and research. This section provides an overview of E-Health's evolution and significance in modern healthcare (Anitha et al., 2023; Pramila et al., 2023; Subha et al., 2023).

Historical Perspective

E-Health emerged in the late 20th century as a result of the digital revolution, primarily focusing on digitizing medical records with electronic health records (EHRs), aiming to improve record-keeping, accessibility, and information exchange among healthcare providers (Chakraborty et al., 2022).

Important Components of E-Health

- **Electronic Health Records (EHRs):** EHRs are digital versions of patients' medical histories, treatment plans, and test results. They enable healthcare professionals to access and update patient information efficiently. Moreover, EHRs facilitate the sharing of patient data among different providers, promoting seamless and coordinated care (Chakraborty et al., 2022).
- **Telemedicine and Telehealth:** E-Health includes telemedicine and telehealth services, which leverage telecommunications technology to provide remote medical consultations, monitoring, and healthcare delivery. These services have become invaluable, especially in rural or underserved areas, and during emergencies like the COVID-19 pandemic (Agrawal et al., 2023; Reddy, Gaurav, et al., 2023; Reddy, Reddy, et al., 2023).
- **Mobile Health (mHealth):** The widespread adoption of smartphones has given rise to mHealth applications and wearable devices. These tools empower individuals to monitor their health, receive medical advice, and track vital signs, fostering proactive and preventive healthcare.
- **Health Information Exchange (HIE):** HIE platforms enable the secure sharing of patient data among healthcare organizations, improving care coordination and reducing redundant tests and procedures.
- **Healthcare Analytics:** E-Health incorporates data analytics to extract valuable insights from healthcare data. Machine learning and artificial intelligence play a significant role in analyzing large datasets for predictive analytics, disease detection, and treatment optimization (Boopathi, 2013, 2022; Boopathi et al., 2021; Boopathi & Sivakumar, 2013).

Figure 1. Various components of e-healthcare

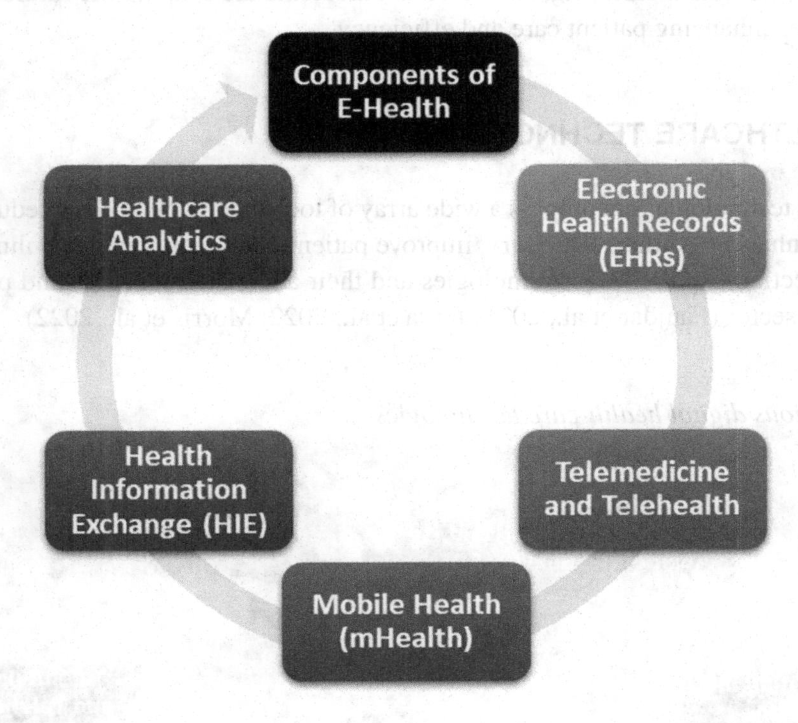

Benefits of E-Health:

E-Health offers a multitude of advantages (Nutbeam, 2021):

- **Improved Patient Care:** EHRs provide comprehensive patient histories, enabling more accurate diagnoses and treatment plans. Telemedicine expands access to healthcare services, particularly for remote or vulnerable populations.
- **Enhanced Efficiency:** Digital records streamline administrative processes, reduce paperwork, and minimize errors. Telemedicine appointments save time and reduce waiting room congestion.
- **Cost Savings:** By reducing duplicate tests and improving care coordination, E-Health can lead to cost savings for healthcare systems and patients alike.
- **Data-Driven Insights:** Healthcare analytics unlock the potential of big data, enabling data-driven decision-making, early disease detection, and personalized medicine.
- **Patient Empowerment:** mHealth applications and wearable devices empower individuals to take an active role in managing their health and well-being.

Challenges

E-Health offers significant potential but also presents challenges like data security, privacy, interoperability issues, and the need for robust regulatory frameworks to regulate healthcare data collection and use (Lupton & Leahy, 2019). E-Health has revolutionized healthcare delivery by offering innovative solutions for patient care, efficiency, and data-driven decision-making. Emerging technologies like artificial intelligence, remote monitoring, and precision medicine are shaping the future of E-Health and Digital Healthcare, enhancing patient care and efficiency.

DIGITAL HEALTHCARE TECHNOLOGIES

Digital healthcare technologies encompass a wide array of tools, methods, and procedures that leverage digitalization to enhance healthcare delivery, improve patient outcomes, and streamline administrative processes. This section explores key technologies and their associated methods and procedures in the digital healthcare sector (Faujdar et al., 2021; Jarva et al., 2022; Morris et al., 2022).

Figure 2. The various digital health-care technologies

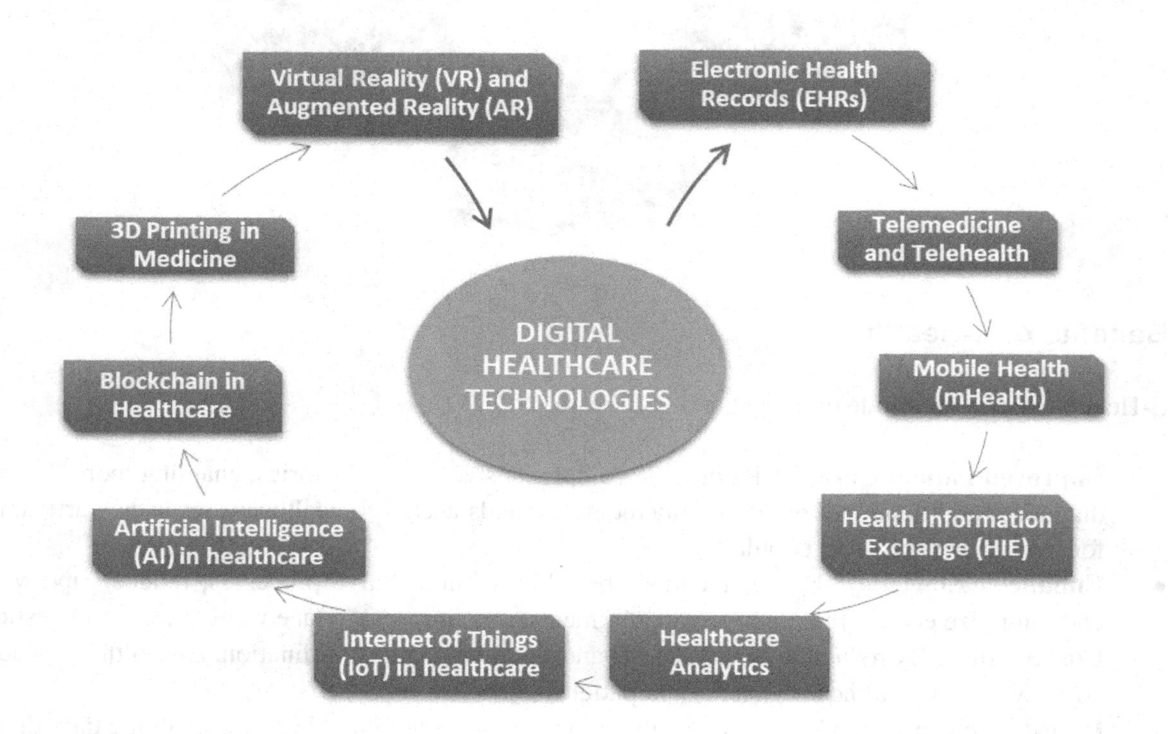

Electronic Health Records (EHRs)

Methods: Data Digitization: Conversion of paper-based patient records into electronic format. Data Entry: Inputting patient information, medical histories, and clinical notes into EHR systems. Interoperability: Establishing standards to enable data sharing among different EHR systems and healthcare providers (Durairaj et al., 2023; Ramudu et al., 2023; Ravisankar et al., 2023; Satav, Hasan, et al., 2024).

Procedures: Patient Record Management: Storing, retrieving, and updating patient information electronically. Clinical Decision Support: Implementing alerts and recommendations based on patient data to assist healthcare providers in making informed decisions.

Telemedicine and Telehealth

Methods: Video Conferencing: Facilitating remote consultations between patients and healthcare providers. Remote Monitoring: Using wearable devices and sensors to track vital signs and health metrics. Store-and-Forward: Capturing patient data, such as images or videos, and transmitting it to specialists for evaluation (Pramila et al., 2023; Sengeni et al., 2023).

Procedures: Teleconsultations: Conducting medical examinations, diagnoses, and follow-ups via video or audio calls. Remote Patient Monitoring: Continuous tracking of patients with chronic conditions to provide timely interventions. Telemedicine Infrastructure: Setting up secure and reliable communication systems for telehealth services.

Mobile Health (mHealth)

Methods: Mobile Applications: Developing smartphone apps for health monitoring, medication reminders, and access to medical information. Wearable Devices: Creating wearable sensors and trackers for monitoring physical activity, heart rate, sleep patterns, and more. Mobile Data Collection: Using mobile devices to collect patient data, such as surveys and symptom tracking.

Procedures: Self-Monitoring: Encouraging patients to use mobile apps and wearables for daily health tracking. Remote Data Transmission: Transmitting collected data to healthcare providers for analysis and feedback. Medication Adherence: Sending medication reminders and educational content to patients' mobile devices.

Health Information Exchange (HIE)

Methods: Data Standardization: Establishing uniform data formats and coding systems for consistent information exchange. Secure Data Sharing: Implementing encryption and authentication measures to protect patient data during transmission (Boopathi, 2023a, 2023b; Karthik et al., 2023; Pramila et al., 2023).

Procedures: Inter-Provider Data Sharing: Facilitating the exchange of patient information among different healthcare organizations. Cross-Border Data Exchange: Enabling secure data sharing across geographical boundaries to support global healthcare.

Healthcare Analytics

Methods: Data Mining: Identifying patterns, trends, and correlations within large healthcare datasets. Machine Learning: Employing algorithms to make predictions, classify diseases, and optimize treatment plans. Natural Language Processing (NLP): Extracting insights from unstructured clinical notes and medical literature (Anitha et al., 2023; Subha et al., 2023).

Procedures: Predictive Analytics: Developing models to forecast disease outbreaks, patient readmissions, and healthcare resource utilization. Clinical Research: Analyzing vast datasets to discover new treatment options and potential drug interactions. Healthcare Dashboards: Creating visualizations and dashboards for monitoring key performance indicators in healthcare facilities.

Internet of Things (IoT) in healthcare

Methods: Sensor Integration: Deploying IoT sensors and devices to monitor patients' health in real-time. Data Transmission: Sending sensor data to healthcare systems for analysis and alerts (Durairaj et al., 2023; Karthik et al., 2023; Subha et al., 2023).

Procedures: Remote Patient Monitoring: Utilizing IoT devices to track vital signs, medication adherence, and environmental factors. Early Warning Systems: Detecting anomalies in patient data and triggering alerts for timely interventions.

Artificial Intelligence (AI) in Healthcare

Methods: Deep Learning: Training neural networks for complex tasks like medical image analysis and natural language understanding. Predictive Modeling: Building AI models to predict patient outcomes, disease progression, and treatment responses (Gowri et al., 2023; Pramila et al., 2023; Sengeni et al., 2023).

Procedures: Medical Imaging Analysis: Using AI to assist radiologists in diagnosing conditions from X-rays, MRIs, and CT scans. Personalized Treatment Plans: Developing AI-driven treatment recommendations based on patient data and medical literature.

Blockchain in Healthcare

Methods: Distributed Ledger Technology: Creating a secure, tamper-proof record of patient transactions and data access. Smart Contracts: Automating healthcare processes such as insurance claims and consent management (Pramila et al., 2023; Boopathi, 2023a).

Procedures: Data Security and Privacy: Ensuring that patient data is stored securely and can only be accessed with appropriate permissions. Supply Chain Management: Using blockchain to track the authenticity and integrity of pharmaceuticals and medical devices.

3D Printing in Medicine

Methods: 3D Printing Technology: Utilizing 3D printers to produce custom implants, prosthetics, and anatomical models. Medical Imaging Data Conversion: Converting medical scans into 3D-printable files (Boopathi, Khare, et al., 2023; Palaniappan et al., 2023; Senthil et al., 2023).

Procedures: Patient-Specific Implants: Creating implants and prosthetics tailored to an individual's anatomy. Surgical Planning: Using 3D-printed models to plan complex surgeries and practice procedures in advance.

Virtual Reality (VR) and Augmented Reality (AR) in Healthcare

Methods: VR Simulation: Building realistic healthcare scenarios for training healthcare professionals. AR Visualization: Overlaying digital information onto the physical world for medical procedures.

Procedures: Medical Training: Allowing medical students and professionals to practice surgeries, procedures, and diagnoses in a virtual environment. Using AR to superimpose patient data and surgical plans onto a surgeon's field of view during surgery.

These technologies, methods, and procedures collectively drive the digitization and transformation of healthcare, improving patient care, enhancing efficiency, and enabling data-driven decision-making in the modern healthcare landscape. However, it's crucial to address challenges such as data security, interoperability, and regulatory compliance to ensure the successful integration and utilization of digital healthcare technologies.

THE ROLE OF DATA IN HEALTHCARE TRANSFORMATION

Data is at the heart of the ongoing transformation in healthcare. It serves as the lifeblood of decision-making, innovation, and optimization in the industry. This section delves into the significant role of data in driving healthcare transformation (Dhanya et al., 2023; Pramila et al., 2023; Ramudu et al., 2023). The role of data in healthcare transformation is shown in Figure 3.

Figure 3. Role of data in healthcare transformation

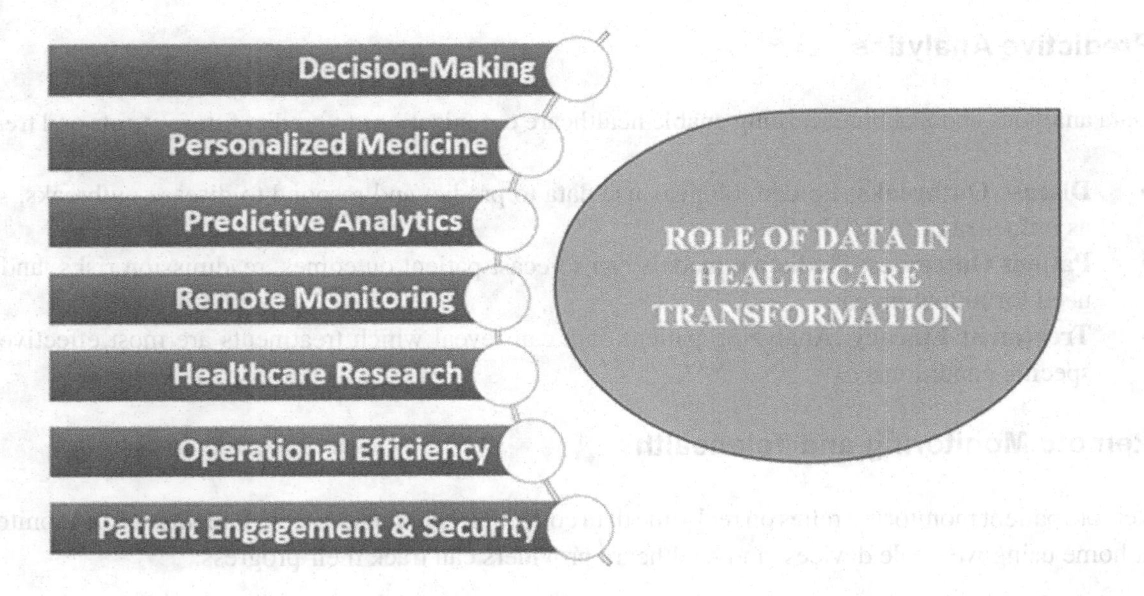

Informed Decision-Making

Data-driven decision-making is fundamental to healthcare transformation. It empowers healthcare providers, administrators, and policymakers to make informed choices about patient care, resource allocation, and strategy development. Here's how data contributes:

- **Clinical Decision Support**: Healthcare professionals can access patient records, treatment guidelines, and research findings to make accurate diagnoses and treatment decisions.
- **Population Health Management**: Data analytics enable the identification of at-risk populations and the development of preventive interventions.
- **Resource Allocation**: Hospitals and healthcare systems use data to optimize resource allocation, such as staffing levels and inventory management.

Personalized Medicine

The era of personalized medicine relies heavily on data. By analyzing an individual's genetic, clinical, and lifestyle data, healthcare providers can tailor treatments and interventions to a patient's unique characteristics (Anitha et al., 2023; Boopathi, Khare, et al., 2023; Subha et al., 2023):

- **Genomic Data**: Genomic sequencing data help identify genetic predispositions to diseases and determine optimal treatment approaches.
- **Health History**: Patient histories, including past treatments and responses, inform the choice of therapies and interventions.
- **Real-Time Monitoring**: Continuous monitoring of patient data allows for immediate adjustments to treatment plans.

Predictive Analytics

Data analytics and machine learning enable healthcare organizations to predict future events and trends:

- **Disease Outbreaks**: Epidemiologists use data to predict and respond to disease outbreaks, such as influenza or COVID-19.
- **Patient Outcomes**: Predictive models can forecast patient outcomes, readmission risks, and the need for follow-up care.
- **Treatment Efficacy**: Analyzing patient data can reveal which treatments are most effective for specific conditions.

Remote Monitoring and Telehealth

Remote patient monitoring relies on real-time data collection and transmission. Patients can be monitored at home using wearable devices, and healthcare providers can track their progress:

- **Vital Signs**: Data from wearable sensors provide continuous monitoring of vital signs, such as heart rate, blood pressure, and glucose levels.
- **Medication Adherence**: Data can verify that patients are taking medications as prescribed, helping prevent complications.

Healthcare Research and Innovation

Data fuels medical research and innovation, leading to the development of new drugs, treatments, and technologies:

- **Clinical Trials**: Clinical trial data are essential for evaluating the safety and efficacy of new therapies.
- **Drug Discovery**: Computational methods and data analysis accelerate drug discovery by identifying potential drug candidates.
- **Medical Imaging**: Advanced imaging techniques generate vast amounts of data for diagnosing and monitoring diseases.

Operational Efficiency

Data optimization extends to healthcare operations, improving efficiency and reducing costs:

- **Healthcare Analytics**: Analyzing operational data helps identify bottlenecks, reduce wait times, and optimize workflows.
- **Resource Allocation**: Data informs decisions about staffing levels, supply chain management, and facility utilization.

Patient Engagement and Education

Data-driven patient engagement platforms provide patients with access to their health data, educational resources, and tools for self-management:

- **Patient Portals**: Patients can access their medical records, test results, and educational materials online.
- **Mobile Apps**: Health apps and wearables offer patients real-time data about their health and fitness, encouraging healthier behaviors.

Regulatory Compliance and Security

Data plays a critical role in ensuring compliance with healthcare regulations and safeguarding patient privacy:

- **HIPAA Compliance**: Healthcare organizations must adhere to regulations like the Health Insurance Portability and Accountability Act (HIPAA) to protect patient data.
- **Cybersecurity**: Robust cybersecurity measures are essential to prevent data breaches and protect sensitive healthcare information.

The data is a transformative force in healthcare, driving improvements in patient care, research, operations, and innovation (Durairaj et al., 2023; Karthik et al., 2023; Pramila et al., 2023; Ramudu et al., 2023). As technology advances and data analytics capabilities grow, healthcare transformation will continue to be shaped by the effective collection, analysis, and utilization of healthcare data. However, it is imperative that healthcare stakeholders also address data privacy, security, and ethical considerations to ensure that patients' trust and well-being are protected in this data-driven era.

APPLICATIONS OF MACHINE LEARNING IN HEALTHCARE

Predictive Analytics for Disease Detection

Predictive analytics powered by machine learning has emerged as a ground-breaking application within the healthcare sector. It enables early disease detection, risk assessment, and timely interventions, ultimately improving patient outcomes and reducing healthcare costs (Boopathi, Alqahtani, et al., 2023; Gnanaprakasam et al., 2023; Haribalaji et al., 2014). This section delves into the significance of predictive analytics in disease detection and its diverse applications, as illustrated in Figure 4.

Early Disease Detection

Predictive analytics algorithms can analyze a patient's medical history, genetic data, and lifestyle factors to identify early warning signs of diseases. For example:

- **Cancer Detection**: Machine learning models can analyze imaging data (e.g., mammograms, CT scans) to detect early-stage cancers, leading to more successful treatment outcomes.
- **Diabetes Risk Assessment**: Predictive models consider factors such as family history, BMI, and blood sugar levels to assess an individual's risk of developing diabetes.

Disease Progression Monitoring

Once a disease is diagnosed, predictive analytics can help monitor its progression and adjust treatment plans accordingly (Boopathi, 2023a; Subha et al., 2023):

- **Cardiovascular Health**: Algorithms can predict the risk of heart attacks or strokes by analyzing patient data, including blood pressure, cholesterol levels, and lifestyle factors.
- **Neurodegenerative Diseases**: Machine learning models can track the progression of conditions like Alzheimer's disease or Parkinson's disease based on cognitive assessments and biomarker data.

Personalized Treatment Plans

Predictive analytics tailors treatment plans to individual patient characteristics, optimizing therapy choices and dosages:

- **Drug Response Prediction**: Algorithms predict how a patient will respond to a specific medication based on genetic and clinical data, minimizing adverse effects.

Figure 4. Predictive analytics for disease detection

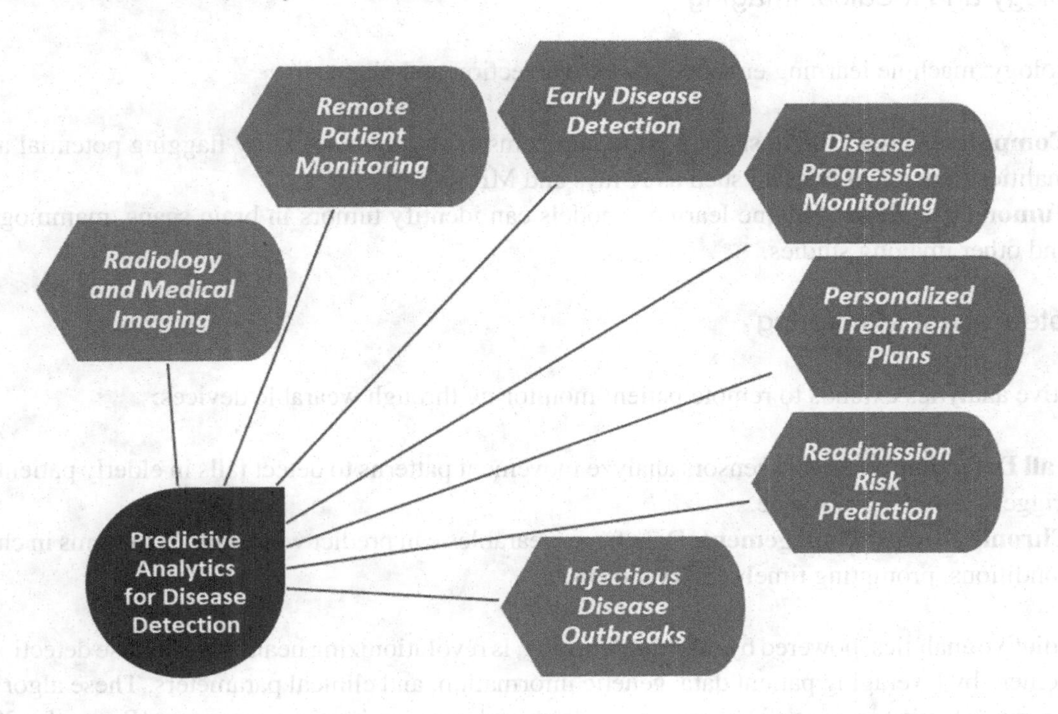

- **Cancer Treatment**: Machine learning assists in determining the most effective cancer treatments by analyzing genetic mutations and drug sensitivities.

Readmission Risk Prediction

Hospitals use predictive models to identify patients at risk of readmission after discharge, allowing for targeted interventions:

- **Heart Failure**: Algorithms consider factors like patient age, comorbidities, and lab results to predict the likelihood of heart failure readmissions.
- **Pneumonia**: Predictive analytics helps healthcare providers focus on patients most at risk of pneumonia-related readmissions.

Infectious Disease Outbreaks

Machine learning can assist in monitoring and predicting infectious disease outbreaks:

- **Epidemic Tracking**: Algorithms analyze global health data to detect outbreaks early, aiding in containment and resource allocation.
- **Flu Forecasting**: Predictive models can forecast flu activity based on historical data and real-time symptom reports.

Radiology and Medical Imaging

In radiology, machine learning enhances disease detection and diagnosis:

- **Computer-Aided Diagnosis (CAD)**: Algorithms assist radiologists by flagging potential abnormalities in medical images, such as X-rays and MRIs.
- **Tumor Detection**: Machine learning models can identify tumors in brain scans, mammograms, and other imaging studies.

Remote Patient Monitoring

Predictive analytics extends to remote patient monitoring through wearable devices:

- **Fall Detection**: Wearable sensors analyze movement patterns to detect falls in elderly patients and trigger alerts for assistance.
- **Chronic Disease Management**: Data from wearables can predict worsening symptoms in chronic conditions, prompting timely interventions.

Predictive analytics, powered by machine learning, is revolutionizing healthcare disease detection and management by leveraging patient data, genetic information, and clinical parameters. These algorithms enable early detection, personalized treatment plans, and improved patient outcomes (Boopathi, 2023b; Maguluri et al., 2023; Samikannu et al., 2022). As technology advances, predictive analytics will play a crucial role in preventive healthcare and disease management, but ethical considerations, data privacy, and model transparency remain important.

Personalized Medicine

Personalized medicine, also known as precision medicine, is a medical approach that customizes treatments and interventions to each patient's unique genetic makeup, clinical history, and lifestyle factors (Pramila et al., 2023; Reddy, Gaurav, et al., 2023; Sengeni et al., 2023). Machine learning plays a critical role in this application by analyzing complex data sets to develop highly customized treatment plans. Here are some key aspects of personalized medicine and treatment planning:

- **Genomic Data Analysis**: Machine learning algorithms analyze a patient's genetic information to identify specific genetic markers, mutations, and variations associated with diseases or drug responses.
- **Risk Assessment**: Models use genetic and clinical data to assess a patient's risk of developing certain diseases, such as cancer, diabetes, or heart disease.
- **Drug Response Prediction**: Machine learning predicts how a patient will respond to various medications, allowing healthcare providers to choose the most effective and least harmful treatment options.
- **Clinical Decision Support**: Algorithms provide healthcare professionals with real-time recommendations and guidelines based on a patient's individual data, improving treatment decision-making.

- **Treatment Optimization**: Machine learning helps optimize treatment plans by considering factors like drug interactions, side effects, and the patient's overall health.

Remote Monitoring and Telemedicine

Remote monitoring and telemedicine have gained significant importance, especially in situations where in-person healthcare visits are challenging or not feasible (Anitha et al., 2023; Boopathi, Khare, et al., 2023; Gowri et al., 2023). Machine learning enhances these applications by enabling continuous data analysis and real-time patient monitoring:

- **Wearable Devices**: Machine learning algorithms analyze data from wearable devices, such as smartwatches and fitness trackers, to monitor vital signs, activity levels, and detect anomalies that may indicate health issues.
- **Chronic Disease Management**: Patients with chronic conditions, like diabetes or hypertension, can use connected devices that transmit data to healthcare providers. Machine learning identifies trends and alerts providers to intervene when necessary.
- **Teleconsultations**: Telemedicine appointments benefit from machine learning-enhanced diagnostic tools, which provide immediate insights based on patient symptoms and data.
- **Emergency Triage**: In emergency situations, machine learning algorithms can assist in prioritizing patients based on their condition and urgency, helping healthcare professionals make informed decisions.
- **Medication Adherence**: Machine learning can track patients' medication adherence and provide reminders, ensuring they take their prescribed medications as directed.
- **Mental Health Monitoring**: Telehealth platforms enhanced by machine learning can monitor mental health patients through natural language processing, identifying concerning patterns in text or voice interactions.

Machine learning is revolutionizing healthcare by enabling personalized medicine and remote monitoring/telemedicine, enhancing treatment and care access. As algorithms evolve and data becomes available, these applications will improve patient outcomes and healthcare delivery. However, addressing privacy, security, and regulatory considerations is crucial for responsible and ethical use of these technologies.

Healthcare Chatbots and Virtual Assistants

Healthcare chatbots and virtual assistants, utilizing machine learning and NLP, are revolutionizing patient interactions, improving access to healthcare information, and enhancing administrative processes (Boopathi, 2023a; Durairaj et al., 2023; Ramudu et al., 2023). Here's how they are making an impact:

- **Patient Engagement**: Chatbots provide patients with 24/7 access to information and support, helping them schedule appointments, receive medication reminders, and answer basic health-related questions.
- **Triage and Symptom Assessment**: Virtual assistants can triage patients by asking questions about their symptoms and medical history, helping prioritize those in need of immediate care.

- **Appointment Scheduling**: Chatbots streamline appointment scheduling by allowing patients to book, reschedule, or cancel appointments through automated systems.
- **Medication Management**: Virtual assistants remind patients to take their medications, provide dosing instructions, and answer medication-related queries.
- **Telehealth Support**: During telemedicine consultations, chatbots can assist healthcare providers by retrieving patient data, presenting medical history, and offering real-time information.
- **Administrative Tasks**: Virtual assistants handle administrative tasks, such as verifying insurance information and processing billing inquiries, reducing administrative burden on staff.

Healthcare Fraud Detection

Machine learning plays a crucial role in identifying fraudulent activities within healthcare systems, saving billions of dollars annually and ensuring fair distribution of resources (Boopathi, Khare, et al., 2023, 2023; Gowri et al., 2023). Here's how it contributes to healthcare fraud detection:

- **Anomaly Detection**: Machine learning models continuously monitor healthcare transactions and claims data to detect anomalies or patterns indicative of fraud or abuse. Unusual billing patterns, repetitive claims, or inconsistencies can trigger alerts.
- **Predictive Modelling**: Predictive analytics models use historical data to forecast potentially fraudulent activities, enabling proactive prevention measures.
- **Behaviour Analysis**: Machine learning algorithms analyse the behavior of healthcare providers, patients, and payers to identify suspicious patterns, such as unnecessary procedures or unusual billing.
- **Claim Validation**: Natural language processing (NLP) can extract and analyze unstructured text in medical records and claim documents to verify the accuracy of claims and identify discrepancies.
- **Network Analysis**: Machine learning identifies relationships between healthcare providers and patients that may suggest collusion or fraud rings.
- **Real-Time Monitoring**: Real-time fraud detection systems use machine learning to flag and investigate suspicious activities as they occur, preventing fraudulent claims from being paid.

Healthcare chatbots and virtual assistants enhance patient engagement, reduce administrative burden, and improve healthcare access. Machine learning-driven fraud detection systems protect system integrity, directing resources towards genuine patient care. These applications contribute to healthcare system efficiency and sustainability, showcasing the transformative power of machine learning in administration and patient support (Dhanya et al., 2023; Nishanth et al., 2023; Satav, Hasan, et al., 2024; Satav, Lamani, et al., 2024).

PRACTICAL STRATEGIES FOR IMPLEMENTATION

The practical strategies for implementation of ml in digital health care applications are illustrated in Figure 5.

Figure 5. Practical strategies for implementation of ML in digital healthcare applications

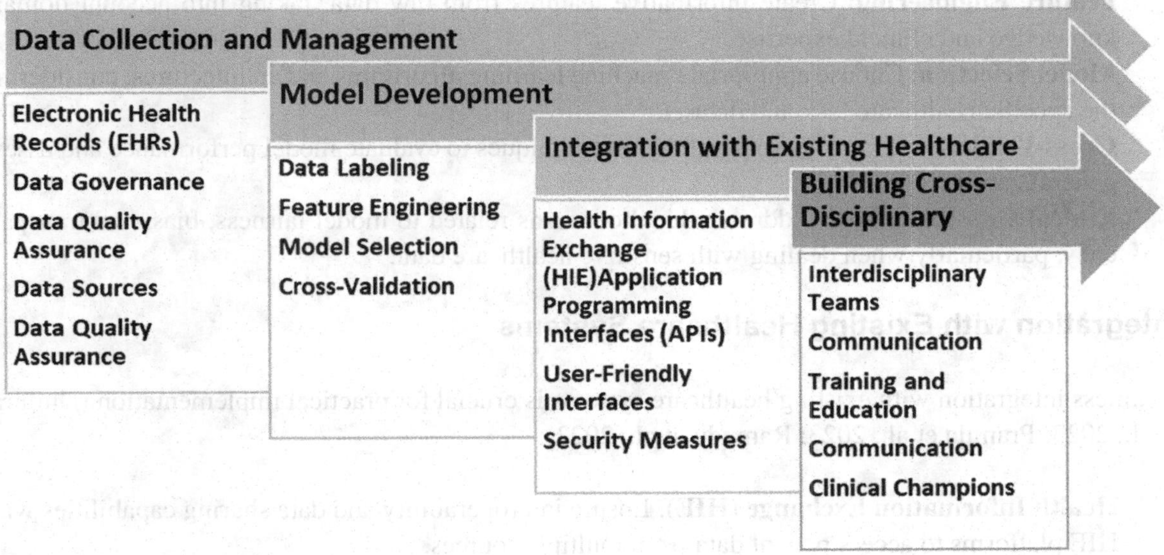

Data Collection and Management

Efficient data collection and management are fundamental to successful machine learning implementation in healthcare (Karthik et al., 2023; Maguluri et al., 2023; Syamala et al., 2023):

- **Electronic Health Records (EHRs)**: Integrate machine learning into EHR systems to ensure access to comprehensive patient data. Implement data standards and validation processes to maintain data accuracy.
- **Data Governance**: Establish data governance frameworks that address data privacy, security, and compliance with regulations like HIPAA. Define roles and responsibilities for data management.
- **Data Quality Assurance**: Implement data quality assurance practices, including data cleansing, normalization, and validation, to ensure the reliability of input data.
- **Data Sources**: Identify and incorporate relevant data sources, including clinical data, genomic data, wearable device data, and external data (e.g., population health data, medical literature).
- **Scalable Infrastructure**: Invest in scalable data storage and processing infrastructure, such as cloud-based solutions, to handle large healthcare datasets effectively.

Model Development and Training

Developing and training machine learning models for healthcare applications requires careful planning and execution (Jeevanantham et al., 2022; Karthik et al., 2023; Koshariya et al., 2023; Samikannu et al., 2022):

- **Data Labeling**: Annotate and label datasets for supervised learning tasks, ensuring the availability of high-quality training data.
- **Feature Engineering**: Create informative features from raw data, taking into account domain knowledge and clinical expertise.
- **Model Selection**: Choose appropriate machine learning algorithms and architectures, considering the specific healthcare task and dataset.
- **Cross-Validation**: Employ cross-validation techniques to evaluate model performance and assess generalizability.
- **Ethical Considerations**: Address ethical concerns related to model fairness, bias, and transparency, particularly when dealing with sensitive healthcare data.

Integration with Existing Healthcare Systems

Seamless integration with existing healthcare systems is crucial for practical implementation (Durairaj et al., 2023; Pramila et al., 2023; Ramudu et al., 2023):

- **Health Information Exchange (HIE)**: Ensure interoperability and data sharing capabilities with HIE platforms to access patient data from multiple sources.
- **Application Programming Interfaces (APIs)**: Develop standardized APIs that enable healthcare systems to communicate with machine learning models and receive predictions or recommendations.
- **User-Friendly Interfaces**: Design user-friendly interfaces for healthcare professionals to interact with machine learning tools within their existing workflows.
- **Security Measures**: Implement robust security measures, including encryption and access controls, to protect patient data and machine learning models from cyber threats.

Building Cross-Disciplinary Teams

Collaboration across diverse disciplines is essential for successful machine learning implementation in healthcare (Boopathi & Kanike, 2023; Maheswari et al., 2023; Syamala et al., 2023; Zekrifa et al., 2023):

- **Interdisciplinary Teams**: Assemble cross-disciplinary teams that include data scientists, healthcare professionals, software engineers, and domain experts.
- **Communication**: Foster effective communication among team members to bridge the gap between technical and clinical perspectives.
- **Training and Education**: Provide ongoing training and education to ensure that healthcare professionals understand the capabilities and limitations of machine learning models.
- **Clinical Champions**: Identify clinical champions who can advocate for and guide the integration of machine learning into clinical practice.
- **Ethical Expertise**: Involve ethicists and legal experts to address ethical and legal considerations in healthcare data usage and model deployment.

By adhering to these practical strategies, healthcare organizations can maximize the benefits of machine learning in improving patient care, optimizing operations, and driving healthcare innovation while mitigating potential challenges and risks. Collaboration and careful planning are key to realizing the full potential of machine learning in healthcare.

CASE STUDIES

Case Study: Early Detection of Diabetic Retinopathy

- **Problem**: Diabetic retinopathy is a leading cause of blindness among diabetic patients. Early detection is crucial for timely intervention.
- **Solution**: Google's DeepMind developed a machine learning model that analyzes retinal images to detect diabetic retinopathy. The model was trained on a dataset of 128,000 images.
- **Success**: The model achieved high accuracy in diagnosing diabetic retinopathy, enabling early detection and intervention. It is now used in clinical settings to assist ophthalmologists in screening patients.
- **Lessons Learned**: Collaborations between tech companies and healthcare providers can yield powerful tools for disease detection. However, ensuring the privacy and security of patient data is a paramount concern.

Case Study: Predictive Analytics for Patient Readmissions

- **Problem**: Hospital readmissions are costly and can indicate gaps in patient care. Reducing readmissions is a priority for healthcare systems.
- **Solution**: A major U.S. hospital implemented a predictive analytics system that used machine learning to assess patient data, including demographics, medical history, and vital signs, to predict readmission risk.
- **Success**: By identifying high-risk patients, the hospital reduced readmission rates and optimized resource allocation. The system led to cost savings and improved patient outcomes.
- **Lessons Learned**: Predictive analytics can help hospitals allocate resources more effectively and reduce costs, but successful implementation requires collaboration between data scientists and healthcare professionals.

Case Study: Natural Language Processing for Radiology Reports

- **Problem**: Radiology reports contain valuable diagnostic information but are often unstructured and time-consuming to analyze.
- **Solution**: A healthcare organization implemented natural language processing (NLP) to extract key information from radiology reports. Machine learning models were trained to identify diagnoses, findings, and recommendations.
- **Success**: NLP significantly reduced the time required to review radiology reports. It improved accuracy in identifying critical findings and enhanced communication between radiologists and referring physicians.

- **Lessons Learned**: NLP can streamline information extraction from unstructured clinical text, improving efficiency and quality of care. Close collaboration with radiologists is essential for model training and validation.

Case Study: Remote Monitoring and Heart Failure Prediction

- **Problem**: Patients with heart failure require continuous monitoring to prevent exacerbations and readmissions.
- **Solution**: A healthcare provider implemented a remote monitoring system that collected data from patients' wearable devices, including heart rate, activity levels, and weight. Machine learning models were used to predict heart failure exacerbations.
- **Success**: The remote monitoring system allowed for early intervention in cases of heart failure exacerbation, reducing hospital readmissions and improving patient quality of life.
- **Lessons Learned**: Remote monitoring, coupled with predictive analytics, can enhance patient care and reduce healthcare costs. However, it requires robust data infrastructure and patient engagement.

The case studies demonstrate the potential of machine learning in healthcare, enhancing diagnosis, treatment, and resource allocation, emphasizing data privacy, interdisciplinary collaboration, and the necessity for continuous validation and refinement of machine learning models.

Challenges and Ethical Considerations

Data Privacy and Security

- **Challenge**: Healthcare data is highly sensitive, containing personal information and medical histories. Ensuring the privacy and security of this data is paramount (Pramila et al., 2023; Ramudu et al., 2023).
- **Ethical Considerations**: Healthcare organizations must implement robust encryption, access controls, and cybersecurity measures to safeguard patient data. Consent mechanisms and transparency in data handling should also be in place. Balancing data accessibility for research with data protection is a constant ethical challenge.

Bias and Fairness in Healthcare Algorithms

- **Challenge**: Machine learning algorithms may inadvertently perpetuate or amplify biases present in healthcare data, leading to unfair treatment or diagnosis disparities.
- **Ethical Considerations**: Ethical machine learning practices involve actively identifying and mitigating bias in algorithms. Diverse and representative training data, fairness metrics, and ongoing monitoring are essential to address bias in healthcare AI.

Regulatory Compliance

- **Challenge**: Healthcare is subject to strict regulations and compliance requirements, such as HIPAA in the United States. Implementing AI solutions while adhering to these regulations is a complex challenge.
- **Ethical Considerations**: Healthcare organizations must ensure that machine learning solutions comply with all relevant regulations and standards. Ethical considerations include data sharing, informed consent, and transparency in data usage.

Informed Consent and Patient Autonomy

- **Challenge**: Patients may not fully understand the implications of sharing their data or receiving AI-based healthcare recommendations. Informed consent processes need to strike a balance between comprehensibility and completeness.
- **Ethical Considerations**: Ethical AI in healthcare involves clear and understandable explanations to patients about how their data will be used and how AI will impact their care. Patients should have the autonomy to opt-in or opt-out of AI-driven care.

Accountability and Transparency

- **Challenge**: Understanding and interpreting machine learning models can be complex, making it challenging to assign responsibility when something goes wrong.
- **Ethical Considerations**: Transparency in AI algorithms and decision-making processes is crucial. Healthcare providers should be able to explain and justify AI-driven recommendations. Establishing clear lines of accountability for AI decisions is an ethical imperative.

Resource Allocation and Equity

- **Challenge**: Machine learning in healthcare can optimize resource allocation but may inadvertently exacerbate healthcare disparities if not implemented carefully.
- **Ethical Considerations**: Ethical considerations involve ensuring that AI applications promote equity in healthcare. This includes addressing the needs of underserved populations and avoiding discriminatory practices.

Continual Monitoring and Evaluation

- **Challenge**: Machine learning models in healthcare must evolve and adapt to changing patient populations and medical practices.
- **Ethical Considerations**: Continuous monitoring, validation, and improvement of AI models are ethical obligations. Models must be regularly evaluated for performance, fairness, and safety to ensure they align with the evolving needs of healthcare.

Ethical considerations are crucial in the development and deployment of machine learning applications in healthcare to build trust, protect patient rights, and ensure responsible AI use. This requires interdisciplinary collaboration among data scientists, healthcare professionals, ethicists, and legal experts to uphold ethical standards in healthcare AI.

FUTURE TRENDS AND INNOVATIONS

Emerging Technologies in Healthcare

- **Quantum Computing**: Quantum computing holds the potential to revolutionize drug discovery, genomics, and healthcare simulations by solving complex problems at unprecedented speeds (Boopathi, 2023b; Durairaj et al., 2023; Karthik et al., 2023).
- **Nanomedicine**: Nanotechnology enables precise drug delivery, diagnostics, and imaging at the molecular level, offering targeted therapies with minimal side effects (Boopathi, Umareddy, et al., 2023; Boopathi & Davim, 2023a, 2023b).
- **Blockchain**: Blockchain technology enhances the security and integrity of health records, supply chain management, and clinical trials, ensuring trust and transparency in healthcare.
- **Edge Computing**: Edge computing enables real-time data processing at the device level, improving the speed and efficiency of remote patient monitoring and telemedicine.

AI-Driven Drug Discovery

- **AI-Generated Compounds**: Machine learning models can design and suggest novel drug compounds, accelerating drug discovery and reducing development costs.
- **Drug Repurposing**: AI identifies existing drugs with potential applications in new disease areas, expediting treatment options for various conditions.
- **Clinical Trial Optimization**: Predictive analytics and AI-driven patient recruitment streamline clinical trials, enhancing trial success rates and efficiency.
- **Personalized Medicine**: AI tailors drug treatments to individual patient characteristics, optimizing therapeutic outcomes and minimizing adverse effects.

Precision Health and Genomic Medicine

- **Genomic Sequencing**: Widespread genomic sequencing allows for personalized treatment plans based on an individual's genetic makeup, leading to targeted therapies and precision medicine.
- **Biobanks**: Growing biobanks of genetic and health data enable large-scale research on disease genetics and biomarkers, driving advances in precision health.
- **Microbiome Research**: Understanding the gut microbiome's impact on health can lead to personalized dietary and treatment recommendations.
- **AI and Genomic Analysis**: Machine learning aids in interpreting vast genomic datasets, identifying disease risks, and predicting treatment responses.

Telehealth and Remote Monitoring

- **Virtual Reality (VR) and Augmented Reality (AR)**: VR and AR technologies enhance telehealth by creating immersive patient-doctor interactions and remote surgical simulations.
- **Wearable Health Tech**: Wearables evolve to provide more accurate health data, enabling continuous monitoring of vital signs and chronic conditions.
- **5G Connectivity**: The rollout of 5G networks facilitates high-speed, low-latency data transmission for real-time telemedicine and remote surgery.

Behavioural Health and Mental Health Tech

- **Digital Therapeutics**: Mobile apps and digital platforms deliver evidence-based interventions for mental health conditions, offering scalable and accessible solutions.
- **AI-Enhanced Mental Health Screening**: AI-powered chatbots and sentiment analysis tools can assist in early detection and monitoring of mental health issues.
- **Predictive Analytics**: Machine learning models analyze patient data to predict mental health risks, allowing for timely interventions.

Bioinformatics and Computational Biology

- **Structural Biology**: Computational methods and AI facilitate the understanding of protein structures and their interactions, aiding in drug design and disease mechanisms.
- **Single-Cell Omics**: Advancements in single-cell RNA sequencing and data analysis reveal cellular heterogeneity, advancing our understanding of diseases at the single-cell level.
- **Drug-Drug Interactions**: AI models predict potential drug interactions, reducing adverse effects and improving treatment safety.

Emerging technologies like AI-driven drug discovery, precision health, and genomic medicine are revolutionizing healthcare by offering more effective and personalized treatments, transforming diagnosis, treatment, and patient care (Maguluri et al., 2023; Maheswari et al., 2023). Telehealth, remote monitoring, and behavioral health tech aim to improve mental health outcomes and access to care, but interdisciplinary collaboration, ethical considerations, and regulatory adaptation are crucial for patient and healthcare system benefits.

SUMMARY

The integration of machine learning and emerging technologies in healthcare is paving the way for transformative possibilities, enhancing patient care, streamlining clinical workflows, and driving groundbreaking medical discoveries.

- Tailoring treatments to the individual based on their genetic makeup, medical history, and real-time health data, leading to more effective and personalized care.

- Revolutionizing the pharmaceutical industry by accelerating the development of novel drugs and repurposing existing ones.
- Expanding access to healthcare, improving patient engagement, and providing timely interventions, especially in underserved areas.
- Maintaining a strong ethical foundation to ensure patient data privacy, mitigate biases, and address the implications of AI in healthcare.
- Fostering collaboration between data scientists, healthcare professionals, ethicists, and policymakers to navigate the complex challenges and opportunities presented by these technologies.

Healthcare is set to become patient-centric, data-driven, and accessible, but stakeholders must address ethical, privacy, and security concerns to ensure responsible deployment. With careful planning, regulation, and ethical principles, the future of healthcare holds great promise for improving individual and community well-being worldwide.

REFERENCES

Agrawal, A. V., Pitchai, R., Senthamaraikannan, C., Balaji, N. A., Sajithra, S., & Boopathi, S. (2023). Digital Education System During the COVID-19 Pandemic. In Using Assistive Technology for Inclusive Learning in K-12 Classrooms (pp. 104–126). IGI Global. doi:10.4018/978-1-6684-6424-3.ch005

Anitha, C., Komala, C., Vivekanand, C. V., Lalitha, S., & Boopathi, S. (2023). Artificial Intelligence driven security model for Internet of Medical Things (IoMT). *IEEE Explore*, 1–7.

Boopathi, S. (2013). *Experimental study and multi-objective optimization of near-dry wire-cut electrical discharge machining process* [PhD Thesis]. http://hdl.handle.net/10603/16933

Boopathi, S. (2022). Experimental investigation and multi-objective optimization of cryogenic Friction-stir-welding of AA2014 and AZ31B alloys using MOORA technique. *Materials Today. Communications*, *33*, 104937. doi:10.1016/j.mtcomm.2022.104937

Boopathi, S. (2023a). Internet of Things-Integrated Remote Patient Monitoring System: Healthcare Application. In *Dynamics of Swarm Intelligence Health Analysis for the Next Generation* (pp. 137–161). IGI Global. doi:10.4018/978-1-6684-6894-4.ch008

Boopathi, S. (2023b). Securing Healthcare Systems Integrated With IoT: Fundamentals, Applications, and Future Trends. In Dynamics of Swarm Intelligence Health Analysis for the Next Generation (pp. 186–209). IGI Global.

Boopathi, S., Alqahtani, A. S., Mubarakali, A., & Panchatcharam, P. (2023). Sustainable developments in near-dry electrical discharge machining process using sunflower oil-mist dielectric fluid. *Environmental Science and Pollution Research International*, 1–20. doi:10.100711356-023-27494-0 PMID:37199846

Boopathi, S., & Davim, J. P. (2023a). Applications of Nanoparticles in Various Manufacturing Processes. In *Sustainable Utilization of Nanoparticles and Nanofluids in Engineering Applications* (pp. 1–31). IGI Global. doi:10.4018/978-1-6684-9135-5.ch001

Boopathi, S., & Davim, J. P. (2023b). *Sustainable Utilization of Nanoparticles and Nanofluids in Engineering Applications*. IGI Global. doi:10.4018/978-1-6684-9135-5

Boopathi, S., & Kanike, U. K. (2023). Applications of Artificial Intelligent and Machine Learning Techniques in Image Processing. In *Handbook of Research on Thrust Technologies' Effect on Image Processing* (pp. 151–173). IGI Global. doi:10.4018/978-1-6684-8618-4.ch010

Boopathi, S., Khare, R., KG, J. C., Muni, T. V., & Khare, S. (2023). Additive Manufacturing Developments in the Medical Engineering Field. In Development, Properties, and Industrial Applications of 3D Printed Polymer Composites (pp. 86–106). IGI Global.

Boopathi, S., Myilsamy, S., & Sukkasamy, S. (2021). *Experimental Investigation and Multi-Objective Optimization of Cryogenically Cooled Near-Dry Wire-Cut EDM Using TOPSIS Technique*. IJAMT Preprint.

Boopathi, S., & Sivakumar, K. (2013). Experimental investigation and parameter optimization of near-dry wire-cut electrical discharge machining using multi-objective evolutionary algorithm. *International Journal of Advanced Manufacturing Technology*, *67*(9–12), 2639–2655. doi:10.100700170-012-4680-4

Boopathi, S., Umareddy, M., & Elangovan, M. (2023). Applications of Nano-Cutting Fluids in Advanced Machining Processes. In *Sustainable Utilization of Nanoparticles and Nanofluids in Engineering Applications* (pp. 211–234). IGI Global. doi:10.4018/978-1-6684-9135-5.ch009

Chakraborty, I., Vigneswara Ilavarasan, P., & Edirippulige, S. (2022). E-Health Startups' Framework for Value Creation and Capture: Some Insights from Systematic Review. *Proceedings of the International Conference on Cognitive and Intelligent Computing: ICCIC 2021,* 1, 141–152. 10.1007/978-981-19-2350-0_13

Dhanya, D., Kumar, S. S., Thilagavathy, A., Prasad, D., & Boopathi, S. (2023). Data Analytics and Artificial Intelligence in the Circular Economy: Case Studies. In Intelligent Engineering Applications and Applied Sciences for Sustainability (pp. 40–58). IGI Global.

Durairaj, M., Jayakumar, S., Karpagavalli, V., Maheswari, B. U., Boopathi, S., & ... (2023). Utilization of Digital Tools in the Indian Higher Education System During Health Crises. In *Multidisciplinary Approaches to Organizational Governance During Health Crises* (pp. 1–21). IGI Global. doi:10.4018/978-1-7998-9213-7.ch001

Faujdar, D. S., Singh, T., Kaur, M., Sahay, S., & Kumar, R. (2021). Stakeholders' perceptions of the implementation of a patient-centric digital health application for primary healthcare in India. *Healthcare Informatics Research*, *27*(4), 315–324. doi:10.4258/hir.2021.27.4.315 PMID:34788912

Ganapathy, K., Das, S., Reddy, S., Thaploo, V., Nazneen, A., Kosuru, A., & Shankar Nag, U. (2021). Digital health care in public private partnership mode. *Telemedicine Journal and e-Health*, *27*(12), 1363–1371. doi:10.1089/tmj.2020.0499 PMID:33819433

Gnanaprakasam, C., Vankara, J., Sastry, A. S., Prajval, V., Gireesh, N., & Boopathi, S. (2023). Long-Range and Low-Power Automated Soil Irrigation System Using Internet of Things: An Experimental Study. In Contemporary Developments in Agricultural Cyber-Physical Systems (pp. 87–104). IGI Global.

Gowri, N. V., Dwivedi, J. N., Krishnaveni, K., Boopathi, S., Palaniappan, M., & Medikondu, N. R. (2023). Experimental investigation and multi-objective optimization of eco-friendly near-dry electrical discharge machining of shape memory alloy using Cu/SiC/Gr composite electrode. *Environmental Science and Pollution Research International*, *30*(49), 1–19. doi:10.100711356-023-26983-6 PMID:37126160

Haribalaji, V., Boopathi, S., & Balamurugan, S. (2014). Effect of Welding Processes on Mechanical and Metallurgical Properties of High Strength Low Alloy (HSLA) Steel Joints. *International Journal of Innovation and Scientific Research*, *12*(1), 170–179.

Jarva, E., Oikarinen, A., Andersson, J., Tuomikoski, A.-M., Kääriäinen, M., Meriläinen, M., & Mikkonen, K. (2022). Healthcare professionals' perceptions of digital health competence: A qualitative descriptive study. *Nursing Open*, *9*(2), 1379–1393. doi:10.1002/nop2.1184 PMID:35094493

Jeevanantham, Y. A., Saravanan, A., Vanitha, V., Boopathi, S., & Kumar, D. P. (2022). Implementation of Internet-of Things (IoT) in Soil Irrigation System. *IEEE Explore*, 1–5.

Karthik, S., Hemalatha, R., Aruna, R., Deivakani, M., Reddy, R. V. K., & Boopathi, S. (2023). Study on Healthcare Security System-Integrated Internet of Things (IoT). In Perspectives and Considerations on the Evolution of Smart Systems (pp. 342–362). IGI Global.

Koshariya, A. K., Kalaiyarasi, D., Jovith, A. A., Sivakami, T., Hasan, D. S., & Boopathi, S. (2023). AI-Enabled IoT and WSN-Integrated Smart Agriculture System. In *Artificial Intelligence Tools and Technologies for Smart Farming and Agriculture Practices* (pp. 200–218). IGI Global. doi:10.4018/978-1-6684-8516-3.ch011

Kruszyńska-Fischbach, A., Sysko-Romańczuk, S., Napiórkowski, T. M., Napiórkowska, A., & Kozakiewicz, D. (2022). Organizational e-health readiness: How to prepare the primary healthcare providers' services for digital transformation. *International Journal of Environmental Research and Public Health*, *19*(7), 3973. doi:10.3390/ijerph19073973 PMID:35409656

Lupton, D., & Leahy, D. (2019). Reimagining digital health education: Reflections on the possibilities of the storyboarding method. *Health Education Journal*, *78*(6), 633–646. doi:10.1177/0017896919841413

Maguluri, L. P., Ananth, J., Hariram, S., Geetha, C., Bhaskar, A., & Boopathi, S. (2023). Smart Vehicle-Emissions Monitoring System Using Internet of Things (IoT). In Handbook of Research on Safe Disposal Methods of Municipal Solid Wastes for a Sustainable Environment (pp. 191–211). IGI Global.

Maheswari, B. U., Imambi, S. S., Hasan, D., Meenakshi, S., Pratheep, V., & Boopathi, S. (2023). Internet of Things and Machine Learning-Integrated Smart Robotics. In Global Perspectives on Robotics and Autonomous Systems: Development and Applications (pp. 240–258). IGI Global. doi:10.4018/978-1-6684-7791-5.ch010

Morris, B. B., Rossi, B., & Fuemmeler, B. (2022). The role of digital health technology in rural cancer care delivery: A systematic review. *The Journal of Rural Health*, *38*(3), 493–511. doi:10.1111/jrh.12619 PMID:34480506

Nishanth, J., Deshmukh, M. A., Kushwah, R., Kushwaha, K. K., Balaji, S., & Sampath, B. (2023). Particle Swarm Optimization of Hybrid Renewable Energy Systems. In *Intelligent Engineering Applications and Applied Sciences for Sustainability* (pp. 291–308). IGI Global. doi:10.4018/979-8-3693-0044-2.ch016

Nutbeam, D. (2021). From health education to digital health literacy–building on the past to shape the future. *Global Health Promotion*, *28*(4), 51–55. doi:10.1177/17579759211044079 PMID:34719292

Palaniappan, M., Tirlangi, S., Mohamed, M. J. S., Moorthy, R. S., Valeti, S. V., & Boopathi, S. (2023). Fused Deposition Modelling of Polylactic Acid (PLA)-Based Polymer Composites: A Case Study. In Development, Properties, and Industrial Applications of 3D Printed Polymer Composites (pp. 66–85). IGI Global.

Pramila, P., Amudha, S., Saravanan, T., Sankar, S. R., Poongothai, E., & Boopathi, S. (2023). Design and Development of Robots for Medical Assistance: An Architectural Approach. In Contemporary Applications of Data Fusion for Advanced Healthcare Informatics (pp. 260–282). IGI Global.

Ramudu, K., Mohan, V. M., Jyothirmai, D., Prasad, D., Agrawal, R., & Boopathi, S. (2023). Machine Learning and Artificial Intelligence in Disease Prediction: Applications, Challenges, Limitations, Case Studies, and Future Directions. In Contemporary Applications of Data Fusion for Advanced Healthcare Informatics (pp. 297–318). IGI Global.

Ravisankar, A., Sampath, B., & Asif, M. M. (2023). Economic Studies on Automobile Management: Working Capital and Investment Analysis. In Multidisciplinary Approaches to Organizational Governance During Health Crises (pp. 169–198). IGI Global.

Reddy, M. A., Gaurav, A., Ushasukhanya, S., Rao, V. C. S., Bhattacharya, S., & Boopathi, S. (2023). Bio-Medical Wastes Handling Strategies During the COVID-19 Pandemic. In Multidisciplinary Approaches to Organizational Governance During Health Crises (pp. 90–111). IGI Global. doi:10.4018/978-1-7998-9213-7.ch006

Reddy, M. A., Reddy, B. M., Mukund, C., Venneti, K., Preethi, D., & Boopathi, S. (2023). Social Health Protection During the COVID-Pandemic Using IoT. In *The COVID-19 Pandemic and the Digitalization of Diplomacy* (pp. 204–235). IGI Global. doi:10.4018/978-1-7998-8394-4.ch009

Samikannu, R., Koshariya, A. K., Poornima, E., Ramesh, S., Kumar, A., & Boopathi, S. (2022). Sustainable Development in Modern Aquaponics Cultivation Systems Using IoT Technologies. In *Human Agro-Energy Optimization for Business and Industry* (pp. 105–127). IGI Global.

Satav, S. D., Hasan, D. S., Pitchai, R., Mohanaprakash, T. A., Sultanuddin, S. J., & Boopathi, S. (2024). Next Generation of Internet of Things (NGIoT) in Healthcare Systems. In Practice, Progress, and Proficiency in Sustainability (pp. 307–330). IGI Global. doi:10.4018/979-8-3693-1186-8.ch017

Satav, S. D., Lamani, D., G, H. K., Kumar, N. M. G., Manikandan, S., & Sampath, B. (2024). Energy and Battery Management in the Era of Cloud Computing. In Practice, Progress, and Proficiency in Sustainability (pp. 141–166). IGI Global. doi:10.4018/979-8-3693-1186-8.ch009

Sengeni, D., Padmapriya, G., Imambi, S. S., Suganthi, D., Suri, A., & Boopathi, S. (2023). Biomedical Waste Handling Method Using Artificial Intelligence Techniques. In *Handbook of Research on Safe Disposal Methods of Municipal Solid Wastes for a Sustainable Environment* (pp. 306–323). IGI Global. doi:10.4018/978-1-6684-8117-2.ch022

Senthil, T., Puviyarasan, M., Babu, S. R., Surakasi, R., Sampath, B., & Associates. (2023). Industrial Robot-Integrated Fused Deposition Modelling for the 3D Printing Process. In Development, Properties, and Industrial Applications of 3D Printed Polymer Composites (pp. 188–210). IGI Global.

Siriwardhana, Y., Gür, G., Ylianttila, M., & Liyanage, M. (2021). The role of 5G for digital healthcare against COVID-19 pandemic: Opportunities and challenges. *Ict Express*, *7*(2), 244–252. doi:10.1016/j.icte.2020.10.002

Subha, S., Inbamalar, T., Komala, C., Suresh, L. R., Boopathi, S., & Alaskar, K. (2023). A Remote Health Care Monitoring system using internet of medical things (IoMT). *IEEE Explore*, 1–6.

Syamala, M., Komala, C., Pramila, P., Dash, S., Meenakshi, S., & Boopathi, S. (2023). Machine Learning-Integrated IoT-Based Smart Home Energy Management System. In *Handbook of Research on Deep Learning Techniques for Cloud-Based Industrial IoT* (pp. 219–235). IGI Global. doi:10.4018/978-1-6684-8098-4.ch013

Tebeje, T. H., & Klein, J. (2021). Applications of e-health to support person-centered health care at the time of COVID-19 pandemic. *Telemedicine Journal and e-Health*, *27*(2), 150–158. doi:10.1089/tmj.2020.0201 PMID:32746750

Veeranjaneyulu, R., Boopathi, S., Kumari, R. K., Vidyarthi, A., Isaac, J. S., & Jaiganesh, V. (2023). Air Quality Improvement and Optimisation Using Machine Learning Technique. *IEEE- Explore*, 1–6.

Veeranjaneyulu, R., Boopathi, S., Narasimharao, J., Gupta, K. K., Reddy, R. V. K., & Ambika, R. (2023). Identification of Heart Diseases using Novel Machine Learning Method. *IEEE- Explore*, 1–6.

Zekrifa, D. M. S., Kulkarni, M., Bhagyalakshmi, A., Devireddy, N., Gupta, S., & Boopathi, S. (2023). Integrating Machine Learning and AI for Improved Hydrological Modeling and Water Resource Management. In *Artificial Intelligence Applications in Water Treatment and Water Resource Management* (pp. 46–70). IGI Global. doi:10.4018/978-1-6684-6791-6.ch003

LIST OF ABBREVIATIONS

EHR - Electronic Health Records
NLP - Natural Language Processing
HIPAA - Health Insurance Portability and Accountability Act
HIE - Health Information Exchange
API - Application Programming Interface
VR - Virtual Reality
AR - Augmented Reality
5G - Fifth Generation (of wireless technology)
AI - Artificial Intelligence
CAD - Computer-Aided Diagnosis

Chapter 13
Unsupervised Learning Techniques for Vibration-Based Structural Health Monitoring Systems Driven by Data:
A General Overview

Francesco Colace
https://orcid.org/0000-0003-2798-5834
University of Salerno, Italy

Brij B. Gupta
Asia University, Taichung, Taiwan & Lebanese American University, Beirut, Lebanon

Angelo Lorusso
University of Salerno, Italy

Alfredo Troiano
University of Salerno, Italy

Domenico Santaniello
https://orcid.org/0000-0002-5783-1847
University of Salerno, Italy

Carmine Valentino
University of Salerno, Italy

ABSTRACT

Structural damage detection is a crucial issue for the safety of civil buildings, which are subject to gradual deterioration over time and at risk from sudden seismic events. To prevent irreparable damage, the scientific community has directed its attention toward developing innovative methods for structural health monitoring (SHM), which can provide a timely and reliable assessment of structural conditions. In this domain, the significance of unsupervised learning approaches has grown considerably, as they enable the identification of structural irregularities solely based on data obtained from intact structures to train statistical models. Despite the importance of studies on unsupervised learning methods for structural health monitoring, no reviews are specifically dedicated to this topic, considering the application part. The review of studies, therefore, made it possible to highlight the progress achieved in this field and identify areas where improvements could still be made to develop increasingly accurate and effective methods for structural damage detection.

DOI: 10.4018/978-1-6684-9999-3.ch013

INTRODUCTION

Civil engineers consistently face the challenge of ensuring the stability of civil buildings, given that these constructions endure many degradation factors, such as extended usage, adverse environmental conditions, and the potential for seismic events. To safeguard against the potential loss of life and property resulting from sudden or gradual damage, they have devised approaches and regulations for structural assessment and evaluation(Lorusso & Celenta, 2023a). These methods encompass visual inspection and non-destructive damage estimation techniques. Nevertheless, these conventional approaches may come with high expenses, time demands, and potential hazards for operators(Dhruva Kumar et al., 2023). For this reason, the scientific community is focusing on the development of SHM, which can provide a real-time assessment of the state of structures, thanks to continuous advances in the hardware part of the systems, such as the ongoing development of the IoT paradigm through real-time monitoring(Deivasigamani et al., 2013; Kumar & Kota, 2023). Undoubtedly, the field of civil engineering has greatly profited from the advancement of sensor technology and machine learning (ML)(Padmapoorani et al., 2023). SHM research has also reaped the rewards of these progressions, as an ever-growing array of SHM approaches now incorporates deep learning (DL) frameworks(Bezas et al., 2020). Specifically, extensive research efforts have been directed toward monitoring the health of facilities using unsupervised learning algorithms. These methodologies rely on analyzing data obtained from intact structures to develop statistical models, eliminating the need for supplementary information regarding the structures' condition(Meribout et al., 2021). Within this context, novelty detection using vibration data is the widely acknowledged approach for unsupervised learning in SHM (Arcadius Tokognon et al., 2017). Recent scientific literature has extensively explored ML-based SHM techniques, encompassing applications of DL (DL) and reinforcement learning, with a special emphasis on vibration analysis. Moreover, researchers have also investigated the effectiveness of CNNs in SHM methods(De Simone et al., 2022). However, to date, there are no systematic literature reviews on unsupervised learning SHM methods taking into account possible applications, which is one of the most promising research areas for the implementation of an effective and practical structural damage detection system to support predictive monitoring(Lorusso & Guida, 2022). Some of the models studied introduced a statistical pattern recognition approach divided into phases, i.e., operational assessment, data acquisition, feature selection, and the increase of statistical models for diversifying these features. Despite the primary focus of the analyzed studies being on the last two stages, several approaches emphasise feature selection rather than pattern development (Sarah et al., 2019)t. Certainly, the main objective of statistical models is to identify possible harm by ascertaining its presence, position, and scope. In recent times, there have been noteworthy advancements in the field of SHM, with particular emphasis on vibration-based and visual-based methods (Fig.1) (Lorusso & Celenta, 2023b)(Martens et al., 2023). The visual-based approach leverages state-of-the-art computer vision techniques and unmanned aerial vehicles (drones) to conduct visual inspections autonomously. These innovative developments have contributed to the progress and effectiveness of SHM methodologies. On the other hand, vibration-based SHM relies on gathering vibration data from structures to identify structural damage that may prove challenging to detect through conventional inspection methods. In the field of SHM, there are two main categories of vibration-based monitoring: model-based and data-based(G. Hou et al., 2022). Model-based techniques require specific skills to develop accurate physical models of structures, calibrated using data collected from the structure itself. System recognition and model refinement techniques are fundamental to the functioning of model-based techniques. However, such procedures can be time-consuming and costly, especially for complex structures(Duan et al., 2014). Moreover, with the growing intricacy of the structure, the likelihood of modelling errors may also elevate. On the other hand, data-centric SHM involves using either supervised or unsupervised learning methodologies to train statistical models that can effectively detect damage and anomalies by analyzing critical engineered elements. These data-driven approaches play a crucial role in enhancing the accuracy and reliability of structural health assessment(De Simone & Guida, 2020)(Ampadi Ramachandran et al., 2023). Within supervised learning methodologies, the training data encompass diverse structural conditions, encompassing both impaired and intact states, and are suitably annotated. Instances of supervised SHM approaches consist of neural networks, decision trees, Support Vector Machine (SVM), and deep neural networks. When considering SHM applications, unsupervised learning offers a viable alternative to supervised learning, particularly in vibration-based systems (Rather et al., 2023). Notably, most unsupervised learning methods in SHM are oriented toward detecting novel patterns and anomalies. This approach allows models to learn directly from the normal condition of the structure, which information is usually abundantly available without the need for damaged state data(Galanopoulos et al., 2023). This confers a notable benefit over supervised learning, as it requires a comprehensive dataset containing information about the structure's health and condition. Nevertheless, it is crucial to acknowledge that unsupervised learning generally exhibits lower precision when it comes to detecting and quantifying damage compared to supervised learning(Pawar et al., 2023). In particular, unsupervised learning can detect any change in the system that the model can detect whereas supervised learning is able to train the model to specifically detect the types of damage present in the datasets(Di Lorenzo et al., 2023). Unsupervised learning still represents an interesting and viable option for vibration-based SHM, especially since the acquisition of damage state data can be costly and impractical for most structures. In many researches, damage detection methods based precisely on unsupervised learning have been introduced and implemented. These techniques entail extracting characteristics through signal processing methodologies like transmissibility and adopting multivariate anomaly analysis along with Mahalanobis squared distance (MSD) to assess discordance for decision-making. The fundamental idea behind outlier analysis is to calibrate the data to the normal condition's distribution, allowing examinations that fail the outlier test to be classified as novel or damaged (Sarmadi et al., 2021). Furthermore, various research endeavours have explored outlier analysis methods utilizing Mean Squared Deviation to detect structural damage, incorporating features derived from autoregressive (AR) models. Artificial neural networks, such as auto-associative neural networks (AANN), self-organizing neural networks, and wavelet neural networks, have also demonstrated their effectiveness in damage detection. In certain cases, novelty detection techniques can be evaluated as classifiers for a specific category, while cluster analysis methods, like k-means clustering and fuzzy c-means, can classify data into multiple clusters(Hurtik & Tomasiello, 2019). These diverse approaches contribute to advancing reliable and efficient structural health monitoring strategies. All these models rely on extracting signal features, which aid in detecting anomalies. Nevertheless, it is essential to recognize that these approaches may exhibit lower precision in detecting and quantifying damage compared to supervised learning, as previously mentioned, due to the absence of inherent data regarding a specific irregularity at the outset. The objective of this document is to examine the latest SHM methodologies that make use of unsupervised learning. Specifically, we systematically reviewed approximately 90 studies published between 2013 and 2023. It should be noted that the study carried out is not comprehensive of all applications of SHM methods based on unsupervised learning in this timeframe. Rather, it represents a methodical selection of peer-reviewed articles that allows for a broader overview of the current scenario to lay the foundation for future studies.

Thus, peer-reviewed articles were selected from well-known academic databases, including Google Scholar, Scopus and Web of Science. The search was conducted using relevant keywords such as "IoT and Structural Health Monitoring", "structural damage detection", "damage detection and ML", "structural health monitoring", "structural novelty detection", and "anomaly damage detection". In order to narrow down the initial search outcomes, a two-step screening process was conducted. During the first step, the pertinence of article titles, abstracts, and keywords to the research topic was examined. Only after this initial screening the remaining papers were chosen for a thorough analysis and review of the entire content. The documents were evaluated according to the relevance of the learning mode, method, types of functionality, objectives, type of application, decision-making process, threshold and results. It has to be repeated that the final list of 90 papers is not exhaustive, and indeed, many examples should be integrated into the study, but it represents a skimmed list of reviewed studies that give an overview of the most up-to-date SHM techniques based precisely on unsupervised learning (Fig 2). Following a concise introduction, we outline the current status of SHM studies that employ unsupervised learning techniques, categorized based on the types of ML methods utilized. Additionally, we delve into the validation approaches commonly employed for unsupervised learning SHM methodologies. Finally, we engage in a thorough discussion of the primary findings and constraints observed in the existing literature to evaluate forthcoming challenges

Figure 1. Sensor data acquisition scheme for structural monitoring

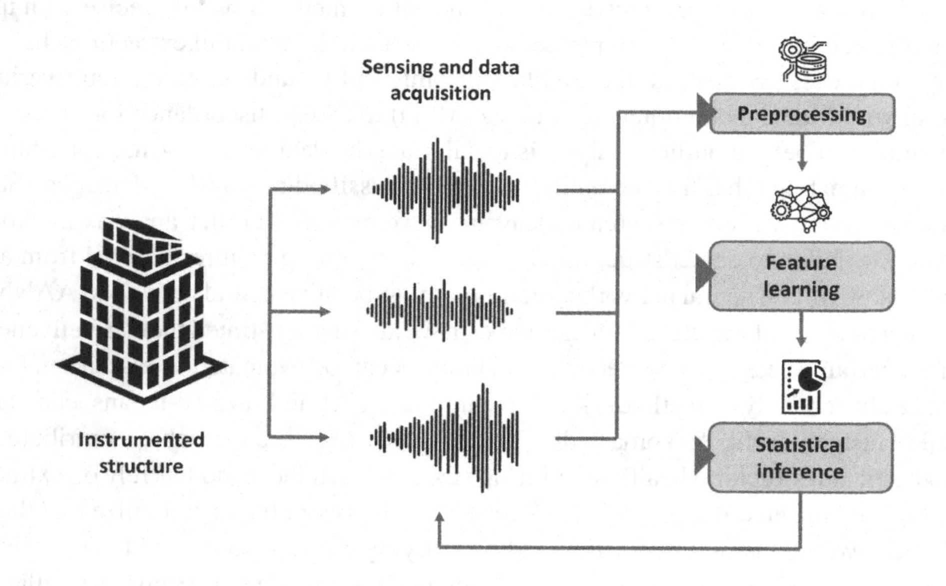

Figure 2. Scheme of review flowchart

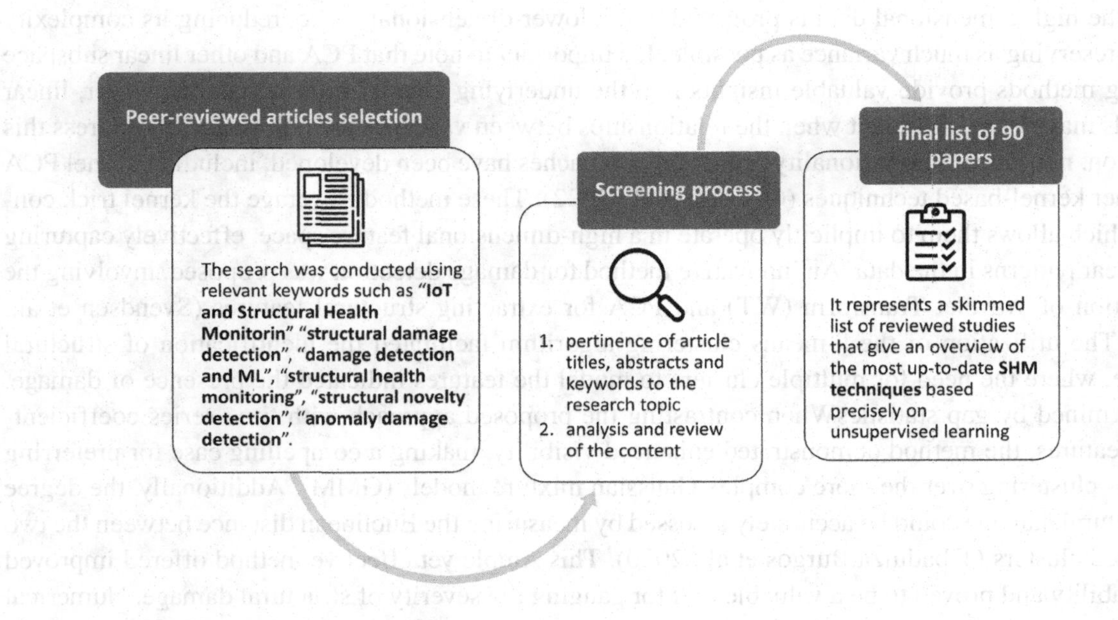

FEATURE EXTRACTION TECHNIQUES

This section of the conducted study will provide an overview of the various methodological approaches of SHM that are mainly based on the use of conventional feature extraction techniques. The analyzed methods will be sorted according to the types of practices employed for feature extraction, which can be further divided into two distinct categories: dimensionality reduction methods and signal processing methods. It is important to note that all techniques used for damage detection include using a novelty detector, which may be univariate or multivariate (H. Wang et al., 2023). An example of a novelty detector is outlier analysis. However, this section will mainly focus on methods that attach more importance to feature extraction. Furthermore, it should be borne in mind that many frameworks adopt a combined approach, using different feature extraction techniques that are performed sequentially. This sequential approach is convenient in cases where ML methods with lower complexity are used.

Dimensionality Reduction Based on Subspace Analysis

Initial measurements often include complex signals from several sources, each with a variable intensity contribution. Addressing this situation requires the application of subspace analysis techniques that allow signals to be linearly combined in an optimal way to explain the underlying data, reducing overall complexity. Among the methodologies widely used to reduce the dimensionality of measurements in the context of SHM, principal component analysis (PCA) stands out. In practical scenarios, PCAserves as a fundamental dimensionality reduction technique by performing singular value decomposition on the normalized data (Aoki et al., 2022). This process identifies principal components, which correspond to the eigenvectors associated with the most significant eigenvalues, representing the directions of maximum variance in the data (C. Zhang et al., 2022). By selecting a subset of these principal components, the high-dimensional data is projected into a lower-dimensional space, reducing its complexity while preserving as much variance as possible. It's important to note that PCA and other linear subspace learning methods provide valuable insights into the underlying structure of the data. However, linear methods may not be sufficient when the relationships between variables are non-linear. To address this limitation, non-linear dimensionality reduction approaches have been developed, including kernel PCA and other kernel-based techniques (Q. Chen et al., 2022). These methods leverage the kernel trick concept, which allows them to implicitly operate in a high-dimensional feature space, effectively capturing non-linear patterns in the data. An innovative method for damage detection was proposed, involving the integration of Wavelet Transform (WT) and PCA for extracting structural features (Svendsen et al., 2022). The utilization of the k-means clustering algorithm facilitated the identification of structural damage, where the need for multiple clusters to model the features indicated the presence of damage, as determined by gap statisticsWhen contrasting the proposed approach with time series coefficient-based features, the method demonstrated enhanced visibility, making a compelling case for preferring k-means clustering over the more complex Gaussian mixture models (GMM). Additionally, the degree of structural damage could be accurately assessed by measuring the Euclidean distance between the two identified clusters (Tibaduiza Burgos et al., 2020). This simple yet effective method offered improved observability and proved to be a valuable tool for gauging the severity of structural damage. Numerical simulations on the reference structure were used to validate the efficacy of the method. PCA was preferred by numerous researchers for its efficiency and practicality in dimensionality reduction over other approaches (C. Guo et al., 2023a). For instance, a combination of transmissibility with PCA proved useful in reducing the number of transmissibility functions by selecting specific projected components in the principal component space. Through PCA mapping, four damage indices were derived to detect structural damage occurrence. The T2 statistic and Q statistic, computed from the data residuals, constituted the first two indices. Subsequently, the combined index and I2 index were established by merging the previous two indices(Zhu et al., 2022). Moreover, a pragmatic strategy was introduced for identifying and localizing damage in cable-stayed bridges, utilizing the discrepancies in a multilinear regression model of cable forces as indicators of damage sensitivity. For novelty detection, Hotelling T2 control charts, computed via PCA of the model discrepancies, were applied. The method effectively identified stays with less than 1% area reductions in a study conducted on a numerically simulated viaduct. Furthermore, damage localization was accomplished by examining the relative variation of the T2 statistic. In a subsequent investigation, multivariate cointegration analysis based on the Johansen test replaced PCA, resulting in the creation of Moving Principal Component Analysis (MPCA), a more suitable variant for continuously monitoring structural conditions. In MPCA, instead of considering the entire signal, PCA is implemented on moving data windows to significantly reduce computational costs (Z. Wang, Bai, et al., 2022). The research investigated the influence of incorporating Multivariate Principal Component Analysis (MPCA) in combination with four regression techniques for the purpose of structural damage detection. The residuals obtained from the regression models were employed as damage indicators and subjected to novelty detection using a confidence interval derived from six standard deviations. Experimental tests conducted on concrete bridges validated the efficacy of the approach, affirming that the integration of MPCA not only enhances damage detection accuracy but also improves computational efficiency. In order to advance real-time bridge monitoring capabilities, a groundbreaking technique for damage detection and localization using Modified Principal Component Analysis with a fixed window was introduced (Z. Wang, Yi, et al., 2022). This innovative approach surpasses conventional MPCA by dynamically determining the moving window's length based on the cumulative contribution ratio's convergent spectrum. Damage indicators are then extracted from the principal components' vectors and eigenvalues. Nevertheless, the study lacks explicit details on the threshold values for these indicators, which are essential for automated decision-making. To address the limitations of PCA in handling measurement uncertainty and missing data, a novel anomaly detection method based on Probabilistic Principal Component Analysis (PPCA) was proposed. PPCA, being a probabilistic variant of PCA that incorporates the Gaussian latent variable model, proves especially effective in handling missing values within the input data matrix(Salehi et al., 2023; K. Yang et al., 2023b). This method employs two anomaly statistics, namely the Q statistic and the T2 statistic, with the residual in the Q statistic aiding in localizing structural damage. An evaluation on a collected dataset demonstrated the method's high success in identifying damage, both in scenarios with and without missing data, outperforming traditional PCA. However, the method may encounter challenges in detecting damage in elements with high redundancy when moderate noise is present. While PCA and its variations are extensively used for dimensionality reduction in SHM, alternative subspace learning methods, such as tensor decomposition, are gaining popularity. In this context, a novel damage detection approach based on tensor analysis was introduced, enabling a better understanding of the interconnections between sensors and their relationship to various parameters. The acquired data is structured in a three-dimensional array with time, position, and frequency as axes. A CANDECOMP/PARAFAC tensor decomposition is then applied to generate three feature extraction matrices(Salehi et al., 2023). These features are utilized to adapt an One-Class Support Vector Machine (OCSVM) model for novelty detection. Another novel damage detection algorithm incorporates residual vectors that exhibit higher resilience to environmental changes, combined with a generalized likelihood ratio test for effective anomaly detection(Z. Yang et al., 2023). Subsequently, a different investigation utilized the Mahalanobis Distance (MD) computed from empirical block Hankel matrices, derived from the structure's acceleration response, as a valuable damage indicator(Lu et al., 2023). The method demonstrated remarkable sensitivity at low damage levels but also exhibited a relatively elevated rate of false positives. To address this, the approach was enhanced by integrating other subspace damage detection techniques, employing Hotelling T2 control charts, which yielded significantly more favorable outcomes. Additionally, a sophisticated algorithm based on the null space of a class kernel harnessed the power of the Foley-Sammon transform (FST), a renowned linear subspace analysis method. The damage index was ingeniously determined by measuring the distance between the new null projection of an observation and the average of all training sample transformations in the kernel space (M. Wang et al., 2023). To accurately estimate the threshold, Extreme Value Theory (EVT) was thoughtfully employed, utilizing a Generalized Pareto Distribution (GPD) with an effective spiking technique. However, it is crucial to underscore that the algorithm's performance greatly hinges on the proper selection of the kernel and its parameters, as any misjudgment in this regard could significantly impact its accuracy, leading to a notable increase in the error rate.

Signal Processing Methods

In the domain of SHM, signal processing methods are utilized to analyze time series data in various domains such as the time domain, frequency domain, and time-frequency domain. The objective is to extract pertinent features from sensor measurements. Among the various methodologies for extracting characteristics in the time domain, the adoption of time series modeling, such as utilizing autoregressive models, is prevalent. In time series modeling, model parameters serve as direct representations of features or are further reduced through dimensionality reduction techniques like PCA (Candon et al., 2022). Alternatively, the reconstruction error, also referred to as the residual, can be utilized as an indicator of irregularities. The selection of the model order presents a significant challenge in time series modeling, as it directly impacts the accuracy of damage detection. Several model order selection techniques based on regression accuracy or model complexity have been proposed, including Mean Squared Error (MSE) or Akaike's Information Criterion (AIC) (Angulo-Saucedo et al., 2022). To evaluate the effectiveness of different models applied to damage detection, the study conducted a comparative analysis of four approaches: Single Variable Regression (SAR), Collinear Regression (COLR), Autoregressive Models (AR), and Autoregressive Models with Exogenous Input (ARX)(C. Chen et al., 2023). To verify the results, a small-scale test rig consisting of a steel frame was utilized. The study revealed that all methods successfully detected damage, and the ARX model outperformed the others in terms of damage localization. An innovative method for identifying, localizing, and quantifying damages was introduced, leveraging parameters and residuals extracted from autoregressive (AR) models. For novelty detection, the Parametric Guarantee Criterion (PAC) and Residual Reliability Criterion (RRC) were utilized as damage indicators, supported by a 95% confidence interval (Cinque et al., 2022). However, the study did not provide specific guidance on integrating both indicators within a unified novelty detection framework. The method's validation was carried out using two datasets from experimental facilities: a three-story LANL laboratory building and an IASC-ASCE reference facility, ensuring comprehensive testing and validation of the proposed approach. In a subsequent investigation, alterations were made to solely extract features from the residuals of autoregressive models, employing Partition-based Kullback-Leibler Divergence (PKLD) as the corruption indicator. These modifications enhanced the efficiency and reliability of the damage detection system under diverse environmental and operating conditions. Another feature extraction technique, akin to the previous one, involved utilizing coefficients and residuals from Autoregressive Moving Average (ARMA) models. Additionally, a hybrid measurement, combining Euclidean Distance Squared and PKLD, along with the Nearest Neighbor Rule, served as the damage indicator. In next times, an ARX model was trained, and a non-graphical method was employed to autoterminate the model order. The damage indicator in this context combined PKLD with Mean Squared Difference (MSD), using residuals from the ARX model (Nasir et al., 2022). Their proposal was validated on a cable-stayed bridge, and the results demonstrated an enhancement in efficiency over previous methods. Over the past decade, time-series modeling has been favored by other researchers for extracting damage-sensitive features. Time-frequency domain representations have gained growing attention in the realm of SHM due to their ability to capture essential damage-sensitive features. In contrast to frequency-domain characteristics, time-frequency-domain characteristics record both frequency and temporal information of the signal. In the realm of feature extraction, the Short-Term Fourier Transform (STFT) and the Wavelet Transform (WT) are both widely utilized techniques (Yeter et al., 2022). However, a novel approach that combines the Synchronized Wavelet Transform (SWT) with fractal modeling has been proposed for damage detection, localization, and quantification (Y. Jiang & Niu, 2022a). The application of SWT helps in reducing signal noise, while the Fractal Dimension (FD) comes into play for detecting system changes by comparing the median absolute deviation between the FD of the training data and new observations. The introduction of the Synchronized Wavelet Transform (SWT) and its integration with fractal modeling presents an exciting avenue for enhancing damage detection methodologies. By effectively mitigating signal noise through SWT and utilizing FD to detect system changes, this approach showcases the potential to improve the accuracy and reliability of damage assessment techniques. The combination of these two methods offers a comprehensive solution for the intricate task of damage localization and quantification, contributing to the advancement of SHM practices. Furthermore, this approach aligns with the current trends in signal processing and pattern recognition, where researchers are continuously seeking innovative ways to enhance damage detection methodologies (Bhattacharya et al., 2022). The utilization of SWT, a refined extension of the conventional Wavelet Transform, demonstrates the commitment to harnessing cutting-edge technology to address real-world challenges. Moreover, the incorporation of fractal modeling further enriches the analysis, allowing for a deeper understanding of system behavior under varying conditions. As research in the field of damage detection and SHM continues to evolve, the proposed approach holds significant promise for its ability to tackle complex scenarios, such as noisy data and varying damage patterns. The use of Fractal Dimension as a key metric for detecting system changes showcases the versatility of this method, as it can adapt to different types of structural damage, providing a robust and adaptive solution (López-Castro et al., 2022). To validate the efficacy of this approach, experimental testing on diverse structures with controlled damage scenarios and real-world data from various engineering applications is essential. Such comprehensive testing will enable researchers and practitioners to gain a deeper insight into the capabilities and limitations of the SWT-fractal modeling combination, paving the way for its widespread adoption in the field of Structural Health Monitoring. This approach has demonstrated significant enhancements in damage detection by employing SWT for noise reduction. An approach for damage detection and localization, employing the Continuous Wavelet Transform (CWT) and a discrete generalized Teager-Kaiser operator, was introduced. Initially, PCA was used for damage detection, followed by Median Absolute Deviation (MAD) Outlier Value Analysis to pinpoint anomalies (Parida et al., 2022). The method's efficacy was demonstrated through two case studies: one involving a numerical model of beams and the other based on an experimental investigation of a wind turbine blade. In both cases, the method effectively located introduced cracks However, it should be emphasized that such an approach requires a large set of sensors, which may limit its applicability in some situations. An innovative approach for damage detection was presented, employing a two-level method based on anomaly detection through Wavelet Transform (WT), Generalized Pareto Distribution (GPD), and Moving Fast Fourier Transform (MFFT) (Cha & Wang, 2018). Moreover, a threshold update strategy was implemented to accommodate variations in traffic volume and facility degradation over time. The anomaly detection process was further complemented by MFFT-based anomaly trending. The results illustrated the method's competence in identifying the majority of anomalies and a decrease in stiffness of main cables. Nonetheless, it was relatively less successful in detecting a decrease in beam stiffness.

Signal Decomposition Methods

A different technique for dealing with non-stationary signals is empirical mode decomposition (EMD). EMD is a method that relies on data-driven approaches to systematically break down a signal into simpler elements known as intrinsic mode functions (IMFs). These IMFs capture distinct oscillation patterns within the signal (Doroudi et al., 2022; D. Zhang et al., 2022). Examining these resulting IMFs offers valuable insights into the signal, encompassing details such as amplitudes and frequencies . The inclusion of Hilbert spectral analysis allows for the extraction of additional signal characteristics, facilitating the creation of a spectrogram through the innovative Hilbert-Huang Transform (HHT) (Momeni & Ebrahimkhanlou, 2021). Numerous research investigations have explored and compared various signal decomposition methods concerning fault and damage diagnosis. Researchers extensively examined the feasibility of EMD has emerged as a powerful tool for identifying and locating damages in numerical beams. By analyzing the response to a moving load and detecting discontinuities in the Intrinsic Mode Functions (IMFs), EMD demonstrates great potential in accurately detecting cracks and structural anomalies. To further improve the interpretability of the results, researchers have found that applying a moving average filter before EMD enhances the insights gained from the analysis. In another insightful study, damage detection in a bridge was explored using data obtained through the Hilbert-Huang Transform (HHT) process under the influence of a moving load (Civera & Surace, 2021). The researchers discovered that the Hilbert fringe spectrum derived from a sensor positioned near the damage location exhibited a noticeable decrease in peak frequency compared to sensors situated farther away. Moreover, changes in the instantaneous phase demonstrated a heightened responsiveness to simulated damage, further validating the efficacy of HHT in capturing structural anomalies. Additionally, a sophisticated damage diagnosis system was introduced, integrating the normalized cumulative energy distribution (NCED) with HHT and Mean Square Deviation (MSD) hypothesis testing. This holistic approach provides a comprehensive and robust mechanism for detecting and quantifying damage in structural systems (Zhi et al., 2021). The combination of NCED, HHT, and MSD offers a more accurate and reliable damage assessment, thereby significantly enhancing the overall performance of the damage diagnosis system. This advanced system represents a significant step forward in the field of structural health monitoring, enabling engineers to detect and evaluate damages with greater precision and confidence. By harnessing the power of HHT and its ability to unravel intricate signal characteristics, coupled with the statistical insights from MSD hypothesis testing, the proposed system demonstrates great potential in enhancing damage diagnosis capabilities (M. Mousavi & Gandomi, 2021). The amalgamation of Hilbert spectral analysis with various signal processing methodologies presents a significant advancement in the field of structural health monitoring and damage assessment. This comprehensive approach not only extends the scope of fault diagnosis but also enriches the understanding of structural behavior under varying conditions. By exploring the unique features captured by HHT and effectively integrating them with established methods like EMD and MSD, researchers can unlock new possibilities for precise and reliable damage detection in critical engineering structures. As this area of research continues to evolve, it opens up exciting opportunities for real-world applications in civil engineering, aerospace, and other industries reliant on robust structural health monitoring systems (Kaloni et al., 2021). Through continuous validation and experimentation on diverse structures and scenarios, the potential benefits of the HHT-based approach can be fully harnessed, contributing to the optimization of structural maintenance, safety, and longevity. This methodology combined four damage indicators and compared the NCEDs between the reference configuration and structures monitored using various techniques, such as the Kolmogorov-Smirnov distance. In spite of its somewhat lower efficiency, the HHT-centered methodology exhibited superior performance compared to a power spectral density-based technique when implemented on a reduced-scale three-story steel frame experiment. The results highlighted the potential of the HHT-based approach in delivering more accurate and robust outcomes in damage detection and structural assessment. The study showcased the advantages of harnessing the unique capabilities of the Hilbert-Huang Transform to capture essential signal characteristics and improve the overall accuracy of damage identification. The proposal put forth for localizing structural damage involved the utilization of multivariate EMD, which incorporates cross-channel information (A. A. Mousavi et al., 2022). The localization indicator relied on the absolute percentage change in energy for each sensor relative to the reference configuration, and an adaptive threshold was established based on the average value across all sensors. Despite certain limitations concerning the number of available sensors and observations, this method successfully detected damage in a numerical model with 10 degrees of freedom. Nevertheless, EMD is confronted with the well-known issue of mode mixing. This drawback arises when the EMD process generates IMFs that blend multiple frequencies instead of isolating them into individual IMFs. In response to this challenge, several modifications have been suggested to address the problem, resulting in the development of ensemble EMD (EEMD), full EEMD with adaptive noise (CEEMDAN), and variational mode decomposition (VMD). Additionally, in the context of a suspension bridge, researchers have successfully employed EMD to segregate thermal strain from the raw strain measurements. This innovative application of EMD enables the precise detection of structural damage, enhancing the bridge's overall health monitoring capabilities (Delgadillo & Casas, 2022). The utilization of EMD to differentiate thermal strain from raw strain measurements in the suspension bridge showcases the versatility and adaptability of this signal processing technique. By effectively isolating the thermal effects, the researchers were able to focus on the true structural strain, allowing for more accurate and reliable damage detection. Consequently, the thermal deformation was employed to create a matrix of indices based on Euclidean distance, which assisted in the process of damage diagnosis. Nevertheless, this method lacks a formal decision-making procedure. To address this, a semi-automated framework for damage diagnosis was devised, incorporating the ensemble empirical mode decomposition (EEMD) in conjunction with a threshold method based on sensitivity analysis (Y. Jiang & Niu, 2022b). The damage index is calculated on the basis of the relative energy variation in the IMFs, while the localization index is obtained by comparing the damage index of the single sensors with the average of all the sensors. However, it should be emphasized that this approach is not fully automated, as it requires user involvement in several stages of the process, including threshold determination, this could necessitate access to past damage data, rendering the method partially supervised. Furthermore, there have been advancements in hybrid methodologies that integrate EEMD with other techniques for SHM, exemplified by the EEMD-AR-ARX approach, which leverages dynamic time warping to extract damage-sensitive features. Despite its advantages, EEMD does have certain limitations, particularly when confronted with white noise of exceedingly low or high amplitude. To tackle this concern, the introduction of Complementary EMD (CEEMD) presented a solution by utilizing pairs of complementary white noise signals for signal decomposition. In a further evolution, Full EMD with Adaptive Noise (CEEMDAN) uses adaptive noise, updated based on the residual of the signal. These methods have proven effective in detecting and locating structural damage, particularly for weak vibration signals (Y. Jiang & Niu, 2022c). The combined diagnostic approach CEEMDAN and artificial neural networks has shown good results in predicting IMF characteristics and assessing damage with estimated damage indices. An alternative to EEMD, which avoids mode mixing and is potentially more efficient, is VMD. This method has proven successful in locating damage in situations where EMD had limitations. The VMD utilizes an analysis of the first IMF's instantaneous frequencies and amplitudes to produce a damage indicator. Experiments on numerical models and composite beams confirmed the effectiveness of the VMD in detecting structural damage through the energy change in the central frequencies of the second mode (D. Zhang et al., 2022). In conclusion, EEMD, CEEMDAN and VMD represent promising methodologies for the diagnosis of structural damage, each with its own characteristics and specific applications. These advanced approaches have the potential to significantly improve facility surveillance and diagnostics, helping to ensure long-term infrastructure safety and reliability.

SHM UNSUPERVISED LEARNING BASED ON ARTIFICIAL NEURAL NETWORKS

In this segment, we provide a brief summary of pertinent research carried out in the last decade concerning the application of artificial neural networks (ANN) in the domain of SHM, with the networks being trained independently of any data related to structural damages (Fig.3). Recent trends show an increasing adoption of advanced approaches based on DL techniques and architectures, going beyond the traditional use of basic ANNs in SHM. Comparable to the techniques for extracting features mentioned before, artificial neural networks are utilized to acquire meaningful data representations, frequently resulting in the reduction of dimensionality. Deeper neural networks can automatically extract relevant features from raw vibration measurements, eliminating the need for extensive preprocessing. Anomaly detection models are then applied to identify significant variations within the monitored system accurately(Dederichs & Øiseth, 2023). These models generally depend on straightforward assessments, taking advantage of the depth of the network and the responsiveness of the acquired features. Regarding SHM, certain neural networks adopt supervised learning techniques during the training process. However, it is essential to highlight that the automatic labeling of faulty data is exclusively done using reference data from normal structural conditions (Azimi et al., 2020). Moreover, these approaches often integrate representation learning strategies and anomaly detection, effectively making them unsupervised learning methods within the SHM context. Common learning objectives consist of input reconstruction employing techniques such as autoencoders, prediction tasks with the assistance of recurrent neural networks, and generative learning utilizing adversarial generative networks (Tomasiello et al., 2021). These diverse learning goals play a crucial role in training neural networks for a wide range of applications and tasks in artificial intelligence and ML.

Figure 3. Example of artificial neural networks fo SHM

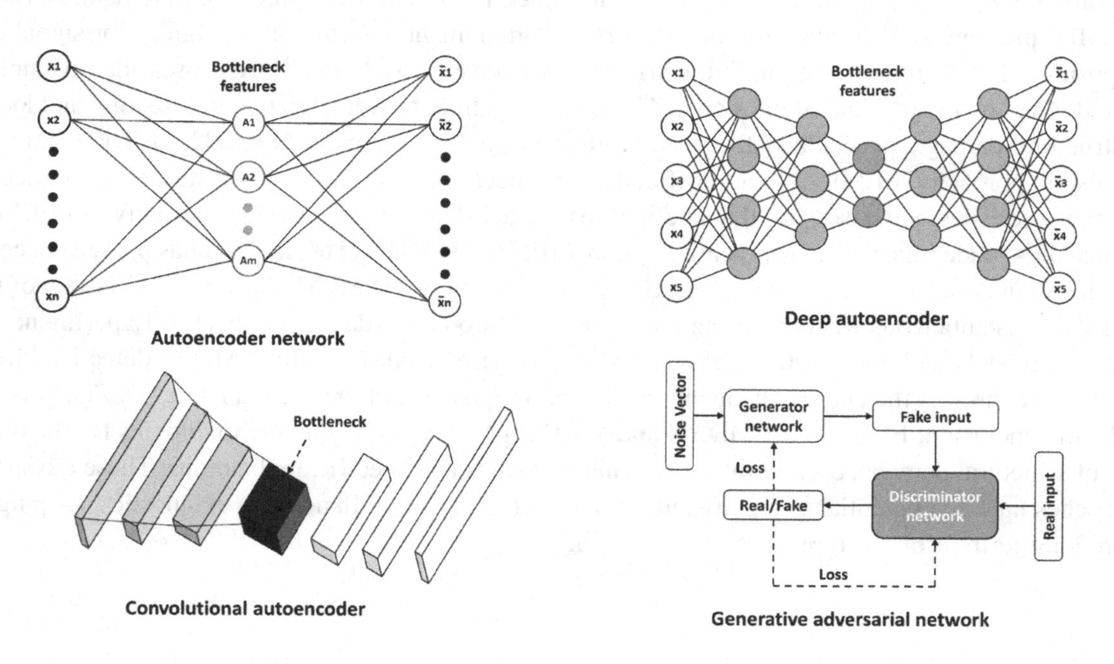

Classical Neural Networks

Classical artificial neural networks (ANNs) consist of a limited number of parameters and have at most two hidden layers. These networks primarily serve as methods for reducing dimensionality, similar to the PCA algorithms discussed earlier (Farhangdoust et al., 2019). An intriguing approach involves the use of Self-Organizing Maps (SOM), which is a type of artificial neural network capable of mapping high-dimensional data onto a lower-dimensional grid while preserving the topological structure of the data(C. Guo et al., 2023b). As an example, a novel structural monitoring algorithm utilizing Self-Organizing Maps (SOM) has been put forward. This algorithm makes use of acceleration data to detect structural damage by computing the Root Mean Squared Error (RMSE) from the topological maps of both the reference and test data. By leveraging the SOM approach, it offers a unique and effective way to identify potential damage in structures based on the analysis of acceleration patterns (Smarsly & Law, 2014). Another technique proposes employing an Artificial Neural Network (ANN) for reconstructing the structural response. In this method, temperature measurements are integrated as an additional input to account for and mitigate the impact of thermal variations. By considering temperature data, the ANN gains a more comprehensive understanding of the structural behavior, enhancing its ability to accurately reconstruct the response under varying environmental conditions (Gomes et al., 2019). Both of these methods exemplify the versatility and potential of advanced algorithms in structural monitoring and damage detection. By incorporating cutting-edge techniques such as SOM and ANN, researchers can develop sophisticated tools for assessing structural integrity and performance under various circumstances(Zaparoli Cunha et al., 2023). These advancements contribute to safer and more resilient infrastructures, where potential damage can be detected and addressed promptly to ensure the longevity and safety of critical structures. As research continues to evolve in this field, we can anticipate even more innovative approaches to tackle complex challenges in structural monitoring and maintenance. The discrepancy between the expected network response and the desired response is used as an indicator of novelty. ANNs can be used for damage prediction and detection in structures. For example, the use of a trained ANN has been proposed to predict future acceleration values based on the structural responses to the stresses generated by passing trains. Subsequently, a Gaussian process was utilized to evaluate the disparity between the forecasted values generated by the network and the actual observed values across different train velocities (Oh et al., 2019a). This advanced statistical technique offers valuable insights into the accuracy and precision of the network's predictions, allowing for a comprehensive assessment of its performance under varying conditions. While an explicit definition of the detection threshold was not provided, it was suggested to select it using Receiver Operating Characteristics (ROC) curves and False Alarm Costs. Nonetheless, it is essential to highlight that ROC curves can only be obtained with access to observed damage data. In a distinct approach to damage detection, an ANN was trained to predict the Maximum Dissimilarity Index (MD) based on specific characteristics sensitive to structural damage. Furthermore, a method was introduced, proposing a damage indicator derived from the forecast error generated by the network. The evaluation of this methodology was carried out using a dataset from a wind turbine, with the cross-covariance analysis of the acceleration response being employed as a distinctive feature (Rafiei & Adeli, 2017). By leveraging the cross-covariance technique, researchers aimed to identify unique patterns in the acceleration data that could be indicative of specific structural behavior or operational conditions. Lastly, the limitations of traditional PCA for data compression have been underscored, as it relies on a linear mapping. In response, a hybrid approach named Hybrid Autoencoder-PCA has been introduced, combining an autoencoder (AE) with PCA(Shang et al., 2021).

his method integrates non-linear connections between high- and low-dimensional feature levels while preserving the linear mapping aspect of PCA. The resulting hybrid network utilizes the reconstruction error as a damage indicator, which is then compared against a threshold derived from the 99th percentile of a reference index (Abdeljaber et al., 2018). Experimental validation on two numerically simulated bridges, using real data, demonstrated the superior accuracy of damage detection compared to linear PCA. Moreover, this innovative approach precisely identified damages in close proximity to the utilized sensors. In contrast to conventional autoencoders (AE), the use of generalized autoencoders (GAE) allows each input instance to reconstruct a group of instances, expanding beyond individual self-reconstruction (V.-H. Dang, 2023). Taking advantage of this concept, a novel framework for SHM was developed, employing a modified GAE network trained to model the power cepstral coefficients extracted from the structural response. The reconstruction error obtained from this GAE network serves as the damage indicator and is compared to a threshold established based on the 99th percentile of reference index values (Li et al., 2023). During the decision-making process, the minimum standardized value (MSD) was utilized in conjunction with the 0.95 quantile of an F-distribution, calculated using the training data. The method showcased heightened accuracy in damage detection when compared to traditional AE and PCA techniques, further demonstrating the promising potential of GAE in SHM applications (Ben Ali et al., 2015). The integration of non-linear connections and the ability of GAE to handle multiple instances for reconstruction makes this approach a promising advancement in Structural Health Monitoring. By exploiting the enhanced capabilities of GAE, this framework showcases improved accuracy in detecting and locating damages within structures. The utilization of power cepstral coefficients as input features also highlights the importance of selecting relevant and meaningful parameters for damage detection (Parida et al., 2022). As research continues to progress in the field of Structural Health Monitoring and predictive maintenance, innovative methodologies like this GAE-based approach contribute to advancing the state-of-the-art. The ability to accurately detect damages and their locations through efficient feature extraction and reconstruction techniques is critical for ensuring the safety and longevity of critical infrastructure. The robustness and superior performance of this GAE framework offer exciting prospects for practical applications in various engineering domains.

Dense and Deep Neural Networks

Several scholars have adopted increasingly complex neural architectures for their models, taking advantage of advances in artificial intelligence and hardware technology. By using deep layering of the networks, it is possible to further reduce or even eliminate the design steps of the pre-networks characteristics, directly employing acceleration measurements as inputs. These sophisticated deep neural networks offer the capability to generate remarkably efficient damage indices, streamlining the use of simpler novelty detection algorithms(Alzubaidi et al., 2023). Dense neural networks, commonly known as standard neural networks, form dense connections between all nodes in adjacent layers, resulting in a network with rich interconnections. The study embarked on an in-depth exploration of two distinct unsupervised learning paradigms: one hinged on dense autoencoders to accomplish proficient representation learning, while the other relied on the time-tested PCAtechnique (L. Hou et al., 2021). The overarching objective was to gauge the magnitude of structural damage by precisely computing the Euclidean distance between the original measurements and the reconstructed output generated by each model. This meticulous investigation sought to ascertain the efficacy of both approaches in accurately quantifying the extent of damage within the structural systems under scrutiny. Moreover, the inclusion of temperature measurements as input parameters in both structures was also explored. Through the utilization of these cutting-edge deep neural networks, scholars have the opportunity to devise remarkably efficient damage indices, streamlining the integration of subsequent novelty detection algorithms(H. Dang et al., 2022). The interconnected layers of dense neural networks endow them with an unparalleled capability to grasp intricate patterns and correlations embedded within the data, rendering them invaluable assets in the realms of structural health monitoring and damage evaluation. By tapping into the power of these advanced networks, researchers can pave the way for more accurate and robust methods to assess and detect structural damage.The utilization of dense autoencoders for representation learning represents a cutting-edge approach in the realm of unsupervised learning(Azimi et al., 2020). These networks excel at capturing and encoding intricate features from the input data, enabling more accurate and reliable damage detection. Simultaneously, the integration of PCA in the investigation allows for a comparison of its performance with that of dense autoencoders, providing valuable insights into the effectiveness of different unsupervised learning methodologies(Rastin et al., 2021). Moreover, the incorporation of temperature measurements as additional input parameters in the structures enhances the richness of the data representation, potentially improving the models' ability to detect and quantify structural damage under varying environmental conditions. As the field of artificial intelligence and DLcontinues to evolve, the exploration of diverse network architectures and learning approaches holds great promise for advancements in structural health monitoring and predictive maintenance (T. Liu et al., 2020). The combination of advanced unsupervised learning models with domain-specific knowledge allows researchers to develop tailored methodologies that address specific challenges in structural assessment. The methods were validated through three case studies, demonstrating the effectiveness of their approach in detecting and locating damage, taking into account thermal variation, especially when input parameters include mode shapes. A novel method leveraging DL has been introduced to tackle the challenge of damage detection in scenarios characterized by vast amounts of high-dimensional data (Liang et al., 2016). This innovative approach incorporates ARMA (Autoregressive Moving Average) coefficients and residuals as crucial features, integrates a deep autoencoder for efficient dimensionality reduction, and employs mean deviation (MD) as the novelty detector, seamlessly integrating these components within a unified framework. The determination of the optimal number of nodes in the deeper levels of the autoencoder is achieved through a final prediction error function(R. Wang et al., 2018). This novel strategy leverages the power of DLtechniques to effectively process and analyze vast datasets, making it particularly well-suited for complex structural health monitoring scenarios. By incorporating ARMA coefficients and residuals, the approach captures essential information related to the dynamic behavior of the structure, enhancing the accuracy and robustness of damage detection (Hosseini et al., 2017). The utilization of a deep autoencoder for dimensionality reduction is a key aspect of this methodology. The autoencoder efficiently compresses the high-dimensional input data into a lower-dimensional representation while retaining the most critical features(H. V. Dang et al., 2021). This reduction in dimensionality not only facilitates the computational efficiency of subsequent analysis but also helps in identifying underlying patterns and relationships within the data. The threshold is established through the application of the generalized extreme value distribution (GEV) with the maximum blocks technique. In a separate investigation that emphasizes automated feature extraction, two distinct deep autoencoder architectures were introduced to directly learn damage-sensitive characteristics from raw acceleration measurements (Mao et al., 2021). The first model extracts features from the "bottleneck" layer, while the second utilizes the reconstruction error as a characteristic. Multiple autoencoders were simultaneously trained, with each sensor having its dedicated autoencoder, and the learned features were assessed against a predefined threshold for damage detection(Seventekidis et al., 2020a). This threshold can be determined based on the structure's significance, and the location of the damage can be identified through the sensor from which the anomalous features were derived. These approaches underwent validation using both the LANL and QUGS three-story frame datasets. A method for extracting damage-sensitive features was proposed using stacked autoencoders. The network took natural frequencies as input and generated a compressed and highly damage-sensitive feature vector from the "bottleneck" layer, often consisting of only two attributes. Although this approach did not include a novelty detection method, they demonstrated the use of expectation maximization-based Gaussian Mixture Models (GMMs) as an example (Entezami et al., 2022). Their model was compared to other representation learning techniques like PCA, AANN, and kernel PCA, showing superior damage detection results, slightly outperforming PCA kernel with a smaller number of feature vector parameters. However, this method lacks the capability to locate and quantify damage. In the investigation of DL approaches for unsupervised learning in SHM, a novel deep and dense autoencoder network was introduced to autonomously extract significant features from raw acceleration data (Seventekidis et al., 2020b). Three key metrics, namely mean squared error (MSE), original signal to reconstructed ratio (ORSR), and Arias intensity, were harnessed to derive these essential features. Subsequently, these metrics were utilized as inputs for an OCSVMmodel, facilitating effective novelty detection. However, it's worth noting that this method requires predefining certain hyperparameters that influence the decision boundary and decision-making process. The methodology's performance was thoroughly validated through two insightful case studies, including one involving a lab-scale steel bridge, where it successfully detected a significant 10% reduction in stiffness(Flah et al., 2021). Nevertheless, it was observed that this approach might not be the most suitable for precise damage localization and quantification tasks. Another notable ensemble-based technique for damage detection and localization in large-scale structures also made use of autoencoder models and the same set of performance metrics (MSE, ORSR, and Arias intensity). This ensemble-based methodology, leveraging the power of multiple autoencoders, offers a robust and comprehensive approach to extract and combine meaningful features from the raw acceleration data (Kanarachos et al., 2017). The utilization of multiple autoencoders enhances the model's ability to learn intricate patterns and relationships within the data, further improving the accuracy and reliability of damage detection. The incorporation of the three key metrics, MSE, ORSR, and Arias intensity, as indicators of damage, provides valuable insights into the structural health and performance. These metrics capture important aspects of the structural response and serve as reliable damage indicators, enabling timely and accurate detection of potential structural anomalies(Azimi & Pekcan, 2020). An autoencoder was trained for each sensor, and binary decision matrices were formed based on threshold exceedance. A sum-of-whole inference method was employed to assess overall damage and location capabilities. The method showed promising performance in damage detection with a good localization ability.

Convolutional Neural Networks (CNNs)

A convolutional neural network (CNN) is an advanced DL architecture that utilizes the convolution operation with learned filters, which assign weights. The distinctive characteristic of CNNs offers the advantage of parameter reduction compared to dense and deep neural networks, making them highly suitable for DLtasks (Lopez-Pacheco et al., 2020). CNNs are widely employed when dealing with structured data arranged in grids that exhibit local spatial correlation, such as images. The unique property of CNNs lies in their ability to leverage shared weights and local connectivity, enabling them to efficiently process grid-like data (Seon Park et al., 2020). By employing convolutional layers, CNNs can extract relevant features from the input data while significantly reducing the number of learnable parameters. This reduction in parameters not only speeds up the training process but also helps prevent overfitting, leading to more robust and generalizable models(Abdeljaber et al., 2018). Consequently, when applying CNNs to vibration data, researchers often propose data organization techniques that optimize the capabilities of CNNs (Iyer et al., 2021). In the domain of data compression for SHM, novel approaches utilizing two deep convolutional autoencoder models were introduced(Jamshidi & El-Badry, 2023). These models serve the dual purpose of anomaly detection in measurements and data compression, and they are built on 1D-CNN autoencoder architectures. In this context, the first model is trained in a supervised manner specifically for anomaly detection, while the second model is responsible for data compression and accurate reconstruction. The effectiveness of these models was thoroughly assessed through testing on a dataset obtained from a suspension bridge (Azuara et al., 2021). The results demonstrated exceptional compression and reconstruction performance, achieving an impressive compression rate of 0.1. This high level of compression efficiency ensures that the essential structural information is preserved while significantly reducing the amount of data needed for storage and transmission (Tran et al., 2023). For real-time structural health monitoring, two densely connected neural networks, leveraging the power of CNNs, were devised. A notable aspect of this framework is the integration of model pruning, a technique that optimizes network efficiency by eliminating redundant connections between neurons without substantially compromising model accuracy(Yuan et al., 2020). This strategic pruning process ensures that the network remains highly efficient while retaining its capability to make accurate predictions. In this context, model prediction errors are harnessed as essential features, and deep support vector domain description (SVDD) is employed to establish a decision boundary that acts as a robust novelty detector. This decision boundary aids in detecting potential structural anomalies and deviations from the baseline behavior. The effectiveness and reliability of this approach were validated through two comprehensive case studies on frame structures, including a real experiment (Suh & Cha, 2018a). The results illustrated the methodology's exceptional real-time damage detection capabilities, demonstrating its potential to significantly improve structural health monitoring systems' efficiency and accuracy. In addition to the aforementioned techniques, an innovative framework for damage detection utilizing a Convolutional Autoencoder (CAE) was introduced(Tang et al., 2019). The CAE is specifically trained on a matrix derived from stacked acceleration measurements, enabling it to learn relevant patterns in the data. The damage indicator is determined by calculating the Euclidean distance between the baseline latent vectors and the unknown structural states after normalization into unitary vectors (Merenda et al., 2019). For novelty detection, a threshold is set, typically at 1.6 or 1.4 standard deviations from the mean, to distinguish between normal and anomalous behavior in the structural response. This CAE-based approach showcases its efficacy in capturing subtle changes in the data, making it a promising technique for real-time damage detection in critical infrastructures. By leveraging the power of convolutional autoencoders and employing the Euclidean distance as the damage indicator, this framework offers a practical and effective approach for damage detection in structural health monitoring scenarios. The setting of the threshold for novelty detection allows for the identification of deviations from the normal behavior, enabling timely and accurate identification of potential structural anomalies.

Other Example of DL Applications

Apart from the aforementioned neural network modeling paradigms, there are several other approaches utilized in unsupervised learning for SHM. Another approach involves a generative model utilizing variational inference, where latent characteristics are considered as stochastic variables with a prior distribution (L. Guo et al., 2021). Throughout the training process, the network gains an understanding of the latent distribution, enabling it to generate new samples when needed. In addressing the issue of damage localization in a bridge, a dedicated approach was introduced, utilizing a one-dimensional convolutional variational autoencoder (CVAE) as an effective dimensionality reduction technique. In this context, the CVAE model is fed with acceleration responses to a moving load as input, generating corresponding output representations(Almeida Cardoso et al., 2019). The localization index is subsequently computed by evaluating the Euclidean distance between the latent features extracted at various time steps throughout the response data. This method demonstrates its potential in accurately pinpointing damage locations by leveraging the learned features and their temporal evolution. This alternative technique leverages the power of variational inference to model the latent characteristics as stochastic variables. By considering prior distributions, the model can better understand the underlying structure of the data and generate new samples that align with the learned distribution (D. Wang et al., 2020). This capability is particularly valuable in scenarios where generating new data samples is necessary, such as during damage localization in structural health monitoring. The introduction of the one-dimensional convolutional variational autoencoder (CVAE) as a dimensionality reduction technique is another significant aspect of this approach. The CVAE effectively compresses the input data, extracting essential features while retaining their probabilistic nature(Sarmadi & Yuen, 2022). This reduction in dimensionality not only enhances the efficiency of subsequent analysis but also aids in detecting structural anomalies more effectively. The customized method for damage localization in a bridge demonstrates the practicality and effectiveness of this approach. By utilizing acceleration responses to a moving load, the model can learn meaningful patterns and relationships that are indicative of potential damage. The computed location index, based on the Euclidean distance between the latent features at different time steps, enables precise localization of the structural anomalies. As the field of generative models and variational inference continues to evolve, we can expect further refinements and enhancements in this approach (B. Zhang et al., 2019). The combination of probabilistic modeling with dimensionality reduction techniques offers promising prospects for more accurate and efficient damage localization in Structural Health Monitoring applications. Moreover, the adaptability of this technique to various types of data and structural systems makes it a valuable tool for ensuring the safety and reliability of critical infrastructure ("Application of Deep Autoencoder Model for Structural Condition Monitoring," 2018). By exploring alternative techniques and leveraging advances in DLand generative modeling, researchers can continue to push the boundaries of Structural Health Monitoring, paving the way for more resilient and sustainable infrastructure worldwide. The primary focus of this method is damage localization, without proposing a threshold strategy solely based on uncorrupted data. Another remarkable model worth mentioning is the graph convolutional network (GCN), an extension of CNNs that operates on graph-structured data (Deng et al., 2020). Unlike standard local connections in CNNs, the GCN leverages predefined or learned node connectivity using a global adjacency matrix (C. Jiang & Chen, 2023). The graph convolutional network (GCN) represents an innovative approach to process data arranged in graph structures. In contrast to traditional CNNs that excel in handling grid-like data, the GCN is designed to accommodate data with complex interconnected relationships, where the nodes represent entities, and edges represent connections between these entities. GCN belongs to the family of graph-based neural networks. It has been pointed out that representing vibration data as an image structure is not the optimal approach since CNNs struggle to learn the spatial correlation of sensors (Lin et al., 2018). In response to the challenge of damage localization in bridges, a specialized method was proposed, introducing the utilization of a one-dimensional convolutional variational autoencoder (CVAE) as an efficient dimensionality reduction technique. The methodology involved processing acceleration responses to a moving load as the model's input, which in turn generated corresponding output. To pinpoint the location of potential damage, the location index was determined by calculating the Euclidean distance between the latent features extracted at different time steps. As an alternative and innovative approach, researchers introduced a spatiotemporal graph convolutional network (GCN) for detecting sensor anomalies(Bao et al., 2019). This advanced architecture was designed to learn and effectively capture both the spatial and temporal dependencies present in the measurements. The methodology combined the strengths of the graph convolutional network (GCN) with trainable adjacency matrices and one-dimensional temporal CNNs. For evaluation purposes, a dataset containing cable forces from a suspension bridge was employed, and the framework demonstrated remarkable capabilities in identifying faulty sensors through the integration of a highly effective novelty detection scheme. In another line of research focusing on capturing intricate temporal patterns from time-varying data, the long-term memory network (LSTM) emerged as a popular choice. Particularly in the domain of SHM data anomaly detection, a two-stage framework leveraging an encoder-decoder LSTM architecture was introduced (HOSKERE et al., 2019). The primary emphasis of this study lay in the monitoring of suspension bridges, where the two-layer LSTM network efficiently processed the raw time series of cable voltages as its input data. This sophisticated approach displayed promising results in terms of anomaly detection and accurate assessment of structural integrity(X. Zhao & Jia, 2020). The integration of GCN with trainable adjacency matrices and one-dimensional temporal CNNs offers a comprehensive and powerful framework for processing graph-structured data with temporal dependencies. By employing trainable adjacency matrices, the model can adaptively learn the node connections from the data, capturing essential patterns and relationships in the graph. Anomaly scores were estimated using the reconstruction error. However, it is important to note that their method does not specifically pinpoint the cause of the anomaly, which could be due to structural damage, sensor malfunctions, or environmental effects (Bukhsh et al., 2021). For damage detection and localization in systems with a large set of sensors, an autoencoder-based spatiotemporal composite neural model was proposed. The model combined elements of CNN and LSTM to provide both signal reconstruction and prediction. Raw accelerations were arranged in a grid structure considering sensor positions and the time domain. Damage indices were derived from latent characteristics and output residuals, and novelty was identified using a threshold based on generalized extreme value theory (EVT). Generative Adversarial Networks (GANs) are sophisticated generative models comprising a generator and a discriminator engaged in a competitive training process (Lorusso et al., 2023). A novel approach for anomaly detection in structural monitoring involved integrating a convolutional GAN with a convolutional autoencoder (CAE) to create a unified model. To enable processing through convolutional layers, vibration data was transformed into images using the Gramian Angular Field method (Lei et al., 2021). In this hybrid model, representative images served as input, and during evaluation, latent features and cumulative sum control maps were extracted to identify anomalous data in a cable-stayed bridge. The model achieved an impressive accuracy of over 94% for all channels considered (Altabey et al., 2021). The study explored the sensitivity of sensor configurations by employing three different GAN models within a comprehensive damage detection framework. While these models shared the same generative architecture, they differed in the discriminators used, including density-based models, CNNs, and long memory recurrent neural networks (LSTMs) (H. Jiang et al., 2022). All GANs received normalized FFT amplitudes as input and generated a discriminatory score used for damage detection. An interesting aspect was the ability to adjust the threshold and adopt adaptive thresholding to efficiently detect recurring anomalous events(Luleci et al., 2023). The extensive testing conducted on various datasets, including QUGS and the IASC-ASCE reference structure, revealed that the LSTM-based GAN model outperformed the others in terms of damage detection performance (L. Zhang et al., 2018). In a subsequent research endeavor, an in-depth exploration was undertaken to investigate the effects of dimensionality reduction on the performance of damage detection methods. Established models were enlisted, and various dimensionality reduction techniques, such as Principal Component Analysis, kernel PCA, and variational autoencoders (AE), were meticulously examined(Entezami & Shariatmadar, 2018). The outcomes of the study indicated that reducing the input vector's dimensions had a noticeable detrimental impact on the overall accuracy of damage detection. However, the researchers did not stop at identifying the challenges posed by dimensionality reduction. Instead, they proactively sought solutions to overcome this limitation (G. Liu et al., 2020). By skillfully incorporating non-linear smoothing methods into the models, they found that the adverse effects of dimensionality reduction could be effectively mitigated. This innovative approach breathed new life into the performance of the models, significantly enhancing their accuracy and making them more robust in detecting structural damages. The study's comprehensive analysis and novel insights offer valuable guidance for future research in the field of SHM and damage detection (Suh & Cha, 2018b). By addressing the crucial issue of dimensionality reduction and intelligently applying non-linear smoothing methods, the research community is better equipped to develop more reliable and efficient SHM models, ultimately ensuring the safety and integrity of critical structures. By leveraging the capabilities of GANs and integrating them with dimensionality reduction techniques, researchers continue to advance the field of structural health monitoring and anomaly detection. These innovative approaches offer valuable insights into detecting anomalies efficiently and accurately, thereby enhancing the safety and reliability of critical infrastructure. As research in this domain evolves, the continuous exploration of cutting-edge methods will undoubtedly lead to further breakthroughs, driving progress in the realm of structural health monitoring and beyond.

NOVELTY DETECTION TECHNIQUES

Within the domain of vibration-based SHM, the predominant focus of unsupervised learning methods lies in the detection of novelty or anomalies. The main objective is to recognize and monitor system variations, particularly those that signal potential structural damage. To achieve this, researchers often combine multiple novelty detectors to either enhance the compression of input features or create an ensemble learning framework. The fundamental principle behind most novelty detection approaches is rooted in the assumption that normal system behavior forms a cohesive single cluster or class. As a result, any data points found outside this cluster are designated as novelties, anomalies, or outliers(Tan et al., 2021). Thus, novelty detection can be seen as a single-class classification problem, with certain methods utilizing Outlier Analysis and One-Class Support Vector Machines (OCSVMs) for this purpose. In the domain of vibration-based SHM, the prevailing focus of unsupervised learning methods revolves around detecting anomalies and novelties to identify and monitor system variations, particularly those that indicate potential structural damage (Z. Wang & Cha, 2021). These approaches often combine multiple novelty detectors to either compress input features or create an ensemble learning framework. The underlying assumption in most novelty detection methods is that normal system behavior constitutes a single cluster or class, with data points lying outside this cluster labeled as novelties, anomalies, or outliers. However, real-world scenarios may present a more complex situation where data exhibits a multi-class structure due to diverse operational conditions, including normal and damaged states. In such cases, employing cluster analysis techniques becomes more appropriate to handle the intricacies of the data. In this section, we will delve into a comprehensive overview of noteworthy research studies from the past decade, focusing specifically on novelty detection within the context of unsupervised learning-based SHM(J. Liu et al., 2020). This exploration will shed light on the advancements made in the field and the practical applications of these techniques in real-world scenarios. Over the years, numerous advancements have been made in the field of SHM, particularly in the domain of unsupervised learning techniques. Novelty detection plays a crucial role in identifying subtle changes in system behavior, providing early warnings for potential structural damage and ensuring the safety and reliability of critical infrastructure(Kurian & Liyanapathirana, 2020). The exploration of both single-class and multi-class approaches highlights the adaptability and versatility of unsupervised learning methods in dealing with various data structures and system conditions. By gaining a comprehensive understanding of these significant studies, researchers can continue to refine and enhance unsupervised learning-based SHM approaches, enabling more accurate and efficient damage detection in real-world applications. The continuous evolution of unsupervised learning techniques paves the way for more sophisticated and reliable structural health monitoring systems, contributing to the preservation and resilience of infrastructure worldwide. These studies have demonstrated the efficacy of using unsupervised learning techniques in detecting and monitoring structural variations, making them valuable contributions to the field of SHM.

Detection of Novelty of Class

In the field of SHM, univariate outlier analysis represents the predominant approach for novelty detection in unsupervised learning. However, there are also other methodologies that go beyond univariate analysis, such as cluster analysis and feature processing using complex DL techniques. In these cases, the decision-making process for novelty detection becomes more sophisticated. In the field of univariate outlier analysis, common statistical significance tests such as the Z test o T- test are frequently employed to determine a significance threshold for the calculated damage index. These tests play a crucial role in identifying outliers or anomalies in the data, enabling a more accurate identification of potential structural damage. Selecting the appropriate significance thresholds is essential for precise and reliable damage detection, and various statistical tests offer flexible and effective means to achieve this goal(Bao & Li, 2021). This threshold can be established using confidence intervals, significance levels, percentiles, or other data statistics, depending on the specific requirements of your application.In the case of multidimensional features, multivariate outlier analysis offers more possibilities to capture complex relationships between the variables of interest. The Mahalanobis measure is often adopted as a metric for multivariate outlier analysis, accounting for correlations between features to provide a more accurate estimate of the distance between data points(Rosafalco et al., 2020). Methods such as Hotelling T2 control charts can then be used to identify outliers by defining appropriate detection thresholds. In addition to outlier analysis, numerous alternative methods for single-class novelty detection have been devised for unsupervised learning within the realm of SHM. These techniques aim to identify and detect anomalies or novelties within the data without relying on labeled samples, making them valuable tools for detecting structural damage or deviations from normal behavior. The development of such methods has enriched the field of SHM by providing diverse approaches to tackle the challenges of detecting and monitoring structural changes without the need for labeled training data. One notable example is the One-Class Support Vector Machine (OCSVM), which employs a learning strategy based on a single dataset to create a decision boundary enclosing the points belonging to the normal class. Similarly, the Support Vector Data Description (SVDD) is based on constructing a hyperspace that includes all data points of the normal class, aiming to minimize the radius of the enclosing hypersphere. In summary, the detection of novelties in unsupervised learning for SHM encompasses a range of approaches, spanning from univariate outlier analysis to multivariate analysis, and utilizing models like OCSVM and SVDD. Selecting the most appropriate method will depend on the specific characteristics of the data and the requirements of the application.

Group Analysis

In the realm of SHM, clustering stands as a fundamental statistical modeling approach aimed at organizing similar observations into distinct groups or clusters. Various techniques for grouping data are commonly classified into distinct classes, encompassing partition-based methods like the well-known K-means algorithm, hierarchy-based approaches, distribution-based models such as Gaussian mixture models, fuzzy theory-based methods like fuzzy c-means, and density-based techniques like peak density methods(Tomasiello et al., 2022). In the field of SHM, clustering plays a pivotal role as a decision-making algorithm, enabling the extraction of damage-sensitive features for effective anomaly detection. Novelty detection involves scrutinizing whether a new observation belongs to an existing cluster or forms a separate cluster, thus providing valuable insights into data uniqueness(Wiemann et al., 2020). Moreover, clustering is sometimes employed for feature extraction and dimensionality reduction, followed by the implementation of single-class novelty detectors. Among the diverse clustering methods available, the K-means algorithm stands out as one of the simplest and most widely adopted techniques in the context of SHM. In this approach, each data point is assigned to one of the 'k' clusters based on its proximity to the nearest centroid. The simplicity and efficiency of the K-means algorithm make it a popular choice for various SHM applications, where it aids in handling large datasets and contributes to identifying structural abnormalities effectively(Bouzenad et al., 2019). This process partitions the data into distinct groups, enabling a deeper understanding of patterns and relationships within the dataset. The application of clustering in SHM holds significant importance, as it allows for the efficient organization of data and the identification of abnormal behavior in structures (Ochôa et al., 2020). By grouping data points into clusters, subtle variations and potential damage indicators can be detected, contributing to the early detection and prevention of structural issues. In the realm of unsupervised learning-based SHM, clustering techniques play a pivotal role in enhancing decision-making processes and enabling efficient feature extraction. The versatility of clustering methods makes them applicable to various types of data and structural monitoring scenarios. As researchers continue to explore and refine clustering algorithms, we can expect further advancements in the field of SHM (Ma & Du, 2020). The continuous evolution of these methodologies contributes to the development of more sophisticated and reliable systems for ensuring the safety and integrity of critical infrastructure worldwide. The appeal of this algorithm lies in its simplicity and efficiency, which make it particularly attractive to researchers engaged in unsupervised learning in the SHM setting (Gholizadeh et al., 2023). However, it is important to note that advanced feature processing is often required in the preliminary stage of the framework before applying clustering. As an illustration, researchers have suggested the utilization of the Rapid Fourier Transform (RFT) technique on dynamic response data to enhance computational efficiency. Furthermore, an additional step in data preprocessing involved the utilization of the k-nearest neighbor algorithm to remove outliers. Subsequently, the K-medians algorithm was employed to identify abnormal conditions based on the extracted characteristics. To assess the effectiveness of this approach, performance evaluation was conducted using the Sydney Harbor Bridge benchmark dataset. Similarly, the K-medians clustering technique was applied to detect stiffness reductions in a cable-stayed bridge (Ardani et al., 2023). The global silhouette index was adopted as a measure of cluster validity in this case, while the Gowda-Diday dissimilarity measure served as a damage indicator. In addition to the widely used K-means algorithm, several variants have been developed to address its limitations. One noteworthy variant is the K-means-- algorithm, which modifies the original K-means method to overcome sensitivity to outliers by eliminating single-member clusters. This adaptation enhances the robustness of the algorithm, making it more suitable for datasets with potential outlier influences. The implementation of the k-nearest neighbor algorithm for outlier removal contributes to data cleansing and enhances the quality of input data for subsequent clustering analysis. By identifying and eliminating outliers, the clustering algorithm can focus on meaningful patterns and relationships in the data, leading to more accurate and reliable damage detection results (Oliveira & Alegre, 2020). The application of K-medians clustering for identifying abnormal conditions and stiffness reductions in structural components demonstrates the versatility of this approach in various SHM scenarios. The selection of appropriate evaluation measures, such as the global silhouette index and Gowda-Diday dissimilarity measure, ensures the robustness and validity of the clustering results, enabling effective damage detection and localization. In a study, this variant of the K-means algorithm was applied for damage detection. The effectiveness of this approach lies in its ability to detect anomalous responses in damaged structures. Another relevant variation is the K-medians algorithm, which distinguishes itself from K-means by employing an actual data point as the cluster center, thus minimizing differences between points within a cluster and enhancing the method's resilience to outliers(Xiong et al., 2023). This particular variant was applied for novelty detection based on distances between all potential medoids. The detection of novelties relied on damage-sensitive characteristics, including medians and interquartile ranges in both the time and frequency domains (derived via FFT). Experimental findings demonstrated that, with carefully optimized hyperparameters, the method exhibited the capability to detect subtle levels of damage even amidst varying environmental and operational conditions. Nevertheless, hyperparameter tuning can pose a challenge in the context of unsupervised learning. The Fuzzy C-Means (FCM) algorithm, a widely-used statistical modeling method, is another variant that organizes observations into groups or clusters based on their similarity. In contrast to the traditional K-means algorithm, which assigns each data point to a single cluster, the Fuzzy C-Means (FCM) algorithm allocates a membership degree (ranging from 0 to 1) to each cluster, enabling points to partially belong to multiple clusters. FCM has gained widespread popularity in recent decades for unsupervised learning in the context of SHM (Turo et al., 2019). An illustrative demonstration of FCM for damage detection involved the integration of the low-frequency response function with fuzzy c-means clustering to identify and monitor structural damage. Additionally, to enhance process efficiency, the adoption of dimensionality reduction techniques, such as PCAor Kernel PCA, has been proposed. The effectiveness of this combined approach was evaluated on a steel truss bridge, where various damage scenarios were simulated by controlling bolt relaxation(K. Yang et al., 2023c). The results showcased the method's capability in accurately detecting and monitoring structural damage. In another study, a monitoring method was introduced that employed symbolic signal analysis and clustering techniques, with a particular emphasis on the superior effectiveness of FCM compared to other clustering methods. Hierarchical clustering, characterized as a greedy algorithm, is utilized to organize data points into clusters, creating a hierarchical structure based on their similarity. This method can be implemented in two modes: bottom-up (agglomeration) mode, where each point starts as a separate cluster and clusters are gradually merged, and top-down (division) mode, where the entire dataset starts as a single cluster and is recursively divided into smaller clusters. The application of FCM in damage detection showcases its versatility and efficacy in the context of SHM. By combining fuzzy c-means clustering with low-frequency response function analysis, the method provides valuable insights into the identification and monitoring of structural damage. Moreover, the integration of dimensionality reduction techniques further enhances the efficiency of the process, making it a powerful tool for real-world applications (Hu & Ma, 2020). The effectiveness of FCM was corroborated by its outperformance against other clustering methods in the context of symbolic signal analysis for structural monitoring. This highlights the importance of selecting appropriate clustering techniques that align with the specific requirements and characteristics of the SHM scenario. An example of using hierarchical clustering for damage detection was demonstrated in a study proposing a hierarchical clustering model for the identification and assessment of structural damage. The methodology was built upon transmissibility analysis as a fundamental input feature, and it incorporated comparison metrics such as cosine similarity and distance similarity. In a separate study, a comprehensive framework for bridge integrity monitoring was introduced, which integrated modal parameters with hierarchical clustering techniques (Y. Jiang & Niu, 2022d). Univariate anomaly detection based on the Gaussian distribution served as the discrimination criterion, and the method's validity was confirmed through laboratory experiments involving both a steel bridge and a pedestrian bridge. However, the challenge of carefully selecting the clustering threshold in the absence of corrupted data was emphasized, as cross-validation becomes complex. Another clustering technique involves the probabilistic Gaussian Mixture Model (GMM), assuming that all data points arise from a combination of a limited number of Gaussian distributions with unknown parameters. In one study, various novelty detection methods, such as Mahalanobis distance squared (MSD), Principal Component Analysis (PCA), self-associative neural networks, and GMM, were compared for damage detection in a bridge, accounting for varying operating and environmental conditions(Sarmadi & Karamodin, 2020). The assessment revealed that utilizing the GMM parameter Mean Square Deviation (MSD) as a damage indicator resulted in fewer errors compared to other methods. In addition to the aforementioned findings, it was emphasized that linear techniques such as PCAand Mean Square Deviation (MSD) encountered challenges in effectively handling non-linear patterns induced by operational and environmental influences, leading to an increased occurrence of false positives. In a separate investigation, a promising approach involved fusing the Gaussian Mixture Model (GMM) with the expectation maximization (EM) algorithm for clustering to detect anomalies. To further enhance the overall system performance, researchers proposed a fusion of genetic algorithm and EM, which demonstrated improved stability of the EM while effectively minimizing type 2 errors (Dardeno et al., 2022). Addressing progressive damages, especially in tie rods with corrosion, a novel damage detection method based on the Gaussian Mixture Model (GMM) was introduced, utilizing eigenfrequencies as crucial features. The core of this method was to compare the likelihood of two GMM hypotheses, each represented by single or double Gaussian densities. Notably, experimental studies involving tie rods confirmed the superiority of the GMM-based approach over the MSD-based method in detecting progressive deterioration phenomena. However, it was observed that the MSD method might be more suitable for scenarios involving sudden damages. These advancements signify significant progress in the domain of SHM, contributing valuable insights and methodologies to identify and mitigate structural anomalies more accurately and efficiently. The utilization of GMM as a robust damage detection tool showcases its potential in various SHM applications (Y. Ren et al., 2022). By leveraging the power of statistical modeling and probability distributions, the GMM-based approach provides a more accurate and reliable means of detecting subtle and progressive damages, such as corrosion in tie rods. Its ability to account for uncertainties and variations in data makes it a valuable asset in SHM systems where accurate and timely detection of structural anomalies is of utmost importance. Moreover, the integration of genetic algorithms and EM further enhances the capabilities of the GMM method, enabling it to adapt and optimize its performance based on specific SHM scenarios. As research in this field progresses, we can anticipate the development of more sophisticated and efficient damage detection techniques that leverage the strengths of advanced statistical methods like GMM (Mei et al., 2019). In the context of structural analysis based on unsupervised learning, there are many variants of clustering algorithms that find application. Among these variants, an example of interest is a method based on the use of genetic algorithms for clustering in the context of bridge health monitoring. In this context, the optimal choice of the number of clusters represents a significant challenge, therefore an algorithm based on concentric hyperspheres is used for the optimization of the number of clusters. The proposed method calculates the damage indicator by determining the minimum Euclidean distance between the newly acquired data points and the centroids of the identified clusters. When compared to other methods like Gaussian mixture models (GMM)-based anomaly analysis or cumulative sum of least squares (MSD), this technique offers certain advantages (Adhikari & Bhalla, 2019). Another frequently used algorithm is spectral clustering (SC), which leverages graph theory and employs eigenvalues of the similarity matrix to detect clusters. Particularly, the variant referred to as kernel spectral clustering (KSC) demonstrates notable benefits in handling linearly non-separable data. This variant is particularly suitable for the analysis of large datasets, since it allows to reduce the computational cost by exploiting suitable kernel functions. An example of KSC use in structural damage monitoring has been proposed, in which an adaptive damage detection method based on this algorithm has been introduced. The effectiveness of this approach has been validated using a reference dataset known as Z-24. In this methodology, the model is calibrated during the structural integrity phase, allowing any anomalies to be identified during the subsequent phases. Another set of clustering algorithms that is relevant in the context of structural damage monitoring is density-based methods (WANG et al., 2019). These methods establish clusters identified by areas of dense observation concentration within the feature space. An instance of utilizing such algorithms for damage detection was presented, where modifications were made to a peak density-based clustering algorithm to adapt it for an unsupervised learning environment to monitor and locate structural damage. In this context, characteristics obtained from the continuous wavelet transform (CWT) and crest factor were employed. During testing, observations with a local density below a predetermined threshold were considered as new data points and potentially indicative of anomalies (L. Zhao et al., 2019). By using an experimental model of a steel structure, the proposed method exhibited superior performance compared to the OCSVMin damage localization. Nevertheless, it should be noted that this approach may entail a high computational cost, and, as a result, recommendations have been provided to enhance its efficiency.

Bayesian Methods

Bayesian analysis represents a powerful statistical approach that leverages Bayes' theorem to update probabilities by incorporating prior information. Emphasizing the concept of probability as a measure of belief, Bayesian methods have become pervasive in tackling uncertainty associated with parameter estimates. In the domain of structural engineering, Bayesian approaches have proven to be instrumental in quantifying uncertainty during the identification, localization, and quantification of damages. In the context of damage detection, Bayesian hypothesis testing emerged as a valuable tool(Rogers et al., 2019). By utilizing model residuals, researchers could effectively compute the Bayes factor, which provided essential insights into the relative strength of evidence supporting various hypotheses(F.-L. Zhang et al., 2023). This Bayesian framework facilitated the rigorous examination of competing scenarios and contributed to more informed decision-making in structural health monitoring applications. Harold Jeffreys' bounds were utilized to evaluate the magnitude of damage severity, offering a robust means of assessing the extent of structural deterioration. Moreover, the Bayes factor was leveraged to estimate the probability of alternative scenarios, providing a measure of associated uncertainty and aiding in decision-making processes related to structural health monitoring(Q. Ren et al., 2023). The integration of Bayesian methods in damage detection contributes to a comprehensive understanding of uncertainties and enhances the reliability of damage assessment in various structural engineering applications. Concerning damage localization and quantification, likelihood and Bayesian inference concepts were employed to quantify uncertainty. A method for updating uncertainty by acquiring new measurements was also presented. Subsequent research aimed to devise a fresh damage indicator by employing the natural excitation technique, taking into account both the real and imaginary aspects of the data(Mao et al., 2023). To capture the connection between the real and imaginary components effectively, the study incorporated a Bayesian sparse learning regression model. This model was trained to predict the imaginary components by utilizing the real components as input variables, enabling a comprehensive understanding of the structural condition(Sajedi & Liang, 2022). The resulting damage indicator demonstrated promising results in detecting anomalies and quantifying damage severity, signifying the potential of the proposed method in SHM applications. The relative change between these components was utilized as a quantifier of the structural condition, and the Bayes factor emerged as a promising indicator to assess the severity of damages(Y.-M. Zhang et al., 2022). By employing this comprehensive approach, researchers aimed to enhance the accuracy and reliability of damage assessment in structural engineering applications, offering valuable insights into the structural health of the system. Regarding unsupervised damage detection, a novel technique was introduced to handle uncertainty arising from diverse factors, including data variability, measurement inaccuracies, and environmental fluctuations. This method encompassed the calculation of the symmetric KL divergence between the transmissibility function (TF) under a reference condition and the TF under an unknown condition, presenting an effective approach for damage detection. To address measurement uncertainty, a statistical threshold estimation process was implemented, combining Bayesian inference with Monte Carlo discordance testing. The results obtained demonstrated satisfactory performance in detecting global damages and assessing their severity. However, it was observed that the method could not precisely pinpoint the exact damage location in certain instances. This discrepancy arose as the sensor positions causing anomalies did not always align precisely with the actual damage location (Entezami et al., 2020). Nonetheless, the approach displayed promising potential for structural health monitoring, providing valuable insights into damage detection and assessment in real-world applications. Future research efforts may focus on refining the method to enhance the localization accuracy, further bolstering its overall effectiveness. In the realm of parameter estimation, Bayesian methodologies have demonstrated their utility as well. For instance, a novel Bayesian technique, leveraging Markov Chain Monte Carlo, was introduced for clustering Gaussian mixtures (GMM), surpassing the limitations of traditional EM (Expectation-Maximization) methods(Eltouny & Liang, 2021). As a novelty detection measure, a cumulative sum of least squares (MSD) method was applied. In a separate investigation, Bayesian optimization was harnessed to construct a probabilistic model using the Kenker Density Maximum-Entropy (KDME) approach for damage localization. This innovative approach eliminated the need for predefining data distributions. Bayesian optimization, a well-established technique in ML for global hyperparameter optimization without assuming the objective function's shape, was employed to train a multivariate KDME. To facilitate KDME training, an initial independent component analysis process was conducted. In the pursuit of an efficient damage detection and localization method, the study formulated a unique damage indicator by combining the joint probabilities of new observations. The threshold for anomaly detection was thoughtfully set using the principles of EVT (Extreme Value Theory), enabling effective identification of potential damage occurrences. The groundbreaking aspect of this research lies in the integration of Bayesian optimization, which resulted in a significant acceleration of the KDME (Kenker Density Maximum-Entropy) model's tuning process(Y.-M. Zhang et al., 2021). The traditional genetic algorithms were outperformed, showcasing the tremendous potential of Bayesian optimization as a powerful and time-saving tool for refining and fine-tuning models in the realm of damage detection and localization. By leveraging the capabilities of Bayesian optimization, researchers and practitioners in the field of SHM can achieve more accurate and reliable outcomes. The optimization process ensures that the KDME model is precisely tailored to the specific structural context, enhancing its overall performance and increasing the system's resilience to uncertainties and environmental variations. Ultimately, this integration holds promise for the practical implementation of SHM technologies, fostering safer and more sustainable infrastructures.

Other Example of Appilications

In orther to the ML methods discussed earlier for detecting system variations, researchers have explored various alternative approaches, including resilient regression, collective learning, and experiential ML. Resilient regression is a valuable technique employed to estimate the parameters of mathematical models while minimizing the impact of outliers. Further studies might explore the shift from resilient linear regression to resilient non-linear regression. Ensemble learning, on the other hand, is a ML technique that combines multiple models to enhance prediction stability through diversity. Several researchers have adopted ensemble learning to improve the performance of novelty detectors in damage detection. For example, a series of models for anomaly detection in facility health monitoring was introduced to address the challenges posed by inclusive outliers. Model ensembles were formed through the bootstrap sampling technique, and the ensemble output was computed by averaging the model predictions. The Monte Carlo simulation thresholding method was applied, resulting in comparable outcomes with a notable reduction in computational resources. In an alternative method that utilizes ensemble learning, the aim is to take advantage of the computational efficiency of ensemble techniques while minimizing the impact of environmental variations on structural health monitoring. The proposed sequential learning framework involves three variations of multi-level detection. At each level, a group of nearest neighbors is identified using the distance participation factor. The final level produces local MSD values that act as damage indicators, and a threshold based on the theory of extrema is established. The method was validated through the use of two experimental case studies, successfully detecting damages even in the presence of marked environmental variations and exhibiting a lower error rate than several traditional techniques. Diverse model learning (DML) is an approach closely related to ensemble learning, wherein multiple statistical models are utilized to analyze or predict a given dataset. It was compared against two other conventional autoregressive methods. Using DML, an attempt was made to capture the inherent dynamics of an undamaged structure by employing a series of conventional models, using vectors of estimated parameters and Gaussian probability density functions. The performance evaluation was conducted using ROC curves to assess the accuracy of each model. In the realm of SHM, certain methodologies prioritize the adoption of flexible nonparametric techniques, avoiding the imposition of pre-established assumptions on the data. An exemplar of such an approach is Empirical Machine Learning (EML), which relies solely on observed data and their relative distances to construct effective models. To serve as a damage indicator, a novel method emerged, incorporating the multiplication of empirical local density by the minimum distance of each sample from all other samples. This innovative non-parametric novelty detection approach draws inspiration from the density peak clustering technique, known for its efficiency in identifying anomalies and structural deviations. During rigorous experimental evaluations, this non-parametric approach showcased superior performance when compared to various other non-parametric novelty detection methods, excelling in terms of both damage detection accuracy and computational efficiency. By embracing EML and harnessing the capabilities of nonparametric methodologies, researchers and practitioners in the SHM field can unlock the potential to achieve robust and reliable outcomes. The absence of rigid assumptions allows the model to adapt seamlessly to diverse and intricate structural behaviors, making it a promising avenue for advancing damage detection capabilities and promoting the safety and integrity of critical infrastructures.

FUTURE CHALLENGES AND DISCUSSIONS

Within the realm of SHM, the application of unsupervised learning provides a pragmatic focus on vibrations. Nonetheless, despite its promise, the extensive adoption of this methodology in the industry faces multiple hurdles and limitations. Unsupervised learning in SHM presents various challenges due to its fundamental concept, where the absence of predefined classes makes it difficult to differentiate between damaged and intact structural conditions. This lack of labeled data and predefined categories adds complexity to the process of detecting and identifying structural damage. Therefore, these barriers need to be addressed in order to promote the widespread use of unsupervised learning in vibration-based monitoring applications. In the literature reviewed, several future research areas were identified that aim to overcome these challenges and improve the effectiveness and applicability of unsupervised learning in SHM (K. Yang et al., 2023a).

Incorporating cross-validation to optimize parameters in unsupervised learning for SHM proves to be complex since there are no predefined impaired and unimpaired classes. Selecting an optimal threshold for outlier analysis, determining the appropriate "ν" parameter in OCSVM, or specifying the number of clusters in clustering methods often requires making engineering judgment assumptions. However, these assumptions may not always guarantee the most optimal selection for the SHM model (Entezami et al., 2020). This challenge arises from the need to strike a balance between sensitivity and specificity, making the choice of parameters crucial for accurate and reliable damage detection in real-world applications. In novelty detection, achieving a precise trade-off between the rates of false positives and false negatives is essential, but this balance can be influenced by the data size and the lack of representative observations for various normal or overlapping conditions with the unknown impaired class (Gulgec et al., 2019). Certain models may not readily lend themselves to parameter selection, but post-test sensitivity analysis might unveil more effective models. To tackle these challenges, several strategies have been proposed to establish more robust thresholds, such as leveraging Monte Carlo sampling techniques or employing the theory of extrema (EVT). Nevertheless, these methods might still demand the choice of extra parameters, like sample size or the maximum block window size. Certain research endeavors have aimed to offer recommendations for threshold determination, relying on ROC curve findings derived from post-test case analyses. At the same time, many of the methods proposed for the SHM are specific to particular structural typologies and may encounter difficulties in their process of generalization to other situations. Some approaches have addressed this limitation by proposing adaptive thresholds, which allow the novelty detector to identify future damage scenarios in systems that have already undergone changes. Threshold selection strategy remains an open and promising research area aimed at improving the accuracy and reliability of SHM models.

Environmental and operational variations present a significant hurdle in developing SHM techniques, especially when it comes to unsupervised learning-based approaches. The detection of deviations, which aims to identify departures from normal system conditions, can result from structural damage or changes in environmental factors like temperature fluctuations or shifts in operating conditions. In recent years, a significant emphasis has been placed on the development of robust damage-specific features and anomaly detection methods that can better withstand the influence of environmental fluctuations. These efforts are driven by the need to enhance the reliability and accuracy of structural health monitoring systems, particularly when dealing with real-world scenarios where external factors can introduce uncertainties and noise in the collected data. By improving the resilience of these techniques, researchers aim to achieve more precise and effective damage detection, leading to safer and more reliable structural assessment in various applications. Some researchers have proposed probabilistic strategies that account for uncertainty, encompassing environmental changes and measurement noise (Sajedi & Liang, 2022). To counteract the effects and decrease false positives, it becomes imperative to expand the training dataset to include a diverse range of environmental conditions (Oh et al., 2019b). Additionally, integrating non-vibration data, like temperature, wind velocity, and applied forces, can improve the model's performance and reduce uncertainty in pinpointing structural damage locations. Recent literature emphasizes the necessity of addressing the impact of environmental and operational variations when developing unsupervised learning-based SHM systems. Future research in this domain should concentrate on several key areas of investigation. Firstly, leveraging technological advancements in sensors and the Internet of Things will enable the acquisition of long-term monitoring data, thereby minimizing associated uncertainties. Secondly, utilizing unsupervised learning methods, like DL, will enable a more accurate and sophisticated modeling of the complex and non-stationary relationships inherent in monitoring data. Finally, incorporating environmental surveillance systems, like temperature, moisture, and wind velocity sensors, and amalgamating data from different origins will lead to a more comprehensive and accurate information repository, thus enabling improved detection of structural damage.

Despite the wide range of unsupervised learning approaches to SHM, there is a clear lack of a standardized methodology for direct comparison between them. In ML, such as computer vision or human sensing, there are established benchmarks such as ImageNet and Human3.6m, which allow for comparative evaluation of models. However, in the context of SHM, direct comparisons between previously proposed methods are limited, as authors often use different metrics and data partitions. Others prefer to split the data by allocating 75% of the unimpaired state period to training and the remaining 25% to testing (C. Guo et al., 2023a). Also, there are variations that include including specific months in the training set or using different test sets. It is equally important to underline that the sharing of the model source code, a practice that would facilitate the comparison and the creation of a sort of "zoo" of models, is not yet widespread in SHM research. As a result, in order to expedite the advancement of ML models in the realm of SHM, it is beneficial to adopt universally recognized standards that offer a timely and extensive evaluation of the present condition of the discipline..

According to the analysis of existing literature, it is evident that several datasets are accessible for verifying the proposed techniques in the realm of structural monitoring using unsupervised learning. However, it is important to underline that most of these data sets are not able to fully represent real situations of damaged structures. Several methods heavily rely on computer-generated simulations and controlled laboratory experiments, which offer valuable insights but may have limitations when applied to real-world scenarios. Although controlled trials in laboratory settings represent an improvement over purely simulated data, they might not fully capture the complexities and variations present in actual operational conditions. In many cases, datasets obtained from real damaged structures lack comprehensive information about the precise environmental and loading conditions during the occurrence of the damage (R. Wang et al., 2018). This limitation can impact the robustness and generalizability of the detection techniques when applied to real-life situations with dynamic and unpredictable conditions, such as vehicular traffic and varying external loads. Therefore, there is a need to collect datasets that encompass structural states with realistic operating conditions, considering associated uncertainties, and that are extensive enough to facilitate the training of DLmodels. This would represent a promising research area for developing advanced methods in the field of structural monitoring based on unsupervised learning.

The interest and attention towards DL-based structural monitoring methods has undergone a remarkable growth over the last few years. However, one of the major hurdles we face is the requirement for large amounts of training data to power these complex models. The limited availability of experimental data represents a real challenge in the field, especially for supervised learning approaches, where the need for a large number of labeled samples can be problematic. On the other hand, the use of unsupervised learning methods offers a potential advantage as they allow you to collect large amounts of training data without requiring explicit labels. This paves the way for the development of more complex and profound algorithms that can extract meaningful information from the intrinsic characteristics of the data. With the advent of big data benchmarks, future research is expected to increasingly focus on unsupervised learning-based structural monitoring methods, fully exploiting the potential of deep neural networks and driving further advances in the field.

The application of unsupervised learning in SHM systems faces significant challenges that require scientific insights and innovative developments. Despite the progress made, there are inherent limitations that hinder the uptake of such methods in industry. One of the main challenges is the ability of unsupervised learning-based SHM models to generalize beyond specific structures, as these models are often adapted to a particular structural context and struggle to transfer to other structures with different characteristics. To overcome this limitation, it is essential to focus on the development of methodologies that allow domain adaptation and knowledge transfer between different structures, making use of advanced techniques such as transfer learning and self-supervised learning. Furthermore, it is crucial to address the heterogeneity of data and operating conditions present in SHM applications, developing robust models that can handle environmental variability and unwanted disturbances. This requires the development of innovative algorithms and the use of DL-based approaches, which make it possible to extract complex structural features and understand the intrinsic relationships present in the monitored data. Furthermore, it is of paramount importance to have large and diverse training datasets to ensure the effectiveness of unsupervised learning-based SHM models. Therefore, it is necessary to promote data sharing and standardization, as well as to encourage collaboration between the scientific community and industry to collect real data from damaged structures and realistic operating conditions. Only through an interdisciplinary approach and a continuous commitment will it be possible to overcome the current limitations and favor the practical application of SHM methods based on unsupervised learning in a wide range of structural contexts.

CONCLUSION

SHM is a crucial area that allows real-time and autonomous assessment of structural conditions, ensuring the security and integrity of critical infrastructures. In recent years, there has been a significant interest in studying unsupervised learning approaches in the context of Vibration-Based SHM, with the aim of bridging the gap between cutting-edge research and practical industrial implementations. This comprehensive study aims to provide an in-depth and up-to-date analysis of the latest applications of unsupervised learning techniques in SHM, covering a decade's worth of advancements. Various sophisticated methodologies are examined, ranging from classical feature extraction methods like PCAto state-of-the-art DLmodels, such as Autoencoders (AE) and Generative Adversarial Networks (GANs). In particular, the study explores specialized techniques tailored for novelty detection and cluster analysis, specifically designed to tackle the unique challenges posed by SHM scenarios. Rigorous evaluations of unsupervised learning models are conducted using established benchmarks commonly employed in SHM research. However, the study also highlights significant challenges encountered in implementing unsupervised learning in SHM. These challenges include accurately establishing optimal detection thresholds, effectively handling environmental variations that may impact the data, and ensuring model generalization across diverse structural contexts. By presenting a detailed analysis of the latest developments and complexities of unsupervised learning in SHM, this study aims to accelerate progress in ML models and pave the way for more efficient and reliable SHM technologies applicable to real-world scenarios. Ultimately, the integration of advanced unsupervised learning techniques promises to enhance the overall safety and longevity of critical infrastructures, benefiting society as a whole.

REFERENCES

Abdeljaber, O., Avci, O., Kiranyaz, M. S., Boashash, B., Sodano, H., & Inman, D. J. (2018). 1-D CNNs for structural damage detection: Verification on a structural health monitoring benchmark data. *Neurocomputing*, *275*, 1308–1317. doi:10.1016/j.neucom.2017.09.069

Adhikari, S., & Bhalla, S. (2019). Modified Dual Piezo Configuration for Improved Structural Health Monitoring Using Electro-Mechanical Impedance (EMI) Technique. *Experimental Techniques*, *43*(1), 25–40. doi:10.100740799-018-0249-y

Almeida Cardoso, R., Cury, A., Barbosa, F., & Gentile, C. (2019). Unsupervised real-time SHM technique based on novelty indexes. *Structural Control and Health Monitoring*, *26*(7), e2364. doi:10.1002tc.2364

Altabey, W. A., Noori, M., Wang, T., Ghiasi, R., Kuok, S.-C., & Wu, Z. (2021). Deep Learning-Based Crack Identification for Steel Pipelines by Extracting Features from 3D Shadow Modeling. *Applied Sciences (Basel, Switzerland)*, *11*(13), 6063. doi:10.3390/app11136063

Alzubaidi, L., Bai, J., Al-Sabaawi, A., Santamaría, J., Albahri, A. S., Al-dabbagh, B. S. N., Fadhel, M. A., Manoufali, M., Zhang, J., Al-Timemy, A. H., Duan, Y., Abdullah, A., Farhan, L., Lu, Y., Gupta, A., Albu, F., Abbosh, A., & Gu, Y. (2023). A survey on deep learning tools dealing with data scarcity: Definitions, challenges, solutions, tips, and applications. *Journal of Big Data*, *10*(1), 46. doi:10.118640537-023-00727-2

Ampadi Ramachandran, R., Barão, V. A. R., Ozevin, D., & Sukotjo, C., Pai. P, S., & Mathew, M. (. (2023). Early predicting tribocorrosion rate of dental implant titanium materials using random forest machine learning models. *Tribology International*, *187*, 108735. doi:10.1016/j.triboint.2023.108735 PMID:37720691

Angulo-Saucedo, G. A., Leon-Medina, J. X., Pineda-Muñoz, W. A., Torres-Arredondo, M. A., & Tibaduiza, D. A. (2022). Damage Classification Using Supervised Self-Organizing Maps in Structural Health Monitoring. *Sensors (Basel)*, *22*(4), 1484. doi:10.339022041484 PMID:35214386

Aoki, E., do Cabo, C. T., Mao, Z., Ding, Y., & Ozharar, S. (2022). Vibration-Based Status Identification of Power Transmission Poles. *IFAC-PapersOnLine*, *55*(27), 214–217. doi:10.1016/j.ifacol.2022.10.514

Application of deep autoencoder model for structural condition monitoring. (2018). *Journal of Systems Engineering and Electronics*, *29*(4), 873. doi:10.21629/JSEE.2018.04.22

Arcadius Tokognon, C., Gao, B., Tian, G. Y., & Yan, Y. (2017). Structural Health Monitoring Framework Based on Internet of Things: A Survey. *IEEE Internet of Things Journal*, *4*(3), 619–635. doi:10.1109/JIOT.2017.2664072

Ardani, S., Akintunde, E., Linzell, D., Eftekhar Azam, S., & Alomari, Q. (2023). Evaluating pod-based unsupervised damage identification using controlled damage propagation of out-of-service bridges. *Engineering Structures*, *286*, 116096. doi:10.1016/j.engstruct.2023.116096

Azimi, M., Eslamlou, A., & Pekcan, G. (2020). Data-Driven Structural Health Monitoring and Damage Detection through Deep Learning: State-of-the-Art Review. *Sensors (Basel)*, *20*(10), 2778. doi:10.339020102778 PMID:32414205

Azimi, M., & Pekcan, G. (2020). Structural health monitoring using extremely compressed data through deep learning. *Computer-Aided Civil and Infrastructure Engineering*, *35*(6), 597–614. doi:10.1111/mice.12517

Azuara, G., Ruiz, M., & Barrera, E. (2021). Damage Localization in Composite Plates Using Wavelet Transform and 2-D Convolutional Neural Networks. *Sensors (Basel)*, *21*(17), 5825. doi:10.339021175825 PMID:34502715

Bao, Y., & Li, H. (2021). Machine learning paradigm for structural health monitoring. *Structural Health Monitoring*, *20*(4), 1353–1372. doi:10.1177/1475921720972416

Bao, Y., Tang, Z., Li, H., & Zhang, Y. (2019). Computer vision and deep learning–based data anomaly detection method for structural health monitoring. *Structural Health Monitoring*, *18*(2), 401–421. doi:10.1177/1475921718757405

Ben Ali, J., Fnaiech, N., Saidi, L., Chebel-Morello, B., & Fnaiech, F. (2015). Application of empirical mode decomposition and artificial neural network for automatic bearing fault diagnosis based on vibration signals. *Applied Acoustics*, *89*, 16–27. doi:10.1016/j.apacoust.2014.08.016

Bezas, K., Komianos, V., Koufoudakis, G., Tsoumanis, G., Kabassi, K., & Oikonomou, K. (2020). Structural Health Monitoring in Historical Buildings: A Network Approach. *Heritage*, *3*(3), 796–818. doi:10.3390/heritage3030044

Bhattacharya, S., Yadav, N., Ahmad, A., Melandsø, F., & Habib, A. (2022). Multiple Damage Detection in PZT Sensor Using Dual Point Contact Method. *Sensors (Basel)*, *22*(23), 9161. doi:10.339022239161 PMID:36501870

Bouzenad, A. E., El Mountassir, M., Yaacoubi, S., Dahmene, F., Koabaz, M., Buchheit, L., & Ke, W. (2019). A Semi-Supervised Based K-Means Algorithm for Optimal Guided Waves Structural Health Monitoring: A Case Study. *Inventions (Basel, Switzerland)*, *4*(1), 17. doi:10.3390/inventions4010017

Bukhsh, Z. A., Jansen, N., & Saeed, A. (2021). Damage detection using in-domain and cross-domain transfer learning. *Neural Computing & Applications*, *33*(24), 16921–16936. doi:10.100700521-021-06279-x

Candon, M., Levinski, O., Ogawa, H., Carrese, R., & Marzocca, P. (2022). A nonlinear signal processing framework for rapid identification and diagnosis of structural freeplay. *Mechanical Systems and Signal Processing*, *163*, 107999. doi:10.1016/j.ymssp.2021.107999

Cha, Y.-J., & Wang, Z. (2018). Unsupervised novelty detection–based structural damage localization using a density peaks-based fast clustering algorithm. *Structural Health Monitoring*, *17*(2), 313–324. doi:10.1177/1475921717691260

Chen, C., Tang, L., Lu, Y., Wang, Y., Liu, Z., Liu, Y., Zhou, L., Jiang, Z., & Yang, B. (2023). Reconstruction of long-term strain data for structural health monitoring with a hybrid deep-learning and autoregressive model considering thermal effects. *Engineering Structures*, *285*, 116063. doi:10.1016/j.engstruct.2023.116063

Chen, Q., Cao, J., & Xia, Y. (2022). Physics-Enhanced PCA for Data Compression in Edge Devices. *IEEE Transactions on Green Communications and Networking*, *6*(3), 1624–1634. doi:10.1109/TGCN.2022.3171681

Cinque, D., Saccone, M., Capua, R., Spina, D., Falcolini, C., & Gabriele, S. (2022). Experimental Validation of a High Precision GNSS System for Monitoring of Civil Infrastructures. *Sustainability (Basel)*, *14*(17), 10984. doi:10.3390u141710984

Civera, M., & Surace, C. (2021). A Comparative Analysis of Signal Decomposition Techniques for Structural Health Monitoring on an Experimental Benchmark. *Sensors (Basel)*, *21*(5), 1825. doi:10.339021051825 PMID:33807884

Dang, H., Tatipamula, M., & Nguyen, H. X. (2022). Cloud-Based Digital Twinning for Structural Health Monitoring Using Deep Learning. *IEEE Transactions on Industrial Informatics*, *18*(6), 3820–3830. doi:10.1109/TII.2021.3115119

Dang, H. V., Tran-Ngoc, H., Nguyen, T. V., Bui-Tien, T., De Roeck, G., & Nguyen, H. X. (2021). Data-Driven Structural Health Monitoring Using Feature Fusion and Hybrid Deep Learning. *IEEE Transactions on Automation Science and Engineering*, *18*(4), 2087–2103. doi:10.1109/TASE.2020.3034401

Dang, V.-H. (2023). Development of Structural Damage Detection Method Working with Contaminated Vibration Data via Autoencoder and Gradient Boosting. *Periodica Polytechnica. Civil Engineering*. Advance online publication. doi:10.3311/PPci.22373

Dardeno, T. A., Bull, L. A., Mills, R. S., Dervilis, N., & Worden, K. (2022). Modelling variability in vibration-based PBSHM via a generalised population form. *Journal of Sound and Vibration*, *538*, 117227. doi:10.1016/j.jsv.2022.117227

De Simone, M. C., & Guida, D. (2020). *Experimental Investigation on Structural Vibrations by a New Shaking Table*. doi:10.1007/978-3-030-41057-5_66

De Simone, M. C., Lorusso, A., & Santaniello, D. (2022). Predictive maintenance and Structural Health Monitoring via IoT system. *2022 IEEE Workshop on Complexity in Engineering (COMPENG)*, 1–4. 10.1109/COMPENG50184.2022.9905441

Dederichs, A. C., & Øiseth, O. (2023). Experimental comparison of automatic operational modal analysis algorithms for application to long-span road bridges. *Mechanical Systems and Signal Processing*, *199*, 110485. doi:10.1016/j.ymssp.2023.110485

Deivasigamani, A., Daliri, A. H., Wang, C., & John, S. (2013). A Review of Passive Wireless Sensors for Structural Health Monitoring. *Modern Applied Science*, *7*(2). Advance online publication. doi:10.5539/mas.v7n2p57

Delgadillo, R. M., & Casas, J. R. (2022). Bridge damage detection via improved completed ensemble empirical mode decomposition with adaptive noise and machine learning algorithms. *Structural Control and Health Monitoring*, *29*(8). Advance online publication. doi:10.1002tc.2966

Deng, L., Chu, H.-H., Shi, P., Wang, W., & Kong, X. (2020). Region-Based CNN Method with Deformable Modules for Visually Classifying Concrete Cracks. *Applied Sciences (Basel, Switzerland)*, *10*(7), 2528. doi:10.3390/app10072528

Dhruva Kumar, D., Fang, C., Zheng, Y., & Gao, Y. (2023). Semi-supervised transfer learning-based automatic weld defect detection and visual inspection. *Engineering Structures*, *292*, 116580. doi:10.1016/j.engstruct.2023.116580

Di Lorenzo, D., Champaney, V., Marzin, J. Y., Farhat, C., & Chinesta, F. (2023). Physics informed and data-based augmented learning in structural health diagnosis. *Computer Methods in Applied Mechanics and Engineering*, *414*, 116186. doi:10.1016/j.cma.2023.116186

Doroudi, R., Hosseini Lavassani, S. H., Shahrouzi, M., & Dadgostar, M. (2022). Identifying the dynamic characteristics of super tall buildings by multivariate empirical mode decomposition. *Structural Control and Health Monitoring*, *29*(11). Advance online publication. doi:10.1002tc.3075

Duan, J., Chen, X. M., Qi, H., & Li, Y. G. (2014). An Integrated Simulation System for Building Structures. *Applied Mechanics and Materials*, *580–583*, 3127–3133. . doi:10.4028/www.scientific.net/AMM.580-583.3127

Eltouny, K. A., & Liang, X. (2021). Bayesian-optimized unsupervised learning approach for structural damage detection. *Computer-Aided Civil and Infrastructure Engineering*, *36*(10), 1249–1269. doi:10.1111/mice.12680

Entezami, A., & Shariatmadar, H. (2018). An unsupervised learning approach by novel damage indices in structural health monitoring for damage localization and quantification. *Structural Health Monitoring*, *17*(2), 325–345. doi:10.1177/1475921717693572

Entezami, A., Shariatmadar, H., & De Michele, C. (2022). Non-parametric empirical machine learning for short-term and long-term structural health monitoring. *Structural Health Monitoring*, *21*(6), 2700–2718. doi:10.1177/14759217211069842

Entezami, A., Shariatmadar, H., & Mariani, S. (2020). Fast unsupervised learning methods for structural health monitoring with large vibration data from dense sensor networks. *Structural Health Monitoring*, *19*(6), 1685–1710. doi:10.1177/1475921719894186

Farhangdoust, S., Tashakori, S., Baghalian, A., Mehrabi, A., & Tansel, N. I. (2019). Prediction of damage location in composite plates using artificial neural network modeling. In K.-W. Wang, H. Sohn, H. Huang, & J. P. Lynch (Eds.), Sensors and Smart Structures Technologies for Civil, Mechanical, and Aerospace Systems 2019 (p. 20). SPIE. doi:10.1117/12.2517422

Flah, M., Nunez, I., Ben Chaabene, W., & Nehdi, M. L. (2021). Machine Learning Algorithms in Civil Structural Health Monitoring: A Systematic Review. *Archives of Computational Methods in Engineering*, *28*(4), 2621–2643. doi:10.100711831-020-09471-9

Galanopoulos, G., Milanoski, D., Eleftheroglou, N., Broer, A., Zarouchas, D., & Loutas, T. (2023). Acoustic emission-based remaining useful life prognosis of aeronautical structures subjected to compressive fatigue loading. *Engineering Structures*, *290*, 116391. doi:10.1016/j.engstruct.2023.116391

Gholizadeh, S., Leman, Z., & Baharudin, B. T. H. T. (2023). State-of-the-art ensemble learning and unsupervised learning in fatigue crack recognition of glass fiber reinforced polyester composite (GFRP) using acoustic emission. *Ultrasonics*, *132*, 106998. doi:10.1016/j.ultras.2023.106998 PMID:37001339

Gomes, G. F., Mendez, Y. A. D., da Silva Lopes Alexandrino, P., da Cunha, S. S. Jr, & Ancelotti, A. C. Jr. (2019). A Review of Vibration Based Inverse Methods for Damage Detection and Identification in Mechanical Structures Using Optimization Algorithms and ANN. *Archives of Computational Methods in Engineering*, *26*(4), 883–897. doi:10.100711831-018-9273-4

Gulgec, N. S., Takáč, M., & Pakzad, S. N. (2019). Convolutional Neural Network Approach for Robust Structural Damage Detection and Localization. *Journal of Computing in Civil Engineering*, *33*(3), 04019005. Advance online publication. doi:10.1061/(ASCE)CP.1943-5487.0000820

Guo, C., Jiang, L., Yang, F., Yang, Z., & Zhang, X. (2023). An intelligent impact load identification and localization method based on autonomic feature extraction and anomaly detection. *Engineering Structures*, *291*, 116378. doi:10.1016/j.engstruct.2023.116378

Guo, L., Li, R., & Jiang, B. (2021). A Cascade Broad Neural Network for Concrete Structural Crack Damage Automated Classification. *IEEE Transactions on Industrial Informatics*, *17*(4), 2737–2742. doi:10.1109/TII.2020.3010799

Hoskere, V., Narazaki, Y., Spencer, B. F., & Smith, M. D. (2019). Deep Learning-based Damage Detection of Miter Gates Using Synthetic Imagery from Computer Graphics. *Structural Health Monitoring*. Advance online publication. doi:10.12783hm2019/32463

Hosseini, M.-P., Tran, T. X., Pompili, D., Elisevich, K., & Soltanian-Zadeh, H. (2017). Deep Learning with Edge Computing for Localization of Epileptogenicity Using Multimodal rs-fMRI and EEG Big Data. *2017 IEEE International Conference on Autonomic Computing (ICAC)*, 83–92. 10.1109/ICAC.2017.41

Hou, G., Li, L., Xu, Z., Chen, Q., Liu, Y., & Mu, X. (2022). A Visual Management System for Structural Health Monitoring Based on Web-BIM and Dynamic Multi-source Monitoring Data-driven. *Arabian Journal for Science and Engineering*, *47*(4), 4731–4748. doi:10.100713369-021-06268-1 PMID:36032406

Hou, L., Chen, H., Zhang, G., & Wang, X. (2021). Deep Learning-Based Applications for Safety Management in the AEC Industry: A Review. *Applied Sciences (Basel, Switzerland)*, *11*(2), 821. doi:10.3390/app11020821

Hu, J., & Ma, F. (2020). Statistical modelling for high arch dam deformation during the initial impoundment period. *Structural Control and Health Monitoring*, *27*(12). Advance online publication. doi:10.1002tc.2638

Hurtik, P., & Tomasiello, S. (2019). A review on the application of fuzzy transform in data and image compression. *Soft Computing*, *23*(23), 12641–12653. doi:10.100700500-019-03816-8

Iyer, S., Velmurugan, T., Gandomi, A. H., Noor Mohammed, V., Saravanan, K., & Nandakumar, S. (2021). Structural health monitoring of railway tracks using IoT-based multi-robot system. *Neural Computing & Applications*, *33*(11), 5897–5915. doi:10.100700521-020-05366-9

Jamshidi, M., & El-Badry, M. (2023). Structural damage severity classification from time-frequency acceleration data using convolutional neural networks. *Structures*, *54*, 236–253. doi:10.1016/j.istruc.2023.05.009

Jiang, C., & Chen, N.-Z. (2023). Graph Neural Networks (GNNs) based accelerated numerical simulation. *Engineering Applications of Artificial Intelligence*, *123*, 106370. doi:10.1016/j.engappai.2023.106370

Jiang, H., Wan, C., Yang, K., Ding, Y., & Xue, S. (2022). Continuous missing data imputation with incomplete dataset by generative adversarial networks–based unsupervised learning for long-term bridge health monitoring. *Structural Health Monitoring*, *21*(3), 1093–1109. doi:10.1177/14759217211021942

Jiang, Y., & Niu, G. (2022). An iterative frequency-domain envelope-tracking filter for dispersive signal decomposition in structural health monitoring. *Mechanical Systems and Signal Processing*, *179*, 109329. doi:10.1016/j.ymssp.2022.109329

Kaloni, S., Singh, G., & Tiwari, P. (2021). Nonparametric damage detection and localization model of framed civil structure based on local gravitation clustering analysis. *Journal of Building Engineering*, *44*, 103339. doi:10.1016/j.jobe.2021.103339

Kanarachos, S., Christopoulos, S.-R. G., Chroneos, A., & Fitzpatrick, M. E. (2017). Detecting anomalies in time series data via a deep learning algorithm combining wavelets, neural networks and Hilbert transform. *Expert Systems with Applications*, *85*, 292–304. doi:10.1016/j.eswa.2017.04.028

Kumar, P., & Kota, S. R. (2023). IoT enabled diagnosis and prognosis framework for structural health monitoring. *Journal of Ambient Intelligence and Humanized Computing*, *14*(8), 11301–11318. doi:10.100712652-023-04646-1

Kurian, B., & Liyanapathirana, R. (2020). *Machine Learning Techniques for Structural Health Monitoring*., doi:10.1007/978-981-13-8331-1_1

Lei, X., Sun, L., & Xia, Y. (2021). Lost data reconstruction for structural health monitoring using deep convolutional generative adversarial networks. *Structural Health Monitoring*, *20*(4), 2069–2087. doi:10.1177/1475921720959226

Li, L., Morgantini, M., & Betti, R. (2023). Structural damage assessment through a new generalized autoencoder with features in the quefrency domain. *Mechanical Systems and Signal Processing*, *184*, 109713. doi:10.1016/j.ymssp.2022.109713

Liang, Y., Wu, D., Liu, G., Li, Y., Gao, C., Ma, Z. J., & Wu, W. (2016). Big data-enabled multiscale serviceability analysis for aging bridges☆. *Digital Communications and Networks*, *2*(3), 97–107. doi:10.1016/j.dcan.2016.05.002

Lin, Z., Pan, H., Gui, G., & Yan, C. (2018). Data-driven structural diagnosis and conditional assessment: from shallow to deep learning. In H. Sohn (Ed.), *Sensors and Smart Structures Technologies for Civil, Mechanical, and Aerospace Systems 2018* (p. 38). SPIE. doi:10.1117/12.2296964

Liu, G., Li, L., Zhang, L., Li, Q., & Law, S. S. (2020). Sensor faults classification for SHM systems using deep learning-based method with Tsfresh features. *Smart Materials and Structures*, *29*(7), 075005. doi:10.1088/1361-665X/ab85a6

Liu, J., Chen, S., Bergés, M., Bielak, J., Garrett, J. H., Kovačević, J., & Noh, H. Y. (2020). Diagnosis algorithms for indirect structural health monitoring of a bridge model via dimensionality reduction. *Mechanical Systems and Signal Processing*, *136*, 106454. doi:10.1016/j.ymssp.2019.106454

Liu, T., Xu, H., Ragulskis, M., Cao, M., & Ostachowicz, W. (2020). A Data-Driven Damage Identification Framework Based on Transmissibility Function Datasets and One-Dimensional Convolutional Neural Networks: Verification on a Structural Health Monitoring Benchmark Structure. *Sensors (Basel)*, *20*(4), 1059. doi:10.339020041059 PMID:32075311

López-Castro, B., Haro-Baez, A. G., Arcos-Aviles, D., Barreno-Riera, M., & Landázuri-Avilés, B. (2022). A Systematic Review of Structural Health Monitoring Systems to Strengthen Post-Earthquake Assessment Procedures. *Sensors (Basel)*, *22*(23), 9206. doi:10.339022239206 PMID:36501906

Lopez-Pacheco, M., Morales-Valdez, J., & Yu, W. (2020). Frequency domain CNN and dissipated energy approach for damage detection in building structures. *Soft Computing*, *24*(20), 15821–15840. doi:10.100700500-020-04912-w

Lorusso, A., & Celenta, G. (2023a). *Internet of Things in the Construction Industry: A General Overview*. doi:10.1007/978-3-031-31066-9_65

Lorusso, A., & Celenta, G. (2023b). *Structural Dynamics of Steel Frames with the Application of Friction Isolators*. doi:10.1007/978-3-031-34721-4_28

Lorusso, A., & Guida, D. (2022). *IoT System for Structural Monitoring*. doi:10.1007/978-3-031-05230-9_72

Lorusso, A., Messina, B., & Santaniello, D. (2023). *The Use of Generative Adversarial Network as Graphical Support for Historical Urban Renovation*. doi:10.1007/978-3-031-13588-0_64

Lu, Y., Tang, L., Chen, C., Zhou, L., Liu, Z., Liu, Y., Jiang, Z., & Yang, B. (2023). Reconstruction of structural long-term acceleration response based on BiLSTM networks. *Engineering Structures*, *285*, 116000. doi:10.1016/j.engstruct.2023.116000

Luleci, F., Necati Catbas, F., & Avci, O. (2023). CycleGAN for undamaged-to-damaged domain translation for structural health monitoring and damage detection. *Mechanical Systems and Signal Processing*, *197*, 110370. doi:10.1016/j.ymssp.2023.110370

Ma, G., & Du, Q. (2020). Optimization on the intellectual monitoring system for structures based on acoustic emission and data mining. *Measurement*, *163*, 107937. doi:10.1016/j.measurement.2020.107937

Mao, J., Su, X., Wang, H., & Li, J. (2023). Automated Bayesian operational modal analysis of the long-span bridge using machine-learning algorithms. *Engineering Structures*, *289*, 116336. doi:10.1016/j.engstruct.2023.116336

Mao, J., Wang, H., & Spencer, B. F. Jr. (2021). Toward data anomaly detection for automated structural health monitoring: Exploiting generative adversarial nets and autoencoders. *Structural Health Monitoring*, *20*(4), 1609–1626. doi:10.1177/1475921720924601

Martens, J., Blut, T., & Blankenbach, J. (2023). Cross domain matching for semantic point cloud segmentation based on image segmentation and geometric reasoning. *Advanced Engineering Informatics*, *57*, 102076. doi:10.1016/j.aei.2023.102076

Mei, H., Haider, M., Joseph, R., Migot, A., & Giurgiutiu, V. (2019). Recent Advances in Piezoelectric Wafer Active Sensors for Structural Health Monitoring Applications. *Sensors (Basel)*, *19*(2), 383. doi:10.339019020383 PMID:30669307

Merenda, P., & Fedele, C. (2019). A Real-Time Decision Platform for the Management of Structures and Infrastructures. *Electronics (Basel)*, *8*(10), 1180. doi:10.3390/electronics8101180

Meribout, M., Mekid, S., Kharoua, N., & Khezzar, L. (2021). Online monitoring of structural materials integrity in process industry for I4.0: A focus on material loss through erosion and corrosion sensing. *Measurement*, *176*, 109110. doi:10.1016/j.measurement.2021.109110

Momeni, H., & Ebrahimkhanlou, A. (2021, November 1). Applications of High-Dimensional Data Analytics in Structural Health Monitoring and Non-Destructive Evaluation: Thermal Videos Processing Using Tensor-Based Analysis. *Volume 13: Safety Engineering, Risk, and Reliability Analysis; Research Posters*. doi:10.1115/IMECE2021-71878

Mousavi, A. A., Zhang, C., Masri, S. F., & Gholipour, G. (2022). Structural damage detection method based on the complete ensemble empirical mode decomposition with adaptive noise: A model steel truss bridge case study. *Structural Health Monitoring*, *21*(3), 887–912. doi:10.1177/14759217211013535

Mousavi, M., & Gandomi, A. H. (2021). Structural health monitoring under environmental and operational variations using MCD prediction error. *Journal of Sound and Vibration*, *512*, 116370. doi:10.1016/j.jsv.2021.116370

Nasir, V., Ayanleye, S., Kazemirad, S., Sassani, F., & Adamopoulos, S. (2022). Acoustic emission monitoring of wood materials and timber structures: A critical review. *Construction & Building Materials*, *350*, 128877. doi:10.1016/j.conbuildmat.2022.128877

Ochôa, P. A., Groves, R. M., & Benedictus, R. (2020). Effects of high-amplitude low-frequency structural vibrations and machinery sound waves on ultrasonic guided wave propagation for health monitoring of composite aircraft primary structures. *Journal of Sound and Vibration*, *475*, 115289. doi:10.1016/j.jsv.2020.115289

Oh, B. K., Glisic, B., Kim, Y., & Park, H. S. (2019). Convolutional neural network-based wind-induced response estimation model for tall buildings. *Computer-Aided Civil and Infrastructure Engineering*, *34*(10), 843–858. doi:10.1111/mice.12476

Oliveira, S., & Alegre, A. (2020). Seismic and structural health monitoring of Cabril dam. Software development for informed management. *Journal of Civil Structural Health Monitoring*, *10*(5), 913–925. doi:10.100713349-020-00425-0

Padmapoorani, P., Senthilkumar, S., & Mohanraj, R. (2023). Machine Learning Techniques for Structural Health Monitoring of Concrete Structures: A Systematic Review. *Civil Engineering (Shiraz)*, *47*(4), 1919–1931. doi:10.100740996-023-01054-5

Parida, L., Moharana, S., Ferreira, V. M., Giri, S. K., & Ascensão, G. (2022). A Novel CNN-LSTM Hybrid Model for Prediction of Electro-Mechanical Impedance Signal Based Bond Strength Monitoring. *Sensors (Basel)*, *22*(24), 9920. doi:10.339022249920 PMID:36560293

Pawar, A., Jolly, A., Pandey, V., Chaurasiya, P. K., Verma, T. N., & Meshram, K. (2023). Artificial intelligence algorithms for prediction of cyclic stress ratio of soil for environment conservation. *Environmental Challenges*, *12*, 100730. doi:10.1016/j.envc.2023.100730

Rafiei, M. H., & Adeli, H. (2017). A novel machine learning-based algorithm to detect damage in high-rise building structures. *Structural Design of Tall and Special Buildings*, *26*(18), e1400. doi:10.1002/tal.1400

Rastin, Z., Ghodrati Amiri, G., & Darvishan, E. (2021). Unsupervised Structural Damage Detection Technique Based on a Deep Convolutional Autoencoder. *Shock and Vibration*, *2021*, 1–11. doi:10.1155/2021/6658575

Rather, A. I., Mirgal, P., Banerjee, S., & Laskar, A. (2023). Application of Acoustic Emission as Damage Assessment Technique for Performance Evaluation of Concrete Structures: A Review. *Practice Periodical on Structural Design and Construction*, *28*(3), 03123003. Advance online publication. doi:10.1061/PPSCFX.SCENG-1256

Ren, Q., Li, H., Li, M., Kong, T., & Guo, R. (2023). Bayesian incremental learning paradigm for online monitoring of dam behavior considering global uncertainty. *Applied Soft Computing*, *143*, 110411. doi:10.1016/j.asoc.2023.110411

Ren, Y., Ye, Q., Xu, X., Huang, Q., Fan, Z., Li, C., & Chang, W. (2022). An anomaly pattern detection for bridge structural response considering time-varying temperature coefficients. *Structures*, *46*, 285–298. doi:10.1016/j.istruc.2022.10.020

Rogers, T. J., Worden, K., Fuentes, R., Dervilis, N., Tygesen, U. T., & Cross, E. J. (2019). A Bayesian non-parametric clustering approach for semi-supervised Structural Health Monitoring. *Mechanical Systems and Signal Processing*, *119*, 100–119. doi:10.1016/j.ymssp.2018.09.013

Rosafalco, L., Manzoni, A., Mariani, S., & Corigliano, A. (2020). Fully convolutional networks for structural health monitoring through multivariate time series classification. *Advanced Modeling and Simulation in Engineering Sciences*, *7*(1), 38. doi:10.118640323-020-00174-1

Sajedi, S., & Liang, X. (2022). Deep generative Bayesian optimization for sensor placement in structural health monitoring. *Computer-Aided Civil and Infrastructure Engineering*, *37*(9), 1109–1127. doi:10.1111/mice.12799

Salehi, H., Gorodetsky, A., Solhmirzaei, R., & Jiao, P. (2023). High-dimensional data analytics in civil engineering: A review on matrix and tensor decomposition. *Engineering Applications of Artificial Intelligence*, *125*, 106659. doi:10.1016/j.engappai.2023.106659

Sarah, J., Hejazi, F., Rashid, R. S. M., & Ostovar, N. (2019). A Review of Dynamic Analysis in Frequency Domain for Structural Health Monitoring. *IOP Conference Series. Earth and Environmental Science*, *357*(1), 012007. doi:10.1088/1755-1315/357/1/012007

Sarmadi, H., Entezami, A., Saeedi Razavi, B., & Yuen, K. (2021). Ensemble learning-based structural health monitoring by Mahalanobis distance metrics. *Structural Control and Health Monitoring*, *28*(2). Advance online publication. doi:10.1002tc.2663

Sarmadi, H., & Karamodin, A. (2020). A novel anomaly detection method based on adaptive Mahalanobis-squared distance and one-class kNN rule for structural health monitoring under environmental effects. *Mechanical Systems and Signal Processing*, *140*, 106495. doi:10.1016/j.ymssp.2019.106495

Sarmadi, H., & Yuen, K.-V. (2022). Structural health monitoring by a novel probabilistic machine learning method based on extreme value theory and mixture quantile modeling. *Mechanical Systems and Signal Processing*, *173*, 109049. doi:10.1016/j.ymssp.2022.109049

Seon Park, H., Hwan An, J., Jun Park, Y., & Kwan Oh, B. (2020). Convolutional neural network-based safety evaluation method for structures with dynamic responses. *Expert Systems with Applications*, *158*, 113634. doi:10.1016/j.eswa.2020.113634

Seventekidis, P., Giagopoulos, D., Arailopoulos, A., & Markogiannaki, O. (2020). Structural Health Monitoring using deep learning with optimal finite element model generated data. *Mechanical Systems and Signal Processing*, *145*, 106972. doi:10.1016/j.ymssp.2020.106972

Shang, Z., Sun, L., Xia, Y., & Zhang, W. (2021). Vibration-based damage detection for bridges by deep convolutional denoising autoencoder. *Structural Health Monitoring*, *20*(4), 1880–1903. doi:10.1177/1475921720942836

Smarsly, K., & Law, K. H. (2014). Decentralized fault detection and isolation in wireless structural health monitoring systems using analytical redundancy. *Advances in Engineering Software*, *73*, 1–10. doi:10.1016/j.advengsoft.2014.02.005

Suh, G., & Cha, Y.-J. (2018). Deep faster R-CNN-based automated detection and localization of multiple types of damage. In H. Sohn (Ed.), *Sensors and Smart Structures Technologies for Civil, Mechanical, and Aerospace Systems 2018* (p. 27). SPIE., doi:10.1117/12.2295954

Svendsen, B. T., Frøseth, G. T., Øiseth, O., & Rønnquist, A. (2022). A data-based structural health monitoring approach for damage detection in steel bridges using experimental data. *Journal of Civil Structural Health Monitoring*, *12*(1), 101–115. doi:10.100713349-021-00530-8

Tan, X., Sun, X., Chen, W., Du, B., Ye, J., & Sun, L. (2021). Investigation on the data augmentation using machine learning algorithms in structural health monitoring information. *Structural Health Monitoring*, *20*(4), 2054–2068. doi:10.1177/1475921721996238

Tang, Z., Chen, Z., Bao, Y., & Li, H. (2019). Convolutional neural network-based data anomaly detection method using multiple information for structural health monitoring. *Structural Control and Health Monitoring*, *26*(1), e2296. doi:10.1002tc.2296

Tibaduiza Burgos, D. A., Gomez Vargas, R. C., Pedraza, C., Agis, D., & Pozo, F. (2020). Damage Identification in Structural Health Monitoring: A Brief Review from its Implementation to the Use of Data-Driven Applications. *Sensors (Basel)*, *20*(3), 733. doi:10.339020030733 PMID:32013073

Tomasiello, S., Loia, V., & Khaliq, A. (2021). A granular recurrent neural network for multiple time series prediction. *Neural Computing & Applications*, *33*(16), 10293–10310. doi:10.100700521-021-05791-4

Tomasiello, S., Pedrycz, W., & Loia, V. (2022). On Fractional Tikhonov Regularization: Application to the Adaptive Network-Based Fuzzy Inference System for Regression Problems. *IEEE Transactions on Fuzzy Systems*, *30*(11), 4717–4727. doi:10.1109/TFUZZ.2022.3157947

Tran, N. C., Wang, J., Vu, T. H., Tai, T.-C., & Wang, J.-C. (2023). Anti-aliasing convolution neural network of finger vein recognition for virtual reality (VR) human–robot equipment of metaverse. *The Journal of Supercomputing*, *79*(3), 2767–2782. doi:10.100711227-022-04680-4 PMID:36035635

Turo, T., Neumann, V., & Krobot, Z. (2019). Health and Usage Monitoring System Assestment. *2019 International Conference on Military Technologies (ICMT)*, 1–4. 10.1109/MILTECHS.2019.8870072

Wang, D., Zhang, Y., Pan, Y., Peng, B., Liu, H., & Ma, R. (2020). An Automated Inspection Method for the Steel Box Girder Bottom of Long-Span Bridges Based on Deep Learning. *IEEE Access : Practical Innovations, Open Solutions*, *8*, 94010–94023. doi:10.1109/ACCESS.2020.2994275

Wang, H., Barone, G., & Smith, A. (2023). A novel multi-level data fusion and anomaly detection approach for infrastructure damage identification and localisation. *Engineering Structures*, *292*, 116473. doi:10.1016/j.engstruct.2023.116473

Wang, M., He, M., Liang, Z., Wu, D., Wang, Y., Qing, X., & Wang, Y. (2023). Fatigue damage monitoring of composite laminates based on acoustic emission and digital image correlation techniques. *Composite Structures*, *321*, 117239. doi:10.1016/j.compstruct.2023.117239

Wang, R., Li, L., & Li, J. (2018). A Novel Parallel Auto-Encoder Framework for Multi-Scale Data in Civil Structural Health Monitoring. *Algorithms*, *11*(8), 112. doi:10.3390/a11080112

Wang, X., Cai, J., & Zhou, Z. (2019). A Lamb wave signal reconstruction method for high-resolution damage imaging. *Chinese Journal of Aeronautics*, *32*(5), 1087–1099. doi:10.1016/j.cja.2019.03.001

Wang, Z., Bai, L., Zhang, Y., Zhao, K., Wu, J., & Fu, W. (2022). Spatial variation, sources identification and risk assessment of soil heavy metals in a typical Torreya grandis cv. Merrillii plantation region of southeastern China. *The Science of the Total Environment*, *849*, 157832. doi:10.1016/j.scitotenv.2022.157832 PMID:35932857

Wang, Z., & Cha, Y.-J. (2021). Unsupervised deep learning approach using a deep auto-encoder with a one-class support vector machine to detect damage. *Structural Health Monitoring*, *20*(1), 406–425. doi:10.1177/1475921720934051

Wang, Z., Yi, T.-H., Yang, D.-H., Li, H.-N., & Liu, H. (2022). Bridge Performance Warning Based on Two-Stage Elimination of Environment-Induced Frequency. *Journal of Performance of Constructed Facilities*, *36*(6), 04022056. Advance online publication. doi:10.1061/(ASCE)CF.1943-5509.0001760

Wiemann, M., Bonekemper, L., & Kraemer, P. (2020). Methods to enhance the automation of operational modal analysis. *Vibroengineering PROCEDIA*, *31*, 46–51. doi:10.21595/vp.2020.21443

Xiong, Q., Yuan, C., He, B., Xiong, H., & Kong, Q. (2023). GTRF: A general deep learning framework for tuples recognition towards supervised, semi-supervised and unsupervised paradigms. *Engineering Applications of Artificial Intelligence*, *124*, 106500. doi:10.1016/j.engappai.2023.106500

Yang, K., Kim, S., & Harley, J. B. (2023). Guidelines for effective unsupervised guided wave compression and denoising in long-term guided wave structural health monitoring. *Structural Health Monitoring*, *22*(4), 2516–2530. doi:10.1177/14759217221124689

Yang, Z., Yang, H., Tian, T., Deng, D., Hu, M., Ma, J., Gao, D., Zhang, J., Ma, S., Yang, L., Xu, H., & Wu, Z. (2023). A review on guided-ultrasonic-wave-based structural health monitoring: From fundamental theory to machine learning techniques. *Ultrasonics*, *133*, 107014. doi:10.1016/j.ultras.2023.107014 PMID:37178485

Yeter, B., Garbatov, Y., & Soares, C. G. (2022). Review on Artificial Intelligence-aided Life Extension Assessment of Offshore Wind Support Structures. *Journal of Marine Science and Application*, *21*(4), 26–54. doi:10.100711804-022-00298-3

Yuan, F.-G., Zargar, S. A., Chen, Q., & Wang, S. (2020). Machine learning for structural health monitoring: challenges and opportunities. In D. Zonta, H. Sohn, & H. Huang (Eds.), *Sensors and Smart Structures Technologies for Civil, Mechanical, and Aerospace Systems 2020* (p. 2). SPIE. doi:10.1117/12.2561610

Zaparoli Cunha, B., Droz, C., Zine, A.-M., Foulard, S., & Ichchou, M. (2023). A review of machine learning methods applied to structural dynamics and vibroacoustic. *Mechanical Systems and Signal Processing*, *200*, 110535. doi:10.1016/j.ymssp.2023.110535

Zhang, B., Zhang, S., & Li, W. (2019). Bearing performance degradation assessment using long short-term memory recurrent network. *Computers in Industry*, *106*, 14–29. doi:10.1016/j.compind.2018.12.016

Zhang, C., Mousavi, A. A., Masri, S. F., Gholipour, G., Yan, K., & Li, X. (2022). Vibration feature extraction using signal processing techniques for structural health monitoring: A review. *Mechanical Systems and Signal Processing*, *177*, 109175. doi:10.1016/j.ymssp.2022.109175

Zhang, D., Zhu, A., Hou, W., Liu, L., & Wang, Y. (2022). Vision-Based Structural Modal Identification Using Hybrid Motion Magnification. *Sensors (Basel)*, *22*(23), 9287. doi:10.339022239287 PMID:36501990

Zhang, F.-L., Gu, D.-K., Li, X., Ye, X.-W., & Peng, H. (2023). Structural damage detection based on fundamental Bayesian two-stage model considering the modal parameters uncertainty. *Structural Health Monitoring*, *22*(4), 2305–2324. doi:10.1177/14759217221114262

Zhang, L., Zhou, G., Han, Y., Lin, H., & Wu, Y. (2018). Application of Internet of Things Technology and Convolutional Neural Network Model in Bridge Crack Detection. *IEEE Access : Practical Innovations, Open Solutions*, *6*, 39442–39451. doi:10.1109/ACCESS.2018.2855144

Zhang, Y.-M., Wang, H., Bai, Y., Mao, J.-X., & Xu, Y.-C. (2022). Bayesian dynamic regression for reconstructing missing data in structural health monitoring. *Structural Health Monitoring*, *21*(5), 2097–2115. doi:10.1177/14759217211053779

Zhang, Y.-M., Wang, H., Wan, H.-P., Mao, J.-X., & Xu, Y.-C. (2021). Anomaly detection of structural health monitoring data using the maximum likelihood estimation-based Bayesian dynamic linear model. *Structural Health Monitoring*, *20*(6), 2936–2952. doi:10.1177/1475921720977020

Zhao, L., Huang, X., Zhang, Y., Tian, Y., & Zhao, Y. (2019). A Vibration-Based Structural Health Monitoring System for Transmission Line Towers. *Electronics (Basel)*, *8*(5), 515. doi:10.3390/electronics8050515

Zhao, X., & Jia, M. (2020). A novel unsupervised deep learning network for intelligent fault diagnosis of rotating machinery. *Structural Health Monitoring*, *19*(6), 1745–1763. doi:10.1177/1475921719897317

Zhi, L., Hu, F., Li, Q., & Hu, Z. (2021). Identification of modal parameters from non-stationary responses of high-rise buildings. *Advances in Structural Engineering*, *24*(15), 3519–3533. doi:10.1177/13694332211033959

Zhu, Y.-C., Xiong, W., & Song, X.-D. (2022). Structural performance assessment considering both observed and latent environmental and operational conditions: A Gaussian process and probability principal component analysis method. *Structural Health Monitoring*, *21*(6), 2531–2546. doi:10.1177/14759217211062099

Chapter 14
Convergence of Data Science–AI–Green Chemistry–Affordable Medicine:
Transforming Drug Discovery

B. Rebecca
Department of Computer Science and Engineering (Data Science), Marri Laxman Reddy Institute of Technology and Management, Hyderabad, India

K. Pradeep Mohan Kumar
Department of Computing Technologies, SRM Institute of Science and Technology, Chennai, India

S. Padmini
https://orcid.org/0000-0001-8133-3453
Department of Computing Technologies, SRM Institute of Science and Technology, Chennai, India

Bipin Kumar Srivastava
Department of Applied Sciences, Galgotias College of Engineering and Technology, India

Shubhajit Halder
https://orcid.org/0000-0001-7958-4667
Department of Chemistry, Hislop College, Nagpur, India

Sampath Boopathi
https://orcid.org/0000-0002-2065-6539
Mechanical Engineering, Muthayammal Engineering College, India

ABSTRACT

The drug discovery and design process has been significantly transformed by the integration of data science, artificial intelligence (AI), green chemistry principles, and affordable medicine. AI techniques enable rapid analysis of vast datasets, predicting molecular interactions, optimizing drug candidates, and identifying potential therapeutics. Green chemistry practices promote sustainability and efficiency, resulting in environmentally friendly and cost-effective production processes. The goal is to develop affordable medicines that are not only efficacious but also accessible to a wider population. This chapter explores case studies and emerging trends to highlight the transformation of the pharmaceutical industry and innovation in drug discovery.

DOI: 10.4018/978-1-6684-9999-3.ch014

INTRODUCTION

The field of drug discovery and design has embarked on a remarkable journey of transformation, catalyzed by the powerful convergence of data science, artificial intelligence (AI), green chemistry principles, and the overarching objective of providing affordable medicine to a broader populace. This chapter embarks on an exploration of the synergistic interplay among these factors and their profound impact on the landscape of pharmaceutical research and development. In the contemporary era, scientific advancements have led to an explosion of molecular data, which, when harnessed effectively, can unlock a treasure trove of insights into disease mechanisms, molecular interactions, and potential therapeutic interventions (Blakemore et al., 2018). Enter data science and AI—tools that have proven indispensable in deciphering the intricate language of molecules. These technologies empower researchers to sift through vast datasets, predict molecular behaviors, and guide the design of novel drug candidates with unprecedented precision and efficiency.

The echoes of this revolution reverberate across the spectrum of drug discovery, from virtual screening to de novo drug design. Our journey will encompass the applications of data science and AI that have revolutionized lead optimization, compound selection, and even the prediction of a drug candidate's pharmacokinetic and toxicity profiles. Through illuminating case studies, we will showcase how these tools have not only accelerated the discovery process but have also influenced the economics of drug development (Glicksberg et al., 2019). As we delve deeper into the chapters that follow, the spotlight will shift to green chemistry—an ethos that echoes the growing societal call for sustainability and environmental stewardship. Green chemistry's principles resonate strongly with the pharmaceutical industry, where the optimization of production processes to minimize waste, reduce hazardous materials, and conserve energy aligns seamlessly with the broader goals of ethical pharmaceutical manufacturing (Arshad et al., 2016). We will explore how these principles have reshaped synthetic routes, solvent selections, and manufacturing strategies, resulting in not only ecologically friendly processes but also in cost savings that can trickle down to the end consumer.

However, the narrative doesn't end there. An underpinning theme woven throughout this exploration is the pursuit of affordable medicine. As the cost of healthcare continues to challenge global economies, the need for cost-effective drug discovery and development becomes more pressing. In tandem with the advancements in data science and green chemistry, the chapter will underscore the importance of these methodologies in reducing the economic barriers to healthcare access (Jensen et al., 2015). Through the intricate dance of data science, AI, green chemistry, and affordability considerations, the pharmaceutical landscape is undergoing a renaissance of innovation. This chapter's voyage through case studies, trends, and future prospects seeks to illuminate the interwoven tapestry of these elements, offering a glimpse into a world where cutting-edge science converges with ethical responsibility to transform drug discovery, production, and distribution.

The pharmaceutical industry stands on the threshold of a new era—one marked by the seamless integration of data science, artificial intelligence (AI), green chemistry, and the quest for affordable medicine. The symbiotic relationship among these elements holds the potential to reshape drug discovery, addressing critical challenges faced by both researchers and patients alike (Bountra et al., 2017). Traditionally, drug discovery has been a lengthy and resource-intensive process. The identification of potential drug candidates often involved a trial-and-error approach, leading to high attrition rates and exorbitant costs. Enter data science and AI—the dynamic duo that has revolutionized the field. The exponential growth of molecular data, coupled with computational prowess, has unleashed the ability to analyze intricate molecular interactions, predict compound behaviors, and navigate the vast chemical space with unparalleled speed. This transformation has not only accelerated the early stages of drug discovery but has also paved the way for rational design, minimizing costly dead-ends and streamlining the development pipeline.

In this context, the significance of data science and AI cannot be overstated. By leveraging machine learning algorithms, researchers can discern hidden patterns within datasets that human intuition might miss. These insights translate into the identification of novel targets, the optimization of lead compounds, and the anticipation of potential side effects—a holistic approach that not only accelerates drug development but also mitigates risks and reduces financial burdens (Yang et al., 2022). Green chemistry enters the stage as a natural partner in this transformation. As public awareness of environmental issues grows, the pharmaceutical industry faces increasing pressure to adopt sustainable practices. Green chemistry principles emphasize the design of efficient and eco-friendly processes, minimizing waste, energy consumption, and the use of hazardous materials. Beyond aligning with global sustainability goals, the implementation of green chemistry practices directly influences cost-effectiveness. By reducing resource consumption and waste generation, green chemistry fosters economically viable and ecologically responsible drug production (Moingeon et al., 2022).

Yet, the convergence doesn't end with streamlined discovery and sustainable manufacturing—it extends to the heart of healthcare accessibility. The escalating costs of medicines have become a societal concern, particularly when considering life-saving treatments. This chapter underscores the imperative of affordable medicine, examining how the integration of data science, AI, and green chemistry can lead to cost-efficient drug development and, consequently, more accessible healthcare solutions (Haghi et al., 2022). As data-driven optimizations minimize development expenses and green chemistry practices reduce production costs, the cumulative effect can contribute to the democratization of healthcare access—a shared aspiration on a global scale. The narrative presented in this chapter resonates deeply with the challenges faced by the pharmaceutical industry, researchers, healthcare providers, and patients. By delving into the intricate web of data science, AI, green chemistry, and affordability, this exploration aims to illuminate the transformative potential of their convergence. Ultimately, this convergence not only drives scientific progress but also underscores the industry's commitment to ethical responsibility, sustainable practices, and equitable healthcare access (Stevens et al., 2020).

The integration of data science, artificial intelligence (AI), green chemistry principles, and the pursuit of affordable medicine has emerged as a dynamic and transformative research frontier within the realm of drug discovery. This interdisciplinary approach brings together diverse expertise to address the multifaceted challenges of creating effective, sustainable, and accessible pharmaceutical solutions.

Research efforts have extensively explored the application of data science and AI in various stages of drug discovery. Machine learning algorithms trained on vast datasets of molecular properties and interactions aid in the identification of potential drug candidates. These algorithms can predict the binding affinity, pharmacokinetics, and toxicity of compounds, enabling researchers to focus on those with the highest likelihood of success. This approach expedites lead optimization and minimizes the attrition rates commonly associated with costly late-stage failures (Vert, 2023).

Data-driven virtual screening has revolutionized compound selection. By leveraging AI models that predict the activity of molecules against specific targets, researchers can rapidly sift through vast compound libraries. Moreover, AI-driven de novo drug design generates novel molecular structures with desired properties, effectively expanding the chemical space accessible to researchers. This amalgamation of data science and AI empowers scientists to identify novel drug candidates with a higher probability of success (Hema et al., 2023; Myilsamy et al., 2021; Selvakumar et al., 2023; Subha et al., 2023a).

Green chemistry principles have been incorporated into drug discovery to address the environmental impact of pharmaceutical production. Researchers are exploring novel synthetic routes that minimize waste, utilize renewable resources, and reduce energy consumption. These eco-friendly approaches not only align with sustainability goals but also contribute to cost savings by optimizing processes and minimizing the need for expensive reagents (Peña-Guerrero et al., 2021).

One of the driving forces behind the integration of these elements is the goal of making medicines more affordable. Data science and AI-driven optimizations expedite drug discovery timelines, reducing the overall cost of development. Green chemistry practices, in addition to being environmentally responsible, contribute to lower production costs. The synergy of these efforts has the potential to drive down the final cost of medications, making them more accessible to a wider population. Researchers and institutions worldwide are collaborating to harness the potential of this convergence. Open-source datasets, machine learning tools, and collaborative platforms are emerging as cornerstones of this effort. International organizations, governments, and pharmaceutical companies are recognizing the importance of accessible healthcare and are investing in initiatives that prioritize affordable medicine through innovation (Ferrero et al., 2020; Workman et al., 2019). As this integrated approach gains momentum, researchers are also exploring the ethical implications and societal impact. Balancing the benefits of rapid drug discovery with patient safety, ensuring equitable distribution, and addressing potential biases in AI algorithms are among the critical considerations.

Thus, research at the intersection of data science, AI, green chemistry, and affordable medicine in drug discovery is paving the way for a new era of pharmaceutical innovation. By leveraging the strengths of each field, researchers are streamlining drug development, promoting sustainability, and addressing the imperative of healthcare accessibility. As ongoing studies continue to unravel the potential of this convergence, the prospect of more effective, affordable, and sustainable medications comes into sharper focus.

DATA SCIENCE AND AI IN DRUG DISCOVERY

In the modern landscape of drug discovery, the marriage of data science and artificial intelligence (AI) has revolutionized the way researchers approach the identification and development of potential drug candidates. This dynamic duo's ability to analyze vast quantities of molecular data, predict intricate molecular interactions, optimize compound properties, and expedite the drug discovery process has reshaped the field. Here, we delve into the integral role that data science and AI play in transforming drug discovery, focusing on their applications in analyzing molecular data, predictive modeling, virtual screening, and de novo drug design (Figure 1).

Analysing Molecular Data

In the realm of modern drug discovery, the transformational power of data science emerges as a beacon of progress. At the core of this transformation lies the ability to unravel the intricate tapestry of molecular data—genomic, proteomic, and chemical—once daunting in its complexity. This convergence of data science and molecular biology enables researchers to distill this wealth of information into insights that drive scientific breakthroughs and pave the way for novel therapeutic interventions (Burki, 2019).

Data Pre-Processing

The journey commences with data pre-processing a meticulous process of cleaning and organizing raw data to ensure its accuracy and reliability. Inaccurate or noisy data can lead to misleading conclusions, making pre-processing a crucial step. Data science techniques come to the fore, automating tasks such as removing duplicates, handling missing values, and correcting errors. Through data pre-processing, the quality of the dataset is elevated, setting the stage for meaningful analysis (Peck et al., 2020).

Dimensionality Reduction

Molecular datasets often manifest as high-dimensional spaces, where each data point corresponds to numerous variables. Dimensionality reduction techniques are the vanguard against the curse of dimensionality an issue where data becomes sparse and unwieldy in high-dimensional spaces. Techniques like Principal Component Analysis (PCA) and t-distributed Stochastic Neighbor Embedding (t-SNE) reduce the dataset's dimensions while preserving its salient features. This reduction not only simplifies visualization but also aids in identifying critical patterns that might otherwise remain obscured (Savage, 2021).

Figure 1. Data science and AI in drug discovery

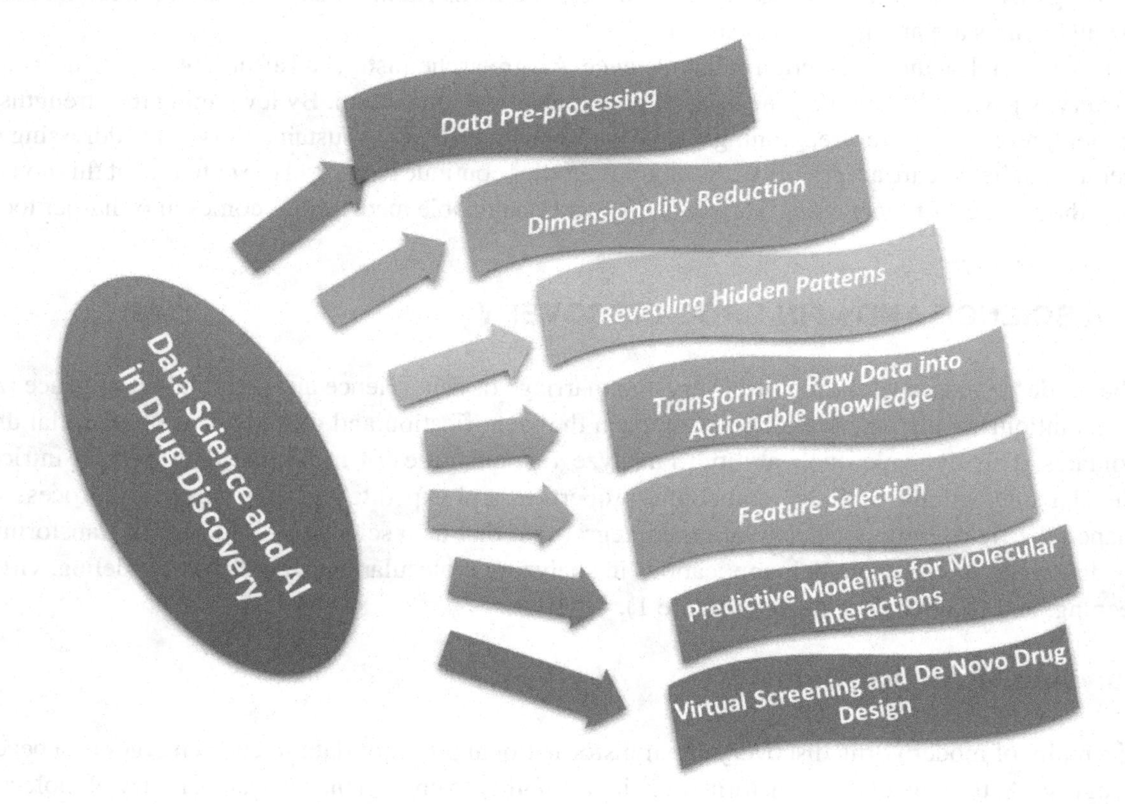

Feature Selection

Feature selection is the art of identifying the most relevant variables that contribute to a specific outcome. In molecular data, not all features are equally influential in determining a drug's efficacy or interactions. Data science algorithms sift through the variables, highlighting those with the most predictive power. This focused approach streamlines subsequent analyses and improves the interpretability of the results (Boopathi et al., 2023; Boopathi, Pandey, et al., 2023a; Mohanty et al., 2023a; Venkateswaran et al., 2023a).

Revealing Hidden Patterns and Correlations

As the data undergoes preprocessing, dimensionality reduction, and feature selection, hidden patterns and correlations begin to emerge. Data points that were previously indistinguishable now exhibit distinct relationships. AI algorithms excel in identifying these intricate connections mapping how genes influence protein interactions or unveiling chemical structures that contribute to specific properties. These revelations form the bedrock of hypothesis generation, guiding researchers to experimentally validate newfound insights (Boopathi, Pandey, et al., 2023a; Koshariya, Kalaiyarasi, et al., 2023a; Koshariya, Khatoon, et al., 2023a; Veeranjaneyulu et al., 2023).

Transforming Raw Data Into Actionable Knowledge

The true triumph of data-driven drug discovery lies in its ability to transform raw data into actionable knowledge. AI algorithms, armed with a refined dataset, traverse the landscape of possibilities. They uncover links between genetic mutations and disease predisposition, elucidate signaling pathways that govern cellular responses, and predict the likelihood of a compound binding to its target. This knowledge guides researchers' decisions, making the drug discovery process more focused, efficient, and fruitful.

Thus, the synergy of data science and molecular data has redefined the boundaries of drug discovery. Through data preprocessing, dimensionality reduction, feature selection, and AI-driven analysis, the once overwhelming complexity of molecular datasets is distilled into insights that shape our understanding of diseases and therapeutic interventions. This integration empowers researchers to navigate the intricate molecular landscape, unlocking the secrets hidden within and advancing the frontiers of medicine (Boopathi et al., 2023, 2023; Domakonda et al., 2022).

Predictive Modeling for Molecular Interactions and Compound Properties

Predictive modeling is at the heart of data science and AI's impact on drug discovery. Machine learning algorithms are trained on diverse datasets that include molecular structures, binding affinities, and biological activities. Once trained, these models can predict how a molecule will interact with a target protein, estimate its pharmacokinetic properties, and even forecast potential toxicity. These predictions guide researchers in selecting compounds with the greatest likelihood of success, significantly reducing the hit-to-lead time (Becker et al., 2022).

Virtual Screening and De Novo Drug Design

Virtual screening, powered by AI, has transformed the way researchers evaluate compounds for their potential as drug candidates. In silico screening rapidly assesses large libraries of compounds against target proteins, identifying those with the highest binding affinities. AI-driven de novo drug design pushes the envelope further by generating novel molecular structures that possess desired properties. These AI-generated molecules open new avenues for creativity, enabling the exploration of chemical space that might have been overlooked using traditional methods (Pramila et al., 2023; Ramudu et al., 2023).

Case Studies: Successful Applications

Numerous case studies underscore the efficacy of data science and AI in drug discovery. For instance, Atomwise's AI-driven approach identified existing drugs that could potentially be repurposed for Ebola treatment, saving time and resources. BenevolentAI's AI platform facilitated the discovery of a potential treatment for amyotrophic lateral sclerosis (ALS). In the realm of de novo drug design, Exscientia's AI-designed molecule for treating obsessive-compulsive disorder (OCD) entered clinical trials in record time. These examples highlight the transformative potential of data science and AI across various stages of drug discovery (Anitha et al., 2023a; Gunasekaran & Boopathi, 2023; Reddy, Gaurav, et al., 2023; Sampath et al., 2023; Sengeni et al., 2023a).

In summary, the integration of data science and AI is transforming drug discovery by leveraging the power of data analytics, predictive modeling, virtual screening, and de novo drug design. The ability to analyze molecular data comprehensively and make informed predictions accelerates the identification of promising drug candidates. As case studies continue to showcase successful applications, the pharmaceutical landscape is poised for continued innovation, efficiency, and the discovery of novel treatments that hold the promise of improved patient outcomes.

GREEN CHEMISTRY PRINCIPLES IN PHARMACEUTICAL PRODUCTION

In an era of heightened environmental awareness, the pharmaceutical industry stands poised to embrace a paradigm shift toward sustainability through the integration of green chemistry principles. These principles, designed to minimize the ecological footprint of chemical processes, have found a crucial application in drug manufacturing. This section delves into the essence of green chemistry, its relevance to pharmaceutical production, sustainable synthesis routes, waste reduction, and illustrative examples of greener production methods (Becker et al., 2022; Sharma et al., 2020).

Green Chemistry Principles

Green chemistry embodies a set of twelve principles that promote the design and execution of chemical processes with minimal environmental impact. These principles advocate for the reduction or elimination of hazardous substances, energy efficiency, and the utilization of renewable resources. They serve as guiding beacons for industries, including pharmaceuticals, to achieve sustainable practices while maintaining scientific rigor and economic viability.

Relevance to Drug Manufacturing

The pharmaceutical industry inherently involves chemical synthesis on various scales, often entailing complex reactions, multiple steps, and the use of diverse reagents. The integration of green chemistry principles is pivotal to reduce the generation of hazardous byproducts, minimize energy consumption, and mitigate the release of harmful chemicals into the environment. This alignment with sustainability dovetails seamlessly with societal expectations and regulatory pressures for more eco-friendly production (Dunn et al., 2010; Sharma et al., 2020).

Sustainable Synthesis Routes and Solvent Selection in Green Chemistry

In the pursuit of environmentally responsible drug manufacturing, green chemistry champions sustainable synthesis routes and prudent solvent selection. These practices embody a shift from conventional methods, advocating for minimized waste generation, reduced energy consumption, and diminished ecological impact (Figure 2). Catalytic processes and the adoption of green solvents stand at the forefront of this transformation, not only enhancing the ecological profile of drug production but also optimizing efficiency and cost-effectiveness (Gupta & Mahajan, 2015; Roschangar et al., 2015).

Minimizing Waste Through Catalytic Processes

Catalytic processes lie at the heart of sustainable synthesis routes. Unlike stoichiometric reactions that require large amounts of reagents, catalysts facilitate reactions with minimal waste generation. They promote the conversion of reactants into products more efficiently and selectively, often requiring lower reaction temperatures. This not only reduces waste but also conserves energy. Catalysts can be fine-tuned to yield specific products, thus enabling targeted synthesis and minimizing byproduct formation.

Choice of Green Solvents

The choice of solvents is equally instrumental in greening chemical processes. Green solvents, distinguished by their reduced impact on human health and the environment, replace traditional organic solvents that contribute to pollution and waste. Water, a universal solvent, is a prime example of a green solvent due to its low toxicity and abundant availability (Boopathi, 2021; Fowziya et al., 2023; Venkateswaran et al., 2023b). Supercritical fluids, such as supercritical carbon dioxide, offer tunable properties and can replace volatile organic solvents. Ionic liquids, liquid salts with unique properties, are gaining attention for their solvent-like behavior and minimal environmental impact (Dunn et al., 2010; Sharma et al., 2020).

Enhancing Ecological Profile and Efficiency

Adopting sustainable synthesis routes and green solvents synergistically enhances the ecological profile of drug manufacturing. Reduced waste generation and minimized use of hazardous substances alleviate the burden on ecosystems and public health. Additionally, these practices often lead to energy savings due to lower reaction temperatures and shorter reaction times. As a result, the overall carbon footprint of the manufacturing process is reduced, aligning with sustainability goals (Boopathi et al., 2023; Boopathi & Davim, 2023; Boopathi & Kanike, 2023a; Sampath et al., 2023).

Figure 2. Sustainable synthesis routes and solvent selection in green chemistry

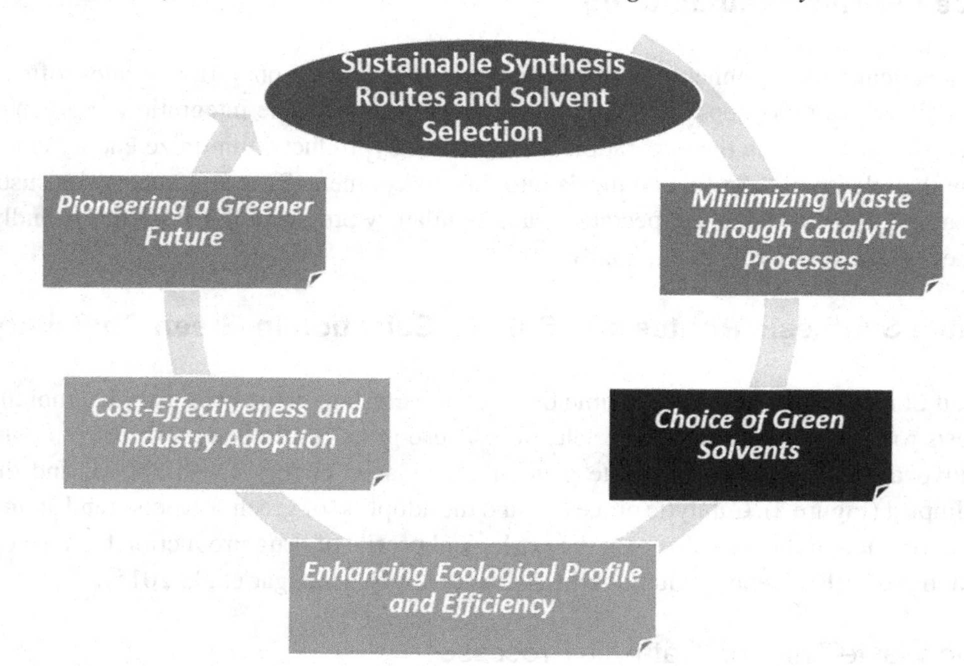

Cost-Effectiveness and Industry Adoption

The adoption of green chemistry practices in synthesis routes and solvent selection has implications for cost-effectiveness. While the initial investment in developing greener processes might be higher, the long-term benefits in terms of reduced waste disposal costs, energy savings, and compliance with regulatory requirements are substantial. Furthermore, the optimization of manufacturing processes through green practices can enhance operational efficiency and contribute to cost savings over the product's lifecycle (Hanumanthakari et al., 2023; Reddy, Gaurav, et al., 2023; Sengeni et al., 2023a).

Pioneering a Greener Future

The shift toward sustainable synthesis routes and green solvents is not just an industry trend but a transformative movement. Pharmaceutical companies are recognizing the strategic value of sustainable practices. Beyond the ecological benefits, these practices align with societal expectations for responsible production, enhance corporate social responsibility, and foster a positive public image.

Thus, sustainable synthesis routes and solvent selection underscore the transformative potential of green chemistry in drug manufacturing. By minimizing waste, conserving energy, and adopting green solvents, the pharmaceutical industry pioneers a greener and more efficient future (Kumara et al., 2023a; Nishanth et al., 2023; Vanitha et al., 2023a; Venkateswaran et al., 2023a). This shift not only aligns with environmental responsibility but also advances cost-effectiveness, operational efficiency, and industry reputation, ultimately reshaping the landscape of drug production in a more sustainable direction.

Minimizing Waste and Reducing Environmental Impact

Waste reduction is a cornerstone of green chemistry. By implementing methods such as atom economy and retrosynthetic analysis, chemists optimize reactions to maximize the incorporation of reactants into the final product, minimizing byproducts. This focus on efficiency inherently reduces waste. Additionally, greener production processes lead to lower emissions of greenhouse gases, toxic byproducts, and hazardous waste, further reducing the industry's environmental footprint (Judson et al., 2009).

Examples of Greener Production Methods

In the pursuit of greener pharmaceutical production, numerous examples highlight the application of green chemistry principles. Flow chemistry, characterized by continuous reactions in microreactors, minimizes reagent waste, enhances safety, and enables the use of hazardous reagents with greater control. Microwave-assisted reactions expedite reaction rates, saving energy and time. Additionally, the use of renewable feedstocks, such as biomass-derived chemicals, offers an alternative to petrochemical-based starting materials (BOOPATHI, 2022; Boopathi & Sivakumar, 2012; Dass james & Boopathi, 2016; Sampath et al., 2021).

Thus, green chemistry principles offer a transformative path to sustainable pharmaceutical manufacturing. The integration of green practices, such as sustainable synthesis routes, solvent selection, and waste reduction, aligns the pharmaceutical industry with environmental responsibility while maintaining economic viability. As the industry continues to evolve, the implementation of greener production methods becomes not just an ethical imperative but a strategic advantage in a world that values both innovation and sustainability.

CHALLENGES AND OPPORTUNITIES IN AFFORDABLE MEDICINE

In the pursuit of equitable healthcare, the challenges and opportunities surrounding the accessibility and affordability of medicines hold paramount significance. The intersection of socioeconomic factors, healthcare systems, and pharmaceutical industry dynamics creates a complex landscape that demands innovative solutions. This section delves into the pervasive issues of accessibility and affordability in healthcare and pharmaceuticals, emphasizes the importance of cost-effective drug development and production, and underscores the pivotal role that technology plays in addressing these challenges (Dunn et al., 2010; Gupta & Mahajan, 2015; Roschangar et al., 2015).

Accessibility and Affordability Issues in Healthcare and Pharmaceuticals

Global disparities in healthcare access remain a stark reality. Many individuals, particularly in low- and middle-income countries, face barriers to essential medications due to their high cost. Life-saving treatments can often be financially out of reach for those who need them the most. Additionally, even in developed nations, segments of the population struggle to afford necessary medications, leading to adverse health outcomes and strained healthcare systems. These issues underscore the urgency of finding solutions that ensure that essential medicines are accessible to all.

Importance of Cost-Effective Drug Development and Production

The cost of developing and bringing a new drug to market is substantial, encompassing years of research, clinical trials, and regulatory processes. These costs contribute significantly to the final price of medications. Cost-effective drug development and production are not only crucial for the pharmaceutical industry's sustainability but also for the well-being of patients. Reducing the financial burden of drug development and production can have a direct impact on the affordability of medications, making them more accessible to a wider population.

The Role of Technology in Addressing Affordability Challenges

Technology has emerged as a formidable ally in the quest for affordable medicine. Data science and artificial intelligence (AI) streamline drug discovery, shortening development timelines and reducing costs. Computational modeling, aided by AI, facilitates the prediction of compound properties, toxicity, and interactions, helping researchers select the most promising candidates efficiently. Furthermore, advancements in manufacturing technologies, such as continuous processing and 3D printing of pharmaceuticals, hold the promise of reducing production costs and waste. Telemedicine and digital health platforms extend healthcare access to remote and underserved areas, enhancing patient outcomes and reducing healthcare expenditure (Boopathi, 2022; Boopathi et al., 2023; Gowri et al., 2023; Janardhana et al., 2023; Mohanty et al., 2023b; Sampath, 2021).

Collaboration and Policy Initiatives

Addressing the challenges of affordability requires collaboration among governments, pharmaceutical companies, healthcare providers, and non-governmental organizations. Governments can implement pricing policies, patent regulations, and incentives that promote fair pricing and innovation. Collaborative initiatives, such as public-private partnerships, can facilitate the development of affordable treatments for neglected diseases. Open access to research and sharing of knowledge are integral to fostering innovation and reducing duplication of efforts.

Hence, the challenges of accessibility and affordability in healthcare and pharmaceuticals are multifaceted, deeply intertwined with societal, economic, and technological factors. While these challenges are significant, they also present opportunities for transformative change. The convergence of innovative technologies, policy initiatives, and collaborative efforts holds the potential to bridge the gap between the availability of essential medicines and the individuals who need them, ultimately realizing the vision of affordable healthcare for all.

INTEGRATION OF DATA SCIENCE, AI, AND GREEN CHEMISTRY FOR AFFORDABLE DRUG DEVELOPMENT

In the dynamic landscape of drug development, the convergence of data science, artificial intelligence (AI), and green chemistry principles emerges as a potent catalyst for innovation, affordability, and sustainability. This section delves into the symbiotic relationship between these fields, showcasing how data-driven insights optimize green chemistry practices, streamline drug discovery and development workflows, and provide tangible examples of the combined impact on cost and efficiency (Rogers & Jensen, 2019; Watson, 2012; Xie et al., 2020).

Leveraging Data-Driven Insights to Optimize Green Chemistry Practices

The integration of data science and AI brings a new dimension to green chemistry practices. Data-driven insights enable researchers to predict chemical reactions, assess reaction pathways, and evaluate the environmental impact of different synthetic routes. By analyzing the molecular interactions, predictive modeling can guide the selection of greener solvents, energy-efficient reactions, and catalytic processes. This optimized design reduces waste, minimizes hazardous byproducts, and enhances the efficiency of synthesis, directly aligning with green chemistry principles (Boopathi et al., 2023; Boopathi, Pandey, et al., 2023b; Domakonda et al., 2022; Kumara et al., 2023b; Rahamathunnisa et al., 2023; Samikannu et al., 2022; Vanitha et al., 2023b; Vennila et al., 2022).

Streamlining Drug Discovery and Development Workflows

The collaboration between data science, AI, and green chemistry enhances the entire drug discovery and development pipeline. Virtual screening, driven by AI algorithms, rapidly evaluates compound libraries for potential drug candidates. Simultaneously, green chemistry principles guide the selection of compounds with favorable environmental profiles. This tandem approach minimizes costly iterations by aligning chemical properties with desired biological activities. Moreover, AI-assisted de novo drug design introduces novel molecules with optimized properties, further enriching the compound pool for development (Anastas & Eghbali, 2010; Floresta et al., 2022; Paulick et al., 2022).

Case Studies Showcasing the Combined Impact on Cost and Efficiency

Numerous case studies underscore the transformative synergy of data science, AI, and green chemistry. For instance, researchers have utilized AI algorithms to predict reaction outcomes, leading to the discovery of greener synthetic pathways that yield higher yields with reduced waste. In drug repurposing, data-driven insights have identified existing drugs with potential therapeutic applications, bypassing extensive lead optimization phases. The combined effect of these approaches not only accelerates drug development timelines but also minimizes costs by optimizing processes and resource utilization (Boopathi et al., 2018).

Advancements in Personalized Medicine

Personalized medicine, driven by molecular data analysis, AI, and green chemistry principles, holds great promise for affordability. Tailoring treatments to individual patient profiles minimizes trial and error, reducing adverse effects and optimizing therapeutic outcomes. Data-driven insights identify patient-specific genetic markers and molecular signatures, enabling the design of precision therapies. Moreover, green chemistry practices contribute to sustainable production of personalized medications, ensuring affordability and accessibility.

A Procedural Approach

The integration of data science, artificial intelligence (AI), and green chemistry in drug development offers a powerful framework for enhancing affordability, efficiency, and sustainability (Bender & Cortés-Ciriano, 2021; Benke & Benke, 2018; Livingston et al., 2020). To operationalize this integration, a step-by-step procedural approach can be followed, spanning from early-stage drug discovery to final manufacturing target for drug development. Identify the molecular pathways and interactions relevant to the target (Figure 3).

Data Collection and Generation: Gather relevant molecular data, including genomics, proteomics, and chemical structures. Acquire existing datasets from public repositories or generate new data through experimentation.

Data Preprocessing and Integration: Clean and preprocess the collected data, handling missing values and outliers. Integrate diverse datasets to create a comprehensive molecular profile.

Data Analysis and Feature Selection: Apply data science techniques for dimensionality reduction (e.g., PCA, t-SNE) to reveal patterns. Select relevant features based on predictive power using AI algorithms.

Figure 3. Procedure for integration of data science, AI, and green chemistry for affordable drug development

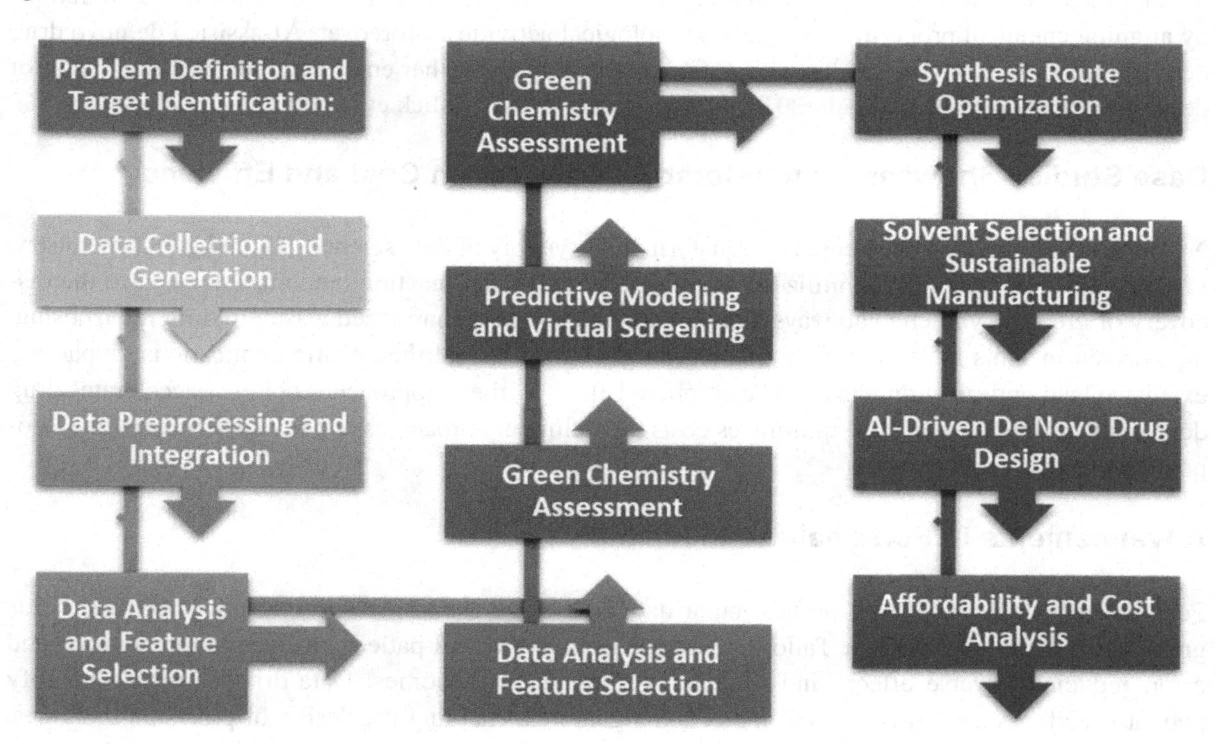

Problem Definition and Target Identification: Define the medical need and therapeutic

Predictive Modeling and Virtual Screening: Utilize AI algorithms to build predictive models for molecular interactions, compound properties, and drug-target interactions. Implement virtual screening to rapidly evaluate compound libraries for potential drug candidates.

Green Chemistry Assessment: Analyze the environmental impact of potential synthesis routes using life cycle assessment (LCA) methods. Evaluate the toxicity and environmental implications of chosen compounds and solvents.

Synthesis Route Optimization: Apply predictive modeling to predict reaction outcomes and identify optimal synthesis pathways. Utilize catalytic processes to minimize waste and enhance reaction efficiency.

Solvent Selection and Sustainable Manufacturing: Choose green solvents (e.g., water, supercritical fluids) to replace traditional organic solvents. Optimize manufacturing processes to reduce energy consumption and waste generation.

AI-Driven De Novo Drug Design: Employ AI algorithms to generate novel molecular structures with desired properties. Validate generated molecules for drug-likeness, bioavailability, and synthetic feasibility.

Affordability and Cost Analysis: Evaluate the cost-effectiveness of the optimized synthesis routes and manufacturing processes. Compare the projected costs with traditional methods to assess the economic benefits.

The integration of data science, AI, and green chemistry marks a pivotal juncture in the trajectory of drug development. This dynamic collaboration optimizes synthesis routes, enhances compound selection, and streamlines processes, culminating in a more affordable and sustainable approach (Boopathi, 2023b, 2023a; Reddy, Reddy, et al., 2023; Subha et al., 2023b). As case studies highlight successful applications, the pharmaceutical industry stands poised to embrace this synergy, transcending traditional limitations and driving innovation toward a future where breakthrough medications are not just accessible but also aligned with ethical and environmental imperatives.

PERSONALIZED MEDICINE AND AFFORDABILITY

In the landscape of healthcare, personalized medicine has emerged as a transformative paradigm that recognizes the unique genetic, molecular, and clinical attributes of individuals. This approach tailors' medical interventions to the specific characteristics of each patient, optimizing treatment outcomes. However, as personalized medicine gains momentum, questions arise about its affordability and accessibility (Floresta et al., 2022; Gruson et al., 2019; Paulick et al., 2022). This section delves into the intricacies of tailoring treatments through personalized medicine approaches, the integration of patient data and molecular insights, and the delicate balance between achieving precision and managing costs (Figure 4).

Tailoring Treatments Through Personalized Medicine Approaches

Personalized medicine acknowledges that every patient is distinct, possessing unique genetic predispositions, biomarkers, and responses to treatments. This understanding prompts a departure from the one-size-fits-all approach to a tailored therapeutic strategy. Molecular profiling, such as genetic testing, proteomic analysis, and omics data, informs physicians about the patient's molecular landscape. Armed with this information, clinicians can choose interventions that are more likely to be effective and have fewer adverse effects, fostering better patient outcomes (Boopathi, 2023a; Karthik et al., 2023; Pramila et al., 2023; Ramudu et al., 2023).

Figure 4. Personalized medicine and affordability approaches

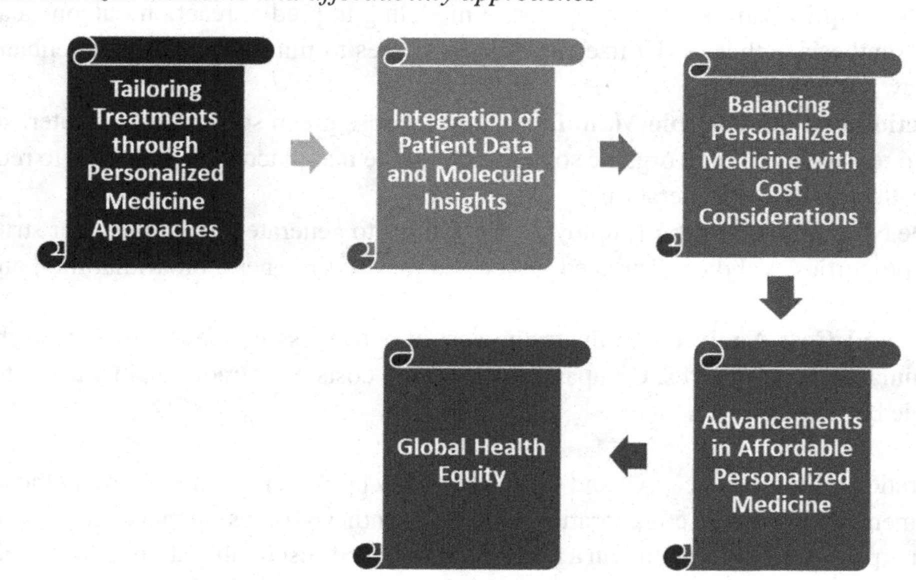

Integration of Patient Data and Molecular Insights

Central to personalized medicine is the integration of patient data and molecular insights into clinical decision-making. Electronic health records, genetic sequencing, and biomarker analysis contribute to a comprehensive patient profile. Data science and AI play a pivotal role in deciphering this data deluge, identifying patterns and correlations that guide treatment strategies. Molecular insights enable the identification of targetable pathways, aiding in the selection of therapies tailored to the patient's unique molecular signature (Pramila et al., 2023; Ramudu et al., 2023).

Balancing Personalized Medicine With Cost Considerations

While personalized medicine offers remarkable benefits, the pursuit of precision must be harmonized with cost considerations. The development and implementation of personalized therapies can be resource-intensive, from genetic testing to targeted drug development. Striking a balance between achieving optimal clinical outcomes and managing costs requires careful evaluation of the value proposition. Economic models, cost-effectiveness analyses, and reimbursement policies play vital roles in ensuring that personalized treatments are not only efficacious but also financially viable.

Advancements in Affordable Personalized Medicine

Advancements in technology and data analytics have the potential to make personalized medicine more affordable. As the cost of genomic sequencing continues to decrease, genetic profiling becomes more accessible, democratizing the foundation of personalized treatment. AI-driven platforms streamline the interpretation of complex molecular data, reducing the time and expertise needed for analysis. Furthermore, the integration of green chemistry practices in pharmaceutical production contributes to cost-effective and sustainable manufacturing of personalized therapies.

Global Health Equity

Addressing the affordability of personalized medicine also has implications for global health equity. While personalized treatments can be costly, innovative collaborations, research incentives, and policies can ensure that these interventions reach underserved populations. Integrating cost-effectiveness considerations into clinical guidelines can optimize resource allocation, ensuring that personalized treatments are accessible to those who stand to benefit the most (Anitha et al., 2023b; Boopathi et al., 2023; Pramila et al., 2023; Sengeni et al., 2023b; Subha et al., 2023b).

Hence, personalized medicine holds immense potential to revolutionize healthcare by aligning treatments with individual molecular profiles. However, ensuring its affordability is a multifaceted challenge that requires a delicate balance between precision and economic feasibility. Through the integration of patient data, molecular insights, and advancements in technology, the vision of accessible and affordable personalized medicine becomes attainable, ultimately improving patient outcomes and transforming the landscape of healthcare.

GLOBAL COLLABORATION AND REGULATORY CONSIDERATIONS

In the pursuit of affordable and sustainable medicine, global collaboration and regulatory frameworks play pivotal roles. Collaborative efforts among researchers, industries, and governments are essential for driving innovation, sharing knowledge, and addressing healthcare challenges (Haghi et al., 2022; Moingeon et al., 2022; Stevens et al., 2020). Moreover, establishing regulatory guidelines for the incorporation of AI and green chemistry in drug development ensures safety, efficacy, and environmental responsibility. International initiatives further promote the shared goal of making medicines accessible and sustainable for all. This section explores these crucial aspects in depth.

Collaborative Efforts Among Researchers, Industries, and Governments

Addressing the complex challenges of drug development, affordability, and sustainability requires a united front. Collaboration among researchers from academia, industry, and governments fosters knowledge exchange, sharing of best practices, and pooling of resources. Joint research projects accelerate advancements in data science, AI, and green chemistry. Public-private partnerships facilitate the development of innovative therapies, ensuring that diverse expertise converges to tackle intricate healthcare dilemmas.

Regulatory Framework for Incorporating AI and Green Chemistry

As AI and green chemistry become integral to drug development, establishing a regulatory framework is imperative. Regulatory agencies need to adapt to the evolving landscape, ensuring that AI-driven predictions and green chemistry practices meet rigorous standards for safety and efficacy. Guidelines must address data quality, model validation, and the ecological impact of manufacturing processes. Collaborative efforts between regulatory bodies, researchers, and industries can harmonize standards globally while encouraging innovation.

International Initiatives for Promoting Affordable and Sustainable Medicine

International initiatives are pivotal in the pursuit of affordable and sustainable medicine. Organizations like the World Health Organization (WHO) and the United Nations (UN) facilitate dialogue, resource sharing, and the establishment of global health priorities. Initiatives such as the Access to Medicines Index encourage pharmaceutical companies to prioritize affordability and accessibility in drug pricing strategies. International agreements, like the Paris Agreement on Climate Change, emphasize sustainable manufacturing practices, aligning environmental goals with pharmaceutical production (Bountra et al., 2017; Moingeon et al., 2022; Peña-Guerrero et al., 2021).

Promoting Technology Transfer and Capacity Building

Global collaboration involves not only the sharing of knowledge but also the transfer of technology and capacity building. High-income countries can support lower-income regions by sharing expertise, technologies, and best practices in drug development and manufacturing. Training programs, workshops, and technology transfer initiatives empower local scientists and researchers, enhancing their capability to contribute to affordable medicine solutions.

Balancing Global Health Equity

In the pursuit of affordable and sustainable medicine, a critical consideration is ensuring global health equity. Collaborative efforts must encompass strategies that prioritize underserved populations and neglected diseases. Initiatives such as the Medicines Patent Pool (MPP) facilitate access to patented medicines in low-income countries through voluntary licensing agreements, contributing to more equitable healthcare access.

In conclusion, global collaboration and regulatory considerations are paramount in shaping the landscape of affordable and sustainable medicine. As researchers, industries, and governments unite their efforts, innovative solutions become more achievable. Regulatory frameworks that adapt to evolving technologies ensure that advancements are aligned with safety and environmental responsibility. International initiatives drive collective action toward making healthcare accessible to all, fostering a world where affordable and sustainable medicine is not a distant dream but a shared reality.

FUTURE TRENDS AND EMERGING TECHNOLOGIES

The horizon of drug discovery and development is illuminated by the promise of future trends and emerging technologies that span data science, artificial intelligence (AI), and green chemistry. These trends hold the potential to reshape drug discovery workflows, enhance affordability, and drive transformative change. However, alongside these advancements, considerations of ethics and societal impact are paramount. This section delves into the potential trends, their implications, and the ethical dimensions of this evolving landscape (Anastas & Eghbali, 2010; Livingston et al., 2020; Paulick et al., 2022; Xie et al., 2020).

Advancements in Data Science, AI, and Green Chemistry

The future promises a confluence of innovations that amplify the capabilities of data science, AI, and green chemistry. In data science, predictive models will evolve to encompass more intricate molecular interactions, resulting in improved compound screening and target prediction accuracy. AI algorithms will become more adept at generating novel molecular structures, fostering unparalleled creativity in drug design. In green chemistry, sustainable synthesis routes will become more standardized, reducing the environmental impact of pharmaceutical production.

Implications for Drug Discovery and Affordability

The integration of these advancements will revolutionize drug discovery and affordability. Data-driven insights will guide researchers toward more viable candidates, expediting the drug development pipeline. AI-driven de novo drug design will introduce a new era of innovation, facilitating the creation of molecules tailored for specific therapeutic needs. Green chemistry practices will become an integral aspect of pharmaceutical manufacturing, contributing to both cost-effectiveness and environmental responsibility (Boopathi, Pandey, et al., 2023b; Boopathi & Kanike, 2023b; Ramudu et al., 2023; Sengeni et al., 2023b; Venkateswaran et al., 2023b). These combined implications will lead to shorter development timelines, reduced costs, and more accessible medicines.

Ethical Considerations and Societal Impacts

As technology evolves, ethical considerations come to the forefront. AI-driven predictions and algorithms raise concerns about transparency, accountability, and bias. Ensuring the ethical use of AI in decision-making, especially in matters of patient health, becomes crucial. Additionally, the societal impact of affordable medicine initiatives needs careful assessment. Balancing the need for affordable drugs with sustainable industry growth is a delicate ethical consideration. Collaborative efforts among stakeholders will be necessary to navigate these complex ethical dimensions (Anitha et al., 2023b; Boopathi et al., 2022; Koshariya, Kalaiyarasi, et al., 2023b; Koshariya, Khatoon, et al., 2023b; Ramudu et al., 2023; Vanitha et al., 2023b).

Shaping Research and Regulatory Landscape

Future trends will inevitably shape the research and regulatory landscape. Regulatory agencies will adapt to accommodate the integration of AI and green chemistry, outlining guidelines that ensure safety, efficacy, and sustainability. Research priorities will be guided by the pursuit of innovative AI-driven drug discovery and environmentally responsible manufacturing. Multidisciplinary collaborations, involving scientists, regulators, ethicists, and patient advocates, will be essential in shaping an inclusive and responsible future.

Global Impact and Access to Healthcare

The fusion of data science, AI, and green chemistry transcends geographical boundaries, offering the potential for global impact. Collaborative efforts will advance healthcare access in underserved regions through technology transfer and capacity building. As technology becomes more democratized, AI-driven drug discovery tools can empower researchers worldwide, leveling the playing field and enhancing global health equity (Floresta et al., 2022; Gruson et al., 2019; Paulick et al., 2022).

The future trends in data science, AI, and green chemistry promise transformative changes in drug discovery and affordability. These advancements carry the potential to reshape industry practices, enhance patient outcomes, and address pressing societal challenges. However, as we tread into this future, ethical considerations remain a guiding compass, ensuring that these innovations are harnessed responsibly and equitably, ultimately shaping a healthcare landscape that is both visionary and conscientious.

CONCLUSION

The convergence of data science, artificial intelligence (AI), green chemistry, and affordability in drug discovery and development presents a paradigm shift that holds transformative potential for the pharmaceutical industry and global healthcare. This chapter has explored the intricate interplay of these fields, unveiling a multidimensional approach that addresses challenges, fosters innovation, and aligns with ethical and environmental imperatives.

From the exploration of data-driven insights optimizing green chemistry practices to the streamlining of drug discovery workflows through AI-driven virtual screening and de novo drug design, the integration of these disciplines offers a comprehensive framework for achieving breakthroughs in efficiency, cost-effectiveness, and sustainable manufacturing. The case studies highlighted the tangible impact of this synergy, showcasing successful applications that accelerate drug development, reduce costs, and minimize environmental footprints. The considerations of personalized medicine further underline the significance of this integration. By leveraging molecular data analysis, AI-driven predictions, and green chemistry principles, the vision of tailoring treatments to individual profiles becomes attainable, optimizing patient outcomes while maintaining affordability and accessibility.

As this chapter draws to a close, it is evident that the collaboration of researchers, industries, governments, and regulatory bodies is essential for harnessing the full potential of this integration. Ethical considerations remain a cornerstone, guiding the responsible use of AI, data privacy, and equitable access to innovative therapies. Looking forward, the future trends and emerging technologies offer glimpses into a landscape where data science, AI, and green chemistry continue to evolve, propelling drug discovery towards unprecedented heights of precision, efficiency, and sustainability. By embracing these advancements, the pharmaceutical industry can pave the way for a future where affordable medicines are not just a possibility but a global reality, transforming lives and ushering in a new era of healthcare for all.

REFERENCES

Anastas, P., & Eghbali, N. (2010). Green chemistry: Principles and practice. *Chemical Society Reviews*, *39*(1), 301–312. doi:10.1039/B918763B PMID:20023854

Anitha, C., Komala, C., Vivekanand, C. V., Lalitha, S., & Boopathi, S. (2023a). Artificial Intelligence driven security model for Internet of Medical Things (IoMT). *IEEE Explore*, (pp. 1–7). IEEE.

Arshad, Z., Smith, J., Roberts, M., Lee, W. H., Davies, B., Bure, K., Hollander, G. A., Dopson, S., Bountra, C., & Brindley, D. (2016). Open access could transform drug discovery: A case study of JQ1. *Expert Opinion on Drug Discovery*, *11*(3), 321–332. doi:10.1517/17460441.2016.1144587 PMID:26791045

Becker, J., Manske, C., & Randl, S. (2022). Green chemistry and sustainability metrics in the pharmaceutical manufacturing sector. *Current Opinion in Green and Sustainable Chemistry*, *33*, 100562. doi:10.1016/j.cogsc.2021.100562

Bender, A., & Cortés-Ciriano, I. (2021). Artificial intelligence in drug discovery: What is realistic, what are illusions? Part 1: Ways to make an impact, and why we are not there yet. *Drug Discovery Today*, *26*(2), 511–524. doi:10.1016/j.drudis.2020.12.009 PMID:33346134

Benke, K., & Benke, G. (2018). Artificial intelligence and big data in public health. *International Journal of Environmental Research and Public Health*, *15*(12), 2796. doi:10.3390/ijerph15122796 PMID:30544648

Blakemore, D. C., Castro, L., Churcher, I., Rees, D. C., Thomas, A. W., Wilson, D. M., & Wood, A. (2018). Organic synthesis provides opportunities to transform drug discovery. *Nature Chemistry*, *10*(4), 383–394. doi:10.103841557-018-0021-z PMID:29568051

Boopathi, S. (2021). Improving of Green Sand-Mould Quality using Taguchi Technique. *Journal of Engineering Research*.

Boopathi. S. (2022). Effects of Cryogenically-treated Stainless Steel on Eco-friendly Wire Electrical Discharge Machining Process. Preprint : Springer.

Boopathi, S. (2022). Performance Improvement of Eco-Friendly Near-Dry wire-Cut Electrical Discharge Machining Process Using Coconut Oil-Mist Dielectric Fluid. *World Scientific: Journal of Advanced Manufacturing Systems*.

Boopathi, S. (2023a). Internet of Things-Integrated Remote Patient Monitoring System: Healthcare Application. In *Dynamics of Swarm Intelligence Health Analysis for the Next Generation* (pp. 137–161). IGI Global. doi:10.4018/978-1-6684-6894-4.ch008

Boopathi, S. (2023b). Securing Healthcare Systems Integrated With IoT: Fundamentals, Applications, and Future Trends. In Dynamics of Swarm Intelligence Health Analysis for the Next Generation (pp. 186–209). IGI Global.

Boopathi, S., & Davim, J. P. (2023). Applications of Nanoparticles in Various Manufacturing Processes. In *Sustainable Utilization of Nanoparticles and Nanofluids in Engineering Applications* (pp. 1–31). IGI Global. doi:10.4018/978-1-6684-9135-5.ch001

Boopathi, S., & Kanike, U. K. (2023). Applications of Artificial Intelligent and Machine Learning Techniques in Image Processing. In *Handbook of Research on Thrust Technologies' Effect on Image Processing* (pp. 151–173). IGI Global. doi:10.4018/978-1-6684-8618-4.ch010

Boopathi, S., & Khare, R. KG, J. C., Muni, T. V., & Khare, S. (2023). Additive Manufacturing Developments in the Medical Engineering Field. In Development, Properties, and Industrial Applications of 3D Printed Polymer Composites (pp. 86–106). IGI Global.

Boopathi, S., Kumar, P. K. S., Meena, R. S., Sudhakar, M., & Associates. (2023). Sustainable Developments of Modern Soil-Less Agro-Cultivation Systems: Aquaponic Culture. In Human Agro-Energy Optimization for Business and Industry (pp. 69–87). IGI Global.

Boopathi, S., Pandey, B. K., & Pandey, D. (2023a). Advances in Artificial Intelligence for Image Processing: Techniques, Applications, and Optimization. In Handbook of Research on Thrust Technologies' Effect on Image Processing (pp. 73–95). IGI Global.

Boopathi, S., Saranya, A., Raghuraman, S., & Revanth, R. (2018). Design and Fabrication of Low Cost Electric Bicycle. *International Research Journal of Engineering and Technology*, *5*(3), 146–147.

Boopathi, S., & Sivakumar, K. (2012). Experimental Analysis of Eco-friendly Near-dry Wire Electrical Discharge Machining Process. [UGC]. *Archives des Sciences*, *65*(10), 334–346.

Boopathi, S., Sureshkumar, M., & Sathiskumar, S. (2022). Parametric Optimization of LPG Refrigeration System Using Artificial Bee Colony Algorithm. *International Conference on Recent Advances in Mechanical Engineering Research and Development*, (pp. 97–105). IEEE.

Boopathi, S., Umareddy, M., & Elangovan, M. (2023). Applications of Nano-Cutting Fluids in Advanced Machining Processes. In *Sustainable Utilization of Nanoparticles and Nanofluids in Engineering Applications* (pp. 211–234). IGI Global. doi:10.4018/978-1-6684-9135-5.ch009

Bountra, C., Lee, W., & Lezaun, J. (2017). *A new pharmaceutical commons: Transforming drug discovery*. Oxford Martin Policy Paper Oxford.

Burki, T. (2019). Pharma blockchains AI for drug development. *Lancet*, *393*(10189), 2382. doi:10.1016/S0140-6736(19)31401-1 PMID:31204669

Dass, A., & Boopathi, S. (2016). Experimental Study of Eco-friendly Wire-Cut Electrical Discharge Machining Processes. *International Journal of Innovative Research in Science, Engineering and Technology, 5*.

Domakonda, V. K., Farooq, S., Chinthamreddy, S., Puviarasi, R., Sudhakar, M., & Boopathi, S. (2022). Sustainable Developments of Hybrid Floating Solar Power Plants: Photovoltaic System. In Human Agro-Energy Optimization for Business and Industry (pp. 148–167). IGI Global.

Dunn, P. J., Wells, A. S., & Williams, M. T. (2010). Future trends for green chemistry in the pharmaceutical industry. *Green Chemistry in the Pharmaceutical Industry*, *16*, 333–355. doi:10.1002/9783527629688.ch16

Ferrero, E., Brachat, S., Jenkins, J. L., Marc, P., Skewes-Cox, P., Altshuler, R. C., Gubser Keller, C., Kauffmann, A., Sassaman, E. K., Laramie, J. M., Schoeberl, B., Borowsky, M. L., & Stiefl, N. (2020). Ten simple rules to power drug discovery with data science. *PLoS Computational Biology*, *16*(8), e1008126. doi:10.1371/journal.pcbi.1008126 PMID:32853229

Floresta, G., Zagni, C., Gentile, D., Patamia, V., & Rescifina, A. (2022). Artificial intelligence technologies for COVID-19 de novo drug design. *International Journal of Molecular Sciences*, *23*(6), 3261. doi:10.3390/ijms23063261 PMID:35328682

Fowziya, S., Sivaranjani, S., Devi, N. L., Boopathi, S., Thakur, S., & Sailaja, J. M. (2023). Influences of nano-green lubricants in the friction-stir process of TiAlN coated alloys. *Materials Today: Proceedings*. doi:10.1016/j.matpr.2023.06.446

Glicksberg, B. S., Li, L., Chen, R., Dudley, J., & Chen, B. (2019). Leveraging big data to transform drug discovery. *Bioinformatics and Drug Discovery*, 91–118.

Gowri, N. V., Dwivedi, J. N., Krishnaveni, K., Boopathi, S., Palaniappan, M., & Medikondu, N. R. (2023). Experimental investigation and multi-objective optimization of eco-friendly near-dry electrical discharge machining of shape memory alloy using Cu/SiC/Gr composite electrode. *Environmental Science and Pollution Research International*, *30*(49), 1–19. doi:10.100711356-023-26983-6 PMID:37126160

Gruson, D., Helleputte, T., Rousseau, P., & Gruson, D. (2019). Data science, artificial intelligence, and machine learning: Opportunities for laboratory medicine and the value of positive regulation. *Clinical Biochemistry*, *69*, 1–7. doi:10.1016/j.clinbiochem.2019.04.013 PMID:31022391

Gunasekaran, K., & Boopathi, S. (2023). Artificial Intelligence in Water Treatments and Water Resource Assessments. In *Artificial Intelligence Applications in Water Treatment and Water Resource Management* (pp. 71–98). IGI Global. doi:10.4018/978-1-6684-6791-6.ch004

Gupta, P., & Mahajan, A. (2015). Green chemistry approaches as sustainable alternatives to conventional strategies in the pharmaceutical industry. *RSC Advances*, *5*(34), 26686–26705. doi:10.1039/C5RA00358J

Haghi, M., Benis, A., & Deserno, T. M. (2022). Accident & Emergency Informatics and One Digital Health. *Yearbook of Medical Informatics, 31* (01), 040–046.

Hanumanthakari, S., Gift, M. M., Kanimozhi, K., Bhavani, M. D., Bamane, K. D., & Boopathi, S. (2023). Biomining Method to Extract Metal Components Using Computer-Printed Circuit Board E-Waste. In *Handbook of Research on Safe Disposal Methods of Municipal Solid Wastes for a Sustainable Environment* (pp. 123–141). IGI Global. doi:10.4018/978-1-6684-8117-2.ch010

Hema, N., Krishnamoorthy, N., Chavan, S. M., Kumar, N., Sabarimuthu, M., & Boopathi, S. (2023). A Study on an Internet of Things (IoT)-Enabled Smart Solar Grid System. In *Handbook of Research on Deep Learning Techniques for Cloud-Based Industrial IoT* (pp. 290–308). IGI Global. doi:10.4018/978-1-6684-8098-4.ch017

Janardhana, K., Singh, V., Singh, S. N., Babu, T. R., Bano, S., & Boopathi, S. (2023). Utilization Process for Electronic Waste in Eco-Friendly Concrete: Experimental Study. In Sustainable Approaches and Strategies for E-Waste Management and Utilization (pp. 204–223). IGI Global.

Jensen, A. J., Molina, D. M., & Lundbäck, T. (2015). CETSA: A target engagement assay with potential to transform drug discovery. *Future Medicinal Chemistry*, *7*(8), 975–978. doi:10.4155/fmc.15.50 PMID:26062395

Judson, R., Richard, A., Dix, D. J., Houck, K., Martin, M., Kavlock, R., Dellarco, V., Henry, T., Holderman, T., Sayre, P., Tan, S., Carpenter, T., & Smith, E. (2009). The toxicity data landscape for environmental chemicals. *Environmental Health Perspectives*, *117*(5), 685–695. doi:10.1289/ehp.0800168 PMID:19479008

Karthik, S., Hemalatha, R., Aruna, R., Deivakani, M., Reddy, R. V. K., & Boopathi, S. (2023). Study on Healthcare Security System-Integrated Internet of Things (IoT). In Perspectives and Considerations on the Evolution of Smart Systems (pp. 342–362). IGI Global.

Koshariya, A. K., Kalaiyarasi, D., Jovith, A. A., Sivakami, T., Hasan, D. S., & Boopathi, S. (2023a). AI-Enabled IoT and WSN-Integrated Smart Agriculture System. In *Artificial Intelligence Tools and Technologies for Smart Farming and Agriculture Practices* (pp. 200–218). IGI Global. doi:10.4018/978-1-6684-8516-3.ch011

Koshariya, A. K., Khatoon, S., Marathe, A. M., Suba, G. M., Baral, D., & Boopathi, S. (2023a). Agricultural Waste Management Systems Using Artificial Intelligence Techniques. In *AI-Enabled Social Robotics in Human Care Services* (pp. 236–258). IGI Global. doi:10.4018/978-1-6684-8171-4.ch009

Kumara, V., Mohanaprakash, T., Fairooz, S., Jamal, K., Babu, T., & Sampath, B. (2023a). Experimental Study on a Reliable Smart Hydroponics System. In *Human Agro-Energy Optimization for Business and Industry* (pp. 27–45). IGI Global. doi:10.4018/978-1-6684-4118-3.ch002

Livingston, A., Trout, B. L., Horvath, I. T., Johnson, M. D., Vaccaro, L., Coronas, J., Babbitt, C. W., Zhang, X., Pradeep, T., Drioli, E., & others. (2020). Challenges and directions for green chemical engineering—Role of nanoscale materials. *Sustainable Nanoscale Engineering*, 1–18.

Mohanty, A., Jothi, B., Jeyasudha, J., Ranjit, P., Isaac, J. S., & Boopathi, S. (2023a). Additive Manufacturing Using Robotic Programming. In *AI-Enabled Social Robotics in Human Care Services* (pp. 259–282). IGI Global. doi:10.4018/978-1-6684-8171-4.ch010

Moingeon, P., Kuenemann, M., & Guedj, M. (2022). Artificial intelligence-enhanced drug design and development: Toward a computational precision medicine. *Drug Discovery Today*, *27*(1), 215–222. doi:10.1016/j.drudis.2021.09.006 PMID:34555509

Myilsamy, S., Boopathi, S., & Yuvaraj, D. (2021). A study on cryogenically treated molybdenum wire electrode. *Materials Today: Proceedings*, *45*(9), 8130–8135. doi:10.1016/j.matpr.2021.02.049

Nishanth, J., Deshmukh, M. A., Kushwah, R., Kushwaha, K. K., Balaji, S., & Sampath, B. (2023). Particle Swarm Optimization of Hybrid Renewable Energy Systems. In *Intelligent Engineering Applications and Applied Sciences for Sustainability* (pp. 291–308). IGI Global. doi:10.4018/979-8-3693-0044-2.ch016

Paulick, K., Seidel, S., Lange, C., Kemmer, A., Cruz-Bournazou, M. N., Baier, A., & Haehn, D. (2022). Promoting sustainability through next-generation biologics drug development. *Sustainability (Basel)*, *14*(8), 4401. doi:10.3390u14084401

Peck, R. W., Shah, P., Vamvakas, S., & van der Graaf, P. H. (2020). Data science in clinical pharmacology and drug development for improving health outcomes in patients. *Clinical Pharmacology and Therapeutics*, *107*(4), 683–686. doi:10.1002/cpt.1803 PMID:32202650

Peña-Guerrero, J., Nguewa, P. A., & García-Sosa, A. T. (2021). Machine learning, artificial intelligence, and data science breaking into drug design and neglected diseases. *Wiley Interdisciplinary Reviews. Computational Molecular Science*, *11*(5), e1513. doi:10.1002/wcms.1513

Pramila, P., Amudha, S., Saravanan, T., Sankar, S. R., Poongothai, E., & Boopathi, S. (2023). Design and Development of Robots for Medical Assistance: An Architectural Approach. In Contemporary Applications of Data Fusion for Advanced Healthcare Informatics (pp. 260–282). IGI Global.

Rahamathunnisa, U., Sudhakar, K., Murugan, T. K., Thivaharan, S., Rajkumar, M., & Boopathi, S. (2023). Cloud Computing Principles for Optimizing Robot Task Offloading Processes. In *AI-Enabled Social Robotics in Human Care Services* (pp. 188–211). IGI Global. doi:10.4018/978-1-6684-8171-4.ch007

Ramudu, K., Mohan, V. M., Jyothirmai, D., Prasad, D., Agrawal, R., & Boopathi, S. (2023). Machine Learning and Artificial Intelligence in Disease Prediction: Applications, Challenges, Limitations, Case Studies, and Future Directions. In Contemporary Applications of Data Fusion for Advanced Healthcare Informatics (pp. 297–318). IGI Global.

Reddy, M. A., Gaurav, A., Ushasukhanya, S., Rao, V. C. S., Bhattacharya, S., & Boopathi, S. (2023). Bio-Medical Wastes Handling Strategies During the COVID-19 Pandemic. In Multidisciplinary Approaches to Organizational Governance During Health Crises (pp. 90–111). IGI Global. doi:10.4018/978-1-7998-9213-7.ch006

Reddy, M. A., Reddy, B. M., Mukund, C., Venneti, K., Preethi, D., & Boopathi, S. (2023). Social Health Protection During the COVID-Pandemic Using IoT. In *The COVID-19 Pandemic and the Digitalization of Diplomacy* (pp. 204–235). IGI Global. doi:10.4018/978-1-7998-8394-4.ch009

Rogers, L., & Jensen, K. F. (2019). Continuous manufacturing–the Green Chemistry promise? *Green Chemistry*, *21*(13), 3481–3498. doi:10.1039/C9GC00773C

Roschangar, F., Sheldon, R. A., & Senanayake, C. H. (2015). Overcoming barriers to green chemistry in the pharmaceutical industry–the Green Aspiration Level™ concept. *Green Chemistry*, *17*(2), 752–768. doi:10.1039/C4GC01563K

Samikannu, R., Koshariya, A. K., Poornima, E., Ramesh, S., Kumar, A., & Boopathi, S. (2022). Sustainable Development in Modern Aquaponics Cultivation Systems Using IoT Technologies. In *Human Agro-Energy Optimization for Business and Industry* (pp. 105–127). IGI Global.

Sampath, B. (2021). *Sustainable Eco-Friendly Wire-Cut Electrical Discharge Machining: Gas Emission Analysis*.

Sampath, B., Sasikumar, C., & Myilsamy, S. (2023). Application of TOPSIS Optimization Technique in the Micro-Machining Process. In Trends, Paradigms, and Advances in Mechatronics Engineering (pp. 162–187). IGI Global.

Sampath, B., Sureshkumar, T., Yuvaraj, M., & Velmurugan, D. (2021). Experimental Investigations on Eco-Friendly Helium-Mist Near-Dry Wire-Cut EDM of M2-HSS Material. *Materials Research Proceedings*, *19*, 175–180.

Savage, N. (2021). *Tapping into the drug discovery potential of AI.* Nature. Com. doi:10.1038/d43747-021-00045-7

Selvakumar, S., Adithe, S., Isaac, J. S., Pradhan, R., Venkatesh, V., & Sampath, B. (2023). A Study of the Printed Circuit Board (PCB) E-Waste Recycling Process. In Sustainable Approaches and Strategies for E-Waste Management and Utilization (pp. 159–184). IGI Global.

Sengeni, D., Padmapriya, G., Imambi, S. S., Suganthi, D., Suri, A., & Boopathi, S. (2023a). Biomedical Waste Handling Method Using Artificial Intelligence Techniques. In *Handbook of Research on Safe Disposal Methods of Municipal Solid Wastes for a Sustainable Environment* (pp. 306–323). IGI Global. doi:10.4018/978-1-6684-8117-2.ch022

Sharma, S., Das, J., Braje, W. M., Dash, A. K., & Handa, S. (2020). A glimpse into green chemistry practices in the pharmaceutical industry. *ChemSusChem*, *13*(11), 2859–2875. doi:10.1002/cssc.202000317 PMID:32212245

Stevens, R., Taylor, V., Nichols, J., Maccabe, A. B., Yelick, K., & Brown, D. (2020). Ai for science: Report on the department of energy (doe) town halls on artificial intelligence (ai) for science. Argonne National Lab. (ANL), Argonne, IL (United States).

Subha, S., Inbamalar, T., Komala, C., Suresh, L. R., Boopathi, S., & Alaskar, K. (2023a). A Remote Health Care Monitoring system using internet of medical things (IoMT). *IEEE Explore*, (pp. 1–6). IEEE.

Vanitha, S., Radhika, K., & Boopathi, S. (2023a). Artificial Intelligence Techniques in Water Purification and Utilization. In *Human Agro-Energy Optimization for Business and Industry* (pp. 202–218). IGI Global. doi:10.4018/978-1-6684-4118-3.ch010

Veeranjaneyulu, R., Boopathi, S., Kumari, R. K., Vidyarthi, A., Isaac, J. S., & Jaiganesh, V. (2023). Air Quality Improvement and Optimisation Using Machine Learning Technique. *IEEE- Explore*, 1–6.

Venkateswaran, N., Vidhya, K., Ayyannan, M., Chavan, S. M., Sekar, K., & Boopathi, S. (2023a). A Study on Smart Energy Management Framework Using Cloud Computing. In 5G, Artificial Intelligence, and Next Generation Internet of Things: Digital Innovation for Green and Sustainable Economies (pp. 189–212). IGI Global. doi:10.4018/978-1-6684-8634-4.ch009

Vennila, T., Karuna, M., Srivastava, B. K., Venugopal, J., Surakasi, R., & Sampath, B. (2022). New Strategies in Treatment and Enzymatic Processes: Ethanol Production From Sugarcane Bagasse. In Human Agro-Energy Optimization for Business and Industry (pp. 219–240). IGI Global.

Vert, J.-P. (2023). How will generative AI disrupt data science in drug discovery? *Nature Biotechnology*, *41*(6), 1–2. doi:10.103841587-023-01789-6 PMID:37156917

Watson, W. J. (2012). How do the fine chemical, pharmaceutical, and related industries approach green chemistry and sustainability? *Green Chemistry*, *14*(2), 251–259. doi:10.1039/C1GC15904F

Workman, P., Antolin, A. A., & Al-Lazikani, B. (2019). Transforming cancer drug discovery with Big Data and AI. *Expert Opinion on Drug Discovery*, *14*(11), 1089–1095. doi:10.1080/17460441.2019.1637414 PMID:31284790

Xie, W., Li, T., Tiraferri, A., Drioli, E., Figoli, A., Crittenden, J. C., & Liu, B. (2020). Toward the next generation of sustainable membranes from green chemistry principles. *ACS Sustainable Chemistry & Engineering*, *9*(1), 50–75. doi:10.1021/acssuschemeng.0c07119

Yang, H., Feng, D., & Baumgartner, R. (2022). AI and Machine Learning in Drug Discovery. In Data Science, AI, and Machine Learning in Drug Development (pp. 63–93). Chapman and Hall/CRC. doi:10.1201/9781003150886-4

ABBREVIATION LIST

AI: Artificial Intelligence
LCA: Life Cycle Assessment
PCA: Principal Component Analysis
t-SNE: t-Distributed Stochastic Neighbor Embedding

Chapter 15
Intelligent Machines, IoT, and AI in Revolutionizing Agriculture for Water Processing

Krishnagandhi Pachiappan
Department of Electrical and Electronics Engineering, Nandha Engineering College, India

K. Anitha
Department of Computing Technologies, School of Computing, SRM Institute of Science and Technology, India

R. Pitchai
https://orcid.org/0000-0002-3759-6915
Department of Computer Science and Engineering, B.V. Raju Institute of Technology, India

S. Sangeetha
Department of Computer Science Engineering, Karpagam College of Engineering, India

T. V. V. Satyanarayana
https://orcid.org/0000-0002-8764-9982
Department of Electronics and Communication Engineering, Mohan Babu University, India

Sampath Boopathi
https://orcid.org/0000-0002-2065-6539
Mechanical Engineering, Muthayammal Engineering College, India

ABSTRACT

Modern agriculture faces numerous challenges, ranging from rising global food demand to water scarcity. To address these issues, the incorporation of intelligent machines, the Internet of Things (IoT), and artificial intelligence (AI) in agricultural water processing has become critical. This chapter investigates these technologies' transformative potential for optimizing water usage, increasing crop yields, and ensuring sustainable agricultural practices. It delves into the key concepts and applications, emphasizing the advantages and disadvantages of this novel approach. Farmers can make data-driven decisions, automate irrigation processes, and adapt to changing environmental conditions by leveraging AI and IoT-enabled systems, ultimately contributing to a more efficient and environmentally friendly agricultural sector.

DOI: 10.4018/978-1-6684-9999-3.ch015

INTRODUCTION

The integration of IoT and AI technologies in agriculture has significantly improved water processing, addressing challenges such as inefficient water usage, suboptimal crop yields, and environmental degradation. This technology enables farmers to make data-driven decisions in real-time, transforming the industry and ensuring sustainable water resource use for billions worldwide (Kumar et al., 2023; Sankar et al., 2023). Modern agriculture is confronted with a myriad of challenges that demand innovative solutions. The ever-increasing global demand for food, coupled with the escalating scarcity of water resources, has pushed the agricultural sector to seek novel approaches to optimize resource utilization, enhance crop yields, and ensure the long-term sustainability of farming practices. In this context, the integration of intelligent machines, the Internet of Things (IoT), and artificial intelligence (AI) has emerged as a critical imperative. These cutting-edge technologies hold the transformative potential to revolutionize water processing in agriculture, ushering in an era of data-driven, efficient, and eco-friendly farming practices (Agrawal et al., 2024; Zekrifa et al., 2023).

The integration of IoT and AI in agriculture is revolutionizing the industry by providing real-time data on soil moisture, weather conditions, and crop health. This data helps farmers optimize irrigation, reduce water wastage, and improve crop quality. It also enables proactive responses to environmental factors like droughts and pests, promoting sustainability. This chapter explores the applications, benefits, and challenges of these intelligent machines in modern farming practices (Koshariya, Kalaiyarasi, et al., 2023; Koshariya, Khatoon, et al., 2023). The integration of IoT and AI technologies is revolutionizing the agricultural industry by transforming traditional farming techniques, crop cultivation, livestock management, and resource allocation. This new era of precision agriculture aims to address inefficiencies, resource wastage, and unpredictable outcomes, making agriculture more sustainable, efficient, and productive than ever before (Gnanaprakasam et al., 2023). Modern agriculture finds itself at a crossroads. As the world's population continues to grow, the demand for food production soars. However, this surge in demand is juxtaposed with the looming specter of water scarcity, driven by climate change and over-extraction of groundwater. The traditional practices of agriculture are increasingly unsustainable, relying heavily on intuition and historical knowledge to manage water resources. In this context, technology becomes not just an option but a necessity to meet the global food security challenge while conserving precious water resources.

IoT technology connects physical objects like sensors and machinery to the internet, enabling real-time data collection and transmission. In agriculture, IoT sensors monitor variables like soil moisture levels, temperature, humidity, and nutrient content, enabling data-driven decisions and optimizing operations (Boopathi, Kumar, et al., 2023; Gnanaprakasam et al., 2023). IoT sensors offer real-time insights into crop conditions, enabling precise adjustments in irrigation, fertilizer application, and pest control. Livestock health monitoring alerts farmers to potential issues, while IoT-enabled tracking ensures food safety and traceability. AI in agriculture enhances IoT by processing sensor data, transforming it into actionable insights. Machine learning algorithms can identify patterns, predict crop diseases, optimize planting schedules, and automate tasks like weed control and harvesting (Jeevanantham et al., 2022). Intelligent machines, ranging from autonomous drones for aerial monitoring to self-driving tractors for precision farming, have emerged as indispensable tools in the modern agricultural landscape. These machines enable farmers to monitor and manage their fields with unprecedented accuracy and efficiency. They can analyze soil conditions, assess crop health, and precisely apply water and nutrients. By harnessing the power of automation and data analytics, farmers can optimize resource utilization and minimize waste, all while reducing the labor-intensive aspects of farming. The Internet of Things (IoT) has ushered in a new era of connectivity, allowing farmers to collect real-time data from sensors deployed throughout their fields, irrigation systems, and reservoirs. These sensors provide a constant stream of information on soil moisture levels, weather conditions, and equipment performance. With this wealth of data, farmers can make informed decisions about irrigation, adapting to changing environmental conditions and ensuring that water is applied precisely where and when it is needed, reducing both water consumption and environmental impact (Satav, Lamani, et al., 2024).

The integration of IoT and AI technologies is revolutionizing agriculture by improving efficiency, sustainability, and resource optimization. This digital transformation helps farmers tackle climate change and population growth challenges. Traditional practices often lead to water wastage and soil degradation, while climate change poses challenges like prolonged droughts and extreme weather events. The integration of IoT and AI technologies is a revolutionary shift in water management and crop cultivation. IoT technology uses sensors in agricultural fields to monitor factors like soil moisture, temperature, humidity, and plant health, providing farmers with real-time data and a comprehensive view of their fields' conditions (Kumar et al., 2023; Kumara et al., 2023; Zekrifa et al., 2023).

AI and machine learning algorithms can process vast data from IoT devices, making predictions, detecting patterns, and providing actionable insights. In agriculture, AI can predict optimal irrigation schedules, identify crop diseases early, and assess climate change impact on yields. This empowers farmers to make data-driven decisions, optimize water usage, reduce resource waste, and enhance crop quality. AI can also help mitigate climate change effects (Kumar et al., 2023). The integration of IoT and AI in agriculture is transforming traditional practices into precision agriculture, ensuring efficient water usage and greater crop care. Real-world case studies will showcase the transformative potential of these intelligent machines, paving the way for a new era of sustainability, efficiency, and productivity.

Artificial intelligence (AI) has emerged as a powerful tool for processing and interpreting the vast amounts of data generated by intelligent machines and IoT sensors. AI algorithms can analyze historical data, predict future trends, and recommend optimal courses of action. In agriculture, AI can be harnessed to create predictive models for crop growth, disease detection, and pest control. By leveraging AI-driven insights, farmers can make data-driven decisions, further enhancing water management efficiency and crop yields. In this chapter, we delve into the transformative potential of intelligent machines, IoT, and AI in the context of water processing for agriculture. We explore the key concepts, practical applications, benefits, and challenges of integrating these technologies into agricultural practices (B et al., 2024; Satav, Hasan, et al., 2024). By leveraging AI and IoT-enabled systems, farmers are poised to make data-driven decisions, automate irrigation processes, and adapt to changing environmental conditions, ultimately contributing to a more efficient and eco-friendly agricultural sector that addresses the pressing challenges of our time.

SIGNIFICANCE OF INTELLIGENT MACHINES, IOT, AND AI IN AGRICULTURE

The integration of Intelligent Machines, IoT, and AI in agriculture offers transformative benefits and addresses industry challenges (Jeevanantham et al., 2022; Koshariya, Kalaiyarasi, et al., 2023; Maguluri et al., 2023).

Precision Agriculture

- **Optimized Resource Management:** Intelligent Machines equipped with IoT sensors and AI algorithms enable precise resource management. Farmers can tailor irrigation, fertilization, and pest control to the specific needs of each crop, minimizing resource waste.
- **Increased Crop Yields:** By providing real-time data on soil conditions and crop health, these technologies enhance crop yields. AI can predict disease outbreaks and optimize planting schedules, contributing to higher productivity.

Water Conservation

- **Efficient Irrigation:** IoT-enabled sensors monitor soil moisture levels and weather conditions, ensuring that irrigation is applied only when necessary. This conserves water resources and reduces environmental impacts.
- **Drought Mitigation:** AI can predict drought conditions based on historical data and real-time weather information, enabling proactive water management strategies to mitigate the effects of droughts.

Sustainability

- **Reduced Environmental Impact:** IoT and AI assist in reducing the use of chemical fertilizers and pesticides. Targeted application minimizes pollution and lowers the environmental footprint of agriculture.
- **Preservation of Ecosystems:** Intelligent Machines support sustainable farming practices that protect ecosystems, including wetlands and watersheds, by preventing overuse of water resources.

Data-Driven Decision-Making

- **Real-Time Insights:** IoT sensors continuously collect data, and AI processes this data rapidly. Farmers can make informed decisions on when to plant, irrigate, and harvest, leading to better outcomes.
- **Risk Mitigation:** AI can analyze historical data and predict potential crop diseases or pest infestations. Farmers can take preventive measures, reducing the risk of crop losses.

Economic Benefits

- **Cost Reduction:** By optimizing resource use and crop management, IoT and AI can lower production costs, increasing profitability for farmers.
- **Market Competitiveness:** Precision agriculture makes it possible for farmers to produce high-quality crops efficiently, enhancing their competitiveness in the market.

Adaptation to Climate Change

- **Climate-Responsive Farming:** IoT and AI help farmers adapt to changing weather patterns and shifting climate conditions by providing timely information and adaptive strategies.
- **Resilience:** The ability to respond quickly to climate-related challenges, such as extreme heat or prolonged droughts, increases the resilience of agricultural systems.

Food Security

- **Increased Production:** Higher crop yields and reduced resource waste contribute to greater food production, aiding in global food security efforts.
- **Consistent Quality:** Precision agriculture ensures consistent crop quality, meeting the demands of an increasingly discerning consumer base.

The integration of Intelligent Machines, IoT, and AI in agriculture is a significant shift towards sustainable, efficient, and data-driven practices. These technologies address water scarcity, climate change, and food security while supporting economic growth. As they evolve, they have the potential to revolutionize food production, making it more resilient and environmentally friendly.

CHAPTER OBJECTIVES

- **Introduce the Current Challenges in Modern Agriculture:** Begin by outlining the key challenges faced by modern agriculture, including increasing global food demand, water scarcity, and the need for sustainable practices.
- **Explore the Role of Intelligent Machines in Agriculture:** Examine the significance of intelligent machines, such as drones and automated tractors, in revolutionizing farming practices and optimizing resource utilization.
- **Discuss the Internet of Things (IoT) in Agricultural Water Management:** Investigate how IoT technologies enable real-time data collection from sensors placed in fields, reservoirs, and irrigation systems to enhance water management.
- **Examine the Integration of Artificial Intelligence (AI) in Agriculture:** Analyze the application of AI algorithms for data analysis, prediction, and decision-making in agriculture, with a focus on water-related processes.
- **Highlight Key Concepts and Terminologies:** Define and explain essential terms and concepts related to intelligent machines, IoT, and AI in the context of agricultural water processing.
- **Present Case Studies and Practical Applications:** Showcase real-world examples and case studies where AI and IoT have been successfully implemented in agriculture to optimize water usage and improve crop yields.
- **Evaluate Benefits and Challenges:** Discuss the advantages and potential drawbacks of integrating AI and IoT in agricultural water management, including cost-effectiveness, environmental impact, and data security concerns.

- **Demonstrate Data-Driven Decision-Making:** Illustrate how farmers can use data from IoT-enabled systems and AI algorithms to make informed decisions about irrigation, fertilization, and crop protection.
- **Address Adaptability to Changing Environmental Conditions:** Explore how AI and IoT systems can help farmers adapt to climate change and other environmental factors by providing real-time insights and recommendations.
- **Promote Sustainable Agriculture:** Emphasize the role of these technologies in achieving sustainability goals, including water conservation, reduced environmental impact, and long-term agricultural viability.
- **Discuss Future Trends and Innovations:** Provide insights into emerging trends and innovations in the field of AI, IoT, and intelligent machines in agriculture, and their potential impact on the industry.
- **Conclusion and Future Prospects:** Summarize the key takeaways from the chapter and discuss the future prospects of AI and IoT in transforming water processing for agriculture.

WATER RESOURCES IN AGRICULTURE

Importance of Water in Agriculture

Water is a vital resource in agriculture, essential for crop growth, livestock sustainability, food security, and environmental sustainability, with its significance extending across various sectors (Boopathi, 2021b; Boopathi, Lewise, et al., 2022; Boopathi, Thillaivanan, et al., 2022; Boopathi & Sivakumar, 2014).

- **Hydration:** Water is a primary component of plant cells, and its uptake through roots is essential for plant growth and development. Adequate water supply ensures proper hydration, leading to healthy crops and higher yields.
- **Photosynthesis:** Water is a key ingredient in the photosynthesis process, where plants convert sunlight into energy and produce oxygen. Without water, photosynthesis cannot occur, and crop productivity suffers.
- **Nutrient Uptake:** Water serves as a medium for the transport of essential nutrients from the soil to plant roots. It plays a vital role in the uptake of minerals and fertilizers, ensuring nutrient availability for plants.
- **Thermal Buffer:** Water has a high heat capacity, which means it can absorb and release heat slowly. This property helps moderate temperature fluctuations in agricultural ecosystems, protecting crops from extreme heat or cold.
- **Soil Moisture:** Adequate soil moisture content is essential for maintaining soil structure and fertility. Properly hydrated soil promotes root growth, prevents erosion, and supports beneficial soil microorganisms.
- **Fungus and Disease Prevention:** Regular irrigation can help prevent soil-borne diseases and fungal infections that thrive in dry conditions.
- **Animal Hydration:** Water is essential for the health and well-being of livestock. Providing clean and ample water ensures that animals are adequately hydrated, promoting their growth and productivity.

- **Resilience:** Adequate water resources and efficient irrigation systems make agricultural systems more resilient to droughts. Water management strategies can help mitigate the adverse effects of water scarcity.
- **Crop Production:** Water availability directly affects crop yields. Insufficient water can lead to crop failures and food shortages, highlighting its role in ensuring food security.
- **Ecosystem Health:** Sustainable water use in agriculture preserves natural ecosystems, including wetlands, rivers, and watersheds. Responsible water management helps protect biodiversity and aquatic habitats.
- **Water Quality:** Proper irrigation practices can prevent water contamination, ensuring that agricultural runoff does not harm nearby water bodies.
- **Agricultural Economy:** Agriculture is a significant contributor to the global economy. Efficient water use in agriculture improves economic outcomes for farmers and supports rural communities.

Water is vital for agriculture, impacting crop growth, livestock health, food security, and environmental sustainability. Efficient use is crucial to address water scarcity and climate change, while sustainable water management practices are essential for global population stability.

Challenges in Water Management

Water management in agriculture faces challenges such as resource scarcity and environmental concerns, necessitating effective solutions for sustainable farming usage (Gunasekaran & Boopathi, 2023; Vanitha et al., 2023; Zekrifa et al., 2023).

- **Water Scarcity:** Many regions experience water scarcity due to factors such as over-extraction of groundwater, droughts, and competition for water resources among various sectors.
- **Inefficient Irrigation:** Traditional irrigation methods, such as flood and furrow irrigation, are often inefficient, leading to water wastage. Transitioning to more efficient irrigation systems is a challenge for many farmers.
- **Water Pollution:** Agricultural runoff can introduce pollutants like pesticides, fertilizers, and sediment into water bodies, causing water pollution and harming aquatic ecosystems.
- **Groundwater Depletion:** Excessive groundwater pumping for irrigation can lead to aquifer depletion, causing land subsidence and long-term water availability concerns.
- **Climate Change:** Changing precipitation patterns and increased temperatures due to climate change can disrupt water availability, making it challenging to plan irrigation and manage water resources effectively.
- **Infrastructure and Technology Gaps:** Access to modern irrigation infrastructure and technology can be limited in some regions, hindering the adoption of efficient water management practices.
- **Lack of Data and Information:** Accurate and timely data on soil moisture, weather conditions, and crop water needs are essential for effective water management. Many farmers lack access to such data.

Sustainable Water Usage in Farming

Various strategies and practices can be employed to address water challenges and promote sustainable farming practices (Boopathi, Arigela, et al., 2022; Koshariya, Kalaiyarasi, et al., 2023).

- **Precision Agriculture:** Implement precision agriculture techniques that use data from IoT sensors and AI to precisely target water and nutrients to crops, reducing waste and optimizing yields.
- **Drip and Sprinkler Irrigation:** Transition from flood and furrow irrigation to more efficient methods like drip and sprinkler irrigation, which deliver water directly to the root zone of plants, minimizing wastage.
- **Rainwater Harvesting:** Capture and store rainwater for agricultural use during dry periods, reducing reliance on groundwater and surface water sources.
- **Crop Rotation and Cover Crops:** Employ crop rotation and cover cropping practices to improve soil health and water retention, reducing the need for excessive irrigation.
- **Water-Efficient Crop Varieties:** Choose crop varieties that are drought-resistant and have lower water requirements.
- **Regulatory Measures:** Implement and enforce regulations to prevent over-extraction of groundwater and reduce water pollution from agricultural activities.
- **Education and Extension Services:** Provide farmers with access to training and information on sustainable water management practices and technologies.
- **Research and Innovation:** Invest in research and innovation to develop water-efficient farming practices, drought-resistant crops, and improved irrigation technologies.
- **Monitoring and Data Collection:** Expand the availability of real-time data and monitoring systems to help farmers make informed decisions about water use.
- **Community Collaboration:** Encourage collaboration among farmers, local communities, and government agencies to develop and implement sustainable water management plans.

Sustainable water usage in agriculture is vital for long-term water availability, environmental conservation, food security, and responsible water resource management by addressing challenges and adopting sustainable practices.

INTELLIGENT MACHINES IN AGRICULTURE

Intelligent agriculture machines utilize advanced technologies like automation, robotics, IoT, and AI to improve efficiency, precision, and sustainability in farming operations (Chandrika et al., 2023; Dhanya et al., 2023; Nishanth et al., 2023). The text provides an overview of intelligent machines in agriculture, as illustrated in Figure 1.

- **Automation and Robotics:** Intelligent machines can automate tasks such as planting, harvesting, weeding, and sorting. Robotics in agriculture includes autonomous tractors, drones, and robotic arms for various applications.
- **IoT Sensors:** IoT sensors are deployed in fields to collect data on soil moisture, temperature, humidity, and crop health. This real-time data helps farmers make informed decisions.

Figure 1. Essential components of intelligent machines in agriculture

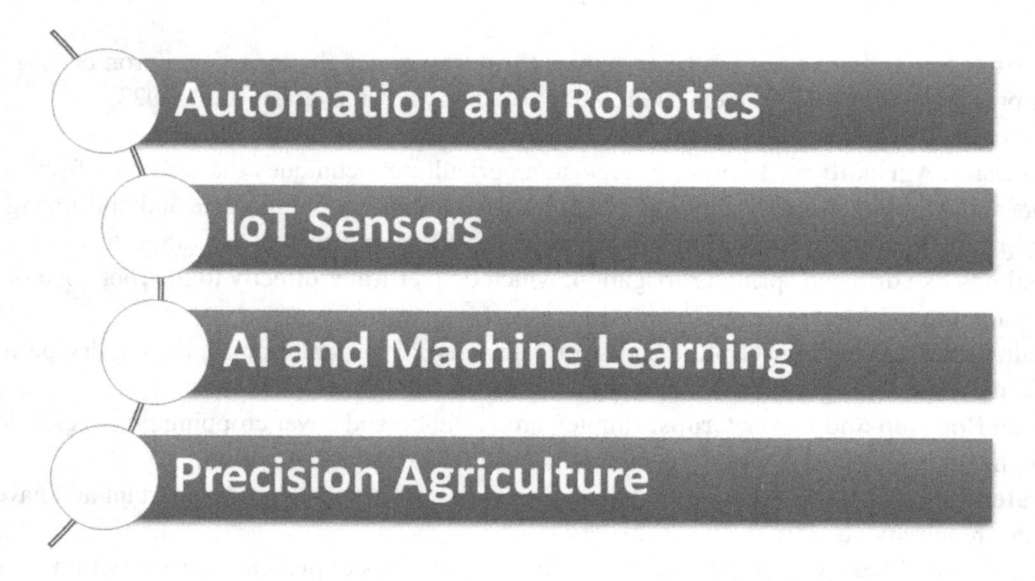

- **AI and Machine Learning:** AI algorithms and machine learning models analyze the data collected by IoT sensors. They can predict crop diseases, optimize irrigation, and make recommendations for planting and harvesting.
- **Precision Agriculture:** Intelligent machines enable precision agriculture, where every action in the field is optimized based on data, resulting in reduced resource wastage and improved yields.

Application of Intelligent Machines in Agriculture

Intelligent machines are revolutionizing traditional farming practices in agriculture (Koshariya, Kalaiyarasi, et al., 2023; Sankar et al., 2023), with a wide range of applications including (Figure 2)

Precision Irrigation

- **IoT Sensors:** Soil moisture sensors and weather data help determine when and how much to irrigate.
- **Automated Irrigation Systems:** Smart irrigation systems adjust water flow and distribution based on real-time data, reducing water waste.

Crop Monitoring and Management

- **Satellite Imagery:** Drones and satellites equipped with AI can monitor crop health and detect issues like pests, diseases, and nutrient deficiencies.
- **Predictive Analytics:** AI algorithms analyze historical data to predict crop diseases and prescribe treatment in advance.

Figure 2. Application of Intelligent machines in agriculture

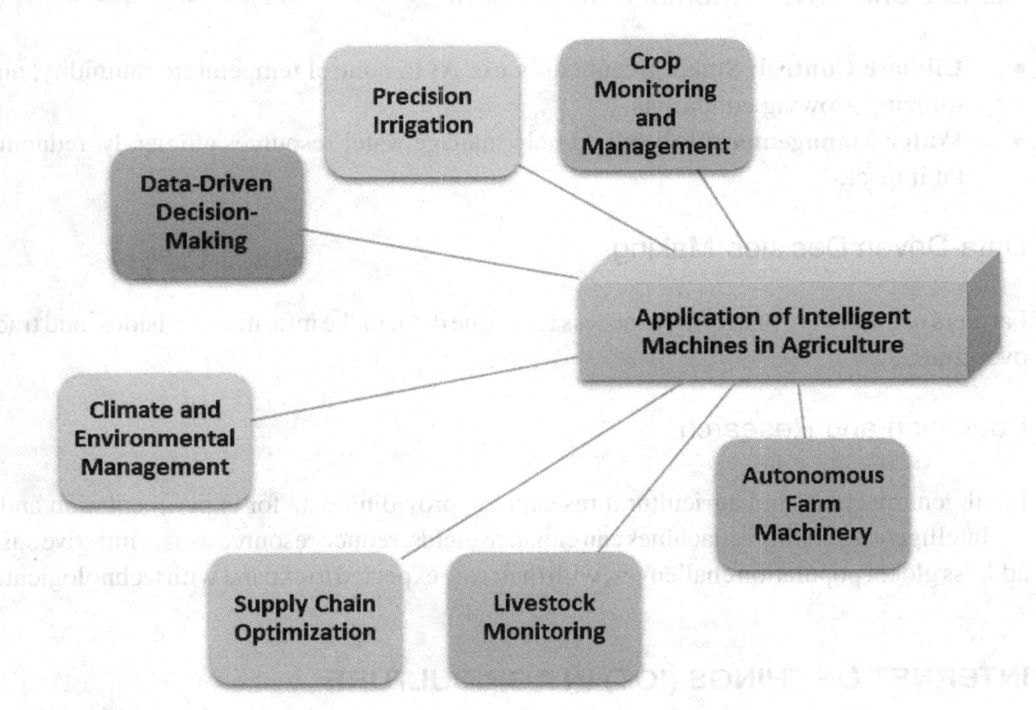

Autonomous Farm Machinery

- **Autonomous Tractors:** Self-driving tractors can plow, plant, and harvest crops with precision, reducing labor requirements.
- **Weeding Robots:** Robots equipped with AI can identify and remove weeds without the need for herbicides.

Livestock Monitoring

Sensors can monitor the health and behavior of livestock, providing early detection of diseases and optimizing feeding regimes.

Supply Chain Optimization

- **IoT in Logistics:** IoT devices track the location and condition of agricultural products during transportation, ensuring freshness and minimizing spoilage.
- **Sorting and Grading:** Intelligent machines can sort and grade fruits, vegetables, and other agricultural products based on quality and size.

Climate and Environmental Management

- **Climate Control:** Smart greenhouses use AI to control temperature, humidity, and lighting, optimizing growing conditions.
- **Water Management:** IoT and AI help manage water resources efficiently, reducing environmental impact.

Data-Driven Decision-Making

Farmers use software platforms to access real-time data, make informed decisions, and track performance over time.

Education and Research

Intelligent machines aid agricultural research by providing data for experimentation and analysis.

Intelligent agriculture machines can enhance yields, reduce resource usage, improve sustainability, and address global population challenges, with their role expected to expand with technological advancements.

INTERNET OF THINGS (IOT) IN AGRICULTURE

The Internet of Things (IoT) in agriculture is a network of interconnected devices, sensors, and equipment that collects, transmits, and analyzes data, enabling real-time information collection about fields, livestock, and equipment, enhancing decision-making and resource management (Boopathi, 2023; Karthik et al., 2023; Koshariya, Kalaiyarasi, et al., 2023; Maguluri et al., 2023; Samikannu et al., 2022).

IoT Applications in Water Management

The Internet of Things (IoT) has significantly influenced water management in agriculture through the implementation of smart irrigation systems.

Soil Moisture Monitoring

IoT sensors, installed in the ground, measure soil moisture levels, aiding farmers in determining irrigation timing and minimizing water wastage, thereby ensuring optimal crop moisture.

Weather Monitoring

IoT-connected weather stations collect data on temperature, humidity, wind speed, and precipitation, enabling farmers to anticipate weather patterns and make irrigation decisions.

Irrigation Control

IoT-based smart irrigation systems utilize real-time data from soil moisture sensors and weather forecasts to automatically adjust water flow and distribution, thereby minimizing over-irrigation and saving water.

Water Quality Monitoring

These sensors measure parameters such as pH, turbidity, and chemical concentrations in irrigation water sources. Farmers use this data to ensure water quality and prevent contamination of crops.

Flow Monitoring

IoT-enabled flow meters measure the volume of water delivered through irrigation systems. This data helps farmers track water usage and identify potential leaks or inefficiencies.

Remote Monitoring and Control

Farmers can monitor and control irrigation systems remotely through mobile apps, allowing them to make adjustments even when off-site.

Enhancing Efficiency with IoT

IoT technology is revolutionizing agriculture by providing real-time data and automation capabilities (Boopathi, 2023; Hema et al., 2023; Samikannu et al., 2022).

- **Resource Optimization:** IoT sensors and data analytics help farmers optimize resource use, such as water, fertilizers, and pesticides. This reduces waste and improves cost-efficiency.
- **Predictive Analytics:** IoT data, combined with AI and machine learning, enables predictive analytics. Farmers can anticipate crop diseases, pest infestations, and weather-related challenges, allowing for proactive responses.
- **Energy Efficiency:** IoT-connected equipment can be optimized for energy efficiency, reducing the environmental footprint of farming operations.
- **Labor Savings:** Automation and remote monitoring reduce the need for manual labor in tasks like irrigation and livestock monitoring, saving time and labor costs.
- **Precision Agriculture:** IoT supports precision agriculture practices, where every action in the field is data-driven and optimized. This leads to increased crop yields and improved crop quality.
- **Environmental Sustainability:** By reducing resource wastage and using data to make eco-conscious decisions, IoT contributes to the environmental sustainability of agriculture.

IoT technology enhances water management and agriculture efficiency by providing real-time data, automation, and predictive capabilities, enabling farmers to make informed decisions, conserve resources, and enhance sustainability and productivity.

ARTIFICIAL INTELLIGENCE (AI) IN AGRICULTURE

Role of AI in Modern Agriculture

AI is revolutionizing farming by utilizing machine learning and deep learning technologies to analyze vast data, enabling informed and precise decision-making in modern agriculture (Gunasekaran & Boopathi, 2023; Koshariya, Kalaiyarasi, et al., 2023; Ramudu et al., 2023; Sengeni et al., 2023; Vanitha et al., 2023; Zekrifa et al., 2023).

- **Data Analysis:** AI algorithms analyze data from sources like IoT sensors, satellites, and drones to provide insights into soil conditions, weather patterns, crop health, and more.
- **Predictive Analytics:** AI models predict crop diseases, pest infestations, and optimal planting and harvesting times based on historical and real-time data.
- **Resource Optimization:** AI helps optimize resource use, including water, fertilizers, and pesticides, reducing waste and costs.
- **Automation:** AI-driven automation, such as autonomous tractors and drones, can perform tasks like planting, harvesting, and monitoring, improving efficiency and reducing labor requirements.
- **Crop Monitoring:** AI-powered image recognition and computer vision systems analyze images from drones and satellites to monitor crop growth, detect anomalies, and assess overall health.
- **Decision Support:** AI provides farmers with decision support systems that offer recommendations for crop management, irrigation, and pest control.

AI for Water Processing and Irrigation

AI significantly impacts water processing and irrigation practices in agriculture.

- **Precise Irrigation:** AI algorithms analyze data from IoT sensors to determine the precise irrigation needs of crops. This information ensures that water is applied when and where it is needed, reducing water wastage.
- **Weather Predictions:** AI can process weather data and provide accurate short-term and long-term forecasts. Farmers use this information to optimize irrigation schedules and mitigate the impact of weather extremes like droughts or heavy rains.
- **Drought Management:** AI helps in early detection and management of drought conditions by analyzing historical data and real-time information. It provides strategies for mitigating the effects of water scarcity.
- **Irrigation System Optimization:** AI can optimize the functioning of irrigation systems, adjusting water flow and distribution based on real-time data, soil conditions, and crop requirements.
- **Water Quality Monitoring:** AI-driven sensors can continuously monitor water quality in irrigation sources, ensuring that the water is free from contaminants that could harm crops.

AI-driven Crop Management

AI is being utilized in crop management to enhance crop yields, reduce resource usage, and ensure crop health (Veeranjaneyulu, Boopathi, Kumari, et al., 2023; Veeranjaneyulu, Boopathi, Narasimharao, et al., 2023; Venkateswaran et al., 2023).

- **Disease and Pest Detection:** AI models analyze images and sensor data to detect signs of crop diseases, pests, or nutrient deficiencies. Early detection allows for targeted interventions.
- **Crop Health Assessment:** AI monitors crop health by analyzing multispectral and hyperspectral data from drones and satellites, identifying stressed areas that may require special attention.
- **Nutrient Management:** AI-driven systems assess nutrient levels in the soil and recommend precise fertilization strategies, reducing excess use of fertilizers.
- **Crop Planning:** AI algorithms consider various factors like soil types, climate, and market demand to help farmers make informed decisions about what crops to plant and where.
- **Harvest Optimization:** AI provides insights into the optimal timing for harvesting crops to maximize quality and yield.

AI is revolutionizing agriculture by enhancing data-driven, precise, and efficient management of resources, crop monitoring, and adaptation to environmental changes, leading to improved sustainability and productivity in the sector.

Integration of Intelligent Machines, IoT, and AI for Water Processing

The integration of Intelligent Machines, IoT, and AI in water processing in agriculture can optimize water usage, enhance crop yields, and promote sustainability, as demonstrated in Figure 3.

Data Collection with IoT

IoT sensors are used in agricultural fields to collect real-time data on soil moisture, temperature, humidity, and weather conditions. They also monitor water sources like rivers, reservoirs, and wells, providing information on water availability and quality (Dhanya et al., 2023; Pramila et al., 2023; Ramudu et al., 2023).

Data Transmission and Management

IoT sensors transmit the collected data wirelessly to a centralized system or cloud-based platform, ensuring data accessibility from anywhere. The data is stored securely and made available for analysis.

AI-driven Data Analysis

AI algorithms preprocess the raw data, cleaning and formatting it for analysis. AI uses machine learning and deep learning techniques to analyze the data. This analysis includes predicting soil moisture trends, assessing crop health, and identifying potential water-related issues. AI provides actionable insights, such as when and how much to irrigate, based on the analysis of historical and real-time data. It can also offer recommendations for crop selection, disease management, and pest control.

Intelligent Machines and Automation

Intelligent machines, such as automated irrigation systems, receive instructions from AI algorithms. They adjust irrigation schedules and water distribution in real-time based on the data and recommendations, ensuring precise and efficient water use. Robots equipped with AI can perform tasks like weeding, planting, and harvesting with precision, reducing the need for manual labor and resource waste.

Figure 3. Integration of Intelligent Machines, IoT, and AI for Water Processing

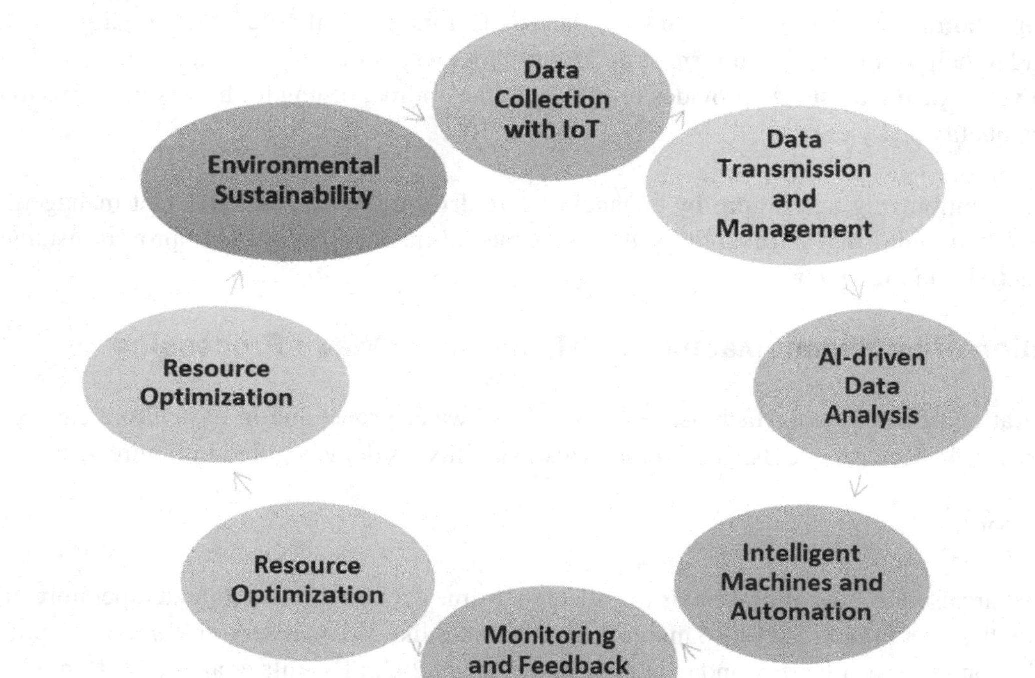

Monitoring and Feedback Loop

IoT sensors continue to collect data throughout the crop growth cycle, allowing for ongoing assessment and adjustment of water management strategies. The data collected during the growing season is fed back into AI models, enhancing their accuracy and enabling adaptive decision-making as environmental conditions change.

Resource Optimization

The integration of intelligent machines, IoT, and AI ensures that water is used efficiently, minimizing wastage and reducing the environmental impact. AI can optimize the allocation of water resources based on crop needs, soil conditions, and weather forecasts, ensuring that water is distributed where it is most needed (Boopathi, Kumar, et al., 2023; Boopathi, Pandey, et al., 2023; Domakonda et al., 2022; Samikannu et al., 2022; Vennila et al., 2022).

Environmental Sustainability

The precise management of water resources reduces the environmental impact of agriculture by minimizing runoff and preventing water pollution. Responsible water management through intelligent machines, IoT, and AI helps protect natural ecosystems such as wetlands and watersheds.

The integration of these technologies enhances water processing in agriculture, promotes sustainable farming practices, increases crop yields, and reduces resource consumption by creating a closed-loop system where data drives decisions and machines execute actions.

Synergy Between Intelligent Machines, IoT, and AI

The integration of Intelligent Machines, IoT, and AI is revolutionizing various industries, including agriculture, by fostering a robust ecosystem that utilizes data, automation, and advanced analytics for efficiency, sustainability, and innovation (Anitha et al., 2023; Boopathi, 2021b; Boopathi, Pandey, et al., 2023; Boopathi & Kanike, 2023; Vanitha et al., 2023). Here's how the synergy between these technologies unfolds

IoT sensors collect data from agricultural ecosystems, such as soil moisture, temperature, humidity, crop health, and weather conditions, and transmit it in real-time to centralized systems or cloud-based platforms. AI algorithms analyze these data, identifying patterns and anomalies, and making predictions like weather forecasts and crop disease predictions. AI-driven decision support systems provide actionable recommendations, such as optimal irrigation schedules and maintenance needs. Automation is a key component of intelligent machines, which can perform tasks autonomously based on AI instructions.

Intelligent machines, like autonomous tractors and drones, utilize AI systems to perform tasks efficiently and accurately. They use real-time data from IoT sensors to adapt to changing conditions and optimize their actions. This integration also allows for resource optimization, such as real-time irrigation schedule adjustments based on soil moisture data, minimizing water wastage.

The integration of IoT, AI, and intelligent machines in agriculture promotes sustainable practices by optimizing resource usage, reducing waste, and preventing pesticide and fertilizer overuse. This reduces the environmental footprint and encourages agricultural research and innovation. This holistic approach empowers farmers and stakeholders to meet global population demands while minimizing environmental impacts and optimizing resource utilization. It also addresses climate change and food security challenges.

Smart Irrigation Systems

Smart irrigation systems utilize advanced technologies like sensors, controllers, and automation to optimize agricultural fields and landscapes by efficiently delivering water to plants at the right time.

Components of Smart Irrigation Systems

- **IoT Sensors:** Soil moisture sensors, weather stations, and other IoT devices collect real-time data on soil conditions, weather forecasts, and crop needs.
- **Controllers:** Centralized or cloud-based controllers receive data from sensors and make irrigation decisions based on preset parameters and algorithms.
- **Automation:** Automated valves and pumps control the flow of water to specific zones or areas within the irrigation system.

How Smart Irrigation Systems Work

- **Data Collection:** IoT sensors continuously monitor soil moisture levels, weather conditions, and other relevant data points in the field.
- **Data Analysis:** The collected data is sent to the controller, which uses AI and data analytics to process the information and make irrigation decisions.
- **Irrigation Optimization:** Based on the analyzed data, the controller determines when and how much water to apply to the crops.
- **Automated Water Delivery:** Automated valves and pumps open and close to deliver the right amount of water to the designated zones or areas.
- **Remote Monitoring:** Farmers and agronomists can monitor and adjust the irrigation system remotely through mobile apps or web-based interfaces.

Benefits of Smart Irrigation Systems

- **Water Efficiency:** Smart irrigation systems use data-driven decisions to optimize water use, reducing wastage and conserving valuable water resources.
- **Energy Savings:** By reducing water pump usage and minimizing over-irrigation, these systems also save energy and reduce operational costs.
- **Improved Crop Yields:** Providing the right amount of water at the right time enhances crop health and can lead to increased yields.
- **Environmental Sustainability:** Efficient irrigation practices reduce the environmental impact of agriculture, such as groundwater depletion and water pollution.
- **Cost Savings:** Smart irrigation systems can lower water and energy costs, making farming more economically viable.

Data Analytics for Water Management

Data analytics is crucial in agriculture, enabling informed decisions about irrigation, resource allocation, and sustainability, thereby enhancing the efficiency of agricultural operations (Dhanya et al., 2023; Pramila et al., 2023; Ramudu et al., 2023).

Figure 4. Data analytics for water management

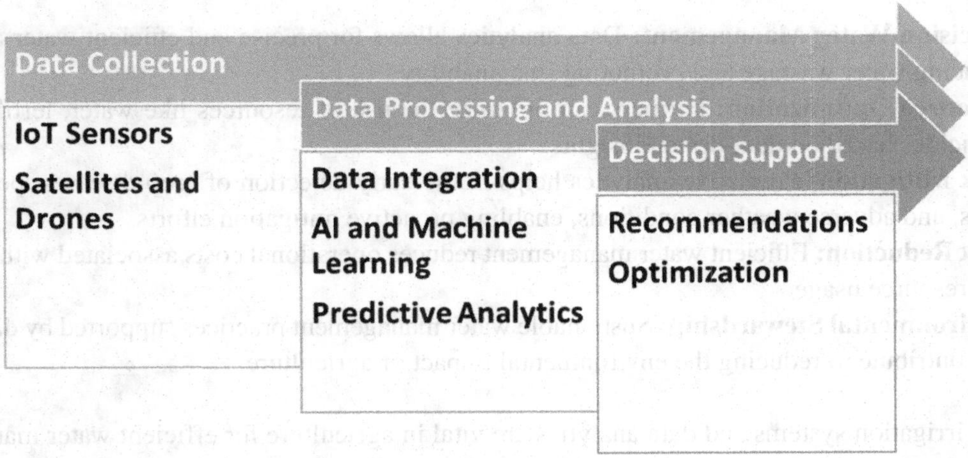

Data Collection

- **IoT Sensors:** IoT sensors collect data on soil moisture, temperature, humidity, and weather conditions in real-time.
- **Satellites and Drones:** Remote sensing technologies, such as satellites and drones, provide high-resolution imagery and data on crop health and land conditions.

Data Processing and Analysis

- **Data Integration:** Data from various sources are integrated into a centralized platform or database for analysis.
- **AI and Machine Learning:** AI and machine learning algorithms process the data, identifying patterns and trends related to water usage, soil conditions, and crop health.
- **Predictive Analytics:** Predictive models are developed to forecast irrigation needs, detect potential crop stress, and assess the impact of weather events.

Decision Support

- **Recommendations:** Data analytics generates actionable insights and recommendations for farmers and agricultural stakeholders.
- **Optimization:** Data-driven decisions are made to optimize irrigation schedules, resource allocation, and crop management practices.

Benefits of Data Analytics for Water Management

- **Precision Water Management:** Data analytics allows for precise and efficient water allocation, reducing water wastage and promoting sustainability.
- **Resource Optimization:** Farmers can optimize the use of resources like water, fertilizers, and pesticides based on data-driven insights.
- **Risk Mitigation:** Predictive analytics helps in the early detection of crop diseases, pest infestations, and adverse weather conditions, enabling proactive mitigation efforts.
- **Cost Reduction:** Efficient water management reduces operational costs associated with irrigation and resource usage.
- **Environmental Stewardship:** Sustainable water management practices supported by data analytics contribute to reducing the environmental impact of agriculture.

Smart irrigation systems and data analytics are vital in agriculture for efficient water management, cost reduction, and environmental sustainability by conserving resources and boosting crop yields.

BENEFITS AND CHALLENGES

Benefits of Integration

Enhanced Resource Efficiency

The integration of intelligent machines, IoT, and AI enhances water optimization, reducing waste and conserving resources like water, while also enhancing resource allocation efficiency and minimizing environmental impact (Figure 5).

Improved Crop Yields and Quality

Data analytics and AI-driven insights enhance crop management, leading to higher yields and improved quality. Precision agriculture practices address specific crop needs, resulting in healthier plants and increased productivity (Boopathi, 2021a; Veeranjaneyulu, Boopathi, Kumari, et al., 2023).

Reduced Environmental Impact

Smart irrigation, resource management, and early disease detection promote sustainable agriculture and water conservation, reducing environmental strain on aquifers and ecosystems, while also promoting water conservation.

Proactive Pest and Disease Management

AI aids in early detection of disease outbreaks and pest infestations, reducing crop losses and minimizing chemical use through targeted pesticide application.

Economic Benefits

Precision agriculture practices enhance farmers' profitability by reducing resource consumption, labor costs, and increasing yields, while also enhancing market competitiveness through efficient crop production.

Figure 5. Benefits and Challenges

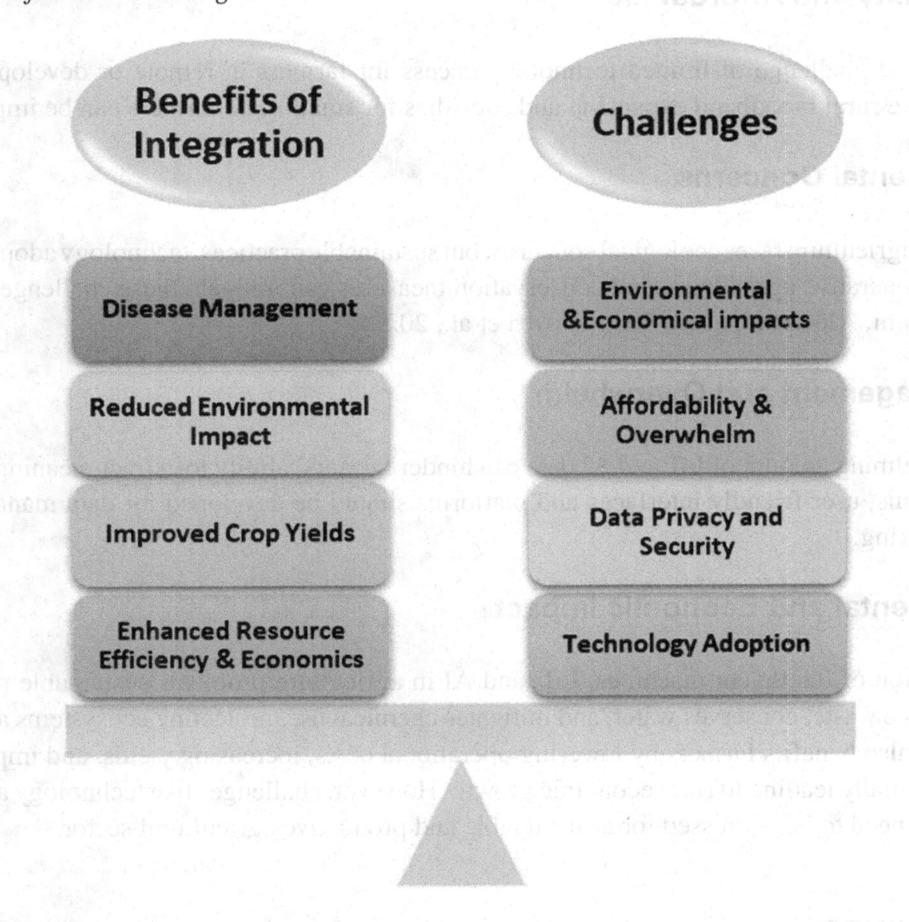

CHALLENGES AND SOLUTIONS

Technology Adoption

Farmers face challenges in adopting new technologies due to cost, training, and infrastructure limitations. Government incentives, subsidies, and support programs can facilitate technology adoption, while training and education programs can help farmers learn effectively.

Data Privacy and Security:

The collection and sharing of sensitive farm data raises privacy and security concerns. Solutions include robust data protection measures, secure communication channels, clear data ownership agreements, and industry-wide standardized protocols.

Accessibility and Affordability

To address the challenge of limited technology access for farmers in remote or developing regions, initiatives like rural broadband expansion and subsidies for small-scale farmers can be implemented.

Environmental Concerns

Large-scale agriculture faces ecological concerns, but sustainable practices, technology adoption, organic farming, regenerative agriculture, and conservation measures can mitigate these challenges (Boopathi, 2022; Boopathi, Alqahtani, et al., 2023; Gowri et al., 2023).

Data Management and Overwhelm

The overwhelming amount of IoT and AI data can hinder farmers' ability to extract meaningful insights. To address this, user-friendly interfaces and platforms should be developed for data management and decision-making.

Environmental and Economic Impact

The integration of intelligent machines, IoT, and AI in agriculture promotes sustainable practices, reduces resource waste, conserves water, and mitigates chemical use, protecting ecosystems and reducing pollution. It also benefits farmers by lowering operational costs, increasing yields, and improving crop quality, potentially leading to rural economic growth. However, challenges like technology adoption and data privacy need to be addressed for a sustainable and productive agricultural sector.

CASE STUDIES

IBM Watson Decision Platform for Agriculture

IBM Watson Decision Platform for Agriculture is an AI-powered solution that aids farmers in making informed decisions. It uses IoT sensors to collect real-time data on weather, soil conditions, and crop health, and uses AI algorithms to suggest irrigation, pest control, and crop management. The platform has led to increased crop yields, reduced water usage, and improved resource efficiency, as seen in a US cotton farmer (Gnanaprakasam et al., 2023; Koshariya, Kalaiyarasi, et al., 2023).

John Deere's See & Spray Precision Farming

John Deere's See & Spray Precision Farming system uses AI and computer vision to apply herbicides to weeds in real time, reducing chemical usage. The system uses cameras and AI algorithms to identify weeds, directing a spray nozzle to target only these. Farmers report significant reductions in herbicide usage, cost savings, and reduced environmental harm.

The Climate Corporation's Climate Field View

Climate Field View is a digital platform that uses data analytics and IoT technology to provide farmers with insights into their fields' performance. It uses IoT sensors and satellite data to collect weather, soil conditions, and crop growth information, allowing AI to provide personalized recommendations for planting, fertilizing, and irrigating. This has led to improved crop yields and resource efficiency.

Netafim's Precision Irrigation

Netafim, a global leader in smart drip and micro-irrigation solutions, uses IoT sensors and AI to create precision irrigation systems. These systems use soil moisture sensors, weather data, and crop data to determine optimal irrigation schedules and water requirements. Farmers have reported significant water savings, improved crop health, and increased yields. Case studies show that integrating intelligent machines, IoT, and AI in agriculture improves resource efficiency, increases crop yields, reduces environmental impact, and enhances economic sustainability.

CONCLUSION

The chapter explores the use of Intelligent Machines, IoT, and AI in agriculture, specifically in water processing and management. It highlights the importance of water and the challenges associated with it. The technologies are used in irrigation, data analysis, and crop management to enhance water processing, reduce waste, and improve resource efficiency, ultimately promoting sustainable water usage. The chapter underscores the potential of these technologies in agriculture.

The chapter explores the use of intelligent machines, IoT, and AI in agriculture, highlighting its potential for optimizing water usage, automating farming, and making data-driven decisions. It highlights benefits like increased resource efficiency, improved crop yields, and reduced environmental impact, while also addressing challenges like technology adoption and data privacy. Real-world case studies demonstrate successful implementations.

The chapter discusses the future trends in agriculture, focusing on AI, robotics, sustainable practices, and data-driven decision-making. It highlights the integration of intelligent machines, IoT, and AI in water processing and agriculture, highlighting their transformative shift towards more efficient and advanced farming practices.

REFERENCES

Agrawal, A. V., Shashibhushan, G., Pradeep, S., Padhi, S. N., Sugumar, D., & Boopathi, S. (2024). Synergizing Artificial Intelligence, 5G, and Cloud Computing for Efficient Energy Conversion Using Agricultural Waste. In Practice, Progress, and Proficiency in Sustainability (pp. 475–497). IGI Global. doi:10.4018/979-8-3693-1186-8.ch026

Anitha, C., Komala, C., Vivekanand, C. V., Lalitha, S., & Boopathi, S. (2023). Artificial Intelligence driven security model for Internet of Medical Things (IoMT). *IEEE Explore*, 1–7.

B, M. K., K, K. K., Sasikala, P., Sampath, B., Gopi, B., & Sundaram, S. (2024). Sustainable Green Energy Generation From Waste Water. In *Practice, Progress, and Proficiency in Sustainability* (pp. 440–463). IGI Global. doi:10.4018/979-8-3693-1186-8.ch024

Boopathi, S. (2021a). Improving of Green Sand-Mould Quality using Taguchi Technique. *Journal of Engineering Research*, in–Press.

Boopathi, S. (2021b). *Pollution monitoring and notification: Water pollution monitoring and notification using intelligent RC boat.*

Boopathi, S. (2022). An investigation on gas emission concentration and relative emission rate of the near-dry wire-cut electrical discharge machining process. *Environmental Science and Pollution Research International*, *29*(57), 86237–86246. doi:10.100711356-021-17658-1 PMID:34837614

Boopathi, S. (2023). Securing Healthcare Systems Integrated With IoT: Fundamentals, Applications, and Future Trends. In Dynamics of Swarm Intelligence Health Analysis for the Next Generation (pp. 186–209). IGI Global.

Boopathi, S., Alqahtani, A. S., Mubarakali, A., & Panchatcharam, P. (2023). Sustainable developments in near-dry electrical discharge machining process using sunflower oil-mist dielectric fluid. *Environmental Science and Pollution Research International*, •••, 1–20. doi:10.100711356-023-27494-0 PMID:37199846

Boopathi, S., Arigela, S. H., Raman, R., Indhumathi, C., Kavitha, V., & Bhatt, B. C. (2022). Prominent Rule Control-based Internet of Things: Poultry Farm Management System. *IEEE Explore*, 1–6.

Boopathi, S., & Kanike, U. K. (2023). Applications of Artificial Intelligent and Machine Learning Techniques in Image Processing. In *Handbook of Research on Thrust Technologies' Effect on Image Processing* (pp. 151–173). IGI Global. doi:10.4018/978-1-6684-8618-4.ch010

Boopathi, S., Kumar, P. K. S., Meena, R. S., Sudhakar, M., & Associates. (2023). Sustainable Developments of Modern Soil-Less Agro-Cultivation Systems: Aquaponic Culture. In Human Agro-Energy Optimization for Business and Industry (pp. 69–87). IGI Global.

Boopathi, S., Lewise, K. A. S., Subbiah, R., & Sivaraman, G. (2022). Near-dry wire-cut electrical discharge machining process using water–air-mist dielectric fluid: An experimental study. *Materials Today: Proceedings*, *50*(5), 1885–1890. doi:10.1016/j.matpr.2021.08.077

Boopathi, S., Pandey, B. K., & Pandey, D. (2023). Advances in Artificial Intelligence for Image Processing: Techniques, Applications, and Optimization. In Handbook of Research on Thrust Technologies' Effect on Image Processing (pp. 73–95). IGI Global.

Boopathi, S., & Sivakumar, K. (2014). Study of water assisted dry wire-cut electrical discharge machining. [SCI]. *Indian Journal of Engineering and Materials Sciences*, *21*, 75–82.

Boopathi, S., Thillaivanan, A., Mohammed, A. A., Shanmugam, P., & VR, P. (2022). Experimental investigation on Abrasive Water Jet Machining of Neem Wood Plastic Composite. *IOP: Functional Composites and Structures, 4*, 025001.

Chandrika, V., Sivakumar, A., Krishnan, T. S., Pradeep, J., Manikandan, S., & Boopathi, S. (2023). Theoretical Study on Power Distribution Systems for Electric Vehicles. In *Intelligent Engineering Applications and Applied Sciences for Sustainability* (pp. 1–19). IGI Global. doi:10.4018/979-8-3693-0044-2.ch001

Dhanya, D., Kumar, S. S., Thilagavathy, A., Prasad, D., & Boopathi, S. (2023). Data Analytics and Artificial Intelligence in the Circular Economy: Case Studies. In Intelligent Engineering Applications and Applied Sciences for Sustainability (pp. 40–58). IGI Global.

Domakonda, V. K., Farooq, S., Chinthamreddy, S., Puviarasi, R., Sudhakar, M., & Boopathi, S. (2022). Sustainable Developments of Hybrid Floating Solar Power Plants: Photovoltaic System. In Human Agro-Energy Optimization for Business and Industry (pp. 148–167). IGI Global.

Gnanaprakasam, C., Vankara, J., Sastry, A. S., Prajval, V., Gireesh, N., & Boopathi, S. (2023). Long-Range and Low-Power Automated Soil Irrigation System Using Internet of Things: An Experimental Study. In Contemporary Developments in Agricultural Cyber-Physical Systems (pp. 87–104). IGI Global.

Gowri, N. V., Dwivedi, J. N., Krishnaveni, K., Boopathi, S., Palaniappan, M., & Medikondu, N. R. (2023). Experimental investigation and multi-objective optimization of eco-friendly near-dry electrical discharge machining of shape memory alloy using Cu/SiC/Gr composite electrode. *Environmental Science and Pollution Research International*, *30*(49), 1–19. doi:10.100711356-023-26983-6 PMID:37126160

Gunasekaran, K., & Boopathi, S. (2023). Artificial Intelligence in Water Treatments and Water Resource Assessments. In *Artificial Intelligence Applications in Water Treatment and Water Resource Management* (pp. 71–98). IGI Global. doi:10.4018/978-1-6684-6791-6.ch004

Hema, N., Krishnamoorthy, N., Chavan, S. M., Kumar, N., Sabarimuthu, M., & Boopathi, S. (2023). A Study on an Internet of Things (IoT)-Enabled Smart Solar Grid System. In *Handbook of Research on Deep Learning Techniques for Cloud-Based Industrial IoT* (pp. 290–308). IGI Global. doi:10.4018/978-1-6684-8098-4.ch017

Jeevanantham, Y. A., Saravanan, A., Vanitha, V., Boopathi, S., & Kumar, D. P. (2022). Implementation of Internet-of Things (IoT) in Soil Irrigation System. *IEEE Explore*, 1–5.

Karthik, S., Hemalatha, R., Aruna, R., Deivakani, M., Reddy, R. V. K., & Boopathi, S. (2023). Study on Healthcare Security System-Integrated Internet of Things (IoT). In Perspectives and Considerations on the Evolution of Smart Systems (pp. 342–362). IGI Global.

Koshariya, A. K., Kalaiyarasi, D., Jovith, A. A., Sivakami, T., Hasan, D. S., & Boopathi, S. (2023). AI-Enabled IoT and WSN-Integrated Smart Agriculture System. In *Artificial Intelligence Tools and Technologies for Smart Farming and Agriculture Practices* (pp. 200–218). IGI Global. doi:10.4018/978-1-6684-8516-3.ch011

Koshariya, A. K., Khatoon, S., Marathe, A. M., Suba, G. M., Baral, D., & Boopathi, S. (2023). Agricultural Waste Management Systems Using Artificial Intelligence Techniques. In *AI-Enabled Social Robotics in Human Care Services* (pp. 236–258). IGI Global. doi:10.4018/978-1-6684-8171-4.ch009

Kumar, P., Sampath, B., Kumar, S., Babu, B. H., & Ahalya, N. (2023). Hydroponics, Aeroponics, and Aquaponics Technologies in Modern Agricultural Cultivation. In IGI:Trends, Paradigms, and Advances in Mechatronics Engineering (pp. 223–241). IGI Global.

Kumara, V., Mohanaprakash, T., Fairooz, S., Jamal, K., Babu, T., & Sampath, B. (2023). Experimental Study on a Reliable Smart Hydroponics System. In *Human Agro-Energy Optimization for Business and Industry* (pp. 27–45). IGI Global. doi:10.4018/978-1-6684-4118-3.ch002

Maguluri, L. P., Ananth, J., Hariram, S., Geetha, C., Bhaskar, A., & Boopathi, S. (2023). Smart Vehicle-Emissions Monitoring System Using Internet of Things (IoT). In Handbook of Research on Safe Disposal Methods of Municipal Solid Wastes for a Sustainable Environment (pp. 191–211). IGI Global.

Nishanth, J., Deshmukh, M. A., Kushwah, R., Kushwaha, K. K., Balaji, S., & Sampath, B. (2023). Particle Swarm Optimization of Hybrid Renewable Energy Systems. In *Intelligent Engineering Applications and Applied Sciences for Sustainability* (pp. 291–308). IGI Global. doi:10.4018/979-8-3693-0044-2.ch016

Pramila, P., Amudha, S., Saravanan, T., Sankar, S. R., Poongothai, E., & Boopathi, S. (2023). Design and Development of Robots for Medical Assistance: An Architectural Approach. In Contemporary Applications of Data Fusion for Advanced Healthcare Informatics (pp. 260–282). IGI Global.

Ramudu, K., Mohan, V. M., Jyothirmai, D., Prasad, D., Agrawal, R., & Boopathi, S. (2023). Machine Learning and Artificial Intelligence in Disease Prediction: Applications, Challenges, Limitations, Case Studies, and Future Directions. In Contemporary Applications of Data Fusion for Advanced Healthcare Informatics (pp. 297–318). IGI Global.

Samikannu, R., Koshariya, A. K., Poornima, E., Ramesh, S., Kumar, A., & Boopathi, S. (2022). Sustainable Development in Modern Aquaponics Cultivation Systems Using IoT Technologies. In *Human Agro-Energy Optimization for Business and Industry* (pp. 105–127). IGI Global.

Sankar, K. M., Booba, B., & Boopathi, S. (2023). Smart Agriculture Irrigation Monitoring System Using Internet of Things. In *Contemporary Developments in Agricultural Cyber-Physical Systems* (pp. 105–121). IGI Global. doi:10.4018/978-1-6684-7879-0.ch006

Satav, S. D., Hasan, D. S., Pitchai, R., Mohanaprakash, T. A., Sultanuddin, S. J., & Boopathi, S. (2024). Next Generation of Internet of Things (NGIoT) in Healthcare Systems. In Practice, Progress, and Proficiency in Sustainability (pp. 307–330). IGI Global. doi:10.4018/979-8-3693-1186-8.ch017

Satav, S. D., & Lamani, D. G, H. K., Kumar, N. M. G., Manikandan, S., & Sampath, B. (2024). Energy and Battery Management in the Era of Cloud Computing. In Practice, Progress, and Proficiency in Sustainability (pp. 141–166). IGI Global. doi:10.4018/979-8-3693-1186-8.ch009

Sengeni, D., Padmapriya, G., Imambi, S. S., Suganthi, D., Suri, A., & Boopathi, S. (2023). Biomedical Waste Handling Method Using Artificial Intelligence Techniques. In *Handbook of Research on Safe Disposal Methods of Municipal Solid Wastes for a Sustainable Environment* (pp. 306–323). IGI Global. doi:10.4018/978-1-6684-8117-2.ch022

Vanitha, S., Radhika, K., & Boopathi, S. (2023). Artificial Intelligence Techniques in Water Purification and Utilization. In *Human Agro-Energy Optimization for Business and Industry* (pp. 202–218). IGI Global. doi:10.4018/978-1-6684-4118-3.ch010

Veeranjaneyulu, R., Boopathi, S., Kumari, R. K., Vidyarthi, A., Isaac, J. S., & Jaiganesh, V. (2023). Air Quality Improvement and Optimisation Using Machine Learning Technique. *IEEE- Explore*, 1–6.

Veeranjaneyulu, R., Boopathi, S., Narasimharao, J., Gupta, K. K., Reddy, R. V. K., & Ambika, R. (2023). Identification of Heart Diseases using Novel Machine Learning Method. *IEEE- Explore*, 1–6.

Venkateswaran, N., Vidhya, R., Naik, D. A., Raj, T. M., Munjal, N., & Boopathi, S. (2023). Study on Sentence and Question Formation Using Deep Learning Techniques. In *Digital Natives as a Disruptive Force in Asian Businesses and Societies* (pp. 252–273). IGI Global. doi:10.4018/978-1-6684-6782-4.ch015

Vennila, T., Karuna, M., Srivastava, B. K., Venugopal, J., Surakasi, R., & Sampath, B. (2022). New Strategies in Treatment and Enzymatic Processes: Ethanol Production From Sugarcane Bagasse. In Human Agro-Energy Optimization for Business and Industry (pp. 219–240). IGI Global.

Zekrifa, D. M. S., Kulkarni, M., Bhagyalakshmi, A., Devireddy, N., Gupta, S., & Boopathi, S. (2023). Integrating Machine Learning and AI for Improved Hydrological Modeling and Water Resource Management. In *Artificial Intelligence Applications in Water Treatment and Water Resource Management* (pp. 46–70). IGI Global. doi:10.4018/978-1-6684-6791-6.ch003

LIST OF ABBREVIATIONS

IoT - Internet of Things
AI - Artificial Intelligence
AIoT - Artificial Intelligence of Things
AI-driven - Artificial Intelligence-driven
CRISPR - Clustered Regularly Interspaced Short Palindromic Repeats
GPS - Global Positioning System
CEA - Controlled Environment Agriculture
5G - Fifth Generation (wireless technology)
QR - Quick Response (code)
IBM - International Business Machines Corporation

Chapter 16
A Mixture Model for Fruit Ripeness Identification in Deep Learning

Bingjie Xiao
Auckland University of Technology, New Zealand

Minh Nguyen
Auckland University of Technology, New Zealand

Wei Qi Yan
Auckland University of Technology, New Zealand

ABSTRACT

Visual object detection is a foundation in the field of computer vision. Since the size of visual objects in an images is various, the speed and accuracy of object detection are the focus of current research projects in computer vision. In this book chapter, the datasets consist of fruit images with various maturity. Different types of fruit are divided into the classes "ripe" and "overripe" according to the degree of skin folds. Then the object detection model is employed to automatically classify different ripeness of fruits. A family of YOLO models are representative algorithms for visual object detection. The authors make use of ConvNeXt and YOLOv7, which belong to the CNN network, to locate and detect fruits, respectively. YOLOv7 employs the bag-of-freebies training method to achieve its objectives, which reduces training costs and enhances detection accuracy. An extended E-ELAN module, based on the original ELAN, is proposed within YOLOv7 to increase group convolution and improve visual feature extraction. In contrast, ConvNeXt makes use of a standard neural network architecture, with ResNet-50 serving as the baseline. The authors compare the proposed models, which result in an optimal classification model with best precision of 98.9%.

DOI: 10.4018/978-1-6684-9999-3.ch016

INTRODUCTION

In the field of computer vision (Gowdra, 2021), digital cameras are utilized to emulate biological vision, enabling computers to process the contents of images or videos in a manner akin to human perception (Pan, & Yan, 2020). The object detection (Qi, Nguyen, & Yan, 2022) task (Zhang, Wang, Liu, & Xiong 2022). in computer vision primarily focuses on identifying visual objects within entire images, which includes detecting both the object and its location (Zhao, & Yan, 2021). In the realm of visual object detection (Shi, Li, & Yamaguchi 2020)., models such as CNN (Liu, Yan,&Yang, 2018), R-CNN, Fast R-CNN, Faster R-CNN(Al-Sarayreh,et. al., 2019), and the YOLO (Zhijun, et. al., 2021) series have successfully located and classified (Liu, Nouaze, Touko Mbouembe, & Kim 2020) fruit images (Gowdra, et. al., 2021). Building upon this foundation, the YOLOv7 (Liu, & Yan, 2023) model has improved the speed and accuracy of visual object detection (Yao et. al., 2021).

In recent years, artificial intelligence has been widely employed in various fields (Wang, & Yan, 2021). In view of the lack of labor in fruit picking and subsequent fruit quality classification (Xia, Nguyen, & Yan, 2022) that requires a lot of human labors (Xia, Nguyen, & Yan, 2023). In this book chapter, we propose an automatic fruit recognition algorithm based on YOLOv7 and ConvNext (Tian, 2022) models. The application of the above is mainly to build a deep learning model that can distinguish different fruit categories (Bazame, 2021 (apples and pears) for the same kind of fruit to distinguish the category level according to the degree of skin folds (Kang, & Chen, 2020). The high-precision fruit (apple, pear) detection and recognition (Wang, &Yan, 2021) system based on deep learning can be harnessed in daily life or in the wild to detect and locate fruit targets (Fu, Nguyen, & Yan, 2022). Using deep learning algorithms, it can realize fruit target detection and recognition in the form of pictures, videos, cameras, etc. In addition, it supports results visualization and export of image or video inspection results (Bhargava, & Bansal, 2021).

Visual object detection is characterized by using location and classification (Liu, Sun, Gu, & Deng, 2022). In a two-dimensional image, target detection can locate the position of an apple in the picture, and distinguish the current apple type as "ripe apple". Firstly, we preprocess the dataset, then input the backbone network to extract features, and take use of ELAN attention. The module acts on the corresponding channel of the feature map to obtain effective features for fruit recognition; then the model performs feature fusion to obtain semantic information and locate the feature map of the information. Finally, accurate detection results are obtained through classification and prediction frame regression calculations (Gokhale, Chavan, & Sonawane, 2023).

In this book chapter, we employ anchor boxes to label fruits and their maturity levels (Xiao, Nguyen, Yan, 2023). We leverage ConvNeXt (Qi, Nguyen, & Yan, 2022) and YOLOv7 models to obtain an optimal model for fruit ripeness classification. ConvNeXt (Hassanien, Singh, Puig, & Abdel-Nasser, 2022) optimizes the technology and parameters of the original CNN to achieve state-of-the-art performance. A characteristic of ConvNeXt is that it does not consider the visual features; it simply inputs the image as a patch and sends it to the deep learning network model for training and testing (Feng, Tan, Li, & Xie, 2022). Conversely, YOLOv7 focuses on optimizing modules and methods without increasing training costs. YOLOv7 serializes or parallelizes network layers into a convolutional group to reduce computations and enhance training speed (Junos, Mohd Khairuddin, Thannirmalai, & Dahari, 2022).

Agricultural harvesting is a labor-intensive process. Utilizing a visual object detection model to classify fruits is the motivation of this book chapter. The visual object detection pipeline is illustrated in Figure 1. The dataset is input into the model for training, and a predicted bounding box is subsequently output. As demonstrated in Figure 1, YOLO model (KIVRAK, & GÜRBÜZ, 2022) is use of the entire image as input, employs a CNN network for end-to-end design, and effectively returns the position and class label of the bounding box at the output layer (Zhang et al., 2022). In our experiments, the YOLOv7 model accurately detects and classifies fruits.

This study demonstrates the use of YOLOv7 model (Kuznetsova, Maleva, & Soloviev, 2021), ConvNeXt, and their transfer learning to detect fruits, which can accurately classify fruit types and their maturity levels (Lee & Kim, 2020). Simultaneously, we also created our own datasets using mobile phones to increase the influence of the environment on experimental results.

The contribution of this book chapter is that we created our own dataset. We take advantage of YOLOv7, ConvNext, and improved models to locate and classify fruits. The model can realize the fruit detection task and achieve high precision.

In the second section of this book chapter, we will discuss the development of the proposed YOLOv7 model. The third section will include the experimental details and outcomes. In the fourth section, we will showcase the training results of the YOLOv7 model and summarize the advantages and disadvantages of the proposed model. In this book chapter, following the section on related work, the proposed methods will be elaborated upon. The result analysis will be explained, leading to the final conclusion of this book chapter.

Figure 1. The pipeline of YOLO model in visual object detection

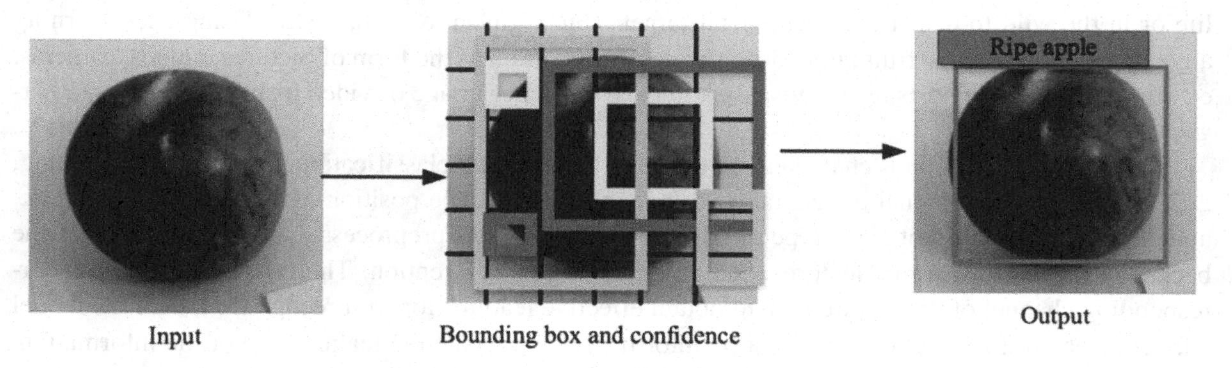

RELATED WORK

A model with YOLOv4-based convolution block was proposed to add an attention mechanism, which judges the maturity of apples by distinguishing colors. For the actual situation in the orchard, the size of fruit trees in the orchard, the influence of branches and leaves on fruits, actual size and color of the apples (Gongal, Karkee, & Amatya, 2018) are all issues that need to be considered in the real object detection (Pan, Liu, & Yan, Zhou, 2021). The YOLOv4 model proposed by Lu et. al. achieved an accuracy 86.2% for a number of fruits, which is about 3% higher than the original YOLOv4 model (Lu et. al., 2022).

Ou et. al. proposed an improved FSOne-YOLOv7 model for the detection of passion fruit (Ou et. al., 2023). ShuffleOne as a new backbone network and slim-neck as an improved YOLOv7 network of the neck network are employed for passion fruit detection in complex natural environments. The FSOne-

YOLOv7 model takes advantage of gradient weighted class activation mapping to enhance the feature extraction and fusion capabilities. Ou et. al. achieved an average accuracy 94.5%. The improved model can better extract features, thereby improving detection speed.

Zhou, et al. also studied how visual inspection can replace manual picking of dragon fruits (Zhou, Zhang, Wang, 2023). Zhou et. al. proposed a PSP-Ellipse method based on YOLOv7 to implement classification. The PSP-Ellipse method detects the endpoints of dragon fruit in the picture by segmenting the detection target and using an ellipse fitting algorithm, and then takes use of ResNet to implement the classification task. In the PSP-Ellipse endpoint detection task, the model achieved an accuracy of 92% for dragon fruit.

Another experiment that has great significance is studied. (Wu, et. al., 2022). Previous agricultural-related detections were based on fruit color and shape classification. However, traditional agricultural detection is prone to false detection in complex natural environments, and the model lacks robustness. Wu et. al. studied target detection based on complex environments, made use of the module characteristics of YOLOv7 data enhancement, and established an improved DA-YOLOv7 model. The DA-YOLOv7 model strengthens the generalization ability of the model in complex environments, which is adopted for the detection of Camellia oleifera under the interference of side light, backlight, slight occlusion and heavy occlusion.

A visual object detection method was proposed to solve fruit counting problem. SSD was employed with MobileNet and Faster R-CNN with Inception V2 for multi-fruit object tracking based on Gaussian estimation. Vasconez et. al. achieved 90% accuracy using SSD model and 93% accuracy using Faster R-CNN (Vasconez et. al., 2020).

A nighttime dataset was collected which demonstrated that YOLOv4 model achieved F1 score 0.968 and average precision 0.983 with images from various orchards, varieties and lighting conditions for real-time mango detection (Koirala et al., 2019).

YOLOv2 was designed for visual object location prediction (Sozzi, et. al, 2022). YOLOv3 continues the idea of YOLOv2. The FPN structure is adopted in YOLOv3 to improve the accuracy of corresponding multi-scale target detection (Liu, et. al., 2022). YOLOv5 (Wang et. al., 2022) was basically modified based on the structure of YOLOv3(Wang, Jin, Wang, & Xu, 2022). YOLOv5 was use of CSPDarknet (Cross Stage Partial Networks) as the backbone to extract visual features from the input image (Wang, & He, 2021).

The difficulty of multi-target tracking in model training lies in the fact that real-life targets are occluded, blurred and deformed. In this project, fruit recognition from digital images is also affected by the natural environment, such as lighting, overlapping, scales change and other factors that affect the final results (Yang et. al., 2022). The E-ELAN module of YOLOv7 can enhance the ability of network and guide different modules. E-ELAN can enhance the net ability without changing the gradient.

Hussain et al. also proposed YOLOv7 for visual object detection (Hussain et. al., 2022). YOLOv7 NAS can implement iterative search by mining the optimal scale factor according to the resolution, width, and depth, as well as the number of feature pyramids. Reparameterization can assist the gradient propagation path to reintegrate the parameters of the model, so that the head module can be applied to fruit detection.

Swin Transformer (Ruiz, et. al., 2022) takes advantage of hierarchical feature maps which are similar to convolutional neural networks (Gowdra, 2021). After the image is downsampled by 4 times, 8 times, or 16 times in the size of the feature map, the backbone builds tasks such as target detection and instance segmentation on this basis. The concept of Windows Multi-Head Self-Attention (W-MSA) was employed in Swin Transformer. For example, in the 4 times downsampling or 8 times downsampling, the feature map is segmented into multiple disjoint regions, and each self-attention is only performed within each window. Transformer can effectively reduce the amount of calculations if the shallow feature map is large. ConvNeXt is based on ResNet, the process of transforming ResNet into ConvNet is similar to the construction process of Transformer. ConvNeXt maintains the simplicity of CNN neural networks (An, &Yan, 2021) while following the structure of Swin Transformer model.

METHODOLOGY

YOLO series models are a one-stage network structure. There is only one neural network in the whole process, which is able to achieve end-to-end structure. After an image is input into YOLO model, the image is segmented into s × s grids, each grid can be processed to obtain bounding boxes and the confidence score of each box. In Figure 2, the blue, yellow, and red boxes are the bounding boxes. Each grid predicts the conditional class probabilities, the confidence of each bounding box is multiplied by using probability. The result contains class information and accuracy of the bounding box prediction. Finally, we set the threshold, filter out the low scores, and cast the rest to non-maximum suppression, and then get the prediction box.

Figure 2. Prediction of bounding boxes

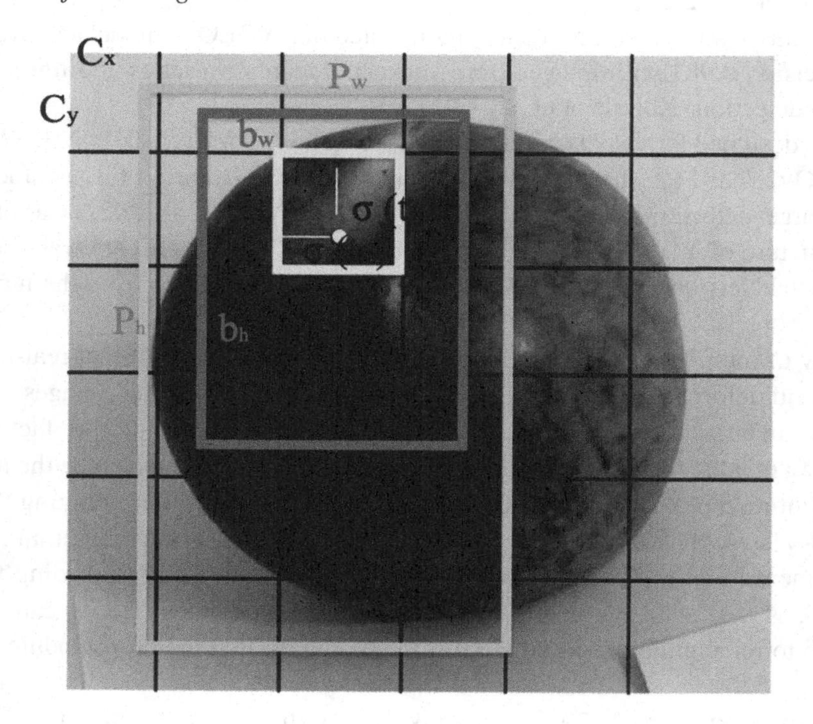

In Figure 2, b_x, b_y, b_w, and b_h show the value of the predicted bounding box, C_x and C_y represent the distance from the upper left corner of the current grid to the upper left corner of the image. P_w and P_h are the width and height of anchor box, respectively; σ is the sigmoid function. t_x, t_y, t_w, t_h, and t_0 are the parameters which are employed to calculate the bounding box and confidence. The centre of the predicted box is the point inside the yellow box as shown in Figure 2, then a group of equations for calculating is listed as follows:

$$b_x = \sigma(t_x) + C_x \tag{1}$$

$$b_y = \sigma(t_y) + C_y \tag{2}$$

$$b_x = P_w e^{t_w} \tag{3}$$

$$b_x = P_h e^{t_h} \tag{4}$$

The confidence and maximum suppression values are:

$$Confidence\ score = P(Object) \times IoU_{truth\_pred} \tag{5}$$

$$Class - specific\ confidence\ scores = Confidence \times P(Class \mid Object) \tag{6}$$

Neck in YOLO is mainly applied to generate feature pyramids. The feature pyramid will enhance the object detection at hierarchical scales, the same object having different sizes and scales will be recognized. CSPNet backbone solves the gradient duplication problem of network optimization in the backbone, large-scale convolutional neural network frameworks integrate the gradient changes into the feature map from beginning to end, thus reduce the parameter amount and FLOPS value of the model, and ensure the inference speed and accuracy, and decrease the size of the proposed model. Head is mainly employed to the final part which applies anchor boxes to the feature map and generate the final output vector with class probabilities, object scores and bounding boxes.

In YOLOv7, a decoupled training-time and inference-time architecture was proposed for training a multi-branch model, converting it into a single-channel model and deploying it. In this way, the high performance of the multi-branch architecture and the fast inference advantage of the single-branch model are realized.

In Figure 3, YOLOv7 merges all Conv and BN layers, and converts the fused Conv layer into a 3×3 Conv layer. A 1×1 Conv layer is converted into a 3×3 Conv layer by using the center weight that equals to the 1×1 Conv layer which merges the branch 3×3 Conv layer. Finally, the weights and bias of the convolution kernels of all branches are added to form a new 3×3 Conv layer. YOLOv7 finally constitutes an identity mapping branch, that is, a RepVGG block. The mapping network of YOLOv7 is similar to the residual network of ResNet, that is, adding a branch at a specific layer.

Figure 3. Reparameterization process of RepVGG

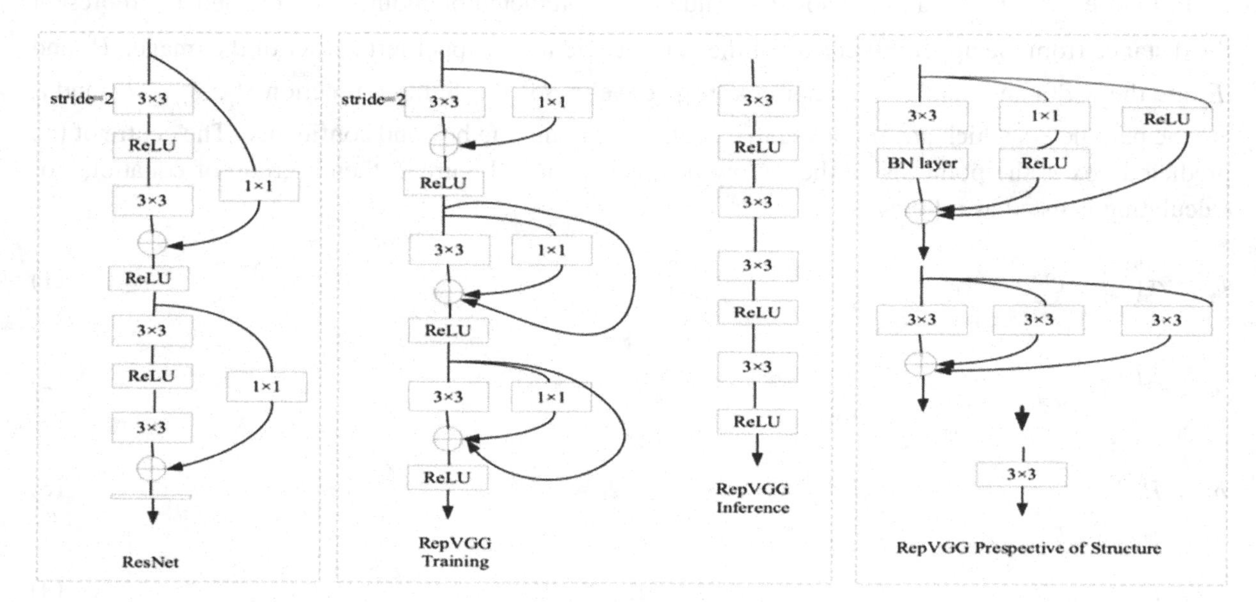

The parameter fusion process is:

$$W'_{i,:,:,:} = \frac{\gamma_i}{\sigma_i} W_{i,:,:,:} \quad (7)$$

$$b'_i = -\frac{\mu_i \gamma_i}{\sigma_i} + \beta_i \quad (8)$$

where μ_i, σ_i, γ_i, β_i are the mean, variance, scale factor and offset factor of BN, respectively, W_i is the original convolution weight. Eq.(9) for each calculation in Conv is,

$$Conv(x) = W \times X \quad (9)$$

$$BN(x) = \gamma_i \left(\frac{x - \mu_i}{\sqrt{\sigma_i^2 + \varepsilon}} \right) \quad (10)$$

where ε is equal to the minimum. The fused result of Conv and BN is,

$$\left(\frac{W \times x - \mu_i}{\sqrt{\sigma_i^2 + \varepsilon}} \right) + \beta_i = \frac{\gamma_i}{\sqrt{\sigma_i^2 + \varepsilon}} W_i X - \frac{\mu_i \gamma_i}{\sqrt{\sigma_i^2 + \varepsilon}} + \beta_i \quad (11)$$

Ignored the minimum value ε, the new convolution is,

$$Conv\left(x\right) = W'x + \beta' \tag{12}$$

Compared with the calculations after fusion, it is essentially a linear operation of convolution.

In Figure 4, scale has always been one of the characteristics of the YOLO model. YOLOv7 adopts the composite model scaling method, modifies the depth factor and calculates the proportion of the corresponding change in the transfer layer. The optimal state of the model can be maintained by scaling the model and the width corresponding to the depth scaling factor [20, 38].

Figure 4. Stitch-based model scaling

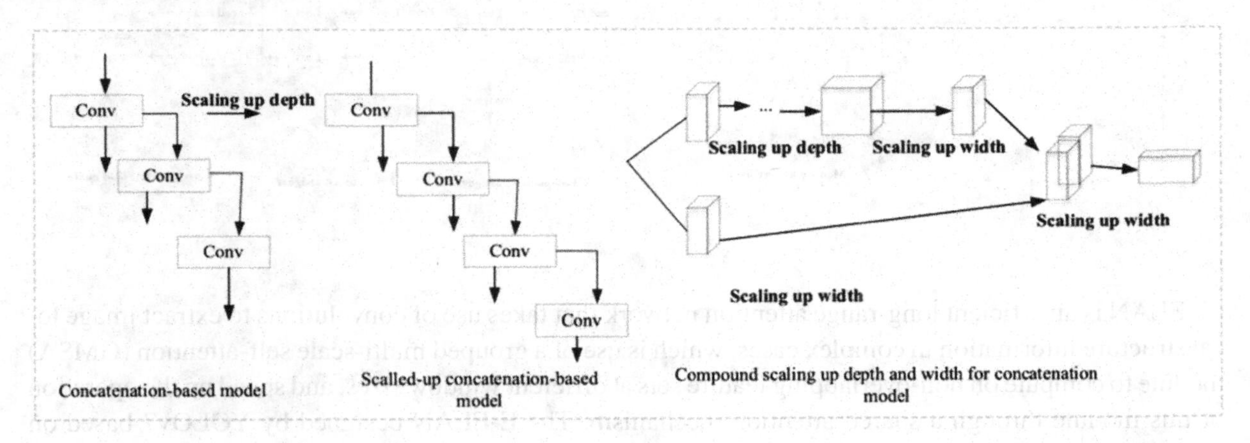

YOLOv7 model scales the stack in the Neck module, which is use of a composite scaling method to scale the depth and width of the entire model to obtain the weight YOLOv7-X. E-ELAN is applied to the weights YOLOv7-E6. The weights of YOLOv7, YOLOv7-X and YOLOv7-E6 are use of SiLu as the activation function,

$$SiLu\left(x\right) = x \times sigmoid\left(x\right) \tag{13}$$

The SiLU function is the abbreviation of sigmoid weighted linear unit, which is adopted as the activation function. Unlike other activation functions (e.g., sigmoid, tanh), the activation function SiLU is not monotonically increasing. The SiLU function is self-stabilizing and acting as an implicit regulariser on the weights at the global minimum with zero derivative, inhibiting the learning of a large number of weights. The weight YOLOv7-tiny is an edge GPU-oriented architecture. Leaky tunes the zero-gradient problem for negative values by giving the negative input x a tiny linear component.

In Figure 5, YOLOv7 separates the auxiliary and dominant heads, and performs label assignment with the respective predictions and ground truths. Deep supervision information is employed for adding additional supervision information to the model so as to improve the performance of the model. The leading head represents the feature map responsible for the final output, and the auxiliary head represents the additional training branch added for auxiliary training.

Figure 5. Auxiliary head and leading head perform label assignment process using prediction results and ground-truth values

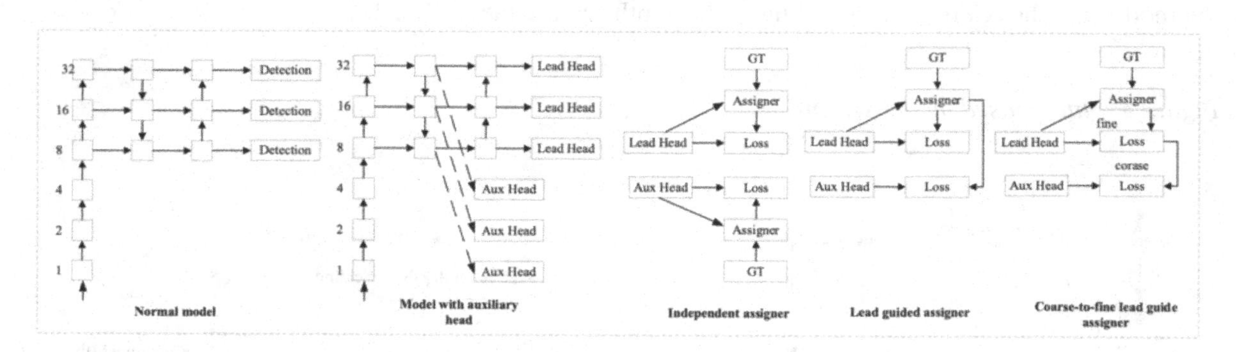

ELAN is an efficient long-range attention network that takes use of convolutions to extract image local structure information in complex cases, which is use of a grouped multi-scale self-attention (GMSA) module to compute on non-overlapping feature sets at different window sizes, and speed up the operation of this module through a shared attention mechanism. The E-ELAN designed by YOLOv7 based on ELAN has the ability to expand, shuffle, and merge cardinality to achieve the ability and continuously enhance the network learning ability without destroying the original gradient path.

All ConvNeXt models are the existing structures and methods, the transformation process is similar to Transformer's construction process. The starting point of ConvNeXt is ResNet, which takes use of enhanced training methods to improve the performance of the ResNet-50 model. The ConvNeXt network structure is composed of macro design, ResNeXt, inverted bottleneck, large kernel size, and various micro designs with layers as the smallest granularity.

The ConvNeXt network adjusts the stacking times of each stage of ResNet from (3, 4, 6, 3) to (3, 3, 9, 3), which increases the accuracy with the cost of increasing calculation scales. The stem layer in Swin Transformer network is a convolutional layer with a convolution kernel size of 4 and a stride of 4. The stem layer of ResNet50 consists of a convolutional layer with a kernel size of 7 and a stride of 2 plus a maximum pooling layer with a kernel size of 3 and a stride of 2. As a combination of Transformer networks and ResNet models, ConvNeXt replaces the stem layer with the same convolution layer as the Swin Transformer network with a convolution kernel size of 4 and a step size of 4, and its accuracy has a small improvement.

Compared with classical ResNet network, the ResNeXt network has achieved a balance between FLOPs and accuracy. ResNeXt takes use of group-wise convolution in the middle of the convolution block to make the convolution block form a parallel structure, while the volume of the ResNet network increases, the block is similar to the structure of bottleneck "thick at both ends and thin in the middle".

In Figure 6, compared with ResNeX and ResNet, the ConvNeXt network makes use of depth-wise convolution to form a convolution block, which greatly reduces the parameter scale of the network while sacrificing a part of the accuracy.

Figure 6. ResNet and ResNeXt blocks

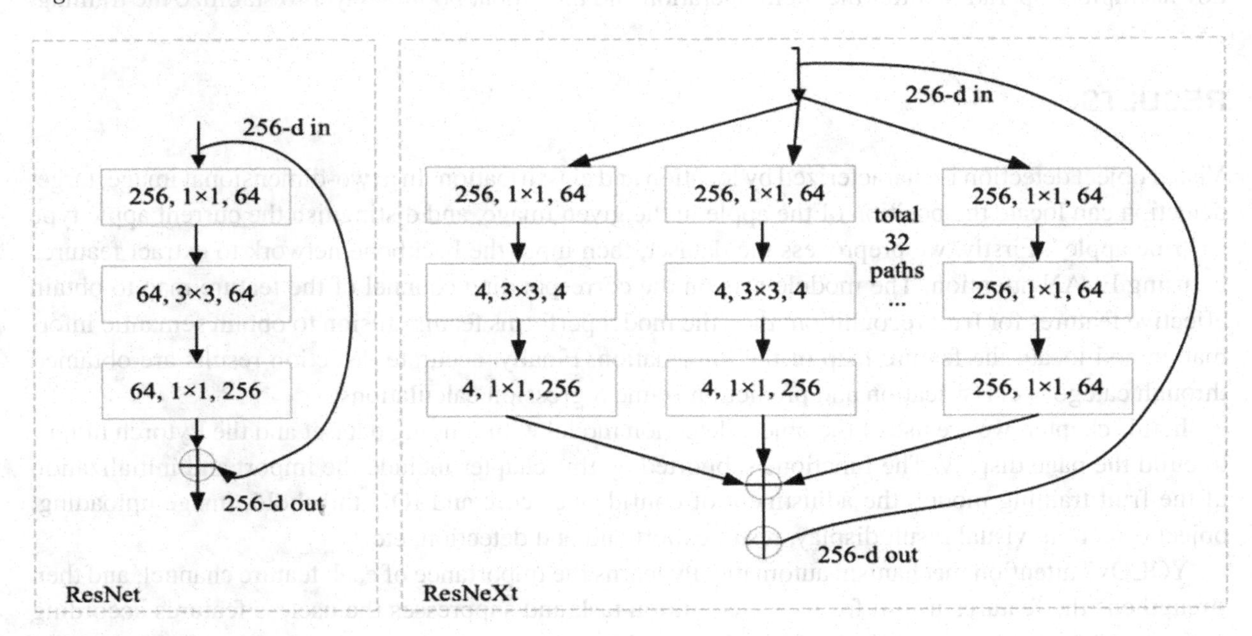

The number of output feature channels of the stem layer in the Swin Transformer network is 96, while the output of the stem layer of the ResNet network is only 64 dimensions. In order to be consistent with the Swin Transformer network, the ConvNeXt network increases the number of output dimensions to make it the same as the Swin-T network, which greatly improves the accuracy of the network, but at the same time inevitably increases the parameter scale of the model.

The ConvNeXt network was designed with a similar inverted bottleneck structure, which can partially reduce the parameter size of the model and improve the overall performance of the model while slightly improving the accuracy.

In the ConvNeXt network, the convolution kernel size of depthwise conv is changed from 3×3 to 7×7 like Swin Transformer, which saturates the accuracy. The current mainstream convolutional neural network has a 3×3 window size. However, a 3×3 window will result in a smaller receptive field, and ConvNeXt can increase the receptive field by using a large convolution kernel, and to a certain extent, more information can be obtained.

ConvNeX replaces the regular activation function ReLU with GELU with fewer activation functions. In a convolutional neural network, an activation function is generally connected after each convolutional layer or full connection. It is not every module in ConvNeXt which is followed by an activation function. At the same time, ConvNeXt is use of less Normalization. The normalization layer in the ConvNeXt block only retains the normalization layer after the depthwise convolution. Batch Normalization (BN) can speed up the convergence of network and reduce overfitting in the convolutional neural network. The downsampling operation of ResNet is completed at the beginning of each stage using a 3×3 convolution with a step size of 2 and a 1×1 convolution with a direct step size of 2. ConvNext performs independent downsampling between different stages, using 2 × 2 convolution with a step size of 2 for spatial downsampling. This change will lead to unstable training, so a layer-based normalization is added before the downsampling operation, after the Stem operation and the global pooling layer to stabilize the training.

RESULTS

Visual object detection is characterized by location and classification. In a two-dimensional image, target detection can locate the position of the apple in the given image, and distinguish the current apple type as "ripe apple". Firstly, we preprocess the dataset, then input the backbone network to extract features by using ELAN attention. The module acts on the corresponding channel of the feature map to obtain effective features for fruit recognition; then the model performs feature fusion to obtain semantic information and locate the feature map of the information. Finally, accurate detection results are obtained through category classification and prediction frame regression calculations.

In this chapter, we are use of the object detection model with training dataset and the Pytorch library to build the page display. The functions supported by this chapter include the import and initialization of the fruit training model; the adjustment of confidence score and IOU threshold, image uploading, object detection, visual result display, result export and end detection, etc.

YOLOv7 attention mechanism automatically learns the importance of each feature channel, and then strengthens the features useful for fruit recognition task and suppresses the useless features according to the importance. Aiming at the problem that GIOU cannot accurately express the overlap relationship between fruit recognition frames when the prediction frame overlaps with the target frame. In this book chapter, the original frame regression loss function GIOU is replaced with CIOU, taking into account the height-to-width ratio and the center point of the target frame and prediction frame. relationship, thereby making the fruit prediction frame closer to the real frame and improving the prediction accuracy. Therefore, mean average precision (mAP) is shown as an indicator to evaluate the model.

The IOU threshold is the degree of overlap between the predicted frame and the ground-truth frame. mAP@.5 means that if IoU is set to 0.5, the AP of all pictures in each category is calculated and averaged. mAP@.5:.95 indicates different IoU thresholds, from 0.5 to 0.95, with a step length 0.05. The larger the IOU, the smaller the number of preselected boxes, which leads to a corresponding increase in the ratio. We observed Table 1~ Table 5 that mAP@.5 results are better than mAP@.5:.95. We set the IOU higher to filter out boxes with low confidence scores. Therefore, the identified frame is basically around the target and counted as a positive sample. The object detection is to select the closest positive sample based on a group of positive samples in Figure 7.

In this book chapter, fruit images taken by mobile phones are employed for fruit detection. We took use of LabelMe to annotate the dataset, the dataset has four classes: "Ripe apple", "overripe apple", "ripe pear", "overripe pear", with the bounding boxes. Given the IOU threshold 0.7, we calculate the average precision as the evaluation. We adjusted the weights of YOLOv7 model. Under the same weights, the number of iterations is taken into account on with the accuracy. We trained the model with batch size 64, Adam optimization with an initial learning rate 0.0002. Our dataset has a total of two thousand fruits and their maturity labels.

We chose four weights YOLOv7, YOLOv7-X, YOLOv7-E6 and YOLOv7-tiny for our experiments, and compared the model performance. We loaded the pretrained weights, compared the backbone network with the network parameters including the pretrained weights, and see how many layers are the same. The training process will only load the same number of layers, we observe how the model is trained by adjusting the number of iterations. At the same time, we are use of the ConvNeXt model for training, and take advantage of ConvNeXt pre-training model for transfer learning.

The bag-of-freebies in YOLOv7 model improved the accuracy of fruit detection. The scaling method of YOLOv7 model reduces the loss of visual information. Adaptive image scaling can deepen the model with visual features, ensure that the overall image transformation is consistent, the information of receptive field can be effectively utilized. The replacement of the reparametrized module and the assignment of dynamic label assignments compute the prediction results and ground truth values, which enable the dominant leader to have strong learning ability through the optimization process.

In order to improve the recognition accuracy of fruits with only different local features and similar global features, the ELAN module of YOLOv7 aims at the problem of poor model performance in model scaling. YOLO v7 borrows from ResNeXt, takes use of 1×1 conv for dimensionality reduction, then convolutes separately, and finally adds YOLOv7 re-parameterization method in residual structure and the problem of dynamic label assignment in multiple output layers.

In Tables 1, 2 and 3, YOLOv7 can achieve better fruit positioning. Figure 7(a) shows that though the model cannot accurately determine the category of the fruit, which can still precisely locate the location of the fruit. After the model has learned enough features, Figure 9(b) shows the detection results of the model.

In Table 4, we observed that E-ELAN module with YOLO-E6 weights enables the deep network to converge efficiently by controlling the shortest and longest gradient paths with the same number of iterations. The weight YOLO-tiny takes use of LeakyReLU function to resolve the problem that the parameters cannot be updated after the neural network accepts the input of the outlier range. During the backpropagation process, a large gradient will be generated because the derivatives are multiplied continuously, so the parameters cannot be updated. This leads to the vanishing gradient problem. For the input of LeakyReLU less than 0, the value is negative, so there is a small gradient, which avoids the problem of aliasing in the gradient direction. As the number of epochs increases in Table 5, the number of iterations for weight updating increases, the curve goes from the initial unfitting state to the optimal fitting state.

Figure 7. The images including fruits and the predicted boxes

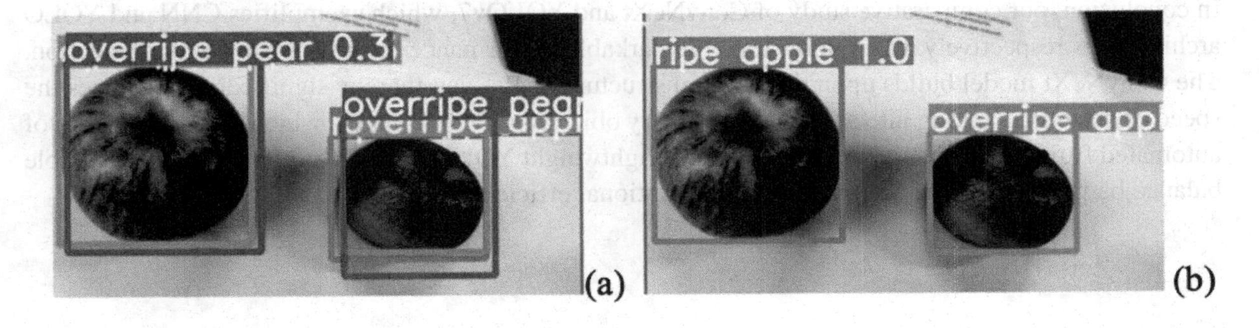

Table 1. The precisions after trained YOLOv7 model

Model	Weights	Epoch	Class Synchronous	AP@.5	AP@.5:.95
YOLOv7	yolov7	10	Ripe apple	0.468	0.427
			Over apple	0.885	0.764
			Ripe pear	0.169	0.115
			Overripe pear	0.209	0.151
		20	Ripe apple	0.427	0.419
			Over apple	0.995	0.932
			Ripe pear	0.673	0.622
			Overripe pear	0.967	0.944
		30	Ripe apple	0.993	0.974
			Over apple	0.996	0.948
			Ripe pear	0.542	0.534
			Overripe pear	0.995	0.932
		50	Ripe apple	0.996	0.992
			Over apple	0.996	0.979
			Ripe pear	0.996	0.981
			Overripe pear	0.996	0.991
		100	Ripe apple	0.996	0.992
			Over apple	0.996	0.988
			Ripe pear	0.996	0.995
			Over apple	0.996	0.990

As a traditional CNN model, ConvNeXt shows better training results. Under the same training parameters, the transfer learning model does not have much advantage. But in Figure 9, transfer learning saves much time if pre-training parameters are frozen. The ConvNeXt model cannot fully capture fruit features. In Tables 5 and Table 6, the ConvNeXt transfer learning model has a slight advantage in detection speed. But in terms of accuracy, the YOLO model is still better.

CONCLUSION

In conclusion, our comparative study of ConvNeXt and YOLOv7, which exemplifies CNN and YOLO architectures respectively, has demonstrated remarkable performance in the domain of fruit detection. The ConvNeXt model builds upon the residual structure of ResNet, thereby significantly enhances the speed of detection. Taking into account the primary objective of our research, which is the realization of automated fruit harvesting, we conclude that the lightweight YOLOv7 model presents a more favorable balance between detection accuracy and computational efficiency.

Table 2. The precisions after trainied YOLOv7-tiny model

Model	Weights	Epoch	Class	AP@.5	AP@.5:.95
YOLOv7	yolov7-tiny	10	Ripe apple	0.660	0.259
			Over apple	0.144	0.041
			Ripe pear	0.075	0.031
			Overripe pear	0.056	0.018
		20	Ripe apple	0.470	0.336
			Over apple	0.730	0.526
			Ripe pear	0.859	0.697
			Overripe pear	0.165	0.113
		30	Ripe apple	0.665	0.608
			Over apple	0.651	0.593
			Ripe pear	0.778	0.691
			Overripe pear	0.492	0.370
		50	Ripe apple	0.995	0.935
			Over apple	0.995	0.905
			Ripe pear	0.995	0.898
			Overripe pear	0.880	0.757
		100	Ripe apple	0.995	0.958
			Over apple	0.995	0.963
			Ripe pear	0.995	0.930
			Over apple	0.995	0.953

Fruit detection from digital images still encounters many complex problems, and the impact of the environment on image quality can easily cause errors in detection. In our experiments, in order to increase the possibility of the detection target, we screened a part of the data that was greatly affected by the environment when making the dataset. Our follow-up experiments will make up for this shortcoming, and study how to use the deep learning model to realize the target detection of fruits in environments such as light and rain.

Table 3. The precisions after trained YOLOv7-X model

Model	Weights	Epoch	Class	AP@.5	AP@.5:.95
YOLOv7	yolov7-X	10	Ripe apple	0.439	0.393
			Over apple	0.843	0.718
			Ripe pear	0.344	0.224
			Overripe pear	0.436	0.396
		20	Ripe apple	0.918	0.906
			Over apple	0.996	0.949
			Ripe pear	0.390	0.352
			Overripe pear	0.517	0.470
		30	Ripe apple	0.996	0.978
			Over apple	0.996	0.960
			Ripe pear	0.996	0.919
			Overripe pear	0.996	0.959
		50	Ripe apple	0.996	0.991
			Over apple	0.997	0.983
			Ripe pear	0.996	0.994
			Overripe pear	0.996	0.989
		100	Ripe apple	0.996	0.992
			Over apple	0.996	0.989
			Ripe pear	0.996	0.996
			Over apple	0.996	0.993

Table 4. The precisions after trained YOLOv7-E6

Model	Weights	Epoch	Class	AP@.5	AP@.5:.95
YOLOv7	yolov7-E6	10	Ripe apple	0.334	0.277
			Over apple	0.437	0.358
			Ripe pear	0.156	0.088
			Overripe pear	0.714	0.538
		20	Ripe apple	0.377	0.342
			Over apple	0.365	0.319
			Ripe pear	0.234	0.204
			Overripe pear	0.589	0.511
		30	Ripe apple	0.993	0.897
			Over apple	0.995	0.921
			Ripe pear	0.227	0.209
			Overripe pear	0.995	0.971
		50	Ripe apple	0.994	0.988
			Over apple	0.996	0.969
			Ripe pear	0.995	0.930
			Overripe pear	0.995	0.923
		100	Ripe apple	0.995	0.991
			Over apple	0.996	0.986
			Ripe pear	0.995	0.995
			Over apple	0.995	0.992

Table 5. The mean average precisions (mAP)

Model	Weights	Epoch	AP@.5	AP@.5:.95	Average inference time(millisecond)
YOLOv7	YOLOv7	10	0.433	0.364	6
		20	0.766	0.730	6
		30	0.882	0.847	6
		50	0.996	0.986	6
		100	0.996	0.991	6
	YOLOv7-tiny	10	0.234	0.087	8
		20	0.556	0.418	8
		30	0.644	0.565	7
		50	0.966	0.874	7
		100	0.995	0.951	6
	YOLOv7-X	10	0.515	0.433	9
		20	0.705	0.669	9
		30	0.996	0.954	9
		50	0.996	0.989	9
		100	0.996	0.993	10
	YOLOv7-E6	10	0.410	0.315	13
		20	0.391	0.344	13
		30	0.803	0.749	13
		50	0.995	0.952	13
		100	0.995	0.991	12

Table 6. The results of ConvNeXt model for fruit detection

Model	Weights	Epoch	AP@.5	AP@.5:.95	Average inference time(millisecond)
ConvNeXt	ConvNext + Mask R-CNN	10	0.848	0.719	5
		20	0.948	0.678	13
		30	0.926	0.669	37
		50	0.844	0.617	58
	ConvNext + Mask R-CNN Transfer Learning	10	0.500	0.701	14
		20	0.487	0.695	4
		30	0.483	0.694	5
		50	0.483	0.695	5

REFERENCES

Al-Sarayreh, M., Reis, M., Yan, W., & Klette, R. (2019) A sequential CNN approach for foreign object detection in hyperspectral images. *International Conference on Information, Communications and Signal*. Springer. 10.1007/978-3-030-29888-3_22

Al-Sarayreha, M. (2020). *Hyperspectral Imaging and Deep Learning for Food Safety*. [PhD Thesis, Auckland University of Technology, New Zealand].

An, N. (2020). *Anomalies Detection and Tracking Using Siamese Neural Networks*. [Master's Thesis, Auckland University of Technology, New Zealand].

An, N., & Yan, W. (2021). Multitarget tracking using Siamese neural networks. *ACM Transactions on Multimedia Computing Communications and Applications*, *17*(2s), 1–16. doi:10.1145/3441656

Bazame, H. C., Molin, J. P., Althoff, D., & Martello, M. (2021). Detection, classification, and mapping of coffee fruits during harvest with computer vision. *Computers and Electronics in Agriculture*, *183*, 106066. doi:10.1016/j.compag.2021.106066

Bhargava, A., & Bansal, A. (2021). Fruits and vegetables quality evaluation using computer vision: A review. *Journal of King Saud University. Computer and Information Sciences*, *33*(3), 243–257. doi:10.1016/j.jksuci.2018.06.002

Feng, J., Tan, H., Li, W., & Xie, M. (2022). Conv2NeXt: Reconsidering Conv NeXt Network Design for Image Recognition. *International Conference on Computers and Artificial Intelligence Technologies (CAIT)* (pp. 53-60). IEEE. 10.1109/CAIT56099.2022.10072172

Fu, Y. (2020) Fruit Freshness Grading Using Deep Learning. Master's Thesis, Auckland University of Technology, New Zealand.

Fu, Y., Nguyen, M., & Yan, W. (2022). *Grading methods for fruit freshness based on deep learning*. Springer Nature Computer Science. doi:10.100742979-022-01152-7

Gokhale, A., Chavan, A., & Sonawane, S. (2023). Leveraging ML techniques for image-based freshness index prediction of fruits and vegetables. *International Conference on Emerging Smart Computing and Informatics (ESCI)* (pp. 1-6). IEEE. 10.1109/ESCI56872.2023.10100260

Gongal, A., Karkee, M., & Amatya, S. (2018). Apple fruit size estimation using a 3D machine vision system. *Information Processing in Agriculture*, *5*(4), 498–503. doi:10.1016/j.inpa.2018.06.002

Gowdra, N. (2021). *Entropy-Based Optimization Strategies for Convolutional Neural Networks*. [PhD Thesis, Auckland University of Technology, New Zealand].

Gowdra, N., Sinha, R., MacDonell, S., & Yan, W. (2021). Maximum Categorical Cross Entropy (MCCE): A noise-robust alternative loss function to mitigate racial bias in Convolutional Neural Networks (CNNs) by reducing overfitting. *Pattern Recognition*.

Hassanien, M. A., Singh, V. K., Puig, D., & Abdel-Nasser, M. (2022). Predicting breast tumor malignancy using deep ConvNeXt radiomics and quality-based score pooling in ultrasound sequences. *Diagnostics (Basel)*, *12*(5), 1053. doi:10.3390/diagnostics12051053 PMID:35626208

Hussain, M., Al-Aqrabi, H., Munawar, M., Hill, R., & Alsboui, T. (2022). Domain feature mapping with YOLOv7 for automated edge-based pallet racking inspections. *Sensors (Basel), 22*(18), 6927. doi:10.339022186927 PMID:36146273

Junos, M. H., Mohd Khairuddin, A. S., Thannirmalai, S., & Dahari, M. (2022). Automatic detection of oil palm fruits from UAV images using an improved YOLO model. *The Visual Computer, 38*(7), 2341–2355. doi:10.100700371-021-02116-3

Kang, H., & Chen, C. (2020). Fast implementation of real-time fruit detection in apple orchards using deep learning. *Computers and Electronics in Agriculture, 168*, 105108. doi:10.1016/j.compag.2019.105108

Kivrak, O., & Gürbüz, M. Z. (2022). Performance comparison of YOLOv3, YOLOv4 and YOLOv5 algorithms: A case study for poultry recognition. *Avrupa Bilim ve Teknoloji Dergisi*, (38), 392–397.

Koirala, A., Walsh, K. B., Wang, Z., & McCarthy, C. (2019). Deep learning for real-time fruit detection and orchard fruit load estimation: Benchmarking of 'MangoYOLO'. *Precision Agriculture, 20*(6), 1107–1135. doi:10.100711119-019-09642-0

Kuznetsova, A., Maleva, T., & Soloviev, V. (2021). *YOLOv5 versus YOLOv3 for apple detection. Cyber-Physical Systems: Modelling and Intelligent Control*. Springer.

Lee, Y. H., & Kim, Y. (2020). Comparison of CNN and YOLO for object detection. *Journal of the Semiconductor & Display Technology, 19*(1), 85–92.

Liu, G., Nouaze, J. C., Touko Mbouembe, P. L., & Kim, J. H. (2020). YOLO-Tomato: A robust algorithm for tomato detection based on YOLOv3. *Sensors (Basel), 20*(7), 2145. doi:10.339020072145 PMID:32290173

Liu, H., Sun, F., Gu, J., & Deng, L. (2022). SF-YOLOv5: A lightweight small object detection algorithm based on improved feature fusion mode. *Sensors (Basel), 22*(15), 5817. doi:10.339022155817 PMID:35957375

Liu, X., Li, G., Chen, W., Liu, B., Chen, M., & Lu, S. (2022). Detection of dense citrus fruits by combining coordinated attention and cross-scale connection with weighted feature fusion. *Applied Sciences (Basel, Switzerland), 12*(13), 6600. doi:10.3390/app12136600

Liu, X., & Yan, W. Q. (2023). *Vehicle-related distance estimation using customized YOLOv7. IVCNZ 2022, Auckland, New Zealand*. Springer Nature Switzerland.

Liu, Z., Yan, W., & Yang, B. (2018) Image denoising based on a CNN model. *International Conference on Control, Automation and Robotics*. IEEE. 10.1109/ICCAR.2018.8384706

Lu, S., Chen, W., Zhang, X., & Karkee, M. (2022). Canopy-attention-YOLOv4-based immature/mature apple fruit detection on dense-foliage tree architectures for early crop load estimation. *Computers and Electronics in Agriculture, 193*, 106696. doi:10.1016/j.compag.2022.106696

Ou, J., Zhang, R., Li, X., & Lin, G. (2023). Research and explainable analysis of a real-time passion fruit detection model based on FSOne-YOLOv7. *Agronomy (Basel), 13*(8), 1993. doi:10.3390/agronomy13081993

Pan, C., Liu, J., Yan, W., & Zhou, Y. (2021). Salient object detection based on visual perceptual saturation and two-stream hybrid networks. *IEEE Transactions on Image Processing*, *30*, 4773–4787. doi:10.1109/TIP.2021.3074796 PMID:33929959

Pan, C., & Yan, W. (2018) A learning-based positive feedback in salient object detection. *International Conference on Image and Vision Computing New Zealand*. IEEE. 10.1109/IVCNZ.2018.8634717

Pan, C., & Yan, W. (2020). Object detection based on saturation of visual perception. *Multimedia Tools and Applications*, *79*(27-28), 19925–19944. doi:10.100711042-020-08866-x

Qi, J., Nguyen, M., & Yan, W. (2022). *Waste classification from digital images using ConvNeXt*. Pacific-Rim Symposium on Image and Video Technology.

Qi, J., Nguyen, M., & Yan, W. (2022) Small visual object detection in smart waste classification using Transformers with deep learning. *International Conference on Image and Vision Computing New Zealand (IVCNZ)*. IEEE.

Ruiz, N., Bargal, S., Xie, C., Saenko, K., & Sclaroff, S. (2022). Finding differences between Transformers and ConvNets using counterfactual simulation testing. *Advances in Neural Information Processing Systems*, *35*, 14403–14418.

Shi, R., Li, T., & Yamaguchi, Y. (2020). An attribution-based pruning method for real-time mango detection with YOLO network. *Computers and Electronics in Agriculture*, *169*, 105214. doi:10.1016/j.compag.2020.105214

Sozzi, M., Cantalamessa, S., Cogato, A., Kayad, A., & Marinello, F. (2022). Automatic bunch detection in white grape varieties using YOLOv3, YOLOv4, and YOLOv5 deep learning algorithms. *Agronomy (Basel)*, *12*(2), 319. doi:10.3390/agronomy12020319

Tian, G., Wang, Z., Wang, C., Chen, J., Liu, G., Xu, H., Lu, Y., Han, Z., Zhao, Y., Li, Z., Luo, X., & Peng, L. (2022). A deep ensemble learning-based automated detection of COVID-19 using lung CT images and Vision Transformer and ConvNeXt. *Frontiers in Microbiology*, *13*, 13. doi:10.3389/fmicb.2022.1024104 PMID:36406463

Vasconez, J. P., Delpiano, J., Vougioukas, S., & Cheein, F. A. (2020). Comparison of convolutional neural networks in fruit detection and counting: A comprehensive evaluation. *Computers and Electronics in Agriculture*, *173*, 105348. doi:10.1016/j.compag.2020.105348

Wang, D., & He, D. (2021). Channel pruned YOLO V5s-based deep learning approach for rapid and accurate apple fruitlet detection before fruit thinning. *Biosystems Engineering*, *210*, 271–281. doi:10.1016/j.biosystemseng.2021.08.015

Wang, L., & Yan, W. (2021) Tree leaves detection based on deep learning. *International Symposium on Geometry and Vision*. Springer. 10.1007/978-3-030-72073-5_3

Wang, L., Zhao, Y., Xiong, Z., Wang, S., Li, Y., & Lan, Y. (2022). Fast and precise detection of litchi fruits for yield estimation based on the improved YOLOv5 model. *Frontiers in Plant Science*, *13*, 13. doi:10.3389/fpls.2022.965425 PMID:36017261

Wang, Z., Jin, L., Wang, S., & Xu, H. (2022). Apple stem/calyx real-time recognition using YOLO-v5 algorithm for fruit automatic loading system. *Postharvest Biology and Technology*, *185*, 111808. doi:10.1016/j.postharvbio.2021.111808

Wu, D., Jiang, S., Zhao, E., Liu, Y., Zhu, H., Wang, W., & Wang, R. (2022). Detection of Camellia oleifera fruit in complex scenes by using YOLOv7 and data augmentation. *Applied Sciences (Basel, Switzerland)*, *12*(22), 11318. doi:10.3390/app122211318

Xia, Y., Nguyen, M., & Yan, W. (2022) A real-time Kiwifruit detection based on improved YOLOv7. *International Conference on Image and Vision Computing New Zealand (IVCNZ)*. IEEE.

Xia, Y., Nguyen, M., & Yan, W. (2023). *Kiwifruit counting using KiwiDetector and KiwiTracker*. IntelliSys.

Xiao, B., Nguyen, M., & Yan, W. (2021) Apple ripeness identification using deep learning. *International Symposium on Geometry and Vision*. Springer. 10.1007/978-3-030-72073-5_5

Xiao, B., Nguyen, M., & Yan, W. (2023). *Apple ripeness identification from digital images using transformers. Multimedia Tools and Applications*. Springer.

Xiao, B., Nguyen, M., & Yan, W. (2023). *Fruit ripeness identification using transformers. Applied Intelligence*. Springer Science and Business Media LLC.

Yan, B., Fan, P., Lei, X., Liu, Z., & Yang, F. (2021). A real-time apple targets detection method for picking robot based on improved YOLOv5. *Remote Sensing (Basel)*, *13*(9), 1619. doi:10.3390/rs13091619

Yang, R., Hu, Y., Yao, Y., Gao, M., & Liu, R. (2022). Fruit target detection based on BCo-YOLOv5 model. *Mobile Information Systems*, *2022*, 2022. doi:10.1155/2022/8457173

Yao, J., Qi, J., Zhang, J., Shao, H., Yang, J., & Li, X. (2021). A real-time detection algorithm for Kiwifruit defects based on YOLOv5. *Electronics (Basel)*, *10*(14), 1711. doi:10.3390/electronics10141711

Zhang, H., Wang, Y., Liu, Y., & Xiong, N. (2022). IFD: An intelligent fast detection for real-time image information in industrial IoT. *Applied Sciences (Basel, Switzerland)*, *12*(15), 7847. doi:10.3390/app12157847

Zhang, Y., Zhang, Y., & Zhang, Y. (2022). Fruit and vegetable disease identification based on updating the activation function for the ConvNeXt model. *International Conference on Electronic Information Technology and Computer Engineering* (pp. 1045-1049). ACM. 10.1145/3573428.3573616

Zhao, K. (2021) *Fruit Detection Using CenterNet*. [Master's Thesis, Auckland University of Technology, New Zealand].

Zhao, K., & Yan, W. (2021) Fruit detection from digital images using CenterNet. *International Symposium on Geometry and Vision*. Springer. 10.1007/978-3-030-72073-5_24

Zhijun, L. I., Shenghui, Y. A. N. G., Deshuai, S. H. I., Xingxing, L. I. U., & Yongjun, Z. H. E. N. G. (2021). Yield estimation method of apple tree based on improved lightweight YOLOv5. *Smart Agriculture*, *3*(2), 100.

Zhou, J., Zhang, Y., & Wang, J. (2023). A dragon fruit picking detection method based on YOLOv7 and PSP-Ellipse. *Sensors (Basel)*, *23*(8), 3803. doi:10.339023083803 PMID:37112144

Chapter 17
YOLO Models for Fresh Fruit Classification From Digital Videos

Yinzhe Xue
Auckland University of Technology, New Zealand

Wei Qi Yan
Auckland University of Technology, New Zealand

ABSTRACT

Identifying food freshness is a very important; it is a part of a long historical actions by humans, because fruit freshness can tell us the information about the quality of foods. With the advancement of machine learning and computer science, which will be broadly employed in factories and markets, instead of manual classification. Recognition of the freshness of food is rapidly being replaced by computers or robots. In this book chapter, the authors conduct the research work on fruit freshness detection, we make use of YOLOv6, YOLOv7, and YOLOv8 in this project to implement fruit classifications based on a variety of digital images, which can improve the efficiency and accuracy of the classification incredibly; after the classification, the output will showcase the result of fruit fresheness classification, namely, fresh, or rotten, etc. They also compare the results of different deep learning models to discover which architecture is the best one in terms of speed and accuracy. At the end of this book chapter, the authors made use of the majority vote method to combine the results of different models to get better accuracy and recall scores. To generate the final result, the authors trained the three models individually, and also propose a majority vote to get a better performance for fresh fruit detection. Compared with the previous work, this method has higher accuracy and a much faster speed. Because this one uses the clustering method to generate the final result, it will be easy for researchers to change the backbone and get a better result in the future.

DOI: 10.4018/978-1-6684-9999-3.ch017

INTRODUCTION

Fruit freshness detection is a very interesting topic in machine vision that is also a very important task for human ordinary lives, because every day we need to know which food is safe to be eaten, which will cause illness or diseases, rotten foods may lead to poisoning, hence, we develop a number of ways to classify fruits and detect as well as predict the freshness.

In this book chapter, we use deep neural networks for freshness and rotten fruit classification, YOLO is a very famous architecture that can be employed for almost all types of fruit classification and freshness detection, meanwhile, we also introduce the potential methods such as Transformers in this project. This project will mainly make use of deep learning methods (like YOLO) to classify the digital images for fresh or rotten fruits, we take advantage of three YOLO models and compare the results.

The focus of this book chapter is to detect fruit freshness or classify fresh and rotten fruits from the input digital images. According to the most advanced YOLO architecture, it will be easy for us to get a high precision and recall for fruit freshness detection compared with the human labor method. Our contributions to this book chapter include: (1) Collecting a large dataset for three classes of fruits (i.e., apple, banana, and orange) (2) Classifying each image with YOLOv6, YOLOv7 or YOLOv8 models (3) Detecting the freshness of the given fruits (4) Proving machine learning methods to detect the freshness of the fruit (5) Seeking an ensemble method to combine the detection result from different architecture, and finding the best clustering weights for different architecture.

The structure of this book chapter is that we show our literature review and discuss the relevant studies of visual object detection and classification in Section 2. Meanwhile, we also introduce the details related to Transformer in deep learning. In Section 3, we introduce our research methods and dataset. In Section 4, we implement the proposed algorithms, collect experimental data and demonstrate our outcomes. Additionally, the limitations of these proposed methods will be detailed. In Section 5, we summarize and analyze the experimental results. We draw the conclusion and state our future work in Section 6.

We collected our dataset through the Kaggle website. In this dataset, we totally have six different types of fruits, there are images for fresh and rotten apples, bananas and oranges, and the number for each group is different. To give more information about the dataset, we will show the distribution of the dataset for this project.

Figure 1. Our dataset: (a) Training dataset; (b)Test dataset

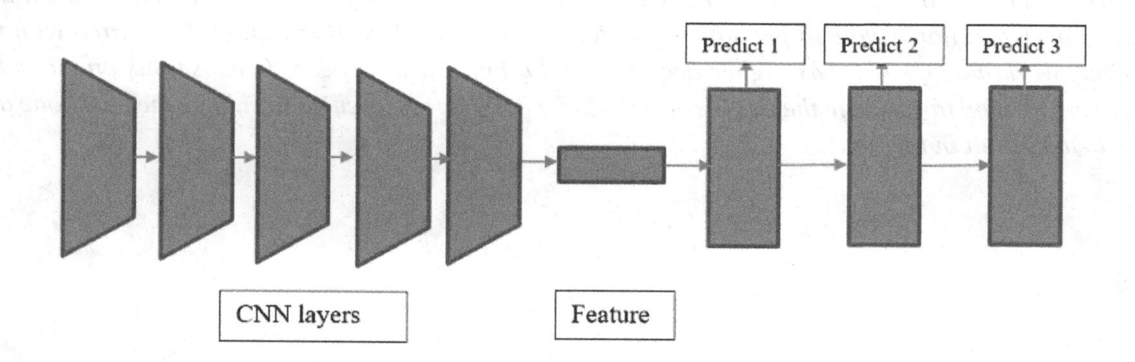

We take advantage of around 2% samples as the test set and the rest of samples will be used as our training set, Fig. 1. Shows the distribution of the samples in our training set and test set. We see that the training and test dataset almost have the same distributions of the samples. We also show the pie chart for the training and testing dataset to offer a more intuitive reflection of the dataset for this project.

Figure 2. Pie charts for training and test dataset

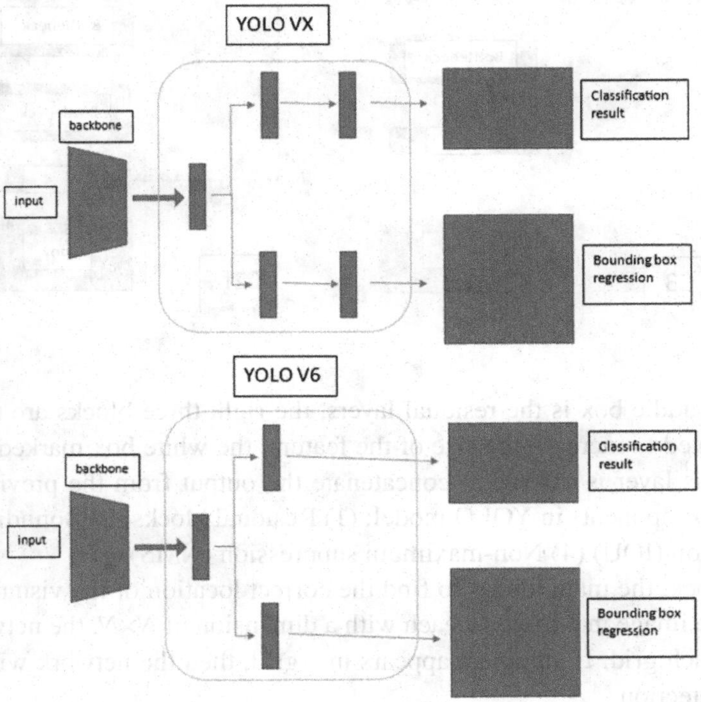

RELATED WORK

YOLO (Fang, 2019; Fang, 2021; Redmon, 2021; Parico, 2021) is the abbreviation of "You Only Look Once", which is a very famous and widely employed in computer vision. There are a lot of advantages of YOLO models, firstly, YOLO is a "lightweight" architecture, which means it will be trained in a very fast way. The trained weights will not consume a large space. Secondly, YOLO can provide visual object detection in real time. The real-time is the network which can make detections on the input image much faster than human reaction, so it feels like the network "promptly" generates the result without delay. Compared to Faster R-CNN or Transformer model, the accuracy of YOLO is not high, the differences of the performance between YOLO and other models are pretty minor.

Figure 3. YOLO architecture

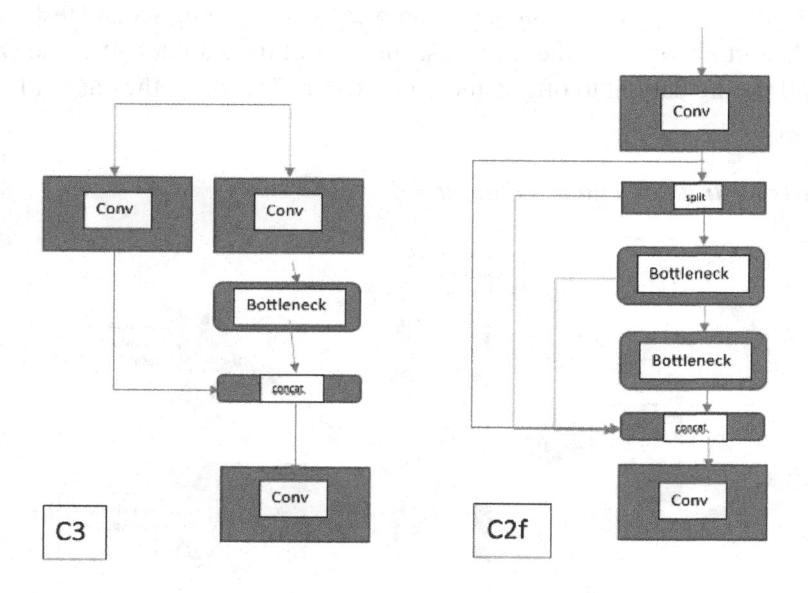

In Figure 3, the middle box is the residual layers, the right three blocks are the upsampling layer, which is accommodated to increase the size of the feature, the white box marked with "Predict" is the concatenate layer, this layer is offered to concatenate the output from the previous layers. There are four very important components in YOLO model: (1) Residual blocks (2) Bounding box regression (3) Intersection over Union (IOU) (4) Non-maximum suppression (NMS).

In the residual block, the main idea is to find the correct location of the visual object. To do this, it firstly splits the given image into regions, each with a dimension of $N \times N$. the network will perform the detection based on each grid. If an object appears in a grid, then the network will mark that grid as a candidate for final detection.

YOLO is the network with a single-box regression to predict the four pieces of information for a bounding box. Each grid cell is responsible for predicting the bounding boxes and their confidence scores. The IOU is equal to 1.0 if the predicted bounding box is as same as the real box. If we assume the blue square presents the prediction, and the red one is the ground truth. This mechanism eliminates bounding boxes that are not equal to the real box.

The red bounding box is the ground truth, we need to conduct object detection, and the blue square is the prediction that the architecture works on the image. On the right side, the green box indicates that our prediction is overlapped with the ground truth (GT).

Compared with the previous architectures, YOLOv6 is use of the same dataset to pre-train, but the backbone has been changed to EfficientRep, and the neck has been changed to Rep-PAN.

YOLOv7 has a higher accuracy for multiple tasks shown in the book chapter. The main difference of YOLOv7 is that YOLOv7 modifies the ELAN architecture (efficient layer aggregation network), The-modified ELAN network is called E-Ellan, which is just a combination of multiple convolution layers and two more concatenation layers.

YOLOv8 was published in 2022, which is the latest version of YOLO models. YOLOv8 is an open-source architecture. Leveraging previous YOLO models, YOLOv8 models are faster and more accurate, while providing a unified framework for training models to perform object detection, instance segmentation, and image classification. YOLOv8 is similar to YOPLOv7, but YOLOv8 replaces C3 architecture to C2f architecture (Lin, 2017), which has two backbones for YOLOv8. Pertaining to YOLOv8, we will take use of different backbones for the different tasks (like detection or classification). YOLOv8 looks like a combination of the previous YOLO architecture, but YOLOv8 is much faster than the previous versions of YOLO models.

Compared to the architecture of C3 and C2f, there are a few changes: (1) The kernel size of the first convolution layer is changed to 6×6 instead of 3×3; (2) Two convolution layers in neck are deleted; (3) The number of blocks in backbone has been changed to 3-6-6-3; (4) A split part is added beforebottleneck; (5) The residual connections are added between the input and output; (6) The kernel parameters are modified; (7) Instead of using the parallel bottlenecklayers, the serial connected layers are employed; (8) The head of YOLOv8 changes to the anchor-free architecture, but the previous YOLO is anchor-based architecture.

Figure 4. Architecture of C3 and C2f

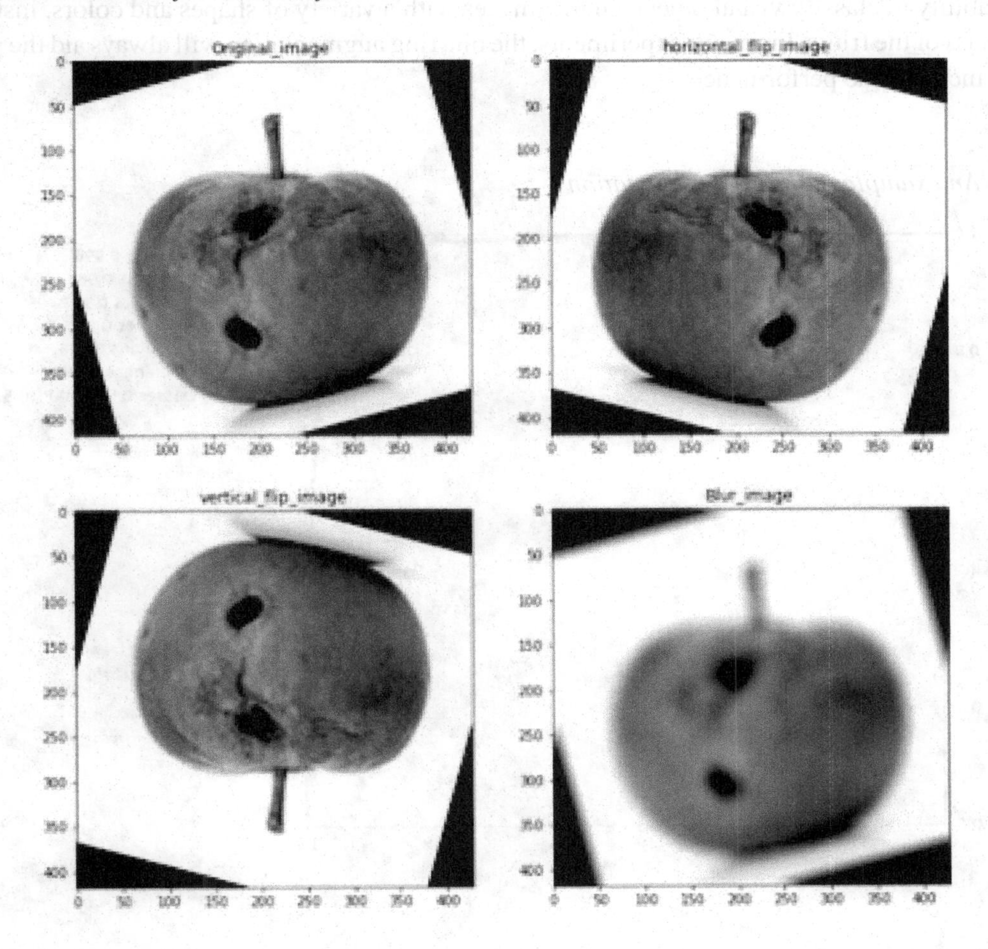

OUR WORK

Data argumentation is a broadly used method to improve the network performance, it adds more types or more samples in the training set based on the existing training dataset, more types of training data will make the proposed model more robust.

Regarding image classification using data arguments, the normal method is always to rotate or flip the original images, because from the previous experiments, these two methods are very simple to implement and can obviously increase the performance of the proposed network. In this project, we also add a blurring method in our data augmentation. Pertaining to this blurring method, a Gaussian kernel is adopted to convolve the original image, then we get a blurred image with the same fruit location and class of the original input. Because we use kernels to process the image, it is also easy for us to undertake different degrees to the image and find out the best one to increase the network performance as much as possible as shown in Figure 5.

In Figure 5, the top left is the original input image, we add three different augmentations in our augmented dataset, namely, horizontal flipping, vertical flipping and image blurring, the two kinds of flips will add more visual appearances to the network, these visual features will show in the cases when we take photos for the object from different angles, we rotate the images (no only 180 degrees or 90 degrees), the step will add more “ambiguous” features to the architecture, then the network model will have the ability to classify visual objects in the images with a variety of shapes and colors, instead only detect a class of the fruits. From our experiments, the blurring augmentation will always aid the proposed model to increase the performance.

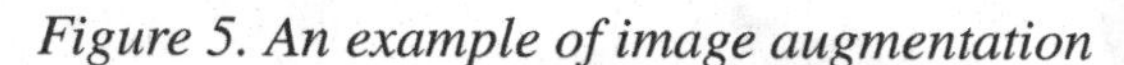

Figure 5. An example of image augmentation

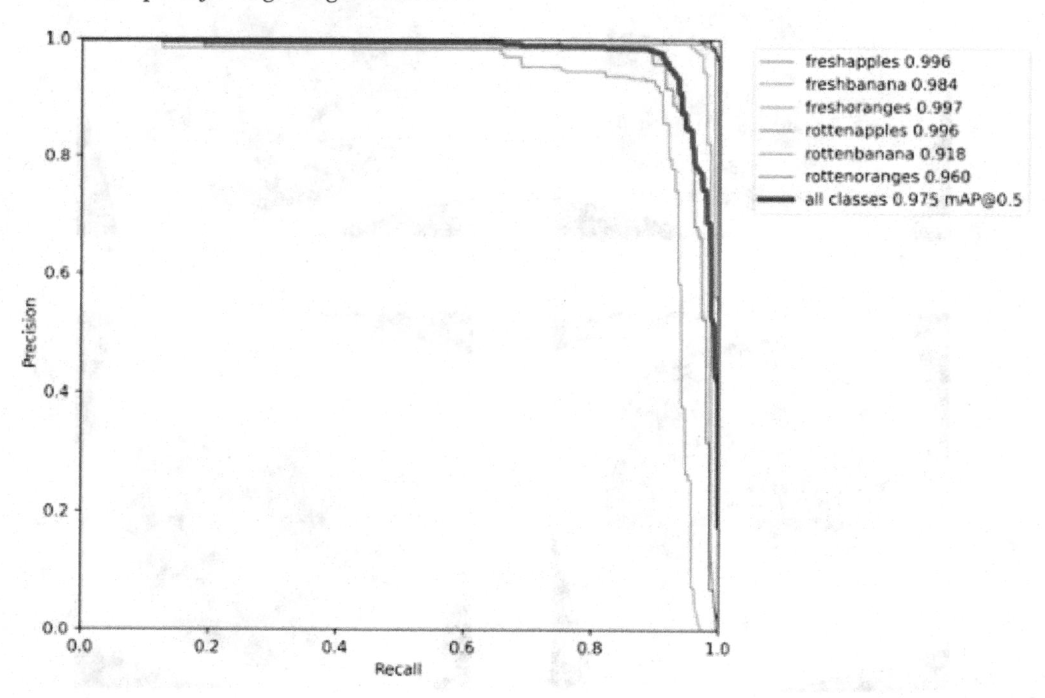

Figure 6. PR curves of YOLO models

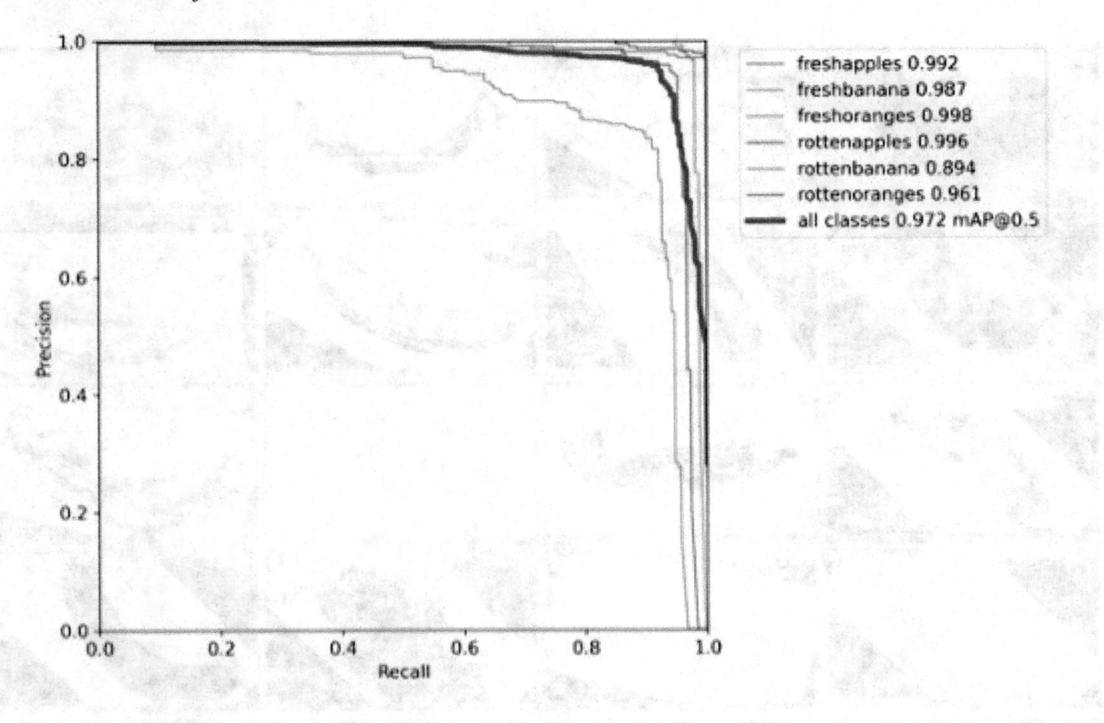

Figure 7. F1 score vs. confidence of YOLO models

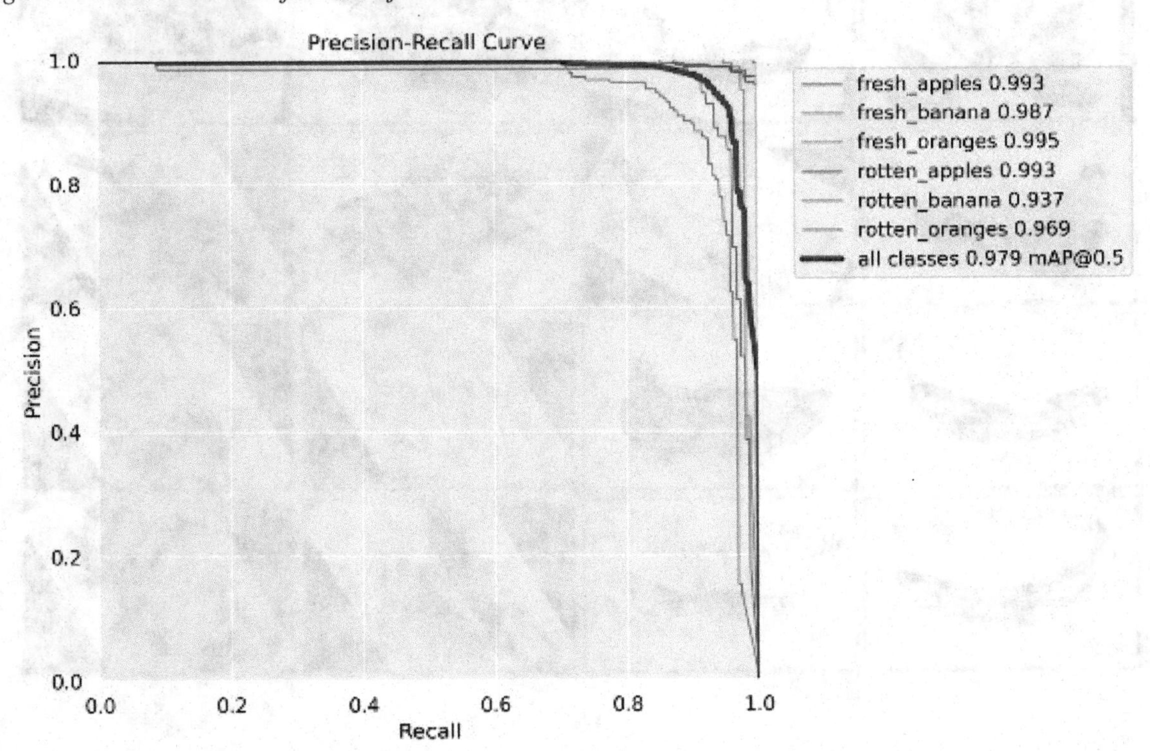

Figure 8. Our classification results using YOLOv8 model

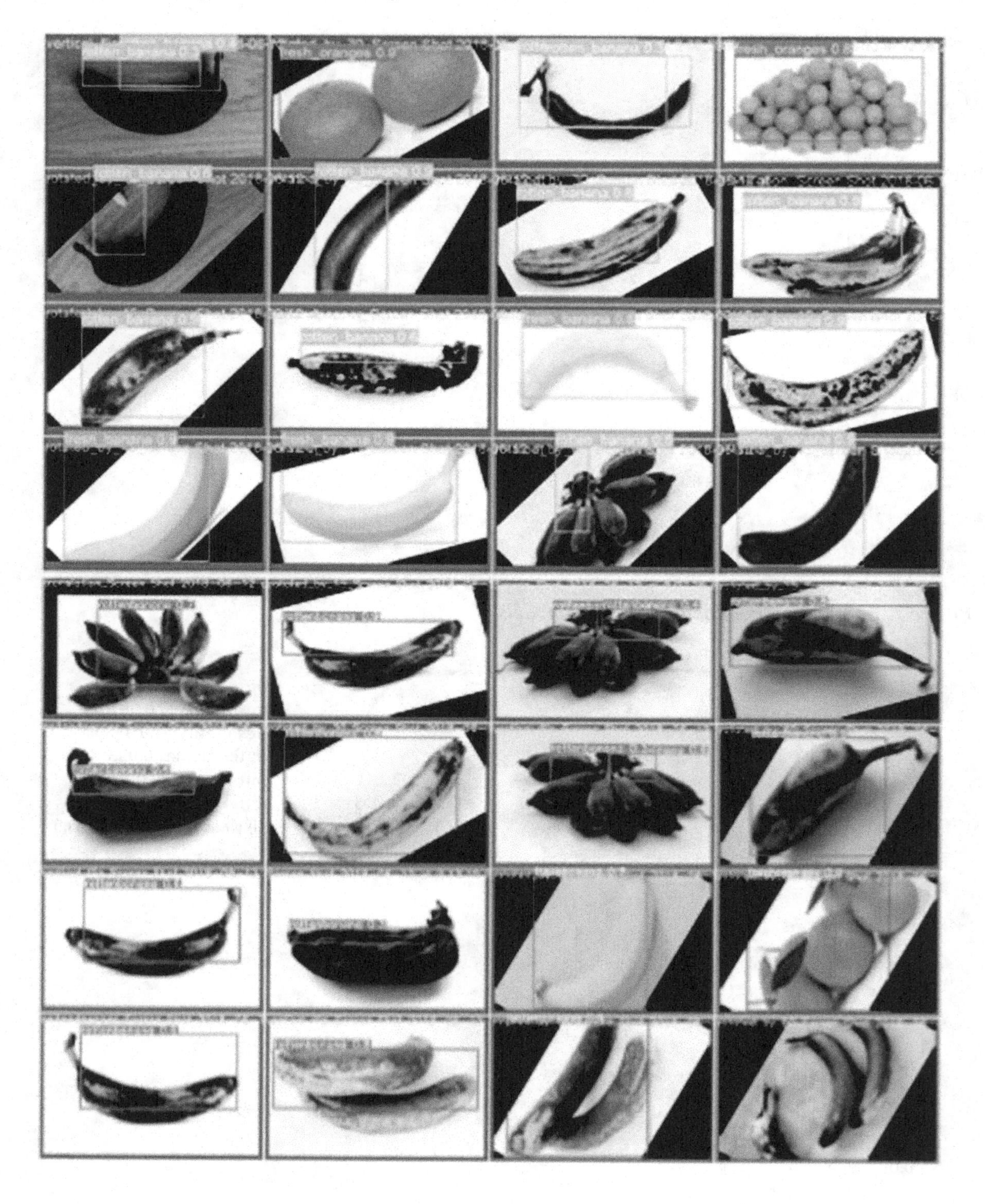

RESULT ANALYSIS

In Figure 7, we show our prediction results generated by YOLOv8 model. By following the given images, the purple bounding box means rotten banana, green bounding boxes refer to fresh oranges, and orange bounding boxes indicate fresh banana. As we see from the test image, the test images also have a rotation around the center, this kind of argumentation methods can give us a better estimation of the classification in the real scene, because in our real world, it is impossible for us or for a machine to take a photo for a fruit from a constant angle, for example, we always take photos for apple from a side perspective, this will increase the workload of human.

To better compare the results, we will show the final prediction results from three different models. Because this is a classification task, we finally make use of precision and recall to estimate the average performance, as we discussed in the previous section, precision means how many objects the network can find with the correct label from the given images, and recall is the score to show the ratio of our correct detection. A good classifier should have a balance between precision and recall; but for this fresh fruit classification task, the results from different architectures are shown in Table 1.

In Table 1, YOLOv6 model has a better result compared with other different architectures, but actually, from the confusion matrix, YOLOv8 model has a better classification accuracy, a large error only occurs for rotten orange. So in the real scene, if we can only make use of a single network to detect fruits from the image dataset, we highly recommend consideringthe YOLOv8 for the first try.

Table 1. The results of training the deep nets

Models	Avg precision	Avg recall
YOLOv6	0.846	0.884
YOLOv7	0.852	0.879
YOLOv8	0.861	0.892

MAJORITY VOTE

Majority vote is a traditional method used to improve the detection performance, because the YOLOv8 architecture is very easy to implement, we are planning to propose it as a base method to get a better result. There are two main reasons why we are use of majority vote as a method in this book chapter. The first reason is that the majority vote is a method that can be employed for "outside" of the model. The second reason is with a majority vote, we can easily determine how many results and which one we will use it to generate the final output, and the majority vote method also has the best adaptability, replacing the backbone (like YOLOv8) that will not influence the architecture, we still can take use of the same code to change the input that we get from the new backbone, this will also help us easily change our architecture in the future. Our results are shown in Table 3.

Table 2. The results for testing the deep nets

Models	All classes IOU	All classes F1	The best threshold	accumulate error rate
YOLOv6	0.975	0.96	0.659	12%
YOLOv7	0.972	0.96	0.284	9%
YOLOv8	0.979	0.96	0.569	14%

Table 3. Comparisons of various models

	Precision	Recall
YOLOv6	0.956	0.943
YOLOv7	0.933	0.92
YOLOv8	0.957	0.941
Majority vote	0.961	0.957

ALATION STUDY

While we train the network models, we make use of the original fruit images, because the network with an augmentation dataset will increase the performance of our proposed model, there are three types of datasets: The first is the original dataset, which includes the images without any augmentation. The second dataset is that we add horizontal flip and vertical flip to the training dataset. The third dataset is that we add flips and blur to the training dataset, the result are shown in Table 4.

Table 4. Ablation experiments

	Train on original dataset		Train on dataset with flip augmentation		Train on dataset with flip and blur	
Architecture	Precision	Recall	Precision	Recall	Precision	Recall
YOLOv6	0.956	0.943	0.955	0.94	0.967	0.96
YOLOv7	0.933	0.92	0.937	0.921	0.952	0.931
YOLOv8	0.957	0.941	0.96	0.934	0.97	0.964

From the results in Table 4, it is obvious that when we add flipping operations to the original dataset, the performance increases but only a little bit. Compared with only the flip augmentation dataset, we add the blurry images to the dataset which enhances the performance a lot, both precision and recall from different architectures ramp up at least one percent, this means compared with the flipped image, blurring images will provide more new information to the model training.

Ablation Study

It is obvious that with the majority vote method, our network can perform better than any individual network. In general, the majority vote method is,

$$voting\left(w_1*net_1,\ w_2*net_2.....w_n*net_n\right) \quad (1)$$

where w_1 to w_n is the weights for different architecture, the net_1 to net_n is the network detection result from different network architecture, the weights will always in range [0,1]. All the other numbers refer to the confidence of the network prediction we use, if a network has a high weight (high confidence), the result is more accurate than the other results. With various weights, we can make the majority vote result become more accurate and robust, we show different weight results if we combine YOLOv6, YOLOv7 and YOLOv8 together.

From Table 5, we easily see that the weights may not be the best strategy to generate the result, given the best performance of our proposed model, relatively large weights will have a better performance, but increasing the weights for a certain network too much will make the majority vote result get close to the single network result.

Table 5. Ablation results with various weights

Series	Weights for YOLOv6 model	Weights for YOLOv7 model	Weights for YOLOv8 model	Majority voting results	
				precision	recall
1	0.33	0.33	0.33	0.961	0.957
2	0.6	0.2	0.2	0.96	0.942
3	0.2	0.6	0.2	0.94	0.92
4	0.2	0.2	0.6	0.97	0.963
5	0.8	0.1	0.1	0.954	0.94
6	0.1	0.1	0.8	0.955	0.942

CONCLUSION

In this book chapter, we summary our work and show our future research directions. For the dataset, we are use of the fruit freshness dataset from Kaggle which includes 6 different types of image samples. In this dataset, fruits are fresh apple, fresh banana, fresh orange, rotten apple, rotten banana, and rotten orange, After getting the dataset, we split them into training and test dataset. Following the normal pre-process steps, we split the dataset with a ratio 8:2, which means we will randomly choose 80% data to train and the rest for the test. Pertaining to the methods for fruit classification, we took use of YOLOv6, YOLOv7, and YOLOv8 to generate the classification result.

Compared to the result from other architectures, during the training and testing, YOLOv8 will have the highest accuracy and the fastest training speed. YOLOv8 is always very easy after we got the result from all YOLO architectures, we make use of the majority voting method to ensemble their result and finally, we got a higher accuracy than the previous work for fresh fruit detection with the ensemble method.

In the future, we are planning to use YOLOv8 as a baseline of the architecture and combine with other models, we believe this will be a great choice. With the development of the detection method created in recent years, a lot of new models, like diffusion model and attention models will always have a very good performance of object detection classification. In our next step, we can also add those architectures in our ensemble group and generate result, we believe such a method can help us get a higher accuracy even we detect with more types of fruits.

REFERENCES

Abadi, M., Agarwal, A., Barham, P., Brevdo, E., Chen, Z., Citro, C., & Ghemawat, S. (2016). TensorFlow: Large-scale machine learning on heterogeneous distributed systems. *arXiv preprint arXiv:1603.04467.*

Abdulrahman, A., & Iqbal, K. (2014) Capturing human body dynamics using RNN based on persistent excitation data generator. *International Symposium on Computer-Based Medical Systems (CBMS),* (pp. 221-226). IEEE. 10.1109/CBMS.2014.145

Anderson, C., Burt, P., & Van Der Wal, G. (1985). Change detection and tracking using pyramid transform techniques. *Cambridge Symposium* (pp. 72-78) International Society for Optics and Photonics. 10.1117/12.950785

Barnea, E., Mairon, R., & Ben-Shahar, O. (2016). Colour-agnostic shape-based 3D fruit detection for crop harvesting robots. *Biosystems Engineering, 146*, 57–70. doi:10.1016/j.biosystemseng.2016.01.013

Baum, L. E., & Sell, G. (1968). Growth transformations for functions on manifolds. *Pacific Journal of Mathematics, 27*(2), 211–227. doi:10.2140/pjm.1968.27.211

Bengio, Y., Simard, P., & Frasconi, P. (1994). Learning long-term dependencies with gradient descent is difficult. *IEEE Transactions on Neural Networks, 5*(2), 157–166. doi:10.1109/72.279181 PMID:18267787

Chatzis, S. P., & Kosmopoulos, D. I. (2011). A variational Bayesian methodology for hidden Markov models utilizing Student's-*t* mixtures. *Pattern Recognition, 44*(2), 295–306. doi:10.1016/j.patcog.2010.09.001

Chen, Y. N., Han, C. C., Wang, C. T., Jeng, B. S., & Fan, K. C. (2006). The application of a convolution neural network on face and license plate detection. *International Conference on Pattern Recognition, 3*, 552–555.

Collins, R. T., Lipton, A. J., Kanade, T., Fujiyoshi, H., Duggins, D., Tsin, Y., & Wixson, L. (2000). *A system for video surveillance and monitoring. Research Report.* Carnegie Mellon University.

Eickeler, S., & Muller, S. (1999). Content-based video indexing of TV broadcast news using hidden Markov models. *IEEE International Conference on Acoustics, Speech, and Signal Processing.*(Vol. 6, pp. 2997-3000). IEEE 10.1109/ICASSP.1999.757471

Fang, W., Lin, W., & Ren, P. (2019). Tinier-YOLO: A real-time object detection method for constrained environments. *IEEE Access : Practical Innovations, Open Solutions, 8*, 1935–1944. doi:10.1109/ACCESS.2019.2961959

Fang, Y. (2021). You only look at one sequence: Rethinking transformer in vision through object detection. *Advances in Neural Information Processing Systems, 34*, 26183–26197.

Fu, R., Zhang, Z., & Li, L. (2016). Using LSTM and GRU neural network methods for traffic flow prediction. *Youth Academic Annual Conference of Chinese Association of Automation (YAC).* IEEE. 10.1109/YAC.2016.7804912

Fu, Y. (2020). *Fruit Freshness Grading Using Deep Learning*. [Master's Thesis, Auckland University of Technology, New Zealand].

Fu, Y., Nguyen, M., & Yan, W. (2022). *Grading methods for fruit freshness based on deep learning*. Springer Nature Computer Science. doi:10.100742979-022-01152-7

Galvez, R., Bandala, A., Dadios, E., et al. (2018) Object detection using convolutional neural networks. *IEEE TENCON*, (pp. 2023-2027).

Gers, F. A., & Schmidhuber, J. (2000). *Recurrent nets that time and count. Neural Networks* (Vol. 3). IEEE – INNS.

Gers, F. A., Schmidhuber, J., & Cummins, F. (2000). Learning to forget: Continual prediction with LSTM. *Neural Computation, 12*(10), 2451–2471. doi:10.1162/089976600300015015 PMID:11032042

Han, X., Gao, Y., Lu, Z., Zhang, Z., & Niu, D. (2015). Research on moving object detection algorithm based on improved three frame difference method and optical flow. *International Conference on Instrumentation and Measurement, Computer, Communication and Control (IMCCC)* (pp. 580-584). IEEE. 10.1109/IMCCC.2015.420

Haritaoglu, I., Harwood, D., & Davis, L. S. (2000). W/sup 4: Real-time surveillance of people and their activities. *IEEE Transactions on Pattern Analysis and Machine Intelligence, 22*(8), 809–830. doi:10.1109/34.868683

Heikkila, M., & Pietikainen, M. (2006). A texture-based method for modeling the background and detecting moving objects. *IEEE Transactions on Pattern Analysis and Machine Intelligence, 28*(4), 657–662. doi:10.1109/TPAMI.2006.68 PMID:16566514

Joseph, R., Divvala, S., Girshick, R., & Farhadi, A. (2021) You only look once: Unified, real-time object detection. *IEEE Conference on Computer Vision and Pattern Recognition*, (pp. 779 – 788). IEEE.

Karpathy, A., Toderici, G., Shetty, S., Leung, T., Sukthankar, R., & Fei-Fei, L. (2014). Large-scale video classification with convolutional neural networks. *IEEE Conference on Computer Vision and Pattern Recognition* (pp. 1725 - 1732). IEEE. 10.1109/CVPR.2014.223

Katagiri, S., & Lee, C. H. (1993). A new hybrid algorithm for speech recognition based on HMM segmentation and learning vector quantization. *IEEE Transactions on Speech and Audio Processing, 1*(4), 421–430. doi:10.1109/89.242488

Khan, R. & Debnath, R. (2015). Multi class fruit classification using efficient object detection and recognition techniques. *International Journal of Image, Graphics and Signal Processing.*

LeCun, Y. (1989). Generalization and network design strategies. *Connectionism in Perspective,* 143 – 155.

LeCun, Y., & Ranzato, M. (2013). Deep learning tutorial. *Tutorials in International Conference on Machine Learning (ICML'13).* IEEE.

Liu, Y., Sun, P., Wergeles, N., & Shang, Y. (2021). A survey and performance evaluation of deep learning methods for small object detection. *Expert Systems with Applications*, *172*, 114602. doi:10.1016/j.eswa.2021.114602

Liu, Z., Yan, W., & Yang, B. (2018) Image denoising based on a CNN model. *International Conference on Control, Automation and Robotics*. IEEE. 10.1109/ICCAR.2018.8384706

Maryam, K., & Reza, K. M. (2012). An analytical framework for event mining in video data. *Artificial Intelligence Review*, *41*(3), 401–413.

Matsugu, M., Mori, K., Mitari, Y., & Kaneda, Y. (2003). Subject independent facial expression recognition with robust face detection using a convolutional neural network. *Neural Networks*, *16*(5), 555–559. doi:10.1016/S0893-6080(03)00115-1 PMID:12850007

Mikolov, T., Kombrink, S., Burget, L., Černocký, J., & Khudanpur, S. (2011). Extensions of recurrent neural network language model. *IEEE International Conference on Acoustics, Speech and Signal Processing (ICASSP),* (pp. 5528 - 5531). IEEE. 10.1109/ICASSP.2011.5947611

Narendra, K. S., & Parthasarathy, K. (1990). Identification and control of dynamical systems using neural networks. *IEEE Transactions on Neural Networks*, *1*(1), 4–27. doi:10.1109/72.80202 PMID:18282820

Noorit, N., & Suvonvorn, N. (2014). Human activity recognition from basic actions using finite state machine. *International Conference on Advanced Data and Information Engineering (DaEng - 2013)* (pp. 379 - 386). Springer. 10.1007/978-981-4585-18-7_43

Pan, C., Liu, J., Yan, W., & Zhou, Y. (2021). Salient object detection based on visual perceptual saturation and two-stream hybrid networks. *IEEE Transactions on Image Processing*, *30*, 4773–4787. doi:10.1109/TIP.2021.3074796 PMID:33929959

Pan, C., & Yan, W. (2018) A learning-based positive feedback in salient object detection. *International Conference on Image and Vision Computing New Zealand.* IEEE. 10.1109/IVCNZ.2018.8634717

Pan, C., & Yan, W. (2020). Object detection based on saturation of visual perception. *Multimedia Tools and Applications*, *79*(27-28), 19925–19944. doi:10.100711042-020-08866-x

Parico, A., & Ahamed, T. (2021). Real time pear fruit detection and counting using YOLOv4 models and deep SORT. *Sensors (Basel)*, *21*(14), 4803. doi:10.339021144803 PMID:34300543

Pascanu, R., Mikolov, T., & Bengio, Y. (2013). On the difficulty of training recurrent neural networks. *ICML*, *28*(3), 1310–1318.

Petrushin, V. A. (2005). Mining rare and frequent events in multi-camera surveillance video using self-organizing maps. *ACM SIGKDD international Conference on Knowledge discovery in Data Mining,* (pp. 794 – 800). ACM.

Popoola, O. P., & Wang, K. (2012). Video-based abnormal human behavior recognition—A review. *IEEE Transactions on Systems, Man, and Cybernetics. Part C, Applications and Reviews*, *42*(6), 865–878. doi:10.1109/TSMCC.2011.2178594

Xia, Y., Nguyen, M., & Yan, W. (2022). A real-time Kiwifruit detection based on improved YOLOv7. *International Conference on Image and Vision Computing New Zealand (IVCNZ).* IEEE.

Xia, Y., Nguyen, M., & Yan, W. (2023) Kiwifruit counting using KiwiDetector and KiwiTracker. *IntelliSys conference*. Auckland University of Technology.

Xiao, B., Nguyen, M., & Yan, W. (2021) Apple ripeness identification using deep learning. *International Symposium on Geometry and Vision*. Springer. 10.1007/978-3-030-72073-5_5

Xiao, B., Nguyen, M., & Yan, W. (2023). *Apple ripeness identification from digital images using transformers. Multimedia Tools and Applications*. Springer Science and Business Media LLC.

Xiao, B., Nguyen, M., & Yan, W. (2023). *Fruit ripeness identification using transformers. Applied Intelligence*. Springer Science and Business Media LLC.

Yan, W. (2019). *Introduction to Intelligent Surveillance: Surveillance Data Capture, Transmission, and Analytics*. Springer London. doi:10.1007/978-3-030-10713-0

Yan, W. (2021). *Computational Methods for Deep Learning: Theoretic, Practice and Applications*. Springer London. doi:10.1007/978-3-030-61081-4

Yu, Y., Zhang, K., Yang, L., & Zhang, D. (2019). Fruit detection for strawberry harvesting robot in non-structural environment based on Mask R-CNN. *Computers and Electronics in Agriculture*, *163*, 104846. doi:10.1016/j.compag.2019.06.001

Zhang, Y., Zhang, W., Yu, J., He, L., Chen, J., & He, Y. (2022). Complete and accurate holly fruits counting using YOLOX object detection. *Computers and Electronics in Agriculture*, *198*, 107062. doi:10.1016/j.compag.2022.107062

Zhao, K. (2021). *Fruit Detection Using CenterNet*. [Master's Thesis, Auckland University of Technology, New Zealand].

Zhao, K., & Yan, W. (2021) Fruit detection from digital images using CenterNet. *International Symposium on Geometry and Vision*. 10.1007/978-3-030-72073-5_24

Zou, Z. (2023) Object detection in 20 years: A survey. *Proceedings of the IEEE*. 10.1109/JPROC.2023.3238524

Compilation of References

Abadi, M., Agarwal, A., Barham, P., Brevdo, E., Chen, Z., Citro, C., & Ghemawat, S. (2016). TensorFlow: Large-scale machine learning on heterogeneous distributed systems. *arXiv preprint arXiv:1603.04467.*

Abdeljaber, O., Avci, O., Kiranyaz, M. S., Boashash, B., Sodano, H., & Inman, D. J. (2018). 1-D CNNs for structural damage detection: Verification on a structural health monitoring benchmark data. *Neurocomputing*, *275*, 1308–1317. doi:10.1016/j.neucom.2017.09.069

Abdulrahman, A., & Iqbal, K. (2014) Capturing human body dynamics using RNN based on persistent excitation data generator. *International Symposium on Computer-Based Medical Systems (CBMS),* (pp. 221-226). IEEE. 10.1109/CBMS.2014.145

Abolfazli, S., Sanaei, Z., Wong, S. Y., Tabassi, A., & Rosen, S. (2015). *Throughput Measurement in 4G Wireless Data Networks: Performance Evaluation and Validation.* Academic Press.

Aboubakar, M., Kellil, M., & Roux, P. (2022). A review of IoT network management: Current status and perspectives. *Journal of King Saud University. Computer and Information Sciences*, *34*(7), 4163–4176. doi:10.1016/j.jksuci.2021.03.006

Abowd, G. D., Dey, A. K., Brown, P. J., Davies, N., Smith, M., & Steggles, P. (1999). Towards a better understanding of context and context-awareness. Lecture Notes in Computer Science (Including Subseries Lecture Notes in Artificial Intelligence and Lecture Notes in Bioinformatics), 1707. doi:10.1007/3-540-48157-5_29

Act, P. (1977). Chapter 37. Retrieved from http://images.policy.mofcom.gov.cn/article/201512/1449024985281.pdf

Adadi, A., & Berrada, M. (2018). Peeking inside the black-box: A survey on explainable artificial intelligence (XAI). *IEEE Access : Practical Innovations, Open Solutions*, *6*, 52138–52160. doi:10.1109/ACCESS.2018.2870052

Adhikari, S., & Bhalla, S. (2019). Modified Dual Piezo Configuration for Improved Structural Health Monitoring Using Electro-Mechanical Impedance (EMI) Technique. *Experimental Techniques*, *43*(1), 25–40. doi:10.100740799-018-0249-y

Adomavicius, G., Mobasher, B., Ricci, F., & Tuzhilin, A. (2011). Context-aware recommender systems. *AI Magazine*, *32*(3), 67–80. Advance online publication. doi:10.1609/aimag.v32i3.2364

Agrawal, A. V., Pitchai, R., Senthamaraikannan, C., Balaji, N. A., Sajithra, S., & Boopathi, S. (2023). Digital Education System During the COVID-19 Pandemic. In Using Assistive Technology for Inclusive Learning in K-12 Classrooms (pp. 104–126). IGI Global. doi:10.4018/978-1-6684-6424-3.ch005

Agrawal, A. V., Shashibhushan, G., Pradeep, S., Padhi, S. N., Sugumar, D., & Boopathi, S. (2024). Synergizing Artificial Intelligence, 5G, and Cloud Computing for Efficient Energy Conversion Using Agricultural Waste. In Practice, Progress, and Proficiency in Sustainability (pp. 475–497). IGI Global. doi:10.4018/979-8-3693-1186-8.ch026

Agrawal, A. V., Magulur, L. P., Priya, S. G., Kaur, A., Singh, G., & Boopathi, S. (2023). Smart Precision Agriculture Using IoT and WSN. In *Handbook of Research on Data Science and Cybersecurity Innovations in Industry 4.0 Technologies* (pp. 524–541). IGI Global. doi:10.4018/978-1-6684-8145-5.ch026

Agrawal, S., & Agrawal, J. (2015). Survey on anomaly detection using data mining techniques. Procedia Computer Science. doi:10.1016/j.procs.2015.08.220

Ahmad, M. B., Muhammad, A. S., Abdullahi, A. A., Tijjani, A., Iliyasu, A. S., Muhammad, I. M., ... Sani, K. M. (2019). Need for security alarm system installation and their challenges faced. *International Journal of New Computer Architectures and their Applications, 9*(3), 68-77.

Ahmadi, H. B., Kusi-Sarpong, S., & Rezaei, J. (2017). Assessing the social sustainability of supply chains using Best Worst Method. *Resources, Conservation and Recycling, 126*, 99–106. doi:10.1016/j.resconrec.2017.07.020

AI Act : A step closer to the first rules on Artificial Intelligence. (2023, Nov. 5). European Parliament.

AI IP Protection Debate. (n.d.). *Legal Service India - Law, Lawyers and Legal Resources.* Retrieved June 30, 2023, from https://www.legalserviceindia.com/legal/article-11428-ai-ip-protection-debate.html

Ajchariyavanich, C., Limpisthira, T., Chanjarasvichai, N., Jareonwatanan, T., Phongphanpanya, W., Wareechuensuk, S., Srichareonkul, S., Tachatanitanont, S., Ratanamahatana, C., Prompoon, N., & ... (2019). Park king: An IoT-based smart parking system. *2019 IEEE International Smart Cities Conference (ISC2)*, 729–734. 10.1109/ISC246665.2019.9071721

Ajith, G., Sudarsaun, J., Arvind, S. D., & Sugumar, R. (2018). IoT based fire deduction and safety navigation system. *International Journal of Innovative Research in Science, Engineering and Technology*, *7*(2), 257–267.

Akerkar, R., & Sajja, P. (2009). *Knowledge-based systems.* Jones & Bartlett Publishers.

Al Amiri, W., Baza, M., Banawan, K., Mahmoud, M., Alasmary, W., & Akkaya, K. (2020). Towards secure smart parking system using blockchain technology. *2020 IEEE 17th Annual Consumer Communications & Networking Conference (CCNC)*, 1–2.

Al Amiri, W., Baza, M., Banawan, K., Mahmoud, M., Alasmary, W., & Akkaya, K. (2019). Privacy-preserving smart parking system using blockchain and private information retrieval. *2019 International Conference on Smart Applications, Communications and Networking (SmartNets)*, 1–6. 10.1109/SmartNets48225.2019.9069783

Al Maruf, M. A., Ahmed, S., Ahmed, M. T., Roy, A., & Nitu, Z. F. (2019). A proposed model of integrated smart parking solution for a city. *2019 International Conference on Robotics, Electrical and Signal Processing Techniques (ICREST)*, 340–345. 10.1109/ICREST.2019.8644414

Alam, T., Tajammul, M., & Gupta, R. (2022). Towards the sustainable development of smart cities through cloud computing. *AI and IoT for Smart City Applications*, 199–222.

Albahri, A. S., Duhaim, A. M., Fadhel, M. A., Alnoor, A., Baqer, N. S., Alzubaidi, L., Albahri, O. S., Alamoodi, A. H., Bai, J., Salhi, A., Santamaría, J., Ouyang, C., Gupta, A., Gu, Y., & Deveci, M. (2023). A systematic review of trustworthy and explainable artificial intelligence in healthcare: Assessment of quality, bias risk, and data fusion. *Information Fusion, 96*, 156–191. doi:10.1016/j.inffus.2023.03.008

Alexandridis, G., Chrysanthi, A., Tsekouras, G. E., & Caridakis, G. (2019). Personalised and content adaptive cultural heritage path recommendation: An application to the Gournia and Çatalhöyük archaeological sites. *User Modeling and User-Adapted Interaction, 29*(1), 201–238. doi:10.100711257-019-09227-6

Al-Farhani, L. H., Alqahtani, Y., Alshehri, H. A., Martin, R. J., Lalar, S., & Jain, R. (2023). IOT and Blockchain-Based Cloud Model for Secure Data Transmission for Smart City. *Security and Communication Networks*, *2023*, 2023. doi:10.1155/2023/3171334

Al-Fuqaha, A., Guizani, M., Mohammadi, M., Aledhari, M., & Ayyash, M. (2015). Internet of Things: A Survey on Enabling Technologies, Protocols, and Applications. *IEEE Communications Surveys and Tutorials*, *17*(4), 2347–2376. Advance online publication. doi:10.1109/COMST.2015.2444095

Alharbi, A., Halikias, G., Yamin, M., & Abi Sen, A. A. (2021). Web-based framework for smart parking system. *International Journal of Information Technology : an Official Journal of Bharati Vidyapeeth's Institute of Computer Applications and Management*, *13*(4), 1495–1502. doi:10.100741870-021-00725-8

Ali, A. S., Radwan, A. G., & Soliman, A. M. (2013). Fractional order Butterworth filter: Active and passive realizations. *IEEE Journal on Emerging and Selected Topics in Circuits and Systems*, *3*(3), 346–354. doi:10.1109/JETCAS.2013.2266753

Ali, M. (1990). Intelligent systems in aerospace. *The Knowledge Engineering Review*, *5*(3), 147–166. doi:10.1017/S0269888900005385

Alkhatib, A. A. (2014). A review on forest fire detection techniques. *International Journal of Distributed Sensor Networks*, *10*(3), 597368. doi:10.1155/2014/597368

Almeida Cardoso, R., Cury, A., Barbosa, F., & Gentile, C. (2019). Unsupervised real-time SHM technique based on novelty indexes. *Structural Control and Health Monitoring*, *26*(7), e2364. doi:10.1002tc.2364

Almiani, M., AbuGhazleh, A., Al-Rahayfeh, A., Atiewi, S., & Razaque, A. (2020). Deep recurrent neural network for IoT intrusion detection system. *Simulation Modelling Practice and Theory*, *101*, 102031. Advance online publication. doi:10.1016/j.simpat.2019.102031

Alotaiby, T., El-Samie, F. E. A., Alshebeili, S. A., & Ahmad, I. (2015). A review of channel selection algorithms for EEG signal processing. *EURASIP Journal on Advances in Signal Processing*, *2015*(1), 1–21. doi:10.118613634-015-0251-9

Alqourabah, H., Muneer, A., & Fati, S. M. (2021). A smart fire detection system using IoT technology with automatic water sprinkler. *International Journal of Electrical & Computer Engineering, 11*(4).

Al-Sarayreh, M., Reis, M., Yan, W., & Klette, R. (2019) A sequential CNN approach for foreign object detection in hyperspectral images. *International Conference on Information, Communications and Signal*. Springer. 10.1007/978-3-030-29888-3_22

Al-Sarayreha, M. (2020). *Hyperspectral Imaging and Deep Learning for Food Safety*. [PhD Thesis, Auckland University of Technology, New Zealand].

Altabey, W. A., Noori, M., Wang, T., Ghiasi, R., Kuok, S.-C., & Wu, Z. (2021). Deep Learning-Based Crack Identification for Steel Pipelines by Extracting Features from 3D Shadow Modeling. *Applied Sciences (Basel, Switzerland)*, *11*(13), 6063. doi:10.3390/app11136063

Al-Turjman, F., & Malekloo, A. (2019). Smart parking in IoT-enabled cities: A survey. *Sustainable Cities and Society*, *49*, 101608. doi:10.1016/j.scs.2019.101608

Alves, B. R., Alves, G. V., Borges, A. P., & Leitão, P. (2019). Experimentation of negotiation protocols for consensus problems in smart parking systems. *Industrial Applications of Holonic and Multi-Agent Systems: 9th International Conference, HoloMAS 2019, Linz, Austria, August 26–29, 2019. Proceedings*, *9*, 189–202.

Alyavina, E., Nikitas, A., & Njoya, E. T. (2022). Mobility as a service (MaaS): A thematic map of challenges and opportunities. *Research in Transportation Business & Management*, *43*, 100783. doi:10.1016/j.rtbm.2022.100783

Alzubaidi, L., Bai, J., Al-Sabaawi, A., Santamaría, J., Albahri, A. S., Al-dabbagh, B. S. N., Fadhel, M. A., Manoufali, M., Zhang, J., Al-Timemy, A. H., Duan, Y., Abdullah, A., Farhan, L., Lu, Y., Gupta, A., Albu, F., Abbosh, A., & Gu, Y. (2023). A survey on deep learning tools dealing with data scarcity: Definitions, challenges, solutions, tips, and applications. *Journal of Big Data, 10*(1), 46. doi:10.118640537-023-00727-2

Amato, F., Moscato, F., Moscato, V., & Sperlì, G. (2021). Smart conversational user interface for recommending cultural heritage points of interest. *CEUR Workshop Proceedings*, 2994.

Ament, J., & Schmelz, J. (2023). Scientific use of Game Engines for Flight Simulation and Human Machine Interface Development. *CEAS Aeronautical Journal.*

Amiri, V., & Nakagawa, K. (2021). Using a linear discriminant analysis (LDA)-based nomenclature system and self-organizing maps (SOM) for spatiotemporal assessment of groundwater quality in a coastal aquifer. *Journal of Hydrology (Amsterdam), 603*, 127082. doi:10.1016/j.jhydrol.2021.127082

Ampadi Ramachandran, R., Barão, V. A. R., Ozevin, D., & Sukotjo, C., Pai. P, S., & Mathew, M. (. (2023). Early predicting tribocorrosion rate of dental implant titanium materials using random forest machine learning models. *Tribology International, 187*, 108735. doi:10.1016/j.triboint.2023.108735 PMID:37720691

An, N. (2020). *Anomalies Detection and Tracking Using Siamese Neural Networks.* [Master's Thesis, Auckland University of Technology, New Zealand].

Anastas, P., & Eghbali, N. (2010). Green chemistry: Principles and practice. *Chemical Society Reviews, 39*(1), 301–312. doi:10.1039/B918763B PMID:20023854

AnceaumeE.Del PozzoA.RieutordT.Tucci-PiergiovanniS. (2022). On Finality in Blockchains. *Leibniz International Proceedings in Informatics, LIPIcs, 217*. doi:10.4230/LIPIcs.OPODIS.2021.6

Anderson, C., Burt, P., & Van Der Wal, G. (1985). Change detection and tracking using pyramid transform techniques. *Cambridge Symposium* (pp. 72-78) International Society for Optics and Photonics. 10.1117/12.950785

Angulo-Saucedo, G. A., Leon-Medina, J. X., Pineda-Muñoz, W. A., Torres-Arredondo, M. A., & Tibaduiza, D. A. (2022). Damage Classification Using Supervised Self-Organizing Maps in Structural Health Monitoring. *Sensors (Basel), 22*(4), 1484. doi:10.339022041484 PMID:35214386

Anitha, C., Komala, C., Vivekanand, C. V., Lalitha, S., & Boopathi, S. (2023). Artificial Intelligence driven security model for Internet of Medical Things (IoMT). *IEEE Explore*, 1–7.

Anitha, C., Komala, C., Vivekanand, C. V., Lalitha, S., & Boopathi, S. (2023a). Artificial Intelligence driven security model for Internet of Medical Things (IoMT). *IEEE Explore*, (pp. 1–7). IEEE.

An, N., & Yan, W. (2021). Multitarget tracking using Siamese neural networks. *ACM Transactions on Multimedia Computing Communications and Applications, 17*(2s), 1–16. doi:10.1145/3441656

Anusha, G., & Rao, V. S. (2018). IoT based ware house fire safety system using ARM7. *Int J Eng Technol, 7*(3.6), 240-242.

Anwar, S., Zain, J. M., Zolkipli, M. F., Inayat, Z., Khan, S., Anthony, B., & Chang, V. (2017). From intrusion detection to an intrusion response system: Fundamentals, requirements, and future directions. *Algorithms, 107*(2), 39. Advance online publication. doi:10.3390/a10020039

Aoki, E., do Cabo, C. T., Mao, Z., Ding, Y., & Ozharar, S. (2022). Vibration-Based Status Identification of Power Transmission Poles. *IFAC-PapersOnLine, 55*(27), 214–217. doi:10.1016/j.ifacol.2022.10.514

Application of deep autoencoder model for structural condition monitoring. (2018). *Journal of Systems Engineering and Electronics, 29*(4), 873. doi:10.21629/JSEE.2018.04.22

Arana-Pulido, V., Cabrera-Almeida, F., Perez-Mato, J., Dorta-Naranjo, B. P., Hernandez-Rodriguez, S., & Jimenez-Yguacel, E. (2018). Challenges of an autonomous wildfire geolocation system based on synthetic vision technology. *Sensors (Basel), 18*(11), 3631. doi:10.339018113631 PMID:30366471

Arcadius Tokognon, C., Gao, B., Tian, G. Y., & Yan, Y. (2017). Structural Health Monitoring Framework Based on Internet of Things: A Survey. *IEEE Internet of Things Journal, 4*(3), 619–635. doi:10.1109/JIOT.2017.2664072

Ardani, S., Akintunde, E., Linzell, D., Eftekhar Azam, S., & Alomari, Q. (2023). Evaluating pod-based unsupervised damage identification using controlled damage propagation of out-of-service bridges. *Engineering Structures, 286*, 116096. doi:10.1016/j.engstruct.2023.116096

Arifin, M. S., Razi, S. A., Haque, A., & Mohammad, N. (2019). A microcontroller based intelligent traffic control system. *American Journal of Embedded Systems and Applications, 7*(1), 21–25. doi:10.11648/j.ajesa.20190701.13

Arshad, Z., Smith, J., Roberts, M., Lee, W. H., Davies, B., Bure, K., Hollander, G. A., Dopson, S., Bountra, C., & Brindley, D. (2016). Open access could transform drug discovery: A case study of JQ1. *Expert Opinion on Drug Discovery, 11*(3), 321–332. doi:10.1517/17460441.2016.1144587 PMID:26791045

Arunprasad, R., & Boopathi, S. (2019). Chapter-4 Alternate Refrigerants for Minimization Environmental Impacts: A Review. In Advances in Engineering Technology (p. 75). AkiNik Publications.

Ashfaq, R. A. R., Wang, X. Z., Huang, J. Z., Abbas, H., & He, Y. L. (2017b). Fuzziness-based semi-supervised learning approach for the intrusion detection system. *Information Sciences, 378*, 484–497. Advance online publication. doi:10.1016/j.ins.2016.04.019

Atlam, H. F., & Wills, G. B. (2020). *IoT Security*. Privacy, Safety and Ethics. doi:10.1007/978-3-030-18732-3_8

Avazov, K., Mukhiddinov, M., Makhmudov, F., & Cho, Y. I. (2021). Fire Detection Method in Smart City Environments Using a Deep-Learning-Based Approach. *Electronics (Basel), 11*(1), 73. doi:10.3390/electronics11010073

Azimi, M., Eslamlou, A., & Pekcan, G. (2020). Data-Driven Structural Health Monitoring and Damage Detection through Deep Learning: State-of-the-Art Review. *Sensors (Basel), 20*(10), 2778. doi:10.339020102778 PMID:32414205

Azimi, M., & Pekcan, G. (2020). Structural health monitoring using extremely compressed data through deep learning. *Computer-Aided Civil and Infrastructure Engineering, 35*(6), 597–614. doi:10.1111/mice.12517

Azuara, G., Ruiz, M., & Barrera, E. (2021). Damage Localization in Composite Plates Using Wavelet Transform and 2-D Convolutional Neural Networks. *Sensors (Basel), 21*(17), 5825. doi:10.339021175825 PMID:34502715

Azwar, H., Murtaz, M., Siddique, M., & Rehman, S. (2018). *Intrusion Detection in secure network for Cybersecurity systems using Machine Learning and Data Mining*. Academic Press.

B, M. K., K, K. K., Sasikala, P., Sampath, B., Gopi, B., & Sundaram, S. (2024). Sustainable Green Energy Generation From Waste Water. In *Practice, Progress, and Proficiency in Sustainability* (pp. 440–463). IGI Global. doi:10.4018/979-8-3693-1186-8.ch024

Baan, A., Allo, M. D. G., & Patak, A. A. (2022). The cultural attitudes of a funeral ritual discourse in the indigenous Torajan, Indonesia. *Heliyon, 8*(2), e08925. doi:10.1016/j.heliyon.2022.e08925 PMID:35198784

Babu, B. S., Kamalakannan, J., Meenatchi, N., Karthik, S., & Boopathi, S. (2022). Economic impacts and reliability evaluation of battery by adopting Electric Vehicle. *IEEE Explore*, 1–6.

Badidi, E. (2022). Edge AI and blockchain for smart sustainable cities: Promise and potential. *Sustainability (Basel), 14*(13), 7609. doi:10.3390u14137609

Bag, D., Ghosh, A., & Saha, S. (2023). An Automatic Approach to Control Wheelchair Movement for Rehabilitation Using Electroencephalogram. In *Design and Control Advances in Robotics* (pp. 105–126). IGI Global.

Bainbridge, D. (2010). *Intellectual Property.* Pearson Education Limited.

BalajiS.BalamuruganB.KumarT. A.RajmohanR.KumarP. P. (2021, March 29). A brief survey on AI based Face Mask Detection System for public places. SSRN. https://ssrn.com/abstract=3814341

Bale, A. S., Hashim, M. F., Bundele, K. S., & Vaishnav, J. (2023). 9 Applications and future trends of Artificial Intelligence and blockchain-powered Augmented Reality. Handbook of Augmented and Virtual Reality, 1, 137.

Ballon, P., Glidden, J., Kranas, P., Menychtas, A., Ruston, S., & Van Der Graaf, S. (2011). Is there a need for a cloud platform for european smart cities. *eChallenges E-2011 Conference Proceedings, IIMC International Information Management Corporation*, 1–7.

Bange, J., Reuder, J., & Platis, A. (2021). Unmanned Aircraft Systems. In *Springer Handbook of Atmospheric Measurements* (pp. 1331–1349). Springer. doi:10.1007/978-3-030-52171-4_49

Bansal, M., Goyal, A., & Choudhary, A. (2022). A comparative analysis of K-nearest neighbor, genetic, support vector machine, decision tree, and long short term memory algorithms in machine learning. *Decision Analytics Journal, 3*, 100071. doi:10.1016/j.dajour.2022.100071

Bao, Y., & Li, H. (2021). Machine learning paradigm for structural health monitoring. *Structural Health Monitoring, 20*(4), 1353–1372. doi:10.1177/1475921720972416

Bao, Y., Tang, Z., Li, H., & Zhang, Y. (2019). Computer vision and deep learning–based data anomaly detection method for structural health monitoring. *Structural Health Monitoring, 18*(2), 401–421. doi:10.1177/1475921718757405

Baranovskiy, N. V., Podorovskiy, A., & Malinin, A. (2021). Parallel implementation of the algorithm to compute forest fire impact on infrastructure facilities of JSC Russian railways. *Algorithms, 14*(11), 333. doi:10.3390/a14110333

Barnea, E., Mairon, R., & Ben-Shahar, O. (2016). Colour-agnostic shape-based 3D fruit detection for crop harvesting robots. *Biosystems Engineering, 146*, 57–70. doi:10.1016/j.biosystemseng.2016.01.013

Barron, L. (2022). Smart cities, connected cars and autonomous vehicles: Design fiction and visions of smarter future urban mobility. *Technoetic Arts : a Journal of Speculative Research, 20*(3), 225–240. doi:10.1386/tear_00092_1

Bartolini, I., Moscato, V., Pensa, R. G., Penta, A., Picariello, A., Sansone, C., & Sapino, M. L. (2016). Recommending multimedia visiting paths in cultural heritage applications. *Multimedia Tools and Applications, 75*(7), 3813–3842. Advance online publication. doi:10.100711042-014-2062-7

Basu, S., & Mamud, S. (2020, September). Comparative study on the effect of order and cut off frequency of Butterworth low pass filter for removal of noise in ECG signal. In *2020 IEEE 1st International Conference for Convergence in Engineering (ICCE)* (pp. 156-160). IEEE. 10.1109/ICCE50343.2020.9290646

Batool, T., Abbas, S., Alhwaiti, Y., Saleem, M., Ahmad, M., Asif, M., & Elmitwal, N. S. (2021). Intelligent model of ecosystem for smart cities using artificial neural networks. *Intelligent Automation & Soft Computing, 30*(2), 513–525. doi:10.32604/iasc.2021.018770

Baum, L. E., & Sell, G. (1968). Growth transformations for functions on manifolds. *Pacific Journal of Mathematics, 27*(2), 211–227. doi:10.2140/pjm.1968.27.211

Bazame, H. C., Molin, J. P., Althoff, D., & Martello, M. (2021). Detection, classification, and mapping of coffee fruits during harvest with computer vision. *Computers and Electronics in Agriculture*, *183*, 106066. doi:10.1016/j.compag.2021.106066

Bechmann, A., & Bowker, G. C. (2019). Unsupervised by any other name: Hidden layers of knowledge production in artificial intelligence on social media. *Big Data & Society*, *6*(1), 2053951718819569. doi:10.1177/2053951718819569

Becker, J., Manske, C., & Randl, S. (2022). Green chemistry and sustainability metrics in the pharmaceutical manufacturing sector. *Current Opinion in Green and Sustainable Chemistry*, *33*, 100562. doi:10.1016/j.cogsc.2021.100562

Bedi, P., Gupta, N., & Jindal, V. (2021). I-SiamIDS: An improved Siam-IDS for handling class imbalance in network-based intrusion detection systems. *Applied Intelligence*, *51*(2), 1133–1151. doi:10.100710489-020-01886-y

Beemsterboer, D. J. C., Hendrix, E. M. T., & Claassen, G. D. H. (2018). On solving the Best-Worst Method in multi-criteria decision-making. *IFAC-PapersOnLine*, *51*(11), 1660–1665. doi:10.1016/j.ifacol.2018.08.218

Belavagi, M. C., & Muniyal, B. (2016). Performance Evaluation of Supervised Machine Learning Algorithms for Intrusion Detection. *Procedia Computer Science*, *89*, 117–123. doi:10.1016/j.procs.2016.06.016

Ben Ali, J., Fnaiech, N., Saidi, L., Chebel-Morello, B., & Fnaiech, F. (2015). Application of empirical mode decomposition and artificial neural network for automatic bearing fault diagnosis based on vibration signals. *Applied Acoustics*, *89*, 16–27. doi:10.1016/j.apacoust.2014.08.016

Benavente-Sánchez, D., Moreno-Molina, J., & Argilés-Herrero, R. (2021). Prospects for remotely piloted aircraft systems in area-wide integrated pest management programmes. *Area-Wide Integrated Pest Management*, 903–916.

Bender, A., & Cortés-Ciriano, I. (2021). Artificial intelligence in drug discovery: What is realistic, what are illusions? Part 1: Ways to make an impact, and why we are not there yet. *Drug Discovery Today*, *26*(2), 511–524. doi:10.1016/j.drudis.2020.12.009 PMID:33346134

Bengio, Y., Simard, P., & Frasconi, P. (1994). Learning long-term dependencies with gradient descent is difficult. *IEEE Transactions on Neural Networks*, *5*(2), 157–166. doi:10.1109/72.279181 PMID:18267787

Benke, K., & Benke, G. (2018). Artificial intelligence and big data in public health. *International Journal of Environmental Research and Public Health*, *15*(12), 2796. doi:10.3390/ijerph15122796 PMID:30544648

Benouaret, I., & Lenne, D. (2016). Personalising the Museum Experience through Context-Aware Recommendations. *Proceedings - 2015 IEEE International Conference on Systems, Man, and Cybernetics, SMC 2015*. 10.1109/SMC.2015.139

Berne Convention for the Protection of Literary and Artistic Works (adopted 9 September 1886) ("Berne Convention")

Bezas, K., Komianos, V., Koufoudakis, G., Tsoumanis, G., Kabassi, K., & Oikonomou, K. (2020). Structural Health Monitoring in Historical Buildings: A Network Approach. *Heritage*, *3*(3), 796–818. doi:10.3390/heritage3030044

Bhadoria, R. S., Pandey, M. K., & Kundu, P. (2021). RVFR: Random vector forest regression model for integrated &enhanced approach in forest fires predictions. *Ecological Informatics*, *66*, 101471. doi:10.1016/j.ecoinf.2021.101471

Bhargava, A., & Bansal, A. (2021). Fruits and vegetables quality evaluation using computer vision: A review. *Journal of King Saud University. Computer and Information Sciences*, *33*(3), 243–257. doi:10.1016/j.jksuci.2018.06.002

Bhaskar, S., Singh, V. B., & Nayak, A. K. (2015).. . *Managing Data in SVM Supervised Algorithm for Data Mining Technology*, *27–30*, 1–4. Advance online publication. doi:10.1109/CSIBIG.2014.7056946

Bhattacharya, S., Yadav, N., Ahmad, A., Melandsø, F., & Habib, A. (2022). Multiple Damage Detection in PZT Sensor Using Dual Point Contact Method. *Sensors (Basel)*, *22*(23), 9161. doi:10.339022239161 PMID:36501870

Bhoi, S. K., Panda, S. K., Padhi, B. N., Swain, M. K., Hembram, B., Mishra, D., ... Khilar, P. M. (2018). Fireds-iot: A fire detection system for smart home based on IoT data analytics. In *2018 International Conference on Information Technology (ICIT)* (pp. 161-165). IEEE. 10.1109/ICIT.2018.00042

Bhosale, D. A., & Mane, V. M. (2016). Comparative study and analysis of network intrusion detection tools. *Proceedings of the 2015 International Conference on Applied and Theoretical Computing and Communication Technology, ICATccT 2015*, 312–315. 10.1109/ICATCCT.2015.7456901

Bikash Chandra Saha, M. S., Deepa, R., Akila, A., & Sai Thrinath, B. V. (2022). IOT based smart energy meter for smart grid. Academic Press.

Blakemore, D. C., Castro, L., Churcher, I., Rees, D. C., Thomas, A. W., Wilson, D. M., & Wood, A. (2018). Organic synthesis provides opportunities to transform drug discovery. *Nature Chemistry*, *10*(4), 383–394. doi:10.103841557-018-0021-z PMID:29568051

Bobadilla, J., Ortega, F., Hernando, A., & Gutiérrez, A. (2013). Recommender systems survey. *Knowledge-Based Systems*, *46*, 109–132. Advance online publication. doi:10.1016/j.knosys.2013.03.012

Bocek, T., Rodrigues, B. B., Strasser, T., & Stiller, B. (2017). Blockchains everywhere-a use-case of blockchains in the pharma supply-chain. *2017 IFIP/IEEE Symposium on Integrated Network and Service Management (IM)*, 772–777. 10.23919/INM.2017.7987376

Bock, F., Di Martino, S., & Origlia, A. (2019). Smart parking: Using a crowd of taxis to sense on-street parking space availability. *IEEE Transactions on Intelligent Transportation Systems*, *21*(2), 496–508. doi:10.1109/TITS.2019.2899149

Boopathi, S. (2013). *Experimental study and multi-objective optimization of near-dry wire-cut electrical discharge machining process* [PhD Thesis]. http://hdl.handle.net/10603/16933

Boopathi, S. (2021). Improving of Green Sand-Mould Quality using Taguchi Technique. *Journal of Engineering Research*.

Boopathi, S. (2021a). Improving of Green Sand-Mould Quality using Taguchi Technique. *Journal of Engineering Research*, in–Press.

Boopathi, S. (2021b). *Pollution monitoring and notification: Water pollution monitoring and notification using intelligent RC boat.*

Boopathi, S. (2022). Performance Improvement of Eco-Friendly Near-Dry wire-Cut Electrical Discharge Machining Process Using Coconut Oil-Mist Dielectric Fluid. *World Scientific: Journal of Advanced Manufacturing Systems.*

Boopathi, S. (2022b). Cryogenically treated and untreated stainless steel grade 317 in sustainable wire electrical discharge machining process: A comparative study. *Environmental Science and Pollution Research*, 1–10.

Boopathi, S. (2023). Securing Healthcare Systems Integrated With IoT: Fundamentals, Applications, and Future Trends. In Dynamics of Swarm Intelligence Health Analysis for the Next Generation (pp. 186–209). IGI Global.

Boopathi, S. (2023a). Deep Learning Techniques Applied for Automatic Sentence Generation. In Promoting Diversity, Equity, and Inclusion in Language Learning Environments (pp. 255–273). IGI Global. doi:10.4018/978-1-6684-3632-5.ch016

Boopathi, S. (2023b). Securing Healthcare Systems Integrated With IoT: Fundamentals, Applications, and Future Trends. In Dynamics of Swarm Intelligence Health Analysis for the Next Generation (pp. 186–209). IGI Global.

Boopathi, S. (2023c). Securing Healthcare Systems Integrated With IoT: Fundamentals, Applications, and Future Trends. In Dynamics of Swarm Intelligence Health Analysis for the Next Generation (pp. 186–209). IGI Global.

Boopathi, S., & Khare, R. KG, J. C., Muni, T. V., & Khare, S. (2023). Additive Manufacturing Developments in the Medical Engineering Field. In Development, Properties, and Industrial Applications of 3D Printed Polymer Composites (pp. 86–106). IGI Global.

Boopathi, S., Arigela, S. H., Raman, R., Indhumathi, C., Kavitha, V., & Bhatt, B. C. (2022). Prominent Rule Control-based Internet of Things: Poultry Farm Management System. *IEEE Explore*, 1–6.

Boopathi, S., Gavaskar, T., Dogga, A. D., Mahendran, R. K., Kumar, A., Kathiresan, G., N., V., Ganesan, M., Ishwarya, K. R., … Ramana, G. V. (2021). *Emergency medicine delivery transportation using unmanned aerial vehicle (Patent Grant)*. Academic Press.

Boopathi, S., Khare, R., KG, J. C., Muni, T. V., & Khare, S. (2023). Additive Manufacturing Developments in the Medical Engineering Field. In Development, Properties, and Industrial Applications of 3D Printed Polymer Composites (pp. 86–106). IGI Global.

Boopathi, S., Kumar, P. K. S., Meena, R. S., Sudhakar, M., & Associates. (2023). Sustainable Developments of Modern Soil-Less Agro-Cultivation Systems: Aquaponic Culture. In Human Agro-Energy Optimization for Business and Industry (pp. 69–87). IGI Global.

Boopathi, S., Pandey, B. K., & Pandey, D. (2023). Advances in Artificial Intelligence for Image Processing: Techniques, Applications, and Optimization. In Handbook of Research on Thrust Technologies' Effect on Image Processing (pp. 73–95). IGI Global.

Boopathi, S., Pandey, B. K., & Pandey, D. (2023a). Advances in Artificial Intelligence for Image Processing: Techniques, Applications, and Optimization. In Handbook of Research on Thrust Technologies' Effect on Image Processing (pp. 73–95). IGI Global.

Boopathi, S., Thillaivanan, A., Mohammed, A. A., Shanmugam, P., & VR, P. (2022). Experimental investigation on Abrasive Water Jet Machining of Neem Wood Plastic Composite. *IOP: Functional Composites and Structures, 4*, 025001.

Boopathi. S. (2022). Effects of Cryogenically-treated Stainless Steel on Eco-friendly Wire Electrical Discharge Machining Process. Preprint : Springer.

Boopathi, S. (2022). An experimental investigation of Quench Polish Quench (QPQ) coating on AISI 4150 steel. *Engineering Research Express*, *4*(4), 045009. doi:10.1088/2631-8695/ac9ddd

Boopathi, S. (2022). Experimental investigation and multi-objective optimization of cryogenic Friction-stir-welding of AA2014 and AZ31B alloys using MOORA technique. *Materials Today. Communications*, *33*, 104937. doi:10.1016/j.mtcomm.2022.104937

Boopathi, S. (2022a). An investigation on gas emission concentration and relative emission rate of the near-dry wire-cut electrical discharge machining process. *Environmental Science and Pollution Research International*, *29*(57), 86237–86246. doi:10.100711356-021-17658-1 PMID:34837614

Boopathi, S. (2023b). Internet of Things-Integrated Remote Patient Monitoring System: Healthcare Application. In *Dynamics of Swarm Intelligence Health Analysis for the Next Generation* (pp. 137–161). IGI Global. doi:10.4018/978-1-6684-6894-4.ch008

Boopathi, S., Alqahtani, A. S., Mubarakali, A., & Panchatcharam, P. (2023). Sustainable developments in near-dry electrical discharge machining process using sunflower oil-mist dielectric fluid. *Environmental Science and Pollution Research International*, 1–20. doi:10.100711356-023-27494-0 PMID:37199846

Boopathi, S., Balasubramani, V., Kumar, R. S., & Singh, G. R. (2021). The influence of human hair on kenaf and Grewia fiber-based hybrid natural composite material: An experimental study. *Functional Composites and Structures*, *3*(4), 045011. doi:10.1088/2631-6331/ac3afc

Boopathi, S., & Davim, J. P. (2023a). Applications of Nanoparticles in Various Manufacturing Processes. In *Sustainable Utilization of Nanoparticles and Nanofluids in Engineering Applications* (pp. 1–31). IGI Global. doi:10.4018/978-1-6684-9135-5.ch001

Boopathi, S., & Davim, J. P. (2023b). *Sustainable Utilization of Nanoparticles and Nanofluids in Engineering Applications*. IGI Global. doi:10.4018/978-1-6684-9135-5

Boopathi, S., & Kanike, U. K. (2023). Applications of Artificial Intelligent and Machine Learning Techniques in Image Processing. In *Handbook of Research on Thrust Technologies' Effect on Image Processing* (pp. 151–173). IGI Global. doi:10.4018/978-1-6684-8618-4.ch010

Boopathi, S., Lewise, K. A. S., Subbiah, R., & Sivaraman, G. (2022). Near-dry wire-cut electrical discharge machining process using water–air-mist dielectric fluid: An experimental study. *Materials Today: Proceedings*, *50*(5), 1885–1890. doi:10.1016/j.matpr.2021.08.077

Boopathi, S., & Myilsamy, S. (2021). Material removal rate and surface roughness study on Near-dry wire electrical discharge Machining process. *Materials Today: Proceedings*, *45*(9), 8149–8156. doi:10.1016/j.matpr.2021.02.267

Boopathi, S., Myilsamy, S., & Sukkasamy, S. (2021). *Experimental Investigation and Multi-Objective Optimization of Cryogenically Cooled Near-Dry Wire-Cut EDM Using TOPSIS Technique*. IJAMT Preprint.

Boopathi, S., Saranya, A., Raghuraman, S., & Revanth, R. (2018). Design and Fabrication of Low Cost Electric Bicycle. *International Research Journal of Engineering and Technology*, *5*(3), 146–147.

Boopathi, S., & Sivakumar, K. (2012). Experimental Analysis of Eco-friendly Near-dry Wire Electrical Discharge Machining Process. [UGC]. *Archives des Sciences*, *65*(10), 334–346.

Boopathi, S., & Sivakumar, K. (2013). Experimental investigation and parameter optimization of near-dry wire-cut electrical discharge machining using multi-objective evolutionary algorithm. *International Journal of Advanced Manufacturing Technology*, *67*(9–12), 2639–2655. doi:10.100700170-012-4680-4

Boopathi, S., & Sivakumar, K. (2014). Study of water assisted dry wire-cut electrical discharge machining. [SCI]. *Indian Journal of Engineering and Materials Sciences*, *21*, 75–82.

Boopathi, S., & Sivakumar, K. (2016). Optimal parameter prediction of oxygen-mist near-dry wire-cut EDM. *Inderscience: International Journal of Manufacturing Technology and Management*, *30*(3–4), 164–178. doi:10.1504/IJMTM.2016.077812

Boopathi, S., Sureshkumar, M., & Sathiskumar, S. (2022). Parametric Optimization of LPG Refrigeration System Using Artificial Bee Colony Algorithm. *International Conference on Recent Advances in Mechanical Engineering Research and Development*, 97–105.

Boopathi, S., Umareddy, M., & Elangovan, M. (2023). Applications of Nano-Cutting Fluids in Advanced Machining Processes. In *Sustainable Utilization of Nanoparticles and Nanofluids in Engineering Applications* (pp. 211–234). IGI Global. doi:10.4018/978-1-6684-9135-5.ch009

Bountra, C., Lee, W., & Lezaun, J. (2017). *A new pharmaceutical commons: Transforming drug discovery*. Oxford Martin Policy Paper Oxford.

Bouzenad, A. E., El Mountassir, M., Yaacoubi, S., Dahmene, F., Koabaz, M., Buchheit, L., & Ke, W. (2019). A Semi-Supervised Based K-Means Algorithm for Optimal Guided Waves Structural Health Monitoring: A Case Study. *Inventions (Basel, Switzerland)*, *4*(1), 17. doi:10.3390/inventions4010017

Brasse, J., Broder, H. R., Förster, M., Klier, M., & Sigler, I. (2023). Explainable artificial intelligence in information systems: A review of the status quo and future research directions. *Electronic Markets*, *33*(1), 26. Advance online publication. doi:10.100712525-023-00644-5

Bro, R., & Smilde, A. K. (2014). Principal component analysis. In Analytical Methods (Vol. 6, Issue 9). doi:10.1039/C3AY41907J

Brown, T. (2019, October 11). *The AI Skills Shortage - ITChronicles.* ITChronicles. https://itchronicles.com/artificial-intelligence/the-ai-skills-shortage/

Brunner, C., Leeb, R., Müller-Putz, G., Schlögl, A., & Pfurtscheller, G. (2008). BCI Competition 2008–Graz data set A. Institute for Knowledge Discovery (of Brain-Computer Interfaces), Graz University of Technology.

Bryatov, S., & Borodinov, A. (2019). Blockchain technology in the pharmaceutical supply chain: Researching a business model based on Hyperledger Fabric. *Proceedings of the International Conference on Information Technology and Nanotechnology (ITNT),* Samara, Russia, *10*, 1613–0073. 10.18287/1613-0073-2019-2416-134-140

BSA The Software Alliance. (2020). Retrieved from https://www.bsa.org/news-events/media/usmca-formalizes-free-flow-of-data-other-tech-issues

Buckley, J., Gaetano, D., McCarthy, K., Loizou, L., O'Flynn, B., & O'mathuna, C. (2014). Compact 433 MHz antenna for wireless smart system applications. *Electronics Letters*, *50*(8), 572–574. doi:10.1049/el.2013.4312

Buczak, A. L., & Guven, E. (2016). A Survey of Data Mining and Machine Learning Methods for Cyber Security Intrusion Detection. *IEEE Communications Surveys and Tutorials*, *18*(2), 1153–1176. doi:10.1109/COMST.2015.2494502

Bui, V., Alaei, A., & Bui, M. (2022). On the integration of AI and IoT systems: A case study of airport smart parking. *Artificial Intelligence-Based Internet of Things Systems*, 419–444.

Bukhsh, Z. A., Jansen, N., & Saeed, A. (2021). Damage detection using in-domain and cross-domain transfer learning. *Neural Computing & Applications*, *33*(24), 16921–16936. doi:10.100700521-021-06279-x

Buolamwini, J., & Gebru, T. (2018, January). Gender shades: Intersectional accuracy disparities in commercial gender classification. *Conference on Fairness, Accountability and Transparency*, 77-91.

Burki, T. (2019). Pharma blockchains AI for drug development. *Lancet*, *393*(10189), 2382. doi:10.1016/S0140-6736(19)31401-1 PMID:31204669

Burns, L. N., & Tucci, L. (2023, March 31). *What is artificial intelligence (AI)? - AI definition and how it works. Enterprise AI.* TechTarget. https://www.techtarget.com/searchenterpriseai/definition/AI-Artificial-Intelligence

Butler, L., Yigitcanlar, T., & Paz, A. (2021). Barriers and risks of Mobility-as-a-Service (MaaS) adoption in cities: A systematic review of the literature. *Cities (London, England)*, *109*, 103036. doi:10.1016/j.cities.2020.103036

Cahyadi, N., Dorand, P., Rozi, N. R. F., Haq, L. A., & Maulana, R. I. (2023). A Literature Review for Understanding the Development of Smart Parking Systems. *Journal of Informatics and Communication Technology*, *5*(1), 46–56.

Cameron, I. (2021). *National security and the European Convention on human rights.* BRILL.

Candon, M., Levinski, O., Ogawa, H., Carrese, R., & Marzocca, P. (2022). A nonlinear signal processing framework for rapid identification and diagnosis of structural freeplay. *Mechanical Systems and Signal Processing*, *163*, 107999. doi:10.1016/j.ymssp.2021.107999

Ćemalović, U. (2022). Status of the World Intellectual Property Organization (WIPO) in the United Nations System and its Competences. *International Organizations, Serbia and Contemporary World*, *1*, 391.

Cervantes, J., Garcia-lamont, F., Rodríguez-mazahua, L., & Lopez, A. (2020). *Neurocomputing A comprehensive survey on support vector machine classification : Applications, challenges, and trends.* doi:10.1016/j.neucom.2019.10.118

Chai, R., Ling, S. H., Hunter, G. P., & Nguyen, H. T. (2012, August). *Toward fewer EEG channels and better feature extractor of non-motor imagery mental tasks classification for a wheelchair thought controller. In 2012 Annual International Conference of the IEEE Engineering in Medicine and Biology Society*. IEEE.

Cha, J., Kim, K. S., Zhang, H., & Lee, S. (2019, November). Analysis on EEG signal with machine learning. In *2019 International Conference on Image and Video Processing, and Artificial Intelligence* (Vol. 11321, pp. 564-569). SPIE. 10.1117/12.2548313

Chakraborty, I., Vigneswara Ilavarasan, P., & Edirippulige, S. (2022). E-Health Startups' Framework for Value Creation and Capture: Some Insights from Systematic Review. *Proceedings of the International Conference on Cognitive and Intelligent Computing: ICCIC 2021,* 1, 141–152. 10.1007/978-981-19-2350-0_13

Chandana, M., Aniruddh, M., Fathima, A., Chandan, G., & Kumari, H. C. (2019). A Study on Smart Parking Solutions Using IoT. *2019 International Conference on Intelligent Sustainable Systems (ICISS)*, 542–548. 10.1109/ISS1.2019.8907944

Chandrika, V., Sivakumar, A., Krishnan, T. S., Pradeep, J., Manikandan, S., & Boopathi, S. (2023). Theoretical Study on Power Distribution Systems for Electric Vehicles. In *Intelligent Engineering Applications and Applied Sciences for Sustainability* (pp. 1–19). IGI Global. doi:10.4018/979-8-3693-0044-2.ch001

Chang, Y., Iakovou, E., & Shi, W. (2020). Blockchain in global supply chains and cross border trade: A critical synthesis of the state-of-the-art, challenges and opportunities. *International Journal of Production Research*, *58*(7), 2082–2099. doi:10.1080/00207543.2019.1651946

Chan, H., Hammad, E., & Kundur, D. (2016). Investigating the impact of intrusion detection system performance on communication latency and power system stability. *Proceedings of the Workshop on Communications, Computation and Control for Resilient Smart Energy Systems, RSES 2016*. 10.1145/2939940.2939946

Chatila, R., & Havens, J. C. (2019). The IEEE global initiative on ethics of autonomous and intelligent systems. *Robotics and Well-Being*, 11-16.

Chatterjee, P., Ghosh, A., & Saha, S. (2023). *A Comprehensive Survey on Rehabilitative Applications of Electroencephalogram in Healthcare*. Cognitive Cardiac Rehabilitation Using IoT and AI Tools. doi:10.4018/978-1-6684-7561-4.ch003

Chatzis, S. P., & Kosmopoulos, D. I. (2011). A variational Bayesian methodology for hidden Markov models utilizing Student's-*t* mixtures. *Pattern Recognition*, *44*(2), 295–306. doi:10.1016/j.patcog.2010.09.001

Chaudhary, S., Taran, S., Bajaj, V., & Sengur, A. (2019). Convolutional neural network based approach towards motor imagery tasks EEG signals classification. *IEEE Sensors Journal*, *19*(12), 4494–4500. doi:10.1109/JSEN.2019.2899645

Cha, Y.-J., & Wang, Z. (2018). Unsupervised novelty detection–based structural damage localization using a density peaks-based fast clustering algorithm. *Structural Health Monitoring*, *17*(2), 313–324. doi:10.1177/1475921717691260

Chen, C., Tang, L., Lu, Y., Wang, Y., Liu, Z., Liu, Y., Zhou, L., Jiang, Z., & Yang, B. (2023). Reconstruction of long-term strain data for structural health monitoring with a hybrid deep-learning and autoregressive model considering thermal effects. *Engineering Structures*, *285*, 116063. doi:10.1016/j.engstruct.2023.116063

Cheng, H., Großschädl, J., Rønne, P. B., & Ryan, P. Y. A. (2021). Lightweight Post-quantum Key Encapsulation for 8-bit AVR Microcontrollers. doi:10.1007/978-3-030-68487-7_2

Chen, Q., Cao, J., & Xia, Y. (2022). Physics-Enhanced PCA for Data Compression in Edge Devices. *IEEE Transactions on Green Communications and Networking*, *6*(3), 1624–1634. doi:10.1109/TGCN.2022.3171681

Chen, Y. N., Han, C. C., Wang, C. T., Jeng, B. S., & Fan, K. C. (2006). The application of a convolution neural network on face and license plate detection. *International Conference on Pattern Recognition*, *3*, 552–555.

Chmiel, W., Dańda, J., Dziech, A., Ernst, S., Kadłuczka, P., Mikrut, Z., Pawlik, P., Szwed, P., & Wojnicki, I. (2016). INSIGMA: An intelligent transportation system for urban mobility enhancement. *Multimedia Tools and Applications*, *75*(17), 10529–10560. doi:10.100711042-016-3367-5

Choudhary, S., & Kesswani, N. (2021). A hybrid classification approach for intrusion detection in IoT network. *Journal of Scientific and Industrial Research*, *80*(9), 809–816.

Choudhury, S., & Bhowal, A. (2015). Comparative analysis of machine learning algorithms along with classifiers for network intrusion detection. *2015 International Conference on Smart Technologies and Management for Computing, Communication, Controls, Energy and Materials (ICSTM).* 10.1109/ICSTM.2015.7225395

Christidis, K., & Devetsikiotis, M. (2016). Blockchains and Smart Contracts for the Internet of Things. *IEEE Access : Practical Innovations, Open Solutions*, *4*, 2292–2303. doi:10.1109/ACCESS.2016.2566339

Christlein, V., Spranger, L., Seuret, M., Nicolaou, A., Kral, P., & Maier, A. (2019). Deep generalized Max pooling. *International Conference on Document Analysis and Recognition (ICDAR).* 10.1109/ICDAR.2019.00177

Cinque, D., Saccone, M., Capua, R., Spina, D., Falcolini, C., & Gabriele, S. (2022). Experimental Validation of a High Precision GNSS System for Monitoring of Civil Infrastructures. *Sustainability (Basel)*, *14*(17), 10984. doi:10.3390u141710984

Ciotti, M., Ciccozzi, M., Terrinoni, A., Jiang, W.-C., Wang, C.-B., & Bernardini, S. (2020). The COVID-19 pandemic. *Critical Reviews in Clinical Laboratory Sciences*, *57*(6), 365–388. doi:10.1080/10408363.2020.1783198 PMID:32645276

CiuriakD.FayR. (2021). The USMCA and Mexico's Prospects under the New North American Trade Regime. *Social Science Research Network, 2*, 45-66. doi:10.2139/ssrn.3771338

Civera, M., & Surace, C. (2021). A Comparative Analysis of Signal Decomposition Techniques for Structural Health Monitoring on an Experimental Benchmark. *Sensors (Basel)*, *21*(5), 1825. doi:10.339021051825 PMID:33807884

Collins, R. T., Lipton, A. J., Kanade, T., Fujiyoshi, H., Duggins, D., Tsin, Y., & Wixson, L. (2000). *A system for video surveillance and monitoring. Research Report.* Carnegie Mellon University.

Cui, W. (2015). *A Scheme of Human Face Recognition in Complex Environments* [Master's Thesis]. Auckland University of Technology.

Cui, W., & Yan, W. (2016). A scheme for face recognition in complex environments. *International Journal of Digital Crime and Forensics*, *8*(1), 26–36. doi:10.4018/IJDCF.2016010102

Da Xu, L., Lu, Y., & Li, L. (2021). Embedding Blockchain Technology Into IoT for Security: A Survey. *IEEE Internet of Things Journal*, *8*(13), 10452–10473. doi:10.1109/JIOT.2021.3060508

Dai, M., Zheng, D., Na, R., Wang, S., & Zhang, S. (2019). EEG classification of motor imagery using a novel deep learning framework. *Sensors (Basel), 19*(3), 551. doi:10.339019030551 PMID:30699946

Dang, H. V., Tran-Ngoc, H., Nguyen, T. V., Bui-Tien, T., De Roeck, G., & Nguyen, H. X. (2021). Data-Driven Structural Health Monitoring Using Feature Fusion and Hybrid Deep Learning. *IEEE Transactions on Automation Science and Engineering, 18*(4), 2087–2103. doi:10.1109/TASE.2020.3034401

Dang, H., Tatipamula, M., & Nguyen, H. X. (2022). Cloud-Based Digital Twinning for Structural Health Monitoring Using Deep Learning. *IEEE Transactions on Industrial Informatics, 18*(6), 3820–3830. doi:10.1109/TII.2021.3115119

Dang, V.-H. (2023). Development of Structural Damage Detection Method Working with Contaminated Vibration Data via Autoencoder and Gradient Boosting. *Periodica Polytechnica. Civil Engineering*. Advance online publication. doi:10.3311/PPci.22373

Dardeno, T. A., Bull, L. A., Mills, R. S., Dervilis, N., & Worden, K. (2022). Modelling variability in vibration-based PBSHM via a generalised population form. *Journal of Sound and Vibration, 538*, 117227. doi:10.1016/j.jsv.2022.117227

Dass, A., & Boopathi, S. (2016). Experimental Study of Eco-friendly Wire-Cut Electrical Discharge Machining Processes. *International Journal of Innovative Research in Science, Engineering and Technology, 5*.

Das, S., & Nene, M. J. (2018). A survey on types of machine learning techniques in intrusion prevention systems. *Proceedings of the 2017 International Conference on Wireless Communications, Signal Processing and Networking, WiSPNET 2017,* 2296–2299. 10.1109/WiSPNET.2017.8300169

De Guimarães, J. C. F., Severo, E. A., Júnior, L. A. F., Da Costa, W. P. L. B., & Salmoria, F. T. (2020). Governance and quality of life in smart cities: Towards sustainable development goals. *Journal of Cleaner Production, 253*, 119926. doi:10.1016/j.jclepro.2019.119926

de Klerk, R., Duarte, A. M., Medeiros, D. P., Duarte, J. P., Jorge, J., & Lopes, D. S. (2019). Usability studies on building early stage architectural models in virtual reality. *Automation in Construction, 103*, 104–116. doi:10.1016/j.autcon.2019.03.009

De Oliveira, L. F. P., Manera, L. T., & Da Luz, P. D. G. (2020). Development of a smart traffic light control system with real-time monitoring. *IEEE Internet of Things Journal, 8*(5), 3384–3393. doi:10.1109/JIOT.2020.3022392

De Simone, M. C., & Guida, D. (2020). *Experimental Investigation on Structural Vibrations by a New Shaking Table.* doi:10.1007/978-3-030-41057-5_66

De Simone, M. C., Lorusso, A., & Santaniello, D. (2022). Predictive maintenance and Structural Health Monitoring via IoT system. *2022 IEEE Workshop on Complexity in Engineering (COMPENG)*, 1–4. 10.1109/COMPENG50184.2022.9905441

De Vries, S., & Mulder, T. (2007). Motor imagery and stroke rehabilitation: A critical discussion. *Journal of Rehabilitation Medicine, 39*(1), 5–13. doi:10.2340/16501977-0020 PMID:17225031

Dederichs, A. C., & Øiseth, O. (2023). Experimental comparison of automatic operational modal analysis algorithms for application to long-span road bridges. *Mechanical Systems and Signal Processing, 199*, 110485. doi:10.1016/j.ymssp.2023.110485

Deivasigamani, A., Daliri, A. H., Wang, C., & John, S. (2013). A Review of Passive Wireless Sensors for Structural Health Monitoring. *Modern Applied Science, 7*(2). Advance online publication. doi:10.5539/mas.v7n2p57

Delgadillo, R. M., & Casas, J. R. (2022). Bridge damage detection via improved completed ensemble empirical mode decomposition with adaptive noise and machine learning algorithms. *Structural Control and Health Monitoring, 29*(8). Advance online publication. doi:10.1002tc.2966

Delgado, A., & Calegari, D. (2023). *Process Mining for Improving Urban Mobility in Smart Cities: Challenges and Application with Open Data*. Academic Press.

Deng, L., Chu, H.-H., Shi, P., Wang, W., & Kong, X. (2020). Region-Based CNN Method with Deformable Modules for Visually Classifying Concrete Cracks. *Applied Sciences (Basel, Switzerland)*, *10*(7), 2528. doi:10.3390/app10072528

Devan, P., & Khare, N. (2020). An efficient XGBoost–DNN-based classification model for network intrusion detection system. *Neural Computing & Applications*, *32*(16), 12499–12514. doi:10.100700521-020-04708-x

DG. (2020). *The ethics of artificial intelligence: Issues and initiatives*. European Parliamentary Research Service Scientific Foresight Unit (STOA) PE 634.452, p:15

Dhanya, D., Kumar, S. S., Thilagavathy, A., Prasad, D., & Boopathi, S. (2023). Data Analytics and Artificial Intelligence in the Circular Economy: Case Studies. In Intelligent Engineering Applications and Applied Sciences for Sustainability (pp. 40–58). IGI Global.

Dhruva Kumar, D., Fang, C., Zheng, Y., & Gao, Y. (2023). Semi-supervised transfer learning-based automatic weld defect detection and visual inspection. *Engineering Structures*, *292*, 116580. doi:10.1016/j.engstruct.2023.116580

Di Lorenzo, D., Champaney, V., Marzin, J. Y., Farhat, C., & Chinesta, F. (2023). Physics informed and data-based augmented learning in structural health diagnosis. *Computer Methods in Applied Mechanics and Engineering*, *414*, 116186. doi:10.1016/j.cma.2023.116186

Diran, D., Van Veenstra, A. F., Timan, T., Tesa, P., & Kirova, M. (2021). Artificial Intelligence in smart cities and urban mobility. *Policy Department for Economic, Scientific and Quality of Life Policies*.

Discher, G., & Rutigliano, N. (2021). USPTO Releases Report on Artificial Intelligence and Intellectual Property Policy. *The Journal of Robotics. Artificial Intelligence and Law*, 4.

Dogo, E. M., Salami, A. F., Aigbavboa, C. O., & Nkonyana, T. (2019). Taking cloud computing to the extreme edge: A review of mist computing for smart cities and industry 4.0 in Africa. *Edge Computing: From Hype to Reality*, 107–132.

Domakonda, V. K., Farooq, S., Chinthamreddy, S., Puviarasi, R., Sudhakar, M., & Boopathi, S. (2022). Sustainable Developments of Hybrid Floating Solar Power Plants: Photovoltaic System. In Human Agro-Energy Optimization for Business and Industry (pp. 148–167). IGI Global.

Dooraki, A. R., & Lee, D.-J. (2021). An innovative bio-inspired flight controller for quad-rotor drones: Quad-rotor drone learning to fly using reinforcement learning. *Robotics and Autonomous Systems*, *135*, 103671. doi:10.1016/j.robot.2020.103671

Dornis, T. W. (2020). Artificial intelligence and innovation: The end of patent law as we know it. *SSRN*, *23*, 97. doi:10.2139srn.3668137

Doroudi, R., Hosseini Lavassani, S. H., Shahrouzi, M., & Dadgostar, M. (2022). Identifying the dynamic characteristics of super tall buildings by multivariate empirical mode decomposition. *Structural Control and Health Monitoring*, *29*(11). Advance online publication. doi:10.1002tc.3075

Duan, J., Chen, X. M., Qi, H., & Li, Y. G. (2014). An Integrated Simulation System for Building Structures. *Applied Mechanics and Materials*, *580–583*, 3127–3133. . doi:10.4028/www.scientific.net/AMM.580-583.3127

Dua, S., & Xian, D. (2011). *Data Mining and Machine Learning in Cybersecurity*. In Auerbach Publications., doi:10.1192/bjp.112.483.211-a

Dubey, C., Kumar, D., Singh, A. K., & Dwivedi, V. K. (2022). Confluence of Artificial Intelligence and Blockchain Powered Smart Contract in Finance System. *2022 International Conference on Computing, Communication, and Intelligent Systems (ICCCIS)*, 125–130. 10.1109/ICCCIS56430.2022.10037701

Dunn, P. J., Wells, A. S., & Williams, M. T. (2010). Future trends for green chemistry in the pharmaceutical industry. *Green Chemistry in the Pharmaceutical Industry*, *16*, 333–355. doi:10.1002/9783527629688.ch16

Du, R. (2022). *Intellectual Property Protection and Growth: Evidence from Post-TRIPS Development of Manufacturing Industries*. University of Chicago.

Durairaj, M., Jayakumar, S., Karpagavalli, V., Maheswari, B. U., Boopathi, S., & ... (2023). Utilization of Digital Tools in the Indian Higher Education System During Health Crises. In *Multidisciplinary Approaches to Organizational Governance During Health Crises* (pp. 1–21). IGI Global. doi:10.4018/978-1-7998-9213-7.ch001

Dutta, A., Chakraborty, A., Kumar, A., Roy, A., Roy, S., Chakraborty, D., Saha, H. N., Sharma, A. S., Mullick, M., Santra, S., & ... (2019). Intelligent traffic control system: Towards smart city. *2019 IEEE 10th Annual Information Technology, Electronics and Mobile Communication Conference (IEMCON)*, 1124–1129.

Eickeler, S., & Muller, S. (1999). Content-based video indexing of TV broadcast news using hidden Markov models. *IEEE International Conference on Acoustics, Speech, and Signal Processing*.(Vol. 6, pp. 2997-3000). IEEE 10.1109/ICASSP.1999.757471

Eltouny, K. A., & Liang, X. (2021). Bayesian-optimized unsupervised learning approach for structural damage detection. *Computer-Aided Civil and Infrastructure Engineering*, *36*(10), 1249–1269. doi:10.1111/mice.12680

Entezami, A., & Shariatmadar, H. (2018). An unsupervised learning approach by novel damage indices in structural health monitoring for damage localization and quantification. *Structural Health Monitoring*, *17*(2), 325–345. doi:10.1177/1475921717693572

Entezami, A., Shariatmadar, H., & De Michele, C. (2022). Non-parametric empirical machine learning for short-term and long-term structural health monitoring. *Structural Health Monitoring*, *21*(6), 2700–2718. doi:10.1177/14759217211069842

Entezami, A., Shariatmadar, H., & Mariani, S. (2020). Fast unsupervised learning methods for structural health monitoring with large vibration data from dense sensor networks. *Structural Health Monitoring*, *19*(6), 1685–1710. doi:10.1177/1475921719894186

Ertel, W. (2017). Introduction to Artificial Intelligence (Undergraduate Topics in Computer Science). Springer.

European Council and European Council. (n.d.). Retrieved from https://www.consilium.europa.eu/en/policies/data-protection/data-protection-regulation/#:~:text=The%20GDPR%20lists%20the%20rights,his%20or%20her%20personal%20data

European Parliament. (2020). *Intellectual property rights for the development of artificial intelligence technologies*. Author.

Fang, W., Tan, X., & Wilbur, D. (2020). Application of intrusion detection technology in network safety based on machine learning. *Safety Science, 124*, 104604. doi:10.1016/j.ssci.2020.104604

Fang, W., Lin, W., & Ren, P. (2019). Tinier-YOLO: A real-time object detection method for constrained environments. *IEEE Access : Practical Innovations, Open Solutions, 8*, 1935–1944. doi:10.1109/ACCESS.2019.2961959

Fang, Y. (2021). You only look at one sequence: Rethinking transformer in vision through object detection. *Advances in Neural Information Processing Systems, 34*, 26183–26197.

Farhangdoust, S., Tashakori, S., Baghalian, A., Mehrabi, A., & Tansel, N. I. (2019). Prediction of damage location in composite plates using artificial neural network modeling. In K.-W. Wang, H. Sohn, H. Huang, & J. P. Lynch (Eds.), Sensors and Smart Structures Technologies for Civil, Mechanical, and Aerospace Systems 2019 (p. 20). SPIE. doi:10.1117/12.2517422

Farina, M. (2023). Intellectual property rights in the era of Italian "artificial" public decisions: time to collapse? *Rivista italiana di informatica e diritto*, *5*(1), 16-16.

Faroudja, A. B., & Izeboudjen, N. (2020). Decision tree based system on chip for forest fires prediction. In International conference on electrical engineering (ICEE) 2020 (pp. 1-4). IEEE.

Faujdar, D. S., Singh, T., Kaur, M., Sahay, S., & Kumar, R. (2021). Stakeholders' perceptions of the implementation of a patient-centric digital health application for primary healthcare in India. *Healthcare Informatics Research*, *27*(4), 315–324. doi:10.4258/hir.2021.27.4.315 PMID:34788912

Fayyad, U., Piatetsky-Shapiro, G., & Smyth, P. (1996). From data mining to knowledge discovery in databases. *AI Magazine*, *17*(3).

Feng, S., & Fan, F. (2019). *A Hierarchical Extraction Method of Impervious Surface Based on NDVI Thresholding Integrated With Multispectral and High-Resolution Remote Sensing Imageries*. Academic Press.

Feng, J., Tan, H., Li, W., & Xie, M. (2022). Conv2NeXt: Reconsidering Conv NeXt Network Design for Image Recognition. *International Conference on Computers and Artificial Intelligence Technologies (CAIT)* (pp. 53-60). IEEE. 10.1109/CAIT56099.2022.10072172

Ferrero, E., Brachat, S., Jenkins, J. L., Marc, P., Skewes-Cox, P., Altshuler, R. C., Gubser Keller, C., Kauffmann, A., Sassaman, E. K., Laramie, J. M., Schoeberl, B., Borowsky, M. L., & Stiefl, N. (2020). Ten simple rules to power drug discovery with data science. *PLoS Computational Biology*, *16*(8), e1008126. doi:10.1371/journal.pcbi.1008126 PMID:32853229

Finck, M. (2020). Legal analysis of international trade law and digital trade (PE 603.517). Briefing requested by the INTA Committee. Brussels.

Fiorucci, M., Khoroshiltseva, M., Pontil, M., Traviglia, A., Del Bue, A., & James, S. (2020). Machine Learning for Cultural Heritage: A Survey. *Pattern Recognition Letters*, *133*, 102–108. doi:10.1016/j.patrec.2020.02.017

Flah, M., Nunez, I., Ben Chaabene, W., & Nehdi, M. L. (2021). Machine Learning Algorithms in Civil Structural Health Monitoring: A Systematic Review. *Archives of Computational Methods in Engineering*, *28*(4), 2621–2643. doi:10.100711831-020-09471-9

Floresta, G., Zagni, C., Gentile, D., Patamia, V., & Rescifina, A. (2022). Artificial intelligence technologies for COVID-19 de novo drug design. *International Journal of Molecular Sciences*, *23*(6), 3261. doi:10.3390/ijms23063261 PMID:35328682

Foss-Solbrekk, K. (2021). Three routes to protecting AI systems and their algorithms under IP law: The good, the bad and the ugly. *Journal of Intellectual Property Law & Practice*, *16*(3), 247–258. doi:10.1093/jiplp/jpab033

Fowziya, S., Sivaranjani, S., Devi, N. L., Boopathi, S., Thakur, S., & Sailaja, J. M. (2023). Influences of nano-green lubricants in the friction-stir process of TiAlN coated alloys. *Materials Today: Proceedings*. doi:10.1016/j.matpr.2023.06.446

Francis, S., Ouseph, A., Durgaprasad, S., Abdulla, H., Lal, A., Likhith, S., Dhanaraj, K., & Harikrishna, M. (2023). Machine Learning based Vacant Space Detection for Smart Parking Solutions. *2023 International Conference on Control, Communication and Computing (ICCC)*, 1–6. 10.1109/ICCC57789.2023.10165557

Freeborn, T., Maundy, B., & Elwakil, A. S. (2015). Approximated fractional order Chebyshev lowpass filters. *Mathematical Problems in Engineering*.

Friha, O., Ferrag, M. A., Shu, L., Maglaras, L., & Wang, X. (2021). Internet of things for the future of smart agriculture: A comprehensive survey of emerging technologies. *IEEE/CAA Journal of Automatica Sinica, 8*(4), 718–752.

Fu, Y. (2020) Fruit Freshness Grading Using Deep Learning. Master's Thesis, Auckland University of Technology, New Zealand.

Fu, Y. (2020). *Fruit Freshness Grading Using Deep Learning*. [Master's Thesis, Auckland University of Technology, New Zealand].

Fu, R., Zhang, Z., & Li, L. (2016). Using LSTM and GRU neural network methods for traffic flow prediction. *Youth Academic Annual Conference of Chinese Association of Automation (YAC)*. IEEE. 10.1109/YAC.2016.7804912

Fu, Y., Nguyen, M., & Yan, W. (2022). *Grading methods for fruit freshness based on deep learning*. Springer Nature Computer Science. doi:10.100742979-022-01152-7

Galanopoulos, G., Milanoski, D., Eleftheroglou, N., Broer, A., Zarouchas, D., & Loutas, T. (2023). Acoustic emission-based remaining useful life prognosis of aeronautical structures subjected to compressive fatigue loading. *Engineering Structures*, *290*, 116391. doi:10.1016/j.engstruct.2023.116391

Galvez, R., Bandala, A., Dadios, E., et al. (2018) Object detection using convolutional neural networks. *IEEE TENCON*, (pp. 2023-2027).

Ganapathy, K., Das, S., Reddy, S., Thaploo, V., Nazneen, A., Kosuru, A., & Shankar Nag, U. (2021). Digital health care in public private partnership mode. *Telemedicine Journal and e-Health*, *27*(12), 1363–1371. doi:10.1089/tmj.2020.0499 PMID:33819433

Gao, X. (2022). *A Method for Face Image Inpainting Based on Generative Adversarial Networks* [Masters Thesis]. Auckland University of Technology, New Zealand.

Gao, X., Nguyen, M., & Yan, W. (2022). A face image inpainting method based on autoencoder and adversarial generative networks. *Pacific-Rim Symposium on Image and Video Technology*.

Gao, X., Nguyen, M., & Yan, W. (2021). Face image inpainting based on generative adversarial network. *International Conference on Image and Vision Computing New Zealand.* 10.1109/IVCNZ54163.2021.9653347

Geethanjali, P., Mohan, Y. K., & Sen, J. (2012) Time domain feature extraction and classification of EEG data for brain computer interface. In *2012 9th International Conference on Fuzzy Systems and Knowledge Discovery* (pp. 1136-1139). IEEE. 10.1109/FSKD.2012.6234336

Gerke, S., Kramer, D. B., & Cohen, I. G. (2019). Ethical and legal challenges of artificial intelligence in cardiology. *AIMed Magazine*, *2*, 12–17.

Gerke, S., Minssen, T., & Cohen, G. (2020). Ethical and legal challenges of artificial intelligence-driven healthcare. In *Artificial intelligence in healthcare* (pp. 295–336). Academic Press. doi:10.1016/B978-0-12-818438-7.00012-5

Géron, A. (2017). Hands-on machine learning with Scikit-Learn and TensorFlow : concepts, tools, and techniques to build intelligent systems. O'Reilly Media.

Gers, F. A., & Schmidhuber, J. (2000). *Recurrent nets that time and count. Neural Networks* (Vol. 3). IEEE – INNS.

Gers, F. A., Schmidhuber, J., & Cummins, F. (2000). Learning to forget: Continual prediction with LSTM. *Neural Computation*, *12*(10), 2451–2471. doi:10.1162/089976600300015015 PMID:11032042

Gharaee, H., & Hosseinvand, H. (2017). A new feature selection IDS based on genetic algorithm and SVM. *2016 8th International Symposium on Telecommunications, IST 2016*, 139–144. 10.1109/ISTEL.2016.7881798

Gholizadeh, S., Leman, Z., & Baharudin, B. T. H. T. (2023). State-of-the-art ensemble learning and unsupervised learning in fatigue crack recognition of glass fiber reinforced polyester composite (GFRP) using acoustic emission. *Ultrasonics*, *132*, 106998. doi:10.1016/j.ultras.2023.106998 PMID:37001339

Ghosh, A., & Saha, S. (2020). Interactive game-based motor rehabilitation using hybrid sensor architecture. In *Handbook of Research on Emerging Trends and Applications of Machine Learning* (pp. 312–337). IGI Global. doi:10.4018/978-1-5225-9643-1.ch015

Ghosh, A., Saha, S., & Ghosh, L. (2023). A rehabilitation framework based on motor imagery induced wheelchair movement using fuzzy vector quantization. *International Journal of Information Technology : an Official Journal of Bharati Vidyapeeth's Institute of Computer Applications and Management*, *15*(6), 1–12. doi:10.100741870-023-01359-8

Gibert, D., Mateu, C., & Planes, J. (2020). The rise of machine learning for detection and classification of malware: Research developments, trends and challenges. *Journal of Network and Computer Applications*, *153*, 102526. doi:10.1016/j.jnca.2019.102526

Ginsburg, J. (2018). Copyright. In *The Oxford handbook of intellectual property law*. Oxford University Press.

Glicksberg, B. S., Li, L., Chen, R., Dudley, J., & Chen, B. (2019). Leveraging big data to transform drug discovery. *Bioinformatics and Drug Discovery*, 91–118.

Gnanaprakasam, C., Vankara, J., Sastry, A. S., Prajval, V., Gireesh, N., & Boopathi, S. (2023). Long-Range and Low-Power Automated Soil Irrigation System Using Internet of Things: An Experimental Study. In Contemporary Developments in Agricultural Cyber-Physical Systems (pp. 87–104). IGI Global.

Gokhale, A., Chavan, A., & Sonawane, S. (2023). Leveraging ML techniques for image-based freshness index prediction of fruits and vegetables. *International Conference on Emerging Smart Computing and Informatics (ESCI)* (pp. 1-6). IEEE. 10.1109/ESCI56872.2023.10100260

Gomes, G. F., Mendez, Y. A. D., da Silva Lopes Alexandrino, P., da Cunha, S. S. Jr, & Ancelotti, A. C. Jr. (2019). A Review of Vibration Based Inverse Methods for Damage Detection and Identification in Mechanical Structures Using Optimization Algorithms and ANN. *Archives of Computational Methods in Engineering*, *26*(4), 883–897. doi:10.100711831-018-9273-4

Gongal, A., Karkee, M., & Amatya, S. (2018). Apple fruit size estimation using a 3D machine vision system. *Information Processing in Agriculture*, *5*(4), 498–503. doi:10.1016/j.inpa.2018.06.002

GOV.UK. (2021). *Government response to call for views on artificial intelligence and intellectual property*. Retrieved from https://www.gov.uk/government/consultations/artificial-intelligence-and-intellectual-property-call-for-views/government-response-to-call-for-views-on-artificial-intelligence-and-intellectual-property

GOV.UK. (2022). *Artificial Intelligence and Intellectual Property: copyright and patents*. Retrieved from https://www.gov.uk/government/consultations/artificial-intelligence-and-ip-copyright-and-patents/artificial-intelligence-and-intellectual-property-copyright-and-patents#copyright

Gowdra, N. (2021). *Entropy-Based Optimization Strategies for Convolutional Neural Networks*. [PhD Thesis, Auckland University of Technology, New Zealand].

Gowdra, N., Sinha, R., MacDonell, S., & Yan, W. (2021). Maximum Categorical Cross Entropy (MCCE): A noise-robust alternative loss function to mitigate racial bias in Convolutional Neural Networks (CNNs) by reducing overfitting. *Pattern Recognition*.

Gowri, N. V., Dwivedi, J. N., Krishnaveni, K., Boopathi, S., Palaniappan, M., & Medikondu, N. R. (2023). Experimental investigation and multi-objective optimization of eco-friendly near-dry electrical discharge machining of shape memory alloy using Cu/SiC/Gr composite electrode. *Environmental Science and Pollution Research International*, *30*(49), 1–19. doi:10.100711356-023-26983-6 PMID:37126160

Gralla, P. (2016, May 11). *Amazon Prime and the racist algorithms*. Computerworld. https://www.computerworld.com/article/3068622/amazon-prime-and-the-racist-algorithms.html

Grigorev, N. A., Savosenkov, A. O., Lukoyanov, M. V., Udoratina, A., Shusharina, N. N., Kaplan, A. Y., Hramov, A. E., Kazantsev, V. B., & Gordleeva, S. (2021). A BCI-based vibrotactile neurofeedback training improves motor cortical excitability during motor imagery. *IEEE Transactions on Neural Systems and Rehabilitation Engineering*, *29*, 1583–1592. doi:10.1109/TNSRE.2021.3102304 PMID:34343094

Grospretre, S., Ruffino, C., & Lebon, F. (2016). Motor imagery and cortico-spinal excitability: A review. *European Journal of Sport Science*, *16*(3), 317–324. doi:10.1080/17461391.2015.1024756 PMID:25830411

Grover, P., Kar, A. K., & Vigneswara Ilavarasan, P. (2018). *Blockchain for Businesses: A Systematic Literature Review*. doi:10.1007/978-3-030-02131-3_29

Gruson, D., Helleputte, T., Rousseau, P., & Gruson, D. (2019). Data science, artificial intelligence, and machine learning: Opportunities for laboratory medicine and the value of positive regulation. *Clinical Biochemistry*, *69*, 1–7. doi:10.1016/j.clinbiochem.2019.04.013 PMID:31022391

Guidi, B., Michienzi, A., & Ricci, L. (2021). Data Persistence in Decentralised Social Applications: The IPFS approach. *2021 IEEE 18th Annual Consumer Communications & Networking Conference (CCNC)*, 1–4. doi:10.1109/CCNC49032.2021.9369473

Guillot, A., & Debarnot, U. (2019). Benefits of motor imagery for human space flight: A brief review of current knowledge and future applications. *Frontiers in Physiology*, *10*, 396. doi:10.3389/fphys.2019.00396 PMID:31031635

Gul, F., Rahiman, W., & Nazli Alhady, S. S. (2019). A comprehensive study for robot navigation techniques. *Cogent Engineering*, *6*(1), 1632046. doi:10.1080/23311916.2019.1632046

Gulgec, N. S., Takáč, M., & Pakzad, S. N. (2019). Convolutional Neural Network Approach for Robust Structural Damage Detection and Localization. *Journal of Computing in Civil Engineering*, *33*(3), 04019005. Advance online publication. doi:10.1061/(ASCE)CP.1943-5487.0000820

Gunasekaran, K., & Boopathi, S. (2023). Artificial Intelligence in Water Treatments and Water Resource Assessments. In *Artificial Intelligence Applications in Water Treatment and Water Resource Management* (pp. 71–98). IGI Global. doi:10.4018/978-1-6684-6791-6.ch004

Guo, L., Boukir, S., Inp, B., & Pessac, F. (2017). *Building an ensemble classifier using ensemble margin. Application to image classification*. Academic Press.

Guo, C., Jiang, L., Yang, F., Yang, Z., & Zhang, X. (2023). An intelligent impact load identification and localization method based on autonomic feature extraction and anomaly detection. *Engineering Structures*, *291*, 116378. doi:10.1016/j.engstruct.2023.116378

Guo, L., Li, R., & Jiang, B. (2021). A Cascade Broad Neural Network for Concrete Structural Crack Damage Automated Classification. *IEEE Transactions on Industrial Informatics*, *17*(4), 2737–2742. doi:10.1109/TII.2020.3010799

Guo, S., & Qi, Z. (2021). A Fuzzy Best-Worst Multi-Criteria Group Decision-Making Method. *IEEE Access : Practical Innovations, Open Solutions*, *9*, 118941–118952. doi:10.1109/ACCESS.2021.3106296

Gupta, P., & Mahajan, A. (2015). Green chemistry approaches as sustainable alternatives to conventional strategies in the pharmaceutical industry. *RSC Advances*, *5*(34), 26686–26705. doi:10.1039/C5RA00358J

Haghi, M., Benis, A., & Deserno, T. M. (2022). Accident & Emergency Informatics and One Digital Health. *Yearbook of Medical Informatics, 31* (01), 040–046.

Halibas, A. S., Reazol, L. B., Delvo, E. G. T., & Tibudan, J. C. (2018). Performance analysis of machine learning classifiers for ASD screening. *2018 International Conference on Innovation and Intelligence for Informatics, Computing, and Technologies, 3ICT 2018*, 1–5. 10.1109/3ICT.2018.8855759

Hamilton, J., de Boer, G., Doddi, A., & Lawrence, D. A. (2022). The DataHawk2 uncrewed aircraft system for atmospheric research. *Atmospheric Measurement Techniques*, *15*(22), 6789–6806. doi:10.5194/amt-15-6789-2022

Hanumanthakari, S., Gift, M. M., Kanimozhi, K., Bhavani, M. D., Bamane, K. D., & Boopathi, S. (2023). Biomining Method to Extract Metal Components Using Computer-Printed Circuit Board E-Waste. In *Handbook of Research on Safe Disposal Methods of Municipal Solid Wastes for a Sustainable Environment* (pp. 123–141). IGI Global. doi:10.4018/978-1-6684-8117-2.ch010

Han, X., Gao, Y., Lu, Z., Zhang, Z., & Niu, D. (2015). Research on moving object detection algorithm based on improved three frame difference method and optical flow. *International Conference on Instrumentation and Measurement, Computer, Communication and Control (IMCCC)* (pp. 580-584). IEEE. 10.1109/IMCCC.2015.420

Haribalaji, V., Boopathi, S., & Balamurugan, S. (2014). Effect of Welding Processes on Mechanical and Metallurgical Properties of High Strength Low Alloy (HSLA) Steel Joints. *International Journal of Innovation and Scientific Research*, *12*(1), 170–179.

Haripriya, L., & Jabbar, M. A. (2018). Role of Machine Learning in Intrusion Detection System: Review. *Proceedings of the 2nd International Conference on Electronics, Communication and Aerospace Technology, ICECA 2018, Iceca*, 925–929. 10.1109/ICECA.2018.8474576

Haritaoglu, I., Harwood, D., & Davis, L. S. (2000). W/sup 4: Real-time surveillance of people and their activities. *IEEE Transactions on Pattern Analysis and Machine Intelligence*, *22*(8), 809–830. doi:10.1109/34.868683

Haritha, T., & Anitha, A. (2023). Asymmetric Consortium Blockchain and Homomorphically Polynomial-Based PIR for Secured Smart Parking Systems. *Computers, Materials & Continua*, *75*(2), 3923–3939. doi:10.32604/cmc.2023.036278

Harper, F. M., & Konstan, J. A. (2015). The movielens datasets: History and context. *ACM Transactions on Interactive Intelligent Systems*, *5*(4), 1–19. Advance online publication. doi:10.1145/2827872

Hasan, S. M. S., Imam, N., Kannan, R., Yoginath, S., & Kurte, K. (2021). *Design Space Exploration of Emerging Memory Technologies for Machine Learning Applications*. doi:10.1109/IPDPSW52791.2021.00075

Hassanien, M. A., Singh, V. K., Puig, D., & Abdel-Nasser, M. (2022). Predicting breast tumor malignancy using deep ConvNeXt radiomics and quality-based score pooling in ultrasound sequences. *Diagnostics (Basel)*, *12*(5), 1053. doi:10.3390/diagnostics12051053 PMID:35626208

Hassan, M. U., Rehmani, M. H., & Chen, J. (2019). Privacy preservation in blockchain based IoT systems: Integration issues, prospects, challenges, and future research directions. *Future Generation Computer Systems*, *97*, 512–529. doi:10.1016/j.future.2019.02.060

Heikkila, M., & Pietikainen, M. (2006). A texture-based method for modeling the background and detecting moving objects. *IEEE Transactions on Pattern Analysis and Machine Intelligence*, *28*(4), 657–662. doi:10.1109/TPAMI.2006.68 PMID:16566514

Hema, N., Krishnamoorthy, N., Chavan, S. M., Kumar, N., Sabarimuthu, M., & Boopathi, S. (2023). A Study on an Internet of Things (IoT)-Enabled Smart Solar Grid System. In *Handbook of Research on Deep Learning Techniques for Cloud-Based Industrial IoT* (pp. 290–308). IGI Global. doi:10.4018/978-1-6684-8098-4.ch017

Hendrawan, H., Sukarno, P., & Nugroho, M. A. (2019). Quality of service (QoS) comparison analysis of snort IDS and Bro IDS application in software-defined network (SDN) architecture. *2019 7th International Conference on Information and Communication Technology, ICoICT 2019*. 10.1109/ICoICT.2019.8835211

Hensher, D. A. (2020). What might Covid-19 mean for mobility as a service (MaaS)? *Transport Reviews*, *40*(5), 551–556. doi:10.1080/01441647.2020.1770487

Hensher, D. A., Mulley, C., Ho, C., Wong, Y., Smith, G., & Nelson, J. D. (2020). *Understanding Mobility as a Service (MaaS): Past, present and future*. Elsevier.

Hilmani, A., Maizate, A., & Hassouni, L. (2020). Automated real-time intelligent traffic control system for smart cities using wireless sensor networks. *Wireless Communications and Mobile Computing*, *2020*, 1–28. doi:10.1155/2020/8841893

Hoffmann, U., Vesin, J. M., & Ebrahimi, T. (2007). Recent advances in brain-computer interfaces. *IEEE International Workshop on Multimedia Signal Processing (MMSP07)*.

Hoskere, V., Narazaki, Y., Spencer, B. F., & Smith, M. D. (2019). Deep Learning-based Damage Detection of Miter Gates Using Synthetic Imagery from Computer Graphics. *Structural Health Monitoring*. Advance online publication. doi:10.12783hm2019/32463

Hosseini, M.-P., Tran, T. X., Pompili, D., Elisevich, K., & Soltanian-Zadeh, H. (2017). Deep Learning with Edge Computing for Localization of Epileptogenicity Using Multimodal rs-fMRI and EEG Big Data. *2017 IEEE International Conference on Autonomic Computing (ICAC)*, 83–92. 10.1109/ICAC.2017.41

Hossen, S., & Janagam, A. (2018). *Analysis of network intrusion detection system with machine learning algorithms (deep reinforcement learning Algorithm)*. Academic Press.

Hou, G., Li, L., Xu, Z., Chen, Q., Liu, Y., & Mu, X. (2022). A Visual Management System for Structural Health Monitoring Based on Web-BIM and Dynamic Multi-source Monitoring Data-driven. *Arabian Journal for Science and Engineering*, *47*(4), 4731–4748. doi:10.100713369-021-06268-1 PMID:36032406

Hou, L., Chen, H., Zhang, G., & Wang, X. (2021). Deep Learning-Based Applications for Safety Management in the AEC Industry: A Review. *Applied Sciences (Basel, Switzerland)*, *11*(2), 821. doi:10.3390/app11020821

Hsu, W. L., Jhuang, J. Y., Huang, C. S., Liang, C. K., & Shiau, Y. C. (2019). Application of Internet of Things in a kitchen fire prevention system. *Applied Sciences (Basel, Switzerland)*, *9*(17), 3520. doi:10.3390/app9173520

Huang, G., Zhu, J., Li, J., Wang, Z., Cheng, L., Liu, L., Li, H., & Zhou, J. (2020). Channel-attention U-Net: Channel attention mechanism for semantic segmentation of esophagus and esophageal cancer. *IEEE Access : Practical Innovations, Open Solutions*, *8*, 122798–122810. doi:10.1109/ACCESS.2020.3007719

Hu, B., & Wang, J. (2020). Deep learning based hand gesture recognition and UAV flight controls. *International Journal of Automation and Computing*, *17*(1), 17–29. doi:10.100711633-019-1194-7

Hui, S., Wang, Z., Hou, X., Wang, X., Wang, H., Li, Y., & Jin, D. (2021). Systematically Quantifying IoT Privacy Leakage in Mobile Networks. *IEEE Internet of Things Journal*, *8*(9), 7115–7125. doi:10.1109/JIOT.2020.3038639

Hu, J., He, D., Zhao, Q., & Choo, K.-K. R. (2019). Parking management: A blockchain-based privacy-preserving system. *IEEE Consumer Electronics Magazine*, *8*(4), 45–49. doi:10.1109/MCE.2019.2905490

Hu, J., & Ma, F. (2020). Statistical modelling for high arch dam deformation during the initial impoundment period. *Structural Control and Health Monitoring*, *27*(12). Advance online publication. doi:10.1002tc.2638

Hundera, N. W., Jin, C., Geressu, D. M., Aftab, M. U., Olanrewaju, O. A., & Xiong, H. (2022). Proxy-based public-key cryptosystem for secure and efficient IoT-based cloud data sharing in the smart city. *Multimedia Tools and Applications*, *81*(21), 29673–29697. doi:10.100711042-021-11685-3

Hunt, E. (2016, March 24). Tay, Microsoft's AI chatbot, gets a crash course in racism from Twitter| Artificial intelligence (AI). *The Guardian*.

Hurtik, P., & Tomasiello, S. (2019). A review on the application of fuzzy transform in data and image compression. *Soft Computing*, *23*(23), 12641–12653. doi:10.100700500-019-03816-8

Hussain, M., Al-Aqrabi, H., Munawar, M., Hill, R., & Alsboui, T. (2022). Domain feature mapping with YOLOv7 for automated edge-based pallet racking inspections. *Sensors (Basel)*, *22*(18), 6927. doi:10.339022186927 PMID:36146273

Ibáñez, J., Serrano, J. I., del Castillo, M. D., Minguez, J., & Pons, J. L. (2015). Predictive classification of self-paced upper-limb analytical movements with EEG. *Medical & Biological Engineering & Computing*, *53*(11), 1201–1210. doi:10.100711517-015-1311-x PMID:25980505

Ibrahim, M., Lee, Y., Kahng, H.-K., Kim, S., & Kim, D.-H. (2022). Blockchain-based parking sharing service for smart city development. *Computers & Electrical Engineering*, *103*, 108267. doi:10.1016/j.compeleceng.2022.108267

Igual, L., & Seguí, S. (2017). *Introduction to Data Science: A Python Approach to Concepts, Techniques and Applications*. In Springer International Publishing. doi:10.1007/978-3-319-50017-1

Ilarri, S., Trillo-Lado, R., & Hermoso, R. (2018). Datasets for context-aware recommender systems: Current context and possible directions. *Proceedings - IEEE 34th International Conference on Data Engineering Workshops, ICDEW 2018*. 10.1109/ICDEW.2018.00011

Information Commissioner's Office. (n.d.). Retrieved from https://ico.org.uk/for-organisations/uk-gdpr-guidance-and-resources/data-protection-principles/a-guide-to-the-data-protection-principles/

Ingley, N. A., & Pawar, M. S. (2015). Latency reduced communication in wireless sensor networks. *2015 International Conference on Communication and Signal Processing, ICCSP 2015*, 1111–1114. 10.1109/ICCSP.2015.7322675

Intellectual property of an AI : issues and challenges - iPleaders. (2020, October 19). https://blog.ipleaders.in/intellectual-property-ai-issues-challenges/

Iqbal, N., Ahmad, S., & Kim, D. H. (2021). Towards mountain fire safety using fire spread predictive analytics and mountain fire containment in iot environment. *Sustainability (Basel)*, *13*(5), 2461. doi:10.3390u13052461

Islam, R., & Shahjalal, M. A. (2019). Predicting DRC Violations Using Ensemble Random Forest Algorithm. *Proceedings of the 56th Annual Design Automation Conference 2019 on - DAC '19*. 10.1145/3316781.3322478

Iyer, S., Velmurugan, T., Gandomi, A. H., Noor Mohammed, V., Saravanan, K., & Nandakumar, S. (2021). Structural health monitoring of railway tracks using IoT-based multi-robot system. *Neural Computing & Applications*, *33*(11), 5897–5915. doi:10.100700521-020-05366-9

Jain, L. C., Seera, M., Lim, C. P., & Balasubramaniam, P. (2014). A review of online learning in supervised neural networks. In *Neural Computing and Applications* (Vol. 25, pp. 3–4). doi:10.100700521-013-1534-4

Jalalzai, M. M., & Busch, C. (2018). Window Based BFT Blockchain Consensus. *2018 IEEE International Conference on Internet of Things (iThings) and IEEE Green Computing and Communications (GreenCom) and IEEE Cyber, Physical and Social Computing (CPSCom) and IEEE Smart Data (SmartData)*, 971–979. 10.1109/Cybermatics_2018.2018.00184

Jamshidi, M., & El-Badry, M. (2023). Structural damage severity classification from time-frequency acceleration data using convolutional neural networks. *Structures*, *54*, 236–253. doi:10.1016/j.istruc.2023.05.009

Janardhana, K., Singh, V., Singh, S. N., Babu, T. R., Bano, S., & Boopathi, S. (2023). Utilization Process for Electronic Waste in Eco-Friendly Concrete: Experimental Study. In Sustainable Approaches and Strategies for E-Waste Management and Utilization (pp. 204–223). IGI Global.

Jan, F., Min-Allah, N., Saeed, S., Iqbal, S. Z., & Ahmed, R. (2022). IoT-based solutions to monitor water level, leakage, and motor control for smart water tanks. *Water (Basel)*, *14*(3), 309. doi:10.3390/w14030309

Jarva, E., Oikarinen, A., Andersson, J., Tuomikoski, A.-M., Kääriäinen, M., Meriläinen, M., & Mikkonen, K. (2022). Healthcare professionals' perceptions of digital health competence: A qualitative descriptive study. *Nursing Open*, *9*(2), 1379–1393. doi:10.1002/nop2.1184 PMID:35094493

Jeevanantham, Y. A., Saravanan, A., Vanitha, V., Boopathi, S., & Kumar, D. P. (2022). Implementation of Internet-of Things (IoT) in Soil Irrigation System. *IEEE Explore*, 1–5.

Jegadeesan, S., Azees, M., Kumar, P. M., Manogaran, G., Chilamkurti, N., Varatharajan, R., & Hsu, C.-H. (2019). An efficient anonymous mutual authentication technique for providing secure communication in mobile cloud computing for smart city applications. *Sustainable Cities and Society*, *49*, 101522. doi:10.1016/j.scs.2019.101522

Jelisavac Trošić, S. (2022). *The World Trade Organization and Serbia's Long Path Towards Accession*. Institute of International Politics and Economics,Faculty of Philosophy of the University of St. Cyril and Methodius.

Jensen, A. J., Molina, D. M., & Lundbäck, T. (2015). CETSA: A target engagement assay with potential to transform drug discovery. *Future Medicinal Chemistry*, *7*(8), 975–978. doi:10.4155/fmc.15.50 PMID:26062395

Jeong, T., Chau, R., Kankalale, D. P., & Hyeran, J. (2019). *Going Deeper or Wider: Throughput Prediction for Cluster Tools with Machine Learning*. Academic Press.

Jiang, L., Cai, Z., Wang, D., & Jiang, S. (2007). Survey of improving K-nearest-neighbor for classification. *Proceedings - Fourth International Conference on Fuzzy Systems and Knowledge Discovery, FSKD 2007*, *1*. 10.1109/FSKD.2007.552

Jiang, C., & Chen, N.-Z. (2023). Graph Neural Networks (GNNs) based accelerated numerical simulation. *Engineering Applications of Artificial Intelligence*, *123*, 106370. doi:10.1016/j.engappai.2023.106370

Jiang, D. (2020). The construction of smart city information system based on the Internet of Things and cloud computing. *Computer Communications*, *150*, 158–166. doi:10.1016/j.comcom.2019.10.035

Jiang, H., Wan, C., Yang, K., Ding, Y., & Xue, S. (2022). Continuous missing data imputation with incomplete dataset by generative adversarial networks–based unsupervised learning for long-term bridge health monitoring. *Structural Health Monitoring*, *21*(3), 1093–1109. doi:10.1177/14759217211021942

Jiang, X., Gao, T., Zhu, Z., & Zhao, Y. (2021). Real-time face mask detection method based on yolov3. *Electronics (Basel)*, *10*(7), 837. doi:10.3390/electronics10070837

Jiang, Y., & Niu, G. (2022). An iterative frequency-domain envelope-tracking filter for dispersive signal decomposition in structural health monitoring. *Mechanical Systems and Signal Processing*, *179*, 109329. doi:10.1016/j.ymssp.2022.109329

Jiao, Y., Weir, J., & Yan, W. (2011). Flame detection in surveillance. *Journal of Multimedia*, *6*(1). Advance online publication. doi:10.4304/jmm.6.1.22-32

Jignesh Chowdary, G., Punn, N. S., Sonbhadra, S. K., & Agarwal, S. (2020). Face mask detection using transfer learning of Inceptionv3. *Lecture Notes in Computer Science*, *12581*, 81–90. doi:10.1007/978-3-030-66665-1_6

Jiménez, D. D. J. G., Olvera, T., Orozco-Rosas, U., & Picos, K. (2021, August). Autonomous object manipulation and transportation using a mobile service robot equipped with an RGB-D and LiDAR sensor. In *Optics and Photonics for Information Processing XV* (Vol. 11841, pp. 92–111). SPIE. doi:10.1117/12.2594025

Jordan, M. I., & Mitchell, T. M. (2015). Machine learning: Trends, perspectives, and prospects. In Science (Vol. 349, Issue 6245). doi:10.1126cience.aaa8415

Jordan, M. I. (2019). Artificial intelligence—The revolution hasn't happened yet. *Harvard Data Science Review*, *1*(1), 1–9.

Joseph, R., Divvala, S., Girshick, R., & Farhadi, A. (2021) You only look once: Unified, real-time object detection. *IEEE Conference on Computer Vision and Pattern Recognition*, (pp. 779 – 788). IEEE.

Joshi, M. V. (2001). *Evaluating Boosting Algorithms to Classify Rare Classes : Comparison and Improvements*. Academic Press.

Judson, R., Richard, A., Dix, D. J., Houck, K., Martin, M., Kavlock, R., Dellarco, V., Henry, T., Holderman, T., Sayre, P., Tan, S., Carpenter, T., & Smith, E. (2009). The toxicity data landscape for environmental chemicals. *Environmental Health Perspectives*, *117*(5), 685–695. doi:10.1289/ehp.0800168 PMID:19479008

Jung, I., Lee, J., & Hwang, K. (2022). Smart parking management system using AI. *Webology*, *19*(1), 4629–4638. doi:10.14704/WEB/V19I1/WEB19307

Junos, M. H., Mohd Khairuddin, A. S., Thannirmalai, S., & Dahari, M. (2022). Automatic detection of oil palm fruits from UAV images using an improved YOLO model. *The Visual Computer*, *38*(7), 2341–2355. doi:10.100700371-021-02116-3

Kaen, A., Park, M. K., & Son, S.-K. (2023). Clinical outcomes of uniportal compared with biportal endoscopic decompression for the treatment of lumbar spinal stenosis: A systematic review and meta-analysis. *European Spine Journal*, *32*(8), 1–9. doi:10.100700586-023-07660-1 PMID:36991184

Kaginalkar, A., Kumar, S., Gargava, P., & Niyogi, D. (2021). Review of urban computing in air quality management as smart city service: An integrated IoT, AI, and cloud technology perspective. *Urban Climate*, *39*, 100972. doi:10.1016/j.uclim.2021.100972

Kalantar, B., Ueda, N., Idrees, M. O., Janizadeh, S., Ahmadi, K., & Shabani, F. (2020). Forest fire susceptibility prediction based on machine learning models with resampling algorithms on remote sensing data. *Remote Sensing (Basel)*, *12*(22), 3682. doi:10.3390/rs12223682

Kaloni, S., Singh, G., & Tiwari, P. (2021). Nonparametric damage detection and localization model of framed civil structure based on local gravitation clustering analysis. *Journal of Building Engineering*, *44*, 103339. doi:10.1016/j.jobe.2021.103339

Kanarachos, S., Christopoulos, S.-R. G., Chroneos, A., & Fitzpatrick, M. E. (2017). Detecting anomalies in time series data via a deep learning algorithm combining wavelets, neural networks and Hilbert transform. *Expert Systems with Applications*, *85*, 292–304. doi:10.1016/j.eswa.2017.04.028

Kang, M., & Jameson, N. J. (2018). Machine Learning: Fundamentals. *Prognostics and Health Management of Electronics*, 85–109. doi:10.1002/9781119515326.ch4

Kang, H., & Chen, C. (2020). Fast implementation of real-time fruit detection in apple orchards using deep learning. *Computers and Electronics in Agriculture*, *168*, 105108. doi:10.1016/j.compag.2019.105108

Karpathy, A., Toderici, G., Shetty, S., Leung, T., Sukthankar, R., & Fei-Fei, L. (2014). Large-scale video classification with convolutional neural networks. *IEEE Conference on Computer Vision and Pattern Recognition* (pp. 1725 - 1732). IEEE. 10.1109/CVPR.2014.223

Karthik, S., Hemalatha, R., Aruna, R., Deivakani, M., Reddy, R. V. K., & Boopathi, S. (2023). Study on Healthcare Security System-Integrated Internet of Things (IoT). In Perspectives and Considerations on the Evolution of Smart Systems (pp. 342–362). IGI Global.

Karthik, K., Teferi, A. B., Sathish, R., Gandhi, A. M., Padhi, S., Boopathi, S., & Sasikala, G. (2023). Analysis of delamination and its effect on polymer matrix composites. *Materials Today: Proceedings*. Advance online publication. doi:10.1016/j.matpr.2023.07.199

Kasongo, S. M., & Sun, Y. (2020). A deep learning method with wrapper-based feature extraction for wireless intrusion detection system. *Computers & Security*, *92*, 101752. Advance online publication. doi:10.1016/j.cose.2020.101752

Katagiri, S., & Lee, C. H. (1993). A new hybrid algorithm for speech recognition based on HMM segmentation and learning vector quantization. *IEEE Transactions on Speech and Audio Processing*, *1*(4), 421–430. doi:10.1109/89.242488

Kaur, G., Tomar, P., & Singh, P. (2018). Design of cloud-based green IoT architecture for smart cities. *Internet of Things and Big Data Analytics Toward Next-Generation Intelligence*, 315–333.

Kaushik, P. (2023). Enhanced cloud car parking system using ML and Advanced Neural Network. *International Journal of Research in Science and Technology*, *13*(1), 73–86. doi:10.37648/ijrst.v13i01.009

Kaya, S. (2020). *An Example of Performance Comparison of Supervised Machine Learning Algorithms Before and After PCA and LDA Application : Breast Cancer Detection*. Academic Press.

Kaya, M., Binli, M. K., Ozbay, E., Yanar, H., & Mishchenko, Y. (2018). A large electroencephalographic motor imagery dataset for electroencephalographic brain computer interfaces. *Scientific Data*, *5*(1), 1–16. doi:10.1038data.2018.211 PMID:30325349

Ke, R., Zhuang, Y., Pu, Z., & Wang, Y. (2020). A smart, efficient, and reliable parking surveillance system with edge artificial intelligence on IoT devices. *IEEE Transactions on Intelligent Transportation Systems*, *22*(8), 4962–4974. doi:10.1109/TITS.2020.2984197

Khamis, R. A., & Matrawy, A. (2020). Evaluation of adversarial training on different types of neural networks in deep learning-based IDSs. *2020 International Symposium on Networks, Computers and Communications, ISNCC 2020*. 10.1109/ISNCC49221.2020.9297344

Khan, R. & Debnath, R. (2015). Multi class fruit classification using efficient object detection and recognition techniques. *International Journal of Image, Graphics and Signal Processing*.

Khan, N. A., Jhanjhi, N., Brohi, S. N., Usmani, R. S. A., & Nayyar, A. (2020). Smart traffic monitoring system using unmanned aerial vehicles (UAVs). *Computer Communications*, *157*, 434–443. doi:10.1016/j.comcom.2020.04.049

Khanna, A., Sah, A., Bolshev, V., Jasinski, M., Vinogradov, A., Leonowicz, Z., & Jasiński, M. (2021). Blockchain: Future of e-governance in smart cities. *Sustainability (Basel)*, *13*(21), 11840. doi:10.3390u132111840

Khan, S. A., Kusi-Sarpong, S., Naim, I., Ahmadi, H. B., & Oyedijo, A. (2021). A best-worst-method-based performance evaluation framework for manufacturing industry. *Kybernetes*.

Khraisat, A., Gondal, I., Vamplew, P., & Kamruzzaman, J. (2019). Survey of intrusion detection systems: Techniques, datasets, and challenges. *Cybersecurity*, *2*(1), 20. Advance online publication. doi:10.118642400-019-0038-7

Kim, J., Jeong, B., Kim, D., Kim, J., Jeong, B., & Kim, D. (2021). A Brief History of Patents. *Patent Analytics: Transforming IP Strategy into Intelligence*, 11-19.

Kim, D. (2022). The Paradox of the DABUS Judgment of the German Federal Patent Court. *GRUR International*, *71*(12), 1162–1166. doi:10.1093/grurint/ikac125

Kim, M. S., Jang, K. M., Che, H., Kim, D. W., & Im, C. H. (2012). Electrophysiological correlates of object-repetition effects: sLORETA imaging with 64-channel EEG and individual MRI. *BMC Neuroscience*, *13*(1), 1–10. doi:10.1186/1471-2202-13-124 PMID:23075055

Kivrak, O., & Gürbüz, M. Z. (2022). Performance comparison of YOLOv3, YOLOv4 and YOLOv5 algorithms: A case study for poultry recognition. *Avrupa Bilim ve Teknoloji Dergisi*, (38), 392–397.

Koirala, A., Walsh, K. B., Wang, Z., & McCarthy, C. (2019). Deep learning for real-time fruit detection and orchard fruit load estimation: Benchmarking of 'MangoYOLO'. *Precision Agriculture*, *20*(6), 1107–1135. doi:10.100711119-019-09642-0

Kolhe, N., & Raza, N. (2013). *Throughput Comparison Results of Proposed Algorithm with Existing Algorithm*. Academic Press.

Koren, Y., Bell, R., & Volinsky, C. (2009). Matrix factorisation techniques for recommender systems. *Computer*, *42*(8), 30–37. Advance online publication. doi:10.1109/MC.2009.263

Koshariya, A. K., Kalaiyarasi, D., Jovith, A. A., Sivakami, T., Hasan, D. S., & Boopathi, S. (2023). AI-Enabled IoT and WSN-Integrated Smart Agriculture System. In *Artificial Intelligence Tools and Technologies for Smart Farming and Agriculture Practices* (pp. 200–218). IGI Global. doi:10.4018/978-1-6684-8516-3.ch011

Koshariya, A. K., Khatoon, S., Marathe, A. M., Suba, G. M., Baral, D., & Boopathi, S. (2023a). Agricultural Waste Management Systems Using Artificial Intelligence Techniques. In *AI-Enabled Social Robotics in Human Care Services* (pp. 236–258). IGI Global. doi:10.4018/978-1-6684-8171-4.ch009

Kotpalliwar, M. V. (2015). *Classification of Attacks Using Support Vector Machine (SVM) on KDDCUP'99 IDS Database*. doi:10.1109/CSNT.2015.185

Kretschmer, M., Meletti, B., & Porangaba, L. H. (2022). Artificial Intelligence and Intellectual Property: Copyright and Patents–a response by the CREATe Centre to the UK Intellectual Property Office's open consultation. *Journal of Intellectual Property Law and Practice*, *17*(3), 321–326. doi:10.1093/jiplp/jpac013

Krishnakumar, K. (2002). *Intelligent Systems for Aerospace Engineering: An Overview*. Von Karman Institute Lecture Series on Intelligent Systems for Aeronautics.

Kruszyńska-Fischbach, A., Sysko-Romańczuk, S., Napiórkowski, T. M., Napiórkowska, A., & Kozakiewicz, D. (2022). Organizational e-health readiness: How to prepare the primary healthcare providers' services for digital transformation. *International Journal of Environmental Research and Public Health*, *19*(7), 3973. doi:10.3390/ijerph19073973 PMID:35409656

Kshirsagar, P., & Akojwar, S. (2016) Optimization of BPNN parameters using PSO for EEG signals. In *International Conference on Communication and Signal Processing 2016 (ICCASP 2016)* (pp. 384-393). Atlantis Press.

Kuhn, M., & Johnson, K. (2013). *Applied predictive modeling*. doi:10.1007/978-1-4614-6849-3

Kumar Singh Gautam, R., & Doegar, E. A. (2018). An Ensemble Approach for Intrusion Detection System Using Machine Learning Algorithms. *Proceedings of the 8th International Conference Confluence 2018 on Cloud Computing, Data Science and Engineering, Confluence 2018*, 61–64. 10.1109/CONFLUENCE.2018.8442693

Kumar, P., Sampath, B., Kumar, S., Babu, B. H., & Ahalya, N. (2023). Hydroponics, Aeroponics, and Aquaponics Technologies in Modern Agricultural Cultivation. In IGI:Trends, Paradigms, and Advances in Mechatronics Engineering (pp. 223–241). IGI Global.

Kumar, A., Bhushan, B., Shristi, S., Chaganti, R., & Soufiene, B. O. (2023). Blockchain-based decentralised management of IoT devices for preserving data integrity. In *Blockchain Technology Solutions for the Security of IoT-Based Healthcare Systems* (pp. 263–286). Elsevier. doi:10.1016/B978-0-323-99199-5.00009-4

Kumara, V., Mohanaprakash, T., Fairooz, S., Jamal, K., Babu, T., & Sampath, B. (2023). Experimental Study on a Reliable Smart Hydroponics System. In *Human Agro-Energy Optimization for Business and Industry* (pp. 27–45). IGI Global. doi:10.4018/978-1-6684-4118-3.ch002

Kumari, V. V., & Varma, P. R. K. (2017). A semi-supervised intrusion detection system using active learning SVM and fuzzy c-means clustering. *2017 International Conference on I-SMAC (IoT in Social, Mobile, Analytics, and Cloud)*, 481-485. 10.1109/I-SMAC.2017.8058397

Kumar, N., Vasilakos, A. V., & Rodrigues, J. J. (2017). A multi-tenant cloud-based DC nano grid for self-sustained smart buildings in smart cities. *IEEE Communications Magazine*, *55*(3), 14–21. doi:10.1109/MCOM.2017.1600228CM

Kumar, P., & Kota, S. R. (2023). IoT enabled diagnosis and prognosis framework for structural health monitoring. *Journal of Ambient Intelligence and Humanized Computing*, *14*(8), 11301–11318. doi:10.100712652-023-04646-1

Kumar, R., Arjunaditya, Singh, D., Srinivasan, K., & Hu, Y.-C. (2022). AI-powered blockchain technology for public health: A contemporary review, open challenges, and future research directions. *Health Care*, *11*(1), 81. PMID:36611541

Kurian, B., & Liyanapathirana, R. (2020). *Machine Learning Techniques for Structural Health Monitoring*., doi:10.1007/978-981-13-8331-1_1

Kuznetsova, A., Maleva, T., & Soloviev, V. (2021). *YOLOv5 versus YOLOv3 for apple detection. Cyber-Physical Systems: Modelling and Intelligent Control*. Springer.

La Diega, G. N. (2018). Against the dehumanisation of decision-making. *J. Intell. Prop. Info. Tech. & Elec. Com. L.*, *9*, 3.

Labellapansa, A., Syafitri, N., Kadir, E. A., Saian, R., Rahman, A. S., & Ahmad, M. B. (2019). Prototype for early detection of fire hazards using fuzzy logic approach and Arduino microcontroller. *International Journal of Advanced Computer Research.*, *9*(44), 276–282. doi:10.19101/IJACR.PID47

Ladak, A. (2023). What would qualify an artificial intelligence for moral standing? *AI and Ethics*, 1–16. doi:10.100743681-023-00260-1

Lan, T., Erdogmus, D., Adami, A., Mathan, S., & Pavel, M. (2007). Channel selection and feature projection for cognitive load estimation using ambulatory EEG. *Computational Intelligence and Neuroscience*, *2007*, 2007. doi:10.1155/2007/74895 PMID:18364990

Lashkari, A. H., Gil, G. D., Saiful, M., Mamun, I., & Ghorbani, A. A. (2017). *Characterization of Tor Traffic using Time based Features*. doi:10.5220/0006105602530262

Latifah, A. L., Shabrina, A., Wahyuni, I. N., & Sadikin, R. (2019). *Evaluation of Random Forest model for forest fire prediction based on climatology over Borneo. In International conference on computer, control, informatics and its applications (IC3INA)*. IEEE.

Lauber-Rönsberg, A., & Hetmank, S. (2019). The concept of authorship and inventorship under pressure: Does artificial intelligence shift paradigms? *Journal of Intellectual Property Law and Practice*, *14*(7), 570–579. doi:10.1093/jiplp/jpz061

Laukemann, J., Hammer, J., Hofmann, J., Hager, G., & Wellein, G. (2018). *Automated Instruction Stream Throughput Prediction for Intel and AMD Microarchitectures. 2018 IEEE/ACM Performance Modeling, Benchmarking, and Simulation of High Performance Computer Systems.* doi:10.1109/PMBS.2018.8641578

Leal Sobral, V. A., Nelson, J., Asmare, L., Mahmood, A., Mitchell, G., Tenkorang, K., Todd, C., Campbell, B., & Goodall, J. L. (2023). A Cloud-Based Data Storage and Visualization Tool for Smart City IoT: Flood Warning as an Example Application. *Smart Cities*, *6*(3), 1416–1434. doi:10.3390martcities6030068

Leblond, P., & Aaronson, S. A. (2019). *A plurilateral "single data area" is the solution to Canada's data trilemma.* Academic Press.

LeCun, Y. (1989). Generalization and network design strategies. *Connectionism in Perspective,* 143 – 155.

LeCun, Y., & Ranzato, M. (2013). Deep learning tutorial. *Tutorials in International Conference on Machine Learning (ICML'13).* IEEE.

Lee, H., & Choi, S. (2003). Pca+ hmm+ svm for eeg pattern classification. In *Seventh International Symposium on Signal Processing and Its Applications, 2003 Proceedings.*, *1*, 541–544.

Lee-Makiyama, H. (2012). Presentation at the *WTO ITA Symposium, Geneva*, Switzerland.

Lee, W.-H., & Chiu, C.-Y. (2020). Design and implementation of a smart traffic signal control system for smart city applications. *Sensors (Basel)*, *20*(2), 508. doi:10.339020020508 PMID:31963229

Lee, Y. H., & Kim, Y. (2020). Comparison of CNN and YOLO for object detection. *Journal of the Semiconductor & Display Technology*, *19*(1), 85–92.

Legislation.gov.uk. Copyright, Designs and Patents Act 1988. Part I Chapter I. Retrieved from https://www.legislation.gov.uk/ukpga/1988/48/part/I/chapter/I#commentary-c13754491

Legislation.gov.uk. UK Public General Acts1988 s.9.

Lei, X., Sun, L., & Xia, Y. (2021). Lost data reconstruction for structural health monitoring using deep convolutional generative adversarial networks. *Structural Health Monitoring*, *20*(4), 2069–2087. doi:10.1177/1475921720959226

Lewis, E., & ... (2021). Smart city software systems and Internet of Things sensors in sustainable urban governance networks. *Geopolitics, History, and International Relations*, *13*(1), 9–19.

Li, D., Wang, L., & Huang, Q. (2019). *A case study of SOS-SVR model for PCB throughput estimation in SMT production lines.* Academic Press.

Liang, D., Liu, Q., Zhao, B., Zhu, Z., & Liu, D. (2019). A Clustering-SVM Ensemble Method for Intrusion Detection System. *2019 8th International Symposium on Next Generation Electronics (ISNE).* 10.1109/ISNE.2019.8896514

Liang, C., Lu, J., & Yan, W. Q. (2022). Human action recognition from digital videos based on Deep Learning. *The International Conference on Control and Computer Vision.* 10.1145/3561613.3561637

Liang, Y., Wu, D., Liu, G., Li, Y., Gao, C., Ma, Z. J., & Wu, W. (2016). Big data-enabled multiscale serviceability analysis for aging bridges☆. *Digital Communications and Networks*, *2*(3), 97–107. doi:10.1016/j.dcan.2016.05.002

Liao, W. D., Yang, D. L., & Hung, M. C. (2010). An intelligent recommendation model with a case study on u-Tour Taiwan of historical momuments and cultural heritage. *Proceedings - International Conference on Technologies and Applications of Artificial Intelligence, TAAI 2010*. 10.1109/TAAI.2010.23

Li, H., & Gupta, M. M. (1995). *Fuzzy logic and intelligent systems* (Vol. 3). Springer Science & Business Media.

Li, H., Yu, L., & He, W. (2019). The impact of GDPR on global technology development. *Journal of Global Information Technology Management*, *22*(1), 1–6. doi:10.1080/1097198X.2019.1569186

Li, L., Morgantini, M., & Betti, R. (2023). Structural damage assessment through a new generalized autoencoder with features in the quefrency domain. *Mechanical Systems and Signal Processing*, *184*, 109713. doi:10.1016/j.ymssp.2022.109713

Li, M., Chen, W., & Zhang, T. (2017). Automatic epileptic EEG detection using DT-CWT-based non-linear features. *Biomedical Signal Processing and Control*, *34*, 114–125. doi:10.1016/j.bspc.2017.01.010

Li, M., Chen, W., & Zhang, T. (2017). Classification of epilepsy EEG signals using DWT-based envelope analysis and neural network ensemble. *Biomedical Signal Processing and Control*, *31*, 357–365. doi:10.1016/j.bspc.2016.09.008

Lim, P. H., & Li, P. (2022). Artificial intelligence and inventorship: Patently much ado in the computer program. *Journal of Intellectual Property Law and Practice*, *17*(4), 376–386. doi:10.1093/jiplp/jpac019

Lin, Y., & Zhang, C. (2021). A Method for Protecting Private Data in IPFS. *2021 IEEE 24th International Conference on Computer Supported Cooperative Work in Design (CSCWD)*, 404–409. 10.1109/CSCWD49262.2021.9437830

Lin, K., Zhao, H., Lv, J., Li, C., Liu, X., Chen, R., & Zhao, R. (2020). Face detection and segmentation based on improved mask R-CNN. *Discrete Dynamics in Nature and Society*, *2020*, 1–11. doi:10.1155/2020/9242917

Lin, Z., Pan, H., Gui, G., & Yan, C. (2018). Data-driven structural diagnosis and conditional assessment: from shallow to deep learning. In H. Sohn (Ed.), *Sensors and Smart Structures Technologies for Civil, Mechanical, and Aerospace Systems 2018* (p. 38). SPIE. doi:10.1117/12.2296964

Liu, G., Li, L., Zhang, L., Li, Q., & Law, S. S. (2020). Sensor faults classification for SHM systems using deep learning-based method with Tsfresh features. *Smart Materials and Structures*, *29*(7), 075005. doi:10.1088/1361-665X/ab85a6

Liu, G., Nouaze, J. C., Touko Mbouembe, P. L., & Kim, J. H. (2020). YOLO-Tomato: A robust algorithm for tomato detection based on YOLOv3. *Sensors (Basel)*, *20*(7), 2145. doi:10.339020072145 PMID:32290173

Liu, H., & Lang, B. (2019). Machine learning and deep learning methods for intrusion detection systems: A survey. *Applied Sciences (Basel, Switzerland)*, *9*(20), 4396. Advance online publication. doi:10.3390/app9204396

Liu, H., Sun, F., Gu, J., & Deng, L. (2022). SF-YOLOv5: A lightweight small object detection algorithm based on improved feature fusion mode. *Sensors (Basel)*, *22*(15), 5817. doi:10.339022155817 PMID:35957375

Liu, J., Chen, S., Bergés, M., Bielak, J., Garrett, J. H., Kovačević, J., & Noh, H. Y. (2020). Diagnosis algorithms for indirect structural health monitoring of a bridge model via dimensionality reduction. *Mechanical Systems and Signal Processing*, *136*, 106454. doi:10.1016/j.ymssp.2019.106454

Liu, M., & Yan, W. (2022). *Masked face recognition in real-time using MobileNetV2*. ACM ICCCV.

Liu, R. W., Liang, M., Nie, J., Lim, W. Y. B., Zhang, Y., & Guizani, M. (2022). Deep learning-powered vessel trajectory prediction for improving smart traffic services in maritime Internet of Things. *IEEE Transactions on Network Science and Engineering*, *9*(5), 3080–3094. doi:10.1109/TNSE.2022.3140529

Liu, T., Xu, H., Ragulskis, M., Cao, M., & Ostachowicz, W. (2020). A Data-Driven Damage Identification Framework Based on Transmissibility Function Datasets and One-Dimensional Convolutional Neural Networks: Verification on a Structural Health Monitoring Benchmark Structure. *Sensors (Basel)*, *20*(4), 1059. doi:10.339020041059 PMID:32075311

Liu, X., Li, G., Chen, W., Liu, B., Chen, M., & Lu, S. (2022). Detection of dense citrus fruits by combining coordinated attention and cross-scale connection with weighted feature fusion. *Applied Sciences (Basel, Switzerland)*, *12*(13), 6600. doi:10.3390/app12136600

Liu, X., & Yan, W. Q. (2023). *Vehicle-related distance estimation using customized YOLOv7. IVCNZ 2022, Auckland, New Zealand*. Springer Nature Switzerland.

Liu, X., Zhang, W., Li, W., Zhang, S., Lv, P., & Yin, Y. (2023). Effects of motor imagery based brain-computer interface on upper limb function and attention in stroke patients with hemiplegia: A randomized controlled trial. *BMC Neurology*, *23*(1), 1–14. doi:10.118612883-023-03150-5 PMID:37003976

Liu, Y., Sun, P., Wergeles, N., & Shang, Y. (2021). A survey and performance evaluation of deep learning methods for small object detection. *Expert Systems with Applications*, *172*, 114602. doi:10.1016/j.eswa.2021.114602

Liu, Y., Wang, S., Khan, M. S., & He, J. (2018). A novel deep hybrid recommender system based on auto-encoder with neural collaborative filtering. *Big Data Mining and Analytics*, *1*(3), 211–221. Advance online publication. doi:10.26599/BDMA.2018.9020019

Liu, Z., Yan, W. Q., & Yang, M. L. (2018). Image denoising based on a CNN model. *International Conference on Control, Automation and Robotics (ICCAR)*. 10.1109/ICCAR.2018.8384706

Livingston, A., Trout, B. L., Horvath, I. T., Johnson, M. D., Vaccaro, L., Coronas, J., Babbitt, C. W., Zhang, X., Pradeep, T., Drioli, E., & others. (2020). Challenges and directions for green chemical engineering—Role of nanoscale materials. *Sustainable Nanoscale Engineering*, 1–18.

Loey, M., Manogaran, G., Taha, M. H., & Khalifa, N. E. (2021). Fighting against COVID-19: A novel deep learning model based on YOLO-V2 with resnet-50 for medical face mask detection. *Sustainable Cities and Society*, *65*, 102600. doi:10.1016/j.scs.2020.102600 PMID:33200063

Long, L. N., Hanford, S. D., Janrathitikarn, O., Sinsley, G. L., & Miller, J. A. (2007). A review of intelligent systems software for autonomous vehicles. *2007 IEEE Symposium on Computational Intelligence in Security and Defense Applications*, 69–76. 10.1109/CISDA.2007.368137

Lo, O., Buchanan, W. J., Griffiths, P., & Macfarlane, R. (2018). Distance measurement methods for improved insider threat detection. *Security and Communication Networks*, *2018*, 1–18. Advance online publication. doi:10.1155/2018/5906368

López-Castro, B., Haro-Baez, A. G., Arcos-Aviles, D., Barreno-Riera, M., & Landázuri-Avilés, B. (2022). A Systematic Review of Structural Health Monitoring Systems to Strengthen Post-Earthquake Assessment Procedures. *Sensors (Basel)*, *22*(23), 9206. doi:10.339022239206 PMID:36501906

López-Larraz, E., Montesano, L., Gil-Agudo, Á., & Minguez, J. (2014). Continuous decoding of movement intention of upper limb self-initiated analytic movements from pre-movement EEG correlates. *Journal of Neuroengineering and Rehabilitation*, *11*(1), 1–15. doi:10.1186/1743-0003-11-153 PMID:25398273

Lopez-Pacheco, M., Morales-Valdez, J., & Yu, W. (2020). Frequency domain CNN and dissipated energy approach for damage detection in building structures. *Soft Computing*, *24*(20), 15821–15840. doi:10.100700500-020-04912-w

Lops, P., De Gemmis, M., Semeraro, G., Musto, C., Narducci, F., & Bux, M. (2009). A semantic content-based recommender system integrating folksonomies for personalised access. *Studies in Computational Intelligence, 229*, 27–47. Advance online publication. doi:10.1007/978-3-642-02794-9_2

Lorusso, A., & Celenta, G. (2023a). *Internet of Things in the Construction Industry: A General Overview*. doi:10.1007/978-3-031-31066-9_65

Lorusso, A., & Celenta, G. (2023b). *Structural Dynamics of Steel Frames with the Application of Friction Isolators*. doi:10.1007/978-3-031-34721-4_28

Lorusso, A., & Guida, D. (2022). *IoT System for Structural Monitoring*. doi:10.1007/978-3-031-05230-9_72

Lorusso, A., Messina, B., & Santaniello, D. (2023). *The Use of Generative Adversarial Network as Graphical Support for Historical Urban Renovation*. doi:10.1007/978-3-031-13588-0_64

Lu, J., Nguyen, M., & Yan, W. Q. (2021). Sign language recognition from digital videos using deep learning methods. *Communications in Computer and Information Science, 1386*, 108–118. doi:10.1007/978-3-030-72073-5_9

Lu, J., Yan, W. Q., & Nguyen, M. (2018). Human behaviour recognition using deep learning. *IEEE International Conference on Advanced Video and Signal Based Surveillance (AVSS)*. 10.1109/AVSS.2018.8639413

Luleci, F., Necati Catbas, F., & Avci, O. (2023). CycleGAN for undamaged-to-damaged domain translation for structural health monitoring and damage detection. *Mechanical Systems and Signal Processing, 197*, 110370. doi:10.1016/j.ymssp.2023.110370

Lupton, D., & Leahy, D. (2019). Reimagining digital health education: Reflections on the possibilities of the storyboarding method. *Health Education Journal, 78*(6), 633–646. doi:10.1177/0017896919841413

Lu, R. R., Zheng, M. X., Li, J., Gao, T. H., Hua, X. Y., Liu, G., Huang, S. H., Xu, J. G., & Wu, Y. (2020). Motor imagery based brain-computer interface control of continuous passive motion for wrist extension recovery in chronic stroke patients. *Neuroscience Letters, 718*, 134727. doi:10.1016/j.neulet.2019.134727 PMID:31887332

Lu, S., Chen, W., Zhang, X., & Karkee, M. (2022). Canopy-attention-YOLOv4-based immature/mature apple fruit detection on dense-foliage tree architectures for early crop load estimation. *Computers and Electronics in Agriculture, 193*, 106696. doi:10.1016/j.compag.2022.106696

Lu, Y., & Li, J. (2014a). Efficient and provably-secure certificate-based key encapsulation mechanism in the standard model. *Jisuanji Yanjiu Yu Fazhan, 51*(7), 1497–1505. doi:10.7544/issn1000-1239.2014.20131604

Lu, Y., & Li, J. (2014b). Efficient constructions of certificate-based key encapsulation mechanism. *International Journal of Internet Protocol Technology, 8*(2/3), 96. doi:10.1504/IJIPT.2014.066374

Lu, Y., Tang, L., Chen, C., Zhou, L., Liu, Z., Liu, Y., Jiang, Z., & Yang, B. (2023). Reconstruction of structural long-term acceleration response based on BiLSTM networks. *Engineering Structures, 285*, 116000. doi:10.1016/j.engstruct.2023.116000

Madiega, T. A. (2021). *Artificial intelligence act. European Parliament*. European Parliamentary Research Service.

Ma, G., & Du, Q. (2020). Optimization on the intellectual monitoring system for structures based on acoustic emission and data mining. *Measurement, 163*, 107937. doi:10.1016/j.measurement.2020.107937

Maguluri, L. P., Ananth, J., Hariram, S., Geetha, C., Bhaskar, A., & Boopathi, S. (2023). Smart Vehicle-Emissions Monitoring System Using Internet of Things (IoT). In Handbook of Research on Safe Disposal Methods of Municipal Solid Wastes for a Sustainable Environment (pp. 191–211). IGI Global.

Mahajan, P., Mahajan, P., Mahajan, S., Mahajan, S., Mahajan, V., Mahale, H., & Dandavate, P. P. (n.d.). Smart parking solutions: Maximizing efficiency with sensor-based system. *Enhancing Productivity in Hybrid Mode: The Beginning of a New Era*, 520.

Maheswari, B. U., Imambi, S. S., Hasan, D., Meenakshi, S., Pratheep, V., & Boopathi, S. (2023). Internet of Things and Machine Learning-Integrated Smart Robotics. In Global Perspectives on Robotics and Autonomous Systems: Development and Applications (pp. 240–258). IGI Global. doi:10.4018/978-1-6684-7791-5.ch010

Majhi, A. K., Dash, S., & Barik, C. K. (2021). Arduino based smart home automation. *ACCENTS Transactions on Information Security.*, *6*(22), 7–12.

Mammone, N., Ieracitano, C., & Morabito, F. C. (2020). A deep CNN approach to decode motor preparation of upper limbs from time–frequency maps of EEG signals at source level. *Neural Networks*, *124*, 357–372. doi:10.1016/j.neunet.2020.01.027 PMID:32045838

Mao, J., Su, X., Wang, H., & Li, J. (2023). Automated Bayesian operational modal analysis of the long-span bridge using machine-learning algorithms. *Engineering Structures*, *289*, 116336. doi:10.1016/j.engstruct.2023.116336

Mao, J., Wang, H., & Spencer, B. F. Jr. (2021). Toward data anomaly detection for automated structural health monitoring: Exploiting generative adversarial nets and autoencoders. *Structural Health Monitoring*, *20*(4), 1609–1626. doi:10.1177/1475921720924601

Marsoof, A., Kariyawasam, K., & Talagala, C. (2022). Crafting Domestic Intellectual Property Law–International Obligations, Flexibilities, and Approaches. In Reframing Intellectual Property Law in Sri Lanka: Lessons from the Developing World and Beyond (pp. 13-26). Singapore: Springer Nature Singapore. https://doi.org/ doi:10.1007/978-981-19-4582-3_213

Martens, J., Blut, T., & Blankenbach, J. (2023). Cross domain matching for semantic point cloud segmentation based on image segmentation and geometric reasoning. *Advanced Engineering Informatics*, *57*, 102076. doi:10.1016/j.aei.2023.102076

Martinet, S. (2018, May 27). *GDPR and Blockchain: Is the New EU Data Protection Regulation a Threat or an Incentive?* Cointelegraph. https://cointelegraph.com/news/gdpr-and-blockchain-is-the-new-eu-data-protection-regulation-a-threat-or-an-incentive

Maryam, K., & Reza, K. M. (2012). An analytical framework for event mining in video data. *Artificial Intelligence Review*, *41*(3), 401–413.

Maseer, Z. K., Yusof, R., Bahaman, N., Mostafa, S. A., Feresa, C. I. K., & Foozy, M. (2021). *Benchmarking of Machine Learning for Anomaly Based Intrusion Detection Systems in the CICIDS2017 Dataset.* doi:10.1109/ACCESS.2021.3056614

Mason, R. O. (1986). Four ethical issues of the information age. *Management Information Systems Quarterly*, *10*(1), 5–12. doi:10.2307/248873

Matsugu, M., Mori, K., Mitari, Y., & Kaneda, Y. (2003). Subject independent facial expression recognition with robust face detection using a convolutional neural network. *Neural Networks*, *16*(5), 555–559. doi:10.1016/S0893-6080(03)00115-1 PMID:12850007

Mattke, J., Hund, A., Maier, C., & Weitzel, T. (2019). How an Enterprise Blockchain Application in the US Pharmaceuticals Supply Chain is Saving Lives. *MIS Quarterly Executive*, *18*(4), 245–261. doi:10.17705/2msqe.00019

Mavlutova, I., Atstaja, D., Grasis, J., Kuzmina, J., Uvarova, I., & Roga, D. (2023). Urban Transportation Concept and Sustainable Urban Mobility in Smart Cities: A Review. *Energies*, *16*(8), 3585. doi:10.3390/en16083585

Mazurek, C., & Stroinski, M. (2022). *Technology pillars for digital transformation of cities based on open software architecture for end2end data streaming*. Academic Press.

Medagliani, P., Leguay, J., Duda, A., Rousseau, F., Duquennoy, S., Raza, S., Ferrari, G., Gonizzi, P., Cirani, S., Veltri, L., & ... (2022). Bringing ip to low-power smart objects: The smart parking case in the CALIPSO project. In *Internet of things applications-from research and innovation to market deployment* (pp. 287–313). River Publishers. doi:10.1201/9781003338628-8

Mehmood, M. S., Shahid, M. R., Jamil, A., Ashraf, R., Mehmood, T., & Mehmood, A. (2019). A Comprehensive Literature Review of Data Encryption Techniques in Cloud Computing and IoT Environment. *2019 8th International Conference on Information and Communication Technologies (ICICT)*, 54–59. 10.1109/ICICT47744.2019.9001945

Mehta, N., & Devarakonda, M. V. (2018). Machine learning, natural language programming, and electronic health records: The next step in the artificial intelligence journey? *The Journal of Allergy and Clinical Immunology*, *141*(6), 2019–2021. doi:10.1016/j.jaci.2018.02.025 PMID:29518424

Mei, H., Haider, M., Joseph, R., Migot, A., & Giurgiutiu, V. (2019). Recent Advances in Piezoelectric Wafer Active Sensors for Structural Health Monitoring Applications. *Sensors (Basel)*, *19*(2), 383. doi:10.339019020383 PMID:30669307

Melibari, W., Baodhah, H., & Akkari, N. (2023). IoT-Based Smart Cities Beyond 2030: Enabling Technologies, Challenges, and Solutions. *2023 1st International Conference on Advanced Innovations in Smart Cities (ICAISC)*, 1–6. 10.1109/ICAISC56366.2023.10085126

Meltzer, J. P. (2018). *The impact of artificial intelligence on international trade*. Center for Technology Innovation at Brookings.

Mendis, C., Renda, A., Amarasinghe, S., & Carbin, M. (2019). Ithemal Accurate, Portable and Fast Basic Block Throughput Estimation using Deep. *Neural Networks*.

Meng, T., Jing, X., Yan, Z., & Pedrycz, W. (2020). A survey on machine learning for data fusion. *Information Fusion*, *57*(2), 115–129. doi:10.1016/j.inffus.2019.12.001

Meng, W., Fei, F., Li, W., & Au, M. H. (2017). Evaluating challenge-based trust mechanism in medical smartphone networks: An empirical study. *2017 IEEE Global Communications Conference, GLOBECOM 2017 - Proceedings*, 1–6. 10.1109/GLOCOM.2017.8254002

Merenda, P., & Fedele, C. (2019). A Real-Time Decision Platform for the Management of Structures and Infrastructures. *Electronics (Basel)*, *8*(10), 1180. doi:10.3390/electronics8101180

Meribout, M., Mekid, S., Kharoua, N., & Khezzar, L. (2021). Online monitoring of structural materials integrity in process industry for I4.0: A focus on material loss through erosion and corrosion sensing. *Measurement*, *176*, 109110. doi:10.1016/j.measurement.2021.109110

Mikolov, T., Kombrink, S., Burget, L., Černocký, J., & Khudanpur, S. (2011). Extensions of recurrent neural network language model. *IEEE International Conference on Acoustics, Speech and Signal Processing (ICASSP)*, (pp. 5528 - 5531). IEEE. 10.1109/ICASSP.2011.5947611

Mishra, K. N., & Chakraborty, C. (2020). A novel approach toward enhancing the quality of life in smart cities using clouds and IoT-based technologies. *Digital Twin Technologies and Smart Cities*, 19–35.

Mishra, P., Pilli, E. S., Varadharajan, V., & Tupakula, U. (2017). Intrusion detection techniques in cloud environment: A survey. *Journal of Network and Computer Applications*, *77*, 18–47. doi:10.1016/j.jnca.2016.10.015

Mishra, A., & Yadav, P. (2020). Anomaly-based IDS to detect attack using various artificial intelligence machine learning algorithms: A review. *2nd International Conference on Data, Engineering and Applications, IDEA 2020*. 10.1109/IDEA49133.2020.9170674

Mitchell, T. M. (1997). Does machine learning really work? *AI Magazine*, *18*(3).

Mi, X., Tang, M., Liao, H., Shen, W., & Lev, B. (2019). The state-of-the-art survey on integrations and applications of the best worst method in decision making: Why, what, what for and what's next? *Omega*, *87*, 205–225. doi:10.1016/j.omega.2019.01.009

Miyakawa, T. (2000). Copyright Legislation in Japan and Recent Trends. A Report to the North American Coordinating Council for Japanese Library Resources Year 2000 Conference, San Diego, CA.

Mobashsher, A. T., Pretorius, A. J., & Abbosh, A. M. (2019). Low-profile vertical polarized slotted antenna for on-road RFID-enabled intelligent parking. *IEEE Transactions on Antennas and Propagation*, *68*(1), 527–532. doi:10.1109/TAP.2019.2939590

Moerland, A. (2022). *Artificial Intelligence and Intellectual Property Law. In The Cambridge Handbook of Private Law and Artificial Intelligence*. Cambridge University Press.

Mohammadi, F., Nazri, G.-A., & Saif, M. (2019). A real-time cloud-based intelligent car parking system for smart cities. *2019 IEEE 2nd International Conference on Information Communication and Signal Processing (ICICSP)*, 235–240.

Mohanty, A., Venkateswaran, N., Ranjit, P., Tripathi, M. A., & Boopathi, S. (2023). Innovative Strategy for Profitable Automobile Industries: Working Capital Management. In Handbook of Research on Designing Sustainable Supply Chains to Achieve a Circular Economy (pp. 412–428). IGI Global.

Mohanty, A., Jothi, B., Jeyasudha, J., Ranjit, P., Isaac, J. S., & Boopathi, S. (2023a). Additive Manufacturing Using Robotic Programming. In *AI-Enabled Social Robotics in Human Care Services* (pp. 259–282). IGI Global. doi:10.4018/978-1-6684-8171-4.ch010

Moingeon, P., Kuenemann, M., & Guedj, M. (2022). Artificial intelligence-enhanced drug design and development: Toward a computational precision medicine. *Drug Discovery Today*, *27*(1), 215–222. doi:10.1016/j.drudis.2021.09.006 PMID:34555509

Momeni, H., & Ebrahimkhanlou, A. (2021, November 1). Applications of High-Dimensional Data Analytics in Structural Health Monitoring and Non-Destructive Evaluation: Thermal Videos Processing Using Tensor-Based Analysis. *Volume 13: Safety Engineering, Risk, and Reliability Analysis; Research Posters*. doi:10.1115/IMECE2021-71878

Morfoulaki, M., & Papathanasiou, J. (2021). Use of the sustainable mobility efficiency index (SMEI) for enhancing the sustainable urban mobility in Greek cities. *Sustainability (Basel)*, *13*(4), 1709. doi:10.3390u13041709

Morley, J., Floridi, L., Kinsey, L., & Elhalal, A. (2020). From what to how: An initial review of publicly available AI ethics tools, methods and research to translate principles into practices. *Science and Engineering Ethics*, *26*(4), 2141–2168. doi:10.100711948-019-00165-5 PMID:31828533

Morris, B. B., Rossi, B., & Fuemmeler, B. (2022). The role of digital health technology in rural cancer care delivery: A systematic review. *The Journal of Rural Health*, *38*(3), 493–511. doi:10.1111/jrh.12619 PMID:34480506

Mousavi, A. A., Zhang, C., Masri, S. F., & Gholipour, G. (2022). Structural damage detection method based on the complete ensemble empirical mode decomposition with adaptive noise: A model steel truss bridge case study. *Structural Health Monitoring*, *21*(3), 887–912. doi:10.1177/14759217211013535

Mousavi, M., & Gandomi, A. H. (2021). Structural health monitoring under environmental and operational variations using MCD prediction error. *Journal of Sound and Vibration*, *512*, 116370. doi:10.1016/j.jsv.2021.116370

Moustafa, N., & Slay, J. (2015). UNSW-NB15: A comprehensive data set for network intrusion detection systems (UNSW-NB15 network data set). *2015 Military Communications and Information Systems Conference, MilCIS 2015 - Proceedings*. 10.1109/MilCIS.2015.7348942

MoustafaN. (2019). UNSW_NB15 dataset. IEEE Dataport. doi:10.21227/8vf7-s525

Mulay, S. A., Devale, P. R., & Garje, G. V. (2010). Intrusion Detection System Using Support Vector Machine and Decision Tree. *International Journal of Computer Applications*, *3*(3), 40–43. doi:10.5120/758-993

Müller-Putz, G. R., Kobler, R. J., Pereira, J., Lopes-Dias, C., Hehenberger, L., Mondini, V., Martínez-Cagigal, V., Srisrisawang, N., Pulferer, H., Batistić, L., & Sburlea, A. I. (2022). Feel your reach: An EEG-based framework to continuously detect goal-directed movements and error processing to gate kinesthetic feedback informed artificial arm control. *Frontiers in Human Neuroscience*, *16*, 841312. doi:10.3389/fnhum.2022.841312 PMID:35360289

Muralidharan, S., & Ko, H. (2019). An InterPlanetary File System (IPFS) based IoT framework. *2019 IEEE International Conference on Consumer Electronics (ICCE)*, 1–2. 10.1109/ICCE.2019.8662002

Myilsamy, S., Boopathi, S., & Yuvaraj, D. (2021). A study on cryogenically treated molybdenum wire electrode. *Materials Today: Proceedings*, *45*(9), 8130–8135. doi:10.1016/j.matpr.2021.02.049

Nallaperuma, D., Nawaratne, R., Bandaragoda, T., Adikari, A., Nguyen, S., Kempitiya, T., De Silva, D., Alahakoon, D., & Pothuhera, D. (2019). Online incremental machine learning platform for big data-driven smart traffic management. *IEEE Transactions on Intelligent Transportation Systems*, *20*(12), 4679–4690. doi:10.1109/TITS.2019.2924883

Nard, C. A. (2019). *The law of patents*. Aspen Publishing.

Narendra, K. S., & Parthasarathy, K. (1990). Identification and control of dynamical systems using neural networks. *IEEE Transactions on Neural Networks*, *1*(1), 4–27. doi:10.1109/72.80202 PMID:18282820

Nasir, V., Ayanleye, S., Kazemirad, S., Sassani, F., & Adamopoulos, S. (2022). Acoustic emission monitoring of wood materials and timber structures: A critical review. *Construction & Building Materials*, *350*, 128877. doi:10.1016/j.conbuildmat.2022.128877

Nawaz, A., Arora, A. S., Ali, W., Saxena, N., Khan, M. S., Yun, C. M., & Lee, M. (2022). Intelligent Human–Machine Interface: An Agile Operation and Decision Support for an ANAMMOX SBR System at a Pilot-Scale Wastewater Treatment Plant. *IEEE Transactions on Industrial Informatics*, *18*(9), 6224–6232. doi:10.1109/TII.2022.3153468

Nishanth, J., Deshmukh, M. A., Kushwah, R., Kushwaha, K. K., Balaji, S., & Sampath, B. (2023). Particle Swarm Optimization of Hybrid Renewable Energy Systems. In *Intelligent Engineering Applications and Applied Sciences for Sustainability* (pp. 291–308). IGI Global. doi:10.4018/979-8-3693-0044-2.ch016

Nogueira, P. R. R., de Paula, S. L., de Lima Santana, S. B., da Silva Pinto, J., Braz, M. I., & de Aquino, L. M. P. (n.d.). *Cidades inteligentes e mobilidade urbana: Atores e práticas na cidade de Recife/PE Smart cities and urban mobility: Actors and practices in the city of Recife/PE*. Academic Press.

Noorit, N., & Suvonvorn, N. (2014). Human activity recognition from basic actions using finite state machine. *International Conference on Advanced Data and Information Engineering (DaEng - 2013)* (pp. 379 - 386). Springer. 10.1007/978-981-4585-18-7_43

Nutbeam, D. (2021). From health education to digital health literacy–building on the past to shape the future. *Global Health Promotion*, *28*(4), 51–55. doi:10.1177/17579759211044079 PMID:34719292

Ochôa, P. A., Groves, R. M., & Benedictus, R. (2020). Effects of high-amplitude low-frequency structural vibrations and machinery sound waves on ultrasonic guided wave propagation for health monitoring of composite aircraft primary structures. *Journal of Sound and Vibration*, *475*, 115289. doi:10.1016/j.jsv.2020.115289

Office of the United States Trade Representative. (n.d.). *Intellectual Property in USMC Agreement*. Retrieved from https://ustr.gov/trade-agreements/free-trade-agreements/united-states-mexico-canada-agreement/agreement-between

Ofner, P., Schwarz, A., Pereira, J., & Müller-Putz, G. R. (2017). Upper limb movements can be decoded from the time-domain of low-frequency EEG. *PLoS One*, *12*(8), e0182578. doi:10.1371/journal.pone.0182578 PMID:28797109

Oh, B. K., Glisic, B., Kim, Y., & Park, H. S. (2019). Convolutional neural network-based wind-induced response estimation model for tall buildings. *Computer-Aided Civil and Infrastructure Engineering*, *34*(10), 843–858. doi:10.1111/mice.12476

Oku, K., Nakajima, S., Miyazaki, J., & Uemura, S. (2006). Context-aware SVM for context-dependent information recommendation. *Proceedings - IEEE International Conference on Mobile Data Management, 2006*. 10.1109/MDM.2006.56

Oliveira, S., & Alegre, A. (2020). Seismic and structural health monitoring of Cabril dam. Software development for informed management. *Journal of Civil Structural Health Monitoring*, *10*(5), 913–925. doi:10.100713349-020-00425-0

Oman, R. (2017). Computer Software as Copyrightable Subject Matter: Oracle v. Google, Legislative Intent, and the Scope of Rights in Digital Works. *Harvard Journal of Law & Technology*, *31*, 639.

Osanai, H., Yamamoto, J., & Kitamura, T. (2023). Extracting electromyographic signals from multi-channel LFPs using independent component analysis without direct muscular recording. *Cell Reports Methods*, *3*(6), 100482. doi:10.1016/j.crmeth.2023.100482 PMID:37426755

Ou, J., Zhang, R., Li, X., & Lin, G. (2023). Research and explainable analysis of a real-time passion fruit detection model based on FSOne-YOLOv7. *Agronomy (Basel)*, *13*(8), 1993. doi:10.3390/agronomy13081993

Oyinloye, D. P., Sen Teh, J., Jamil, N., & Alawida, M. (2021). Blockchain Consensus: An Overview of Alternative Protocols. *Symmetry*, *13*(8), 1363. doi:10.3390ym13081363

Padmapoorani, P., Senthilkumar, S., & Mohanraj, R. (2023). Machine Learning Techniques for Structural Health Monitoring of Concrete Structures: A Systematic Review. *Civil Engineering (Shiraz)*, *47*(4), 1919–1931. doi:10.100740996-023-01054-5

Palaniappan, M., Tirlangi, S., Mohamed, M. J. S., Moorthy, R. S., Valeti, S. V., & Boopathi, S. (2023). Fused Deposition Modelling of Polylactic Acid (PLA)-Based Polymer Composites: A Case Study. In Development, Properties, and Industrial Applications of 3D Printed Polymer Composites (pp. 66–85). IGI Global.

Pamučar, D., Ecer, F., Cirovic, G., & Arlasheedi, M. A. (2020). Application of improved best worst method (BWM) in real-world problems. *Mathematics*, *8*(8), 1342. doi:10.3390/math8081342

Pan, C., Liu, J., Yan, W., & Zhou, Y. (2021). Salient object detection based on visual perceptual saturation and two-stream hybrid networks. *IEEE Transactions on Image Processing*, *30*, 4773–4787. doi:10.1109/TIP.2021.3074796 PMID:33929959

Pan, C., & Yan, W. (2018) A learning-based positive feedback in salient object detection. *International Conference on Image and Vision Computing New Zealand*. 10.1109/IVCNZ.2018.8634717

Pan, C., & Yan, W. (2020). Object detection based on saturation of visual perception. *Multimedia Tools and Applications*, *79*(27-28), 19925–19944. doi:10.100711042-020-08866-x

Panigrahi, R., & Borah, S. (2018). A detailed analysis of CICIDS2017 dataset for designing Intrusion Detection Systems. *International Journal of Engineering and Technology(UAE)*, *7*(24), 479–482.

Parico, A., & Ahamed, T. (2021). Real time pear fruit detection and counting using YOLOv4 models and deep SORT. *Sensors (Basel), 21*(14), 4803. doi:10.339021144803 PMID:34300543

Parida, L., Moharana, S., Ferreira, V. M., Giri, S. K., & Ascensão, G. (2022). A Novel CNN-LSTM Hybrid Model for Prediction of Electro-Mechanical Impedance Signal Based Bond Strength Monitoring. *Sensors (Basel), 22*(24), 9920. doi:10.339022249920 PMID:36560293

Paris Convention for the Protection of Industrial Property (opened for signature 20 March 1883) ("Paris Convention").

Park, K., Song, Y., & Cheong, Y. G. (2018). Classification of attack types for intrusion detection systems using a machine learning algorithm. *Proceedings - IEEE 4th International Conference on Big Data Computing Service and Applications, BigDataService 2018*, 282–286. 10.1109/BigDataService.2018.00050

Park, A., & Li, H. (2021). The effect of blockchain technology on supply chain sustainability performances. *Sustainability (Basel), 13*(4), 1726. doi:10.3390u13041726

Pascanu, R., Mikolov, T., & Bengio, Y. (2013). On the difficulty of training recurrent neural networks. *ICML, 28*(3), 1310–1318.

Patro, S. P., Patel, P., Senapaty, M. K., Padhy, N., & Sah, R. D. (2020). IoT based smart parking system: A proposed algorithm and model. *2020 International Conference on Computer Science, Engineering and Applications (ICCSEA)*, 1–6. 10.1109/ICCSEA49143.2020.9132923

Paulick, K., Seidel, S., Lange, C., Kemmer, A., Cruz-Bournazou, M. N., Baier, A., & Haehn, D. (2022). Promoting sustainability through next-generation biologics drug development. *Sustainability (Basel), 14*(8), 4401. doi:10.3390u14084401

Paul, S., Banerjee, C., & Ghoshal, M. (2018). A CFS–DNN-Based Intrusion Detection System. *Lecture Notes in Electrical Engineering, 462*(March), 159–168. doi:10.1007/978-981-10-7901-6_19

Pavlidis, G. (2018). Apollo - A Hybrid Recommender for Museums and Cultural Tourism. *9th International Conference on Intelligent Systems 2018: Theory, Research and Innovation in Applications, IS 2018 - Proceedings*. 10.1109/IS.2018.8710494

Pawar, A., Jolly, A., Pandey, V., Chaurasiya, P. K., Verma, T. N., & Meshram, K. (2023). Artificial intelligence algorithms for prediction of cyclic stress ratio of soil for environment conservation. *Environmental Challenges, 12*, 100730. doi:10.1016/j.envc.2023.100730

Pearlman, R. (2017). Recognising artificial intelligence (AI) as authors and investors under US intellectual property law. *Rich. J.L. & Tech., 2*.

Peck, R. W., Shah, P., Vamvakas, S., & van der Graaf, P. H. (2020). Data science in clinical pharmacology and drug development for improving health outcomes in patients. *Clinical Pharmacology and Therapeutics, 107*(4), 683–686. doi:10.1002/cpt.1803 PMID:32202650

Peña-Guerrero, J., Nguewa, P. A., & García-Sosa, A. T. (2021). Machine learning, artificial intelligence, and data science breaking into drug design and neglected diseases. *Wiley Interdisciplinary Reviews. Computational Molecular Science, 11*(5), e1513. doi:10.1002/wcms.1513

Pendyala, S., Vanama, S., & Upalanchi, S. (n.d.). *Design of automatic flight control system with safety features for small aircraft*. Academic Press.

Perera, C., Qin, Y., Estrella, J. C., Reiff-Marganiec, S., & Vasilakos, A. V. (2017). Fog computing for sustainable smart cities: A survey. *ACM Computing Surveys, 50*(3), 1–43. doi:10.1145/3057266

Perlmutter, S. (2023). *Copyright Registration Guidance: Works Containing Material Generated by Artificial Intelligence*. U.S. Copyright Office, Library of Congress.

Petrushin, V. A. (2005). Mining rare and frequent events in multi-camera surveillance video using self-organizing maps. *ACM SIGKDD international Conference on Knowledge discovery in Data Mining,* (pp. 794 – 800). ACM.

Philip Chen, C. L., & Zhang, C. Y. (2014). Data-intensive applications, challenges, techniques and technologies: A survey on Big Data. *Information Sciences*, *275*, 314–347. Advance online publication. doi:10.1016/j.ins.2014.01.015

Picht, P. G., & Thouvenin, F. (2023). AI and IP: Theory to Policy and Back Again–Policy and Research Recommendations at the Intersection of Artificial Intelligence and Intellectual Property. *IIC-International Review of Intellectual Property and Competition Law*, 1-25.

Popoola, O. P., & Wang, K. (2012). Video-based abnormal human behavior recognition—A review. *IEEE Transactions on Systems, Man, and Cybernetics. Part C, Applications and Reviews*, *42*(6), 865–878. doi:10.1109/TSMCC.2011.2178594

Pramila, P., Amudha, S., Saravanan, T., Sankar, S. R., Poongothai, E., & Boopathi, S. (2023). Design and Development of Robots for Medical Assistance: An Architectural Approach. In Contemporary Applications of Data Fusion for Advanced Healthcare Informatics (pp. 260–282). IGI Global.

Prasath, V. B. S., Haneen Arafat, A. A., Hassanat, A. B. A., Lasassmeh, O., Tarawneh, A. S., Alhasanat, M. B., & Salman, H. S. E. (2019). Effects of Distance Measure Choice on KNN Classifier Performance. *RE:view*. ArXiv1708.04321v3

Prasetio, B. H., Tamura, H., & Tanno, K. (2019). Generalized Discriminant Methods for Improved X-Vector Back-end Based Stress Speech Recognition. *IEEJ Transactions on Electronics. Information Systems*, *139*(11), 1341–1347.

Preeti, T., Kanakaraddi, S., Beelagi, A., Malagi, S., & Sudi, A. (2021). *Forest fire prediction using machine learning techniques. In International conference on intelligent technologies (CONIT)*. IEEE.

Preuveneers, D., Tsingenopoulos, I., & Joosen, W. (2020). Resource Usage and Performance Trade-offs for Machine Learning Models in Smart Environments. *Sensors (Basel)*, *20*(4), 1176. doi:10.339020041176 PMID:32093354

Protopapadakis, E., Doulamis, N., & Voulodimos, A. (2017). Hybrid meta-filtering system for cultural monument related recommendations. *VISIGRAPP 2017 - Proceedings of the 12th International Joint Conference on Computer Vision, Imaging and Computer Graphics Theory and Applications, 5*. 10.5220/0006347104360443

Puthal, D., & Mohanty, S. P. (2019). Proof of Authentication: IoT-Friendly Blockchains. *IEEE Potentials*, *38*(1), 26–29. doi:10.1109/MPOT.2018.2850541

Qi, J., Nguyen, M., & Yan, W. (2022). *Waste classification from digital images using ConvNeXt*. Pacific-Rim Symposium on Image and Video Technology.

Qi, J., Nguyen, M., & Yan, W. (2022) Small visual object detection in smart waste classification using Transformers with deep learning. *International Conference on Image and Vision Computing New Zealand (IVCNZ)*. IEEE.

Qolomany, B., Mohammed, I., Al-Fuqaha, A., Guizani, M., & Qadir, J. (2020). Trust-based cloud machine learning model selection for industrial IoT and smart city services. *IEEE Internet of Things Journal*, *8*(4), 2943–2958. doi:10.1109/JIOT.2020.3022323

Rabah, K., & ... (2018). Convergence of AI, IoT, big data and blockchain: A review. *Law Institute Journal*, *1*(1), 1–18.

Rabby, M. K. M., Islam, M. M., & Imon, S. M. (2019). A review of IoT application in a smart traffic management system. *2019 5th International Conference on Advances in Electrical Engineering (ICAEE)*, 280–285.

Radhakrishna, A., Yan, W., & Kankanhalli, M. (2006). Modelling intent for home video repurposing. *IEEE MultiMedia*, *13*(1), 46–55. doi:10.1109/MMUL.2006.12

Rafiei, M. H., & Adeli, H. (2017). A novel machine learning-based algorithm to detect damage in high-rise building structures. *Structural Design of Tall and Special Buildings*, *26*(18), e1400. doi:10.1002/tal.1400

Ragavan, K., Venkatalakshmi, K., & Vijayalakshmi, K. (2021). Traffic video-based intelligent traffic control system for smart cities using modified ant colony optimizer. *Computational Intelligence*, *37*(1), 538–558. doi:10.1111/coin.12424

Rahamathunnisa, U., Subhashini, P., Aancy, H. M., Meenakshi, S., Boopathi, S., & ... (2023). Solutions for Software Requirement Risks Using Artificial Intelligence Techniques. In *Handbook of Research on Data Science and Cybersecurity Innovations in Industry 4.0 Technologies* (pp. 45–64). IGI Global.

Rahamathunnisa, U., Sudhakar, K., Murugan, T. K., Thivaharan, S., Rajkumar, M., & Boopathi, S. (2023). Cloud Computing Principles for Optimizing Robot Task Offloading Processes. In *AI-Enabled Social Robotics in Human Care Services* (pp. 188–211). IGI Global. doi:10.4018/978-1-6684-8171-4.ch007

Rahman, M. A., Asyhari, A. T., Leong, L. S., Satrya, G. B., Hai Tao, M., & Zolkipli, M. F. (2020). Scalable machine learning-based intrusion detection system for IoT-enabled smart cities. *Sustainable Cities and Society*, *61*(January), 102324. doi:10.1016/j.scs.2020.102324

Rai, A. (2020). Explainable AI: From black box to glass box. *Journal of the Academy of Marketing Science*, *48*(1), 137–141. doi:10.100711747-019-00710-5

Rajapaksha, P., Farahbakhsh, R., Nathanail, E., & Crespi, N. (2017). iTrip, a framework to enhance urban mobility by leveraging various data sources. *Transportation Research Procedia*, *24*, 113–122. doi:10.1016/j.trpro.2017.05.076

Ramirez-Garcia, X., & García-Valdez, M. (2014). Post-filtering for a restaurant context-aware recommender system. *Studies in Computational Intelligence*, *547*, 695–707. Advance online publication. doi:10.1007/978-3-319-05170-3_49

Ramudu, K., Mohan, V. M., Jyothirmai, D., Prasad, D., Agrawal, R., & Boopathi, S. (2023). Machine Learning and Artificial Intelligence in Disease Prediction: Applications, Challenges, Limitations, Case Studies, and Future Directions. In Contemporary Applications of Data Fusion for Advanced Healthcare Informatics (pp. 297–318). IGI Global.

Rastin, Z., Ghodrati Amiri, G., & Darvishan, E. (2021). Unsupervised Structural Damage Detection Technique Based on a Deep Convolutional Autoencoder. *Shock and Vibration*, *2021*, 1–11. doi:10.1155/2021/6658575

Rather, A. I., Mirgal, P., Banerjee, S., & Laskar, A. (2023). Application of Acoustic Emission as Damage Assessment Technique for Performance Evaluation of Concrete Structures: A Review. *Practice Periodical on Structural Design and Construction*, *28*(3), 03123003. Advance online publication. doi:10.1061/PPSCFX.SCENG-1256

Ravisankar, A., Sampath, B., & Asif, M. M. (2023). Economic Studies on Automobile Management: Working Capital and Investment Analysis. In Multidisciplinary Approaches to Organizational Governance During Health Crises (pp. 169–198). IGI Global.

Reddy, M. A., Gaurav, A., Ushasukhanya, S., Rao, V. C. S., Bhattacharya, S., & Boopathi, S. (2023). Bio-Medical Wastes Handling Strategies During the COVID-19 Pandemic. In Multidisciplinary Approaches to Organizational Governance During Health Crises (pp. 90–111). IGI Global. doi:10.4018/978-1-7998-9213-7.ch006

Reddy, M. A., Reddy, B. M., Mukund, C., Venneti, K., Preethi, D., & Boopathi, S. (2023). Social Health Protection During the COVID-Pandemic Using IoT. In *The COVID-19 Pandemic and the Digitalization of Diplomacy* (pp. 204–235). IGI Global. doi:10.4018/978-1-7998-8394-4.ch009

Redmon, J., Divvala, S., Girshick, R., & Farhadi, A. (2016). You Only Look Once: Unified, real-time object detection. *IEEE Conference on Computer Vision and Pattern Recognition (CVPR),* 779–788. 10.1109/CVPR.2016.91

Reis, C., Ruivo, P., Oliveira, T., & Faroleiro, P. (2020). Assessing the drivers of machine learning business value. *Journal of Business Research, 117*, 232–243. doi:10.1016/j.jbusres.2020.05.053

Ren, Q., Li, H., Li, M., Kong, T., & Guo, R. (2023). Bayesian incremental learning paradigm for online monitoring of dam behavior considering global uncertainty. *Applied Soft Computing, 143*, 110411. doi:10.1016/j.asoc.2023.110411

Ren, Y., Ye, Q., Xu, X., Huang, Q., Fan, Z., Li, C., & Chang, W. (2022). An anomaly pattern detection for bridge structural response considering time-varying temperature coefficients. *Structures, 46*, 285–298. doi:10.1016/j.istruc.2022.10.020

Report on intellectual property rights for the development of artificial intelligence technologies. (2020). https://www.europarl.europa.eu/doceo/document/A-9-2020-0176_EN.html

Resende, P. A. A., & Drummond, A. C. (2018). A Survey of Random Forest Based Methods for Intrusion Detection Systems. *ACM Computing Surveys, 51*(3), 1–36. doi:10.1145/3178582

Rezaei, J. (2015). Best-worst multi-criteria decision-making method. *Omega, 53*, 49–57. doi:10.1016/j.omega.2014.11.009

Ricci, F., Shapira, B., & Rokach, L. (2015). Recommender systems: Introduction and challenges. In Recommender Systems Handbook (2nd ed.). doi:10.1007/978-1-4899-7637-6_1

Richter, M. A., Hagenmaier, M., Bandte, O., Parida, V., & Wincent, J. (2022). Smart cities, urban mobility and autonomous vehicles: How different cities needs different sustainable investment strategies. *Technological Forecasting and Social Change, 184*, 121857. doi:10.1016/j.techfore.2022.121857

Rizk, J. G., Lippi, G., Henry, B. M., Forthal, D. N., & Rizk, Y. (2022). Prevention and treatment of Monkeypox. *Drugs, 82*(9), 957–963. doi:10.100740265-022-01742-y PMID:35763248

Rizky Duto Pamungkas, A., Husna, D., Astha Ekadiyanto, F., Eddy Purnama, I. K., Nurul Hidayati, A., Hery Purnomo, M., Nurtanio, I., Fuad Rachmadi, R., Mardi Susiki Nugroho, S., & Agung Putri Ratna, A. (2021). Designing a Blockchain Data Storage System Using Ethereum Architecture and Peer-to-Peer InterPlanetary File System (IPFS). *2021 the 7th International Conference on Communication and Information Processing (ICCIP)*, 152–157. 10.1145/3507971.3507997

Roberts, G., Zoubir, A. M., & Boashash, B. (1997) Time-frequency classification using a multiple hypotheses test: An application to the classification of humpback whale signals. In *1997 IEEE International Conference on Acoustics, Speech, and Signal Processing* (Vol. 1, pp. 563-566). IEEE. 10.1109/ICASSP.1997.599700

Robertson, D. G. E., & Dowling, J. J. (2003). Design and responses of Butterworth and critically damped digital filters. *Journal of Electromyography and Kinesiology, 13*(6), 569–573. doi:10.1016/S1050-6411(03)00080-4 PMID:14573371

Robinson, R. R. R., & Thomas, C. (2015). *Ranking of machine learning algorithms based on the performance in classifying DDoS attacks. 2015 IEEE Recent Advances in Intelligent Computational Systems*. doi:10.1109/RAICS.2015.7488411

Rodic, L. D. (n.d.). *Smart parking solutions for occupancy sensing*. Academic Press.

Rogers, L., & Jensen, K. F. (2019). Continuous manufacturing–the Green Chemistry promise? *Green Chemistry, 21*(13), 3481–3498. doi:10.1039/C9GC00773C

Rogers, T. J., Worden, K., Fuentes, R., Dervilis, N., Tygesen, U. T., & Cross, E. J. (2019). A Bayesian non-parametric clustering approach for semi-supervised Structural Health Monitoring. *Mechanical Systems and Signal Processing, 119*, 100–119. doi:10.1016/j.ymssp.2018.09.013

Rokach, L., & Maimon, O. (2010). *Data Mining and Knowledge Discovery Handbook*. Springer.

Rosafalco, L., Manzoni, A., Mariani, S., & Corigliano, A. (2020). Fully convolutional networks for structural health monitoring through multivariate time series classification. *Advanced Modeling and Simulation in Engineering Sciences*, *7*(1), 38. doi:10.118640323-020-00174-1

Roschangar, F., Sheldon, R. A., & Senanayake, C. H. (2015). Overcoming barriers to green chemistry in the pharmaceutical industry–the Green Aspiration Level™ concept. *Green Chemistry*, *17*(2), 752–768. doi:10.1039/C4GC01563K

Ruffino, C., Papaxanthis, C., & Lebon, F. (2017). Neural plasticity during motor learning with motor imagery practice: Review and perspectives. *Neuroscience*, *341*, 61–78. doi:10.1016/j.neuroscience.2016.11.023 PMID:27890831

Rui, J., & Othengrafen, F. (2023). Examining the Role of Innovative Streets in Enhancing Urban Mobility and Livability for Sustainable Urban Transition: A Review. *Sustainability (Basel)*, *15*(7), 5709. doi:10.3390u15075709

Ruiz, N., Bargal, S., Xie, C., Saenko, K., & Sclaroff, S. (2022). Finding differences between Transformers and ConvNets using counterfactual simulation testing. *Advances in Neural Information Processing Systems*, *35*, 14403–14418.

Rumelhart, D. E., Hinton, G. E., & Williams, R. J. (1986). Learning representations by back-propagating errors. *Nature*, *323*(6088), 533–536. Advance online publication. doi:10.1038/323533a0

Ruotsalo, T., Haav, K., Stoyanov, A., Roche, S., Fani, E., Deliai, R., Mäkelä, E., Kauppinen, T., & Hyvönen, E. (2013). SMARTMUSEUM: A mobile recommender system for the Web of Data. *Journal of Web Semantics*, *20*, 50–67. Advance online publication. doi:10.1016/j.websem.2013.03.001

Saadeh, M., Sleit, A., Sabri, K. E., & Almobaideen, W. (2018). Hierarchical architecture and protocol for mobile object authentication in the context of IoT smart cities. *Journal of Network and Computer Applications*, *121*, 1–19. doi:10.1016/j.jnca.2018.07.009

Sadique, K. M., Rahmani, R., & Johannesson, P. (2020). *Enhancing Data Privacy in the Internet of Things (IoT)*. Using Edge Computing. doi:10.1007/978-3-030-66763-4_20

Saeed, F., Paul, A., Rehman, A., Hong, W. H., & Seo, H. (2018). IoT-based intelligent modeling of smart home environment for fire prevention and safety. *Journal of Sensor and Actuator Networks*, *7*(1), 11. doi:10.3390/jsan7010011

Saha, S., & Ghosh, A. (2019, December). Rehabilitation using neighbor-cluster based matching inducing artificial bee colony optimization. In *2019 IEEE 16th India council international conference (INDICON)* (pp. 1-4). IEEE. 10.1109/INDICON47234.2019.9028975

Saha, S., Lahiri, R., Konar, A., & Nagar, A. K. (2018, November). Gesture driven remote robot manoeuvre for unstructured environment. In *2018 IEEE Symposium Series on Computational Intelligence (SSCI)* (pp. 1793-1800). IEEE. 10.1109/SSCI.2018.8628867

Said, B., Lathamaheswari, M., Singh, P. K., Ouallane, A. A., Bakhouyi, A., Bakali, A., Talea, M., Dhital, A., & Deivanayagampillai, N. (2022). An Intelligent Traffic Control System Using Neutrosophic Sets, Rough sets, Graph Theory, Fuzzy sets and its Extended Approach: A Literature Review. *Neutrosophic Sets Syst*, *50*, 10–26.

Sajedi, S., & Liang, X. (2022). Deep generative Bayesian optimization for sensor placement in structural health monitoring. *Computer-Aided Civil and Infrastructure Engineering*, *37*(9), 1109–1127. doi:10.1111/mice.12799

Saleem, A. A., Siddiqui, H. U. R., Shafique, R., Haider, A., & Ali, M. (2020). A review on smart IOT based parking system. *Recent Advances on Soft Computing and Data Mining: Proceedings of the Fourth International Conference on Soft Computing and Data Mining (SCDM 2020)*, Melaka, Malaysia, January 22–23, 2020, 264–273.

Salehi, H., Gorodetsky, A., Solhmirzaei, R., & Jiao, P. (2023). High-dimensional data analytics in civil engineering: A review on matrix and tensor decomposition. *Engineering Applications of Artificial Intelligence*, *125*, 106659. doi:10.1016/j.engappai.2023.106659

Samikannu, R., Koshariya, A. K., Poornima, E., Ramesh, S., Kumar, A., & Boopathi, S. (2022). Sustainable Development in Modern Aquaponics Cultivation Systems Using IoT Technologies. In *Human Agro-Energy Optimization for Business and Industry* (pp. 105–127). IGI Global.

Sampath, B. (2021). *Sustainable Eco-Friendly Wire-Cut Electrical Discharge Machining: Gas Emission Analysis.*

Sampath, B., Sasikumar, C., & Myilsamy, S. (2023). Application of TOPSIS Optimization Technique in the Micro-Machining Process. In Trends, Paradigms, and Advances in Mechatronics Engineering (pp. 162–187). IGI Global.

Sampath, B., Pandian, M., Deepa, D., & Subbiah, R. (2022). Operating parameters prediction of liquefied petroleum gas refrigerator using simulated annealing algorithm. *AIP Conference Proceedings*, *2460*(1), 070003. doi:10.1063/5.0095601

Sampath, B., Sureshkumar, T., Yuvaraj, M., & Velmurugan, D. (2021). Experimental Investigations on Eco-Friendly Helium-Mist Near-Dry Wire-Cut EDM of M2-HSS Material. *Materials Research Proceedings*, *19*, 175–180.

Sampath, B., Yuvaraj, T., & Velmurugan, D. (2022). Parametric analysis of mould sand properties for flange coupling casting. *AIP Conference Proceedings*, *2460*(1), 070002. doi:10.1063/5.0095599

Sanghvi, K., Shah, A., Shah, P., Shah, P., & Rathod, S. S. (2021). Smart Parking Solutions for On-Street and Off-Street Parking. *2021 International Conference on Communication Information and Computing Technology (ICCICT)*, 1–6. 10.1109/ICCICT50803.2021.9510099

Sanjaya, S. A., & Adi Rakhmawan, S. (2020). Face mask detection using MobileNetv2 in the era of COVID-19 pandemic. *International Conference on Data Analytics for Business and Industry: Way Towards a Sustainable Economy (ICDABI).* 10.1109/ICDABI51230.2020.9325631

Sankar, K. M., Booba, B., & Boopathi, S. (2023). Smart Agriculture Irrigation Monitoring System Using Internet of Things. In *Contemporary Developments in Agricultural Cyber-Physical Systems* (pp. 105–121). IGI Global. doi:10.4018/978-1-6684-7879-0.ch006

Sant, A., Garg, L., Xuereb, P., & Chakraborty, C. (2021). A Novel Green IoT-Based Pay-As-You-Go Smart Parking System. *Computers, Materials & Continua*, *67*(3), 3523–3544. doi:10.32604/cmc.2021.015265

Santos, M. H., Oliveira, N. M., & D'Amore, R. (2021). From Control Requirements to PIL Test: Development of a Structure to Autopilot Implementation. *IEEE Access : Practical Innovations, Open Solutions*, *9*, 154788–154803. doi:10.1109/ACCESS.2021.3127846

Sarah, J., Hejazi, F., Rashid, R. S. M., & Ostovar, N. (2019). A Review of Dynamic Analysis in Frequency Domain for Structural Health Monitoring. *IOP Conference Series. Earth and Environmental Science*, *357*(1), 012007. doi:10.1088/1755-1315/357/1/012007

Saravanan, A., Venkatasubramanian, R., Khare, R., Surakasi, R., Boopathi, S., Ray, S., & Sudhakar, M. (2022). Policy trends of renewable energy and non-*renewable energy*. Academic Press.

Saravanan, M., Vasanth, M., Boopathi, S., Sureshkumar, M., & Haribalaji, V. (2022). Optimization of Quench Polish Quench (QPQ) Coating Process Using Taguchi Method. *Key Engineering Materials*, *935*, 83–91. doi:10.4028/p-z569vy

Sarker, I. H. (2021). Deep Learning: A Comprehensive Overview on Techniques, Taxonomy, Applications and Research Directions. In SN Computer Science (Vol. 2, Issue 6). doi:10.100742979-021-00815-1

Sarmadi, H., Entezami, A., Saeedi Razavi, B., & Yuen, K. (2021). Ensemble learning-based structural health monitoring by Mahalanobis distance metrics. *Structural Control and Health Monitoring*, *28*(2). Advance online publication. doi:10.1002tc.2663

Sarmadi, H., & Karamodin, A. (2020). A novel anomaly detection method based on adaptive Mahalanobis-squared distance and one-class kNN rule for structural health monitoring under environmental effects. *Mechanical Systems and Signal Processing*, *140*, 106495. doi:10.1016/j.ymssp.2019.106495

Sarmadi, H., & Yuen, K.-V. (2022). Structural health monitoring by a novel probabilistic machine learning method based on extreme value theory and mixture quantile modeling. *Mechanical Systems and Signal Processing*, *173*, 109049. doi:10.1016/j.ymssp.2022.109049

Sasikumar, A., Ravi, L., Kotecha, K., Abraham, A., Devarajan, M., & Vairavasundaram, S. (2023). A Secure Big Data Storage Framework Based on Blockchain Consensus Mechanism With Flexible Finality. *IEEE Access : Practical Innovations, Open Solutions*, *11*, 56712–56725. doi:10.1109/ACCESS.2023.3282322

Satav, S. D., Hasan, D. S., Pitchai, R., Mohanaprakash, T. A., Sultanuddin, S. J., & Boopathi, S. (2024). Next Generation of Internet of Things (NGIoT) in Healthcare Systems. In Practice, Progress, and Proficiency in Sustainability (pp. 307–330). IGI Global. doi:10.4018/979-8-3693-1186-8.ch017

Satav, S. D., Lamani, D. G, H. K., Kumar, N. M. G., Manikandan, S., & Sampath, B. (2024). Energy and Battery Management in the Era of Cloud Computing. In Practice, Progress, and Proficiency in Sustainability (pp. 141–166). IGI Global. doi:10.4018/979-8-3693-1186-8.ch009

Sathya, A. (2022). A Blockchain-Based and IoT-Powered Smart Parking System for Smart Cities. *Communication, Software and Networks Proceedings*, *2022*, 583–590.

Savage, N. (2021). *Tapping into the drug discovery potential of AI*. Nature. Com. doi:10.1038/d43747-021-00045-7

Scherer, R., Schloegl, A., Lee, F., Bischof, H., Janša, J., & Pfurtscheller, G. (2007). The self-paced graz brain-computer interface: Methods and applications. *Computational Intelligence and Neuroscience*, *2007*, 2007. doi:10.1155/2007/79826 PMID:18350133

Schlögl, A., & Supp, G. (2006). Analyzing event-related EEG data with multivariate autoregressive parameters. *Progress in Brain Research*, *159*, 135–147. doi:10.1016/S0079-6123(06)59009-0 PMID:17071228

Schmidhuber, J. (2015). Deep Learning in neural networks: An overview. In Neural Networks (Vol. 61). doi:10.1016/j.neunet.2014.09.003

Schuster, C., Hilfiker, R., Amft, O., Scheidhauer, A., Andrews, B., Butler, J., Kischka, U., & Ettlin, T. (2011). Best practice for motor imagery: A systematic literature review on motor imagery training elements in five different disciplines. *BMC Medicine*, *9*(1), 1–35. doi:10.1186/1741-7015-9-75 PMID:21682867

Schwarz, A., Escolano, C., Montesano, L., & Müller-Putz, G. R. (2020). Analyzing and decoding natural reach-and-grasp actions using gel, water and dry EEG systems. *Frontiers in Neuroscience*, *14*, 849. doi:10.3389/fnins.2020.00849 PMID:32903775

Sedrati, A., Stoyanova, N., Mezrioui, A., Hilali, A., & Benomar, A. (2020). Decentralisation and governance in IoT: Bitcoin and Wikipedia case. *International Journal of Electronic Governance*, *12*(2), 166. doi:10.1504/IJEG.2020.109540

Selvakumar, S., Adithe, S., Isaac, J. S., Pradhan, R., Venkatesh, V., & Sampath, B. (2023). A Study of the Printed Circuit Board (PCB) E-Waste Recycling Process. In Sustainable Approaches and Strategies for E-Waste Management and Utilization (pp. 159–184). IGI Global.

Selvakumar, S., Shankar, R., Ranjit, P., Bhattacharya, S., Gupta, A. S. G., & Boopathi, S. (2023). E-Waste Recovery and Utilization Processes for Mobile Phone Waste. In *Handbook of Research on Safe Disposal Methods of Municipal Solid Wastes for a Sustainable Environment* (pp. 222–240). IGI Global. doi:10.4018/978-1-6684-8117-2.ch016

Sengeni, D., Padmapriya, G., Imambi, S. S., Suganthi, D., Suri, A., & Boopathi, S. (2023). Biomedical Waste Handling Method Using Artificial Intelligence Techniques. In *Handbook of Research on Safe Disposal Methods of Municipal Solid Wastes for a Sustainable Environment* (pp. 306–323). IGI Global. doi:10.4018/978-1-6684-8117-2.ch022

Senthil, T., Puviyarasan, M., Babu, S. R., Surakasi, R., Sampath, B., & Associates. (2023). Industrial Robot-Integrated Fused Deposition Modelling for the 3D Printing Process. In Development, Properties, and Industrial Applications of 3D Printed Polymer Composites (pp. 188–210). IGI Global.

Seon Park, H., Hwan An, J., Jun Park, Y., & Kwan Oh, B. (2020). Convolutional neural network-based safety evaluation method for structures with dynamic responses. *Expert Systems with Applications*, *158*, 113634. doi:10.1016/j.eswa.2020.113634

Seventekidis, P., Giagopoulos, D., Arailopoulos, A., & Markogiannaki, O. (2020). Structural Health Monitoring using deep learning with optimal finite element model generated data. *Mechanical Systems and Signal Processing*, *145*, 106972. doi:10.1016/j.ymssp.2020.106972

Seville, C. (2017). The Emergence and Development of Intellectual Property Law in Western Europe in The Oxford Handbook of Intellectual Property Law, 171-197.

Sha, K., Shi, W., & Watkins, O. (2006). Using wireless sensor networks for fire rescue applications: Requirements and challenges. In *2006 IEEE International Conference on Electro/Information Technology* (pp. 239-244). IEEE.

Shamshad, S., Riaz, F., Riaz, R., Rizvi, S. S., & Abdulla, S. (2022). An Enhanced Architecture to Resolve Public-Key Cryptographic Issues in the Internet of Things (IoT), Employing Quantum Computing Supremacy. *Sensors (Basel)*, *22*(21), 8151. doi:10.339022218151 PMID:36365848

Shang, Z., Sun, L., Xia, Y., & Zhang, W. (2021). Vibration-based damage detection for bridges by deep convolutional denoising autoencoder. *Structural Health Monitoring*, *20*(4), 1880–1903. doi:10.1177/1475921720942836

Shani, G., & Gunawardana, A. (2011). Evaluating Recommendation Systems. In Recommender Systems Handbook. doi:10.1007/978-0-387-85820-3_8

Sharafaldin, I., Lashkari, A. H., & Ghorbani, A. A. (2018). Toward generating a new intrusion detection dataset and intrusion traffic characterization. *ICISSP 2018 - Proceedings of the 4th International Conference on Information Systems Security and Privacy*, 108–116. 10.5220/0006639801080116

Sharma, A., Singh, P. K., & Kumar, Y. (2020). An integrated fire detection system using IoT and image processing technique for smart cities. *Sustainable Cities and Society*, *61*, 102332. doi:10.1016/j.scs.2020.102332

Sharma, S., Das, J., Braje, W. M., Dash, A. K., & Handa, S. (2020). A glimpse into green chemistry practices in the pharmaceutical industry. *ChemSusChem*, *13*(11), 2859–2875. doi:10.1002/cssc.202000317 PMID:32212245

Sheldon, M. R., Fillyaw, M. J., & Thompson, W. D. (1996). The use and interpretation of the Friedman test in the analysis of ordinal-scale data in repeated measures designs. *Physiotherapy Research International*, *1*(4), 221–228. doi:10.1002/pri.66 PMID:9238739

Shen, H., Kankanhalli, M., Srinivasan, S., Yan, W. (2004). Mosaic-based view enlargement for moving objects in motion pictures. *IEEE ICME'04*.

Shen, D., Chen, X., Nguyen, M., & Yan, W. Q. (2018). Flame detection using deep learning. *International Conference on Control, Automation and Robotics (ICCAR).* 10.1109/ICCAR.2018.8384711

Shengdong, M., Zhengxian, X., & Yixiang, T. (2019). Intelligent traffic control system based on cloud computing and big data mining. *IEEE Transactions on Industrial Informatics, 15*(12), 6583–6592. doi:10.1109/TII.2019.2929060

Shen, J., Yan, W., Miller, P., & Zhou, H. (2010) Human localization in a cluttered space using multiple cameras. *IEEE International Conference on Advanced Video and Signal Based Surveillance.* 10.1109/AVSS.2010.60

Shi, R., Li, T., & Yamaguchi, Y. (2020). An attribution-based pruning method for real-time mango detection with YOLO network. *Computers and Electronics in Agriculture, 169*, 105214. doi:10.1016/j.compag.2020.105214

Shouran, M., & Elgamli, E. (2020). Design and implementation of Butterworth filter. *International Journal of Innovative Research in Science, Engineering and Technology, 9*(9).

Shroud, M. A., Eame, M., Elsaghayer, E., Almabrouk, A., & Nassar, Y. (2023). Challenges and opportunities in smart parking sensor technologies. *International Journal of Electrical Engineering and Sustainability*, 44–59.

Shukla, R. G., Agarwal, A., & Shukla, S. (2020). Blockchain-powered smart healthcare system. In *Handbook of research on blockchain technology* (pp. 245–270). Elsevier. doi:10.1016/B978-0-12-819816-2.00010-1

Sinde, R. S., Kaijage, S., & Njau, K. N. (2020). Cluster based wireless sensor network for forests environmental monitoring. *International Journal of Advanced Technology and Engineering Exploration.*, 7(63), 36–47. doi:10.19101/IJATEE.2019.650083

Singh, B. K., Kumar, N., & Tiwari, P. (2019). *Extreme learning machine approach for prediction of forest fires using topographical and metrological data of Vietnam. In Women institute of technology conference on electrical and computer engineering (WITCON ECE).* IEEE.

Singh, D. R. (2008). *Law relating to intellectual property: A complete comprehensive material on intellectual property covering acts, rules, conventions, treaties, agreements, digest of cases and much more.* Universal Law Publishing Company.

Singh, S. K., Pan, Y., & Park, J. H. (2022). Blockchain-enabled secure framework for energy-efficient smart parking in sustainable city environment. *Sustainable Cities and Society, 76*, 103364. doi:10.1016/j.scs.2021.103364

Singh, S., Ahuja, U., Kumar, M., Kumar, K., & Sachdeva, M. (2021). Face mask detection using yolov3 and faster R-CNN models: COVID-19 environment. *Multimedia Tools and Applications, 80*(13), 19753–19768. doi:10.100711042-021-10711-8 PMID:33679209

Siraj, A., & Shukla, V. K. (2020). Framework for personalized car parking system using proximity sensor. *2020 8th International Conference on Reliability, Infocom Technologies and Optimization (Trends and Future Directions) (ICRITO)*, 198–202.

Siriwardhana, Y., Gür, G., Ylianttila, M., & Liyanage, M. (2021). The role of 5G for digital healthcare against COVID-19 pandemic: Opportunities and challenges. *Ict Express, 7*(2), 244–252. doi:10.1016/j.icte.2020.10.002

Slavina, Z. (2023). *AI ethics: Chosen challenges for contemporary societies and technological policymaking* [B.A. Dissertation]. Uniwersytet w Białymstoku.

Smarsly, K., & Law, K. H. (2014). Decentralized fault detection and isolation in wireless structural health monitoring systems using analytical redundancy. *Advances in Engineering Software, 73*, 1–10. doi:10.1016/j.advengsoft.2014.02.005

Somefun, O., Akingbade, K., & Dahunsi, F. (2022). Uniformly damped binomial filters: Five-percent maximum overshoot optimal response design. *Circuits, Systems, and Signal Processing, 41*(6), 3282–3305. doi:10.100700034-021-01931-2

Son, M., & Lee, K. (2018). Distributed Matrix Multiplication Performance Estimator for Machine Learning Jobs in Cloud Computing. *2018 IEEE 11th International Conference on Cloud Computing (CLOUD),* 638–645. 10.1109/CLOUD.2018.00088

Song, C., He, L., Yan, W., & Nand, P. (2019) An improved selective facial extraction model for age estimation. *International Conference on Image and Vision Computing New Zealand.* 10.1109/IVCNZ48456.2019.8960965

Sotres, P., Lanza, J., Sánchez, L., Santana, J. R., López, C., & Muñoz, L. (2019). Breaking vendors and city locks through a semantic-enabled global interoperable internet-of-things system: A smart parking case. *Sensors (Basel), 19*(2), 229. doi:10.339019020229 PMID:30634490

Sozzi, M., Cantalamessa, S., Cogato, A., Kayad, A., & Marinello, F. (2022). Automatic bunch detection in white grape varieties using YOLOv3, YOLOv4, and YOLOv5 deep learning algorithms. *Agronomy (Basel), 12*(2), 319. doi:10.3390/agronomy12020319

Sperlí, G. (2021). A cultural heritage framework using a Deep Learning based Chatbot for supporting tourist journey. *Expert Systems with Applications, 183*, 115277. Advance online publication. doi:10.1016/j.eswa.2021.115277

Spiridonov, V. Y., & Shabiev, S. (2020). Smart urban planning: Modern technologies for ensuring sustainable territorial development. *IOP Conference Series. Materials Science and Engineering, 962*(3), 032034. doi:10.1088/1757-899X/962/3/032034

Srinivas, B., Maguluri, L. P., Naidu, K. V., Reddy, L. C. S., Deivakani, M., & Boopathi, S. (2023). Architecture and Framework for Interfacing Cloud-Enabled Robots. In *Handbook of Research on Data Science and Cybersecurity Innovations in Industry 4.0 Technologies* (pp. 542–560). IGI Global. doi:10.4018/978-1-6684-8145-5.ch027

Srivastava, A., Agarwal, A., & Kaur, G. (2019). Novel Machine Learning Technique for Intrusion Detection in Recent Network-based Attacks. *2019 4th International Conference on Information Systems and Computer Networks, ISCON 2019*, 524–528. 10.1109/ISCON47742.2019.9036172

Stamatoudi, I., & Torremans, P. (Eds.). (2021). *EU copyright law: A commentary*. Edward Elgar Publishing. doi:10.4337/9781786437808

Stevens, R., Taylor, V., Nichols, J., Maccabe, A. B., Yelick, K., & Brown, D. (2020). Ai for science: Report on the department of energy (doe) town halls on artificial intelligence (ai) for science. Argonne National Lab. (ANL), Argonne, IL (United States).

Steyrl, D., Scherer, R., Förstner, O., & Müller-Putz, G. R. (2014). Motor imagery brain-computer interfaces: random forests vs regularized LDA-non-linear beats linear. In *Proceedings of the 6th international brain-computer interface conference* (pp. 241-244). Academic Press.

Stiglic, M., Agatz, N., Savelsbergh, M., & Gradisar, M. (2018). Enhancing urban mobility: Integrating ride-sharing and public transit. *Computers & Operations Research, 90*, 12–21. doi:10.1016/j.cor.2017.08.016

Stonebraker, M. (2010). SQL databases v. NoSQL databases. *Communications of the ACM, 53*(4), 10–11. Advance online publication. doi:10.1145/1721654.1721659

Subha, S., Inbamalar, T., Komala, C., Suresh, L. R., Boopathi, S., & Alaskar, K. (2023). A Remote Health Care Monitoring system using internet of medical things (IoMT). *IEEE Explore*, 1–6.

Subha, S., Inbamalar, T., Komala, C., Suresh, L. R., Boopathi, S., & Alaskar, K. (2023a). A Remote Health Care Monitoring system using internet of medical things (IoMT). *IEEE Explore*, (pp. 1–6). IEEE.

Subramanian, C. (2021). An appraisal on intelligent and smart systems. *AIP Conference Proceedings*, *2316*(1), 020003. doi:10.1063/5.0037534

Suh, G., & Cha, Y.-J. (2018). Deep faster R-CNN-based automated detection and localization of multiple types of damage. In H. Sohn (Ed.), *Sensors and Smart Structures Technologies for Civil, Mechanical, and Aerospace Systems 2018* (p. 27). SPIE., doi:10.1117/12.2295954

Sulistianingsih, D., & Ilyasa, R. M. A. (2022). The Impact of Trips Agreement On The Development Of Intellectual Property Laws In Indonesia. *Indonesia Private Law Review*, *3*(2), 85–98. doi:10.25041/iplr.v3i2.2579

Sun, Y., Kamel, M. S., Wong, A. K. C., & Wang, Y. (2007). *Cost-sensitive boosting for classification of imbalanced data.* doi:10.1016/j.patcog.2007.04.009

Su, X., Sperli, G., Moscato, V., Picariello, A., Esposito, C., & Choi, C. (2019). An Edge Intelligence Empowered Recommender System Enabling Cultural Heritage Applications. *IEEE Transactions on Industrial Informatics*, *15*(7), 4266–4275. Advance online publication. doi:10.1109/TII.2019.2908056

Svendsen, B. T., Frøseth, G. T., Øiseth, O., & Rønnquist, A. (2022). A data-based structural health monitoring approach for damage detection in steel bridges using experimental data. *Journal of Civil Structural Health Monitoring*, *12*(1), 101–115. doi:10.100713349-021-00530-8

Syamala, M., Komala, C., Pramila, P., Dash, S., Meenakshi, S., & Boopathi, S. (2023). Machine Learning-Integrated IoT-Based Smart Home Energy Management System. In *Handbook of Research on Deep Learning Techniques for Cloud-Based Industrial IoT* (pp. 219–235). IGI Global. doi:10.4018/978-1-6684-8098-4.ch013

Syed, A. S., Sierra-Sosa, D., Kumar, A., & Elmaghraby, A. (2021). IoT in Smart Cities: A Survey of Technologies, Practices and Challenges. *Smart Cities*, *4*(2), 429–475. doi:10.3390martcities4020024

Tama, B. A., Comuzzi, M., & Rhee, K. H. (2019). TSE-IDS: A Two-Stage Classifier Ensemble for Intelligent Anomaly-Based Intrusion Detection System. *IEEE Access : Practical Innovations, Open Solutions*, *7*, 94497–94507. doi:10.1109/ACCESS.2019.2928048

Tang, Z., Chen, Z., Bao, Y., & Li, H. (2019). Convolutional neural network-based data anomaly detection method using multiple information for structural health monitoring. *Structural Control and Health Monitoring*, *26*(1), e2296. doi:10.1002tc.2296

Tan, X., Sun, X., Chen, W., Du, B., Ye, J., & Sun, L. (2021). Investigation on the data augmentation using machine learning algorithms in structural health monitoring information. *Structural Health Monitoring*, *20*(4), 2054–2068. doi:10.1177/1475921721996238

Taricani, E., Saris, N., & Park, P. A. (2020). Beyond technology: The ethics of artificial intelligence. *World Complexity Science Academy Journal*, *1*(2), 17. doi:10.46473/WCSAJ27240606/15-05-2020-0017//full/html

Taunk, K., De, S., Verma, S., & Swetapadma, A. (2019, May). A brief review of nearest neighbor algorithm for learning and classification. In 2019 international conference on intelligent computing and control systems (ICCS) (pp. 1255-1260). IEEE. doi:10.1109/ICCS45141.2019.9065747

Tebeje, T. H., & Klein, J. (2021). Applications of e-health to support person-centered health care at the time of COVID-19 pandemic. *Telemedicine Journal and e-Health*, *27*(2), 150–158. doi:10.1089/tmj.2020.0201 PMID:32746750

Teing, Y.-Y., Dehghantanha, A., Choo, K.-K. R., & Yang, L. T. (2017). Forensic investigation of P2P cloud storage services and backbone for IoT networks: BitTorrent Sync as a case study. *Computers & Electrical Engineering*, *58*, 350–363. doi:10.1016/j.compeleceng.2016.08.020

Teran, K., Žibret, G., & Fanetti, M. (2020). Impact of urbanization and steel mill emissions on elemental composition of street dust and corresponding particle characterization. *Journal of Hazardous Materials, 384*, 120963. doi:10.1016/j.jhazmat.2019.120963 PMID:31628063

Thaseen, S., & Kumar, A. (2017). Intrusion detection model using fusion of chi-square feature selection and multi class SVM. *Journal of King Saud University. Computer and Information Sciences*, *29*(4), 462–472. doi:10.1016/j.jksuci.2015.12.004

Thibaud, M., Chi, H., Zhou, W., & Piramuthu, S. (2018). Internet of Things (IoT) in high-risk Environment, Health and Safety (EHS) industries: A comprehensive review. *Decision Support Systems, 108*, 79-95.

Thomas, R. N., & Gupta, R. (2020). A Survey on Machine Learning Approaches and Its Techniques. *2020 IEEE International Students' Conference on Electrical, Electronics and Computer Science, SCEECS 2020*. 10.1109/SCEECS48394.2020.190

Tian, G., Wang, Z., Wang, C., Chen, J., Liu, G., Xu, H., Lu, Y., Han, Z., Zhao, Y., Li, Z., Luo, X., & Peng, L. (2022). A deep ensemble learning-based automated detection of COVID-19 using lung CT images and Vision Transformer and ConvNeXt. *Frontiers in Microbiology*, *13*, 13. doi:10.3389/fmicb.2022.1024104 PMID:36406463

Tibaduiza Burgos, D. A., Gomez Vargas, R. C., Pedraza, C., Agis, D., & Pozo, F. (2020). Damage Identification in Structural Health Monitoring: A Brief Review from its Implementation to the Use of Data-Driven Applications. *Sensors (Basel)*, *20*(3), 733. doi:10.339020030733 PMID:32013073

Tomasiello, S., Loia, V., & Khaliq, A. (2021). A granular recurrent neural network for multiple time series prediction. *Neural Computing & Applications*, *33*(16), 10293–10310. doi:10.100700521-021-05791-4

Tomasiello, S., Pedrycz, W., & Loia, V. (2022). On Fractional Tikhonov Regularization: Application to the Adaptive Network-Based Fuzzy Inference System for Regression Problems. *IEEE Transactions on Fuzzy Systems*, *30*(11), 4717–4727. doi:10.1109/TFUZZ.2022.3157947

Tran, N. C., Wang, J., Vu, T. H., Tai, T.-C., & Wang, J.-C. (2023). Anti-aliasing convolution neural network of finger vein recognition for virtual reality (VR) human–robot equipment of metaverse. *The Journal of Supercomputing*, *79*(3), 2767–2782. doi:10.100711227-022-04680-4 PMID:36035635

Tsai, C. W., Lai, C. F., Chao, H. C., & Vasilakos, A. V. (2015). Big data analytics: A survey. *Journal of Big Data*, *2*(1), 21. Advance online publication. doi:10.118640537-015-0030-3 PMID:26191487

Tsepapadakis, M., & Gavalas, D. (2023). Are you talking to me? An Audio Augmented Reality conversational guide for cultural heritage. *Pervasive and Mobile Computing, 92*, 101797. doi:10.1016/j.future.2019.04.020

Tsikoudis, N., Papadogiannakis, A., & Markatos, E. P. (2016). LEoNIDS: A Low-Latency and Energy-Efficient Network-Level Intrusion Detection System. *IEEE Transactions on Emerging Topics in Computing*, *4*(1), 142–155. doi:10.1109/TETC.2014.2369958

Tupasela, A., & Di Nucci, E. (2020). Concordance as evidence in the Watson for Oncology decision-support system. *AI & Society*, *35*(4), 811–818. doi:10.100700146-020-00945-9

Turki, M., Dammak, B., & Mars, R. (2022). A Private Smart parking solution based on Blockchain and AI. *2022 15th International Conference on Security of Information and Networks (SIN)*, 1–7.

Turo, T., Neumann, V., & Krobot, Z. (2019). Health and Usage Monitoring System Assestment. *2019 International Conference on Military Technologies (ICMT)*, 1–4. 10.1109/MILTECHS.2019.8870072

Tuysuzoglu, G., Moarref, N., & Yaslan, Y. (2016). *Ensemble-Based Classifiers Using Dictionary Learning*. Academic Press.

U.S. Copyright Office. (2023). *Copyright Registration Guidance: Works Containing Material Generated by Artificial Intelligence*. Library of Congress. Retrieved from https://www.govinfo.gov/content/pkg/FR-2023-03-16/pdf/2023-05321.pdf

Übeyli, E. D. (2009). Analysis of EEG signals by implementing eigenvector methods/recurrent neural networks. *Digital Signal Processing*, *19*(1), 134–143. doi:10.1016/j.dsp.2008.07.007

Vandenberk, B., Chew, D., Prasana, D., Gupta, S., & Exner, D. V. (2023). Successes and Challenges of Artificial Intelligence in Cardiology. *Frontiers in Digital Health*, *5*, 1201392. doi:10.3389/fdgth.2023.1201392 PMID:37448836

Vanitha, S., Radhika, K., & Boopathi, S. (2023). Artificial Intelligence Techniques in Water Purification and Utilization. In *Human Agro-Energy Optimization for Business and Industry* (pp. 202–218). IGI Global. doi:10.4018/978-1-6684-4118-3.ch010

Vasconez, J. P., Delpiano, J., Vougioukas, S., & Cheein, F. A. (2020). Comparison of convolutional neural networks in fruit detection and counting: A comprehensive evaluation. *Computers and Electronics in Agriculture*, *173*, 105348. doi:10.1016/j.compag.2020.105348

Veeranjaneyulu, R., Boopathi, S., Kumari, R. K., Vidyarthi, A., Isaac, J. S., & Jaiganesh, V. (2023). Air Quality Improvement and Optimisation Using Machine Learning Technique. *IEEE- Explore*, 1–6.

Veeranjaneyulu, R., Boopathi, S., Narasimharao, J., Gupta, K. K., Reddy, R. V. K., & Ambika, R. (2023). Identification of Heart Diseases using Novel Machine Learning Method. *IEEE- Explore*, 1–6.

Venkateswaran, N., Vidhya, K., Ayyannan, M., Chavan, S. M., Sekar, K., & Boopathi, S. (2023). A Study on Smart Energy Management Framework Using Cloud Computing. In 5G, Artificial Intelligence, and Next Generation Internet of Things: Digital Innovation for Green and Sustainable Economies (pp. 189–212). IGI Global. doi:10.4018/978-1-6684-8634-4.ch009

Venkateswaran, N., Vidhya, R., Naik, D. A., Raj, T. M., Munjal, N., & Boopathi, S. (2023). Study on Sentence and Question Formation Using Deep Learning Techniques. In *Digital Natives as a Disruptive Force in Asian Businesses and Societies* (pp. 252–273). IGI Global. doi:10.4018/978-1-6684-6782-4.ch015

Venkateswarlu, I. B., Kakarla, J., & Prakash, S. (2020). Face mask detection using MobileNet and global pooling block. *IEEE Conference on Information Communication Technology (CICT)*. 10.1109/CICT51604.2020.9312083

Vennila, T., Karuna, M., Srivastava, B. K., Venugopal, J., Surakasi, R., & Sampath, B. (2022). New Strategies in Treatment and Enzymatic Processes: Ethanol Production From Sugarcane Bagasse. In Human Agro-Energy Optimization for Business and Industry (pp. 219–240). IGI Global.

Verma, R., Dhanda, N., & Nagar, V. (2023). *Towards a Secured IoT Communication: A Blockchain Implementation Through APIs*. doi:10.1007/978-981-19-1142-2_53

Vert, J.-P. (2023). How will generative AI disrupt data science in drug discovery? *Nature Biotechnology*, *41*(6), 1–2. doi:10.103841587-023-01789-6 PMID:37156917

Vignesh, S., Arulshri, K., SyedSajith, S., Kathiresan, S., Boopathi, S., & Dinesh Babu, P. (2018). Design and development of ornithopter and experimental analysis of flapping rate under various operating conditions. *Materials Today: Proceedings*, *5*(11), 25185–25194. doi:10.1016/j.matpr.2018.10.320

Vikram, R., Sinha, D., De, D., & Das, A. K. (2021). PAFF: Predictive analytics on forest fire using compressed sensing-based localized Ad Hoc wireless sensor networks. *Journal of Ambient Intelligence and Humanized Computing*, *12*(2), 1647–1665. doi:10.100712652-020-02238-x

Vincent, J. (2018). AI that detects cardiac arrests during emergency calls will be tested across Europe this summer. *The Verge*, 25.

Voigt, P., & Von dem Bussche, A. (2017). *The EU General Data Protection Regulation (GDPR). A Practical Guide.* Springer International Publishing. doi:10.1007/978-3-319-57959-7

Volponi, A., Brotherton, T., & Luppold, R. (2004). Development of an information fusion system for engine diagnostics and health management. *AIAA 1st Intelligent Systems Technical Conference*, 6461.

Vuong, T. P., Loukas, G., & Gan, D. (2015). *Performance evaluation of cyber-physical intrusion detection on a robotic vehicle.* doi:10.1109/CIT/IUCC/DASC/PICOM.2015.313

Waheed, A., & Krishna, P. V. (2020). Comparing biometric and blockchain security mechanisms in smart parking system. *2020 International Conference on Inventive Computation Technologies (ICICT)*, 634–638. 10.1109/ICICT48043.2020.9112483

Wang, L., & Yan, W. (2021) Tree leaves detection based on deep learning. *International Symposium on Geometry and Vision.* Springer. 10.1007/978-3-030-72073-5_3

Wang, B., & Pineau, J. (2016). *Online Bagging and Boosting for Imbalanced Data Streams*. Academic Press.

Wang, C.-Y., Bochkovskiy, A., & Liao, H.-Y. M. (2023). YOLOv7: Trainable Bag-of-Freebies sets new state-of-the-art for real-time object detectors. *IEEE Conference on Computer Vision and Pattern Recognition (CVPR)*, 7464–7475. 10.1109/CVPR52729.2023.00721

Wang, D., & He, D. (2021). Channel pruned YOLO V5s-based deep learning approach for rapid and accurate apple fruitlet detection before fruit thinning. *Biosystems Engineering*, *210*, 271–281. doi:10.1016/j.biosystemseng.2021.08.015

Wang, D., Zhang, Y., Pan, Y., Peng, B., Liu, H., & Ma, R. (2020). An Automated Inspection Method for the Steel Box Girder Bottom of Long-Span Bridges Based on Deep Learning. *IEEE Access : Practical Innovations, Open Solutions*, *8*, 94010–94023. doi:10.1109/ACCESS.2020.2994275

Wang, H., Barone, G., & Smith, A. (2023). A novel multi-level data fusion and anomaly detection approach for infrastructure damage identification and localisation. *Engineering Structures*, *292*, 116473. doi:10.1016/j.engstruct.2023.116473

Wang, H., & Yan, W. Q. (2022). Face detection and recognition from distance based on deep learning. *Advances in Digital Crime, Forensics, and Cyber Terrorism*, 144–160. doi:10.4018/978-1-6684-4558-7.ch006

Wang, J., Kankanhalli, M., Yan, W., & Jain, R. (2003) Experiential sampling for video surveillance. *ACM SIGMM International Workshop on Video surveillance*, 77-86. 10.1145/982452.982462

Wang, J., Yan, W., Kankanhalli, M., Jain, R., & Reinders, M. (2003) Adaptive monitoring for video surveillance. *International Conference on Information, Communications and Signal Processing*.

Wang, L., Zhao, Y., Xiong, Z., Wang, S., Li, Y., & Lan, Y. (2022). Fast and precise detection of litchi fruits for yield estimation based on the improved YOLOv5 model. *Frontiers in Plant Science*, *13*, 13. doi:10.3389/fpls.2022.965425 PMID:36017261

Wang, M., He, M., Liang, Z., Wu, D., Wang, Y., Qing, X., & Wang, Y. (2023). Fatigue damage monitoring of composite laminates based on acoustic emission and digital image correlation techniques. *Composite Structures*, *321*, 117239. doi:10.1016/j.compstruct.2023.117239

Wang, R., Li, L., & Li, J. (2018). A Novel Parallel Auto-Encoder Framework for Multi-Scale Data in Civil Structural Health Monitoring. *Algorithms*, *11*(8), 112. doi:10.3390/a11080112

Wang, W., & Siau, K. (2019). Artificial intelligence, machine learning, automation, robotics, future of work and future of humanity: A review and research agenda. *Journal of Database Management*, *30*(1), 61–79. doi:10.4018/JDM.2019010104

Wang, X., Cai, J., & Zhou, Z. (2019). A Lamb wave signal reconstruction method for high-resolution damage imaging. *Chinese Journal of Aeronautics*, *32*(5), 1087–1099. doi:10.1016/j.cja.2019.03.001

Wang, X., Hu, H.-M., & Zhang, Y. (2019). Pedestrian detection based on spatial attention module for outdoor video surveillance. *IEEE International Conference on Multimedia Big Data (BigMM).* 10.1109/BigMM.2019.00-17

Wang, Z., Bai, L., Zhang, Y., Zhao, K., Wu, J., & Fu, W. (2022). Spatial variation, sources identification and risk assessment of soil heavy metals in a typical Torreya grandis cv. Merrillii plantation region of southeastern China. *The Science of the Total Environment*, *849*, 157832. doi:10.1016/j.scitotenv.2022.157832 PMID:35932857

Wang, Z., & Cha, Y.-J. (2021). Unsupervised deep learning approach using a deep auto-encoder with a one-class support vector machine to detect damage. *Structural Health Monitoring*, *20*(1), 406–425. doi:10.1177/1475921720934051

Wang, Z., Jin, L., Wang, S., & Xu, H. (2022). Apple stem/calyx real-time recognition using YOLO-v5 algorithm for fruit automatic loading system. *Postharvest Biology and Technology*, *185*, 111808. doi:10.1016/j.postharvbio.2021.111808

Wang, Z., & Srinivasan, R. S. (2017). A review of artificial intelligence based building energy use prediction: Contrasting the capabilities of single and ensemble prediction models. *Renewable & Sustainable Energy Reviews*, *75*(October), 796–808. doi:10.1016/j.rser.2016.10.079

Wang, Z., Yi, T.-H., Yang, D.-H., Li, H.-N., & Liu, H. (2022). Bridge Performance Warning Based on Two-Stage Elimination of Environment-Induced Frequency. *Journal of Performance of Constructed Facilities*, *36*(6), 04022056. Advance online publication. doi:10.1061/(ASCE)CF.1943-5509.0001760

Watal, J. (2002). Intellectual property rights in the WTO and developing countries. *Intellectual Property Rights in the WTO and Developing Countries.*

Watson, W. J. (2012). How do the fine chemical, pharmaceutical, and related industries approach green chemistry and sustainability? *Green Chemistry*, *14*(2), 251–259. doi:10.1039/C1GC15904F

Wei, X., Chen, K., Liu, Y., Zhao, X., Ma, L., Ai, Q., & Liu, Q. (2021, November). Control of multiple DOFs robots using motor imagery EEG combined with Huffman coding. In *2021 27th International Conference on Mechatronics and Machine Vision in Practice (M2VIP)* (pp. 344-348). IEEE. 10.1109/M2VIP49856.2021.9665060

What is Intellectual Property ? (n.d.). WIPO - World Intellectual Property Organization. Retrieved June 30, 2023, from https://www.wipo.int/about-ip/en/

White, M. (2002). World Intellectual Property Organization. *Journal of Business & Finance Librarianship*, *8*(1), 71–78. doi:10.1300/J109v08n01_08

Whitley, T., Tomiczek, A., Tripp, C., Ortega, A., Mennu, M., Bridge, J., & Ifju, P. (2020). Design of a small unmanned aircraft system for bridge inspections. *Sensors (Basel)*, *20*(18), 5358. doi:10.339020185358 PMID:32962108

Wickramasinghe, I., & Kalutarage, H. (2021). Naive Bayes: applications, variations and vulnerabilities: a review of literature with code snippets for implementation. *Soft Computing*, *25*(3), 2277–2293. doi:10.100700500-020-05297-6

Wiemann, M., Bonekemper, L., & Kraemer, P. (2020). Methods to enhance the automation of operational modal analysis. *Vibroengineering PROCEDIA*, *31*, 46–51. doi:10.21595/vp.2020.21443

Witten, I. H., Frank, E., Hall, M. A., & Pal, C. J. (2016). Data Mining: Practical Machine Learning Tools and Techniques. doi:10.1016/C2009-0-19715-5

Witten, I. H., Frank, E., Hall, M. A., & Pal, C. J. (2016). *Data Mining: Practical Machine Learning Tools and Techniques*. Morgan Kaufmann.

Woo, S., Park, J., Lee, J.-Y., & Kweon, I. S. (2018). *CBAM: Convolutional block attention module*. ECCV. doi:10.1007/978-3-030-01234-2_1

Workman, P., Antolin, A. A., & Al-Lazikani, B. (2019). Transforming cancer drug discovery with Big Data and AI. *Expert Opinion on Drug Discovery*, *14*(11), 1089–1095. doi:10.1080/17460441.2019.1637414 PMID:31284790

World Health Organization. (2019). Brief model disability survey: result for India, Lao People's Democratic Republic and Tajikistan: executive summary, 2019 (No. WHO/NMH/NVI/19.15). World Health Organization.

Wu, D., Jiang, S., Zhao, E., Liu, Y., Zhu, H., Wang, W., & Wang, R. (2022). Detection of Camellia oleifera fruit in complex scenes by using YOLOv7 and data augmentation. *Applied Sciences (Basel, Switzerland)*, *12*(22), 11318. doi:10.3390/app122211318

Wuest, T., Weimer, D., Irgens, C., & Thoben, K.-D. (2016). Machine learning in manufacturing: Advantages, challenges, and applications. *Production & Manufacturing Research*, *4*(1), 23–45. doi:10.1080/21693277.2016.1192517

Wu, P., Li, H., Zeng, N., & Li, F. (2022). FMD-YOLO: An efficient face mask detection method for COVID-19 prevention and control in public. *Image and Vision Computing*, *117*, 104341. doi:10.1016/j.imavis.2021.104341 PMID:34848910

Xia, Y., Nguyen, M., & Yan, W. (2023) Kiwifruit counting using KiwiDetector and KiwiTracker. *IntelliSys conference*. Auckland University of Technology.

Xiao, B., Nguyen, M., & Yan, W. (2021) Apple ripeness identification using deep learning. *International Symposium on Geometry and Vision*. Springer. 10.1007/978-3-030-72073-5_5

Xiao, B., Nguyen, M., & Yan, W. (2023). *Apple ripeness identification from digital images using transformers. Multimedia Tools and Applications*. Springer.

Xiao, B., Nguyen, M., & Yan, W. (2023). *Fruit ripeness identification using transformers. Applied Intelligence*. Springer Science and Business Media LLC.

Xiao, W., Liu, C., Wang, H., Zhou, M., Hossain, M. S., Alrashoud, M., & Muhammad, G. (2020). Blockchain for secure-GaS: Blockchain-powered secure natural gas IoT system with AI-enabled gas prediction and transaction in smart city. *IEEE Internet of Things Journal*, *8*(8), 6305–6312. doi:10.1109/JIOT.2020.3028773

Xia, W., Wen, Y., Foh, C. H., Niyato, D., & Xie, H. (2015). A Survey on Software-Defined Networking. *IEEE Communications Surveys and Tutorials*, *17*(1), 27–51. doi:10.1109/COMST.2014.2330903

Xia, Y., Nguyen, M., & Yan, W. (2022) A real-time Kiwifruit detection based on improved YOLOv7. *International Conference on Image and Vision Computing New Zealand (IVCNZ)*. IEEE.

Xia, Y., Nguyen, M., & Yan, W. (2022). A real-time Kiwifruit detection based on improved YOLOv7. *International Conference on Image and Vision Computing New Zealand (IVCNZ)*. IEEE.

Xia, Y., Nguyen, M., & Yan, W. (2023). *Kiwifruit counting using KiwiDetector and KiwiTracker*. IntelliSys.

Xie, X. (2019). Principal component analysis. *Wiley Interdisciplinary Reviews*.

Xie, W., Li, T., Tiraferri, A., Drioli, E., Figoli, A., Crittenden, J. C., & Liu, B. (2020). Toward the next generation of sustainable membranes from green chemistry principles. *ACS Sustainable Chemistry & Engineering*, *9*(1), 50–75. doi:10.1021/acssuschemeng.0c07119

Xiong, H., Yao, T., Wang, H., Feng, J., & Yu, S. (2022). A Survey of Public-Key Encryption With Search Functionality for Cloud-Assisted IoT. *IEEE Internet of Things Journal*, *9*(1), 401–418. doi:10.1109/JIOT.2021.3109440

Xiong, Q., Yuan, C., He, B., Xiong, H., & Kong, Q. (2023). GTRF: A general deep learning framework for tuples recognition towards supervised, semi-supervised and unsupervised paradigms. *Engineering Applications of Artificial Intelligence*, *124*, 106500. doi:10.1016/j.engappai.2023.106500

Yadav, R., & Rani, P. (2020). Sensor based smart fire detection and fire alarm system. In *Proceedings of the International Conference on Advances in Chemical Engineering (AdChE)*. 10.2139srn.3724291

Yan, W., & Kankanhalli, M. (2015) Face search in encrypted domain. Pacific-Rim Symposium on Image and Video Technology, 775-790.

Yan, B., Fan, P., Lei, X., Liu, Z., & Yang, F. (2021). A real-time apple targets detection method for picking robot based on improved YOLOv5. *Remote Sensing (Basel)*, *13*(9), 1619. doi:10.3390/rs13091619

Yang, H., Feng, D., & Baumgartner, R. (2022). AI and Machine Learning in Drug Discovery. In Data Science, AI, and Machine Learning in Drug Development (pp. 63–93). Chapman and Hall/CRC. doi:10.1201/9781003150886-4

Yang, G., Feng, W., Jin, J., Lei, Q., Li, X., Gui, G., & Wang, W. (2020). Face mask recognition system with YOLOv5 based on image recognition. *IEEE International Conference on Computer and Communications (ICCC)*. 10.1109/ICCC51575.2020.9345042

Yang, J., Dai, J., Gooi, H. B., Nguyen, H. D., & Paudel, A. (2022). A Proof-of-Authority Blockchain-Based Distributed Control System for Islanded Microgrids. *IEEE Transactions on Industrial Informatics*, *18*(11), 8287–8297. doi:10.1109/TII.2022.3142755

Yang, K., Kim, S., & Harley, J. B. (2023). Guidelines for effective unsupervised guided wave compression and denoising in long-term guided wave structural health monitoring. *Structural Health Monitoring*, *22*(4), 2516–2530. doi:10.1177/14759217221124689

Yang, R., Hu, Y., Yao, Y., Gao, M., & Liu, R. (2022). Fruit target detection based on BCo-YOLOv5 model. *Mobile Information Systems*, *2022*, 2022. doi:10.1155/2022/8457173

Yang, X., Xiong, S., Li, H., He, X., Ai, H., & Liu, Q. (2019). *Research on forest fire helicopter demand forecast based on index fuzzy segmentation and TOPSIS. In 9th international conference on fire science and fire protection engineering (ICFSFPE).* IEEE.

Yang, Z., Yang, H., Tian, T., Deng, D., Hu, M., Ma, J., Gao, D., Zhang, J., Ma, S., Yang, L., Xu, H., & Wu, Z. (2023). A review on guided-ultrasonic-wave-based structural health monitoring: From fundamental theory to machine learning techniques. *Ultrasonics*, *133*, 107014. doi:10.1016/j.ultras.2023.107014 PMID:37178485

Yan, W. Q. (2019). *Introduction to Intelligent Surveillance: Surveillance Data Capture, Transmission, and Analytics*. Springer. doi:10.1007/978-3-030-10713-0

Yan, W. Q. (2021). *Computational Methods for Deep Learning: Theoretic, Practice and Applications*. Springer Nature. doi:10.1007/978-3-030-61081-4

Yan, W., Kankanhalli, M., Wang, J., & Reinders, M. (2003) Experiential sampling for monitoring. *ACM SIGMM Workshop on Experiential Telepresence*, 70-72. 10.1145/982484.982497

Yao, J., Qi, J., Zhang, J., Shao, H., Yang, J., & Li, X. (2021). A real-time detection algorithm for Kiwifruit defects based on YOLOv5. *Electronics (Basel)*, *10*(14), 1711. doi:10.3390/electronics10141711

Yeter, B., Garbatov, Y., & Soares, C. G. (2022). Review on Artificial Intelligence-aided Life Extension Assessment of Offshore Wind Support Structures. *Journal of Marine Science and Application*, *21*(4), 26–54. doi:10.100711804-022-00298-3

Yger, F., Lotte, F., & Sugiyama, M. (2015) Averaging covariance matrices for EEG signal classification based on the CSP: an empirical study. In *2015 23rd European Signal Processing Conference (EUSIPCO)* (pp. 2721-2725). IEEE. 10.1109/EUSIPCO.2015.7362879

Yin, M., Chen, Z., & Zhang, C. (2023). A CNN-transformer network combining CBAM for change detection in high-resolution remote sensing images. *Remote Sensing (Basel)*, *15*(9), 2406. doi:10.3390/rs15092406

Yogheshwaran, M., Praveenkumar, D., Pravin, S., Manikandan, P., & Saravanan, D. S. (2020). IoT based intelligent traffic control system. *International Journal of Engineering Technology Research & Management*, *4*(4), 59–63.

Yongze, S., Wang, J., & Lu, Z. (2019). Asynchronous Parallel Surrogate Optimization Algorithm based on Ensemble Surrogating Model and Stochastic Response Surface Method. *2019 IEEE 5th Intl Conference on Big Data Security on Cloud (BigDataSecurity), IEEE Intl Conference on High Performance and Smart Computing, (HPSC), and IEEE Intl Conference on Intelligent Data and Security (IDS).* 10.1109/BigDataSecurity-HPSC-IDS.2019.00024

Yu, X., Yang, Y., Wang, W., & Zhang, Y. (2021). Whether the sensitive information statement of the IoT privacy policy is consistent with the actual behavior. *2021 51st Annual IEEE/IFIP International Conference on Dependable Systems and Networks Workshops (DSN-W)*, 85–92. 10.1109/DSN-W52860.2021.00025

Yuan, F.-G., Zargar, S. A., Chen, Q., & Wang, S. (2020). Machine learning for structural health monitoring: challenges and opportunities. In D. Zonta, H. Sohn, & H. Huang (Eds.), *Sensors and Smart Structures Technologies for Civil, Mechanical, and Aerospace Systems 2020* (p. 2). SPIE. doi:10.1117/12.2561610

Yudhana, A., Muslim, A., Wati, D. E., Puspitasari, I., Azhari, A., & Mardhia, M. M. (2020). Human emotion recognition based on EEG signal using fast fourier transform and K-Nearest neighbor. *Adv. Sci. Technol. Eng. Syst. J*, *5*(6), 1082–1088. doi:10.25046/aj0506131

Yudkowsky, E., & Bostrom, N. (2011). *The ethics of artificial intelligence. In The Cambridge Handbook of Artificial Intelligence*. Cambridge University Press.

Yu, J., & Zhang, W. (2021). Face mask wearing detection algorithm based on improved YOLO-V4. *Sensors (Basel)*, *21*(9), 3263. doi:10.339021093263 PMID:34066802

Yu, R., Charreyron, S. L., Boehler, Q., Weibel, C., Chautems, C., Poon, C. C., & Nelson, B. J. (2020, May). Modeling electromagnetic navigation systems for medical applications using random forests and artificial neural networks. In *2020 IEEE International Conference on Robotics and Automation (ICRA)* (pp. 9251-9256). IEEE. 10.1109/ICRA40945.2020.9197212

Yu, Y., Guo, L., Huang, J., Zhang, F., & Zong, Y. (2018). A Cross-Layer Security Monitoring Selection Algorithm Based on Traffic Prediction. *IEEE Access : Practical Innovations, Open Solutions*, *6*, 35382–35391. doi:10.1109/ACCESS.2018.2851993

Yu, Y., Zhang, K., Yang, L., & Zhang, D. (2019). Fruit detection for strawberry harvesting robot in non-structural environment based on Mask R-CNN. *Computers and Electronics in Agriculture*, *163*, 104846. doi:10.1016/j.compag.2019.06.001

Zaparoli Cunha, B., Droz, C., Zine, A.-M., Foulard, S., & Ichchou, M. (2023). A review of machine learning methods applied to structural dynamics and vibroacoustic. *Mechanical Systems and Signal Processing*, *200*, 110535. doi:10.1016/j.ymssp.2023.110535

Zekrifa, D. M. S., Kulkarni, M., Bhagyalakshmi, A., Devireddy, N., Gupta, S., & Boopathi, S. (2023). Integrating Machine Learning and AI for Improved Hydrological Modeling and Water Resource Management. In *Artificial Intelligence Applications in Water Treatment and Water Resource Management* (pp. 46–70). IGI Global. doi:10.4018/978-1-6684-6791-6.ch003

Zhang, H., Dai, S., Li, Y., & Zhang, W. (2018). Real-time Distributed-Random-Forest-Based Network Intrusion Detection System Using Apache Spark. *2018 IEEE 37th International Performance Computing and Communications Conference, IPCCC 2018*. 10.1109/PCCC.2018.8711068

Zhang, B., Zhang, S., & Li, W. (2019). Bearing performance degradation assessment using long short-term memory recurrent network. *Computers in Industry*, *106*, 14–29. doi:10.1016/j.compind.2018.12.016

Zhang, C., Mousavi, A. A., Masri, S. F., Gholipour, G., Yan, K., & Li, X. (2022). Vibration feature extraction using signal processing techniques for structural health monitoring: A review. *Mechanical Systems and Signal Processing*, *177*, 109175. doi:10.1016/j.ymssp.2022.109175

Zhang, C., Zhu, L., & Xu, C. (2022). BSDP: Blockchain-based smart parking for digital-twin empowered vehicular sensing networks with privacy protection. *IEEE Transactions on Industrial Informatics*.

Zhang, C., Zhu, L., Xu, C., Zhang, C., Sharif, K., Wu, H., & Westermann, H. (2020). BSFP: Blockchain-enabled smart parking with fairness, reliability and privacy protection. *IEEE Transactions on Vehicular Technology*, *69*(6), 6578–6591. doi:10.1109/TVT.2020.2984621

Zhang, D., Zhu, A., Hou, W., Liu, L., & Wang, Y. (2022). Vision-Based Structural Modal Identification Using Hybrid Motion Magnification. *Sensors (Basel)*, *22*(23), 9287. doi:10.339022239287 PMID:36501990

Zhang, F.-L., Gu, D.-K., Li, X., Ye, X.-W., & Peng, H. (2023). Structural damage detection based on fundamental Bayesian two-stage model considering the modal parameters uncertainty. *Structural Health Monitoring*, *22*(4), 2305–2324. doi:10.1177/14759217221114262

Zhang, H., Wang, Y., Liu, Y., & Xiong, N. (2022). IFD: An intelligent fast detection for real-time image information in industrial IoT. *Applied Sciences (Basel, Switzerland)*, *12*(15), 7847. doi:10.3390/app12157847

Zhang, L., Zhou, G., Han, Y., Lin, H., & Wu, Y. (2018). Application of Internet of Things Technology and Convolutional Neural Network Model in Bridge Crack Detection. *IEEE Access : Practical Innovations, Open Solutions*, *6*, 39442–39451. doi:10.1109/ACCESS.2018.2855144

Zhang, Y. C., & Yu, J. (2013). A study on the fire IOT development strategy. *Procedia Engineering*, *52*, 314–319. doi:10.1016/j.proeng.2013.02.146

Zhang, Y.-M., Wang, H., Bai, Y., Mao, J.-X., & Xu, Y.-C. (2022). Bayesian dynamic regression for reconstructing missing data in structural health monitoring. *Structural Health Monitoring*, *21*(5), 2097–2115. doi:10.1177/14759217211053779

Zhang, Y.-M., Wang, H., Wan, H.-P., Mao, J.-X., & Xu, Y.-C. (2021). Anomaly detection of structural health monitoring data using the maximum likelihood estimation-based Bayesian dynamic linear model. *Structural Health Monitoring*, *20*(6), 2936–2952. doi:10.1177/1475921720977020

Zhang, Y., Zhang, W., Yu, J., He, L., Chen, J., & He, Y. (2022). Complete and accurate holly fruits counting using YOLOX object detection. *Computers and Electronics in Agriculture*, *198*, 107062. doi:10.1016/j.compag.2022.107062

Zhang, Y., Zhang, Y., & Zhang, Y. (2022). Fruit and vegetable disease identification based on updating the activation function for the ConvNeXt model. *International Conference on Electronic Information Technology and Computer Engineering* (pp. 1045-1049). ACM. 10.1145/3573428.3573616

Zhao, K. (2021) *Fruit Detection Using CenterNet*. [Master's Thesis, Auckland University of Technology, New Zealand].

Zhao, K. (2021). *Fruit Detection Using CenterNet*. [Master's Thesis, Auckland University of Technology, New Zealand].

Zhao, K., & Yan, W. (2021) Fruit detection from digital images using CenterNet. *International Symposium on Geometry and Vision*. Springer. 10.1007/978-3-030-72073-5_24

Zhao, J., Geng, X., Zhou, J., Sun, Q., Xiao, Y., Zhang, Z., & Fu, Z. (2019). Attribute mapping and autoencoder neural network based matrix factorisation initialisation for recommendation systems. *Knowledge-Based Systems*, *166*, 132–139. Advance online publication. doi:10.1016/j.knosys.2018.12.022

Zhao, L., Huang, X., Zhang, Y., Tian, Y., & Zhao, Y. (2019). A Vibration-Based Structural Health Monitoring System for Transmission Line Towers. *Electronics (Basel)*, *8*(5), 515. doi:10.3390/electronics8050515

Zhao, X., & Jia, M. (2020). A novel unsupervised deep learning network for intelligent fault diagnosis of rotating machinery. *Structural Health Monitoring*, *19*(6), 1745–1763. doi:10.1177/1475921719897317

Zheng, Y., Mobasher, B., & Burke, R. (2016). CARSKit: A Java-Based Context-Aware Recommendation Engine. *Proceedings - 15th IEEE International Conference on Data Mining Workshop, ICDMW 2015*. 10.1109/ICDMW.2015.222

Zhijun, L. I., Shenghui, Y. A. N. G., Deshuai, S. H. I., Xingxing, L. I. U., & Yongjun, Z. H. E. N. G. (2021). Yield estimation method of apple tree based on improved lightweight YOLOv5. *Smart Agriculture*, *3*(2), 100.

Zhi, L., Hu, F., Li, Q., & Hu, Z. (2021). Identification of modal parameters from non-stationary responses of high-rise buildings. *Advances in Structural Engineering*, *24*(15), 3519–3533. doi:10.1177/13694332211033959

Zhongshen, L. (2007, August). Design and analysis of improved butterworth low pass filter. In *2007 8th International Conference on Electronic Measurement and Instruments* (pp. 1-729). IEEE. 10.1109/ICEMI.2007.4350554

Zhou, D., Fang, J., Song, X., Guan, C., Yin, J., Dai, Y., & Yang, R. (2019). IOU loss for 2D/3D object detection. *International Conference on 3D Vision (3DV)*. 10.1109/3DV.2019.00019

Zhou, J., Zhang, Y., & Wang, J. (2023). A dragon fruit picking detection method based on YOLOv7 and PSP-Ellipse. *Sensors (Basel)*, *23*(8), 3803. doi:10.339023083803 PMID:37112144

Zhou, Y., Saad, Y., Tiago, M. L., & Chelikowsky, J. R. (2006). Self-consistent-field calculations using Chebyshev-filtered subspace iteration. *Journal of Computational Physics*, *219*(1), 172–184. doi:10.1016/j.jcp.2006.03.017

Zhou, Z., Pei, J., Liu, X., Fu, H., & Pardalos, P. M. (2021). Effects of resource occupation and decision authority decentralisation on performance of the IoT-based virtual enterprise in central China. *International Journal of Production Research*, *59*(24), 7357–7373. doi:10.1080/00207543.2020.1806369

Zhuk, A. (2023). Navigating the legal landscape of AI copyright: A comparative analysis of EU, US, and Chinese approaches. *AI and Ethics*, 1–8.

Zhu, Y.-C., Xiong, W., & Song, X.-D. (2022). Structural performance assessment considering both observed and latent environmental and operational conditions: A Gaussian process and probability principal component analysis method. *Structural Health Monitoring*, *21*(6), 2531–2546. doi:10.1177/14759217211062099

Zope, V., Dadlani, T., Matai, A., Tembhurnikar, P., & Kalani, R. (2020). IoT sensor and deep neural network-based wildfire prediction system. In *4th international conference on intelligent computing and control systems (ICICCS)* (pp. 205-208). IEEE. 10.1109/ICICCS48265.2020.9120949

Zou, Z., Chen, K., Shi, Z., Guo, Y., & Ye, J. (2023). Object detection in 20 years: A survey. *Proceedings of the IEEE*, *111*(3), 257–276. doi:10.1109/JPROC.2023.3238524

About the Contributors

Brij B. Gupta working as Director of International Center for AI and Cyber Security Research and Innovations, and Distinguished Professor with the Department of Computer Science and Information Engineering (CSIE), Asia University, Taiwan. In more than 17 years of his professional experience, he published over 500 papers in journals/conferences including 35 books and 12 Patents with over 25,000 citations. He has received numerous national and international awards including Canadian Commonwealth Scholarship (2009), Faculty Research Fellowship Award (2017), MeitY, GoI, IEEE GCCE outstanding and WIE paper awards and Best Faculty Award (2018 & 2019), NIT KKR, respectively. Prof. Gupta was selected for Clarivate Web of Science Highly Cited Researchers in Computer Science (top 0.1% researchers in the world) consecutively in 2022 and 2023. Also, he was also selected in Stanford University's ranking of the world's top 2% scientists consecutively in 2020, 2021, 2022 and 2023. He is also a visiting/adjunct professor with several universities worldwide. He is also an IEEE Senior Member (2017) and also selected as 2021 Distinguished Lecturer in IEEE CTSoc. Dr Gupta is also serving as Member-in-Large, Board of Governors, IEEE Consumer Technology Society (2022-2024). Prof Gupta is also leading IJSWIS, IJSSCI, STE and IJCAC as Editor-in-Chief. Moreover, he is also serving as lead-editor of a Book Series with CRC and IET press. He also served as TPC members in more than 150 international conferences also serving as Associate/Guest Editor of various journals and transactions. His research interests include information security, Cyber physical systems, cloud computing, blockchain technologies, intrusion detection, AI, social media and networking.

Francesco Colace, Ph.D., is full professor of Computer Science at the University of Salerno (Department of Industrial Engineering). He has research experience in Computer Science, Data Mining, Knowledge Management, Computer Networks, Context-Aware, and e-Learning. He is the author of more than 150 papers in the field of Computer Science, of each more than 50 were published in international journals with high impact factor. He worked as a Guest Editor for many journals and is a reviewer for various international scientific journals and conferences. He is a member of The Scientific Committee of the Archaeological Park of Pompeii. He is head of the ICT Center for Cultural Heritage () and is the coordinator of the research group Knowman (). His research activities have had a significant impact on national and international projects. The research activities have had a considerable impact on national and international projects. There have been numerous experiences, including as scientific director, in projects involving national and international research groups and companies operating both nationally and internationally.

***

A.R. Arunarani is with the Department of Computational Intelligence, SRM Institute of Science and Technology, Kattankulathur, Tamil Nadu, 603203, India.

Sampath Boopathi is an accomplished individual with a strong academic background and extensive research experience. He completed his undergraduate studies in Mechanical Engineering and pursued his postgraduate studies in the field of Computer-Aided Design. Dr. Boopathi obtained his Ph.D. from Anna University, focusing his research on Manufacturing and optimization. Throughout his career, Dr. Boopathi has made significant contributions to the field of engineering. He has authored and published over 155 research articles in internationally peer-reviewed journals, highlighting his expertise and dedication to advancing knowledge in his area of specialization. His research output demonstrates his commitment to conducting rigorous and impactful research. In addition to his research publications, Dr. Boopathi has also been granted one patent and has three published patents to his name. This indicates his innovative thinking and ability to develop practical solutions to real-world engineering challenges. With 17 years of academic and research experience, Dr. Boopathi has enriched the engineering community through his teaching and mentorship roles. He has served in various engineering colleges in Tamilnadu, India, where he has imparted knowledge, guided students, and contributed to the overall academic development of the institutions. Dr. Sampath Boopathi's diverse background, ranging from mechanical engineering to computer-aided design, along with his specialization in manufacturing and optimization, positions him as a valuable asset in the field of engineering. His research contributions, patents, and extensive teaching experience exemplify his expertise and dedication to advancing engineering knowledge and fostering innovation.

Mario Casillo received his Ph.D. in Science and technology management from Università degli Studi di Napoli, "Federico II", Napoli, Italy. Currently is an adjunct professor at the University of Salerno. He has research experience in Computer Science, Big Data and Knowledge Management, Smart Cities, and Human-Computer Interaction. He is the author of more than 30 papers in the field of Computer Science. He is a member of the Knowman research group (knowman.unisa.it).

Pratyay Das is making significant strides in machine learning, focusing on robotics. Driven by passion and innovation, they have embarked on an inspiring journey, achieving remarkable results in their project, "Machine Learning Approach for Robot Navigation Using Motor Imagery Signals." Early on, Pratyay was fascinated by technology and its potential to shape the world. Passionate about machine learning and robotics, they pursued a path to explore and contribute to this field, recognizing its transformative power. Pursued a Computer Science degree, aiming to excel in machine learning and robotics. This foundation laid the groundwork for future endeavors, focusing on data analysis and robotic systems. Pratyay's project, "Machine Learning Approach for Robot Navigation Using Motor Imagery Signals," uses EEG motor imagery to predict and differentiate human movements, enabling efficient robot navigation. Developed an innovative framework using machine learning to interpret motor imagery signals, combining neuroscience and robotics for intelligent automation and human-robot interaction. Excels in machine learning, robotics, and innovation, showcasing dedication and expertise. Passionate about machine learning and robotics, seeking collaboration and advancements. They drive technological progress with a desire to make a meaningful impact and seek opportunities to collaborate with like-minded individuals. Pratyay is passionate about machine learning and robotics, focusing on robot navigation using motor imagery signals. Their project demonstrates their ability to merge fields and contribute to innovation. With their dedication and passion, they are poised to shape the future of machine learning and robotics.

Hari Priya G. S. is Assistant Professor, Department of Computer Science, M S Ramaiah College of Arts Science and Commerce.

Rajkumar G. Nadakinamani is a specialist cardiologist, Badr Al Samaa Hospital- Muscat-Oman.

Xinyi Gao is a PhD student, his research interests are deep learning and computer vision.

Ahona Ghosh is a B.Tech., M.Tech. in Computer Science and Engineering and currently is an AICTE Doctoral Fellow in the Department of Computer Science and Engineering, Maulana Abul Kalam Azad University of Technology, West Bengal (In house). Before joining her Ph.D., she was an Assistant Professor in the Department of Computational Science at Brainware University, Barasat, Kolkata. Her research interests include Artificial Intelligence and Human-Computer Interaction.

Anitha Kannan is with the Department of Computing Technologies, School of Computing, SRM Institute of Science and Technology, Kattankulathur, Chennai, India.

Angelo Lorusso is a Ph.D. Student at the University of Salerno (Department of Industrial Engineering), with research experience in Computer Science, the Internet of Things, BIM, and Digital Twin. He is the author of more than 15 papers in Computer Science, some of which were published in international journals. He is a member of the Knowman research group (knowman.unisa.it) and collaborates with the research group on several projects funded by the Italian Ministry of University and by the European Community.

A. Maheshwari is with the Department of Computational Intelligence, SRM Institute of Science and Technology, Kattankulathur, Tamil Nadu, 603203, India.

J. Malathi is with the Department of Computer Science and Business Systems, Sri Sai Ram Engineering College, Chennai, India.

Francesco Marongiu is a Ph.D. Student at the University of Salerno (Department of Industrial Engineering), and has research experience in Computer Science, the Internet of Things, Blockchain, and Cybersecurity. He is the author of more than 15 papers in Computer Science, some of which were published in international journals. He is a member of the Knowman research group (knowman.unisa.it) and collaborates with the research group on several projects funded by the Italian Ministry of University and by the European Community.

Akash Mohanty is with the School of Mechanical Engineering, Vellore Institute of Technology, Vellore, Tamil Nadu, India.

M. Sureshkumar completed his undergraduate in Mechanical Engineering and postgraduate in the field of Engineering Design. He completed his Ph.D. from Anna University, Chennai, Tamil Nādu, India.

Minh Nguyen is an Associate Professor in Computer Science, he is the Chair of the Department of Computer Science, Auckland University of Technology, New Zealand.

Krishnagandhi Pachiappan is with the Department of Electrical and Electronics Engineering, Nandha Engineering College, Erode, Tamil Nadu, India.

Princy Pappachan is employed as an assistant professor at the Center for the Development of Language Teaching and Research (CDLTR) in Asia University, Taiwan. Additionally, she has served on the International Advisory Committee for ICRAMLET22. She is also an International Advisory Member of MIMSE2023. She holds a Ph.D. in Applied Linguistics from the Centre of Applied Linguistics and Translation Studies (CALTS) at the University of Hyderabad (UoH), an M.Phil. in Cognitive Science from the Centre for Cognitive Science and Neuroscience (CNCS) at UoH. Her primary research focus areas are Generative Artificial Intelligence, Cognitive Science, Multilingual Education (MLE) in Sustainable Development Research and English Language Teaching (ELT). She has also presented research papers at various international conferences and published papers in peer-reviewed journals.

R. Pitchai is with the Department of Computer Science and Engineering, B V Raju Institute of Technology, Narsapur, India.

Mosiur Rahaman received the Bachelor of Technology degree and the Master of Technology degree in Computer Science and Engineering from Jawaharlal Nehru Technological University, India. He received subsequently Master of Business administration (M.B.A) degree in Human Resource Management with IT (HRM-IT) from Osmania University, India. He is currently pursuing a Ph.D. degree in Computer Science and Information Engineering department at Asia University, Taichung, Taiwan. He is serving as a lecturer College of Humanities and Social Science Department, Asia University Taiwan from 2020. From 2014 to 2015, he was a assistant professor in the Dept. of Computer Science and Engineering at the Royal Institute of Technology and Science, India. His research interests include Information Security, Blockchain, and Artificial Intelligence, IoT Security, Supply chain, Food Safety. He is also working as program manager and assisting with ASEAN and South Asian Countries - AI R&D Elite Class/Doctoral Class. He worked and cooperated with various R&D projects under MOST, Taiwan. He is also serving as a Session Chairs for FICT 2021, FICT 2022, General Chair for ICRAMLET22 (International Conference) and is an International Advisory Member of MIMSE23.

U. Rahamathunnisa is with the School of Computer Science Engineering and Information Systems, Vellore Institute of Technology, Vellore, India.

B. Rebecca is with the Department of Computer Science and Engineering (Data Science), Marri Laxman Reddy Institute of Technology and Management, Dundigal, Hyderabad, India.

Sriparna Saha received her M.E. and Ph.D. degrees from Electronics and Tele-Communication Engineering department of Jadavpur University, Kolkata, India. She is currently an Assistant Professor in the Department of Computer Science and Engineering of Maulana Abul Kalam Azad University of Technology, West Bengal, India. Prior to that, she was associated as a faculty with Jadavpur University and two other institutions. She has more than 8 years of experience in teaching and research. Her research area includes artificial intelligence, image processing, machine learning and recently she is giving more focus on deep learning. She has over 70 publications in IEEE, Elsevier, Springer, etc. She is the author of a book on Gesture Recognition published by Studies in Computational Intelligence, Springer. She is also the reviewer for many international journals. Her major research proposal is accepted for Start Up Grant under UGC Basic Scientific Research Grant. Not only that, she also got University seed money for setting up her laboratory.

S. Sangeetha is a Professor, Department of Computer Science Engineering, Karpagam College of Engineering, India.

Domenico Santaniello received his Ph.D. in Industrial Engineering from the University of Salerno. Currently is a research fellow (Department of Industrial Engineering). He has research experience in Computer Science, Data Mining, Knowledge Management, Situation Awareness, Machine Learning, the Internet of Things, Natural Learning Processing, and Education. He is the author of more than 60 papers in Computer Science, some of which were published in international journals. He is a member of the Knowman research group (knowman.unisa.it), a reviewer for many international scientific journals and international conferences, and collaborates on several research projects funded by the Italian Ministry of University and the European Community.

T. K. Sethuramalingam is with the Department of Electronics and Communication Engineering, Karpagam College of Engineering, Coimbatore, India.

Manju Sharma is with the Department of Computer Science and Engineering, University Institute of Engineering & Technology, Rohtak, Haryana, IN.

Monika Sharma is with the Department of Computer Science and Engineering, The Technological Institute of Textile and Sciences Bhiwani, IN.

Neerav Sharma is with the Department of Computer Science and Engineering, BITS College, Bhiwani, IN.

Ajeet Singh is an Assistant Professor, School of Computing Science and Engineering, VIT Bhopal University, Bhopal-Indore Highway, Sehore Dist., Madhya Pradesh, 466114, India.

Vikash Singh is with the Department of Civil Engineering, Institute of Engineering and Technology, Lucknow, India.

Anitha Jebamani Soundararaj is with the Department of Information Technology, Sri Sai Ram Engineering College, Chennai, Tamil Nadu 602109, India.

Sreerakuvandana is an Assistant Professor of English, specializing in Cognitive Linguistics at the Department of AI-ML and Cybersecurity at Jain deemed-to-be University. She received her Bachelor's, Master's and Doctorate degree from the University of Hyderabad, specializing in Linguistics and Cognitive Science. Her research areas include Artificial Intelligence, English Language Teaching, Sociolinguistics, and Psycholinguistics. She also has a range of academic publications in the areas of her interest to her credit. She currently serves as a guest editor for the Athens Journal of Education and is an International Advisory Member of MIMSE23 and is a guest editor for the same.

Bipin Kumar Srivastava is with the Department of Applied Sciences, Galgotias College of Engineering and Technology, Greater Noida, India.

K. Sudhakar is with the Department of Computer Science & Engineering Madanapalle Institute of Technology & Science, Madanapalle, India.

G. Sumathy is with the Department of Computational Intelligence, SRM Institute of Science and Technology, Kattankulathur, Tamil Nadu, India.

K. Sundaramoorthy Department of Information Technology, Jerusalem College of Engineering, Velachery, Pallikaranai, India.

Maganti Syamala is with the Department of Computer Science and Engineering, Koneru Lakshmaiah Education Foundation, Vaddeswaram, India.

Satyanarayana T.V.V. is with the Department of Electronics and Communication Engineering, Mohan Babu University, Tirupati, India.

Alfredo Troiano is currently the Chief Technical Officer at Netcom Group S.p.A Napoli and adjunct professor at the University of Salerno (Department of Industrial Engineering). His research is related to the Internet of Things, Smart Agriculture, Context-Aware Computing, Blockchain, and Quantum Computing. He is involved in several research projects funded by the Italian Ministry of University and by the European Community.

R. Udendhran is with the Department of Computational Intelligence, SRM Institute of Science and Technology, Kattankulathur, India.

B. Uma Maheswari is with the Department of Computer Science and Engineering, St. Joseph's College of Engineering, Chennai, India.

Carmine Valentino is a Ph.D. student in the Department of Industrial Engineering at the University of Salerno. His research works are related to Recommender Systems, Context-Aware Recommender Systems, Fake News Detection, Machine Learning, and Situation Awareness. He is the author of about 15 papers in Computer Science, some of which were published in international journals. He is a member of the Knowman research group (knowman.unisa.it) and collaborates with the research group on several projects funded by the Italian Ministry of University and by the European Community.

S. Venkatramulu is with the Department of Computer Science Engineering, Kakatiya Institute of Technology and Science, Warangal.

Bingjie Xiao is a PhD student from Department of Computer Science, Auckland University of Technology, Auckland New Zealand. Her research interest is computer vision and deep learning.

Yinzhe Xue is a master's student, his research interest is deep learning and computer vision.

Wei Qi Yan is the Director of Institute of Robotics & Vision (IoRV), Auckland University of Technology (AUT), IoRV is one of the topstream research institutes at AUT. Dr. Yan's research interests include deep learning, intelligent surveillance, computer vision, multimedia computing, etc. Dr. Yan's expertise is computational mathematics and applied mathematics, computer science and computer engineering, his research distinctions encapsulate the published Springer monographs: Computational Methods for Deep Learning (2021, 2023), Introduction to Intelligent Surveillance (2019), Visual Cryptography for Image Processing and Security (2015), etc. Dr. Yan is the Chair of ACM Multimedia Chapter of New Zealand, a Member of the ACM, a Senior Member of the IEEE, TC members of the IEEE, a Mentor of IEEE ISCAS society. Dr. Yan was a world's top 2% cited scientist listed by Stanford University in 2022.

Winfred Yaokumah is a researcher and senior faculty at the Department of Computer Science of the University of Ghana. He has published several articles in highly rated journals including Information and Computer Security, Information Resources Management Journal, IEEE Xplore, International Journal of Human Capital and Information Technology, International Journal of Human Capital and Information Technology Professionals, International Journal of Technology and Human Interaction, International Journal of e-Business Research, International Journal of Enterprise Information Systems, Journal of Information Technology Research, International Journal of Information Systems in the Service Sector, and Education and Information Technologies. His research interest includes cyber security, cyber ethics, network security, and information systems security and governance. He serves as a member of the International Review Board for the International Journal of Technology Diffusion.

Index

A

Accidental Deaths 76
administrative processes. 278, 282
Affordable Medicine 348-351, 357-358, 364-365
aircraft applications 251-257, 264, 270
aircraft systems 254, 266
apple 401, 410, 422, 429, 431
Artificial Intelligence (AI) 1-2, 4-10, 12-20, 52, 107-110, 112, 115, 118-120, 125, 128, 130, 132-133, 138, 194, 222, 224-225, 228, 230, 232, 234, 241, 244, 250, 259, 268, 284, 304, 316, 318-319, 348-351, 358, 360, 365-366, 373-376, 386, 399, 401
Artificial Intelligence techniques 107, 109-110, 118, 120, 130, 133
autonomous vehicles 9, 169-170, 184, 194, 215-217, 222
Autopilot 251, 253, 256-260
Aviation Safety 257, 264

B

backbone network 401-402, 410-411
banana 422, 429, 431
Best Practices 195, 197-198, 363-364
Best Worst Method 75, 80, 82, 90
Blockchain 18, 52, 55-66, 69, 71, 223-225, 227-228, 230, 232, 234-237, 239, 241-242, 244, 284

C

Chatbot 9, 128-130
chemistry practices 348, 350-351, 356, 358-359, 363-366
chemistry principles 348-351, 354-355, 357-359, 366
cloud computing 33, 169-170, 172-175, 177, 181-184, 186-187, 195-200, 202, 207, 209, 216
computer vision 95, 250, 252, 260, 306, 333, 395, 400-401, 423
confusion matrix 131, 429
Consensus Protocol 56, 74
ConvNeXt 400-402, 404, 408-412
Convolutional Block Attention Module 94, 97-98
Cost-Benefit Analysis 223-224, 239-240
Cultural Heritage 107-110, 118-119, 121-122, 125, 128-130, 132-133

D

Damage Detection 305-315, 317-321, 323-325, 327-333
Data Analytics 7, 121, 129, 169-173, 177, 183, 187, 207, 286-287, 354, 363, 375, 390-392, 395
Data Privacy 2, 14-16, 67, 69, 171, 184, 197, 213-214, 223, 250, 253, 269, 276, 278, 288, 290, 296, 366, 394-395
data protection 2, 9, 15-19, 250, 394
Data Science 348-349, 351-354, 358-363, 365-366
Data-driven Decision Making 205
data-driven decisions 199, 209, 374-376, 395
data-driven innovations 186, 188
data-driven insights 171, 173, 216, 277, 358-359, 365-366
Decentralisation 53, 56, 59-60, 63, 74
Decentralised Storage 66, 68, 74
Decentralised Systems 53-55, 61-63, 71
Decision Support 268, 278, 283, 389, 391
decision-making processes 6, 17, 52, 254, 326, 330
Deep Learning 14, 34, 43, 94-95, 103, 108-109, 112, 115, 117, 125, 138, 163, 251-253, 255-257, 259-261, 263, 265-267, 270, 277, 284, 306, 386-387, 400-401, 413, 421-422
detection algorithm 95, 103, 311
Digital Healthcare 276-278, 280, 282, 285, 293
digital images 403, 413, 421-422
Distributed Ledger 61, 63, 74, 284
drug development 349-351, 357-366
Drug Discovery 17, 19, 276-277, 298-299, 348-354, 358-360, 365-366

E

E-Health 276-278, 280-282
ELAN attention 401, 410
Electric Vehicle 250
Electroencephalography 137
energy consumption 57, 69, 170-171, 186, 196, 216, 350-351, 355
ensemble method 27, 31, 37-38, 40-41, 43, 422, 432
Ensemble Model 27, 31, 41
Environmental Impact 173, 181, 186, 195-199, 213-214, 223-224, 351, 354-355, 357, 359, 365, 376, 389, 392, 395
Ethical Artificial Intelligence 1-2, 10, 15-16, 18
ethical concerns 2, 6, 16-18, 20

F

Face Mask Detection 94-98, 102
Fault Detection 251, 253, 257, 264-265
feature extraction 32, 96-97, 99, 127, 143, 145-146, 149, 151, 154, 156, 163, 309, 311-312, 318-319, 326, 335, 400, 403
filter-based feature 27-28, 35, 40, 43
freshness detection 421-422
Fruit classification 421-422, 429, 431
fuel consumption 171, 197, 252

G

Generative Artificial Intelligence 2, 12-18
GPS devices 174
Green Chemistry 348-351, 354-361, 363-366
greenhouse gas 171

H

Healthcare Transformation 285, 287
holistic approach 170, 186, 314, 350, 389

I

inadequate infrastructure 170, 175, 179
Information Security 28
Information Storage 53
intellectual property 1-7, 10-20, 65
Intellectual Property Rights 1-6, 10-18, 20
Intelligence techniques 107, 109-110, 118, 120, 130, 133
intelligence-driven healthcare 2, 20
Intelligent Machines 50-52, 374-376, 378, 381-384, 387-389, 392, 394-395
intelligent parking systems 195-198, 202-205, 207, 217
Intelligent Systems 18, 251-253, 268
Internet of Things (IoT) 29, 50-51, 53, 58-59, 61, 69, 71, 74, 77-78, 110, 172, 175, 181, 194, 196, 222, 250, 284, 333, 374-376, 384, 399
intrusion detection 27-33, 36, 43
IoT Sensors 177, 223, 284, 375-376, 384, 387-389, 394-395

K

Key Encapsulation Mechanisms 69-70, 74

L

language processing 2, 4, 109, 128, 255, 260, 268, 283, 295, 304
LCA 373
learning techniques 2, 19, 27, 29-30, 32-33, 43, 108, 125, 128, 130, 251-252, 254, 256, 259, 305, 308, 316, 320, 325, 335, 387
Logistics regression 27, 33, 37-38, 40-41, 43
low-frequency EEG 163

M

machine learning (ML) 2, 4-7, 13-14, 17-19, 27-34, 43, 52, 108-109, 112-115, 121, 128, 130, 137-138, 169, 224-225, 231-232, 251-254, 256-257, 259-261, 268, 270, 276-280, 283, 286, 288-296, 298-299, 306, 332, 350-351, 353, 375-376, 386-387, 421-422
malicious actors 52-53, 56
mathematical model 32, 109, 112
ML algorithms 28-29, 40, 112-113, 141, 252, 259
molecular data 349, 351, 353-354, 359, 363, 366
Motor Imagery 137-138
multi-criteria decision-making 78, 81, 90

N

Naive Bayes 27-28, 31-33, 37-38, 40-41, 43, 151
Natural Language Processing 2, 4, 109, 128, 255, 260, 268, 283, 295, 304
Network Security 28
neural networks 30, 32-33, 95, 100, 108-109, 115, 163, 255-256, 260, 263, 284, 307, 315-321, 323-324, 328, 334, 403-404, 422
novelty detection 306-308, 310-312, 318-321, 323-328, 331-332, 335

O

object confidence 97, 100
object detection 95-96, 98, 101-102, 260, 400-403, 405, 410, 422-425, 432
One-Class Support 311, 324, 326
orange 87, 422, 429, 431

P

PCA 32, 141, 143-147, 149-151, 153-154, 156, 158, 160-161, 163, 309-313, 317-320, 324, 327-328, 352, 373
Pearson and Spearman Correlation Coefficient 35, 38
Performance Accuracy 27
Personalized Medicine 286, 290-291, 359, 361-363, 366
pharmaceutical industry 19, 348-350, 354-358, 361, 366-367
Pilot Assistance 253, 267
Policy Considerations 184, 195, 198
Predictive Analytics 79, 169, 181, 187-188, 208, 223, 234, 277, 283, 286, 288-290, 295
Predictive maintenance 252, 256, 264-267, 270, 318-319
property rights 1-6, 10-18, 20

Q

Quality of Life 51, 141, 170-173, 175, 179, 181, 183, 186, 188, 195-196, 198-199, 225-226, 242

R

Random Forest 28, 30, 32-33, 37-38, 40-41, 43, 137-138
Real-time data 169-173, 176-177, 181-182, 196, 207, 209, 212, 223, 225, 234, 266, 286, 375-376, 384-385, 387-389, 394
real-time information 170, 176, 196-197, 224, 384
real-time monitoring 173, 306
real-world scenarios 325, 333-335
Recommender Systems 107-109, 118-120, 124, 130, 132
Remote Monitoring 282-283, 286, 291, 296, 299, 385
resident well-being 196, 217
Robot Navigation 137-138, 163

S

secure transactions 232-233
semantic information 97, 401, 410
signal processing 146, 307, 309, 311, 313-315
small aircraft 251-261, 263-267, 270
Smart Parking 169, 171-173, 175-179, 196, 206, 223-228, 230, 232, 234-236, 239, 241-242, 244
social media 4, 170
space allocation 196, 223, 226, 228, 230
Structural Health Monitoring 305, 307-308, 313-314, 318-319, 321-322, 324-325, 330-331, 333
Supply Chain 55, 63, 65, 79, 81, 284, 383
Support Vector Machine 28, 31, 33, 138, 307, 311, 326
Sustainable Cities 173, 225, 241
Sustainable Farming 380-381, 389
sustainable practices 170, 186, 350, 354, 356, 381, 389, 394-395
Sustainable Production 359
sustainable synthesis 354-357, 365
Sustainable Transportation 170, 172-173, 182, 212

T

technological advancements 6, 10, 18, 170, 333, 384
therapeutic interventions 349, 351, 353
trade secrets 4, 6, 17, 19
Traffic Congestion 171-173, 175, 177, 179, 181, 183-184, 195-198, 203-205, 207, 209-211, 213, 216, 224-228, 232, 241-242, 244
transformative change 358, 365
transportation systems 170-174, 179, 181, 196, 212
t-SNE 352, 373
two-dimensional image 401, 410

U

unsupervised learning 2, 5, 113, 254, 305-308, 316, 318-320, 322, 324-327, 329, 332-335
Urban Mobility 169-175, 177, 179, 181-184, 186, 188, 195, 197-198, 204-205, 207, 209-211, 223-226, 228, 230, 234, 236, 241-242, 244
Urban Planning 169-171, 173, 187-188, 228
User Experience 132, 186, 203-204, 208, 223-224, 230, 235, 239-240, 250

V

vibration-based SHM 306-307, 324, 335
virtual screening 349-351, 354, 359, 366
Visual Object Detection 95-96, 400-403, 410, 422-423

W

waste management 170-171
Water Processing 374-376, 386-389, 395

Y

YOLOv7 94-95, 97-103, 400-403, 405, 407-408, 410-412, 421-422, 424, 431
YOLOv8 421-422, 425, 428-429, 431-432

- [illegible] etc.
- [illegible]
- [illegible]
- [illegible] Mathematics [illegible]
- [illegible] (WCAG)

[illegible]

Request [illegible] in the [illegible]

[illegible]

Printed in the United States
by Baker & Taylor Publisher Services